Mechanics: From Theory to Computation

Springer Science+Business Media, LLC

Mechanics: From Theory to Computation

Essays in Honor of Juan-Carlos Simo

Papers Invited by *Journal of Nonlinear Science* Editors

With 62 Illustrations

 Springer

Library of Congress Cataloging-in-Publication Data
Mechanics : from theory to computation : essays in honor of Juan-Carlos Simo.
  p. cm.
  At head of title: Papers invited by Journal of nonlinear science editors.
  Includes bibliographical references.
  ISBN 978-1-4612-7059-1   ISBN 978-1-4612-1246-1 (eBook)
  DOI 10.1007/978-1-4612-1246-1

  1. Mechanics, Applied.   2. Simo, J.-C. (Juan Carlos), 1952–
I. Simo, J.-C. (Juan-Carlos), 1952– II. Journal of nonlinear science.
  TA350.3 .M43 1999
  620.1—dc21   98-46020

Printed on acid-free paper.

Production managed by Timothy Taylor; manufacturing supervised by Jeffrey Taub.
Photocomposed copy prepared from PDF files by Archetype, Monticello, IL.

9 8 7 6 5 4 3 2 1

ISBN 978-1-4612-7059-1   SPIN 10698180

Juan-Carlos Simo (1952–1994)

Preface

Starting in 1996, a sequence of articles appeared in the *Journal of Nonlinear Science* dedicated to the memory of one of its original editors, Juan-Carlos Simo, Applied Mechanics, Stanford University. Sadly, Juan-Carlos passed away at an early age in 1994. We lost a brilliant colleague and a wonderful person.

These articles are collected in the present volume. Many of them are updated and corrected especially for this occasion. These essays are in areas of scientific interest of Juan-Carlos, including mechanics (particles, rigid bodies, fluids, elasticity, plasticity, etc.), geometry, applied dynamics, and, of course, computation. His interests were extremely broad—he did not see boundaries between computation, mathematics, mechanics, and dynamics, and, in that sense, he ideally reflected the spirit of the journal and many of the most exciting areas of current scientific interest.

Juan-Carlos was one of those select and gifted people who could cross interdisciplinary boundaries with extremely high quality and productive interactions of lasting value. His contributions, ranging from concrete engineering problems to fundamental mathematical theorems in geometric mechanics, are remarkable. In current conferences as well as in scientific books and articles, and over a wide range of subjects, one frequently hears how his ideas as well as specific results are often used and quoted—this is one indication of just how profound and fundamental his work has impacted the community.

On top of his brilliance was a generous and caring soul. He did not engage in those petty academic arguments we too often see that divert lesser people from productivity and kindness. However, his standards were high, and those with the intellectual gift and diligence to rise to them were amply and fairly rewarded.

It is hoped that the articles contained herein honor some of Juan-Carlos' spirit and achievement.

Contents

Dynamical Problems for Geometrically Exact Theories of Nonlinearly Viscoelastic Rods

S.S. Antman
Department of Mathematics and Institute for Physical Science and Technology,
University of Maryland, College Park, MD 20742, USA

Received October 5, 1995
Communicated by Jerrold Marsden and Stephen Wiggins

This paper is dedicated to the memory of Juan-Carlos Simo

Summary. This paper surveys recent results and open problems for the equations of motion for geometrically exact theories of nonlinearly viscoelastic and elastic rods. These rods can deform in space by undergoing not only flexure and torsion, but also extension and shear. The paper begins with a derivation of the governing equations, which for viscoelastic rods form a quasilinear system of hyperbolic–parabolic partial differential equations of high order. It then derives the energy equation and discusses difficulties that can arise in getting useful energy estimates. The paper next treats constitutive assumptions precluding total compression. The paper then discusses the curious asymptotic problems that arise when the inertia of the rod is small relative to that of a rigid body attached to its end. The paper concludes with discussions of traveling waves and shock structure, Hopf bifurcation problems, and problems of control.

1. Introduction

The development of theories of rods has been intertwined with the development of the three-dimensional theory of deformable solids for over 250 years. Although nonlinear problems for the planar equilibrium of elastic rods were correctly formulated and solved by Euler [18] in 1744, two hundred years before the solution of three-dimensional equilibrium problems for nonlinearly elastic bodies, it was not until after the appearance of clean, correct, and simple formulations of three-dimensional continuum mechanics that such formulations of geometrically exact problems for rods became available. Thus only recently have the equations of rods attained a form readily accessible to analysis. And only recently have methods of analysis (and computation—see [32], [33], [34], [35], and [36]) attained a level capable of handling geometrically exact problems for rods with nonlinear constitutive equations. It is the purpose of this paper to describe the analysis of several such dynamical problems for nonlinearly viscoelastic rods.

Since they have but one independent variable, rod theories are analytically far simpler than three-dimensional theories. On the other hand, inverse and semi-inverse problems having but one spatial variable are virtually the only problems of the three-dimensional theory that admit solutions with readily determined qualitative properties, and these problems typically have a much simpler structure than do problems for rods. In mechanical terms, the complexity of rod problems is due to the presence of nonuniform flexural effects.

Theories of rods can be derived systematically by regarding rods as (i) three-dimensional bodies so constrained that their strains depend on only one spatial variable, with the governing equations obtained by the theory of material constraints (which generalizes projection methods of numerical analysis like the method of lines), (ii) thin three-dimensional bodies, with the governing equations obtained as the leading terms of asymptotic expansions in a thickness parameter, or (iii) intrinsically one-dimensional bodies, with their governing equations obtained by invoking the standard balance laws of mechanics. In this paper, we adopt the third approach because it is the simplest and most direct, method (i) is in complete agreement with it, and method (ii) agrees with it to the extent permitted by the level of approximation. See [3], [8], [16], [24], [27], [28], [32], [33], [34], [35], [36], [44]. Almost all of our methods extend to far more complicated theories of rods, which are most easily generated by method (i).

Notation

We employ Gibbs notation for vectors and tensors: Vectors, which are elements of Euclidean 3-space $\mathbb{E}^3$, and vector-valued functions are denoted by lower-case, italic, bold-face symbols $u, v, \ldots$. The three vectors $\{i_1, i_2, i_3\} \equiv \{i, j, k\}$ are assumed to form a fixed right-handed orthonormal basis. The dot product of (vectors) u and v is denoted by $u \cdot v$. Tensors are linear transformations of $\mathbb{E}^3$ into itself. They are denoted by upper-case, italic, bold-face symbols $A, B, \ldots$. The value of tensor A at vector v is denoted $A \cdot v$ (in place of the more usual Av) and the product of A and B is denoted $A \cdot B$ (in place of the more usual AB). The transpose of A is denoted A^*. We write $u \cdot A = A^* \cdot u$. The dyadic product of vectors a and b is denoted ab (in place of the more usual $a \otimes b$). It is the tensor defined by $(ab) \cdot u = (b \cdot u)a$ for all u.

Triples of components of vectors are denoted by lower-case, bold-face roman symbols $\mathbf{u}, \mathbf{v}, \ldots$. We use the same notational scheme for operations involving these entities.

Twice-repeated lower-case Latin indices are summed from 1 to 3 and twice-repeated lower-case Greek indices are summed from 1 to 2.

The (Gâteaux) differential of $u \mapsto f(u)$ at v in the direction h is $\frac{d}{dt} f(v + th)\big|_{t=0}$. When it is linear in h, we denote this differential by $\frac{\partial f}{\partial u}(v) \cdot h$ or $f_u(v) \cdot h$. The function f_u is tensor-valued. The partial derivative of a function f with respect to a scalar argument t is denoted by either f_t or $\partial_t f$. Obvious analogs of these notations will also be used. In particular, $\partial f/\partial \mathbf{u}$ is the matrix of partial derivatives of the components of $\mathbf{f}$ with respect to the components of $\mathbf{u}$. If $\mathbf{f}$ and $\mathbf{g}$ are three-component functions of the triples $\mathbf{u}$ and $\mathbf{v}$, then

$$\frac{\partial(\mathbf{f}, \mathbf{g})}{\partial(\mathbf{u}, \mathbf{v})} \equiv \begin{pmatrix} \partial\mathbf{f}/\partial\mathbf{u} & \partial\mathbf{f}/\partial\mathbf{v} \\ \partial\mathbf{g}/\partial\mathbf{u} & \partial\mathbf{g}/\partial\mathbf{v} \end{pmatrix}$$

is the 6×6 matrix consisting of the four indicated 3×3 submatrices.

2. Formulation of the Governing Equations

We give a brief formulation of the geometrically exact equations of motion of a rod that can suffer flexure, extension, torsion, and shear. (For this purpose we use a Cosserat theory of rods with an orthonormal triad of directors.) For full details and motivations, see [3].

Geometry of Deformation

The *motion of a rod* is defined by three vector-valued functions

$$[0, 1] \times \mathbb{R} \ni (s, t) \mapsto r(s, t), \ d_1(s, t), \ d_2(s, t) \in \mathbb{E}^3 \tag{2.1}$$

with $d_1(s, t)$ and $d_2(s, t)$ orthonormal. $r(\cdot, t)$ may be interpreted as the configuration at time t of any material curve connecting the "ends" of a slender three-dimensional body, e.g., a curve of centroids of cross-sections or a curve of mass centers of cross-sections of the body. $d_1(s, t)$ and $d_2(s; t)$ may be interpreted as characterizing the configuration of the material section (at) s at time t. In particular, $d_1(s, t)$ and $d_2(s, t)$ may be regarded as characterizing the configurations of a pair of orthogonal material lines of the section s. We assume that s is the arc-length parameter of the reference configuration of r, and we scale the length so that $0 \le s \le 1$. We set

$$d_3 \equiv d_1 \times d_2. \tag{2.2}$$

Thus $\{d_k(s, t)\}$ for each (s, t) is a right-handed orthonormal basis for $\mathbb{E}^3$. The orthonormality of these vectors implies that there are vector-valued functions u and w such that

$$\partial_s d_k = u \times d_k, \qquad \partial_t d_k = w \times d_k. \tag{2.3a,b}$$

Since the basis $\{d_k\}$ is natural for the intrinsic description of deformation, we decompose relevant vector-valued functions with respect to it:

$$u = u_k d_k, \qquad r_s = v_k d_k, \qquad w = w_k d_k. \tag{2.4a,b,c}$$

By using the equality of mixed partial derivatives of the d_k, we can prove that

$$w_s = u_t + u \times w = (\partial_t u_k) d_k. \tag{2.5}$$

The triples

$$\mathbf{u} \equiv (u_1, u_2, u_3), \qquad \mathbf{v} \equiv (v_1, v_2, v_3) \tag{2.6}$$

are the *strain* variables corresponding to the motion (2.1) with u_1 and u_2 measuring flexure, u_3 measuring torsion, v_1 and v_2 measuring shear, and v_3 measuring dilatation.

Under our interpretations of the kinematic variables, a rod-theoretic analog of the three-dimensional requirement that the Jacobian of the deformation be positive (so that orientation is preserved) is that there be a function $(u_1, u_2, s) \mapsto V(u_1, u_2, s)$ for which $V(0, 0, s) = 0$, $V(u_1, u_2, s) > 0$ for $u_\alpha u_\alpha > 0$, and $V(\cdot, \cdot, s)$ is convex and homogeneous of degree 1 such that

$$v_3 > V(u_1, u_2, s). \tag{2.7}$$

Thus for each s the surface $v_3 = V(u_1, u_2, s)$ is a cone in (u_1, u_2, v_3)-space. (The form of V depends on the shape of the cross-section; V can often be found explicitly.) A consequence of (2.7) is that

$$v_3 \equiv \mathbf{r}_s \cdot \mathbf{d}_3 > 0, \tag{2.8}$$

so that the local ratio $|\mathbf{r}_s| > 0$ of deformed to reference length of the axis cannot be reduced to zero, and so that a typical section s cannot undergo a total shear in which the plane determined by $\mathbf{d}_1(s, t)$ and $\mathbf{d}_2(s, t)$ is tangent to the curve $\mathbf{r}(\cdot, t)$ at $\mathbf{r}(s, t)$.

We assume that the end $s = 0$ of the rod is welded to the $\{\mathbf{i}, \mathbf{j}\}$-plane, so that

$$\mathbf{r}(0, t) = \mathbf{0}, \qquad \mathbf{d}_k(0, t) = \mathbf{i}_k. \tag{2.9a,b}$$

The presence of shearability means that there is no natural way to clamp an end of the rod, i.e., to prescribe the direction of $\mathbf{r}$ at an end. Those methods that effect such a clamping essentially require feedbacks, and these may lead to some surprising dynamical instabilities [41].

Mechanics

In the configuration at time t, the resultant contact force and contact couple exerted by the material of $[0, s]$ on the material of $[s, 1]$ (for $0 < s \leq 1$) are respectively denoted $-\mathbf{n}(s, t)$ and $-\mathbf{m}(s, t)$. At (s, t) the rod is subjected to a body force of intensity $\mathbf{f}(s, t)$ and a body couple of intensity $\mathbf{l}(s, t)$ per unit reference length at (s, t).

Motivated in part by three-dimensional considerations in which $\mathbf{r}$ is the material curve of mass centers of the cross-sections of the three-dimensional rod, we assume that the linear and angular momenta per unit reference length at (s, t) have the forms $\varepsilon(\rho A)(s)\mathbf{r}_t(s, t)$ and

$$\chi(\varepsilon, s)(\rho \mathbf{I})(s, t) \cdot \mathbf{w}(s, t) \equiv \chi(\varepsilon, s)(\rho I_{pq})(s)w_q(s, t)\mathbf{d}_p(s, t). \tag{2.10}$$

We take ε to be a nonnegative parameter that measures the size of the inertias of the rod relative to the size of the inertias of the rigid body. It is not a thickness parameter. We take χ to be a prescribed function with $\chi(\varepsilon, s) > 0$ for $\varepsilon > 0$ and $0 < s < 1$, and with $\chi(0, s) = 0$ (χ is determined by the cross-sectional shape). The function ρA is prescribed to be positive-valued. It may be interpreted as the scaled mass density per reference length. The $(\rho I_{\gamma\delta})(s)$ are the prescribed components of a positive-definite symmetric 2×2 matrix, which may be thought of as a matrix of scaled mass-moments of inertia of the section s, $\rho I_{\gamma 3} = \rho I_{3\gamma} = 0$, $\rho I_{33} = \rho I_{\gamma\gamma}$, and $\rho \mathbf{I} \equiv \rho I_{pq}\mathbf{d}_p\mathbf{d}_q$. The methods for treating more general forms of (2.10) are identical to those used here.

We assume that a rigid body of mass μ is rigidly attached to the end $s = 1$ of the rod. Its mass center is at $\mathbf{r}(1, t) + \mathbf{c}(t)$, where, because of the rigidity of the attachment, $\mathbf{c}(t)$ has the form $c_k\mathbf{d}_k(1, t)$, with the c_k prescribed constants. This rigid body has inertia tensor $\mathbf{J}(t) = J_{kl}\mathbf{d}_k(1, t)\mathbf{d}_l(1, t)$ about its mass center. Here the J_{kl} are the constant components of a positive-definite matrix.

The classical forms of the equations of motion are

$$\mathbf{n}_s + \mathbf{f} = \varepsilon \rho A \mathbf{r}_{tt}, \tag{2.11}$$

$$\mathbf{m}_s + \mathbf{r}_s \times \mathbf{n} + \mathbf{l} = \chi(\varepsilon, s)[(\rho \mathbf{I})(s, t) \cdot \mathbf{w}(s, t)]_t. \tag{2.12}$$

The boundary conditions at $s = 1$ are the equations of motion of the rigid body:

$$-\mathbf{n}(1, t) = \mu[\mathbf{r}_{tt}(1, t) + \mathbf{c}_{tt}(t)], \tag{2.13}$$

$$-\mathbf{m}(1, t) + \mathbf{c}(t) \times \mathbf{n}(1, t) = [\mathbf{J}(t) \cdot \mathbf{w}(1, t)]_t. \tag{2.14}$$

Let

$$\mathbf{m} \equiv (m_1, m_2, m_3), \quad \mathbf{n} \equiv (n_1, n_2, n_3) \quad \text{where} \quad m_k \equiv \mathbf{m} \cdot \mathbf{d}_k, \quad n_k \equiv \mathbf{n} \cdot \mathbf{d}_k. \tag{2.15}$$

We may call m_1 and m_2 the *bending couples*, m_3 the *twisting couple*, n_1 and n_2 the *shear forces*, and $\mathbf{n} \cdot \mathbf{r}_s/|\mathbf{r}_s|$ the *tension*. These terms are not strictly analogous to those used in structural mechanics, in which it is usually assumed that $\mathbf{d}_3 = \mathbf{r}_s/|\mathbf{r}_s|$.

Constitutive Equations

The rod is called *viscoelastic of differential type of complexity 1* if there are constitutive functions

$$(\mathbf{u}, \mathbf{v}, s) \mapsto \varphi(\mathbf{u}, \mathbf{v}, s),$$

$$(\mathbf{u}, \mathbf{v}, \dot{\mathbf{u}}, \dot{\mathbf{v}}, s) \mapsto \mathbf{m}^D(\mathbf{u}, \mathbf{v}, \dot{\mathbf{u}}, \dot{\mathbf{v}}, s), \quad \mathbf{n}^D(\mathbf{u}, \mathbf{v}, \dot{\mathbf{u}}, \dot{\mathbf{v}}, s) \tag{2.16}$$

with $\mathbf{m}^D(\mathbf{u}, \mathbf{v}, \mathbf{0}, \mathbf{0}, s) = \mathbf{0} = \mathbf{n}^D(\mathbf{u}, \mathbf{v}, \mathbf{0}, \mathbf{0}, s)$ such that

$$\mathbf{m}(s, t) = \frac{\partial \varphi}{\partial \mathbf{u}}(\mathbf{u}(s, t), \mathbf{v}(s, t), s) + \mathbf{m}^D(\mathbf{u}(s, t), \mathbf{v}(s, t), \dot{\mathbf{u}}(s, t), \dot{\mathbf{v}}(s, t), s),$$

$$\mathbf{n}(s, t) = \frac{\partial \varphi}{\partial \mathbf{v}}(\mathbf{u}(s, t), \mathbf{v}(s, t), s) + \mathbf{n}^D(\mathbf{u}(s, t), \mathbf{v}(s, t), \dot{\mathbf{u}}(s, t), \dot{\mathbf{v}}(s, t), s). \tag{2.17}$$

The dots over $\mathbf{u}$ and $\mathbf{v}$ in (2.16) have mnemonic, but not operational, significance: They merely identify the third and fourth arguments of $\mathbf{m}^D$ and $\mathbf{n}^D$ in (2.16). The function φ, which is assumed to be nonnegative, is the *stored-energy function* for the *elastic part* of constitutive functions, and $\mathbf{m}^D$ and $\mathbf{n}^D$ are the *dissipative parts*. The common domain of these constitutive functions is defined by (2.7). These forms of the constitutive equations ensure that the material response is unaffected by rigid motions. For simplicity of exposition, we assume that φ, $\mathbf{m}^D$, $\mathbf{n}^D$ are continuously differentiable.

It is reasonable to require that the stored-energy function φ be convex in $\mathbf{u}$, $\mathbf{v}$, but such an assumption would prohibit the sort of material instabilities associated with coexistent phases (cf. [1], [19], [22], [30]). Since this convexity is not necessary for much of the analysis of problems we treat, we do not make this assumption. We merely require that

$$\frac{\partial \varphi}{\partial u_\alpha}(\mathbf{u}, \mathbf{v}, s)\frac{\partial V}{\partial u_\alpha}(u_1, u_2, s) - \frac{\partial \varphi}{\partial v_3}(\mathbf{u}, \mathbf{v}, s) \to \infty \qquad \text{as } v_3 - V(u_1, u_2, s) \searrow 0, \tag{2.18}$$

$$\varphi(\mathbf{u}, \mathbf{v}, s) \to \infty \qquad \text{as } |\mathbf{u}| + |\mathbf{v}| \to \infty. \tag{2.19}$$

The left-hand side of (2.18) is proportional to the directional derivative of φ along the outer normal to the cone $\{(\mathbf{u}, \mathbf{v}): v_3 = V(u_1, u_2, s)\}$. A growth condition far more stringent than (2.18) is that

$$\varphi(\mathbf{u}, \mathbf{v}, s) \to \infty \qquad \text{as } v_3 - V(u_1, u_2, s) \searrow 0. \tag{2.20}$$

Neither (2.18) nor (2.20) seems capable of ensuring (2.7) for dynamical problems, although (2.20) serves this purpose for equilibrium problems (cf. [3, Chap. 7]). We shall see that condition (4.2) ensures (2.7).

We require that effects of internal friction grow with the strain rates:

$$\frac{\partial (\mathbf{m}^D, \mathbf{n}^D)}{\partial (\dot{\mathbf{u}}, \dot{\mathbf{v}})} \quad \text{is positive-definite.} \tag{2.21}$$

This monotonicity condition ensures that the response is truly dissipative and that the governing equations of motion have a parabolic character. We further assume that infinite strain rates produce infinite resultants:

$$\frac{\mathbf{m}^D(\mathbf{u}, \mathbf{v}, \dot{\mathbf{u}}, \dot{\mathbf{v}}, s) \cdot \dot{\mathbf{u}} + \mathbf{n}^D(\mathbf{u}, \mathbf{v}, \dot{\mathbf{u}}, \dot{\mathbf{v}}, s) \cdot \dot{\mathbf{v}}}{\sqrt{|\dot{\mathbf{u}}|^2 + |\dot{\mathbf{v}}|^2}} \to \infty \qquad \text{as } |\dot{\mathbf{u}}|^2 + |\dot{\mathbf{v}}|^2 \to \infty \tag{2.22}$$

for $(\mathbf{u}, \mathbf{v})$ in a compact set of strains satisfying (2.7). In this case, there are numbers $c, C > 0$ such that

$$\mathbf{m}^D(\mathbf{u}, \mathbf{v}, \dot{\mathbf{u}}, \dot{\mathbf{v}}, s) \cdot \dot{\mathbf{u}} + \mathbf{n}^D(\mathbf{u}, \mathbf{v}, \dot{\mathbf{u}}, \dot{\mathbf{v}}, s) \cdot \dot{\mathbf{v}} \geq c \left(|\dot{\mathbf{u}}| + |\dot{\mathbf{v}}|\right) - C. \tag{2.23}$$

We impose initial conditions on $\mathbf{r}, \mathbf{r}_t, \mathbf{d}_k, \partial_t \mathbf{d}_k$ that we assume are compatible with the boundary conditions at $s = 0$. Then our initial-boundary-value problem for a viscoelastic rod consists of the kinematic relations (2.3)–(2.5), the equations of motion (2.11), (2.12), the constitutive equations (2.17), the boundary conditions (2.9), (2.13), (2.14), and the initial conditions just described. If the natural configuration of the rod is straight, then under very reasonable constitutive conditions, it admits purely longitudinal motions, described by a scalar equation of motion of the form

$$\varepsilon \rho A w_{tt} = \hat{n}_3(w_s, w_{st}, s)_s + f. \tag{2.24}$$

Here w is the component of $\mathbf{r}$ along the axis of motion. The analysis of this equation under general physically reasonable constitutive assumptions is itself a formidable exercise (cf. [1], [10], [17], [21], [23], [26], [30], [48]).

The derivation of rod theories with more kinematic structure is most efficiently carried out by suitably constraining the motions of the three-dimensional bodies they are designed to model [3], [8], [36]. These methods can handle any theory in which the three-dimensional stress is a prescribed function of the past history of the strain. Materials of internal-variable type, e.g., those of many models of large-strain plasticity, have this specific form provided that the ordinary differential equations governing the evolution of the internal variables can be solved in closed form, which is rare. An effective modification of the method of constraints capable of handling general materials of internal-variable type is formulated in [43]. The resulting theory is applied to treat the large dynamical buckling of nonlinearly elastoplastic rods without using quasistaticity. The construction of rod theories that exactly account for incompressibility is very tricky [42]. The theories may have nonlocal effects, and they always have higher derivatives than the models described in this section. For a study of the special but interesting problem of coupled longitudinal and radial motions of incompressible rods, see [46].

3. The Energy Equation

As usual, we begin our analysis of equations of motion with a derivation of the energy equation: We take the dot product of (2.11) with r_t, take the dot product of (2.12) with w, add the resulting equations, integrate the sum with respect to s over $(0, 1)$, and then use the boundary conditions (2.9), (2.13), (2.14) and the kinematic relations (2.4), (2.5) to obtain

$$\int_0^1 [\varepsilon \rho A r_{tt} \cdot r_t + \chi(\rho I \cdot w)_t \cdot w]\, ds$$

$$= \int_0^1 [r_t \cdot n_s + w \cdot (m_s + r_s \times n)]\, ds + \int_0^1 (f \cdot r_t + l \cdot w)\, ds$$

$$= r_t(1, t) \cdot n(1, t) + w(1, t) \cdot m(1, t)$$

$$\quad - \int_0^1 [n_k d_k \cdot (v_l d_l)_t + m \cdot w_s - w \cdot (r_s \times n)]\, ds$$

$$= -\mu r_t(1, t) \cdot [r_{tt}(1, t) + c_{tt}(t)] - \mu w(1, t) \cdot \{c(t) \times [r_{tt}(1, t) + c_{tt}(t)]\}$$

$$\quad - w(1, t) \cdot [J(t) \cdot w(1, t)]_t - \int_0^1 (n_k \partial_t v_k + m_k \partial_t u_k)\, ds + \int_0^1 (f \cdot r_t + l \cdot w)\, ds$$

$$= -\frac{1}{2}\frac{d}{dt}\{\mu |r_t(1, t) + c_t(t)|^2 + w(1, t) \cdot J(t) \cdot w(1, t)\}$$

$$\quad - \frac{d}{dt}\int_0^1 \varphi(\mathbf{u}, \mathbf{v}, s)\, ds + \int_0^1 (f \cdot r_t + l \cdot w)\, ds$$

$$\quad - \int_0^1 [\mathbf{m}^D(\mathbf{u}, \mathbf{v}, \mathbf{u}_t, \mathbf{v}_t, s) \cdot \mathbf{u}_t + \mathbf{n}^D(\mathbf{u}, \mathbf{v}, \mathbf{u}_t, \mathbf{v}_t, s) \cdot \mathbf{v}_t]\, ds. \tag{3.1}$$

Integrating (3.1) from 0 to t we get the *energy equation*:

$$\int_0^t \int_0^1 [\mathbf{m}^D(\mathbf{u}, \mathbf{v}, \mathbf{u}_t, \mathbf{v}_t, s) \cdot \mathbf{u}_t + \mathbf{n}^D(\mathbf{u}, \mathbf{v}, \mathbf{u}_t, \mathbf{v}_t, s) \cdot \mathbf{v}_t]\, ds\, d\tau$$

$$+ \int_0^1 \varphi(\mathbf{u}(s, t), \mathbf{v}(s, t), s)\, ds$$

$$+ \frac{1}{2}\int_0^1 [\varepsilon(\rho A)(s)|r_t(s, t)|^2 + \chi(\varepsilon, s)w(s, t) \cdot (\rho I)(s) \cdot w(s, t)]\, ds$$

$$+ \frac{1}{2}\{\mu |r_t(1, t) + c_t(t)|^2 + w(1, t) \cdot J(t) \cdot w(1, t)\}$$

$$= \int_0^1 \varphi(\mathbf{u}(s, 0), \mathbf{v}(s, 0), s)\, ds + \frac{1}{2}\{\mu |r_t(1, 0) + c_t(0)|^2 + w(1, 0) \cdot J(0) \cdot w(1, 0)\}$$

$$+ \frac{1}{2}\int_0^1 [\varepsilon \rho A |r_t(s, 0)|^2 + \chi w(s, 0) \cdot (\rho I) \cdot w(s, 0)]\, ds$$

$$+ \int_0^t \int_0^1 (f \cdot r_t + l \cdot w)\, ds\, dt. \tag{3.2}$$

Notice that each term on the left-hand side of (3.2) is positive. Elementary inequalities show that the work of the body force f and the body couple l cause no difficulty. Therefore, each term on the left-hand side of (3.2) is bounded above. If the strong condition (2.20) is assumed, then we conclude that for any given t, (2.7) holds for almost every s. This result is not adequate for our needs.

The form of (3.2) shows that while our boundary conditions are complicated, they are innocuous. Suppose that we replace (2.13) with the seemingly simpler condition that $r(1, t)$ is prescribed. Then we obtain an equation like (3.2) with the expression $\int_0^t [n(1, \tau) \cdot r_t(1, \tau) - n(0, \tau) \cdot r_t(0, \tau)] \, d\tau$ appearing on the right-hand side. The n's that appear here have elastic parts $\partial \varphi / \partial v$, which for small $v_3 - V$ behave much worse than φ itself, whose integral appears on the left-hand side of (3.2). Consequently, we cannot get a useful estimate for the left-hand side of the modified energy equation in terms of the data. That this difficulty is real is indicated by the fact that the version of (2.24) with $f = 0$ and with ρA and $\hat{n}_3$ independent of s admits the exact solution $w(s, t) = (1 - t)s$, which describes a total compression at $t = 1$, even though its boundary conditions $w(0, t) = 0$, $w(1, t) = 1 - t$ are ostensibly innocuous. A method for handling this difficulty for scalar problems, employing strengthened constitutive hypotheses and an intricate analysis, is given by [10]. We do not pursue this question here.

If in place of our vectorial equations we had used equivalent componential forms, then the derivation of the energy equation (3.3) would be a formidable exercise. The basic difficulty is due to our introduction of the moving basis d_k, which is essential for a simple description of constitutive equations invariant under rigid motions, but which complicates the equations of motion because it is responsible for the appearance of w in various time-derivatives. (This difficulty recurs unavoidably in treatments of differentiated versions of our governing equations, which must be studied in the treatment of regularity; cf. [12].)

Henceforth we take $f = 0 = l$ to simplify our presentation.

4. Preclusion of Total Compression

The essential ingredient in showing that our boundary-value problem admits well-behaved solutions is to show that (2.7) holds everywhere. For this purpose we assume that frictional effects become infinitely large in a suitable way at a total compression. To describe such a condition we define

$$\delta^\sharp(\mathbf{u}, \mathbf{v}, s) \equiv v_3 - V(u_1, u_2, s), \quad \dot{\delta}^\sharp(\mathbf{u}, \mathbf{v}, \dot{\mathbf{u}}, \dot{\mathbf{v}}, s) \equiv \dot{v}_3 - \frac{\partial V}{\partial u_\alpha}(u_1, u_2, s)\dot{u}_\alpha,$$

$$\delta(s, t) \equiv \delta^\sharp(\mathbf{u}(s, t), \mathbf{v}(s, t), s). \tag{4.1}$$

We assume that there is a positive number η, a constant $C > 0$, and a continuously differentiable function $(0, \eta) \times [0, 1] \ni (\delta, s) \mapsto \Omega(\delta, s)$ with $\Omega(\delta, s) \to \infty$ as $\delta \to 0$ such that

$$\hat{n}_3(\mathbf{u}, \mathbf{v}, \dot{\mathbf{u}}, \dot{\mathbf{v}}, s) \leq -\Omega_\delta(\delta^\sharp(\mathbf{u}, \mathbf{v}, s), s)\dot{\delta}^\sharp(\mathbf{u}, \mathbf{v}, \dot{\mathbf{u}}, \dot{\mathbf{v}}, s) + C$$
$$\text{if } 0 < \delta^\sharp(\mathbf{u}, \mathbf{v}, s) \leq \eta. \tag{4.2}$$

Now we employ (4.2) to obtain a positive lower bound for δ and thereby show that (2.7) must hold everywhere, provided that our initial-boundary-value problem has a solution for which δ is continuous. (We could use the lower bound for δ to obtain the existence of solutions for which δ is continuous without defying logic: The bound on δ tells us where to seek solutions. Versions of (2.21)–(2.23) enable us to find weak solutions there and then show that these solutions are regular.)

We assume that the initial function $s \mapsto \delta(s, 0)$ is continuous and positive. Then without loss of generality we choose the η of (4.2) to satisfy $\eta < \min_s \delta(s, 0)$. To show that $\delta(s, t)$ is positive for all (s, t) it suffices to show this only for all (s, t) for which $\delta(s, t) < \eta$. Thus suppose that there is a ξ in $[0, 1)$ and a $\tau_2 > 0$ such that $\delta(\xi, \tau_2) < \eta$. Since δ is continuous, there is a last time τ_1 at which $\delta(\xi, \tau_1) = \eta$. From (2.11), (2.13), and (4.2) we get

$$d_3(\xi, t) \cdot \int_\xi^1 (\rho A)(s) r_t(s, \tau) \, ds \bigg|_{\tau_1}^t$$

$$= d_3(\xi, t) \cdot \int_{\tau_1}^t n(s, \tau) \, d\tau \bigg|_\xi^1$$

$$= -\int_{\tau_1}^t n_3(\xi, \tau) \, d\tau - \mu d_3(\xi, t) \cdot [r_t(1, \tau) + c_t(\tau)]_{\tau_1}^t$$

$$\geq -\int_{\tau_1}^t \Omega_\delta(\delta(\xi, \tau), \xi) \delta_\tau(\xi, \tau) \, d\tau - C(t - \tau_1) - \mu |[r_t(1, \tau) + c_t(\tau)]_{\tau_1}^t|$$

$$\geq -\Omega(\delta(\xi, t), \xi) - \Omega(\eta, \xi) - C(t - \tau_1) - \mu |[r_t(1, \tau) + c_t(\tau)]_{\tau_1}^t| \qquad (4.3)$$

for $\tau_1 \leq t \leq \tau_2$. From the energy equation (3.2) and the Cauchy–Bunyakovskiĭ–Schwarz inequality we deduce that the leftmost and rightmost terms of (4.3) are bounded by well-behaved functions of t. Thus we conclude that there is a continuous function Γ such that

$$\Omega(\delta(\xi, t), \xi) \leq \Gamma(t). \qquad (4.4)$$

Thus there is a positive-valued continuous function γ such that

$$\delta(s, t) \geq \gamma(t) \qquad (4.5)$$

for all s, t.

Condition (4.2) is unduly restrictive when $\dot{\delta}^\sharp(\mathbf{u}, \mathbf{v}, \dot{\mathbf{u}}, \dot{\mathbf{v}}, s)$ is large. More general constitutive assumptions lacking this defect and methods for treating them are given by [10].

Let us set $\kappa \equiv \sqrt{|\mathbf{u}|^2 + |\mathbf{v}|^2}$, $\lambda \equiv \sqrt{\dot{\mathbf{u}} \cdot \mathbf{u} + \dot{\mathbf{v}} \cdot \mathbf{v}}$. Then, by using an analogous but more complicated argument, we can get an upper bound for κ under the following hypothesis, which is inspired by the representation of isotropic constitutive functions:

For each $c > 0$ for which $\delta \geq c$ there is a continuously differentiable function ζ of κ with $\zeta(\kappa) \to \infty$ as $\kappa \to \infty$, there is a nonnegative number D, and there is a positive number η such that

$$\frac{\hat{\mathbf{m}}(\mathbf{u}, \mathbf{v}, \dot{\mathbf{u}}, \dot{\mathbf{v}}, s) \cdot \mathbf{u} + \hat{\mathbf{n}}(\mathbf{u}, \mathbf{v}, \dot{\mathbf{u}}, \dot{\mathbf{v}}, s) \cdot \mathbf{v}}{\kappa} \geq \zeta'(\kappa)\dot{\kappa} - D\zeta(\kappa) + \eta \lambda \kappa. \qquad (4.6)$$

where $\hat{\mathbf{m}} \equiv \partial \varphi / \partial \mathbf{u} + \mathbf{m}^D$, etc.

For details, see [11], [12]. Reference [12] gives the full existence theory for initial-boundary-value problems like ours. Methods by which these results can be extended to problems of large-strain elastoplasticity of internal-variable type are given by [40].

5. Light Rods with Heavy Attachments

The reduced initial-boundary-value problem

Our interest in this section is in the *reduced* initial-boundary-value problem, obtained by setting $\varepsilon = 0$. In this case, from (2.11) and (2.12) we obtain the integrals

$$n(s, t) = n(1, t), \qquad m(s, t) + r(s, t) \times n(1, t) = m(1, t) + r(1, t) \times n(1, t), \quad (5.1a,b)$$

which are very useful for our study.

For simplicity of exposition, we assume that there are no external forces or couples applied to the rigid body by any agent other than the rod.

Let us comment briefly on elastic rods, for which $\mathbf{m}^D = \mathbf{0} = \mathbf{n}^D$. We immediately see from (2.11), (2.12) with $\varepsilon = 0$ that the rod moves through a family of equilibrium states parametrized by t, which enters through the boundary conditions (2.13), (2.14). In general, however, these equilibrium states are not unique. (An exception is given in [2].) Indeed, for each given $m(1, t), n(1, t)$, there may be several equilibrium states. The collection of these states lies on a 6-dimensional manifold S parametrized by $m(1, t), n(1, t)$, which can be decomposed into (an infinite number of) surfaces S_α on each of which the configuration of the rod is a single-valued function of $m(1, t), n(1, t)$. (In several important cases there are a countable number of surfaces S_α that can be distinguished by the nodal properties of certain kinematic variables.) On each S_α, we can express $m(1, t), n(1, t)$ as single-valued functions of $r(1, t), d_k(1, t)$. (We omit the details.) We assemble all these single-valued representations into a multivalued representation

$$m(1, t) = \tilde{m}(r(1, t), d_k(1, t)), \qquad n(1, t) = \tilde{n}(r(1, t), d_k(1, t)). \quad (5.2)$$

We substitute (5.2) into (2.13) and (2.14) and append (2.3b) to obtain the the following multivalued system of ordinary differential equations for $r(1, \cdot), d_k(1, \cdot), w(1, \cdot)$:

$$[J(t) \cdot w(1, t)]_t + \mu c \times [r_{tt}(1, t) + c_{tt}(t)] + \tilde{m}(r(1, t), d_k(1, t)) = 0, \quad (5.3)$$

$$\mu[r_{tt}(1, t) + c_{tt}(t)] + \tilde{n}(r(1, t), d_k(1, t)) = 0, \quad (5.4)$$

$$\partial_t d_k(1, t) = w(1, t) \times d_k(1, t). \quad (5.5)$$

We use (5.2) to replace S, S_α with corresponding manifolds T, T_α parametrized by $r(1, t), d_k(1, t)$. The boundaries of the T_α are fold sets (where a certain "Jacobian" is zero). Now the dynamics embodied in a single-valued version of (5.3)–(5.5) corresponding to a given T_α might take the configuration to the boundary of T_α. When this happens, I surmise that the configuration jumps with constant $r(1, \cdot), d_k(1, \cdot)$ to another such surface on which the potential energy is a local minimum, i.e., on which the equilibrium configuration is "stable." One can make plausible conjectures about which such surfaces are the targets. Notice that such jumps preserve the continuity of $r(1, \cdot), d_k(1, \cdot)$. (These surfaces for the elastica are graphed in [9].) For problems like this to make sense, the initial conditions for the rod must lie on T_α.

These jumps correspond to snapping motions of the rod. Our process of taking the formal limit as $\varepsilon \to 0$ gives us a way to describe these sudden large motions systematically. This approach differs from standard applications of catastrophe theory not only by being global, but more importantly, by having the evolution described by a complete dynamics.

To have any hope of justifying the limit $\varepsilon \to 0$ mathematically, we introduce the strong dissipative mechanism of viscosity. We restrict our attention to the reduced problem for viscoelastic rods.

Conditions (2.21) and (2.22) support a global implicit-function theorem (based on degree theory) that asserts that the system of finite-dimensional equations

$$\frac{\partial \varphi}{\partial \mathbf{u}}(\mathbf{u}, \mathbf{v}, s) + \mathbf{m}^D(\mathbf{u}, \mathbf{v}, \dot{\mathbf{u}}, \dot{\mathbf{v}}, s) = \mathbf{m}, \qquad \frac{\partial \varphi}{\partial \mathbf{v}}(\mathbf{u}, \mathbf{v}, s) + \mathbf{n}^D(\mathbf{u}, \mathbf{v}, \dot{\mathbf{u}}, \dot{\mathbf{v}}, s) = \mathbf{n} \qquad (5.6)$$

can be solved uniquely for $\dot{\mathbf{u}}$ and $\dot{\mathbf{v}}$ in terms of the other variables:

$$\dot{\mathbf{u}} = \dot{\mathbf{u}}^\sharp(\mathbf{u}, \mathbf{v}, \mathbf{m}, \mathbf{n}, s), \qquad \dot{\mathbf{v}} = \dot{\mathbf{v}}^\sharp(\mathbf{u}, \mathbf{v}, \mathbf{m}, \mathbf{n}, s) \qquad (5.7)$$

with $\dot{\mathbf{u}}^\sharp$ and $\dot{\mathbf{v}}^\sharp$ inheriting the regularity of φ, $\mathbf{m}^D$, $\mathbf{n}^D$ and satisfying the monotonicity condition:

$$\frac{\partial(\dot{\mathbf{u}}^\sharp, \dot{\mathbf{v}}^\sharp)}{\partial(\dot{\mathbf{m}}, \dot{\mathbf{n}})} \quad \text{is positive-definite.} \qquad (5.8)$$

The equivalence of (5.6) and (5.7) implies that (2.17) is equivalent to

$$\mathbf{u}_t(s, t) = \dot{\mathbf{u}}^\sharp\left(\mathbf{u}(s, t), \mathbf{v}(s, t), \mathbf{m}(s, t), \mathbf{n}(s, t), s\right),$$
$$\mathbf{v}_t(s, t) = \dot{\mathbf{v}}^\sharp\left(\mathbf{u}(s, t), \mathbf{v}(s, t), \mathbf{m}(s, t), \mathbf{n}(s, t), s\right). \qquad (5.9)$$

From (2.5), (2.4b), (2.3b) we obtain

$$w(s, t) = \int_0^s \dot{u}_k^\sharp d_k(\xi, t) \, d\xi, \qquad (5.10)$$

$$r_t(s, t) = \int_0^s \{\dot{v}_k^\sharp d_k(\xi, t) + \dot{u}_k^\sharp d_k(\xi, t) \times [r(s, t) - r(\xi, t)]\} \, d\xi, \qquad (5.11)$$

$$\partial_t d_k(s, t) = \left[\int_0^s \dot{u}_l^\sharp d_l(\xi, t) \, d\xi\right] \times d_k(s, t) \qquad (5.12)$$

where the arguments of $\dot{\mathbf{u}}^\sharp$ and $\dot{\mathbf{v}}^\sharp$ are those of (5.9) at (ξ, t). To obtain (5.11), we used the identity $r_{st} = (\partial_t v_k) d_k + w \times r_s = (\partial_t v_k) d_k + (w \times r)_s - u_t \times r$ coming from (2.3b), (2.4b), (2.5).

We now substitute (5.1) into (5.9), (5.11), (5.12). If $m(1, \cdot)$ and $n(1, \cdot)$ were known, then the resulting equations would constitute a system of *ordinary* differential equations for the family $\{\mathbf{u}(s, \cdot), \mathbf{v}(s, \cdot), r(s, \cdot), d_k(s, \cdot): s \in [0, 1]\}$. That this family of functions is parametrized by s is just the remnant of the full partial differential equations valid for $\varepsilon \neq 0$.

By substituting (5.9) into representations for $r_{tt}(1, t)$ and $w_t(1, t)$ and then substituting the resulting representations into the boundary conditions (2.13), (2.14), we obtain two ordinary differential equations for $m(1, \cdot)$ and $n(1, \cdot)$, which (5.8) enables us to put in standard form. We thus obtain a full system of ordinary differential equations for $\{\mathbf{u}(s, \cdot), \mathbf{v}(s, \cdot), r(s, \cdot), d_k(s, \cdot): s \in [0, 1]\}$ and $m(1, \cdot), n(1, \cdot)$. We can exploit the energy equation (3.2) and the bound (4.5) to show that the solution of the initial-value problem for this system cannot blow up in finite time, and therefore must exist for all time.

We now ask whether there are conditions ensuring that the motion of the rigid body for the reduced problem for viscoelastic rods is governed by a system of ordinary differential equations, just as it is for the reduced problem for elastic rods. For simplicity of exposition we restrict our attention to rods whose natural reference configuration is straight and prismatic, so that r_s and d_k are constant in this configuration, with $r_s = i_3 = d_3$ here.

For this purpose, we consider a one-parameter family of initial data:

$$r(s, 0) = si_3 + \eta\bar{r}(s), \qquad r_t(s, 0) = \eta\bar{p}(s),$$
$$u(s, 0) = \eta\bar{u}(s), \qquad w(s, 0) = \eta\bar{w}(s). \tag{5.13}$$

For $\eta = 0$, the solution of the reduced problem, which is unique, is just the reference configuration. Now if the reduced problem subject to initial conditions (5.12) is governed by ordinary differential equations, then so is its derivative with respect to η at $\eta = 0$; this derivative is just the linearization of the reduced problem about the reference configuration. This linearization can be formulated as a system of convolution equations, which can be solved by means of the Laplace transform. An analysis of the solution shows that in contrast to the case for elastic rods, (except in one peculiar circumstance) the reduced problem *cannot* be governed by ordinary differential equations when the rod undergoes a flexural motion, i.e., a motion with the property that for some fixed t, the direction of $r(s, t)$ varies with s.

It is still an open problem to connect these results for viscoelastic rods to those for elastic rods. Those for elastic rods should arise in an asymptotic expansion of the problem for viscoelastic rods when a small viscosity parameter ν goes to zero. It is well known that zero-viscosity limits behave poorly. There is nevertheless hope of obtaining sharp results because we have two parameters, ε and ν, with the smallness of ε showing promise of compensating for the smallness of ν.

For full details on the results of this section, see [9]. It is proved in [48] that the solution of the reduced problem for purely longitudinal motion, governed by (2.24) and a scalar version of (2.13), is the leading term of the regular part of a rigorous asymptotic expansion of the solution in ε. Since this result relies on the Maximum Principle for scalar parabolic equations, it does not carry over to the system treated above. Forced versions of the reduced equations exhibit some very interesting dynamics [45].

6. Traveling Waves

The equations for elastic rods form a quasilinear hyperbolic system, provided that φ is convex in **u** and **v**. It is well known that solutions of such systems typically develop shocks. The Principle of Virtual Power yields the Rankine–Hugoniot conditions for shocks, but these are insufficient to determine unique solutions containing shocks. Additional "entropy" conditions are needed. Various forms of these, not fully understood for systems as complicated as ours, are alternatively generated by (i) postulating thermodynamic restrictions that supplement the Second Law, (ii) formulating mathematical criteria that ensure uniqueness, or (iii) introducing a one-parameter family of dissipative mechanisms strong enough to prevent shocks, and then characterizing admissible solutions of the hyperbolic system as asymptotic limits as the dissipativity goes to zero.

We examine only the last approach because it is closely related to concrete aspects of material response, even though the mathematical steps are merely formal. We take the dissipative mechanism to be that of viscosity, and we introduce a small viscosity parameter η by modifying the constitutive functions of (2.17) thus:

$$\mathbf{m}(s, t) = \frac{\partial \varphi}{\partial \mathbf{u}}(\mathbf{u}, \mathbf{v}, s) + \mathbf{m}^D(\mathbf{u}, \mathbf{v}, \eta \dot{\mathbf{u}}, \eta \dot{\mathbf{v}}, s),$$

$$\mathbf{n}(s, t) = \frac{\partial \varphi}{\partial \mathbf{v}}(\mathbf{u}, \mathbf{v}, s) + \mathbf{n}^D(\mathbf{u}, \mathbf{v}, \eta \dot{\mathbf{u}}, \eta \dot{\mathbf{v}}, s). \tag{6.1}$$

(This disposition of η leads to a much cleaner asymptotic analysis than that in which η enters in other ways, say as in $\mathbf{m}(s, t) = \partial \varphi(\mathbf{u}, \mathbf{v}, s)/\partial \mathbf{u} + \eta \mathbf{m}^D(\mathbf{u}, \mathbf{v}, \dot{\mathbf{u}}, \dot{\mathbf{v}}, s)$, etc.)

To treat the system obtained by substituting (6.1) into (2.11), (2.12), we introduce a stretched variable $\xi = [s - \sigma(t)]/\eta$ where $s = \sigma(t)$ is the equation of a shock path for the hyperbolic system. We set $\tilde{r}(\xi, t) = r(s, t)$ and formally seek solutions in the form of a sum of a regular and a transition-layer expansion in η. The leading term of the latter can be shown (cf. [3], [37]) to satisfy the traveling-wave equations for the equations of viscoelasticity:

$$\tilde{n}_\xi = \sigma'(t)^2 \varepsilon \rho A \tilde{r}_{\xi\xi}, \tag{6.2}$$

$$\tilde{m}_\xi + \tilde{r}_\xi \times \tilde{n} = -\sigma'(t) \chi(\varepsilon, s)[(\rho I) \cdot \tilde{w}]_\xi, \tag{6.3}$$

$$\tilde{r}_\xi = \tilde{v}_k \tilde{d}_k, \qquad \partial_\xi \tilde{d}_k = \tilde{u} \times \tilde{d}_k, \qquad \tilde{w}_\xi = -\sigma'(t) \partial_\xi \tilde{u}_k \tilde{d}_k, \tag{6.4a,b,c}$$

where

$$\tilde{n} = \left[\frac{\partial \varphi}{\partial v_k} (\tilde{u}, \tilde{v}, \sigma(t)) + n_k^D \left(\tilde{u}, \tilde{v}, -\sigma'(t) \tilde{u}_\xi, -\sigma'(t) \tilde{v}_\xi, \sigma(t) \right) \right] \tilde{d}_k, \quad \text{etc.} \tag{6.5}$$

The variables of (6.2)–(6.4) satisfy limiting conditions

$$\tilde{u}(\pm\infty, t) = \lim_{s \to \sigma(t)\pm} \mathbf{u}(s, t), \quad \text{etc.,} \tag{6.6}$$

where $\mathbf{u}$ is a set of components of the solution of the hyperbolic problem. The right-hand side of (6.6) gives the limits from the left and the right of $\mathbf{u}$ at the shock. Equation (6.6) says that these values must constitute singular points for the system (6.2)–(6.5) of ordinary differential equations.

A shock, consisting of a specification of σ and of the limits from the left and the right of the functions entering the hyperbolic problem, is *admissible according to the traveling-wave criterion* if the system (6.2)–(6.6) has a unique solution. To prove that there are "connecting orbits" joining the singular points typically requires a careful topological study. For our problem, such a study would be somewhat simplified by the availability of the integrals

$$\tilde{n} = \sigma'(t)^2 \varepsilon \rho A \tilde{r}_\xi + a(t), \tag{6.7}$$

$$\tilde{m} + \tilde{r} \times a(t) = -\sigma'(t) \chi(\varepsilon, s)(\rho I) \cdot \tilde{w} + b(t), \tag{6.8}$$

where $a(t)$ and $b(t)$ are constant vectors of integration. (Further integrals are available when the equations enjoy certain symmetries; cf. [6]. This paper gives a rather complete treatment of traveling waves for elastic rods, a problem with a character very different from that considered here.)

The work of [7] (cf. [3, Sec. XVI.5]), restricted to pure shearing motions of an incompressible medium (which correspond to very special motions of a rod), shows that there can be infinitely many connecting orbits in the traveling-wave problem. Presumably the same is true for our more complicated system, whose traveling waves have not been analyzed. This means that the dissipative mechanism of viscosity by itself is incapable of delivering a definitive shock structure and therefore incapable of delivering a useful entropy condition. Thus we are faced with the open problem of supplementing viscosity with further dissipative mechanisms that yield a shock structure. (It is likely that different mechanisms yield different shock structures.)

7. Hopf Bifurcation Problems

We study the stability of a naturally straight rod under the combined action of a terminal follower couple and a terminal follower load: We retain the equations of Section 2, except that we replace the boundary conditions (2.13) and (2.14) with

$$n(1, t) = -\lambda d_3(1, t), \qquad m(1, t) = \mu d(1, t) \qquad (7.1)$$

where λ and μ are given constants. (Our problem is one of several possible nonlinear generalizations of the problems of Nikolai and Beck.)

We first look at equilibrium states, for which the material response is elastic. We assume that the constitutive equations have the form

$$u = \hat{u}(m, n, s), \qquad v = \hat{v}(m, n, s) \qquad (7.2)$$

where

$$\hat{u}_\alpha(0, 0, m_3, 0, 0, n_3, s) = 0 = \hat{v}_\alpha(0, 0, m_3, 0, 0, n_3, s), \qquad \alpha = 1, 2. \qquad (7.3)$$

(The uncoupling property (7.3) says that there are no shear and flexural strains when there are no shear forces and bending couples. This property holds for transversely isotropic rods, among others.)

Since we assume that there are no body forces or couples, the equilibrium form of (2.11) yields the integral

$$n(s) = n(1) = -\lambda d_3(1). \qquad (7.4)$$

We substitute (7.2), (7.4) into the componential version of (2.12) and into (2.3a) to get the system

$$\partial_s m_1 - m_2 \hat{u}_3 + m_3 \hat{u}_2 + \lambda [\hat{v}_2 d_3(s) - \hat{v}_3 d_2(s)] \cdot d_3(1) = 0,$$
$$\partial_s m_2 - m_3 \hat{u}_1 + m_1 \hat{u}_3 + \lambda [\hat{v}_3 d_1(s) - \hat{v}_1 d_3(s)] \cdot d_3(1) = 0,$$
$$\partial_s m_3 - m_1 \hat{u}_2 + m_2 \hat{u}_1 + \lambda [\hat{v}_1 d_2(s) - \hat{v}_2 d_1(s)] \cdot d_3(1) = 0, \qquad (7.5)$$

$$\partial_s d_1 = -\hat{u}_2 d_3 + \hat{u}_3 d_2,$$
$$\partial_s d_2 = -\hat{u}_3 d_1 + \hat{u}_1 d_3,$$
$$\partial_s d_3 = -\hat{u}_1 d_2 + \hat{u}_2 d_1, \tag{7.6}$$

where the arguments of the constitutive functions $\hat{u}_k$, $\hat{v}_k$ are

$$m_1, m_2, m_3, -\lambda d_1(s) \cdot d_3(1), -\lambda d_2(s) \cdot d_3(1), -\lambda d_3(s) \cdot d_3(1), s).$$

From (7.1) we read off boundary conditions for m:

$$m_1(1) = 0, \qquad m_2(1) = 0, \qquad m_3(1) = \mu. \tag{7.7}$$

We supplement these with tautological boundary conditions on the d_k:

$$d_k(1) = d_k(1). \tag{7.8}$$

Equations (7.5)–(7.8) constitute an initial-value problem for a system of ordinary differential equations. This problem has a unique solution, which, by virtue of (7.3), has the form

$$m_1 = 0, \qquad m_2 = 0, \qquad m_3 = \mu, \qquad d_3(s) = d_3(1) \tag{7.9a,b,c,d}$$

with d_1 and d_2 satisfying the linear system

$$\partial_s d_1 = \hat{u}_3(0, 0, \mu, 0, 0, -\lambda, s) d_2, \qquad \partial_s d_2 = -\hat{u}_3(0, 0, \mu, 0, 0, -\lambda, s) d_1. \tag{7.10}$$

(When $\hat{u}_3$ is independent of s, we easily get an explicit solution of (7.10).) From (2.9b) and (7.9d) we conclude that $d_3 = k$. Therefore, under our assumptions, *the only equilibrium states are those with straight axes.* (This treatment generalizes that of [5].)

Nikolai [29] and Beck [13] recognized that this uniqueness does not mean that the rod is stable for all values of the load parameters λ and μ: By formally studying the linearized equations for (inextensible, unshearable) elastic rods, they showed that stability is lost by dynamical processes.

Since we are dealing with quasilinear equations, the determination of stability is a more difficult task. It is shown in [25] that if the linearized problem for *viscoelastic* rods meets the usual spectral conditions, then the stability of the straight state is lost by a Hopf bifurcation. (Also see [47].) A source of difficulty in the proof is the complexity of boundary conditions like (7.1) that arise in mechanical problems. The determination of the dependence of the spectrum on the load parameters appears quite formidable, even for simplified versions of our problem (cf. [4], [14], [15]) and might require the use of carefully designed numerical procedures.

8. Control Problems

There is an extensive literature on the boundary control of linear hyperbolic equations. Boundary control is studied because it can be realized by practical devices. The linearity of the equations prevents the severe analytic difficulties attending the blowup of solutions of nonlinear hyperbolic equations.

New magnetoelastic materials developed over the last twenty years are being incorporated into new kinds of control devices, the analysis of which has a character quite different from that described in the preceding paragraph. In particular, there are materials for which the tangent elastic modulus can be varied by a factor of 10 by varying an ambient magnetic field through moderate values (see [31]). Here we briefly discuss some preliminary questions that arise in the control of magnetoelastic rods.

First we study the purely longitudinal motion of a magnetoelastic rod (under zero body force), which is governed by

$$\varepsilon \rho A w_{tt} = \hat{n}_3(w_s, h, s)_s \tag{8.1}$$

(cf. (2.24)) where h, the magnetic intensity, is the control variable. We conceive of each short segment of the rod as being wrapped with its own short solenoid, so that effectively the magnetic field at each material point can be controlled separately. Thus, in contrast to the situation described in the first paragraph of this section, we face a quasilinear hyperbolic equation with a body control.

A typical optimal control problem is the following: Given boundary conditions and initial conditions for (8.1), find an h as a function of s, t with values in a prescribed closed interval that brings the rod to rest in the least amount of time. We make no effort here to solve this problem. (Discrete versions of it with realistic constitutive functions are themselves quite challenging.) Rather, we solve a much simpler control problem, which indicates that there are controls h that prevent the quasilinearity of (8.1) from causing any difficulty:

A feedback control $\tilde{h}(w_s, s)$ that ensures that (8.1) transmits signals faithfully at a fixed speed a is any solution of an equation of the form

$$\hat{n}_3(w_s, h, s) = a^2 w_s + b \tag{8.2}$$

where a and b are any numbers for which (8.2) is solvable. Clearly (8.2) converts (8.1) to a linear wave equation. One can even construct more exotic feedbacks that spread the characteristics of (8.1) and thereby tend to reduce the magnitude of the derivative of the solution.

In principle, the same ideas apply to our elastic and viscoelastic rods: All we need to do is add to the arguments of the constitutive functions of (2.17) a variable $\mathbf{h}$ accounting for rod-theoretic effect of the magnetic field. (For suitable forms for $\mathbf{h}$, see [20].) We could then linearize the principal part of the differential operator, and we could conceivably diagonalize parts of it and make $\mathbf{n}$ parallel to r_s. The fundamental issue here is physical, not mathematical: Are there practical ways to control the field $\mathbf{h}$? (These could generalize the disposition of the solenoids described above.) For this purpose, one might exploit the many recently developed ways of embedding magnetoelastic fibers and solenoids in materials.

It was shown in [39] that given the first branch of a bifurcation diagram for the static buckling of an inextensible nonlinearly elastic beam, one can construct the corresponding constitutive function for the bending couple in terms of the curvature. In general, the resulting constitutive function does not describe any natural material. By using ideas like those associated with (8.2), one can synthesize the constitutive function by adjusting the magnetic field in response to the strain [38].

Acknowledgment

The research reported here was supported in part by the NSF and AFOSR.

References

[1] G. Andrews and J. M. Ball, Asymptotic behaviour and changes of phase in one-dimensional nonlinear viscoelasticity, *J. Diff. Eqs.* **44** (1982), 306–341.

[2] S. S. Antman, The paradoxical asymptotic status of massless springs, *SIAM J. Appl. Math.* **48** (1988), 1319–1334.

[3] S. S. Antman, *Nonlinear Problems of Elasticity*, Springer-Verlag, New York, 1995.

[4] S. S. Antman and H. Koch, Self-sustained oscillations of nonlinearly viscoelastic layers, to appear.

[5] S. S. Antman and C. S. Kenney, Large buckled states of nonlinearly elastic rods under torsion, thrust, and gravity, *Arch. Rational Mech. Anal.* **76** (1981), 289–338.

[6] S. S. Antman and T.-P. Liu, Travelling waves in hyperelastic rods, *Quart. Appl. Math.* **36** (1979), 377–399.

[7] S. S. Antman and R. Malek-Madani, Travelling waves in nonlinearly viscoelastic media and shock structure in elastic media, *Quart. Appl. Math.* **46** (1988), 77–93.

[8] S. S. Antman and R. S. Marlow, Material constraints, Lagrange multipliers, and compatibility, *Arch. Rational Mech. Anal.* **116** (1991), 257–299.

[9] S. S. Antman, R. S. Marlow, and C. P. Vlahacos, The complicated dynamics of heavy rigid bodies attached to light deformable rods, *Quart. Appl. Math.* **56** (1998), 431–460.

[10] S. S. Antman and T. I. Seidman, Quasilinear hyperbolic-parabolic equations of nonlinear viscoelasticity, *J. Diff. Eqs.* **124** (1996), 132–185.

[11] S. S. Antman and T. I. Seidman, Large shearing motions of nonlinearly viscoelastic slabs, *Bull. Tech. Univ. Istanbul* **47** (1994), 41–56.

[12] S. S. Antman and T. I. Seidman, Hyperbolic-parabolic systems governing the spatial motion of nonlinearly viscoelastic rods, to appear.

[13] M. Beck, Die Knicklast des einseitig eingespannten tangential gedrückten Stabes, *Z. Angew. Math. Phys.* **3** (1952), 225–228.

[14] J. Carr and M. Z. M. Malhardeen, Beck's problem, *SIAM J. Appl. Math.* **37** (1979), 261–262.

[15] M.-S. Chen, Hopf bifurcation in Beck's problem, *Nonlin. Anal. T. M. A.* **11** (1987), 1061–1073.

[16] A. Cimetière, G. Geymonat, H. Le Dret, A. Raoult, and Z. Tutek, Asymptotic theory and analysis for displacements and stress distribution in nonlinear elastic straight slender rods, *J. Elasticity* **19** (1988), 111–161.

[17] C. M. Dafermos, The mixed initial-boundary value problem for the equations of nonlinear one-dimensional viscoelasticity, *J. Diff. Eqs.* **6** (1969), 71–86.

[18] L. Euler, *Additamentum I de curvis elasticis, methodus inveniendi lineas curvas maximi minimivi proprietate gaudentes*, Bousquent, Lausanne, 1744, in *Opera Omnia* I, Vol. 24, 231–297.

[19] R. L. Fosdick and R. D. James, The elastica and the problem of pure bending for a non-convex stored energy function, *J. Elasticity* **11** (1981), 165–186.

[20] A. E. Green and P. M. Naghdi, Electromagnetic effects in the theory of rods, *Phil. Trans. R. Soc. Lond. A* **314** (1985), 311–352.

[21] J. M. Greenberg, R. C. MacCamy, and V. J. Mizel, On the existence, uniqueness, and stability of solutions of the equation $\sigma'(u_x)u_{xx} + \lambda u_{txt} = \rho_0 u_{tt}$, *J. Math. Mech.* **17** (1968), 707–728.

[22] R. D. James, The equilibrium and post-buckling behavior of an elastic curve governed by a non-convex energy, *J. Elasticity* **11** (1981), 239–269.

[23] Ya. I. Kanel', On a model system of equations of one-dimensional gas motion (in Russian), *Diff. Urav.* **4** (1969), 721–734; English transl: *Diff. Eqs.* **4** (1969), 374–380.

[24] A. J. Karwowski, Asymptotic models for a long, elastic cylinder, *J. Elasticity* **24** (1990), 229–287.

[25] H. Koch and S. S. Antman, Stability and Hopf bifurcation for fully nonlinear parabolic-hyperbolic equations, to appear.

[26] R. C. MacCamy, Existence, uniqueness and stability of $u_{tt} = \frac{\partial}{\partial x}[\sigma(u_x) + \lambda(u_x)u_{xt}]$, *Indiana Univ. Math. J.* **20** (1970), 231–238.

[27] A. Mielke, Saint-Venant's problem and semi-inverse solutions in nonlinear elasticity, *Arch. Rational Mech. Anal.* **102** (1988), 205–229; Corrigendum, ibid. **110** (1990), 351–352.

[28] A. Mielke, Normal hyperbolicity of center manifolds and Saint-Venant's principle, *Arch. Rational Mech. Anal.* **110** (1990), 353–372.

[29] E. L. Nikolai, On the stability of the straight equilibrium form of a compressed and twisted rod (in Russian), *Izv. Leningr. Politekh. Inst.* **31** (1928), 357–387; reprinted in E. L. Nikolai, *Works on Mechanics* (in Russian), G.I.T.T.L., 1955.

[30] R. L. Pego, Phase transitions in one-dimensional nonlinear viscoelasticity: Admissibility and stability, *Arch. Rational Mech. Anal.* **97** (1987), 353–394.

[31] H. T. Savage and M. L. Spano, Theory and application of highly magnetoelastic Metglas 2605SC, *J. Appl. Phys.* (1982), 8092–8097.

[32] J.-C. Simo, A finite strain beam formulation. The three-dimensional dynamical problem. Part I, *Comp. Meths. Appl. Mech. Eng.* **49** (1985), 55–70.

[33] J.-C. Simo and L. Vu-Quoc, Three-dimensional finite strain rod model. Part I: Computational aspects, *Comp. Meths. Appl. Mech. Eng.* **58** (1986), 79–116.

[34] J.-C. Simo and L. Vu-Quoc, On the dynamics of flexible beams under large overall motions—The plane case: Part I, Part II, *J. Appl. Mech.* **53** (1986), 849–863.

[35] J.-C. Simo and L. Vu-Quoc, On the dynamics in space of rods undergoing large motions—A geometrically exact approach, *Comp. Meths. Appl. Mech. Eng.* **66** (1988), 125–161.

[36] J.-C. Simo and L. Vu-Quoc, A geometrically exact rod model incorporating shear and torsion-warping deformation, *Int. J. Solids Structures* **27** (1991), 371–393.

[37] J. Smoller, *Shock Waves and Reaction-Diffusion Equations*, Springer-Verlag, New York, 1983.

[38] S. S. Antman, Synthesis of nonlinear constitutive functions. Applications to the electromagnetic control of snapping, to appear.

[39] S. S. Antman and C. L. Adler, Design of material properties that yield a prescribed global buckling response, *J. Appl. Mech.* **54** (1987), 263–268.

[40] S. S. Antman and F. Klaus, The shearing of nonlinearly viscoplastic slabs, *Nonlinear Problems in Applied Mathematics*, edited by T. Angell, L. P. Cook, R. Kleinman, and W. E. Olmstead, SIAM, 1996, pp. 20–29.

[41] S. S. Antman and M. Lanza de Cristoforis, Peculiar instabilities due to the clamping of shearable rods, *Int. J. Nonlin. Mech.* **32** (1997), 31–54.

[42] S. S. Antman and F. Schuricht, Incompressibility in rod and shell theories, *Math. Modelling Num. Anal.*, to appear.

[43] T. Frohman, Doctoral Dissertation, Univ. Maryland, in preparation.

[44] L. Trabucho and J. M. Viaño, Mathematical Modelling of Rods, in *Handbook of Numerical Analysis*, Vol. IV, P. G. Ciarlet and J.-L. Lions, eds., North-Holland, 1996.

[45] J. P. Wilber, Doctoral Dissertation, Univ. Maryland, in preparation.

[46] T. Wright, Nonlinear waves in rods: results for incompressible elastic materials, *Stud. Appl. Math.* **72** (1984), 149–160.

[47] C.-Y. Xu and J. E. Marsden, Asymptotic stability of equilibria of nonlinear semiflows with applications to rotating viscoelastic rods, Part 1, *Topol. Methods Nonlin. Anal.* **7** (1996), 271–297.

[48] S. C. Yip, *Asymptotic Analysis of Quasilinear Parabolic-Hyperbolic Equations describing the large Longitudinal Motion of a Light Viscoelastic Bar with a Heavy Attachment*, Doctoral Dissertation, Univ. Maryland, 1997.

The Limits of Hamiltonian Structures in Three-Dimensional Elasticity, Shells, and Rods

Z. Ge,[1] H.P. Kruse,[2] and J.E. Marsden[3]

[1] The Fields Institute for Research in Mathematical Sciences, 185 Columbia Street West, Waterloo, Ontario N2L 5Z5, e-mail: 92504gez@cdf.toronto.edu

[2] Zentrum Mathematik, TU München, Arcisstrasse 21, D-80290 München, Germany, e-mail: kruse@mathematik.tu-muenchen.de

[3] Control and Dynamical Systems, California Institute of Technology 107-81, Pasadena, CA 91125, e-mail: marsden@cds.caltech.edu

Received September 1, 1995 and in revised form October 15, 1995
Communicated by Stephen Wiggins

This paper is dedicated to the memory of Juan-Carlos Simo

Summary. This paper uses Hamiltonian structures to study the problem of the limit of three-dimensional (3D) elastic models to shell and rod models. In the case of shells, we show that the Hamiltonian structure for a three-dimensional elastic body converges, in a sense made precise, to that for a shell model described by a one-director Cosserat surface as the thickness goes to zero. We study limiting procedures that give rise to unconstrained as well as constrained Cosserat director models. The case of a rod is also considered and similar convergence results are established, with the limiting model being a geometrically exact director rod model (in the framework developed by Antman, Simo, and coworkers). The resulting model may or may not have constraints, depending on the nature of the constitutive relations and their behavior under the limiting procedure.

The closeness of Hamiltonian structures is measured by the closeness of Poisson brackets on certain classes of functions, as well as the Hamiltonians. This provides one way of justifying the dynamic one-director model for shells. Another way of stating the convergence result is that there is an almost-Poisson embedding from the phase space of the shell to the phase space of the 3D elastic body, which implies that, in the sense of Hamiltonian structures, the dynamics of the elastic body is close to that of the shell. The constitutive equations of the 3D model and their behavior as the thickness tends to zero dictates whether the limiting 2D model is a constrained or an unconstrained director model.

We apply our theory in the specific case of a 3D Saint Venant-Kirchhoff material and *derive* the corresponding limiting shell and rod theories. The limiting shell model is an interesting Kirchhoff-like shell model in which the stored energy function is explicitly derived in terms of the shell curvature. For rods, one gets (with an additional inextensibility constraint) a one-director Kirchhoff elastic rod model, which reduces to the well-known Euler elastica if one adds an additional single constraint that the director lines up with the Frenet frame.

1. Introduction

The Goals and Main Results

This paper studies the problem of convergence of three-dimensional (3D) elasticity models to corresponding models for shells and rods. In our approach we make use of the Hamiltonian structure as a crucial tool in the analysis.

In the case of shells, we show that the Poisson bracket (applied to certain classes of functions) for a 3D elastic body converges to the Poisson bracket for a shell model described by a one-director Cosserat surface when the thickness goes to zero. Alternatively, we prove, in a sense made precise later, that there is an almost-Poisson embedding from the phase space of the shell to the phase space of the 3D elastic body. We also establish the sense in which the Hamiltonians themselves converge. Taken together, this is what we mean by the convergence of Hamiltonian structures. This convergence implies that, in a certain weak sense, the dynamics of the elastic body is close to that of the shell. This provides one justification of the approximation of a thin body by the Cosserat surface in full generality, at least for time-evolution problems.

Starting with the Saint Venant-Kirchhoff constitutive model as a specific case to illustrate the general theory, we derive by our systematic procedure a Kirchhoff shell model in which the stored energy function is an explicit function of the mean and Gaussian curvature of the shell (see Section 4).

Our method also applies to the case of thin rods, as we show in Sections 5 and 6. In particular, we are able to get rather large classes of geometrically exact rod models. In particular, starting with the Saint Venant-Kirchhoff constitutive model, we derive the Euler-Kirchhoff model for an elastica by our procedures without making any intermediate ad hoc hypotheses.

Related Works and Background

The problem of convergence of the Hamiltonian structure for a 3D ideal fluid with a free boundary to that for the shallow water equation was considered in Ge, Kruse, Marsden, and Scovel [1995]. The approach in the present paper is an outgrowth of this previous work. That paper also discusses the general setting of the problem of convergence of Hamiltonian structures and elaborates on the meaning of the weak convergence. The main difficulty solved was how to deal with the incompressibility constraint, whereas in this paper the main analytic complication we overcome is how to deal with the limiting form of the constitutive relation. We do not consider the incompressible case here, although it can presumably be done by combining the two approaches.

Traditionally, there were two methods to derive an approximation of a thin elastic body (see, for example, Antman [1972, 1995]). The first one is using asymptotic analysis, which usually consists of expanding the solution and the equations using powers of the thickness of the plate. This has been applied to special models (cf. Ciarlet and Miara [1992]). A remarkable recent result is that of Fox, Raoult, and Simo [1993], who showed that for the Saint Venant-Kirchhoff material, the membrane model, the inextensional model, and the von Kármán model are all limits of 3D elasticity as the thickness $2\epsilon \to 0$ by choosing different orders of dependence on ϵ for various components of the loading. It is of interest to generalize these ideas to materials with more general constitutive relations.

The second method, sometimes called the projection-constraint method (cf. Antman [1972]), is a Galerkin-type method and has been extensively used. For example, the shell model described by a one-director Cosserat surface can be derived by such a method. However, this method is not easy to justify by means of asymptotic analysis.

More recently, the theory of Γ-convergence has been applied successfully to time-independent problems in elasticity in Le Dret and Raoult [1995]. It would be of interest to see if these ideas can be applied to time-evolutionary problems.

We should mention explicitly that while we do examine the formal asymptotics of the problem and tie this to the Hamiltonian structure, we do not attempt to prove here the strong convergence of solutions in any sense. However, we think that the techniques of Marsden, Ratiu, and Raugel [1995] will be useful in this problem, and this will be a subject of future investigation.

The Approach in the Present Paper

We take a different approach than previous authors, namely to study the convergence of the Poisson bracket, or alternatively to show that the natural embedding from the phase space of the Cosserat surface to that of a thin body is an almost-Poisson map. One reason for this approach is that we are mainly concerned with the time-evolution problem, not with the equilibrium problem.

Poisson brackets are a useful tool for studying evolution problems since the equations of motion can be written in Poisson bracket form (see, for example, Marsden and Hughes [1983], Simo, Marsden, and Krishnaprasad [1988]). Including them in the asymptotic analysis represents one of our main contributions augmenting previous approaches. We should also mention that the approach to asymptotics by incorporating Hamiltonian structures was central to the important work of Camassa and Holm [1993] and related papers.

Our method combines Hamiltonian structures with the projection-constraint method and asymptotics. In particular, we prove convergence of the equations in a weak sense using a calculation similar to that in the projection method. Our method incorporates the boundary conditions into the Hamiltonian structure in a natural way (as in the so-called natural boundary conditions in variational problems) rather than applying (rescaled) forces. Also, we can work either with the original domain or we can rescale the thin domain into one that is independent of ϵ. John [1971] dealt with plates having no loadings and periodic boundary conditions in the plate directions. We note that Antman and Warner [1967] and John [1971] work with a thin domain that is not rescaled, while most other authors that deal with the asymptotics do rescale the domain. Our method works either with or without rescaling; retaining this flexibility is quite useful.

The type of rescaling done determines, in part, the model that one gets in the limit. One of our main points is that all of this structure can be seen as various assumptions on the Hamiltonian structure. In general terms, we can divide the theories one gets as being constrained or unconstrained. Here, constrained means that there is a relation determined between the director variables and the position variables. In other approaches, the assumptions on the asymptotics that are needed to get the various theories are achieved by means of different scalings in different variables and in the external loadings (cf. Fox, Raoult, and Simo [1993]). In our approach one sees the same thing by simple assumptions on the scalings that appear directly in the Hamiltonian. In particular, with a simple assumption on the scaling of the director variables, one gets the *general* unconstrained director theory.

The Set-up for Shells

For the case of shells, we consider the motion of a thin elastic body with a three-dimensional reference configuration of the form[†]

$$D_\epsilon := \Omega \times [-\epsilon, \epsilon], \tag{1.1}$$

where Ω is a bounded domain in $\mathbb{R}^2$. We let x denote the variables in the domain Ω and let y denote the variable in the transverse direction, so that $-\epsilon \leq y \leq \epsilon$. Let ϕ denote a configuration of the body, that is, ϕ is a map of D_ϵ to $\mathbb{R}^3$ and we consider a three-dimensional Hamiltonian of the form

$$H(\phi, \dot{\phi}, \epsilon) := \iiint_{D_\epsilon} \left(\frac{1}{2} \dot{\phi} \cdot \dot{\phi} + g(x, y, \phi, \phi_x, \phi_y, \epsilon) \right) d^2x \, dy, \tag{1.2}$$

where "$\cdot$" stands for the standard inner product in $\mathbb{R}^3$ and the overdot denotes the time derivative. We write $d^2x \, dy = dV$ for the standard Euclidean volume element. The equations of motion are given by Hamilton's equations for this Hamiltonian or, equivalently, the Euler-Lagrange equations for the corresponding Lagrangian (see, for example, Marsden and Hughes [1994]). Let $v := \phi_y$. We will show that if

$$\frac{\partial g}{\partial v} = O(\epsilon^2), \tag{1.3}$$

then the limiting 2D system as $\epsilon \to 0$ is an unconstrained Cosserat model. Notice that the 3D model is in this case necessarily *anisotropic*. Otherwise, the limiting process will lead

---

[†] Even though this reference configuration has a flat midsurface, we will shortly be taking approximations about a nontrivial configuration with a nonflat midsurface. Thus, we are considering shells here and not necessarily the special case of plates.

in general to constraints on the director and the displacement; that is, the limiting system is a constrained director model. Under the additional assumption that $\partial^2 g/\partial v^2$ does not vanish anywhere, we can use the implicit function theorem to express the director field as a function of the displacement field. The 2D shell models studied in Ciarlet and Lods [1994], Ciarlet, Lods, and Miara [1994], and Ciarlet [1994] fall into this category.

We show that there is an almost-Poisson map from the phase space of the constrained-director model to the phase space of the unconstrained-director model. Thus, both the unconstrained one-director model and the constrained-director model are compatible.

The Hamiltonian structure, namely the Poisson bracket, the energy, and the conserved quantities, has also been important in the long time computation of conservative systems (cf. Ge and Marsden [1988], Ge [1991], and Ge and Scovel [1994]). Simo and his collaborators already show that the Hamiltonian structure plays an important role in the numerical computation of the dynamics of rods and shells (compare Simo et al. [1992]). This structure has also been important in stability theory for elasticity; see, for example, Simo, Posbergh, and Marsden [1990, 1991] and Maddocks [1984, 1991]. We also note that Foltinek [1994] shows how to get integrals for the equations of the Euler elastica using symmetry and momentum maps.

As an application of the methods developed in this paper, we derive the membrane model and various inextensional models (for shells and rods) from the 3D Saint Venant-Kirchhoff model.

Organization of the Paper

In Section 2 we discuss the relation between the canonical 3D Poisson bracket and an ϵ-dependent 2D bracket for one-director shells, where, as above, 2ϵ is the shell thickness. The results in this section are used later in the paper to establish the limit theorems as the thickness tends to zero. In particular, in Section 2.4 we show that the natural embedding from the phase space of the one-director shell model to the 3D model is an almost-Poisson embedding. In Section 2.5 we discuss the boundary conditions through a study of the dynamics of the limiting 2D shell model. An example of unconstrained-director models is discussed.

In Section 3 we introduce the constrained-director model for shells and show that the natural embedding from the phase space of the constrained-director model to that of the one-director model is almost-Poisson. In Section 3.2 we give several examples of constrained-director models. In particular, for linear plate theory, we obtain the membrane model (Ciarlet [1994]) as the constrained-director model. We also show how a geometrically exact shell model can be obtained as the limit of a 3D Saint Venant-Kirchhoff material.

In Section 4 we show how to derive the Kirchhoff shell as a limit of a 3D Saint Venant-Kirchhoff material. We discuss constraints that arise from the limiting procedures and how they enter into the models as well as other constraints such as inextensibility that are imposed as holonomic constraints.

In Sections 5 and 6 we discuss most of these same aspects for rod models. In particular, we show how to obtain the Kirchhoff elastic rod model as a limit of a 3D Saint Venant-Kirchhoff material.

Remarks on Notation

For shells, we use x to denote a pair of variables, say (x_1, x_2), which parameterize the reference middle surface of the shell, and y for a variable transverse to the shell. The placement field is denoted ϕ for the corresponding three-dimensional elastic body and φ for the middle surface of the shell. The director is denoted by w.

For rods, we use z to denote the variables that parameterize the reference central line of the rod and x, y for variables transverse to the rod. The placement field for the corresponding three-dimensional elastic body is denoted ϕ and that for the central line of the rod is denoted φ. The directors are denoted by w_1 and w_2.

2. The Convergence of Three-Dimensional Elasticity to Shells

2.1. Three-Dimensional Elastic Bodies

With the reference configuration D_ϵ as defined by (1.1) above, we introduce further notation

$$\Omega_u := \Omega \times \{\epsilon\},$$

$$\Omega_l := \Omega \times \{-\epsilon\},$$

$$\Omega_s := \partial\Omega \times [-\epsilon, \epsilon],$$

which denote the upper, lower, and lateral boundary of D_ϵ, respectively. The material (or Lagrangian) configuration space of the body is

$$\mathcal{M}_{3d}^\epsilon := \{\phi \mid \phi \colon \Omega \times [-\epsilon, \epsilon] \to \mathbb{R}^3 \text{ is an embedding}\}.$$

For now we do not specify the precise smoothness class in this definition and those to follow. In a specific context these can be dealt with as in Marsden and Hughes [1994]; we hope to return in a later paper to these points in connection with existence theory. The Lagrangian phase space is the tangent bundle of $\mathcal{M}_{3d}^\epsilon$, namely

$$T\mathcal{M}_{3d}^\epsilon = \{(\phi, \dot\phi) \mid \phi \in \mathcal{M}_{3d}^\epsilon \text{ and } \dot\phi \colon \Omega \times [-\epsilon, \epsilon] \to \mathbb{R}^3\}.$$

On $\mathcal{M}_{3d}^\epsilon$ we consider the Riemannian metric defined by

$$\langle\langle(\phi, \dot\phi_1), (\phi, \dot\phi_2)\rangle\rangle_{3d}^\epsilon = \iiint_{D_\epsilon} \dot\phi_1 \cdot \dot\phi_2 \, dV. \tag{2.1.1}$$

We identify $T^*\mathcal{M}_{3d}^\epsilon$ and $T\mathcal{M}_{3d}^\epsilon$ via this metric. We will use the standard canonical cotangent bracket for functionals F_1, F_2 on $T^*\mathcal{M}_{3d}^\epsilon$ and denote it by $\{F_1, F_2\}_{3d\,can}^\epsilon$. Note that this depends on ϵ since the domain does. The induced bracket on $T\mathcal{M}_{3d}^\epsilon$ obtained via the Legendre transformation is denoted $\{F_1, F_2\}_{3d}^\epsilon$.

In this section we consider hyperelastic materials whose corresponding energy functionals are defined on $T\mathcal{M}_{3d}^{\epsilon}$ of the form

$$
\begin{aligned}
F = &\iint_\Omega \int_{-\epsilon}^{\epsilon} f(x, y, \phi, \partial_x\phi, \partial_y\phi, \dot\phi, \epsilon)\, d^2x\, dy \\
&+ \iint_{\Omega_u} \epsilon f_u(x, \phi, \partial_x\phi, \partial_y\phi, \epsilon)\, d^2x \\
&+ \iint_{\Omega_l} \epsilon f_l(x, \phi, \partial_x\phi, \partial_y\phi, \epsilon)\, d^2x \\
&+ \iint_{\Omega_s} f_s(s_1, y, \phi, \partial_s\phi, \partial_y\phi, \epsilon)\, dA
\end{aligned}
\tag{2.1.2}
$$

where $s = (s_1, s_2)$ defines coordinates in a neighborhood of the curve $\partial\Omega$, such that the curve itself is parameterized by s_1, and dA is the area element induced on Ω_s. The functions f, f_u, f_l, f_s depend on ϵ as a parameter generally. Note that, in particular, the elastic energy is of this form, where the first term corresponds to the stored energy and the body force, and the remaining terms correspond to energy terms giving rise to surface forces along the upper, lower, and lateral surfaces respectively.

2.2. Unconstrained Elastic Shells

Now we turn to the approximation by the one-director model described by a Cosserat surface in the limit $\epsilon \to 0$. Intuitively, when $y \in [-\epsilon, \epsilon]$ is small, a configuration $\phi(x, y)$ can be expanded in powers of y:

$$
\phi(x, y) = \phi(x, 0) + \frac{\partial\phi}{\partial y}(x, 0)y + O(y^2).
$$

Denote

$$
\varphi(x) := \phi(x, 0) \qquad \text{and} \qquad w(x) := \frac{\partial\phi}{\partial y}(x, 0).
$$

For the approximation, we consider the space of maps affine in y, namely those of the form $\varphi(x) + yw(x)$, which we can identify with the space of pairs

$$
\mathcal{M}_{2d} = \{(\varphi, w) \mid \varphi\colon \Omega \to \mathbb{R}^3 \text{ is an embedding and } w\colon \Omega \to \mathbb{R}^3\}.
$$

Here w is usually called the director field, as in Naghdi [1972]. The tangent bundle of $\mathcal{M}_{2d}$ is given by

$$
T\mathcal{M}_{2d} = \{(\varphi, w, \dot\varphi, \dot w) \mid (\varphi, w) \in \mathcal{M}_{2d} \text{ and } \dot\varphi, \dot w\colon \Omega \to \mathbb{R}^3\}.
$$

If we take a higher order truncation of ϕ, we obtain a model with more than one director field. As we will see, an appropriate ϵ-dependent Poisson bracket on this space will be an approximation to the 3D Poisson bracket. When one is dealing with the Hamiltonian formulation of elasticity, one must include boundary conditions in a standard way (see John [1971] and Marsden and Hughes [1994]). Those conditions do not play an important role in this section but will be important in Sections 3 and 4.

Define a map

$$A_\epsilon: T\mathcal{M}_{2d} \to T\mathcal{M}_{3d}^\epsilon$$

by

$$A_\epsilon(\varphi(x), w(x), \dot\varphi(x), \dot w(x)) = (\varphi(x) + yw(x), \dot\varphi(x) + y\dot w(x)), \qquad y \in [-\epsilon, \epsilon].$$

If we are given a functional of the form (2.1.2), which includes the energy for 3D elasticity, the corresponding approximate functional is obtained by substituting $\phi(x, y) = \varphi(x) + yw(x)$, $\dot\phi(x, y) = \dot\varphi(x) + y\dot w(x)$ in (2.1.2), that is,

$$F \circ A_\epsilon$$

$$= \iint_\Omega \int_{-\epsilon}^\epsilon f(x, y, \varphi(x) + yw(x), \partial_x\varphi(x) + y\partial_x w(x), w(x), \dot\varphi(x) + y\dot w(x), \epsilon)\, d^2x\, dy$$

$$+ \iint_{\Omega_u} \epsilon f_u(x, \varphi(x) + \epsilon w(x), \partial_x\varphi(x) + \epsilon\partial_x w(x), w(x))\, d^2x$$

$$+ \iint_{\Omega_l} \epsilon f_l(x, \varphi(x) \dot- \epsilon w(x), \partial_x\varphi(x) - \epsilon\partial_x w(x), w(x))\, d^2x$$

$$+ \iint_{\Omega_s} f_s(s_1, y, \varphi(s_1) + yw(s_1), \partial_s\varphi(s_1) + y\partial_s w(s_1), w(s_1))\, dA.$$

For example, letting $U = \dot\varphi + y\dot w$, the approximate functional associated with the kinetic energy functional is given by the induced kinetic energy functional

$$\frac{1}{2} \langle\!\langle (\dot\varphi, \dot w), (\dot\varphi, \dot w) \rangle\!\rangle_{2d}^\epsilon := \frac{1}{2} \langle\!\langle U, U \rangle\!\rangle_{3d}^\epsilon$$

$$= \iint_\Omega \left(\epsilon(\dot\varphi \cdot \dot\varphi) + \frac{\epsilon^3}{3}(\dot w \cdot \dot w) \right) d^2x. \qquad (2.2.1)$$

Now we compute the Poisson bracket for 2D elasticity. First we identify $T\mathcal{M}_{2d}$ with $T^*\mathcal{M}_{2d}$ via the Riemannian metric (2.2.1) and obtain the ϵ-dependent Legendre transformation

$$p_1 = 2\epsilon\dot\varphi, \qquad p_2 = \frac{2\epsilon^3}{3}\dot w.$$

Here p_1, p_2 are the generalized conjugate momenta to φ, w respectively. We use, as before, the canonical bracket on $T^*\mathcal{M}_{2d}$ and denote it by $\{F_1, F_2\}_{2d\,\mathrm{can}}$. The bracket induced on $T\mathcal{M}_{2d}$ via the Legendre transform will be denoted by $\{F_1, F_2\}_{2d}^\epsilon$. A short calculation shows that this Poisson bracket is given by

$$\{F_1, F_2\}_{2d}^\epsilon = \frac{1}{2\epsilon} D_\varphi F_1 \frac{\delta F_2}{\delta\dot\varphi} + \frac{3}{2\epsilon^3} D_w F_1 \frac{\delta F_2}{\delta\dot w} - \frac{1}{2\epsilon} D_\varphi F_2 \frac{\delta F_1}{\delta\dot\varphi} - \frac{3}{2\epsilon^3} D_w F_2 \frac{\delta F_1}{\delta\dot w}. \qquad (2.2.2)$$

Notice that this bracket contains singular terms of order $1/\epsilon^3$. In particular, one cannot take the limit $\epsilon = 0$.

A possible way to resolve this singularity is to introduce a change of variable $w \mapsto \tilde w = w\epsilon$; then, in the rescaled coordinates, the bracket has the form

$$\{F_1, F_2\}_{2d}^\epsilon = \frac{1}{2\epsilon} D_\varphi F_1 \frac{\delta F_2}{\delta\dot\varphi} + \frac{3}{2\epsilon} D_{\tilde w} F_1 \frac{\delta F_2}{\delta\dot{\tilde w}} - \frac{1}{2\epsilon} D_\varphi F_2 \frac{\delta F_1}{\delta\dot\varphi} - \frac{3}{2\epsilon} D_{\tilde w} F_2 \frac{\delta F_1}{\delta\dot{\tilde w}}. \qquad (2.2.3)$$

This change of coordinates corresponds to a change of variables $y \mapsto \bar{y} = y/\epsilon$ in the reference configuration of the 3D model. After this rescaling, the singularity in the Poisson tensor has been resolved (up to an overall scale factor). However, the Hamiltonian will have the form

$$\iint_{\Omega} \left[f\left(x, 0, \varphi, \frac{\partial \varphi}{\partial x}, \frac{1}{\epsilon}\bar{w}, \frac{1}{\epsilon}\dot{\bar{w}}\right) + O(\epsilon)\right] d^2x, \tag{2.2.4}$$

which in general does not converge as $\epsilon \to 0$ unless one makes special assumptions on the behavior of the function f "at infinity." In Section 2.6 we will consider Hamiltonians that do converge after rescaling as $\epsilon \to 0$; restriction of the Poisson bracket to such functions leads to unconstrained-director models.

Another way to eliminate the singularity is to introduce holonomic constraints on the director vector field, which is the content of Section 3.

2.3. Approximation of Poisson Brackets

Let F_i, $i = 1, 2$, be two functionals of the form (2.1.2). In this section we prove the following theorem:

Theorem 2.1. *Let F_1 and F_2 be two functionals of the form (2.1.2) and $F_i \circ A_\epsilon$ their pull-backs via the embedding A_ϵ, which we will call the approximation embedding. Let $(\phi, \dot{\phi}) \in T\mathcal{M}_{3d}^\epsilon$, and let $\varphi(x) = \phi(x, 0)$, $w = \frac{\partial}{\partial y}\phi(x, 0)$, $\dot{\varphi}(x) = \dot{\phi}(x, 0)$, $\dot{w} = \frac{\partial}{\partial y}\dot{\phi}(x, 0)$. Then*

$$\{F_1, F_2\}_{3d}^\epsilon(\phi, \dot{\phi}) = \{F_1 \circ A_\epsilon, F_2 \circ A_\epsilon\}_{2d}^\epsilon(\varphi, w, \dot{\varphi}, \dot{w}) + O(\epsilon^3).$$

Proof. The derivatives of F_1 and F_2 are (the index i is omitted for the moment)

$$D_\phi F \cdot \delta\phi = \iiint_{\Omega \times [-\epsilon, \epsilon]} \left(\frac{\partial f}{\partial \phi}\delta\phi + \frac{\partial f}{\partial u}\partial_x\delta\phi + \frac{\partial f}{\partial v}\partial_y\delta\phi\right) d^2x\, dy$$

$$+ \epsilon \iint_{\Omega_u} \left(\frac{\partial f_u}{\partial \phi}\delta\phi + \frac{\partial f_u}{\partial u}\partial_x\delta\phi + \frac{\partial f_u}{\partial v}\partial_y\delta\phi\right) d^2x$$

$$+ \epsilon \iint_{\Omega_l} \left(\frac{\partial f_l}{\partial \phi}\delta\phi + \frac{\partial f_l}{\partial u}\partial_x\delta\phi + \frac{\partial f_l}{\partial v}\partial_y\delta\phi\right) d^2x$$

$$+ \iint_{\Omega_s} \left(\frac{\partial f_s}{\partial \phi}\delta\phi + \frac{\partial f_s}{\partial m}\partial_s\delta\phi + \frac{\partial f_s}{\partial v}\partial_y\delta\phi\right) dA,$$

where $u = \partial_x\phi$, $v = \partial_y\phi$, $m = \partial_s\phi$, and

$$D_{\dot{\phi}}F \cdot \delta\dot{\phi} = \iiint_{\Omega \times [-\epsilon, \epsilon]} \left(\frac{\partial f}{\partial \dot{\phi}} \cdot \delta\dot{\phi}\right) d^2x\, dy.$$

Thus, the 3D bracket is given by

$$
\begin{aligned}
\{F_1, F_2\}_{3d}^{\epsilon} = &\iiint_{\Omega\times[-\epsilon,\epsilon]} \left(\frac{\partial f_1}{\partial \phi}\frac{\partial f_2}{\partial \dot{\phi}} + \frac{\partial f_1}{\partial u}\,\partial_x\frac{\partial f_2}{\partial \dot{\phi}} + \frac{\partial f_1}{\partial v}\,\partial_y\frac{\partial f_2}{\partial \dot{\phi}} \right) d^2x\, dy \\
&- \iiint_{\Omega\times[-\epsilon,\epsilon]} \left(\frac{\partial f_2}{\partial \phi}\frac{\partial f_1}{\partial \dot{\phi}} + \frac{\partial f_2}{\partial u}\,\partial_x\frac{\partial f_1}{\partial \dot{\phi}} + \frac{\partial f_2}{\partial v}\,\partial_y\frac{\partial f_1}{\partial \dot{\phi}} \right) d^2x\, dy \\
&+ \epsilon \iint_{\Omega_u} \left(\frac{\partial f_{1,u}}{\partial \phi}\frac{\partial f_2}{\partial \dot{\phi}} + \frac{\partial f_{1,u}}{\partial u}\,\partial_x\frac{\partial f_2}{\partial \dot{\phi}} + \frac{\partial f_{1,u}}{\partial v}\,\partial_y\frac{\partial f_2}{\partial \dot{\phi}} \right) d^2x \\
&- \epsilon \iint_{\Omega_u} \left(\frac{\partial f_{2,u}}{\partial \phi}\frac{\partial f_1}{\partial \dot{\phi}} + \frac{\partial f_{2,u}}{\partial u}\,\partial_x\frac{\partial f_1}{\partial \dot{\phi}} + \frac{\partial f_{2,u}}{\partial v}\,\partial_y\frac{\partial f_1}{\partial \dot{\phi}} \right) d^2x \\
&+ \epsilon \iint_{\Omega_l} \left(\frac{\partial f_{1,l}}{\partial \phi}\frac{\partial f_2}{\partial \dot{\phi}} + \frac{\partial f_{1,l}}{\partial u}\,\partial_x\frac{\partial f_2}{\partial \dot{\phi}} + \frac{\partial f_{1,l}}{\partial v}\,\partial_y\frac{\partial f_2}{\partial \dot{\phi}} \right) d^2x \\
&- \epsilon \iint_{\Omega_l} \left(\frac{\partial f_{2,l}}{\partial \phi}\frac{\partial f_1}{\partial \dot{\phi}} + \frac{\partial f_{2,l}}{\partial u}\,\partial_x\frac{\partial f_1}{\partial \dot{\phi}} + \frac{\partial f_{2,l}}{\partial v}\,\partial_y\frac{\partial f_1}{\partial \dot{\phi}} \right) d^2x \\
&+ \iint_{\Omega_s} \left(\frac{\partial f_{1,s}}{\partial \phi}\frac{\partial f_2}{\partial \dot{\phi}} + \frac{\partial f_{1,s}}{\partial m}\,\partial_s\frac{\partial f_2}{\partial \dot{\phi}} + \frac{\partial f_{1,s}}{\partial v}\,\partial_y\frac{\partial f_2}{\partial \dot{\phi}} \right) dA \\
&- \iint_{\Omega_s} \left(\frac{\partial f_{2,s}}{\partial \phi}\frac{\partial f_1}{\partial \dot{\phi}} + \frac{\partial f_{2,s}}{\partial m}\,\partial_s\frac{\partial f_1}{\partial \dot{\phi}} + \frac{\partial f_{2,s}}{\partial v}\,\partial_y\frac{\partial f_1}{\partial \dot{\phi}} \right) dA. \qquad (2.3.1)
\end{aligned}
$$

In the integrand we expand the solution in powers of y. First note that after making use of the equality $\phi = \varphi + yw + O(y^2)$, we have

$$
\partial_y \frac{\partial f}{\partial \dot{\phi}} = \frac{\partial^2 f}{\partial y \partial \dot{\phi}} + \frac{\partial^2 f}{\partial \phi \partial \dot{\phi}}w + \frac{\partial^2 f}{\partial u \partial \dot{\phi}}\partial_x w + \frac{\partial^2 f}{\partial \dot{\phi}^2}\dot{w} + O(y).
$$

Thus, the Poisson bracket (2.3.1) becomes

$\{F_1, F_2\}_{3d}^{\epsilon}$

$$
\begin{aligned}
= &\, 2\epsilon \iint_{\Omega} \left[\frac{\partial f_1}{\partial \phi}\frac{\partial f_2}{\partial \dot{\phi}} + \frac{\partial f_1}{\partial u}\,\partial_x\frac{\partial f_2}{\partial \dot{\phi}} + \frac{\partial f_1}{\partial v}\left(\frac{\partial^2 f_2}{\partial y \partial \dot{\phi}} + \frac{\partial^2 f_2}{\partial \phi \partial \dot{\phi}}w + \frac{\partial^2 f_2}{\partial u \partial \dot{\phi}}\partial_x w + \frac{\partial^2 f_2}{\partial \dot{\phi}^2}\dot{w} \right) \right] d^2x \\
&- 2\epsilon \iint_{\Omega} \left[\frac{\partial f_2}{\partial \phi}\frac{\partial f_1}{\partial \dot{\phi}} + \frac{\partial f_2}{\partial u}\,\partial_x\frac{\partial f_1}{\partial \dot{\phi}} + \frac{\partial f_2}{\partial v}\left(\frac{\partial^2 f_1}{\partial y \partial \dot{\phi}} + \frac{\partial^2 f_1}{\partial \phi \partial \dot{\phi}}w + \frac{\partial^2 f_1}{\partial u \partial \dot{\phi}}\partial_x w + \frac{\partial^2 f_2}{\partial \dot{\phi}^2}\dot{w} \right) \right] d^2x \\
&+ \epsilon \iint_{\Omega} \left[\frac{\partial f_{1,u}}{\partial \phi}\frac{\partial f_2}{\partial \dot{\phi}} + \frac{\partial f_{1,u}}{\partial u}\,\partial_x\frac{\partial f_2}{\partial \dot{\phi}} + \frac{\partial f_{1,u}}{\partial v}\left(\frac{\partial^2 f_2}{\partial y \partial \dot{\phi}} + \frac{\partial^2 f_2}{\partial \phi \partial \dot{\phi}}w + \frac{\partial^2 f_2}{\partial u \partial \dot{\phi}}\partial_x w + \frac{\partial^2 f_2}{\partial \dot{\phi}^2}\dot{w} \right) \right] d^2x \\
&- \epsilon \iint_{\Omega} \left[\frac{\partial f_{2,u}}{\partial \phi}\frac{\partial f_1}{\partial \dot{\phi}} + \frac{\partial f_{2,u}}{\partial u}\,\partial_x\frac{\partial f_1}{\partial \dot{\phi}} + \frac{\partial f_{2,u}}{\partial v}\left(\frac{\partial^2 f_1}{\partial y \partial \dot{\phi}} + \frac{\partial^2 f_1}{\partial \phi \partial \dot{\phi}}w + \frac{\partial^2 f_1}{\partial u \partial \dot{\phi}}\partial_x w + \frac{\partial^2 f_2}{\partial \dot{\phi}^2}\dot{w} \right) \right] d^2x \\
&+ \epsilon \iint_{\Omega} \left[\frac{\partial f_{1,l}}{\partial \phi}\frac{\partial f_2}{\partial \dot{\phi}} + \frac{\partial f_{1,l}}{\partial u}\,\partial_x\frac{\partial f_2}{\partial \dot{\phi}} + \frac{\partial f_{1,l}}{\partial v}\left(\frac{\partial^2 f_2}{\partial y \partial \dot{\phi}} + \frac{\partial^2 f_2}{\partial \phi \partial \dot{\phi}}w + \frac{\partial^2 f_2}{\partial u \partial \dot{\phi}}\partial_x w + \frac{\partial^2 f_2}{\partial \dot{\phi}^2}\dot{w} \right) \right] d^2x
\end{aligned}
$$

$$-\epsilon \iint_\Omega \left[\frac{\partial f_{2,l}}{\partial \phi} \frac{\partial f_1}{\partial \phi} + \frac{\partial f_{2,l}}{\partial u} \partial_x \frac{\partial f_1}{\partial \dot\phi} + \frac{\partial f_{2,l}}{\partial v} \left(\frac{\partial^2 f_1}{\partial y \partial \dot\phi} + \frac{\partial^2 f_1}{\partial \phi \partial \dot\phi} w + \frac{\partial^2 f_1}{\partial u \partial \dot\phi} \partial_x w + \frac{\partial^2 f_1}{\partial \dot\phi^2} \dot w \right) \right] d^2 x$$

$$+ 2\epsilon \iint_{\partial\Omega} \left[\frac{\partial f_{1,s}}{\partial \phi} \frac{\partial f_2}{\partial \phi} + \frac{\partial f_{1,s}}{\partial m} \partial_s \frac{\partial f_2}{\partial \dot\phi} + \frac{\partial f_{1,s}}{\partial v} \left(\frac{\partial^2 f_2}{\partial y \partial \dot\phi} + \frac{\partial^2 f_2}{\partial \phi \partial \dot\phi} w + \frac{\partial^2 f_2}{\partial u \partial \dot\phi} \partial_x w + \frac{\partial^2 f_2}{\partial \dot\phi^2} \dot w \right) \right] ds_1$$

$$- 2\epsilon \int_{\partial\Omega} \left[\frac{\partial f_{2,s}}{\partial \phi} \frac{\partial f_1}{\partial \phi} + \frac{\partial f_{2,s}}{\partial m} \partial_s \frac{\partial f_1}{\partial \dot\phi} + \frac{\partial f_{2,s}}{\partial v} \left(\frac{\partial^2 f_1}{\partial y \partial \dot\phi} + \frac{\partial^2 f_1}{\partial \phi \partial \dot\phi} w + \frac{\partial^2 f_1}{\partial u \partial \dot\phi} \partial_x w + \frac{\partial^2 f_1}{\partial \dot\phi^2} \dot w \right) \right] ds_1$$

$$+ O(\epsilon^3),$$

where the functions are evaluated at $(x, y, \phi, u, v, \dot\phi) = (x, 0, \varphi, \partial_x\varphi, w, \dot\varphi)$.

Next we compute the 2D brackets of $F_1 \circ A_\epsilon$ and $F_2 \circ A_\epsilon$. For functions of the form introduced earlier, which includes the 2D energy, we have

$$F \circ A_\epsilon = 2\epsilon \iint_\Omega f(x, 0, \varphi(x), \partial_x\varphi(x), w(x), \dot\varphi, \epsilon) \, d^2 x + \epsilon \iint_\Omega f_u(x, \varphi, \partial_x\varphi, w) \, d^2 x$$

$$+ \epsilon \iint_\Omega f_l(x, \varphi, \partial_x\varphi, w) \, d^2 x + 2\epsilon \int_{\partial\Omega} f_s(s_1, 0, \varphi, \partial_s\varphi, w) \, ds_1$$

$$+ \text{ terms of order } \epsilon^2 \text{ independent of } \dot\phi \text{ and } \dot w$$

$$+ \frac{2\epsilon^3}{3} \iint_\Omega \left[\left(\frac{\partial}{\partial y} + w \frac{\partial}{\partial \phi} + \partial_x w \frac{\partial}{\partial u} + \dot w \frac{\partial}{\partial \dot\phi} \right)^2 f(x, 0, \varphi, \partial_x\varphi, \dot\varphi) \right] d^2 x + O(\epsilon^5).$$

Thus, we obtain an asymptotic formula

$$\frac{\partial(F \circ A_\epsilon)}{\partial \dot w} = 4 \frac{\epsilon^3}{3} \left(\frac{\partial}{\partial y} + w \frac{\partial}{\partial \phi} + \partial_x w \frac{\partial}{\partial u} + \dot w \frac{\partial}{\partial \dot\phi} \right) \frac{\partial}{\partial \dot\phi} f(x, 0, \varphi, \partial_x\varphi, w, \dot\varphi) + O(\epsilon^5).$$

Hence, we see that $\{F_1 \circ A_\epsilon, F_2 \circ A_\epsilon\}_{2D}^\epsilon$ is equal to $\{F_1, F_2\}_{3d}^\epsilon$ up to an error of $O(\epsilon^3)$. This proves the theorem. □

Remark. If one rescales the domain from $\Omega \times [-\epsilon, \epsilon]$ to $\Omega \times [-1, 1]$ by the change of variable $y = \epsilon \bar y$, then one should keep in mind that

$$\frac{1}{\epsilon} \frac{\partial \phi}{\partial \bar y} = O(1)$$

in the computations. This rescaling enables us to work with a 3D reference configuration of fixed height. Our method of deriving the 2D model carries over to this situation and leads to the same limiting system up to a change of variables. This is not surprising because the Poisson bracket is a geometric concept independent of the choice of coordinates (i.e., variables).

2.4. An Almost-Poisson Embedding for the Unconstrained Shell

As a corollary of Theorem 2.1, we show that there is an almost-Poisson embedding from the phase space of the 2D shell into that for the 3D elastic body.

Recall that if we have two Poisson manifolds P_1, P_2 with Poisson brackets $\{\cdot, \cdot\}_{P_1}$, $\{\cdot, \cdot\}_{P_2}$, respectively, then a map $A\colon P_1 \to P_2$ is called a Poisson map if

$$\{F_1, F_2\}_{P_2} \circ A = \{F_1 \circ A, F_2 \circ A\}_{P_1}$$

for every pair of functions F_1, F_2 on P_2. If A is a Poisson map, then every Hamiltonian dynamical system on P_2 can be pulled back to P_1. That is, if Θ_F^t and $\Theta_{F \circ A}^t$ denote the Hamiltonian flows of the Hamiltonians F and $F \circ A$, respectively, then

$$A \circ \Theta_{F \circ A}^t = \Theta_F^t \circ A.$$

These concepts can be generalized to almost-Poisson embeddings. We say that A is an $O(\epsilon)$-Poisson map if

$$\{F_1, F_2\}_{P_2} \circ A = \{F_1 \circ A, F_2 \circ A\}_{P_1} + O(\epsilon).$$

In this case, we have the relation (for a proof, see Ge et al. [1995])

$$A \circ \Theta_{F \circ A}^t = \Theta_F^t \circ A + O(\epsilon),$$

interpreted in a suitable weak sense when one is dealing with infinite-dimensional systems, that is, with PDEs.

Now we apply these general ideas to the approximation of a thin elastic body by a shell. It follows from Theorem 2.1 that the map A_ϵ introduced above is an $O(\epsilon^3)$-Poisson embedding:

Corollary 2.2. *For two functionals F_1, F_2 of the form (2.1.2), we have*

$$\{F_1, F_2\}_{3d}^\epsilon \circ A_\epsilon = \{F_1 \circ A_\epsilon, F_2 \circ A_\epsilon\}_{2d}^\epsilon + O(\epsilon^3).$$

2.5. The Role of Boundary Conditions in the Limiting Equations

Let $H\colon T\mathcal{M}_{3d}^\epsilon \to \mathbb{R}$ be given, for example, as in the introduction. In this chapter we write down the differential equations corresponding to the equations

$$\dot{F} = \{F, H\}_{3d}^\epsilon \qquad \text{for all} \quad F\colon T\mathcal{M}_{3d}^\epsilon \to \mathbb{R}, \tag{2.5.1}$$

and

$$\dot{F} = \{F, H \circ A_\epsilon\}_{2d}^\epsilon \qquad \text{for all} \quad F\colon T\mathcal{M}_{2d}^\epsilon \to \mathbb{R}, \tag{2.5.2}$$

respectively.

We are especially interested in the boundary conditions that are implied by (2.5.1) and (2.5.2) for the corresponding differential equations. These are natural boundary conditions, to be distinguished from boundary conditions which are formulated as part of the definition of configuration space (e.g., conditions that say certain parts of the boundary stay fixed throughout a motion).

To formulate the differential equations corresponding to (2.5.1) and (2.5.2), it is convenient to introduce functional derivatives, defined as follows:

The functional derivatives

$$\frac{\delta F}{\delta \phi}(\phi, \dot{\phi}): \Omega \times [-\epsilon, \epsilon] \to \mathbb{R}^3,$$

$$\frac{\delta' F}{\delta \phi}(\phi, \dot{\phi}): \partial(\Omega \times [-\epsilon, \epsilon]) \to \mathbb{R}^3,$$

$$\frac{\delta F}{\delta \dot{\phi}}(\phi, \dot{\phi}): \Omega \times [-\epsilon, \epsilon] \to \mathbb{R}^3,$$

of a function $F: T\mathcal{M}_{3d}^\epsilon \to \mathbb{R}$ at a point $(\phi, \dot{\phi}) \in T\mathcal{M}_{3d}^\epsilon$ are defined by

$$D_\phi F \cdot \delta\phi = \iiint_{\Omega \times [-\epsilon, \epsilon]} \left(\frac{\delta F}{\delta \phi}(\phi, \dot{\phi}) \cdot \delta\phi \right) d^2x \, dy$$

$$+ \iint_{\partial(\Omega \times [-\epsilon, \epsilon])} \left(\frac{\delta' F}{\delta \phi}(\phi, \dot{\phi}) \cdot \delta\phi \right) dA,$$

$$D_{\dot{\phi}} F \cdot \delta\dot{\phi} = \iiint_{\Omega \times [-\epsilon, \epsilon]} \left(\frac{\delta F}{\delta \dot{\phi}}(\phi, \dot{\phi}) \cdot \delta\dot{\phi} \right) d^2x \, dy.$$

Analogously we define the functional derivatives of a function $L: T\mathcal{M}_{2d} \to \mathbb{R}$ at point $(\varphi, \dot{\varphi}, w, \dot{w}) \in T\mathcal{M}_{2d}$:

$$\frac{\delta L}{\delta \varphi}(\varphi, \dot{\varphi}, w, \dot{w}): \Omega \to \mathbb{R}^3,$$

$$\frac{\delta' L}{\delta \varphi}(\varphi, \dot{\varphi}, w, \dot{w}): \partial\Omega \to \mathbb{R}^3,$$

$$\frac{\delta L}{\delta \dot{\varphi}}(\varphi, \dot{\varphi}, w, \dot{w}): \Omega \to \mathbb{R}^3,$$

$$\frac{\delta L}{\delta w}(\varphi, \dot{\varphi}, w, \dot{w}): \Omega \to \mathbb{R}^3,$$

$$\frac{\delta' L}{\delta w}(\varphi, \dot{\varphi}, w, \dot{w}): \partial\Omega \to \mathbb{R}^3,$$

$$\frac{\delta L}{\delta \dot{w}}(\varphi, \dot{\varphi}, w, \dot{w}): \Omega \to \mathbb{R}^3.$$

Equation (2.5.1) is then equivalent to the system of differential equations

$$\frac{d\phi}{dt} = \frac{\delta H}{\delta \dot{\phi}}(\phi, \dot{\phi}),$$

$$\frac{d\dot{\phi}}{dt} = -\frac{\delta H}{\delta \phi}(\phi, \dot{\phi}), \tag{2.5.3}$$

with boundary conditions

$$\frac{\delta' H}{\delta \phi}(\phi, \dot{\phi}) = 0 \qquad \text{on } \partial(\Omega \times [-\epsilon, \epsilon]). \tag{2.5.4}$$

Equation (2.5.2) is equivalent to the system of differential equations

$$\frac{d\varphi}{dt} = \frac{1}{2\epsilon} \frac{\delta(H \circ A_\epsilon)}{\delta\dot{\varphi}}(\varphi, \dot{\varphi}, w, \dot{w}),$$

$$\frac{d\dot{\varphi}}{dt} = -\frac{1}{2\epsilon} \frac{\delta(H \circ A_\epsilon)}{\delta\varphi}(\varphi, \dot{\varphi}, w, \dot{w}),$$

$$\frac{dw}{dt} = \frac{3}{2\epsilon^3} \frac{\delta(H \circ A_\epsilon)}{\delta\dot{w}}(\varphi, \dot{\varphi}, w, \dot{w}),$$

$$\frac{d\dot{w}}{dt} = -\frac{3}{2\epsilon^3} \frac{\delta(H \circ A_\epsilon)}{\delta w}(\varphi, \dot{\varphi}, w, \dot{w}), \tag{2.5.5}$$

with boundary conditions

$$\frac{\delta'(H \circ A_\epsilon)}{\delta\varphi}(\varphi, \dot{\varphi}, w, \dot{w}) = 0,$$

$$\frac{\delta'(H \circ A_\epsilon)}{\delta w}(\varphi, \dot{\varphi}, w, \dot{w}) = 0. \tag{2.5.6}$$

Now let the specific H be given as in the introduction:

$$H = \iiint_{\Omega \times [-\epsilon, \epsilon]} \left(\frac{1}{2}\dot{\phi} \cdot \dot{\phi} + g(x, y, \phi, \phi_x, \phi_y, \epsilon) \right) d^2x \, dy. \tag{2.5.7}$$

We can also consider the case of nonconstant mass density, but since that case is similar, we assume the reference mass density is equal to 1.

The equations (2.5.3) in this case are equivalent to

$$\frac{d^2\phi}{dt^2} = -\frac{\partial g}{\partial\phi} + \text{div}\frac{\partial g}{\partial(u, v)}, \tag{2.5.8}$$

where

$$\frac{\partial g}{\partial(u, v)} := \begin{pmatrix} \dfrac{\partial g}{\partial u_1^1} & \dfrac{\partial g}{\partial u_1^2} & \dfrac{\partial g}{\partial v_1} \\[2mm] \dfrac{\partial g}{\partial u_2^1} & \dfrac{\partial g}{\partial u_2^2} & \dfrac{\partial g}{\partial v_2} \\[2mm] \dfrac{\partial g}{\partial u_3^1} & \dfrac{\partial g}{\partial u_3^2} & \dfrac{\partial g}{\partial v_3} \end{pmatrix}$$

and where

$$u := \partial_x\phi, \quad u^i := \partial_{x_i}\phi, \quad u_j^i := \partial_{x_i}\phi_j,$$
$$v := \partial_y\phi, \quad v_j := \partial_y\phi_j, \quad i = 1, 2, \; j = 1, 2, 3.$$

The divergence is taken row-wise in (2.5.8). The boundary condition (2.5.4) reads as

$$\frac{\partial g}{\partial(u, v)} \cdot n = 0 \qquad \text{on } \partial(\Omega \times [-\epsilon, \epsilon]), \tag{2.5.9}$$

where $n\colon \partial(\Omega \times [-\epsilon, \epsilon]) \to \mathbb{R}^3$ is the vector field of outer unit normal vectors to $\Omega \times [-\epsilon, \epsilon]$ (defined at the smooth points of the boundary). In the special case that the shell is periodic in the x_1- and x_2-directions, the boundary condition (2.5.9) simplifies to

$$\frac{\partial g}{\partial v} = 0 \qquad \text{for } y = \pm\epsilon. \tag{2.5.10}$$

The function $H \circ A_\epsilon\colon T\mathcal{M}_{2d} \to \mathbb{R}$ has the form

$$(H \circ A_\epsilon)(\varphi, \dot{\varphi}, w, \dot{w})$$

$$= \iint_\Omega \Bigg(\epsilon\dot{\varphi}^2 + \frac{\epsilon^3}{3}\dot{w}^2 + 2\epsilon g(x, 0, \varphi, \varphi_x, w, \epsilon)$$

$$+ \frac{\epsilon^3}{3}\left(g_{y,y} + 2g_{y,\phi}w + 2g_{y,u^i}w_{x_i} + g_{\phi,\phi}w^2 + 2g_{\phi,u^i}ww_{x_i} + g_{u^i,u^j}w_{x_i}w_{x_j} \right)$$

$$+ O(\epsilon^5) \Bigg) d^2x$$

where, for example, $g_{\phi,\phi}w^2$ is shorthand for applying the bilinear form

$$\frac{\partial^2 g}{\partial \phi^2}(x, 0, \varphi, \varphi_x, w, \epsilon)$$

to the pair of vectors (w, w) and where the second derivatives under the integral sign have to be evaluated at $(x, 0, \varphi, \varphi_x, w, \dot{w})$. We have

$$\frac{\delta(H \circ A_\epsilon)}{\delta\varphi} = 2\epsilon g_\phi - 2\epsilon \operatorname{div}_x g_{u^i} + \frac{\epsilon^3}{3}(g_{y,y,\phi} + 2g_{y,\phi,\phi}w + 2g_{y,\phi,u^i}w_{x_i}$$

$$+ g_{\phi,\phi,\phi}w^2 + 2g_{\phi,\phi,u^i}ww_{x_i} + g_{\phi,u^i,u^j}w_{x_i}w_{x_j})$$

$$- \frac{\epsilon^3}{3}\operatorname{div}(g_{y,y,u^i_j} + 2g_{y,\phi,u^i_j}w + 2g_{y,u^k,u^i_j}w_{x_k} + g_{\phi,\phi,u^i_j}w^2$$

$$+ 2g_{\phi,u^k,u^i_j}w_{x_k} + g_{u^k,u^l,u^i_j}w_{x_k}w_{x_l}) + O(\epsilon^5), \tag{2.5.11}$$

$$\frac{\delta'(H \circ A_\epsilon)}{\delta\varphi} = 2\epsilon\langle g_u, n\rangle + \frac{\epsilon^3}{3}(g_{y,y,u^i_j} + 2g_{y,\phi,u^i_j}w + 2g_{y,u^k,u^i_j}w_{x_k}$$

$$+ g_{\phi,\phi,u^i_j}w^2 + 2g_{\phi,u^k,u^i_j}w_{x_k} + g_{u^k,u^l,u^i_j}w_{x_k}w_{x_l}) \cdot n + O(\epsilon^5), \tag{2.5.12}$$

$$\frac{\delta(H \circ A_\epsilon)}{\delta w} = 2\epsilon g_v + \frac{\epsilon^3}{3}(g_{y,y,v} + 2g_{y,\phi} + 2g_{y,\phi,v}w + 2g_{y,u^i,v}w_{x_i}$$

$$- 2(g_{y,u^i})_{x_i} + 2g_{\phi,\phi}w + 2g_{\phi,u^i,v}ww_{x_i} + 2g_{\phi,u^i}w_{x_i}$$

$$- 2(g_{\phi,u^i}w)_{x_i} + g_{u^i,u^j,v}w_{x_i}w_{x_j} - 2(g_{u^i,u^j}w_{x_i})_{x_j}) + O(\epsilon^5), \tag{2.5.13}$$

$$\frac{\delta'(H \circ A_\epsilon)}{\delta w} = \frac{2\epsilon^3}{3}\left(\frac{\partial^2 g}{\partial y \partial u} + \frac{\partial^2 g}{\partial \phi \partial u}w + \frac{\partial^2 g}{\partial u^2}\frac{\partial w}{\partial x} \right) \cdot n + O(\epsilon^5). \tag{2.5.14}$$

It is of interest to see how well the function $\varphi + yw$ satisfies the system of differential equations for 3D elasticity if (φ, w) is a solution for the differential equation in the director-model approximation. Since the top and bottom faces are missing in the two-dimensional model, it is especially interesting to see how well the function $\varphi + yw$ satisfies the boundary conditions of our 3D problem at the upper and the lower face of the plate.

2.6. A Simple Example: An Unconstrained Wave Equation

We study the problem posed at the end of the preceding section for the following example. Choose $\Omega := T^2$ and let in (2.5.7)

$$g = \frac{1}{2}(\phi_x^2 + \phi_y^2 - 2\phi\phi_y). \tag{2.6.1}$$

The equations of motion (2.5.8) read

$$\frac{d^2\phi}{dt^2} = \Delta_x\phi + \frac{\partial^2\phi}{\partial y^2}, \tag{2.6.2}$$

with boundary conditions

$$\phi_y = \phi \qquad \text{for } y = \pm\epsilon. \tag{2.6.3}$$

The equations of motion for the corresponding one-director model (2.5.5) are

$$\frac{d^2\varphi}{dt^2} = \varphi_{xx} + w, \tag{2.6.4}$$

$$\frac{d^2w}{dt^2} = w_{xx} + \frac{3}{\epsilon^2}(w - \varphi). \tag{2.6.5}$$

Note the singular term in equation (2.6.5).

To solve these differential equations, we make the Ansatz

$$w := c \cdot \varphi. \tag{2.6.6}$$

Substitute this into equations (2.6.4)–(2.6.5) to get the algebraic equation

$$\frac{3}{\epsilon^2}(c - 1) = c^2 \tag{2.6.7}$$

for the real number c. Now expand c into a power series in ϵ. We get

$$c = 1 + \frac{1}{3}\epsilon^2 + O(\epsilon^3). \tag{2.6.8}$$

Thus,

$$w = \left(1 + \frac{1}{3}\epsilon^2 + O(\epsilon^3)\right) \cdot \varphi. \tag{2.6.9}$$

We see that the boundary conditions (2.6.3) are satisfied to first order in ϵ by the function $\varphi + yw$.

Now we return to the general case. From equations (2.5.5) and (2.5.11)–(2.5.14) we see that in the general case the equations for the director field are

$$\frac{d^2 w}{dt^2} = -\frac{3}{\epsilon^2} \frac{\partial g}{\partial v} + O(1) \tag{2.6.10}$$

which contains a singular term of order $1/\epsilon^2$. If

$$\frac{\partial g}{\partial v} = O(\epsilon^2), \tag{2.6.11}$$

then the singularity in (2.6.10) disappears. We are led to an unconstrained Cosserat model. In this situation the material is anisotropic and mechanical properties depend on its thickness.

2.7. Unconstrained Limiting Models

Here is one way the situation described in the last subsection arises. We start with a specific isotropic 3D elasticity model (Saint Venant-Kirchhoff, for instance) that has an energy density function of the form

$$\frac{1}{2}\dot{\phi} \cdot \dot{\phi} + g_{\text{isotropic}}\left(x, y, \phi, \frac{\partial \phi}{\partial x}, \frac{\partial \phi}{\partial y}\right). \tag{2.7.1}$$

Then consider a scaled Hamiltonian

$$\begin{aligned}
H &= \iint_\Omega \int_{-\epsilon}^{\epsilon} \left(\frac{1}{2}\dot{\phi} \cdot \dot{\phi} + g\left(x, y, \phi, \frac{\partial \phi}{\partial x}, \frac{\partial \phi}{\partial y}, \epsilon\right)\right) d^2x \, dy \\
&:= \iint_\Omega \int_{-\epsilon}^{\epsilon} \left(\frac{1}{2}\dot{\phi} \cdot \dot{\phi} + g_{\text{isotropic}}\left(x, y, \phi, \frac{\partial \phi}{\partial x}, \epsilon \frac{\partial \phi}{\partial y}\right)\right) d^2x \, dy.
\end{aligned} \tag{2.7.2}$$

In particular, if g is quadratic plus higher order terms in $\partial \phi / \partial y$, then it satisfies the hypotheses needed to get an unconstrained limiting model. The Hamiltonian for the limiting model is then given in terms of the variable φ and the scaled variable $\bar{w} = \epsilon w$ by

$$H_{\text{shell}} = \iint_\Omega \left(\frac{1}{2}\left(\dot{\varphi} \cdot \dot{\varphi} + \frac{1}{3}\dot{\bar{w}} \cdot \dot{\bar{w}}\right) + g_{\text{shell}}\left(x, \varphi, \bar{w}, \frac{\partial \varphi}{\partial x}\right)\right) d^2x, \tag{2.7.3}$$

where

$$g_{\text{shell}}\left(x, \varphi, \bar{w}, \frac{\partial \varphi}{\partial x}\right) = g_{\text{isotropic}}\left(x, 0, \varphi, \frac{\partial \varphi}{\partial x}, \bar{w}\right). \tag{2.7.4}$$

Alternatively, we can write

$$H_{\text{shell}} = \lim_{\epsilon \to 0} \frac{1}{2\epsilon}(H \circ A_\epsilon)\left(\varphi, \frac{\bar{w}}{\epsilon}, \dot{\varphi}, \frac{\dot{\bar{w}}}{\epsilon}\right). \tag{2.7.5}$$

Then the limiting system has a single unconstrained director.

Now we look at the convergence of the 3D Poisson bracket. Introduce

$$\{F_1, F_2\}_{shell} = 2\epsilon \{F_1, F_2\}^{\epsilon}_{2d}$$
$$= D_{\varphi}F_1 \frac{\delta F_2}{\delta \dot{\varphi}} + 3D_{\bar{w}}F_1 \frac{\delta F_2}{\delta \dot{\bar{w}}} - D_{\varphi}F_2 \frac{\delta F_1}{\delta \dot{\varphi}} - 3D_{\bar{w}}F_2 \frac{\delta F_1}{\delta \dot{\bar{w}}} . \quad (2.7.6)$$

Corollary 2.3. *For functions F_1, F_2 satisfying (2.6.11), one has*

$$\{F_{1,shell}, F_{2,shell}\}_{shell}(\varphi, \bar{w}, \dot{\varphi}, \dot{\bar{w}}) = \lim_{\epsilon \to 0} \frac{1}{2\epsilon} \left(\{F_1, F_2\}^{\epsilon}_{3d} \circ A_{\epsilon} \right) \left(\varphi, \frac{\bar{w}}{\epsilon}, \dot{\varphi}, \frac{\dot{\bar{w}}}{\epsilon} \right). \quad (2.7.7)$$

In the next section we deal with the situation that condition (2.6.11) is not satisfied. This leads to a special kind of one-director model in which the director field is not independent from the displacement of the shell; instead, it is a function of the displacement and its derivatives. For lack of terminology, we shall call this model a constrained-director model.

3. Constrained-Director Shell Models

We now consider the case when the limiting procedures lead to constraints in the limiting shell model. Such constraints need to be distinguished from any additional constraints one may wish to impose. In this section we consider the simplest examples, and leave the example of the Kirchhoff shell for the next section. We note that constraints can involve derivatives and this naturally leads one from second-order elasticity models to higher order ones. We will see this explicitly in the Kirchhoff rod and shell models.

3.1. The Limit of the Hamiltonian Structure

Consider a 3D elastic body with Hamiltonian of the form

$$H = \iint_{\Omega} \int_{-\epsilon}^{\epsilon} \left(\frac{1}{2}\dot{\phi} \cdot \dot{\phi} + g(x, y, \phi, \partial_x \phi, \partial_y \phi, \epsilon) \right) d^2x \, dy. \quad (3.1.1)$$

We assume that the boundary condition in the x-direction is periodic for simplicity. The elasto-dynamical equation is

$$\frac{d^2 \phi}{dt^2} = -\frac{\partial g}{\partial \phi} + \text{div} \frac{\partial g}{\partial (u, v)}, \qquad \text{where} \quad v = \partial_y \phi$$

with the boundary conditions

$$\frac{\partial g}{\partial v} = 0, \qquad \text{where} \quad y = \pm\epsilon. \quad (3.1.2)$$

Throughout this section we assume that equation (2.6.11) is not satisfied and that

Assumption A. $\partial^2 g/\partial v^2$ does not vanish anywhere.

As before, we take the approximation

$$\phi(x, y) \approx \varphi(x) + y w(x).$$

However, here we require that the above approximation satisfies the boundary condition (3.1.2) up to order ϵ. Under Assumption A, this implies that w can be written as $w(x) = \mathcal{L}(x, \varphi, \partial_x \phi) + O(y)$ approximately, for some function $\mathcal{L}$. Thus, we define

$$w(x) := \mathcal{L}(x, \varphi, \partial_x \varphi).$$

That is, as mentioned before, the director is determined by $\varphi, \partial_x \varphi$. Note that the configuration space of the constrained-director model is

$$\mathcal{M}_{cd} = \{\varphi \mid \varphi: T^2 \to \mathbb{R}^3 \text{ is an embedding}\}.$$

The Poisson bracket on $T\mathcal{M}_{cd}$ is

$$\{F_1, F_2\}_{cd} = \iint_\Omega \left(\frac{\delta F_1}{\delta \varphi} \frac{\delta F_2}{\delta \dot\varphi} - \frac{\delta F_2}{\delta \varphi} \frac{\delta F_1}{\delta \dot\varphi} \right) d^2 x.$$

At first glance, the 3D Poisson bracket will not converge to that of the constrained-director model. Nevertheless, as we shall see, for Hamiltonians of interest to us, such as kinetic energy plus a potential energy, the 3D Poisson brackets *do* converge to that of the constrained-director model.

Introduce the embedding

$$\mathcal{U}_\epsilon: \mathcal{M}_{cd} \to \mathcal{M}_{3d}^\epsilon, \qquad (\mathcal{U}_\epsilon)(x, y) = \varphi(x) + y \mathcal{L}(x, \varphi, \partial \varphi).$$

Let $\mathcal{C}_\epsilon = T\mathcal{U}_\epsilon$ be the tangent map of $\mathcal{U}_\epsilon$. As we shall see, this is an almost-Poisson embedding for Hamiltonians of most interests:

Theorem 3.1. *As $\epsilon \to 0$,*

$$\{F_1, F_2\}_{3d}^\epsilon \circ \mathcal{C}_\epsilon = \frac{1}{2\epsilon} \{F_1 \circ \mathcal{C}_\epsilon, F_2 \circ \mathcal{C}_\epsilon\}_{cd} + O(\epsilon^2)$$

if either

1. *both F_1, F_2 are of form (3.1.1), i.e., "kinetic energy plus potential," where the kinetic energy is the Riemannian metric, or*
2. *F_1 is of the form of*

$$H = \iint_\Omega \int_{-\epsilon}^{\epsilon} m(x, y) \left(\frac{1}{2} \dot\phi \cdot \dot\phi + g(x, y, \phi, \partial_x \phi, \partial_y \phi, \epsilon) \right) d^2 x \, dy$$

where m is the mass density, g satisfies

$$\frac{\partial g(x, y, \varphi + y\mathcal{L}, \partial_x(\varphi + y\mathcal{L}), \mathcal{L})}{\partial v} = O(\epsilon) \tag{3.1.3}$$

at $y = \pm\epsilon$, and F_2 is arbitrary.

Proof. We only need to prove

$$\iint_\Omega \int_{-\epsilon}^\epsilon \left(\frac{\delta F_1}{\delta v} \cdot \partial_y \left(\frac{\delta F_2}{\delta \dot\phi} \right) \right) \circ C_\epsilon \, d^2 x \, dy$$
$$= 2\epsilon \iint_\Omega \left(\frac{\delta F_1}{\delta v} \frac{\delta \mathcal{L}}{\delta \varphi} \frac{\delta F_2}{\delta \dot\phi} \right) \circ C_\epsilon \, d^2 x + O(\epsilon^3) \tag{3.1.4}$$

where $v = \partial\phi/\partial y = \mathcal{L}$. We first prove this for case 1. In this case, one has

$$\left(\partial_y \frac{\delta F_i}{\delta \dot\phi} \right) \circ C_\epsilon = (\partial_y(\dot\phi)) \circ C_\epsilon = \dot{\mathcal{L}}. \tag{3.1.5}$$

On the other hand, since

$$\frac{\delta F_2}{\delta \dot\phi} \circ C_\epsilon = \dot\phi, \qquad \frac{\delta \mathcal{L}}{\delta \varphi} = \partial_\varphi \mathcal{L},$$

one has

$$\frac{\delta F_2}{\delta \dot\phi} \frac{\delta v}{\delta \varphi} = \partial_\varphi \mathcal{L} \cdot \dot\varphi = \dot{\mathcal{L}}. \tag{3.1.6}$$

Now (3.1.4) follows from (3.1.5), (3.1.6). This proves the theorem in case 1.

For the proof of case 2, one simply notices that, in addition to the above relations, the compatibility condition (3.1.3) implies that both sides of (3.1.4) are of order $O(\epsilon^3)$. $\square$

Remark. Note that $\phi(x, y) = \varphi(x) + y\mathcal{L}$ only satisfies the boundary conditions up to order ϵ. Put another way, it only satisfies

$$\frac{\partial g}{\partial v} = O(\epsilon),$$

which is much weaker than condition (1.3). In order to satisfy the boundary conditions (3.1.2) *exactly*, or to satisfy condition (1.3), it is necessary to use a two-director model,

$$\phi(x, y) \approx \varphi(x) + y\mathcal{L}_1 + y^2 \mathcal{L}_2. \tag{3.1.7}$$

Thus the constrained one-director model is only a first-order approximation. Nevertheless, the Hamiltonian of the constrained one-director model, $C_\epsilon^* E$, differs from that of the constrained two-directors model by a term of order $O(\epsilon^2)$. Thus, if one assumes that the boundary condition in the x-direction is periodic, the constrained one-director model is good enough. However, if the boundary condition in the x-direction is not periodic, presumably a constrained two-director model is better for the purpose of approximation, as a boundary layer might develop. But, as this layer is believed to be very thin, one expects that the constrained-director model still provides a good approximation in the interior (see also the remark in John [1971]).

Corollary 3.2. *For functions F_1, F_2 as in the preceding theorem, one has*

$$\lim_{\epsilon \to 0} \frac{1}{2\epsilon}(\{F_1, F_2\}_{3d}^\epsilon \circ C_\epsilon) = \left\{ \lim_{\epsilon \to 0} \frac{1}{2\epsilon} F_1 \circ C_\epsilon, \lim_{\epsilon \to 0} \frac{1}{2\epsilon} F_2 \circ C_\epsilon \right\}_{cd}. \tag{3.1.8}$$

Now we show that the constrained-director model is compatible with the one-director model. Define

$$\mathcal{U}_1: \mathcal{M}_{cd} \to \mathcal{M}_{2d}, \qquad \mathcal{U}_1: \varphi \to (\varphi, \mathcal{L}(x, \varphi, \partial_x \varphi))$$

and set $\mathcal{B} = T\mathcal{U}_1: T\mathcal{M}_{cd} \to T\mathcal{M}_{2d}$. Then $\mathcal{C}_\epsilon$ can be decomposed as $\mathcal{A}_\epsilon \circ \mathcal{B}$; moreover, $\mathcal{B}$ is also an almost-Poisson embedding:

Corollary 3.3. *Let F_1, F_2 be as in Theorem 3.1, then*

$$\{A_\epsilon^* F_1, A_\epsilon^* F_2\}_{2d}^\epsilon \circ \mathcal{B} = \{(C_\epsilon^* F_1), (C_\epsilon^* F_2)\}_{cd} + O(\epsilon^3).$$

3.2. The Wave Equation as a Constrained-Director Model

In general, when the limiting model for a shell has constraints, these constraints must be worked out in each case and the constraint will depend on the constitutive function chosen. We illustrate the procedure first with some simplified examples to illustrate the ideas and then we consider the case of an inextensible Kirchhoff shell.

Example 1.

$$\frac{\partial^2 \phi}{\partial^2 t} = \frac{\partial^2 \phi}{\partial x_1^2} + \frac{\partial^2 \phi}{\partial x_2^2} + \frac{\partial^2 \phi}{\partial y^2}, \qquad x = (x_1, x_2) \in T^2, \ -\epsilon \le y \le \epsilon, \quad (3.2.1)$$

$$\partial_y \phi = 0 \quad \text{for } y = \pm \epsilon. \tag{3.2.2}$$

The director is given by $w = 0$. The limit equation is

$$\partial_t^2 \varphi = (\partial_{x_1}^2 + \partial_{x_2}^2)\varphi, \qquad x \in S^1. \tag{3.2.3}$$

This can also be seen from the fact that (3.2.1)–(3.2.2) has a family of special exact solutions:

$$\phi(x, y) = \bar{\varphi}(x) \ \exp(y),$$

where $\bar{\varphi}$ satisfies equation (3.2.3).

Example 2. Consider the linear wave equation

$$\frac{\partial^2 \phi}{\partial^2 t} = \frac{\partial^2 \phi}{\partial x_1^2} + \frac{\partial^2 \phi}{\partial x_2^2} + \frac{\partial^2 \phi}{\partial y^2}, \qquad x = (x_1, x_2) \in T^2, \ -\epsilon \le y \le \epsilon \tag{3.2.4}$$

with the boundary conditions

$$\partial_y \phi = \phi \quad \text{for } y = \pm \epsilon, \tag{3.2.5}$$

which has the Hamiltonian

$$H(\phi, \dot{\phi}) = \int_{S^1} \int_{-\epsilon}^\epsilon \frac{1}{2}((\partial_t \phi)^2 + (\partial_x \phi)^2 + (\partial_y \phi)^2 - 2\phi\phi_y) \, d^2x \, dy. \tag{3.2.6}$$

The director is given by $\mathcal{L} = \varphi$. Substitute

$$\phi \approx (1 + y)\varphi$$

into equation (3.2.6) to obtain the Hamiltonian (modulo a multiplication by a constant) for the constrained-director model

$$H_{cd} = \int_{T^2} \frac{1}{2} ((\partial_t \varphi)^2 + (\partial_x \varphi)^2 - \varphi^2) \, d^2x.$$

The corresponding Hamiltonian system is

$$\partial_t^2 \varphi = (\partial_{x_1}^2 + \partial_{x_2}^2) \varphi + \varphi, \qquad x \in T^2. \tag{3.2.7}$$

This is the correct limit equation, as the original equation (3.2.4)–(3.2.5) has a family of special exact solutions:

$$\phi(x, y) = \bar{\varphi}(x) \, \exp(y),$$

where $\bar{\varphi}$ satisfies equation (3.2.7).

Example 3. Consider the following system of hyperbolic equations in linear elasticity on the domain $T^2 \times [-\epsilon, \epsilon]$ with points denoted by (x_1, x_2, y):

$$\partial_t^2 \phi_i = \partial_j \{\lambda \, e_{pp}(\vec{\phi}) \, \delta_{ij} + 2\mu \, e_{ij}(\vec{\phi})\}, \tag{3.2.8}$$

$$\lambda e_{pp}(\vec{\phi}) \, \delta_{i3} + 2\mu \, e_{i3}(\vec{\phi}) = 0, \qquad \text{for } y = \pm\epsilon, \ i = 1, 2, 3, \tag{3.2.9}$$

where $\vec{\phi} = (\phi_1, \phi_2, \phi_3)$, $e_{ij}(\vec{\phi}) = \frac{1}{2}(\partial_i \phi_j + \partial_j \phi_i)$ is the linearized strain tensor, λ, μ the Lamé constants. Here we used $\partial_1 := \partial/\partial x_1$, $\partial_2 := \partial/\partial x_2$, and $\partial_3 := \partial/\partial y$. Note that these equations describe a linearized Saint Venant-Kirchhoff model.

The Hamiltonian of this system is

$$\iiint_{T^2 \times [-\epsilon, \epsilon]} \{\partial_t \vec{\phi} \cdot \partial_t \vec{\phi} + \lambda \, e_{pp}(\vec{\phi}) \, e_{qq}(\vec{\phi}) + 2\mu \, (e_{ij}(\vec{\phi}))^2\} \, d^2x \, dy.$$

The equations considered in Ciarlet and Miara [1992] correspond to (and are more general than) the stationary case of equations (3.2.8) and (3.2.9).

As in previous examples, we take an approximation by a constrained-director model

$$\vec{\phi} \approx \vec{\varphi} + y\vec{w},$$

where $\vec{\varphi} = (\varphi_1, \varphi_2, \varphi_3)$, $\vec{w}$ are vector-valued functions on $\mathbb{R}^3$. The director is obtained from (3.2.9)

$$\vec{w} = \left(-\partial_1 \varphi_3, -\partial_2 \varphi_3, -\frac{\lambda}{2\mu + \lambda} (\partial_1 \varphi_1 + \partial_2 \varphi_2) \right).$$

The corresponding e_{ij}'s are

$$e_{11} = \partial_1\varphi_1 - y\partial_{11}^2\varphi_3,$$

$$e_{12} = \frac{1}{2}(\partial_2\varphi_1 + \partial_1\varphi_2 - 2y\partial_{12}^2\varphi_3),$$

$$e_{22} = \partial_2\varphi_2 - y\partial_{22}^2\varphi_3,$$

$$e_{13} = -\frac{\lambda}{2(2\mu+\lambda)}(\partial_{11}^2\varphi_1 + \partial_{12}^2\varphi_2)y,$$

$$e_{23} = -\frac{\lambda}{2\mu+\lambda}(\partial_{12}^2\varphi_1 + \partial_{22}^2\varphi_2)y,$$

$$e_{33} = -\frac{\lambda}{2\mu+\lambda}(\partial_1\varphi_1 + \partial_2\varphi_2).$$

Thus, the corresponding Hamiltonian is

$$2\epsilon \iint_{T^2} \partial_t\vec{\varphi} \cdot \partial_t\vec{\varphi}\, d^2x$$

plus the potential energy

$$\iint_{T^2}\int_{-\epsilon}^{\epsilon} \left\{ \lambda(\partial_1\varphi_1 - y\partial_{11}^2\varphi_3 + \partial_2\varphi_2 - y\partial_{22}^2\varphi_3 - \frac{\lambda}{2\mu+\lambda}(\partial_1\varphi_1 + \partial_2\varphi_2))^2 \right.$$

$$+ 2\mu((\partial_1\varphi_1 - y\partial_{11}^2\varphi_3)^2 + \frac{1}{2}(\partial_2\varphi_1 + \partial_1\varphi_2 - 2y\partial_{12}^2\varphi_3)^2)$$

$$\left. + (\partial_2\varphi_2 - y\partial_{22}^2\varphi_3)^2 + \left(\frac{\lambda}{2\mu+\lambda}(\partial_1\varphi_1 + \partial_2\varphi_2)\right)^2 \right\} d^2x\, dy + O(\epsilon^3)$$

$$= 2\epsilon \iint_{T^2} \left\{ \frac{4\lambda\mu^2}{(\lambda+2\mu)^2}(\partial_1\varphi_1 + \partial_2\varphi_2)^2 + 2\mu((\partial_1\varphi_1)^2 \right.$$

$$\left. + \frac{1}{2}(\partial_2\varphi_1 + \partial_1\varphi_2)^2 + (\partial_2\varphi_2)^2) \right\} d^2x + O(\epsilon^3).$$

Since the boundary condition in the x-direction is periodic, the higher order terms can be ignored, so the limit Hamiltonian is

$$\lim_{\epsilon\to 0}\frac{1}{\epsilon}(H\circ A_\epsilon) = \iint_{T^2} \left\{ \sum_{i=1}^{2}\partial_t\varphi_i \cdot \partial_t\varphi_i + \frac{2\mu\lambda}{\lambda+2\mu}(\partial_1\varphi_1 + \partial_2\varphi_2)^2 + 2\mu((\partial_1\varphi_1)^2 \right.$$

$$\left. + \mu(\partial_2\varphi_1 + \partial_1\varphi_2)^2 + 2\mu(\partial_2\varphi_2)^2) \right\} d^2x$$

$$= \iint_{T^2} \left\{ \sum_{i=1}^{2}\partial_t\varphi_i \cdot \partial_t\varphi_i + \frac{2\mu\lambda}{\lambda+2\mu}e_{pp}(\vec{\varphi})e_{qq}(\vec{\varphi}) + 2\mu(e_{ij}(\vec{\varphi}))^2 \right\} d^2x.$$

The limiting 2D equation is

$$\partial_t^2 \varphi_i = \partial_j \left\{ \frac{2\mu\lambda}{\lambda + 2\mu} e_{pp}(\vec{\varphi})\, \delta_{ij} + 2\mu\, e_{ij}(\vec{\varphi}) \right\}, \qquad i = 1, 2,$$

which corresponds to a membrane model.

4. The Kirchhoff Shell as a Limit of a 3D Saint Venant-Kirchhoff Material

4.1. Saint Venant-Kirchhoff Materials

Material frame invariance implies that the stored energy density of a hyperelastic material is of the form

$$g(x, y, \phi, D\phi) = W(D\phi^T D\phi) \tag{4.1.1}$$

(compare Marsden and Hughes [1994], Ch. 3). Let

$$E := \frac{1}{2}(D\phi^T D\phi - 1) \tag{4.1.2}$$

denote the Lagrangian strain tensor. A Saint Venant-Kirchhoff material is defined by its constitutive equations

$$S = \lambda \operatorname{tr}(E)\mathrm{Id} + \mu E, \tag{4.1.3}$$

where S is the stress tensor and the numbers $\lambda/2$, $\mu/2$ are the Lamé constants (compare Fox, Raoult, and Simo [1993]). Saint Venant-Kirchhoff materials as defined above are actually hyperelastic with stored energy density given by

$$W(D\phi^T D\phi) = \frac{\mu}{4} \left(\sum_{i,j=1}^{3} (z_i \cdot z_j - \delta_{ij})^2 \right) + \frac{\lambda}{8} \left(\sum_{i=1}^{3} (\|z_i\|^2 - 1) \right)^2,$$

where $z_i := \partial\phi/\partial x_i$, $i = 1, 2, 3$, and x_1, x_2, x_3 are Cartesian coordinates in $\mathbb{R}^3$. To see this, one has to check that

$$S = 2\nabla W(D\phi^T D\phi), \tag{4.1.4}$$

which is a straightforward computation (cf. Marsden and Hughes [1994]). Using the Saint Venant-Kirchhoff constitutive equations in modeling a given elastic material can be interpreted as using a truncation of the real constitutive equations of this material (cf. Fox, Raoult, and Simo [1993]).

4.2. The Kirchhoff Membrane Model

In this section we apply our general methods to derive constrained-director models outlined above to Saint Venant-Kirchhoff materials.

Following our general procedure we rewrite the coordinates in $\mathbb{R}^3$ as x_1, x_2, y and use the notation $u^i = \partial\phi/\partial x_i$, $i = 1, 2$, and $v = \partial\phi/\partial y$. We make the Ansatz

$$\phi(x, y) = \varphi(x) + yw(x) \tag{4.2.1}$$

for the configuration of the 3D body and determine the director field w from the condition

$$\frac{\partial g}{\partial v}(\varphi + yw) = O(\epsilon). \tag{4.2.2}$$

This equation is satisfied if

$$u^1 \cdot w = O(\epsilon),$$
$$u^2 \cdot w = O(\epsilon),$$

and if

$$(2\mu + \lambda)(\|w\|^2 - 1) + \lambda \left(\sum_{\alpha=1}^{2} \|u^\alpha\|^2 - 1 \right) = O(\epsilon).$$

After solving for w in terms of $\partial_{x_1}\varphi$ and $\partial_{x_2}\varphi$, inserting the result into W, and dropping higher order terms, we arrive at a first-order shell model with potential energy density

$$\begin{aligned}
W_m(s_1, s_2) &= \frac{\mu}{4} \left(\sum_{\alpha,\beta=1}^{2} (s_\alpha \cdot s_\beta - \delta_{\alpha,\beta})^2 \right) + \frac{\lambda}{8} \left(\sum_{\alpha=1}^{2} (\|s_\alpha\|^2 - 1) \right)^2 \\
&\quad - \frac{\lambda^2}{8(2\mu + \lambda)} \left(\sum_{\alpha=1}^{2} (\|s_\alpha\|^2 - 1) \right)^2 \\
&= \frac{\mu}{4} \left(\sum_{\alpha,\beta=1}^{2} (s_\alpha \cdot s_\beta - \delta_{\alpha,\beta})^2 \right) + \frac{2\mu\lambda}{8(2\mu + \lambda)} \left(\sum_{\alpha=1}^{2} (\|s_\alpha\|^2 - 1) \right)^2,
\end{aligned}$$

where $s_\alpha = \partial_{x_\alpha}\varphi$ for $i = 1, 2$. Note that $W_m \geq 0$ and $W_m = 0$ if and only if

$$\begin{aligned}
s_1 \cdot s_2 &= 0, \\
\|s_1\|^2 - 1 &= 0, \\
\|s_2\|^2 - 1 &= 0.
\end{aligned}$$

We see that application of our asymptotic procedure to a 3D Saint Venant-Kirchhoff material leads to a 2D membrane model that does not contain any director field.

4.3. The Inextensible Kirchhoff Shell

In the last subsection we derived a 2D membrane model with potential energy W_m. Now consider the equilibrium problem. Assume that the equilibrium state has a limit as $\epsilon \to 0$; then the limit equilibrium has potential energy either $W_m > 0$ or $W_m = 0$. In the former case it is determined by the Euler-Lagrangian equation and the boundary conditions, so we can stop here. However, if $W_m = 0$, then the Euler-Lagrangian equation and the boundary conditions alone do not determine the limit equilibrium, and we need to consider a higher order approximation, which will be the aim of this subsection.

Again, we write

$$\phi(x_1, x_2, y) = \varphi(x_1, x_2) + w(x_1, x_2)y, \tag{4.3.1}$$

where w is the director field. But now we introduce the inextensibility constraint that $\partial\varphi/\partial x_1$ and $\partial\varphi/\partial x_2$ are orthonormal vectors, which is equivalent to setting $W_m = 0$. The preceding discussion shows that one has to choose the director field w so that

$$\frac{\partial\varphi}{\partial x_1}, \quad \frac{\partial\varphi}{\partial x_2}, \quad \text{and} \quad w \tag{4.3.2}$$

are orthonormal. This means that w is a unit normal vector field to the surface which is parameterized by the map φ. Asking that

$$\det\left(\frac{\partial\varphi}{\partial x_1}, \frac{\partial\varphi}{\partial x_2}, w\right) > 0 \tag{4.3.3}$$

uniquely determines the director field w. Obviously, one has

$$w = \frac{\partial\varphi}{\partial x_1} \times \frac{\partial\varphi}{\partial x_2}. \tag{4.3.4}$$

Substituting the Ansatz $\phi = \varphi + yw$ into the Saint Venant-Kirchhoff potential energy density and dropping higher order terms in ϵ we get the potential energy density for an inextensible shell, given by

$$W_s\left(\frac{\partial\varphi}{\partial x_\alpha}, \frac{\partial w}{\partial x_\alpha}\right) = \frac{\mu}{4}\sum_{\alpha,\beta=1}^{2}\left(\frac{\partial\varphi}{\partial x_\alpha} \cdot \frac{\partial w}{\partial x_\beta} + \frac{\partial\varphi}{\partial x_\beta} \cdot \frac{\partial w}{\partial x_\alpha}\right)^2$$

$$+ \frac{\lambda}{2}\left(\sum_{\alpha=1}^{2}\frac{\partial\varphi}{\partial x_\alpha} \cdot \frac{\partial w}{\partial x_\alpha}\right)^2, \tag{4.3.5}$$

which is the potential energy for the inextensional shell model.

We want to give an interpretation of this energy density in terms of geometric data of the surface Σ parameterized by the map φ. Remember that the first fundamental form

$$E(dx_1)^2 + 2F dx_1 dx_2 + G(dx_2)^2 \tag{4.3.6}$$

of the surface Σ is defined by

$$E := \frac{\partial\varphi}{\partial x_1} \cdot \frac{\partial\varphi}{\partial x_1},$$

$$F := \frac{\partial\varphi}{\partial x_1} \cdot \frac{\partial\varphi}{\partial x_2},$$

$$G := \frac{\partial\varphi}{\partial x_2} \cdot \frac{\partial\varphi}{\partial x_2},$$

and that its second fundamental form

$$e(dx_1)^2 + 2f dx_1 dx_2 + g(dx_2)^2 \tag{4.3.7}$$

is defined by

$$e := -\frac{\partial\varphi}{\partial x_1} \cdot \frac{\partial n}{\partial x_1},$$

$$f := -\frac{\partial\varphi}{\partial x_1} \cdot \frac{\partial n}{\partial x_2}$$

$$= -\frac{\partial\varphi}{\partial x_2} \cdot \frac{\partial n}{\partial x_1},$$

$$g := -\frac{\partial\varphi}{\partial x_2} \cdot \frac{\partial n}{\partial x_2}.$$

Alternatively, the second fundamental form Π_p at a point $p \in \Sigma$ is the quadratic form associated to (4.3.7) which takes the value $\Pi_p(v) = ev_1^2 + 2fv_1v_2 + gv_2^2$ at $v = v_1\partial\varphi/\partial x_1 + v_2\partial\varphi/\partial x_2 \in T_p\Sigma$. The mean curvature H of the surface Σ, which is half the sum of the principle curvatures k_1, k_2 at every point of the surface, can then be expressed as

$$H = \frac{1}{2}\frac{eG - 2fF + gE}{EG - F^2}. \tag{4.3.8}$$

The Gaussian curvature $K = k_1k_2$ can be written as

$$K = \frac{eg - f^2}{EG - F^2}. \tag{4.3.9}$$

By choice of the map φ, the first fundamental form is just the identity matrix and the formulas for the mean curvature and the Gaussian curvature reduce to

$$H = \frac{1}{2}(e + g) \tag{4.3.10}$$

and

$$K = eg - f^2. \tag{4.3.11}$$

We can therefore rewrite the stored energy density of the inextensible shell as

$$W_s(H, K) = (4\mu + 2\lambda)H^2 - 2\mu K. \tag{4.3.12}$$

We see that under the additional assumption of inextensibility our method leads to an inextensible 2D shell model without any director variables, the energy density of which can be expressed in terms of mean and Gaussian curvature only. This is in sharp contrast to the membrane model derived earlier where the energy density depends only on the first fundamental form.

5. The Convergence of 3D Elasticity to Rod Models

5.1. Limits of Hamiltonian Structures

The methods in the previous sections also apply to thin rods. For a general introduction to the theory of rods, see Antman [1972], and for the Hamiltonian formulation of rod theories, see Simo, Marsden, and Krishnaprasad [1988] and references therein. We indicate in this section how to carry out this program.

Consider the motion of a rod with reference configuration given by

$$R_\epsilon = \{(x, y, z) \mid x^2 + y^2 \le \epsilon^2, 0 \le z \le l\} \subset \mathbb{R}^3$$

which consists of embeddings

$$\phi: R_\epsilon \to \mathbb{R}^3. \tag{5.1.1}$$

Expanding ϕ at $x = y = 0$, we obtain

$$\varphi(z) = \phi(0, 0, z), \; w_1(z) = \frac{\partial \phi}{\partial x}(0, 0, z), \; w_2(z) = \frac{\partial \phi}{\partial y}(0, 0, z), \tag{5.1.2}$$

where w_1, w_2 are the director fields.

We introduce an approximation embedding $\mathcal{K}_\epsilon$ as the tangent lift of the map

$$(\varphi(z), w_1(z), w_2(z)) \to \varphi(z) + x w_1 + y w_2. \tag{5.1.3}$$

If we are given a functional on the 3D phase space (typically kinetic plus potential energy) of the form

$$F = \iiint_{R_\epsilon} f(\phi, \partial\phi, \dot\phi, \epsilon) \, dx \, dy \, dz, \tag{5.1.4}$$

the induced functional for the rod is obtained by substituting $\phi(x, y, z) = \varphi(z) + x w_1(z) + y w_2(z)$ in the 3D energy, namely

$$F \circ \mathcal{K}_\epsilon = \iiint_{R_\epsilon} f(\varphi(z) + x w_1(z) + y w_2(z), \partial(\varphi + x w_1 + y w_2), \dot\varphi + x \dot w_1 + y \dot w_2) \, dx \, dy \, dz.$$

Suppose that the kinetic energy for the 3D elastic body is

$$\frac{1}{2} \langle (\phi, \dot\phi_1), (\phi, \dot\phi_2) \rangle_{3d} = \frac{1}{2} \iiint_{R_\epsilon} \dot\phi_1 \cdot \dot\phi_2 \, dV, \tag{5.1.5}$$

then the induced kinetic energy for the rod is

$$\frac{\pi \epsilon^2}{2} \int_0^l \dot\varphi \cdot \dot\varphi \, dz + \frac{\pi \epsilon^4}{4} \int_0^l (\dot w_1 \cdot \dot w_1 + \dot w_2 \cdot \dot w_2) \, dz. \tag{5.1.6}$$

We use the corresponding Riemannian metric to identify the tangent bundle and cotangent bundle for the rod, and we obtain the Poisson bracket

$$\{F_1, F_2\}^\epsilon_{1d} = \int_0^l \frac{1}{\pi \epsilon^2} \left(\frac{\delta F_1}{\delta \varphi} \frac{\delta F_2}{\delta \dot\varphi} - \frac{\delta F_2}{\delta \varphi} \frac{\delta F_1}{\delta \dot\varphi} \right) dz$$

$$+ \int_0^l \frac{2}{\pi \epsilon^4} \left(\frac{\delta F_1}{\delta w_1} \frac{\delta F_2}{\delta \dot w_1} + \frac{\delta F_1}{\delta w_2} \frac{\delta F_2}{\delta \dot w_2} - \frac{\delta F_2}{\delta w_1} \frac{\delta F_1}{\delta \dot w_1} - \frac{\delta F_2}{\delta w_2} \frac{\delta F_1}{\delta \dot w_2} \right) dz.$$

Theorem 5.1. *If F_1, F_2 are two functionals of the form (5.1.5), and $F_1 \circ \mathcal{K}_\epsilon$, $F_2 \circ \mathcal{K}_\epsilon$ their counterparts for the rod, then*

$$\{F_1, F_2\}^\epsilon_{3d} = \{F_1 \circ \mathcal{K}_\epsilon, F_2 \circ \mathcal{K}_\epsilon\}^\epsilon_{1d} + O(\epsilon^3).$$

5.2. Constrained and Unconstrained Limiting Models

As in the previous sections, by studying the dynamics of the limiting model, one obtains either a unconstrained-director model or a constrained-director model, depending on different assumptions on the Hamiltonian functional.

For example, consider the Hamiltonian

$$H = \iiint_{R_\epsilon} \left(\frac{1}{2} \dot{\phi} \cdot \dot{\phi} + g(x, y, z, \phi, \partial\phi, \epsilon) \right) dx\, dy\, dz. \tag{5.2.1}$$

The corresponding limiting equations have the form

$$\frac{\partial^2 \varphi}{\partial t^2} = O(1),$$

$$\frac{\partial^2 w_1}{\partial t^2} = \frac{1}{\epsilon^2} \frac{\partial g}{\partial v_1} + O(1),$$

$$\frac{\partial^2 w_2}{\partial t^2} = \frac{1}{\epsilon^2} \frac{\partial g}{\partial v_2} + O(1),$$

where $v_1 = \partial\varphi/\partial x$ and $v_2 = \partial\varphi/\partial y$. If we assume that

$$\frac{\partial g}{\partial v_1} = O(\epsilon^2), \qquad \frac{\partial g}{\partial v_2} = O(\epsilon^2), \tag{5.2.2}$$

then we are led to unconstrained two-director models in the limit $\epsilon \to 0$.

As with shells, this situation arises if one starts with a specific isotropic 3D elasticity model with an energy density of the form

$$\frac{1}{2} \dot{\phi} \cdot \dot{\phi} + g_{\text{isotropic}} \left(x, y, z, \phi, \frac{\partial\phi}{\partial x}, \frac{\partial\phi}{\partial y}, \frac{\partial\phi}{\partial z} \right). \tag{5.2.3}$$

Then, consider a scaled Hamiltonian

$$\begin{aligned}
H &= \iiint_{R_\epsilon} \left(\frac{1}{2} \dot{\phi} \cdot \dot{\phi} + g \left(x, y, z, \phi, \frac{\partial\phi}{\partial x}, \frac{\partial\phi}{\partial y}, \frac{\partial\phi}{\partial z}, \epsilon \right) \right) dx\, dy\, dz \\
&:= \iiint_{R_\epsilon} \left(\frac{1}{2} \dot{\phi} \cdot \dot{\phi} + g_{\text{isotropic}} \left(x, y, z, \phi, \epsilon\frac{\partial\phi}{\partial x}, \epsilon\frac{\partial\phi}{\partial y}, \frac{\partial\phi}{\partial z} \right) \right) dx\, dy\, dz. \tag{5.2.4}
\end{aligned}$$

In particular, if g is quadratic plus higher order terms in $\partial\phi/\partial x$ and $\partial\phi/\partial y$, then it satisfies the hypotheses needed to get an unconstrained limiting model. The Hamiltonian for the limiting model then is given in terms of the variables $\phi(x, y, z) = \varphi(z) + x w_1 + y w_2$ and the scaled variables $\bar{w}_1 = \epsilon w_1$ and $\bar{w}_2 = \epsilon w_2$

$$\begin{aligned}
H_{\text{rod}} = \int_0^l \Big(\frac{1}{2} \left(\dot{\varphi} \cdot \dot{\varphi} + \dot{\bar{w}}_1 \cdot \dot{\bar{w}}_1 + \dot{\bar{w}}_2 \cdot \dot{\bar{w}}_2 \right) \\
+ g_{\text{rod}} \left(z, \varphi, \bar{w}_1, \bar{w}_2, \frac{\partial\varphi}{\partial z}, \frac{\partial\bar{w}_1}{\partial z}, \frac{\partial\bar{w}_2}{\partial z} \right) \Big) dz. \tag{5.2.5}
\end{aligned}$$

where

$$g_{rod}\left(z, \varphi, \bar{w}_1, \bar{w}_2, \frac{\partial\varphi}{\partial z}, \frac{\partial\bar{w}_1}{\partial z}, \frac{\partial\bar{w}_2}{\partial z}\right) = g_{isotropic}\left(0, 0, z, \varphi, \bar{w}_1, \bar{w}_2, \frac{\partial\varphi}{\partial z}\right). \quad (5.2.6)$$

Then the limiting system has two unconstrained directors.

If, on the other hand, $\partial g/\partial v_1 \neq O(\epsilon^2)$, $\partial g/\partial v_2 \neq O(\epsilon^2)$, which has to be evaluated for $\phi = \varphi + xw_1 + yw_2$, then one in general obtains constrained two-director models in the limit.

Suppose that

$$\det\left(\frac{\partial^2 g}{\partial v_1 \partial v_2}\right) \neq 0. \quad (5.2.7)$$

Then one can solve for w_1 and w_2 in terms of φ:

$$w_1 = \mathcal{L}_1(\varphi), \qquad w_2 = \mathcal{L}_2(\varphi) \quad (5.2.8)$$

from the equations

$$\left.\frac{\partial g}{\partial v_1}\right|_{\phi=\varphi(z)+xw_1+yw_2} = O(\epsilon), \qquad \left.\frac{\partial g}{\partial v_2}\right|_{\phi=\varphi(z)+xw_1+yw_2} = O(\epsilon). \quad (5.2.9)$$

In this case the configuration space is the set

$$\mathcal{M}_{1cd} = \{\varphi: \mathbb{R} \to \mathbb{R}^3\}$$

and the Poisson bracket is given by

$$\{F_1, F_2\}_{1cd} = \int\left(\frac{\delta F_1}{\delta\varphi}\frac{\delta F_2}{\delta\dot\varphi} - \frac{\delta F_2}{\delta\varphi}\frac{\delta F_1}{\delta\dot\varphi}\right) dz.$$

We introduce the map $\mathcal{I}_\epsilon: TM_{1cd} \mapsto TM_{3d}$ as the tangent lift of the map

$$\varphi(z) \mapsto \varphi(z) + x\mathcal{L}_1 + y\mathcal{L}_2.$$

The map $\mathcal{I}_\epsilon$ is an almost-Poisson embedding, in the sense that

$$\{F_1 \circ \mathcal{I}_\epsilon, F_2 \circ \mathcal{I}_\epsilon\}_{3d}^\epsilon = \{F_1, F_2\}_{1d}^\epsilon \circ \mathcal{I}_\epsilon + O(\epsilon^3)$$

where F_1 is of the form

$$\iiint_{R_\epsilon}\left(\frac{1}{2}m\dot\phi\cdot\dot\phi + g\left(x, y, z, \phi, \frac{\partial\phi}{\partial x}, \frac{\partial\phi}{\partial y}, \frac{\partial\phi}{\partial z}, \epsilon\right)\right) dx\, dy\, dz$$

where m is the mass density, g satisfies

$$\left.\frac{\partial g}{\partial v_1}\right|_{\phi=\varphi+x\mathcal{L}_1+y\mathcal{L}_2} = O(\epsilon^2), \qquad \left.\frac{\partial g}{\partial v_2}\right|_{\phi=\varphi+x\mathcal{L}_1+y\mathcal{L}_2} = O(\epsilon^2),$$

and F_2 is arbitrary.

6. The Kirchhoff Elastica as a Limit of a 3D Saint Venant-Kirchhoff Material

6.1. Introduction

In this section we apply our method to 3D Saint Venant-Kirchhoff materials. As outlined above, the stored energy density of such a material is given by $g(x, y, z, \phi, D\phi) = W(D\phi^T D\phi)$, where

$$W(D\phi^T D\phi) = \frac{\mu}{4} \left(\sum_{i,j=1}^{3} (z_i \cdot z_j - \delta_{ij})^2 \right) + \frac{\lambda}{8} \left(\sum_{i=1}^{3} (\|z_i\|^2 - 1) \right)^2. \qquad (6.1.1)$$

Here we used $z_1 := \partial\phi/\partial x$, $z_2 := \partial\phi/\partial y$, $z_3 := \partial\phi/\partial z$ and Cartesian coordinates x, y, z in $\mathbb{R}^3$. Introducing an additional inextensibility constraint will lead to an unconstrained one-director Kirchhoff rod model, the stored energy density of which depends only on curvature and torsion of the rod and the derivative of the director variable along the rod.

In case one starts with a 2D Saint Venant-Kirchhoff material instead of a 3D one, the asymptotic procedure will lead to the well-known Euler elastica as a limit model.

6.2. A Limiting Director Model for Rods

To apply our general theory, we make the Ansatz

$$\phi(x, y, z) = \varphi(z) + x w_2(z) + y w_3(z) \qquad (6.2.1)$$

for a configuration of the 3D material and determine the director fields w_2 and w_3 from the equations

$$\frac{\partial g}{\partial v_1} = O(\epsilon),$$

$$\frac{\partial g}{\partial v_2} = O(\epsilon).$$

(We changed the numbering of the director fields slightly to make the discussion of inextensible rods in the next section more transparent.) To determine w_2 and w_3 so that these equations are satisfied, we use the following lemma:

Lemma 6.1. *Let* $h(\varphi_z) = \|\varphi_z\|^2 - 1$. *If*

$$-\frac{\lambda h(\varphi_z)}{2(\mu + \lambda)} > -1,$$

then the equations

$$\frac{\partial W}{\partial z_1}(\bar{w}_2, \bar{w}_3, z_3) = O(\epsilon) \qquad (6.2.2)$$

$$\frac{\partial W}{\partial z_2}(\bar{w}_2, \bar{w}_3, z_3) = O(\epsilon) \qquad (6.2.3)$$

have solutions that satisfy

$$\bar{w}_2 \cdot \varphi_z = 0,$$

$$\bar{w}_3 \cdot \varphi_z = 0,$$

$$\|\bar{w}_2\|^2 - 1 = -\frac{\lambda h(\varphi_z)}{2(\mu + \lambda)},$$

$$\|\bar{w}_3\|^2 - 1 = -\frac{\lambda h(\varphi_z)}{2(\mu + \lambda)}.$$

Proof. We write

$$W = \frac{\mu}{4} \left(\sum_{\alpha, \beta = 1}^{2} (z_\alpha \cdot z_\beta - \delta_{\alpha\beta})^2 \right) + \frac{\lambda}{8} \left(\sum_{\alpha=1}^{2} (\|z_\alpha\|^2 - 1) \right)^2$$

$$+ \frac{\mu}{2} \left(\sum_{\alpha}^{2} (z_\alpha \cdot z_3)^2 \right) + \frac{2\mu + \lambda}{8} \left(\|z_3\|^2 - 1 \right)^2$$

$$+ \frac{\lambda}{4} (\|z_3\|^2 - 1) \left(\sum_{\alpha}^{2} (\|z_\alpha\|^2 - 1) \right).$$

We only consider (z_1, z_2, z_3) for which W is minimal with respect to (z_1, z_2). For W to be minimal, z_1, z_2, z_3 have to be orthogonal. Denote $Y_1 := \|z_1\|^2 - 1$, $Y_2 := \|z_2\|^2 - 1$ and $\bar{h} := \|z_3\|^2 - 1$, then

$$W = \frac{\mu}{4}(Y_1^2 + Y_2^2) + \frac{\lambda}{8}(Y_1 + Y_2)^2 + \frac{2\mu + \lambda}{8}\bar{h}^2 + \frac{\lambda}{4}\bar{h}(Y_1 + Y_2).$$

Thus, $\partial W / \partial Y_1 = \partial W / \partial Y_2 = 0$ is equivalent to

$$\frac{\mu}{2} Y_1 + \frac{\lambda}{4}(Y_1 + Y_2) + \frac{\lambda}{4}\bar{h} = 0,$$

$$\frac{\mu}{2} Y_2 + \frac{\lambda}{4}(Y_1 + Y_2) + \frac{\lambda}{4}\bar{h} = 0.$$

In case $Y_1 \geq -1$, the solution to these equations is

$$Y_1 = Y_2 = -\frac{\lambda \bar{h}}{2(\mu + \lambda)}. \tag{6.2.4}$$

The corresponding potential energy density is

$$W_{\min} = \left(\frac{2\mu + \lambda}{8} \right) \bar{h}^2 - \frac{1}{8(\lambda + \mu)} \lambda^2 \bar{h}^2. \tag{6.2.5}$$

This lemma shows that to solve equations (6.2.2) and (6.2.3), one has to choose the director fields w_2 and w_3 orthogonal to φ_z and both of squared length equal to

$$1 - \frac{\lambda h(\varphi_z)}{2(\mu + \lambda)}.$$

The stored energy density of the limit rod model is then

$$g_{rod} = \left(\frac{2\mu + \lambda}{8}\right) h^2 - \frac{1}{8(\lambda + \mu)} \lambda^2 h^2, \tag{6.2.6}$$

or

$$\lim_{\epsilon \to 0} \frac{2}{\epsilon^2} V \circ K_\epsilon = \left(\frac{2\mu + \lambda}{8}\right) h(\varphi_z)^2 - \frac{\lambda^2}{8(\lambda + \mu)} h(\varphi_z)^2. \tag{6.2.7}$$

We see that the limiting procedure applied to a Saint Venant-Kirchhoff material leads to an unconstrained rod model, the energy of which depends only on the first derivatives of the placement field. In the next section we show that one arrives at a limit model of a very different type if one introduces an inextensibility constraint, which is equivalent to setting (6.2.7) to zero (see also the discussion in Section 4.3).   □

6.3. An Inextensible Elastica with a Director

To derive a theory for an inextensible rod, we make the (inextensibility) assumption

$$\bar{h}(z) = \|\partial_z \varphi\|^2 - 1 = 0, \tag{6.3.1}$$

which is equivalent to setting the potential energy (6.2.7) to zero. Let $w_1 := \partial_z \varphi$. From the discussion in the preceding section we see that the director fields w_2, w_3 of the rod should satisfy the following equations:

$$\begin{aligned} w_2 \cdot w_3 &= 0, \\ w_2 \cdot \varphi_z &= 0, \\ w_3 \cdot \varphi_z &= 0, \\ \|w_2\|^2 - 1 &= 0, \\ \|w_3\|^2 - 1 &= 0. \end{aligned}$$

If we assume that

$$\det(w_1, w_2, w_3) > 0, \tag{6.3.2}$$

then the director fields w_2, w_3 are determined by these constraints up to a rotation about w_1. In particular, we have $w_3 = \varphi_z \times w_2$. After calculation one finds (neglecting higher order terms in ϵ)

$$\begin{aligned} V &= \frac{2\mu + \lambda}{8} \pi \epsilon^4 \int \left(\langle \varphi_z, (w_2)_z \rangle^2 + \langle \varphi_z, (w_3)_z \rangle^2 \right) dz \\ &+ \frac{\mu}{8} \pi \epsilon^4 \int \langle w_2, w_1 \times (w_2)_z \rangle^2 dz. \end{aligned}$$

Because $\langle \varphi_{zz}, \varphi_z \rangle = 0$ and w_2 and w_3 are orthonormal, one has

$$\varphi_{zz} = \langle \varphi_{zz}, w_2 \rangle w_2 + \langle \varphi_{zz}, w_3 \rangle w_3. \tag{6.3.3}$$

Using the equations

$$\langle \varphi_z, (w_i)_z \rangle = \partial_z \langle \varphi_z, w_i \rangle - \langle \varphi_{zz}, w_i \rangle \qquad (i = 2, 3)$$

and the fact that φ_z, w_2, w_3 are orthonormal in equation (6.3.3), we get

$$\varphi_{zz} = -\langle \varphi_z, (w_2)_z \rangle w_2 - \langle \varphi_z, (w_3)_z \rangle w_3$$

and

$$\|\varphi_{zz}\|^2 = \langle \varphi_z, (w_2)_z \rangle^2 + \langle \varphi_z, (w_3)_z \rangle^2.$$

Writing $a := \|\varphi_{zz}\|$ for the curvature function of the curve $z \mapsto \varphi(z)$ and neglecting higher order terms in ϵ then yields

$$V = \frac{2\mu + \lambda}{8} \pi \epsilon^4 \int a^2 \, dz + \frac{\mu}{8} \pi \epsilon^4 \int \langle w_2, w_1 \times (w_2)_z \rangle^2 \, dz.$$

If $a \neq 0$, we may choose the director field w_2 pointwise collinear to the field φ_{zz}:

We introduce the Frenet frame v_1, v_2, v_3, where $v_1(z) = \partial_z \varphi(z)$, $v_2(z)$ is a unit vector in the direction of the curvature vector of φ at $\varphi(z)$, and $v_3(z) = v_1(z) \times v_2(z)$ denotes the binormal vector of φ at $\varphi(z)$. The Frenet equations

$$\frac{dv_1}{dz} = a v_2,$$

$$\frac{dv_2}{dz} = -a v_1 - c v_3,$$

$$\frac{dv_3}{dz} = c v_2$$

hold, where $a(z)$ is the curvature of φ at $\varphi(z)$ and $c(z)$ is the torsion (see, for example, do Carmo [1976]).

Now we can write

$$w_1 = v_1,$$
$$w_2 = A v_2,$$
$$w_3 = A v_3,$$

where $A(z) \in SO(2)$ describes a rotation at $w_1(z)$ by an angle $\Theta(z)$. We plug the Ansatz

$$\phi(x, y, z) = \varphi(z) + x w_2(z) + y w_3(z) \tag{6.3.4}$$

into the (kinetic + potential) energy density of the 3D Saint Venant-Kirchhoff material and integrate over the reference configuration to get the energy H for the inextensible rod. After calculation, one finds that

$$H = K + V, \tag{6.3.5}$$

where

$$K = \int \left[\frac{1}{2}\pi\epsilon^2\dot\varphi^2 + \frac{\pi}{4}\epsilon^4\dot\Theta^2 + \frac{\pi}{4}\epsilon^4\{\langle v_1, v_2 \times \dot v_2\rangle + \langle v_1, v_3 \times \dot v_3\rangle\}\dot\Theta \right.$$
$$\left. + \frac{\pi}{8}\epsilon^4\{\|\dot v_2\|^2 + \|\dot v_3\|^2\} \right] dz$$

$$= \int \left[\frac{1}{2}\pi\epsilon^2\dot\varphi^2 + \frac{\pi}{4}\epsilon^4\dot\Theta^2 + \frac{\pi}{2a^2}\epsilon^4\langle\varphi_z \times \varphi_{zz}, \dot\varphi_{zz}\rangle\dot\Theta + \frac{\pi}{4a^2}\epsilon^4\|\dot\varphi_{zz}\|^2 \right.$$
$$\left. - \frac{\pi}{4a^4}\epsilon^4\langle\varphi_{zz}, \dot\varphi_{zz}\rangle^2 + \frac{\pi}{8a^2}\epsilon^4\|\varphi_{zz} \times \dot\varphi_z\|^2 + \frac{\pi}{4a^2}\epsilon^4\langle\varphi_z \times \varphi_{zz}, \varphi_z \times \dot\varphi_{zz}\rangle \right] dz,$$

and (neglecting terms of higher order in ϵ)

$$V = \int \left[\frac{\pi\epsilon^4}{8}\mu(c + \Theta_z)^2 + \frac{\pi\epsilon^4}{8}(2\mu + \lambda)a^2 \right] dz. \qquad (6.3.6)$$

In Landau and Lifshitz [1959] the quantity Θ_z is called the *torsion angle*, in case the rod is unbent.

The function H may be interpreted in different ways. On one hand, one can think of it as the energy function for a two-director rod model, where the directors are constrained to be orthonormal and to be orthogonal to the tangent vector of the centerline of the rod. On the other hand, one might interpret H as the energy function of an *unconstrained* one-director rod model with variables φ (position of centerline of the rod) and Θ (twist of the normal cross-sections to the centerline of the rod). Here Θ is interpreted as the director variable.

One might also think of H as the energy function of a *three-director* rod model, following Maddocks [1984], who introduces two directors d_1, d_2: $[0, 1] \to \mathbb{R}^3$ that are orthonormal and orthogonal to the tangent vector of the axis r: $[0, 1] \to \mathbb{R}^3$ of the rod. He imposes the additional constraint $\|r'\| = 1$, puts $d_1 := r'$, and regards the resulting rod model as a three-director rod model with directors d_1, d_2, d_3.

To understand how the energy function behaves under rescaling, let

$$\epsilon^\alpha \bar z := z,$$
$$\epsilon^\beta \bar\varphi := \varphi,$$
$$\epsilon^\gamma \bar\theta := \theta,$$
$$\epsilon^\delta \bar t := t,$$
$$\epsilon^\rho \bar\mu := \mu,$$
$$\epsilon^\sigma \bar\lambda := \lambda.$$

After rescaling, the potential energy is given by

$$\bar V = \int \left[\frac{\pi}{4}\epsilon^{4-\alpha+2\gamma+\rho}\bar\mu(\bar\Theta_{\bar z})^2 - \frac{\pi}{2}\epsilon^{4-\beta+\gamma+\rho}\bar\mu\bar c\bar\Theta_{\bar z} \right.$$
$$\left. + \frac{\pi}{4}\epsilon^{4+\alpha-2\beta+\rho}\bar\mu\bar c^2 + \frac{\pi}{2}\epsilon^{4+\alpha-2\beta+\rho}\bar\mu\bar a^2 + \frac{\pi}{4}\epsilon^{4+\alpha-2\beta+\sigma}\bar\lambda\bar a^2 \right] d\bar z$$

where $\bar{a}$ denotes the curvature of the curve $\bar{\varphi}$ and $\bar{c}$ its torsion. The rescaled kinetic energy is given by

$$
\bar{K} = \int \Bigg[\frac{1}{2}\epsilon^{2+\alpha+2\beta-2\delta}\dot{\bar{\varphi}}^2 + \frac{\pi}{4}\epsilon^{4+\alpha+2\gamma-2\delta}\dot{\bar{\Theta}}^2
$$
$$
+ \frac{\pi}{2\bar{a}^2}\epsilon^{4-4\alpha+5\beta+\gamma-2\delta}\langle \bar{\varphi}_{\bar{z}} \times \bar{\varphi}_{\bar{z}\bar{z}}, \dot{\bar{\varphi}}_{\bar{z}\bar{z}}\rangle \dot{\bar{\Theta}}
$$
$$
+ \frac{\pi}{4\bar{a}^2}\epsilon^{4-3\alpha+4\beta-2\delta}\|\dot{\bar{\varphi}}_{\bar{z}\bar{z}}\|^2 - \frac{\pi}{4\bar{a}^4}\epsilon^{4-7\alpha+6\beta-2\delta}\langle \bar{\varphi}_{\bar{z}\bar{z}}, \dot{\bar{\varphi}}_{\bar{z}\bar{z}}\rangle^2
$$
$$
+ \frac{\pi}{8\bar{a}^2}\epsilon^{4-5\alpha+6\beta-2\delta}\|\bar{\varphi}_{\bar{z}\bar{z}} \times \dot{\bar{\varphi}}_{\bar{z}}\|^2
$$
$$
+ \frac{\pi}{4\bar{a}^2}\epsilon^{4-5\alpha+6\beta-2\delta}\langle \dot{\bar{\varphi}}_{\bar{z}} \times \bar{\varphi}_{\bar{z}\bar{z}}, \bar{\varphi}_{\bar{z}} \times \dot{\bar{\varphi}}_{\bar{z}\bar{z}}\rangle \Bigg] d\bar{z}.
$$

Here, $\dot{\bar{\varphi}}$ and $\dot{\bar{\Theta}}$ are the derivatives of $\bar{\varphi}$ and $\bar{\Theta}$ with respect to the rescaled time variable $\bar{t}$. If one chooses β arbitrarily and sets

$$
\alpha := 2\beta - 1,
$$
$$
\gamma := \beta - 1,
$$
$$
\delta := \frac{1}{2}(4\beta + 1),
$$
$$
\rho := -3,
$$
$$
\sigma := -3,
$$

then the equations for potential and kinetic energy become

$$
\bar{V} = \int \Bigg[\frac{\pi}{4}\bar{\mu}\bar{\Theta}_{\bar{z}}^2 - \frac{\pi}{2}\bar{\mu}\bar{c}\bar{\Theta}_{\bar{z}} + \frac{\pi}{4}\bar{\mu}\bar{c}^2 + \frac{\pi}{2}\bar{\mu}\bar{a}^2 + \frac{\pi}{4}\bar{\lambda}\bar{a}^2 \Bigg] d\bar{z},
$$

and

$$
\bar{K} = \int \Bigg[\frac{1}{2}\dot{\bar{\varphi}}^2 + \frac{\pi}{4}\dot{\bar{\Theta}}^2 + \frac{\pi}{2\bar{a}^2}\epsilon^{6-6\beta}\langle \bar{\varphi}_{\bar{z}} \times \bar{\varphi}_{\bar{z}\bar{z}}, \dot{\bar{\varphi}}_{\bar{z}\bar{z}}\rangle \dot{\bar{\Theta}} + \frac{\pi}{4\bar{a}^2}\epsilon^{6-6\beta}\|\dot{\bar{\varphi}}_{\bar{z}\bar{z}}\|^2
$$
$$
- \frac{\pi}{4\bar{a}^4}\epsilon^{10-12\beta}\langle \bar{\varphi}_{\bar{z}\bar{z}}, \dot{\bar{\varphi}}_{\bar{z}\bar{z}}\rangle^2 + \frac{\pi}{8\bar{a}^2}\epsilon^{8-8\beta}\|\bar{\varphi}_{\bar{z}\bar{z}} \times \dot{\bar{\varphi}}_{\bar{z}}\|^2
$$
$$
+ \frac{\pi}{4\bar{a}^2}\epsilon^{8-8\beta}\langle \dot{\bar{\varphi}}_{\bar{z}} \times \bar{\varphi}_{\bar{z}\bar{z}}, \bar{\varphi}_{\bar{z}} \times \dot{\bar{\varphi}}_{\bar{z}\bar{z}}\rangle \Bigg] d\bar{z}.
$$

In particular, if we choose $\beta := 1/2$, then $\alpha = 0$ and the kinetic energy has the form

$$
\bar{K} = \int \Bigg[\frac{1}{2}\dot{\bar{\varphi}}^2 + \frac{\pi}{4}\dot{\bar{\Theta}}^2 + \frac{\pi}{2\bar{a}^2}\epsilon^3\langle \bar{\varphi}_{\bar{z}} \times \bar{\varphi}_{\bar{z}\bar{z}}, \dot{\bar{\varphi}}_{\bar{z}\bar{z}}\rangle \dot{\bar{\Theta}} + \frac{\pi}{4\bar{a}^2}\epsilon^3\|\dot{\bar{\varphi}}_{\bar{z}\bar{z}}\|^2
$$
$$
- \frac{1\pi}{4\bar{a}^4}\epsilon^4\langle \bar{\varphi}_{\bar{z}\bar{z}}, \dot{\bar{\varphi}}_{\bar{z}\bar{z}}\rangle^2 + \frac{\pi}{8\bar{a}^2}\epsilon^4\|\bar{\varphi}_{\bar{z}\bar{z}} \times \dot{\bar{\varphi}}_{\bar{z}}\|^2 + \frac{\pi}{4\bar{a}^2}\epsilon^4\langle \dot{\bar{\varphi}}_{\bar{z}} \times \bar{\varphi}_{\bar{z}\bar{z}}, \bar{\varphi}_{\bar{z}} \times \dot{\bar{\varphi}}_{\bar{z}\bar{z}}\rangle \Bigg] d\bar{z}.
$$

Thus, after neglecting terms of higher order in ϵ, the kinetic energy is given by

$$
\bar{K} = \int \Bigg[\frac{1}{2}\dot{\bar{\varphi}}^2 + \frac{\pi}{4}\dot{\bar{\Theta}}^2 \Bigg] d\bar{z}.
$$

Maddocks [1984] discusses the equilibrium problem for a three-director rod model in Euclidean three space and Caflisch and Maddocks [1984] discuss the dynamics of a *planar* rod model. Our model is a dynamic *three*-dimensional one-director model and we provide a justification for the expression of the kinetic energy. The justification for such expressions seems to be a general problem in the direct approach to the theory of Cosserat continua (compare Antman [1995], p. 262), whereas we *derive* our energy functional by an asymptotic procedure from the 3D theory.

6.4. The Euler-Kirchhoff Elastica

Introducing the additional holonomic constraint $\Theta = 0$ in the one-director model just derived yields a Kirchhoff elastica model whose stored energy density depends only on the curvature and torsion of the rod, namely

$$\bar{V} = \int \left[\frac{\pi}{8} \bar{\mu} \bar{c}^2 + \frac{\pi}{8} (2\bar{\mu} + \bar{\lambda}) \bar{a}^2 \right] d\bar{z}.$$

Note that if one starts from a 2D Saint-Venant Kirchhoff material, one arrives at the classical planar Euler elastica as a limit model without having to impose any extra constraint. The stored energy density of this rod model is (modulo a constant factor) given by the square of the curvature:

$$V_E = \int a^2 \, dz$$

(see Love [1944]).

One can also combine the constraint $\Theta = 0$ and the assumption that the rod is planar in the following way. We say a rod is planar if it is invariant under a one-parameter group of translations in $\mathbb{R}^3$, in which case $c = 0$. Now fix a one-parameter group of translations in $\mathbb{R}^3$. The set

$$\{(\varphi, \dot{\varphi}, \Theta, \dot{\Theta}), \varphi, \dot{\varphi} \text{ are planar (with respect to the group of translations)}, \Theta = \dot{\Theta} = 0\}$$

is an invariant submanifold of the Hamiltonian system of the one-director rod model derived in the last subsection. This can be seen as follows. The Hamiltonian function is invariant under translations in φ, so if at $t = 0$, $\varphi, \dot{\varphi}$ are planar (with respect to the one-parameter group of translations), they will remain so for $t > 0$ and hence $c = 0$. Then the Hamiltonian function splits as the sum of two functions; the first only involves $\Theta, \dot{\Theta}$, and the second only involves $\varphi, \dot{\varphi}$. Moreover, $\Theta = 0$ is a solution of the first Hamiltonian system.

The elastica has received much attention in recent years from the viewpoint of infinite-dimensional Hamiltonian systems as well as finite-dimensional Hamiltonian systems. For example, the elastica occurs as a 'soliton' solution of the vortex filament equation, which is an infinite-dimensional Hamiltonian system closely related to the nonlinear Schrödinger equation (see Langer and Perline [1991]). Here, "soliton solution" means that the elastica moves like a heavy rigid body under the dynamics of the vortex filament equations. The link between the equations for a top and the equations for the elastica can be found in Love [1944], and this link can be used to integrate the equations; the link has also been exploited in, for example, Mielke and Holmes [1988].

In Foltinek [1994], the integrability of the elastica equation is discussed from the viewpoint of finite-dimensional Hamiltonian systems with symmetries. In this approach, as in Love [1944] and Mielke and Holmes [1988], the arc length parameter of the rod is interpreted as a time-like variable. Integrals of motion can then be derived using the theory of reduction of Hamiltonian systems with symmetry, and methods of (singular) reduction are used to analyze the problem.

An instance for the ubiquity of the elastica can be found in Abresch [1987], who shows that planar λ_1-curvature lines (where $\lambda_1 < \lambda_2$ denote the principal curvatures) on a torus with constant mean curvature are solutions of the elastica equations. (As in the work of Langer and Perline, there is a connection to vortex theory: Constant mean curvature tori can be classified with the help of the sinh-Gordon equation, which also describes vortices in Plasma Physics.)

Acknowledgments

The authors wish to especially thank the late Juan-Carlos Simo, who inspired and helped initiate this work. We also thank John Maddocks and Annie Raoult for useful comments. Zhong Ge and Hans-Peter Kruse thank Jerry Marsden and the University of California for their hospitality during their visits.

The authors wish to acknowledge the following research support: the Ministry of Colleges and Universities of Ontario and the Natural Sciences and Engineering Research Council of Canada (ZG), partial support by the Humboldt Foundation during a stay at the Department of Mathematics, University of California, Berkeley CA 94720 and by the DFG under contract Sch 233/3-1 (HPK) and partial support by the Fields Institute (JEM).

References

Abresch, U. [1987] Constant mean curvature tori in terms of elliptic functions. *J. Reine Angew. Math.* **374**, 169–192.

Antman, S. S. [1972], The theory of rods, *Handbuch der Physik*, Band VIa/2, S. Flügge and C. Truesdell, eds., Springer-Verlag, Berlin, 641–703.

Antman, S. S. [1995], *Nonlinear Problems of Elasticity*, Applied Mathematical Sciences, **107**, Springer-Verlag, New York.

Antman, S. S. and W. H. Warner [1967] Dynamical theory of hyperelastic rods. *Arch. Ratl. Mech. Anal.* **23**, 135–162.

Caflisch, R. and J. H. Maddocks [1984] Nonlinear dynamical theory of the elastica. *Proc. R. Soc. Edin.* **99A**, 1–23.

Camassa, R. and D. Holm [1993] An integrable shallow water equation with peaked solitons, *Phys. Rev. Lett.*, **71**, 1661–1664.

Ciarlet, P. G. [1980], A justification of the von Kármán equations. *Arch. Ratl. Mech. Anal.* **73**, 349–389.

Ciarlet, P. G. [1994] Mathematical shell theory: recent developments and open problems, in *Duration and Change: Fifty years at Oberwolfach*, M. Artin, H. Kraft, and R. Remmert, eds., Springer-Verlag, New York, 159–176.

Ciarlet, P. G. and V. Lods [1994] Analyse asymptotique des coques linéairement élastiques. III. Une justification du modèle de W. T. Koiter. *C. R. Acad. Sci. Paris* **319** 299–304.

Ciarlet, P. G., V. Lods, and B. Miara [1994] Analyse asymptotique des coques linéairement élastiques. II. Coques "en flexion". *C. R. Acad. Sci. Paris* **319**, 95–100, 1994.

Ciarlet, P. G. and B. Miara [1992], Two dimensional shallow shell equations. *Comm. Pure Appl. Math.* **XLV**, 327–360.

Destuynder, P. [1985], A classification of thin shell theories. *Acta Appl. Math.* **4**, 15–63.

do Carmo, M. [1976], *Differential Geometry of Curves and Surfaces*, Prentice-Hall, Englewood Cliffs, N.J.

Foltinek, K. [1994] The Hamilton theory of elastica. *Amer. J. Math.* **116**, 1479–1488.

Fox, D., A. Raoult, and J.-C. Simo [1992] Modèles asymptotiques invariants pour des structures minces élastiques. *C. R. Acad. Sci. Paris* **315**, 235–240.

Fox, D., A. Raoult, and J.-C. Simo [1993] A justification of nonlinear properly invariant plate theories. *Arch. Ratl. Mech. Anal.* **124**, 157–199.

Ge, Z. [1991] Equivariant symplectic difference schemes and generating functions, *Physica D* **49**, 376–386.

Ge, Z., H. P. Kruse, J. E. Marsden and C. Scovel [1995] Poisson Brackets in the Shallow Water Approximation. *Canad. Appl. Math. Quart.* **3**, 277–302.

Ge, Z. and J. E. Marsden [1988] Lie-Poisson integrators and Lie-Poisson Hamilton-Jacobi theory, *Phys. Lett.* A **133**, 134–139.

Ge, Z. and C. Scovel [1994] A Hamiltonian truncation of the shallow water equation. *Lett. Math. Phys.* **31**, 1–13.

John, F. [1971], Refined interior equations for the elastic shells. *Comm. Pure Appl. Math.* **24**, 584–675.

Kato, T. [1985] *Abstract Differential Equations and Nonlinear Mixed Problems*. Lezioni Fermiane, Scuola Normale Superiore, Accademia Nazionale dei Lincei.

Koiter, W. T. [1970], On the foundation of the linear theory of thin elastic shells. *Proc. Kon. Nederl. Akad. Wetensch.* **B69**, 1–54.

Landau, L. D. and E. M. Lifshitz [1959], *Theory of Elasticity*, Addison-Wesley, Reading, MA.

Langer, J. and R. Perline [1991] Poisson geometry of the filament equation. *J. Nonlin. Sci.* **1**, 71–94.

Le Dret, H. and A. Raoult [1995] The nonlinear membrane model as a variational limit of nonlinear three-dimensional elasticity. *J. Math. Pure Appl.* **74**, 549–578.

Love, A. E. H. [1944] *A Treatise on the Mathematical Theory of Elasticity*. Dover, New York.

Maddocks, J. [1984] Stability of nonlinearly elastic rods. *Arch. Ratl. Mech. Anal.* **85**, 311–354.

Maddocks, J. [1991] On the stability of relative equilibria. *IMA J. Appl. Math.* **46**, 71–99.

Marsden, J. E. and T. J. R. Hughes [1994] *Mathematical Foundations of Elasticity*. Dover, New York; reprint of [1983] Prentice-Hall edition.

Marsden, J. E., T. S. Ratiu, and G. Raugel [1995] Equations d'Euler dans une coque sphérique mince (The Euler equations in a thin spherical shell), *C. R. Acad. Sci. Paris* **321**, 1201–1206.

Mielke, A. and P. Holmes [1988] Spatially complex equilibria of buckled rods. *Arch. Ratl. Mech. Anal.*, **101**, 319–348.

Naghdi, P. [1972], The theory of shells and plates. *Handbuch der Physik* Band VIa/2, S. Flügge and C. Truesdell, eds., Springer-Verlag, Berlin, 425–640.

Shi, Y. and J. E. Hearst [1994] The Kirchhoff elastic rod, the nonlinear Schrödinger equation, and DNA supercoiling. *J. Chem. Phys.* **101**, 5186–5200.

Simo, J.-C., M. S. Rifai, and D. D. Fox [1992], On a stress resultant geometrically exact shell models. Part VI: Conserving algorithms for nonlinear dynamics. *Comp. Meth. Appl. Mech. Eng.* **34**, 117–164.

Simo, J.-C., J. E. Marsden, and P. S. Krishnaprasad [1988] The Hamiltonian structure of nonlinear elasticity: The material, spatial, and convective representations of solids, rods, and plates, *Arch. Ratl. Mech. Anal.* **104**, 125–183.

Simo, J.-C., T. A. Posbergh, and J. E. Marsden [1990] Stability of coupled rigid body and geometrically exact rods: block diagonalization and the energy-momentum method, *Phys. Rep.* **193**, 280–360.

Simo, J.-C., T. A. Posbergh, and J. E. Marsden [1991] Stability of relative equilibria II: Three dimensional elasticity, *Arch. Ratl. Mech. Anal.* **115**, 61–100.

The Membrane Shell Model in Nonlinear Elasticity: A Variational Asymptotic Derivation

H. Le Dret[1] and A. Raoult[2]

[1] Laboratoire d'Analyse Numérique, Université Pierre et Marie Curie, 75252 Paris Cedex 05, France
[2] Laboratoire de Modélisation et Calcul, Université Joseph Fourier, BP 53, 38041 Grenoble Cedex 9, France

Received September 1, 1995
Communicated by Jerrold Marsden and Stephen Wiggins

This paper is dedicated to the memory of Juan-Carlos Simo

Summary. We consider a shell-like three-dimensional nonlinearly hyperelastic body and we let its thickness go to zero. We show, under appropriate hypotheses on the applied loads, that the deformations that minimize the total energy weakly converge in a Sobolev space toward deformations that minimize a nonlinear shell membrane energy. The nonlinear shell membrane energy is obtained by computing the Γ-limit of the sequence of three-dimensional energies.

1. Introduction

The purpose of this article is to derive nonlinear membrane shell models from genuine three-dimensional nonlinear elasticity by means of a rigorous convergence result. It is a sequel to a previous article concerned with planar membranes; see Le Dret and Raoult [1995].

J.-C. Simo's profound interest in the large deformation theory of thin structures is at the origin of numerous works on the derivation, analysis, and approximation of models for shells and rods that respect the fundamental requirement of continuum mechanics, frame-indifference. These models (see, for instance, Simo and Vu-Quoc [1988, 1991], Simo and Fox [1989], Simo, Fox, and Rifai [1990a, 1990b], Fox and Simo [1992], Simo and Tarnow [1994]) rely on a kinematic assumption on the possible deformed configurations of the body. In this framework, shells are one-director Cosserat structures. The shell model constructed in Simo and Fox [1989] is fully nonlinear, frame-indifferent, and couples membrane, bending and shearing effects together. Recent results on the numerical approximation and on the construction of such a model are given in Carrive-Bédouani, Le Tallec, and Mouro [1995] and Carrive-Bédouani [1995]. Important contributions to

the formulation of classical nonlinear shell theory using the Cosserat hypothesis as well
as thorough and comprehensive analyses of these models can be found in, e.g., Ericksen
and Truesdell [1958], Naghdi [1972], Green and Naghdi [1974], and Antman [1976a,
1976b, 1995].

A complementary approach to thin elastic structures theory is that of formal asymptotic expansions in powers of the thickness pioneered by Friedrichs and Dressler [1961]
and Goldenveizer [1963] and later recast in a modern functional framework by Ciarlet
and Destuynder [1979a, 1979b], Ciarlet [1980], and in the case of shells, Destuynder
[1980]. This approach also led to numerous developments, see Ciarlet [1990] for a bib-
liography. A one-year stay of the second author at the Division of Applied Mechanics
at Stanford University provided the opportunity to try and bridge the two approaches.
More specifically, the goal was to investigate whether the asymptotic expansion method
applied to thin bodies could lead to frame-indifferent limit models.

This objective is attained in Fox, Raoult, and Simo [1993] where, for simplicity, the
case of plate-like—rather than shell-like—bodies is treated and where the nonlinear
material is the Saint Venant-Kirchhoff material. It is shown that a hierarchy of two-
dimensional models can be derived from the nonlinear system of three-dimensional
elasticity by a formal asymptotic expansion. The type of limit model thus obtained
depends on the order of magnitude of the external loads. The first two models in the
hierarchy are a nonlinear membrane plate model and a nonlinear inextensional bending model for smaller loads. Both models are quasilinear and frame-indifferent. The
membrane model is also obtained by Karwowski [1993]. By lowering again the order of
magnitude of the loads, one recovers semilinear plate models that had been previously
derived by Ciarlet and Destuynder [1979b], Ciarlet [1980] for the von Kármán equations, and Raoult [1988] in the dynamical case by analogous formal expansions. Note
that these models are no longer frame-indifferent.

Although establishing the grounds for an asymptotic justification of invariant plate
models, the method of Fox, Raoult, and Simo [1993] is purely formal. It was the purpose
of the work by Le Dret and Raoult [1993, 1995] to provide a rigorous proof of convergence to a nonlinear membrane model. There, as in Fox, Raoult, and Simo [1993],
only plate-like bodies are considered. The assumptions on the external loads are those
of Fox, Raoult, and Simo [1993], but the results are obtained for a general hyperelastic material. The main mathematical tool is Γ-convergence theory, a systematic way
of analyzing the convergence of minimizers of a sequence of problems of the Calculus of Variations. Ideas from Γ-convergence theory had been previously introduced
in the context of lower-dimensional theories in nonlinear elasticity by Acerbi, Buttazzo, and Percivale [1991] for nonlinearly elastic strings. Their method was extended
to nonlinear planar membranes in a preprint by Percivale [1991] recently drawn to the
authors' attention. Note that in the case of strings the limit model is one-dimensional
and thus convexity arguments can be used that are not sufficient in the two-dimensional
case.

Let us briefly recall the limit membrane model obtained in Le Dret and Raoult [1993,
1995]; see also Percivale [1991]. Starting from a three-dimensional stored energy function W defined on three-dimensional deformation gradients, i.e., 3×3 matrices, the limit
membrane energy density, which is defined on membrane deformation gradients, i.e.,
3×2 matrices, is constructed in two steps. First W is minimized with respect to the third

column of its matrix argument, then the resulting function is quasiconvexified. It is worth mentioning that, except in some very special cases, this quasiconvexification step cannot be skipped. In the case of the Saint Venant-Kirchhoff density, the quasiconvexification is carried out explicitly in Le Dret and Raoult [1995]. A quite surprising consequence of this calculation is that the formal limit membrane energy of Fox, Raoult, and Simo [1993] only coincides with the rigorous limit energy on a compact subset of the set of 3×2 matrices.

The present article extends the analysis of Le Dret and Raoult [1995] to the case of shells. Other recent works in the field of asymptotic justification and analysis of lower-dimensional linear or nonlinear shell models include Sanchez-Palencia [1989a, 1989b, 1990], Ciarlet and Lods [1994], Ciarlet, Lods, and Miara [1994], Miara [1994].

An overview of the article is as follows. Section 2 is devoted to introducing the geometrical notation for a shell with midsurface $\tilde{S}$ and thickness 2ε in its reference configuration $\tilde{\Omega}_\varepsilon$. We assume that the shell is made of a hyperelastic homogeneous material with stored energy function W. In Section 3, we state the equilibrium problem for the shell as an energy minimization problem over a set of admissible deformations included in the Sobolev space $W^{1,p}(\tilde{\Omega}_\varepsilon; \mathbb{R}^3)$. To study the asymptotic behavior of the corresponding energy minimizers when $\varepsilon \to 0$, we define an equivalent minimization problem set on a straight cylindrical domain $\Omega = \omega \times]-1, 1[$ of $\mathbb{R}^3$ independent of ε. This is achieved by transporting the deformations and the external loads through a chart and rescaling them. As opposed to the planar case, the geometry of the shell intervenes in the expression of the rescaled hyperelastic energies through the Jacobian matrix of the change of coordinates. This matrix depends on ε and appears notably inside the argument of the stored energy density.

In Section 4, we give our first convergence result expressed in terms of rescaled displacements. We determine the Γ-limit of the sequence of rescaled energies. The construction of the limit energy density extends that of the planar case. Note however that the nontrivial geometry of the shell causes the limit energy to depend on the point $x \in \omega$ even for a homogeneous three-dimensional material.

In Section 5, we translate the Γ-convergence result of Section 4 in terms of deformations and show that the minimizing deformations weakly converge in $W^{1,p}(\Omega; \mathbb{R}^3)$ toward deformations that minimize a limit nonlinear shell energy. The limit deformations depend on two space variables only (they are identified with functions defined on the transported midsurface ω). The limit elastic energy of a deformation $\bar{\varphi}$ depends on its first derivatives and thus does not incorporate bending effects associated with curvature or shear effects. Consequently, it is a membrane energy. Let us emphasize the fact that, contrarily to methods relying on Cosserat assumptions, our analysis provides an exact formula for deriving the limit energy, hence the membrane constitutive law, from the three-dimensional energy W. Furthermore, we give an intrinsic formulation of the membrane minimization problem and an intrinsic expression of the nonlinear membrane energy by transporting the obtained result back on the reference surface $\tilde{S}$. The stored energy depends on a deformation $\tilde{\varphi}$ defined on $\tilde{S}$ only through its gradient (for a definition of this gradient, see Section 5). It depends on the current point of $\tilde{S}$, but only through the normal vector to $\tilde{S}$ at this point.

In Section 6, we study how the limit membrane shell energy inherits the invariance properties of the three-dimensional energy. In particular, it is shown that frame-indifference is preserved and that the shell energy depends on the deformation only through the deformed metric. Moreover, if W has a global zero minimum at $F = I$, the corresponding membrane shell energy is zero for compressive states. Isotropy is also preserved. In this case, the membrane stored energy does no longer depend on the normal vector to the reference surface $\tilde{S}$ and depends on the deformation only through the principal stretches.

The second author is indebted to J.-C. Simo for inviting her to spend a sabbatical year at the Division of Applied Mechanics at Stanford University in 1990–1991. Her work owes much to J.-C. Simo's brilliance and enthusiasm.

2. Geometrical Preliminaries

The summation convention is assumed throughout this article, unless otherwise specified. Greek indices take their values in the set $\{1, 2\}$ and Latin indices take their values in the set $\{1, 2, 3\}$. Let (e_1, e_2, e_3) be the canonical orthonormal basis of the Euclidean space $\mathbb{R}^3$. The norm of a vector of $\mathbb{R}^3$ will be denoted by $\|u\|$, the scalar product of two vectors of $\mathbb{R}^3$ by $u \cdot v$, and their vector product by $u \wedge v$. In the sequel, we will identify $\mathbb{R}^2$ with the plane spanned by the vectors e_1 and e_2. Accordingly and depending on the context, x will denote a generic point of $\mathbb{R}^2$ or $\mathbb{R}^3$. Let M_3 be the space of real 3×3 matrices endowed with the usual Euclidean norm $\|F\| = \sqrt{\operatorname{tr}(F^T F)}$. For any $z_i \in \mathbb{R}^3, i = 1, 2, 3$, we denote by $(z_1 \mid z_2 \mid z_3)$ the matrix whose ith column consists of the components of z_i in the canonical basis.

We assume that the midsurface $\tilde{S}$ is a bounded, two-dimensional, C^2-submanifold of $\mathbb{R}^3$, which, for simplicity, admits an atlas consisting of one chart only. Let ψ be this chart. It is thus a C^2-mapping from a bounded, open set $\omega \subset \mathbb{R}^2$ into $\mathbb{R}^3$ which is a global diffeomorphism between ω and $\tilde{S}$. We assume that ω has a Lipschitz boundary and that ψ admits an extension to $\bar{\omega}$ into a $C^2(\bar{\omega}; \mathbb{R}^3)$-function.

Let $a_\alpha(x) = \frac{\partial \psi}{\partial x_\alpha}(x)$ be the covariant basis vectors of the tangent plane $T_{\psi(x)}\tilde{S}$, associated with the chart ψ. These vectors are linearly independent on ω. We assume furthermore that there exists $\delta > 0$ such that

$$\|a_1(x) \wedge a_2(x)\| \geq \delta \qquad \text{on } \bar{\omega}. \tag{1}$$

We then define $a_3(x) = \frac{a_1(x) \wedge a_2(x)}{\|a_1(x) \wedge a_2(x)\|} \in C^1(\bar{\omega}; S^2)$, which is a unit normal vector to $T_{\psi(x)}\tilde{S}$. If no confusion may arise from it, we will write $a_3(\tilde{x})$ for $a_3(x)$ at point $\tilde{x} = \psi(x)$, since it is important to remember that a_3 is actually chart-independent (modulo multiplication by -1). The contravariant basis vectors are defined by the relations $a^\alpha(x) \in T_{\psi(x)}\tilde{S}$, $a^\alpha(x) \cdot a_\beta(x) = \delta_{\alpha\beta}$, and $a^3(x) = a_3(x)$.

Using this notation, we let $A(x) = (a_1(x) \mid a_2(x) \mid a_3(x))$. This matrix is everywhere nonsingular on $\bar{\omega}$ and its inverse is given by $A^{-1}(x) = (a^1(x) \mid a^2(x) \mid a^3(x))^T$. Note that due to our choice of unit normal vector, $\det A(x) = \|a_1(x) \wedge a_2(x)\| \geq \delta > 0$ on $\bar{\omega}$. Thus, there exists a constant C such that

$$\forall x \in \bar{\omega}, \qquad \|A^{-1}(x)\| \leq C. \tag{2}$$

The function $\det A(x)$ also satisfies

$$\det A(x) = \| \operatorname{cof} A(x)e_3 \| = \sqrt{a(x)} \tag{3}$$

where a is the determinant of the metric on $\tilde{S}$ expressed in the chart ψ.

For $\varepsilon > 0$, we consider the set $\tilde{\Omega}_\varepsilon$ defined by

$$\tilde{\Omega}_\varepsilon = \{\tilde{y} \in \mathbb{R}^3; \exists \tilde{x} \in \tilde{S}, \tilde{y} = \tilde{x} + \eta a_3(\tilde{x}) \text{ with } |\eta| < \varepsilon\}. \tag{4}$$

This set is the reference configuration of a shell of thickness 2ε. Due to our regularity hypothesis on $\tilde{S}$ there exists a C^1-orthogonal projection mapping $\tilde{\Pi}: \tilde{\Omega}_\varepsilon \to \tilde{S}$ if ε is small enough, which will be understood thereafter. Any $\tilde{y}$ in $\tilde{\Omega}_\varepsilon$ can be uniquely decomposed as $\tilde{y} = \tilde{\Pi}(\tilde{y}) + [(\tilde{y} - \tilde{\Pi}(\tilde{y})) \cdot a_3(\tilde{\Pi}(\tilde{y}))] a_3(\tilde{\Pi}(\tilde{y}))$. With this notation, $(x_1, x_2) = \psi^{-1}(\tilde{\Pi}(\tilde{y}))$ and $x_3 = (\tilde{y} - \tilde{\Pi}(\tilde{y})) \cdot a_3(\tilde{\Pi}(\tilde{y}))$ define the natural curvilinear coordinate system in $\tilde{\Omega}_\varepsilon$ that is associated with the chart ψ of the midsurface. If

$$\Omega_\varepsilon = \{x \in \mathbb{R}^3; (x_1, x_2) \in \omega, |x_3| < \varepsilon\}, \tag{5}$$

then the C^1-diffeomorphism $\Psi: \Omega_\varepsilon \to \tilde{\Omega}_\varepsilon$ defined by

$$\Psi(x) = \psi(x_1, x_2) + x_3 a_3(x_1, x_2) \tag{6}$$

is the inverse of this change of coordinates. Its gradient is the matrix

$$\nabla \Psi(x) = A(x_1, x_2) + x_3(\partial_1 a_3(x_1, x_2) \mid \partial_2 a_3(x_1, x_2) \mid 0). \tag{7}$$

Naturally, this gradient is everywhere nonsingular as soon as ε is small enough. In the context of nonlinear elasticity, the mapping Ψ^{-1} can also be viewed as a change of reference configuration for the shell (Ψ is orientation preserving).

3. The Three-Dimensional and Rescaled Problems

We assume that the shells are made of the same hyperelastic homogeneous material whose stored energy function is denoted by W. The function $W: M_3 \to \mathbb{R}$ is continuous and satisfies the growth and coercivity hypotheses

$$\begin{cases} \exists C > 0, \exists p \in]1, +\infty[, \forall F \in M_3, |W(F)| \leq C(1 + \|F\|^p), \\ \exists \alpha > 0, \exists \beta \geq 0, \forall F \in M_3, W(F) \geq \alpha\|F\|^p - \beta, \\ \forall F, F' \in M_3, \quad |W(F) - W(F')| \leq C(1 + \|F\|^{p-1} + \|F'\|^{p-1})\|F - F'\|. \end{cases} \tag{8}$$

Assumption $(8)_3$ was not needed in the case of planar membranes considered in Le Dret and Raoult [1995]. It is however quite natural. In particular, if W is quasiconvex $(8)_1$ implies $(8)_3$ (cf. Marcellini [1985]). Assumption $(8)_3$ also holds true if W is continuously differentiable, and its derivative grows as $\|F\|^{p-1}$ at infinity.

Let $S_\varepsilon^\pm = \omega \times \{\pm\varepsilon\}$ and define $\tilde{S}_\varepsilon^\pm = \Psi(S_\varepsilon^\pm)$ to be the top and bottom surfaces of the shell. For simplicity, we assume that the shells are solely submitted to the action of dead loading surface traction densities $\tilde{g}^\varepsilon \in L^q(\tilde{S}_\varepsilon^\pm; \mathbb{R}^3)$ with $1/p + 1/q = 1$. Taking body forces and lateral forces into account is straightforward. An example of live loads is detailed in the appendix. Let $\Gamma_\varepsilon = \partial \omega \times]-\varepsilon, \varepsilon[$ and $\tilde{\Gamma}_\varepsilon = \Psi(\Gamma_\varepsilon)$ be the lateral surface of $\tilde{\Omega}_\varepsilon$. We assume that the deformations of the shells satisfy a boundary condition of place on $\tilde{\Gamma}_\varepsilon$. The equilibrium problem may be formulated as a minimization problem:

$$\text{Find } \tilde{\phi}^\varepsilon \in \tilde{\Phi}_\varepsilon \text{ such that } \tilde{I}_\varepsilon(\tilde{\phi}^\varepsilon) = \inf_{\tilde{\varphi} \in \tilde{\Phi}_\varepsilon} \tilde{I}_\varepsilon(\tilde{\varphi}), \tag{9}$$

where the total energy $\tilde{I}_\varepsilon$ is

$$\tilde{I}_\varepsilon(\tilde{\varphi}) = \int_{\tilde{\Omega}_\varepsilon} W(\nabla \tilde{\varphi}) \, dx - \int_{\tilde{S}_\varepsilon} \tilde{g}^\varepsilon \cdot \tilde{\varphi} \, d\tilde{\sigma}_\varepsilon, \tag{10}$$

$d\tilde{\sigma}_\varepsilon$ is the surface element on $\tilde{S}_\varepsilon$, and the set of admissible deformations is

$$\tilde{\Phi}_\varepsilon = \{\tilde{\varphi} \in W^{1,p}(\tilde{\Omega}_\varepsilon; \mathbb{R}^3); \tilde{\varphi}(x) = x \text{ on } \tilde{\Gamma}_\varepsilon\}. \tag{11}$$

See Wang and Truesdell [1973], Marsden and Hughes [1983], or Ciarlet [1988], among others, for general references on three-dimensional nonlinear elasticity. A key ingredient in existence proofs using the direct method of the calculus of variations is the sequential weak lower semi-continuity of the energy functional $\tilde{I}_\varepsilon$ on $W^{1,p}(\tilde{\Omega}_\varepsilon; \mathbb{R}^3)$. Under assumptions (8), it is known that the energy functional $\tilde{I}_\varepsilon$ in problem (9) is sequentially weakly lower semi-continuous on $W^{1,p}(\tilde{\Omega}_\varepsilon; \mathbb{R}^3)$ if and only if the function W is quasiconvex, i.e.,

$$\forall F \in M_3, \forall \theta \in W_0^{1,\infty}(O; \mathbb{R}^3), \int_O W(F + \nabla\theta(x)) \, dx \geq (\text{meas } O) W(F), \tag{12}$$

where O is any bounded domain of $\mathbb{R}^3$; see Morrey [1952], Acerbi and Fusco [1984], Dacorogna [1989], and the references therein. Problem (9) was solved in the more physical case $W(F) = +\infty$ if $\det F \leq 0$ and $W(F) \to +\infty$ when $\det F \to 0^+$ by Ball [1977], under an assumption of polyconvexity of W, a notion more restrictive than quasiconvexity, plus appropriate growth and coercivity assumptions. For our purposes here, it is not desirable to assume at the onset that W is quasiconvex or polyconvex. There are two reasons for this. First, the zero thickness limit model we obtain always involves a quasiconvexification, which has to be effected whether W is quasiconvex or not. Second, we do not want to rule out important examples, such as the Saint Venant-Kirchhoff stored energy function, which is neither polyconvex nor quasiconvex; see Raoult [1986]. Consequently, we *do not* assume that W is quasiconvex and problem (9) may well not possess any solutions. Naturally, if it does have solutions that are thus actual equilibrium deformations of the bodies, our results apply to these deformations.

Let us thus be given a diagonal minimizing sequence $\tilde{\phi}^\varepsilon$ for the sequence of energies $\tilde{I}_\varepsilon$ over the sets $\tilde{\Phi}_\varepsilon$. More specifically, we assume that

$$\tilde{\phi}^\varepsilon \in \tilde{\Phi}_\varepsilon, \quad \tilde{I}_\varepsilon(\tilde{\phi}^\varepsilon) \leq \inf_{\tilde{\varphi} \in \tilde{\Phi}_\varepsilon} \tilde{I}_\varepsilon(\tilde{\varphi}) + \varepsilon h(\varepsilon), \tag{13}$$

where h is a positive function such that $h(\varepsilon) \to 0$ when $\varepsilon \to 0$. Such a sequence always exists and, if the minimization problems have solutions, $\tilde{\phi}^\varepsilon$ may be chosen to be such a solution.

As in the case of a planar membrane, we assume that $\|\tilde{g}^{\varepsilon}\|_{L^q(\tilde{S}^{\pm}_{\varepsilon};\mathbb{R}^3)} \le C\varepsilon$ where the constant C does not depend on ε. If we also considered body force densities or lateral traction densities, we would assume them to be of the order of 1 so that all force resultants would be of the order of ε (in particular, the weight of the material is allowed). This is essential in order to obtain a membrane model in the limit; see Le Dret and Raoult [1995] and Fox, Raoult, and Simo [1993] for a discussion of this observation.

We first rewrite problem (9) in the curvilinear coordinate system or, equivalently, we consider Ω_ε as a new reference configuration. Note that this configuration is not homogeneous anymore. If $\tilde{\varphi}$ is a deformation of the shell in its first reference configuration, we thus define for almost all $x \in \Omega_\varepsilon$,

$$\varphi(x) = \tilde{\varphi}(\Psi(x)), \tag{14}$$

and the set of admissible deformations becomes

$$\Phi_\varepsilon = \{\varphi \in W^{1,p}(\Omega_\varepsilon; \mathbb{R}^3); \varphi(x) = \Psi(x) \text{ on } \Gamma_\varepsilon\}. \tag{15}$$

Similarly, we set for almost all $x \in S^{\pm}_\varepsilon$

$$g^\varepsilon(x) = \tilde{g}^\varepsilon(\Psi(x)), \tag{16}$$

and by defining $I_\varepsilon(\varphi) = \tilde{I}_\varepsilon(\tilde{\varphi})$ to be the energy in the new reference configuration, i.e.,

$$\begin{aligned}
I_\varepsilon(\varphi) &= \int_{\Omega_\varepsilon} W(\nabla\varphi(x)\nabla\Psi(x)^{-1})\det(\nabla\Psi(x))\,dx \\
&\quad - \int_{S^{\pm}_\varepsilon} g^\varepsilon(x) \cdot \varphi(x)\|\operatorname{cof}\nabla\Psi(x)e_3\|\,d\sigma_\varepsilon,
\end{aligned} \tag{17}$$

we obtain

$$\phi^\varepsilon \in \Phi_\varepsilon, \qquad I_\varepsilon(\phi^\varepsilon) \le \inf_{\varphi \in \Phi_\varepsilon} I_\varepsilon(\varphi) + \varepsilon h(\varepsilon). \tag{18}$$

All these definitions may also be rewritten in terms of displacements $v(x) = \varphi(x) - \Psi(x)$.

We now are in a position to rescale the problem. Let $\Omega = \Omega_1$, $\Gamma = \Gamma_1$, and $S^\pm = S^{\pm}_1$ and define a rescaling operator Θ_ε by $(\Theta_\varepsilon\varphi)(x_1, x_2, x_3) = \varphi(x_1, x_2, \varepsilon x_3)$. Let $\phi(\varepsilon) = \Theta_\varepsilon\phi^\varepsilon$, $\Psi(\varepsilon)(x) = \Theta_\varepsilon\Psi$, and $g(\varepsilon)(x) = \Theta_\varepsilon g^\varepsilon$. The rescaled displacement $u(\varepsilon) = \phi(\varepsilon) - \Psi(\varepsilon)$ belongs to $V = W^{1,p}_\Gamma(\Omega; \mathbb{R}^3)$. We rescale the energies by setting $I(\varepsilon)(\varphi) = \varepsilon^{-1}I_\varepsilon(\Theta^{-1}_\varepsilon\varphi)$, i.e.,

$$\begin{aligned}
I(\varepsilon)(\varphi) &= \int_\Omega W\left(\left(\partial_1\varphi \mid \partial_2\varphi \mid \frac{\partial_3\varphi}{\varepsilon}\right)A(\varepsilon)^{-1}\right)\det A(\varepsilon)\,dx \\
&\quad - \int_{S^\pm} \varepsilon^{-1}g(\varepsilon) \cdot \varphi\|\operatorname{cof}A(\varepsilon)e_3\|\,d\sigma,
\end{aligned} \tag{19}$$

with the notation

$$A(\varepsilon)(x) = \nabla\Psi(x_1, x_2, \varepsilon x_3) = A(x_1, x_2) + \varepsilon x_3(\partial_1 a_3(x_1, x_2) \mid \partial_2 a_3(x_1, x_2) \mid 0). \tag{20}$$

In terms of the rescaled displacements, the rescaled energy reads:

$$J(\varepsilon)(v) = \int_\Omega W\left(\left(\partial_1 v \,|\, \partial_2 v \,|\, \frac{\partial_3 v}{\varepsilon}\right) A(\varepsilon)^{-1} + I\right) \det A(\varepsilon)\, dx$$
$$- \int_{S^\pm} \varepsilon^{-1} g(\varepsilon) \cdot (\Psi(\varepsilon) + v) \|\operatorname{cof} A(\varepsilon) e_3\|\, d\sigma. \tag{21}$$

It is immediate that

$$J(\varepsilon)(u(\varepsilon)) \le \inf_{v \in V} J(\varepsilon)(v) + h(\varepsilon). \tag{22}$$

For notational brevity, we also introduce the rescaled elastic energy

$$E(\varepsilon)(v) = \int_\Omega W\left(\left(\partial_1 v \,|\, \partial_2 v \,|\, \frac{\partial_3 v}{\varepsilon}\right) A(\varepsilon)^{-1} + I\right) \det A(\varepsilon)\, dx, \tag{23}$$

and the rescaled virtual work of the applied loads

$$L(\varepsilon)(v) = \int_{S^\pm} \varepsilon^{-1} g(\varepsilon) \cdot (\Psi(\varepsilon) + v) \|\operatorname{cof} A(\varepsilon) e_3\|\, d\sigma. \tag{24}$$

The assumed bound on $\tilde{g}^\varepsilon$ ensures that $\|\varepsilon^{-1} g(\varepsilon)\|_{L^q(S^\pm;\mathbb{R}^3)} \le C$. We may thus assume that there exists $g \in L^q(S^\pm; \mathbb{R}^3)$ such that

$$\varepsilon^{-1} g(\varepsilon) \longrightarrow g \qquad \text{in } L^q(S^\pm; \mathbb{R}^3)$$

by extracting a subsequence, if necessary. Examples of such loadings are for instance dead loadings normal to the reference configurations of the shell, of the form $g^\varepsilon(x) = \varepsilon h^+(x_1, x_2) a_3(x_1, x_2)$ if $x_3 = \varepsilon$ and $g^\varepsilon(x) = \varepsilon h^-(x_1, x_2) a_3(x_1, x_2)$ if $x_3 = -\varepsilon$. See the Appendix for a concise treatment of pressure loads.

4. Computation of the Γ-Limit of the Rescaled Energies

We use Γ-convergence theory to determine the asymptotic behavior of the rescaled displacements $u(\varepsilon)$ when $\varepsilon \to 0$. In the sequel, the thickness parameter ε will take its values in a sequence $\varepsilon_n \to 0$. Since the results do not depend on the sequence in question, and also for notational brevity, we will simply use the notation ε. Let us recall that a sequence of functions G_ε from a metric space X into $\bar{\mathbb{R}}$ is said to Γ-converge toward G_0 for the topology of X if the following two conditions are satisfied for all $x \in X$:

$$\begin{cases} \forall x_\varepsilon \to x,\ \liminf G_\varepsilon(x_\varepsilon) \ge G_0(x), \\ \exists\, y_\varepsilon \to x,\ G_\varepsilon(y_\varepsilon) \to G_0(x). \end{cases}$$

If the sequence G_ε Γ-converges, its Γ-limit is alternatively given by

$$G_0(x) = \min\{\liminf G_\varepsilon(x_\varepsilon);\ x_\varepsilon \to x\}.$$

In addition, the set of functions from X into $\bar{\mathbb{R}}$ has a sequential compactness property with respect to Γ-convergence in the sense that any sequence $G_\varepsilon: X \to \bar{\mathbb{R}}$ admits a

Γ-convergent subsequence. The main interest of Γ-convergence is that if the minimizers of G_ε stay in a compact set of X for all ε, then their limit points are minimizers of G_0; see De Giorgi and Franzoni [1975], Attouch [1984], and Dal Maso [1993].

We extend the energies to $L^p(\Omega; \mathbb{R}^3)$ by setting

$$\forall v \in L^p(\Omega; \mathbb{R}^3), \qquad J^*(\varepsilon)(v) = \begin{cases} J(\varepsilon)(v) & \text{if } v \in V, \\ +\infty & \text{otherwise.} \end{cases} \tag{25}$$

Let us now proceed to compute the Γ-limit of the sequence $J^*(\varepsilon)$ for the strong topology of $L^p(\Omega; \mathbb{R}^3)$. Let $M_{3,2}$ be the space of 3×2 real matrices endowed with the usual Euclidean norm

$$\|\bar{F}\| = \sqrt{\text{tr}(\bar{F}^T \bar{F})}.$$

We denote by $(z_1 \mid z_2)$ the matrix of $M_{3,2}$ whose αth column is composed of the components of $z_\alpha \in \mathbb{R}^3$ in the canonical basis. For all $\bar{F} = (z_1 \mid z_2) \in M_{3,2}$ and $z \in \mathbb{R}^3$, we also denote by $(\bar{F} \mid z)$ the matrix whose first two columns are z_1 and z_2 and whose third column is z.

Let us introduce a function $W_0 : \bar{\omega} \times M_{3,2} \to \mathbb{R}$

$$W_0(x, \bar{F}) = \inf_{z \in \mathbb{R}^3} W((\bar{F} \mid z)A^{-1}(x)). \tag{26}$$

Due to the coercivity assumption $(8)_2$, it is clear that this function is well defined. Besides, since W is continuous, the infimum is attained. Let us briefly state a few properties of W_0: The function W_0 is continuous on $\bar{\omega} \times M_{3,2}$ and satisfies the growth and coercivity estimates

$$\begin{cases} \exists C' > 0, \forall \bar{F} \in M_{3,2}, \forall x \in \bar{\omega}, |W_0(x, \bar{F})| \leq C'(1 + \|\bar{F}\|^p), \\ \exists \alpha' > 0, \exists \beta' \geq 0, \forall \bar{F} \in M_{3,2}, \forall x \in \bar{\omega}, W_0(x, \bar{F}) \geq \alpha' \|\bar{F}\|^p - \beta'. \end{cases} \tag{27}$$

See Le Dret and Raoult [1995] for a proof in the planar case. We use here in addition the continuity of A and A^{-1} on $\bar{\omega}$.

Let $QW_0 = \sup\{Z : \bar{\omega} \times M_{3,2} \to \mathbb{R}, Z \text{ quasiconvex}, Z \leq W_0\}$ be the quasiconvex envelope of W_0; see Dacorogna [1982] for the definition and properties of quasiconvex functions and quasiconvex envelopes. Recall that a function Z of x and $\bar{F}$ is quasiconvex if it satisfies

$$\forall x_0 \in \bar{\omega}, \forall \bar{F} \in M_3, \forall \theta \in W_0^{1,\infty}(O; \mathbb{R}^3),$$

$$\int_O Z(x_0, \bar{F} + \nabla\theta(x)) \, dx \geq (\text{meas } O) Z(x_0, \bar{F}), \tag{28}$$

where O is any bounded domain of $\mathbb{R}^2$. This is the same definition as (12) in the 3×2 case with the variable x_0 frozen. We introduce the space

$$V_M = \{v \in V; \partial_3 v = 0\}, \tag{29}$$

which we call the space of membrane displacements. This space is canonically isomorphic to $W_0^{1,p}(\omega; \mathbb{R}^3)$ and we let $\bar{v}$ denote the element of $W_0^{1,p}(\omega; \mathbb{R}^3)$ that is associated with $v \in V_M$ through this isomorphism. The expression of the Γ-limit of the sequence $J^*(\varepsilon)$ is given in the following theorem.

Theorem 1. *The sequence $J^*(\varepsilon)$ Γ-converges for the strong topology of $L^p(\Omega; \mathbb{R}^3)$ when $\varepsilon \to 0$. Let $J^*(0)$ be its Γ-limit. For all $v \in L^p(\Omega; \mathbb{R}^3)$, $J^*(0)(v)$ is given by*

$$J^*(0)(v) = \begin{cases} 2 \int_\omega QW_0(x, (a_1 + \partial_1 \bar{v} \mid a_2 + \partial_2 \bar{v})) \sqrt{a}\, dx_1\, dx_2 \\ \quad - \int_\omega \mathcal{G} \cdot (\psi + \bar{v}) \sqrt{a}\, dx_1\, dx_2 & \text{if } v \in V_M, \quad (30) \\ +\infty & \text{otherwise,} \end{cases}$$

where $\mathcal{G}(x_1, x_2) = g(x_1, x_2, 1) + g(x_1, x_2, -1)$.

For clarity, we break the proof of Theorem 1 into a series of lemmas. We will return to the mechanical interpretation of Theorem 1 in the next section. Let us first give a few simple convergence results for the various geometrical quantities associated with the shells.

Lemma 2. *The matrix $A(\varepsilon)$ satisfies*

$$\det A(\varepsilon) = \| \operatorname{cof} A(\varepsilon) e_3 \| \to \sqrt{a} \text{ in } C^0(\bar{\Omega}), \qquad A(\varepsilon)^{-1} \to A^{-1} \text{ in } C^0(\bar{\Omega}; M_3) \quad (31)$$

and the rescaled chart $\Psi(\varepsilon)$ satisfies

$$\Psi(\varepsilon) \to \Psi(0) \text{ in } C^0(\bar{\Omega}; \mathbb{R}^3) \tag{32}$$

where $\Psi(0)(x_1, x_2, x_3) = \psi(x_1, x_2)$.

We now extract a Γ-convergent subsequence, still denoted $J^*(\varepsilon)$, and call $J^*(0)$ its Γ-limit. The uniqueness of $J^*(0)$ will make the extraction of this subsequence superfluous *a posteriori*.

Lemma 3. *Let $v(\varepsilon) \in L^p(\Omega; \mathbb{R}^3)$ be a sequence such that $J^*(\varepsilon)(v(\varepsilon)) \leq C < +\infty$ where C does not depend on ε. Then $v(\varepsilon)$ is uniformly bounded in V, $\varepsilon^{-1} \partial_3 v(\varepsilon)$ is uniformly bounded in $L^p(\Omega; \mathbb{R}^3)$, and the limit points of $v(\varepsilon)$ for the weak topology of V belong to V_M.*

Proof. Consider a sequence $v(\varepsilon) \in L^p(\Omega; \mathbb{R}^3)$ such that

$$J^*(\varepsilon)(v(\varepsilon)) \leq C < +\infty. \tag{33}$$

The definition (25) of $J^*(\varepsilon)$ implies first that $v(\varepsilon) \in V$ for all $\varepsilon > 0$.

The following inequality is an easy consequence of Lemma 2:

$$\| FA(\varepsilon)^{-1} + I \|^p \geq c_1 \| F \|^p - c_2, \tag{34}$$

where $c_1 > 0$ and c_2 do not depend either on ε or on x. Furthermore, it is clear that for all $\varepsilon \leq 1$, $\| (z_1 \mid z_2 \mid \varepsilon^{-1} z_3) \| \geq \| (z_1 \mid z_2 \mid z_3) \|$. It follows then from the coercivity of the function W, estimate (34), and Lemma 2 that there exists constants $c_3 > 0$ and c_4

such that

$$J^*(\varepsilon)(v(\varepsilon)) \geq c_3 \|\nabla v(\varepsilon)\|_{L^p(\Omega; M_3)}^p - c_4 - \|\varepsilon^{-1}g(\varepsilon)\|_{L^q(S^\pm; \mathbb{R}^3)} \|v(\varepsilon)\|_{W^{1,p}(\Omega; \mathbb{R}_3)}. \quad (35)$$

Therefore, Poincaré's inequality implies the desired uniform bound for $v(\varepsilon)$.

On the other hand, since $\|(z_1 \mid z_2 \mid \varepsilon^{-1}z_3)\| \geq \varepsilon^{-1}\|z_3\|$, it follows from inequalities (33) and (34) that $\|\partial_3 v(\varepsilon)\|_{L^p(\Omega; \mathbb{R}^3)} \leq c_5\varepsilon$, so that $\partial_3 v(\varepsilon) \to 0$ strongly in $L^p(\Omega; \mathbb{R}^3)$. If we let v denote any limit point of the sequence $v(\varepsilon)$ for the weak topology of $W_\Gamma^{1,p}(\Omega; \mathbb{R}^3)$, it follows at once that $\partial_3 v = 0$, hence v belongs to V_M. □

Corollary 4. *If $v \in L^p(\Omega; \mathbb{R}^3)$ but $v \notin V_M$, then $J^*(0)(v) = +\infty$.*

Proof. Indeed, if $J^*(0)(v) < +\infty$, there exists a sequence $v(\varepsilon)$ that converges strongly to v in $L^p(\Omega; \mathbb{R}^3)$ and such that $J^*(\varepsilon)(v(\varepsilon)) \to J^*(0)(v)$. By Lemma 3, $v(\varepsilon) \rightharpoonup v$ in V and $v \in V_M$. □

We thus only have to compute the value of the Γ-limit for displacements in V_M. We first establish a bound from below for the Γ-limit functional.

Proposition 5. *For all $v \in V_M$, we have that*

$$J^*(0)(v) \geq 2 \int_\omega QW_0(x, (a_1 + \partial_1\bar{v} \mid a_2 + \partial_2\bar{v}))\sqrt{a}\,dx_1dx_2$$

$$- \int_\omega \mathcal{G} \cdot (\psi + \bar{v})\sqrt{a}\,dx_1dx_2. \quad (36)$$

Proof. Consider any $v \in V_M$. Since $J(\varepsilon)(v)$ is obviously bounded from above independently of ε, it follows that $J^*(0)(v) < +\infty$. By the definition of Γ-convergence, there exists a sequence $v(\varepsilon)$ such that $v(\varepsilon) \to v$ strongly in $L^p(\Omega; \mathbb{R}^3)$ and $J^*(\varepsilon)(v(\varepsilon)) \to J^*(0)(v)$, so that $v(\varepsilon) \in V$. Moreover, by Lemma 3, $v(\varepsilon) \rightharpoonup v$ weakly in V, hence its trace on $S^\pm$ converges strongly in $L^p(S^\pm; \mathbb{R}^3)$. Since $\varepsilon^{-1}g(\varepsilon) \rightharpoonup g$ weakly in $L^q(S^\pm; \mathbb{R}^3)$, we thus have

$$L(\varepsilon)(v(\varepsilon)) \to L(0)(v) = \int_{S^\pm} g \cdot (\Psi(0) + v)\sqrt{a}\,d\sigma. \quad (37)$$

Let us examine the asymptotic behavior of the rescaled elastic energy. Let $\varphi(\varepsilon) = \Psi(\varepsilon) + v(\varepsilon)$ be the deformation associated with the displacement $v(\varepsilon)$. For any $F = (z_1 \mid z_2 \mid z_3) \in M_3$ and $x = (x_1, x_2, x_3) \in \bar{\Omega}$, we can write

$$W((z_1 \mid z_2 \mid \varepsilon^{-1}z_3)A(\varepsilon)^{-1}(x)) = W((z_1 \mid z_2 \mid \varepsilon^{-1}z_3)A^{-1}(x_1, x_2)) + R(x, \varepsilon, F) \quad (38)$$

where, due to hypothesis $(8)_3$,

$$|R(x, \varepsilon, F)| \leq C\left(1 + \left\|\left(z_1 \mid z_2 \mid \frac{z_3}{\varepsilon}\right)A(\varepsilon)^{-1}(x)\right\|^{p-1}\right.$$

$$+ \left\|\left(z_1 \mid z_2 \mid \frac{z_3}{\varepsilon}\right)A^{-1}(x_1, x_2)\right\|^{p-1}\right)$$

$$\times \left\|\left(z_1 \mid z_2 \mid \frac{z_3}{\varepsilon}\right)[A(\varepsilon)^{-1}(x) - A^{-1}(x_1, x_2)]\right\|. \quad (39)$$

Since the matrix $A(\varepsilon)(x)$ is of the form $A(\varepsilon)(x_1, x_2, x_3) = A(x_1, x_2) + \varepsilon x_3 B(x_1, x_2)$, it follows that for ε small enough

$$A(\varepsilon)^{-1}(x_1, x_2, x_3) = A^{-1}(x_1, x_2)(I + \varepsilon S(\varepsilon, x)), \tag{40}$$

where $S(\varepsilon, \cdot)$ is bounded in $C^0(\bar{\Omega}; M_3)$ uniformly with respect to ε. Consequently,

$$|R(x, \varepsilon, F)| \leq C\varepsilon \left(1 + \left\|\left(z_1 |z_2| \frac{z_3}{\varepsilon}\right)\right\|^p\right) \tag{41}$$

where C does not depend on x and ε. If we replace F by $\nabla\varphi(\varepsilon)$ in (41) and integrate it on Ω against the weight $\det A(\varepsilon)$, we thus obtain by Lemmas 2 and 3

$$\int_\Omega |R(x, \varepsilon, \nabla\varphi(\varepsilon)(x))| \det A(\varepsilon) \, dx \leq C\varepsilon. \tag{42}$$

We now infer from equation (38) and the definition (26) of W_0 that

$$
\begin{aligned}
E(\varepsilon)(v(\varepsilon)) &\geq \int_\Omega W_0((x_1, x_2), (\partial_1\varphi(\varepsilon) \mid \partial_2\varphi(\varepsilon))) \det A(\varepsilon) \, dx \\
&\quad + \int_\Omega R(x, \varepsilon, \nabla\varphi(\varepsilon)(x)) \det A(\varepsilon) \, dx \\
&\geq \int_\Omega QW_0((x_1, x_2), (\partial_1\varphi(\varepsilon) \mid \partial_2\varphi(\varepsilon))) \det A(\varepsilon) \, dx \\
&\quad + \int_\Omega R(x, \varepsilon, \nabla\varphi(\varepsilon)(x)) \det A(\varepsilon) \, dx.
\end{aligned} \tag{43}
$$

Therefore, estimate (42) implies that

$$
\begin{aligned}
\lim_{\varepsilon \to 0} E(\varepsilon)(v(\varepsilon)) &\geq \liminf_{\varepsilon \to 0} \int_\Omega QW_0((x_1, x_2), (\partial_1\varphi(\varepsilon) \mid \partial_2\varphi(\varepsilon))) \det A(\varepsilon) \, dx \\
&= \liminf_{\varepsilon \to 0} \int_\Omega QW_0((x_1, x_2), (\partial_1\varphi(\varepsilon) \mid \partial_2\varphi(\varepsilon))) \det A \, dx \tag{44}
\end{aligned}
$$

since $\det A(\varepsilon) \to \det A$ in $C^0(\bar{\Omega})$. Let $G: W^{1,p}(\Omega; \mathbb{R}^3) \to \mathbb{R}$ be defined by

$$G(\varphi) = \int_\Omega QW_0((x_1, x_2), (\partial_1\varphi \mid \partial_2\varphi)) \det A(x_1, x_2) \, dx. \tag{45}$$

We define a function $Z: \Omega \times M_3 \to \mathbb{R}$ by

$$Z(x, (z_1 \mid z_2 \mid z_3)) = QW_0((x_1, x_2), (z_1 \mid z_2)) \det A(x_1, x_2)$$

so that $G(\varphi) = \int_\Omega Z(x, \nabla\varphi(x)) \, dx$. Since QW_0 is quasiconvex, it is easy to see that Z is also quasiconvex; see Le Dret and Raoult [1995]. Moreover, Z is continuous, bounded below, and satisfies the growth condition $(8)_1$ since QW_0 satisfies $(27)_1$. Therefore, the function G is sequentially weakly lower semi-continuous on $W^{1,p}(\Omega; \mathbb{R}^3)$; see Acerbi and Fusco [1984], Meyers [1965], and Dacorogna [1989]. Consequently, as $\varphi(\varepsilon) \rightharpoonup \varphi = \Psi(0) + v$ in $W^{1,p}(\Omega; \mathbb{R}^3)$,

$$
\begin{aligned}
\lim_{\varepsilon \to 0} E(\varepsilon)(v(\varepsilon)) &\geq \liminf_{\varepsilon \to 0} G(\varphi(\varepsilon)) \geq G(\varphi) \\
&= 2 \int_\omega QW_0((x_1, x_2), (a_1 + \partial_1\bar{v} \mid a_2 + \partial_2\bar{v}))\sqrt{a(x_1, x_2)} \, dx_1 \, dx_2, \tag{46}
\end{aligned}
$$

and the proof is complete. $\qquad\square$

Let us now turn to proving the reverse inequality. We first recall a technical lemma; see Dal Maso [1993] and Le Dret and Raoult [1995].

Lemma 6. *Let $X \hookrightarrow Y$ be two Banach spaces such that X is reflexive and compactly embedded in Y. Consider a functional $G: X \to \mathbb{R}$ such that for all $v \in X$, $G(v) \geq g(\|v\|_X)$ where g is such that $g(t) \to +\infty$ as $t \to +\infty$. Let $G^*: Y \to \mathbb{R}$ be defined by $G^*(v) = G(v)$ if $v \in X$, $G^*(v) = +\infty$ otherwise. Let $\Gamma\text{-}G$ denote the sequential lower semi-continuous envelope of G for the weak topology of X and let $\Gamma\text{-}G^*$ denote the lower semi-continuous envelope of G^* for the strong topology of Y. Then $\Gamma\text{-}G^* = (\Gamma\text{-}G)^*$.*

Proposition 7. *For all $v \in V_M$, the following estimate holds true:*

$$J^*(0)(v) \leq 2 \int_\omega QW_0((x_1, x_2), (a_1 + \partial_1\bar{v} \mid a_2 + \partial_2\bar{v}))\sqrt{a}\,dx_1\,dx_2$$
$$- \int_\omega \mathcal{G} \cdot (\psi + \bar{v})\sqrt{a}\,dx_1\,dx_2. \tag{47}$$

Proof. Let us consider $v \in V_M$. For all $w \in W_0^{1,p}(\omega; \mathbb{R}^3)$, we define a displacement

$$v(\varepsilon)(x) = \bar{v}(x_1, x_2) + \varepsilon x_3 w(x_1, x_2), \tag{48}$$

and the associated deformation $\varphi(\varepsilon) = \Psi(\varepsilon) + v(\varepsilon) = \bar{\varphi} + \varepsilon x_3(a_3 + w)$, with $\bar{\varphi} = \psi + \bar{v}$. Obviously, $v(\varepsilon) \to v$ strongly in $W^{1,p}(\Omega; \mathbb{R}^3)$. Let us examine the limit behavior of the sequence $J^*(\varepsilon)(v(\varepsilon))$. By the dominated convergence theorem and the growth estimate, it is clear that

$$E(\varepsilon)(v(\varepsilon)) = \int_\Omega W((\partial_1\varphi(\varepsilon) \mid \partial_2\varphi(\varepsilon) \mid a_3 + w)A(\varepsilon)^{-1}) \det A(\varepsilon)\,dx$$
$$\to 2 \int_\omega W((\partial_1\bar{\varphi} \mid \partial_2\bar{\varphi} \mid a_3 + w)A^{-1}) \det A\,dx_1\,dx_2 \tag{49}$$

when $\varepsilon \to 0$. Consequently,

$$J^*(\varepsilon)(v(\varepsilon)) \longrightarrow 2 \int_\omega W((\partial_1\bar{\varphi} \mid \partial_2\bar{\varphi} \mid a_3 + w)A^{-1}) \det A\,dx_1\,dx_2$$
$$- \int_\omega \mathcal{G} \cdot (\psi + \bar{v})\sqrt{a}\,dx_1\,dx_2. \tag{50}$$

As this is true for all $w \in W_0^{1,p}(\omega; \mathbb{R}^3)$, it follows from the definition of Γ-convergence that

$$J^*(0)(v) \leq \inf_{w \in W_0^{1,p}(\omega;\mathbb{R}^3)} \left\{ 2 \int_\omega W((\partial_1\bar{\varphi} \mid \partial_2\bar{\varphi} \mid a_3 + w)A^{-1}) \det A\,dx_1\,dx_2 \right\}$$
$$- \int_\omega \mathcal{G} \cdot (\psi + \bar{v})\sqrt{a}\,dx_1\,dx_2. \tag{51}$$

We remark that in inequality (51), the infimum over $W_0^{1,p}(\omega; \mathbb{R}^3)$ can be replaced by the infimum over $L^p(\omega; \mathbb{R}^3)$, by the density of $W_0^{1,p}(\omega; \mathbb{R}^3)$ in $L^p(\omega; \mathbb{R}^3)$, and by the dominated convergence theorem. The function $g: \omega \times \mathbb{R}^3 \to \mathbb{R}$,

$$g(x, z) = W((\partial_1 \bar{\varphi} \mid \partial_2 \bar{\varphi} \mid a_3 + z)A^{-1})$$

is a Carathéodory function. Hence, the measurable selection lemma (cf. Ekeland and Temam [1974]) shows that there exists a measurable function w_0 such that

$$W_0(x, (\partial_1 \bar{\varphi}(x) \mid \partial_2 \bar{\varphi}(x))) = W((\partial_1 \bar{\varphi}(x) \mid \partial_2 \bar{\varphi}(x) \mid a_3(x) + w_0(x))A^{-1}(x)) \quad (52)$$

for almost all $x \in \omega$. Due to the coercivity estimate, $w_0 \in L^p(\omega; \mathbb{R}^3)$ and thus

$$\inf_{w \in L^p(\omega;\mathbb{R}^3)} \left\{ \int_\omega W((\partial_1 \bar{\varphi} \mid \partial_2 \bar{\varphi} \mid a_3 + w)A^{-1}) \det A\, dx \right\}$$
$$\leq \int_\omega W_0(x, (\partial_1 \bar{\varphi} \mid \partial_2 \bar{\varphi})) \det A\, dx. \quad (53)$$

Let $G: W_0^{1,p}(\omega; \mathbb{R}^3) \to \mathbb{R}$ be defined by

$$G(\bar{v}) = 2 \int_\omega W_0(x, (\partial_1 \bar{\varphi} \mid \partial_2 \bar{\varphi}))\sqrt{a}\, dx_1\, dx_2 - \int_\omega \mathcal{G} \cdot (\psi + \bar{v})\sqrt{a}\, dx_1\, dx_2, \quad (54)$$

with $\bar{\varphi} = \psi + \bar{v}$. It follows from (53) that for all $v \in V_M$

$$J^*(0)(v) \leq G(\bar{v}). \quad (55)$$

Let G^* be defined on $L^p(\Omega; \mathbb{R}^3)$ by $G^*(v) = G(\bar{v})$ if $v \in V_M$, $G^*(v) = +\infty$ otherwise. Corollary 4 and (55) then imply that for all $v \in L^p(\Omega; \mathbb{R}^3)$

$$J^*(0)(v) \leq G^*(v). \quad (56)$$

Since $J^*(0)$ is lower semi-continuous on $L^p(\Omega; \mathbb{R}^3)$, it is smaller than the lower semi-continuous envelope of G^*. It is known (see Acerbi and Fusco [1984]) that the sequential weak lower semi-continuous envelope $\Gamma\text{-}G$ of G on $W_0^{1,p}(\omega; \mathbb{R}^3)$ is given by

$$\Gamma\text{-}G(\bar{v}) = 2 \int_\omega QW_0(x, (\partial_1 \bar{\varphi} \mid \partial_2 \bar{\varphi}))\sqrt{a}\, dx_1\, dx_2 - \int_\omega \mathcal{G} \cdot (\psi + \bar{v})\sqrt{a}\, dx_1\, dx_2. \quad (57)$$

Therefore, Lemma 6 with $X = V_M$, $Y = L^p(\Omega; \mathbb{R}^3)$, and $g(t) = \alpha(t^p - 1)$ implies that

$$J^*(0) \leq \Gamma\text{-}G^* = (\Gamma\text{-}G)^*, \quad (58)$$

which proves the proposition. $\square$

Proof of Theorem 1. Use Corollary 4 for the case $v \notin V_M$ and Propositions 5 and 7 for the case $v \in V_M$. $\square$

5. The Limit Nonlinear Membrane Shell Model

We now use Theorem 1 to characterize the asymptotic behavior of diagonal minimizing sequences of rescaled deformations $\phi(\varepsilon)$ satisfying $I(\varepsilon)(\phi(\varepsilon)) \leq \inf_{\varphi \in \Phi(\varepsilon)} I(\varepsilon)(\varphi) + h(\varepsilon)$ where h is a positive function such that $h(\varepsilon) \to 0$ when $\varepsilon \to 0$ and the sets of admissible deformations are

$$\Phi(\varepsilon) = \{\varphi \in W^{1,p}(\Omega; \mathbb{R}^3); \varphi(x) = \Psi(\varepsilon)(x) \text{ on } \Gamma\}.$$

We introduce the space of membrane shell deformations as $\Phi_M = \{\varphi \in W^{1,p}(\Omega; \mathbb{R}^3),$ $\partial_3 \varphi = 0$ in Ω, $\varphi = \psi$ on $\Gamma\}$, which is isomorphic to the space $\bar{\Phi} = \{\bar{\varphi} \in W^{1,p}(\omega; \mathbb{R}^3),$ $\bar{\varphi} = \psi$ on $\partial\omega\}$. We use the same notational device as for displacements to denote this isomorphism.

Theorem 8. *The sequence $\phi(\varepsilon)$ is relatively weakly compact in $W^{1,p}(\Omega; \mathbb{R}^3)$. Its limit points ϕ belong to Φ_M and are identified with elements $\bar{\phi}$ of $\bar{\Phi}$, solutions of the minimization problem $\bar{I}(0)(\bar{\phi}) = \inf_{\bar{\varphi} \in \bar{\Phi}} \bar{I}(0)(\bar{\varphi})$, where the membrane shell energy $\bar{I}(0)$ is given by*

$$\bar{I}(0)(\bar{\varphi}) = 2 \int_\omega Q W_0(x, \nabla\bar{\varphi}(x)) \sqrt{a(x)} \, dx_1 \, dx_2$$
$$- \int_\omega \mathcal{G}(x) \cdot \bar{\varphi}(x) \sqrt{a(x)} \, dx_1 \, dx_2. \tag{59}$$

Moreover, $I(\varepsilon)(\phi(\varepsilon)) \to \bar{I}(0)(\bar{\phi})$ for all weakly convergent subsequences.

Proof. See Le Dret and Raoult [1995], using the classical argument of De Giorgi. □

Comments. (i) The limit energy depends on the deformation only through its first derivatives. In this sense, it is a membrane model with no bending or shear effects. Even if the three-dimensional material is homogenous in its reference configuration, the limit model exhibits a dependence on the point x. See Theorem 9 below for a more precise description of this dependence. Note that the limit minimization problem has a solution.

(ii) If the function $Q W_0$ is smooth enough, the Euler-Lagrange equations for the limit problem assume the form

$$-\frac{2}{\sqrt{a}} \partial_\beta \left[\left(\frac{\partial Q W_0}{\partial \bar{F}}(x, \nabla\bar{\phi}) \right)_{i\beta} \sqrt{a} \right] = \mathcal{G}_i \text{ in } \omega, \quad \bar{\phi}(x_1, x_2) = \psi(x_1, x_2) \text{ on } \partial\omega. \tag{60}$$

System (60) is a system of three second-order quasilinear partial differential equations in the three unknowns $\bar{\phi}_i$.

See Le Dret and Raoult [1995] for more comments.

Theorem 8 gives information on the asymptotic behavior of the actual deformations $\tilde{\phi}^\varepsilon$ of the shell in its given reference configuration $\tilde{\Omega}_\varepsilon$, by reading it through the chart Ψ. However, we could have worked as well with a chart Ψ' associated with any other chart ψ' for $\tilde{S}$. More specifically, let O', e'_1, e'_2, e'_3 be another Cartesian frame in $\mathbb{R}^3$, ω' a bounded open subset of the plane (O', e'_1, e'_2), and ψ': $\omega' \to \tilde{S}$ a chart for $\tilde{S}$ that satisfies the same hypotheses as ψ. With the deformation $\tilde{\phi}^\varepsilon$, we associate a new rescaled deformation $\phi'(\varepsilon)(x') = \tilde{\phi}^\varepsilon(\psi'(x'_1, x'_2) + \varepsilon x'_3 a'_3(x'_1, x'_2))$. Since $\lambda = \psi^{-1} \circ \psi'$ is a C^2-diffeomorphism between ω' and ω, it is fairly clear that if $\bar{\phi} \in \bar{\Phi}_M$ is associated with a limit point of $\phi(\varepsilon)$, then $\bar{\phi}' = \bar{\phi} \circ \lambda$ is associated with a limit point of $\phi'(\varepsilon)$. Applying Theorem 8 in both charts, we see that $\bar{I}(0)(\bar{\phi}) = \bar{I}'(0)(\bar{\phi}')$ with obvious notation. This observation is confirmed by a direct computation. Indeed, let $W'_0 \colon \bar{\omega}' \times M_{3,2} \to \mathbb{R}$ be defined by

$$W'_0((x'_1, x'_2), \bar{F}) = \inf_{z \in \mathbb{R}^3} W((\bar{F} \mid z) A'^{-1}(x'_1, x'_2))$$

where $A'(x'_1, x'_2) = (\partial_1 \psi' \mid \partial_2 \psi' \mid a'_3)(x'_1, x'_2)$. It is a simple matter to check that since the normal vectors are chart-independent, i.e., $a'_3(x'_1, x'_2) = \pm a_3(x_1, x_2)$ whenever $(x_1, x_2) = \lambda(x'_1, x'_2)$, then $W_0((x_1, x_2), \nabla\bar{\varphi}(x_1, x_2)) = W'_0((x'_1, x'_2), \nabla\bar{\varphi}'(x'_1, x'_2))$. Hence, the same holds true for the quasiconvex envelopes and thus for the energies themselves.

The above remarks demonstrate the intrinsic character of the limit minimization problem. It is nonetheless of prime importance to give an expression of the limit nonlinear shell problem in the original reference configuration of the shell, particularly if this configuration has special properties, for example a natural configuration, a homogeneous configuration, or an isotropic configuration. This is the object of the remainder of this section.

With any deformation $\bar{\varphi} \in \bar{\Phi}$ we thus associate a deformation of the shell in its reference configuration $\tilde{\varphi} = \bar{\varphi} \circ \psi^{-1} \in \tilde{\Phi}$ where

$$\tilde{\Phi} = \{\tilde{\varphi} \in W^{1,p}(\tilde{S}; \mathbb{R}^3); \tilde{\varphi}(\tilde{x}) = \tilde{x} \text{ on } \partial\tilde{S}\}.$$

Let $\tilde{\Pi}$ be the orthogonal projection on $\tilde{S}$, which is well defined in a tubular neighborhood of $\tilde{S}$. We extend the deformation to this tubular neighborhood by setting $\varphi(\tilde{x}) = \tilde{\varphi}(\tilde{\Pi}(\tilde{x}))$ and for $\tilde{x} \in \tilde{S}$, we let $D\tilde{\varphi}(\tilde{x}) = \nabla\varphi(\tilde{x})$. Therefore, $D\tilde{\varphi}(\tilde{x})$ is the 3×3 matrix of the components of $\nabla\varphi(\tilde{x})$ in the canonical Cartesian basis (e_1, e_2, e_3). We will call this matrix the deformation gradient. We denote by $d\tilde{\sigma}$ the area element on $\tilde{S}$.

For all unit vectors $e \in S^2$, we choose a bounded open set $O_e \subset e^\perp$ and denote by Π_e the orthogonal projection on $e^\perp$. For all $\chi \in W_0^{1,\infty}(O_e; \mathbb{R}^3)$, we let $\chi_e(y) = \chi(\Pi_e(y))$ and for all $y \in O_e$, we define $D_{e^\perp}\chi(y) = \nabla\chi_e(y)$ which is again a 3×3 matrix. Then we have:

Theorem 9. *Let $\bar{\phi}$ be a shell deformation associated with a minimizer $\bar{\phi}$ of the limit energy in the chart ψ, as in Theorem 8. Then $\tilde{\phi}$ is a solution of the minimization problem*

$$\tilde{I}_{\tilde{S}}(\tilde{\phi}) = \inf_{\tilde{\varphi} \in \tilde{\Phi}} \tilde{I}_{\tilde{S}}(\tilde{\varphi}). \tag{61}$$

The membrane shell energy $\tilde{I}_{\tilde{S}}$ is given by

$$\tilde{I}_{\tilde{S}}(\tilde{\varphi}) = 2 \int_{\tilde{S}} W_m(a_3(\tilde{x}), D\tilde{\varphi}(\tilde{x})) \, d\tilde{\sigma} - \int_{\tilde{S}} \tilde{\mathcal{G}} \cdot \tilde{\varphi} \, d\tilde{\sigma}, \tag{62}$$

where the elastic membrane stored energy function of the material W_m: $S^2 \times M_3 \to \mathbb{R}$ *is defined by*

$$W_m(e, F) = \inf_{\chi \in W_0^{1,\infty}(O_e; \mathbb{R}^3)} \left[\frac{1}{\text{meas } O_e} \int_{O_e} \left[\inf_{z \in \mathbb{R}^3} W(F + z \otimes e + D_{e^\perp} \chi(y)) \right] dy \right] \quad (63)$$

and $\tilde{\mathcal{G}}(\tilde{x}) = \mathcal{G}(\psi^{-1}(\tilde{x}))$.

Proof. Recall that by Theorem 8, $\bar{\phi}$ minimizes the energy

$$\bar{I}(0)(\bar{\phi}) = 2 \int_\omega Q W_0(x, \nabla \bar{\phi}) \sqrt{a} \, dx_1 \, dx_2 - \int_\omega \mathcal{G} \cdot \bar{\phi} \sqrt{a} \, dx_1 \, dx_2.$$

We start from Dacorogna's representation formula for the quasiconvex envelope of W_0 (cf. Dacorogna [1982, 1989]), which states that

$$Q W_0(x_0, \bar{F}) = \inf_{\tilde{\chi} \in W_0^{1,\infty}(O; \mathbb{R}^3)} \left\{ \frac{1}{\text{meas } O} \int_O W_0(x_0, \bar{F} + \nabla \tilde{\chi}(\bar{y})) \, d\bar{y} \right\}, \quad (64)$$

where O is a bounded open subset of $\mathbb{R}^2$ (this infimum does not depend on the choice of O). We choose $O = (D\psi(x_0))^{-1}(O_{a_3(\psi(x_0))})$. Due to the definition of W_0, we thus have

$$Q W_0(x_0, \bar{F}) = \inf_{\tilde{\chi} \in W_0^{1,\infty}(O; \mathbb{R}^3)} \left\{ \frac{1}{\text{meas } O} \int_O \inf_{z \in \mathbb{R}^3} W((\bar{F} + \nabla \tilde{\chi}(\bar{y}) \mid z)A^{-1}(x_0)) \, d\bar{y} \right\}. \quad (65)$$

We now remark that

$$(\bar{F} + \nabla \tilde{\chi}(\bar{y}) \mid z)A^{-1}(x_0) = (\bar{F} \mid 0)A^{-1}(x_0) + (0 \mid z)A^{-1}(x_0)$$
$$+ (\nabla \tilde{\chi}(\bar{y}) \mid 0)A^{-1}(x_0), \quad (66)$$

and we consider each of these three terms separately. First, if $\bar{F}$ is a gradient, i.e., $\bar{F} = \nabla \bar{\varphi}(x_0)$, then $(\bar{F} \mid 0)A^{-1}(x_0) = D\tilde{\varphi}(\psi(x_0))$. Indeed, if we let $\varphi(x_1, x_2, x_3) = \bar{\varphi}(x_1, x_2)$, then $\varphi(x) = \tilde{\varphi}(\bar{\Pi}(\Psi(x)))$. Consequently, $(\nabla \bar{\varphi}(x_0) \mid 0) = \nabla(\tilde{\varphi}(\bar{\Pi}(\Psi(x_0)))\nabla \Psi(x_0) = D\tilde{\varphi}(\psi(x_0))A(x_0)$.

Second, letting $y = D\psi(x_0)\bar{y}$ and $\chi(y) = \tilde{\chi}(\bar{y})$, we likewise note that

$$(\nabla \tilde{\chi}(\bar{y}) \mid 0)A^{-1}(x_0) = D_{a_3(\psi(x_0))^\perp} \chi(y).$$

Moreover, χ belongs to $W_0^{1,\infty}(O_{a_3(\psi(x_0))}; \mathbb{R}^3)$.

Finally, it is easily checked that $(0 \mid z)A^{-1}(x_0) = z \otimes a_3(\psi(x_0))$. Replacing these expressions into (65), we obtain

$$\forall x_0 \in \omega, \qquad Q W_0(x_0, \nabla \bar{\varphi}(x_0)) = W_m(a_3(\psi(x_0)), D\tilde{\varphi}(\psi(x_0))) \quad (67)$$

from which Theorem 9 follows at once by the change of variables $x \mapsto \tilde{x} = \psi(x)$. $\quad \square$

Remarks. (i) The elastic membrane stored energy function depends on two variables: a unit vector and a matrix. The membrane shell energy is obtained by replacing the unit vector by the normal vector and the matrix by the deformation gradient. Note that deformation gradients always satisfy $D\tilde{\varphi}(\tilde{x})a_3(\tilde{x}) = 0$. Thus, expression (63) is only useful for couples (e, F) such that $Fe = 0$. In the planar case, the normal vector is constant and we recover the result of Le Dret and Raoult [1995]. The fact that the energy depends on the surface only through its normal vector is not directly apparent in expression (59) in terms of QW_0.

(ii) The limit energy (62) corresponds to the membrane part of the energy for inextensible one-director Cosserat shells obtained by Simo and Fox [1989]. However, let us point out that we do not make any *a priori* kinematic assumptions. Moreover, our analysis provides a convergence result and at the same time an exact formula for the constitutive law of the shell. Our model is a pure membrane model since it does not include shear and flexural effects.

(iii) Note that $W_m(e, F) = W_m(-e, F)$ which is due to the fact that the energy does not depend on the orientation of the midsurface.

(iv) Definition (63) does not depend on the choice of O_e in $e^\perp$.

(v) Theorem 8 may be reformulated in terms of convergence of the deformations rescaled in the reference configuration, i.e., defining $\tilde{\phi}(\varepsilon)(\tilde{x}) = \tilde{\phi}^\varepsilon(\tilde{\Pi}(\tilde{x}) + \varepsilon[(\tilde{x} - \tilde{\Pi}(\tilde{x})) \cdot a_3(\tilde{\Pi}(\tilde{x}))]a_3(\tilde{\Pi}(\tilde{x})))$ on $\tilde{\Omega}_1$ (assuming this is well defined; otherwise we just rescale on a thinner domain) then $\tilde{\phi}(\varepsilon) \rightharpoonup \tilde{\phi}$ in $W^{1,p}(\tilde{\Omega}_1; \mathbb{R}^3)$ where $\tilde{\phi}$ is a solution of problem (61)–(62).

6. Properties of the Nonlinear Membrane Shell Energy

In the Γ-convergence analysis, we have ignored the fact that the stored energy function W of the three-dimensional bodies has to satisfy material frame-indifference, since this was irrelevant for the convergence proof. In this section, we will investigate the consequences of material frame-indifference for the nonlinear membrane shell energy W_m as well as the consequences of material symmetry assumptions.

First, recall that the principle of material frame-indifference states that, to be legitimate from the standpoint of continuum mechanics, a stored energy function W has to satisfy

$$\forall F \in M_3, \forall R \in SO(3), \ W(RF) = W(F), \tag{68}$$

(see, e.g., Ciarlet [1988], Wang and Truesdell [1973], or Marsden and Hughes [1983]).

Theorem 10. *Let the stored energy function W satisfy the principle of material frame-indifference (68). Then, the nonlinear membrane shell energy W_m is frame-indifferent as well, in the sense that*

$$\forall e \in S^2, \ \forall F \in M_3, \ \forall R \in SO(3), \ W_m(e, RF) = W_m(e, F), \tag{69}$$

and there exists a function $\mathcal{W}_m \colon S^2 \times \mathbb{S}_3^\geq \to \mathbb{R}$, where $\mathbb{S}_3^\geq$ is the set of 3×3 positive semidefinite symmetric matrices, such that

$$\forall e \in S^2, \ \forall F \in M_3, \ W_m(e, F) = \mathcal{W}_m(e, F^T F). \tag{70}$$

Proof. Let $F \in M_3$ and $R \in SO(3)$ be arbitrary matrices. Since for all $e \in S^2$, $y \in O_e$, $\chi \in W_0^{1,\infty}(O_e; \mathbb{R}^3)$, and $z \in \mathbb{R}^3$,

$$
\begin{aligned}
W(RF + z \otimes e + D_{e^\perp}\chi(y)) &= W\left(R(F + R^T z \otimes e + R^T D_{e^\perp}\chi(y))\right) \\
&= W\left(F + (R^T z) \otimes e + D_{e^\perp}(R^T \chi)(y)\right), \quad (71)
\end{aligned}
$$

definition (63) shows that (69) holds true. The existence of the representation function W_m in formula (70) is then classical. $\qquad\square$

Remarks. (i) The representation formula (70) is given for an arbitrary couple (e, F) in $S^2 \times M_3$. Since deformation gradients $D\tilde{\varphi}(\tilde{x}) = F(\tilde{x})$ always satisfy $F(\tilde{x})a_3(\tilde{x}) = 0$, the associated strain tensors $C(\tilde{x}) = F(\tilde{x})^T F(\tilde{x})$ also satisfy $C(\tilde{x})a_3(\tilde{x}) = 0$. Thus, formula (70) is only useful for couples (e, C) such that $Ce = 0$.

(ii) If F is the gradient of a smooth enough shell deformation $\tilde{\phi}$, the matrix $F(\tilde{x})^T F(\tilde{x})$ represents the metric of the deformed surface at point $\tilde{\phi}(\tilde{x})$. The membrane energy thus only depends on this metric, which is consistent with the intuition that the stress state in an elastic membrane depends only on the stretching that the deformed surface undergoes.

We now show that, due to frame-indifference, if the three-dimensional stored energy function has a global minimum at $F = I$, the corresponding nonlinear shell energy is constant under compression. This means that is is possible to crumple a membrane shell without using any energy. This phenomenon was first noticed in the case of nonlinear strings by Acerbi, Buttazzo, and Percivale [1991], then in the case of planar membranes by Percivale [1991] for isotropic materials and Le Dret and Raoult [1993, 1995] for general materials. The proof given in the latter article does not extend to the case of shells. The proof we provide below is at the same time simpler and more general. We note $v_i(F)$, $i = 1, 2, 3$, the singular values of F numbered in increasing order.

Corollary 11. *Assume that the three-dimensional stored energy function W is such that $W(I) = 0$ and $W(F) \geq 0$ for all $F \in M_3$. Then, $W_m(e, F) = 0$ for all $F \in M_3$ such that $Fe = 0$ and $v_3(F) \leq 1$.*

Proof. Fix $e \in S^2$. Since $W \geq 0$, it follows immediately that $W_m(e, F) \geq 0$ for all $F \in M_3$.

Let $F \in M_3$ be such that $Fe = 0$ and $v_3(F) \leq 1$. Let $U = \sqrt{F^T F}$. We can choose an orthonormal basis e, f_2, f_3 of eigenvectors of U, where e is associated with the eigenvalue 0 and f_i, $i = 2, 3$, are associated with the eigenvalues $v_i(F)$, $i = 2, 3$. It follows from (70) and the polar factorization theorem that $W_m(e, F) = W_m(e, U)$. It thus suffices to prove that $W_m(e, U) = 0$.

We consider the 1-periodic functions $\theta_i : \mathbb{R} \to \mathbb{R}$, $i = 2, 3$, defined by their restriction to $[0, 1[$

$$
\theta_i(t) = \begin{cases} (1 - v_i(F))t & \text{if } 0 \leq t \leq \frac{1+v_i(F)}{2}, \\ (-1 - v_i(F))(t - 1) & \text{if } \frac{1+v_i(F)}{2} \leq t < 1. \end{cases}
$$

Note that since $v_i(\bar{F}) \in [0, 1]$, $\frac{1+v_i(\bar{F})}{2} \in [0, 1]$ and the functions θ_i are well defined and belong to $W^{1,\infty}(\mathbb{R})$. We let for $y \in \mathbb{R}^3$

$$\chi_n(y) = \sum_{i=2}^{3} \frac{1}{n} \theta_i (ny \cdot f_i) f_i. \tag{72}$$

Therefore,

$$\nabla \chi_n(y) = \sum_{i=2}^{3} \theta_i'(ny \cdot f_i) f_i \otimes f_i$$

$$= \sum_{i=2}^{3} (h_i^n(y) - v_i(F)) f_i \otimes f_i \tag{73}$$

where h_i^n only takes the values ± 1.

Without loss of generality, we may assume that meas $O_e = 1$. We introduce a smooth cut-off function $0 \leq \rho_n \leq 1$ defined on O_e and such that $\rho_n(y) = 1$ if $d(y, \partial O_e) \geq 1/n$, $\rho_n(y) = 0$ if $y \in \partial O_e$ and that $\|\nabla \rho_n\| \leq 2n$. Since $\rho_n \chi_n \in W_0^{1,\infty}(O_e; \mathbb{R}^3)$, we can use it in definition (63). It follows from this definition that for all measurable functions $h: O_e \to \{-1, 1\}$,

$$W_m(e, U) \leq \int_{O_e} W(U + h(y)e \otimes e + D_{e^\perp}(\rho_n \chi_n)(y)) \, dy. \tag{74}$$

Since $(\rho_n \chi_n)_e(y) = \chi_n(\Pi_e(y))$ for all $y \in \mathbb{R}^3$ such that $d(\Pi_e(y), \partial O_e) \geq 1/n$, we see that $D_{e^\perp}((\rho_n \chi_n))(y) = \nabla \chi_n(y)$ for all $y \in O_e$ such that $d(y, \partial O_e) \geq 1/n$. Therefore, since $U = \sum_{i=2}^{3} v_i(F) f_i \otimes f_i$, we obtain

$$W_m(e, U) \leq \int_{d(y, \partial O_e) \geq 1/n} W(h(y)e \otimes e + h_2^n(y) f_2 \otimes f_2 + h_3^n(y) f_3 \otimes f_3) \, dy$$

$$+ \int_{d(y, \partial O_e) < 1/n} W(U + h(y)e \otimes e + D_{e^\perp}(\rho_n \chi_n)(y)) \, dy. \tag{75}$$

Fix n. We choose $h(y)$ so that $(h(y)e \otimes e + h_2^n(y) f_2 \otimes f_2 + h_3^n(y) f_3 \otimes f_3) \in SO(3)$ for almost all y, which is obviously possible. With this choice, the first integral vanishes by frame-indifference. It is clear that the integrand of the second term is bounded independently of n. Since meas$\{d(y, \partial O_e) < 1/n\} \to 0$ as $n \to +\infty$, we obtain $W(e, U) \leq 0$. $\qquad \square$

We now investigate the consequences of isotropy on the membrane energy. Recall first that an elastic material is said to be isotropic if

$$\forall F \in M_3, \ \forall R \in SO(3), \ W(FR) = W(F). \tag{76}$$

We show below that isotropy added to the principle of material indifference implies that the shell energy W_m does not depend on the normal vector.

Theorem 12. *Assume that the stored energy function W is isotropic (76). Then, the nonlinear membrane shell energy W_m is isotropic as well, in the sense that*

$$\forall e \in S^2, \forall F \in M_3, \forall R \in SO(3), \qquad W_m(e, F) = W_m(R^T e, FR). \tag{77}$$

If W furthermore satisfies the principle of material indifference, there exists a symmetric function $w_m: (\mathbb{R}_+)^2 \to \mathbb{R}$ that

$$\forall e \in S^2, \forall F \in M_3, Fe = 0, \qquad W_m(e, F) = w_m(v_2(F), v_3(F)). \tag{78}$$

Proof. We may assume without loss of generality that for all $e \in S^2$ and all $R \in SO(3)$, $RO_e = O_{Re}$. If $\chi \in W_0^{1,\infty}(O_e; \mathbb{R}^3)$, the function χ_R defined by $\chi_R(y) = \chi(Ry)$ belongs to $W_0^{1,\infty}(O_{R^T e}; \mathbb{R}^3)$. Since for all $F \in M_3$, $e \in S^2$, $y \in O_e$, $\chi \in W_0^{1,\infty}(O_e; \mathbb{R}^3)$, and $z \in \mathbb{R}^3$,

$$
\begin{aligned}
W(FR + z \otimes (R^T e) + D_{(R^T e)^\perp}\chi_R(y)) &= W((F + z \otimes e + D_{e^\perp}\chi(Ry))R) \\
&= W(F + z \otimes e + D_{e^\perp}\chi(Ry)), \tag{79}
\end{aligned}
$$

definition (63) shows that (77) holds true.

Assume now that W satisfies the principle of material indifference. By Theorem 10, $W_m(e, F) = W_m(e, F^T F) = W_m(R^T e, R^T F^T F R)$. In particular, for all $C \in \mathbb{S}_3^{\geq}$ and all $R \in SO(3)$ such that $R^T e = e$,

$$W_m(e, C) = W_m(e, R^T C R). \tag{80}$$

Consider now two matrices C and C' of $\mathbb{S}_3$ such that $Ce = C'e = 0$ and have the same eigenvalues. Proceeding as in Gurtin [1981], we see that there exists $R \in SO(3)$ with $R^T e = e$ such that $R^T C R = C'$. Consequently, there exists a function $w: S^2 \times (\mathbb{R}_+)^2 \to \mathbb{R}$, symmetric with respect to the last two arguments, such that for all C with $Ce = 0$

$$W_m(e, C) = w(e, \lambda_2(C)^{1/2}, \lambda_3(C)^{1/2}), \tag{81}$$

where $\lambda_2(C)$, $\lambda_3(C)$ are the largest eigenvalues of C.

Let us now prove that w does not depend on e. Consider thus two unit vectors e and e' and let $R \in SO(3)$ be such that $e' = R^T e$. For all F such that $Fe = 0$, we have

$$W_m(e, F) = w(e, v_2(F), v_3(F)) \tag{82}$$

by (70) and (81). On the other hand, by (77),

$$W_m(e, F) = W_m(e', FR). \tag{83}$$

Since $F Re' = 0$ we also have

$$W_m(e', FR) = w(e', v_2(FR), v_3(FR)) \tag{84}$$

by (70) and (81) again. Since $v_i(FR) = v_i(F)$ for $i = 2, 3$, we conclude that for all $(v_2, v_3) \in (\mathbb{R}_+)^2$,

$$w(e, v_2, v_3) = w(e', v_2, v_3) = w_m(v_2, v_3), \tag{85}$$

which defines w_m and completes the proof. $\qquad\square$

Remarks. (i) In this case, the shell energy (62) assumes the form

$$\tilde{I}_{\tilde{S}}(\tilde{\varphi}) = 2 \int_{\tilde{S}} w_m(v_2(D\tilde{\varphi}(\tilde{x})), v_3(D\tilde{\varphi}(\tilde{x}))) \, d\tilde{\sigma} - \int_{\tilde{S}} \tilde{\mathcal{G}} \cdot \tilde{\varphi} \, d\tilde{\sigma}. \tag{86}$$

This formula applies to any surface, particularly to planar ones as in Le Dret and Raoult [1995]. Consequently, if the planar membrane energy is explicitly known, the shell energy is also explicitly determined without further computations. This is the case for the Saint Venant-Kirchhoff material

$$W(F) = \frac{\mu}{4} \text{tr}(F^T F - I)^2 + \frac{\lambda}{8} (\text{tr} \, (F^T F - I))^2$$

where $\mu > 0$ and $\lambda \geq 0$ are the Lamé moduli. We thus obtain according to Le Dret and Raoult [1995]:

$$W_m(F) = \frac{E}{8} \left[v_3(F)^2 - 1 \right]_+^2 + \frac{E}{8(1 - v^2)} \left[v_2(F)^2 + v v_3(F)^2 - (1 + v) \right]_+^2$$

$$+ \frac{E}{8(1 - v^2)(1 - 2v)} \left[v(v_2(F)^2 + v_3(F)^2) - (1 + v) \right]_+^2, \tag{87}$$

where $[t]_+^2$ stands for $([t]_+)^2$ and E and v are the Young modulus and the Poisson coefficient.

Appendix. A Live Loading Case: The Pressure Load

Let us briefly show how the case of a quite realistic live loading can be handled. We assume that the upper and lower surfaces $\tilde{S}_\varepsilon^\pm$ of the shell are submitted to uniform hydrostatic pressures $\pi_\varepsilon^\pm$ instead of prescribed dead loads. This means that the Cauchy stress vector on the deformed upper surface satisfies $Tn^+ = -\pi_\varepsilon^+ n^+$, where T is the Cauchy stress tensor, n^+ is the outer unit normal vector to the deformed upper surface and $\pi_\varepsilon^+ \in \mathbb{R}$, and similarly on the lower deformed surface. To be consistent with the order of magnitude of the loads we chose previously, we assume that $\pi_\varepsilon^\pm = \varepsilon \pi^\pm$, where $\pi^\pm$ do not depend on ε. Let $\Delta \pi = \pi^+ - \pi^-$.

It is shown in Ball [1977] (see also Sewell [1967]) that the corresponding equilibrium problem may be formulated as an energy minimization problem as follows. Let $\tilde{\pi}_\varepsilon \in C^1\left(\overline{\tilde{\Omega}_\varepsilon}\right)$ be such that $\tilde{\pi}_\varepsilon(\tilde{x}) = \pi_\varepsilon^+$ for $\tilde{x} \in \tilde{S}_\varepsilon^+$ and $\tilde{\pi}_\varepsilon(\tilde{x}) = \pi_\varepsilon^-$ for $\tilde{x} \in \tilde{S}_\varepsilon^-$. Then the pressure load equilibrium problem is, at least formally, equivalent to minimizing the energy

$$\tilde{I}_\varepsilon(\tilde{\varphi}) = \int_{\tilde{\Omega}_\varepsilon} W(\nabla \tilde{\varphi}) \, d\tilde{x} + P_\varepsilon(\tilde{\varphi}), \tag{88}$$

over the set $\tilde{\Phi}_\varepsilon$ of admissible deformations, where

$$P_\varepsilon(\tilde{\varphi}) = \int_{\tilde{\Omega}_\varepsilon} \left[\tilde{\pi}_\varepsilon(\tilde{x}) \det \nabla \tilde{\varphi}(\tilde{x}) + \frac{1}{3} \nabla \tilde{\pi}_\varepsilon(\tilde{x}) \cdot (\text{adj} \, \nabla \tilde{\varphi}(\tilde{x}) \tilde{\varphi}(\tilde{x})) \right] d\tilde{x}, \tag{89}$$

and adj F is the transpose of the cofactor matrix of F. We are at liberty to choose here $\tilde{\pi}_\varepsilon(\tilde{x}) = \frac{1}{2\varepsilon}[(\varepsilon + x_3(\tilde{x}))\pi_\varepsilon^+ + (\varepsilon - x_3(\tilde{x}))\pi_\varepsilon^-]$, with obvious notation.

For simplicity, we assume that the exponent p is strictly larger than 3, which trivially ensures that the energy (88) is bounded from below and has the same coercivity properties as before. This assumption also implies that there is no distinction between the distributional and the algebraic determinants and adjugates of the deformation gradients, which we thus all denote with a lowercase initial (see Ball [1977], Müller [1990], for a discussion of this question). Performing the same change of variables and rescaling as in Section 3, we are thus led to the computation of the Γ-limit of the sequence of functionals:

$$J(\varepsilon)(v) = E(\varepsilon)(v) + P(\varepsilon)(v), \tag{90}$$

where

$$P(\varepsilon)(v) = \int_\Omega \left[\varepsilon\pi \det\left(\partial_1\varphi \mid \partial_2\varphi \mid \frac{\partial_3\varphi}{\varepsilon} \right) + \frac{\Delta\pi}{6} e_3 \right. $$
$$\left. \cdot \left(\text{adj}\left(\partial_1\varphi \mid \partial_2\varphi \mid \frac{\partial_3\varphi}{\varepsilon} \right) \varphi \right) \right] dx, \tag{91}$$

with $\pi(x) = \frac{1}{2}[(1+x_3)\pi^+ + (1-x_3)\pi^-]$ and $\varphi = \Psi(\varepsilon) + v$ as usual. A simple algebraic calculation shows that

$$e_3 \cdot \left(\text{adj}\left(\partial_1\varphi \mid \partial_2\varphi \mid \frac{\partial_3\varphi}{\varepsilon} \right) \varphi \right) = (\partial_1\varphi \wedge \partial_2\varphi) \cdot \varphi, \tag{92}$$

so that the energy contribution of the pressure load reduces to

$$P(\varepsilon)(v) = \int_\Omega \left[\varepsilon\pi \det\left(\partial_1\varphi \mid \partial_2\varphi \mid \frac{\partial_3\varphi}{\varepsilon} \right) + \frac{\Delta\pi}{6}(\partial_1\varphi \wedge \partial_2\varphi) \cdot \varphi \right] dx. \tag{93}$$

Since we have assumed that $p > 3$, it is not difficult to see that Lemma 3 still holds true. In Propositions 5 and 7, we thus consider sequences $v(\varepsilon) \in V$ such that $v(\varepsilon) \rightharpoonup v$ with $v \in V_M$ and $(\partial_1 v(\varepsilon) \mid \partial_2 v(\varepsilon) \mid \varepsilon^{-1}\partial_3 v(\varepsilon))$ is bounded in $L^p(\Omega; M_3)$. For such sequences, we have

$$P(\varepsilon)(v(\varepsilon)) \to \frac{\Delta\pi}{3} \int_\omega (\partial_1(\psi + \bar{v}) \wedge \partial_2(\psi + \bar{v})) \cdot (\psi + \bar{v}) \, dx_1 \, dx_2. \tag{94}$$

Indeed, $\det(\partial_1\varphi(\varepsilon) \mid \partial_2\varphi(\varepsilon) \mid \varepsilon^{-1}\partial_3\varphi(\varepsilon))$ is bounded in $L^{p/3}(\Omega)$, $\partial_1\varphi(\varepsilon) \wedge \partial_2\varphi(\varepsilon) \rightharpoonup \partial_1\varphi \wedge \partial_2\varphi$ in $L^{p/2}(\Omega; \mathbb{R}^3)$ by the weak continuity of null Lagrangians; see Ball [1977], Ball, Currie, and Olver [1981], and $\varphi(\varepsilon) \to \varphi$ in $C^0(\bar{\Omega}; \mathbb{R}^3)$ by the Rellich-Kondrachov theorem (recall that Ω is Lipschitz). Hence, we obtain the Γ-limit

$$J^*(0)(v) = \begin{cases} 2\int_\omega QW_0(x, (a_1 + \partial_1\bar{v} \mid a_2 + \partial_2\bar{v}))\sqrt{a}\, dx_1\, dx_2 \\ \quad + \frac{\Delta\pi}{3} \int_\omega ((a_1 + \partial_1\bar{v}) \wedge (a_2 + \partial_2\bar{v})) \cdot (\psi + \bar{v})\, dx_1 dx_2 & \text{if } v \in V_M, \\ +\infty & \text{otherwise.} \end{cases} \tag{95}$$

The limit energy expressed in terms of the deformations $\bar{\varphi} \in \bar{\Phi}$ then reads:

$$\bar{I}(0)(\bar{\varphi}) = 2\int_\omega QW_0(x, \nabla\bar{\varphi})\sqrt{a}\, dx_1\, dx_2 + \frac{\Delta\pi}{3}\int_\omega (\partial_1\bar{\varphi} \wedge \partial_2\bar{\varphi}) \cdot \bar{\varphi}\, dx_1\, dx_2. \tag{96}$$

As before, we may express this energy on the surface $\tilde{S}$ itself. This yields:

$$\tilde{I}_{\tilde{S}}(\tilde{\varphi}) = 2 \int_{\tilde{S}} W_m(a_3(\tilde{x}), D\tilde{\varphi}(\tilde{x}))\, d\tilde{\sigma} + \frac{\Delta\pi}{3} \int_{\tilde{S}} (\operatorname{cof} D\tilde{\varphi}(\tilde{x}) a_3(\tilde{x})) \cdot \tilde{\varphi}\, d\tilde{\sigma}. \quad (97)$$

The term $\tilde{P}(\tilde{\varphi}) = \frac{\Delta\pi}{3} \int_{\tilde{S}} (\operatorname{cof} D\tilde{\varphi}(\tilde{x}) a_3(\tilde{x})) \cdot \tilde{\varphi}\, d\tilde{\sigma}$ corresponds to a pressure of amount $\Delta\pi$ applied on the deformed surface. Indeed, the Euler-Lagrange equations for the limit problem involve the term

$$D\tilde{P}(\tilde{\varphi})\tilde{v} = \Delta\pi \int_{\tilde{S}} (\operatorname{cof} D\tilde{\varphi}(\tilde{x}) a_3(\tilde{x})) \cdot \tilde{v}\, d\tilde{\sigma}$$

for all test functions $\tilde{v}$ that vanish on $\partial\tilde{S}$. If we assume that the deformation $\tilde{\varphi}$ is smooth enough, this may be rewritten as an integral over the deformed surface

$$D\tilde{P}(\tilde{\varphi})\tilde{v} = \Delta\pi \int_{\tilde{\varphi}(\tilde{S})} n_{\tilde{\varphi}} \cdot (\tilde{v} \circ \tilde{\varphi}^{-1})\, d\tilde{\sigma}_{\tilde{\varphi}}$$

where $n_{\tilde{\varphi}}(y)$ is the unit normal vector to $\tilde{\varphi}(\tilde{S})$ in the direction of $\partial_1\tilde{\varphi} \wedge \partial_2\tilde{\varphi}(\psi^{-1}(\tilde{\varphi}^{-1}(y)))$.

Acknowledgment

This work is part of the HCM program "Shells: Mathematical Modeling and Analysis, Scientific Computing" of the Commission of the European Communities (contract ERBCHRXCT940536).

References

[1] E. Acerbi, G. Buttazzo, D. Percivale [1991], A variational definition for the strain energy of an elastic string, *J. Elasticity*, 25, pp. 137–148.
[2] E. Acerbi, N. Fusco [1984], Semicontinuity problems in the calculus of variations, *Arch. Ratl. Mech. Anal.*, 86, pp. 125–145.
[3] S. S. Antman [1976a], Ordinary differential equations of one-dimensional nonlinear elasticity I: Foundations of the theories of nonlinearly elastic rods and shells, *Arch. Ratl. Mech. Anal.*, 61, pp. 307–351.
[4] S. S. Antman [1976b], Ordinary differential equations of one-dimensional nonlinear elasticity II: Existence and regularity theory for conservative problems, *Arch. Ratl. Mech. Anal.*, 61, pp. 353–393.
[5] S. S. Antman [1995], *Nonlinear Problems of Elasticity*, Applied Mathematical Sciences 107, Springer-Verlag, New York.
[6] H. Attouch [1984], *Variational Convergence for Functions and Operators*, Pitman, Boston.
[7] J. M. Ball [1977], Convexity conditions and existence theorems in nonlinear elasticity, *Arch. Ratl. Mech. Anal.*, 63, pp. 337–403.
[8] J. M. Ball, J. C. Currie, P. J. Olver [1981], Null Lagrangians, weak continuity, and variational problems of arbitrary order *J. Funct. Anal.*, 41, pp. 135–174.
[9] M. Carrive-Bédouani [1995], Modélisation intrinsèque et analyse numérique d'un problème de coque mince en grands déplacements, Doctoral Dissertation, Université Paris-Dauphine.
[10] M. Carrive-Bédouani, P. Le Tallec, J. Mouro [1995], Approximations par éléments finis d'un modèle de coque géométriquement exact, INRIA Research Report #2504.

[11] P. G. Ciarlet [1980], A justification of the von Kármán equations, *Arch. Ratl. Mech. Anal.*, 73, pp. 349–389.

[12] P. G. Ciarlet [1988], *Mathematical Elasticity. Volume I: Three-Dimensional Elasticity*, North-Holland, Amsterdam.

[13] P. G. Ciarlet [1990], *Plates and Junctions in Elastic Multi-Structures: An Asymptotic Analysis*, Masson/Springer-Verlag, Paris.

[14] P. G. Ciarlet, P. Destuynder [1979a], A justification of the two-dimensional plate model, *J. Mécanique*, 18, pp. 315–344.

[15] P. G. Ciarlet, P. Destuynder [1979b], A justification of a nonlinear model in plate theory, *Comput. Methods Appl. Mech. Eng.*, 17/18, pp. 227–258.

[16] P. G. Ciarlet, V. Lods [1994], Analyse asymptotique des coques linéairement élastiques. I. Coques "membranaires", *C. R. Acad. Sci. Paris*, 318, Série I, pp. 863–868.

[17] P. G. Ciarlet, V. Lods, B. Miara [1994], Analyse asymptotique des coques linéairement élastiques. II. Coques "en flexion", *C. R. Acad. Sci. Paris*, 319, Série I, pp. 95–100.

[18] B. Dacorogna [1982], Quasiconvexity and relaxation of nonconvex variational problems, *J. Funct. Anal.*, 46, pp. 102–118.

[19] B. Dacorogna [1989], *Direct Methods in the Calculus of Variations*, Applied Mathematical Sciences 78, Springer-Verlag, Berlin.

[20] G. Dal Maso [1993], *An Introduction to Γ-Convergence*, Progress in Nonlinear Differential Equations and their Applications, Birkäuser, Basel.

[21] E. De Giorgi, T. Franzoni [1975], Su un tipo di convergenza variazionale, *Atti. Accad. Naz. Lincei*, 58, pp. 842–850.

[22] P. Destuynder [1980], *Sur une justification des modèles de plaques et de coques par les méthodes asymptotiques*, Doctoral Dissertation, Université Pierre et Marie Curie.

[23] I. Ekeland, R. Temam [1974], *Analyse convexe et problèmes variationnels*, Dunod, Paris.

[24] J. L. Ericksen, C. Truesdell [1958], Exact theory of stress and strain in rods and shells, *Arch. Ratl. Mech. Anal.*, 1, pp. 295–323.

[25] D. D. Fox, J.-C. Simo [1992], A drill rotation formulation for geometrically exact shells, *Comput. Methods Appl. Mech. Eng.*, 98, pp. 329–343.

[26] D. D. Fox, A. Raoult, J.-C. Simo [1993], A justification of nonlinear properly invariant plate theories, *Arch. Ratl. Mech. Anal.*, 124, pp. 157–199.

[27] K. O. Friedrichs, R. F. Dressler [1961], A boundary-layer theory for elastic plates, *Comm. Pure Appl. Math.*, 14, pp. 1–33.

[28] A. L. Goldenveizer [1963], Derivation of an approximate theory of shells by means of asymptotic integration of the equations of the theory of elasticity, *Prikl. Mat. Mech.*, 27, pp. 593–608.

[29] A. E. Green, P. M. Naghdi [1974], On the derivation of shell theories by direct approach, *J. Appl. Mech.*, pp. 173–176.

[30] M. E. Gurtin [1981], *An Introduction to Continuum Mechanics*, Academic Press, New York.

[31] A. Karwowski [1993], Dynamical models for plates and membranes. An asymptotic approach, *J. Elasticity*, 32, pp. 93–153.

[32] H. Le Dret, A. Raoult [1993], Le modèle de membrane non linéaire comme limite variationnelle de l'élasticité non linéaire tridimensionnelle, *C. R. Acad. Sci. Paris*, 317, Série I, pp. 221–226.

[33] H. Le Dret, A. Raoult [1995], The nonlinear membrane model as variational limit of nonlinear three-dimensional elasticity, *J. Math. Pures Appl.*, 75, pp. 551–580.

[34] P. Marcellini [1985], Approximation of quasiconvex functions, and lower semicontinuity of multiple integrals, *Manuscripta Math.*, 51, pp. 1–28.

[35] J. E. Marsden, T. J. R. Hughes [1983], *Mathematical Foundations of Elasticity*, Prentice-Hall, Englewood Cliffs.

[36] B. Miara [1994], Analyse asymptotique des coques membranaires non linéairement élastiques, *C. R. Acad. Sci. Paris*, 318, Série I, pp. 689–694.

[37] C. B. Morrey Jr. [1952], Quasiconvexity and the semicontinuity of multiple integrals, *Pacific J. Math.*, 2, pp. 25–53.

[38] N. G. Meyers [1965], Quasiconvexity and lower semicontinuity of multiple variational

integrals of any order, *Trans. A.M.S.*, 119, pp. 125–149.

[39] S. Müller [1990], Det = det. A remark on the distributional determinant, *C. R. Acad. Sci. Paris*, 311, Série I, pp. 13–17.

[40] P. M. Naghdi [1972], *The Theory of Plates and Shells*, in Handbuch der Physik, vol. VIa/2, Springer-Verlag, Berlin.

[41] D. Percivale [1991], The variational method for tensile structures, Politecnico di Torino, Dipartimento di Matematica Research Report #16.

[42] A. Raoult [1986], Non-polyconvexity of the stored energy function of a Saint Venant-Kirchhoff material, *Apl. Mat.*, 6, pp. 417–419.

[43] A. Raoult [1988], Analyse mathématique de quelques modèles de plaques et de poutres élastiques ou élasto-plastiques, Doctoral Dissertation, Université Pierre et Marie Curie, Paris.

[44] E. Sanchez-Palencia [1989a], Statique et dynamique des coques minces. I. Cas de flexion pure non inhibée, *C. R. Acad. Sci. Paris*, 309, Série I, pp. 411–417.

[45] E. Sanchez-Palencia [1989b], Statique et dynamique des coques minces. II. Cas de flexion pure inhibée, *C. R. Acad. Sci. Paris*, 309, Série I, pp. 531–537.

[46] E. Sanchez-Palencia [1990], Passage à la limite de l'élasticité tridimensionnelle à la théorie asymptotique des coques minces, *C. R. Acad. Sci. Paris*, 311, Série II, pp. 909–916.

[47] M. J. Sewell [1967], On configuration-dependent loading, *Arch. Ratl. Mech. Anal.*, 23, pp. 327–351.

[48] J.-C. Simo, D. D. Fox [1989], On a stress resultant geometrically exact shell model. Part I: Formulation and optimal parametrization, *Comput. Methods Appl. Mech. Eng.*, 72, pp. 267–304.

[49] J.-C. Simo, D. D. Fox, M. S. Rifai [1990a], On a stress resultant geometrically exact shell model. Part III: Computational aspects of the nonlinear theory, *Comput. Methods Appl. Mech. Eng.*, 79, pp. 21–70.

[50] J.-C. Simo, D. D. Fox, M. S. Rifai [1990b], On a stress resultant geometrically exact shell model. Part IV: Variable thickness shells with through-the-thickness stretching, *Comput. Methods Appl. Mech. Eng.*, 81, pp. 91–126.

[51] J.-C. Simo, N. Tarnow [1994], A new energy and momentum conserving algorithm for the nonlinear dynamics of shells, *Int. J. Numer. Methods Eng.*, 37, pp. 2527–2549.

[52] J.-C. Simo, L. Vu-Quoc [1988], On the dynamics in space of rods undergoing large overall deformations. A geometrically exact approach, *Comput. Methods Appl. Mech. Eng.*, 66, pp. 125–161.

[53] J.-C. Simo, L. Vu-Quoc [1991], A geometrically exact rod model incorporating shear and torsion-warping deformation, *Int. J. Solids Structures*, 27, pp. 371–393.

[54] C. C. Wang, C. Truesdell [1973], *Introduction to Rational Elasticity*, Noordhoff, Gröningen.

Gravity Waves on the Surface of the Sphere

R. de la Llave[1] and P. Panayotaros[2]

[1] Department of Mathematics and

[2] Department of Physics, The University of Texas at Austin, Austin, TX 78712

Received September 20, 1995
Communicated by Jerrold Marsden and Stephen Wiggins

This paper is dedicated to the memory of Juan-Carlos Simo

Summary. We propose a Hamiltonian model for gravity waves on the surface of a fluid layer surrounding a gravitating sphere. The general equations of motion are nonlocal and can be used as a starting point for simpler models, which can be derived systematically by expanding the Hamiltonian in dimensionless parameters. In this paper, we focus on the small wave amplitude regime. The first-order nonlinear terms can be eliminated by a formal canonical transformation. Similarly, many of the second order terms can be eliminated. The resulting model has the feature that it leaves invariant several finite-dimensional subspaces on which the motion is integrable.

1. Introduction

The goal of this paper is to describe a Hamiltonian formulation for waves on the surface of a fluid layer surrounding a gravitating spherical body. The fluid satisfies hydrodynamic equations inside the layer, and the surface of the layer is moving consistently with the motions of the fluid. The resulting system is a free boundary problem in which the region where the hydrodynamic equations of motion hold is also part of the unknowns.

We will assume that the fluid in the layer is inviscid, incompressible and moving irrotationally. The assumption that the flow is irrotational (potential flow) is very restrictive and in many cases inappropriate. Our model is, however, useful for isolating and studying hydrodynamic wave phenomena where the restoring force is gravitational, such as, for instance, sea waves and atmospheric tides.

In the case where the potential gravity waves take place over a plane, a very elegant Hamiltonian formulation was first introduced by Zakharov [Z] (see also [M]). More recently it has been shown that the formulation leads to a very systematic discussion of approximate gravity wave equations (see [CG]) and to efficient numerical methods [CS]. Using the formalism presented here, similar algorithms could be derived for the sphere.

The theory of Zakharov has also been extended to general inviscid, incompressible free boundary flows in [LMMR]. In that work the Hamiltonian formulation of the potential flow case is derived from the canonical theory of the Euler equations (see [MG], [MW]).

In Sections 2, 3, and 4 we show how to use the potential flow formalism on the sphere and discuss some physical applications and limitations of our model. The main result of these sections is that, indeed, the free boundary problem can be written in a Hamiltonian form. The Poisson bracket does not have the standard form (such brackets are often referred to as noncanonical), but we also find a change of variables that reduces it to standard form. Note that there are already several well-known examples of noncanonical Poisson brackets in hydrodynamics, plasma, magnetism and other areas. The canonical variables in the Hamiltonian formulation of free surface potential flow are the wave amplitude (a function giving the shape of the surface) and the hydrodynamic potential at the surface. These two functions, defined on the sphere, completely determine the velocity inside the layer. This reduction in the dimension of the problem is one of the benefits of the potential flow assumption. On the other hand, the equations of motion for wave amplitude and surface potential are nonlocal. We describe a method that allows us to write the Hamiltonian and the equations of motion in terms of Fourier multiplier operators in Section 5.

The Hamiltonian formalism is particularly convenient for deriving approximate gravity wave equations, for it suffices to consider approximations of the Hamiltonian. These approximations are based on dimensional analysis and are discussed in Section 6, where we identify the dimensionless parameters of the problem and indicate interesting asymptotic regimes. In the present work we will be concerned with a small amplitude regime and we focus on "intermediate depth" waves.

The theory of small amplitude water waves is analogous to the theory of motions of a Hamiltonian system near an elliptic fixed point. The completely quiescent state is the fixed point, and the linear plane waves correspond to the normal modes of the linearised system. The motion of the waves in the linear approximation is governed by the quadratic terms in the Hamiltonian, while the nonlinear evolution arises from cubic and higher order terms of the Hamiltonian. The cubic and higher order terms of the Hamiltonian can be also interpreted as describing the interaction between three waves (cubic terms), four waves (quartic terms) and so on. One of the tools used in the study of motion around an elliptic fixed point is the Poincaré-Birkhoff method of successive canonical changes of coordinates. In each canonical transformation of the method one tries to eliminate the lowest order nonlinear terms of the Hamiltonian. After a number of such transformations the Hamiltonian is reduced to the so-called Birkhoff normal form containing only the resonant nonlinear terms. Analysis of normal form systems can yield extra information on the behavior of the system. In Section 7 we show that for water waves on the surface of the sphere: (a) the cubic (first-order in the nonlinearity) terms of the Hamiltonian can be eliminated, and (b) the only quartic (second-order in the nonlinearity) terms that cannot be eliminated are those of a very reduced class. The canonical transformations are given explicitly as formal time-1 maps of suitable Hamiltonian flows. We will remark, however, that these normal form calculations are more useful for intermediate and large depth water waves. As an application of these calculations we find some finite dimensional manifolds that are invariant under the evolution determined by the quartic normal form

Hamiltonian. On these finite dimensional manifolds we can use the standard methods of dynamical systems to, in particular, identify periodic orbits of the second-order normal form system. These approximate solutions of the full water wave system are traveling and standing waves with amplitude dependent frequency.

Even if at this point these transformations are purely formal, we hope that they can be made rigorous and at least be used to prove good lower bounds for the time of existence of solutions with small initial data. It also seems possible that some of the periodic orbits and quasi-periodic orbits of the linear problem persist in the full nonlinear system. For finite dimensional systems, the persistence of families of periodic orbits in the full nonlinear system was proved first by Lyapunov. More general results of this type are in [Mo], [We]. The persistence of some of the quasi-periodic orbits is also proved in finite dimensions using KAM theory. Since this paper is concerned with developing the formalism, we will postpone these questions to future work.

We also point out that our formalism is well adapted to the development of numerical methods for the problem. For example, to develop finite-dimensional approximations we truncate the Hamiltonian. The truncated system will automatically be Hamiltonian. We plan to discuss the numerical implementations of that scheme in a forthcoming paper.

2. Equations of Motion

We consider a sphere of radius b and, on top of the sphere, a layer of fluid ("sea" or "atmosphere") of thickness ("depth") h. Using the standard spherical coordinates $r =$ radius, $\vartheta =$ polar, $\varphi =$ azimuth, the surface of the sea will be at $r(\vartheta, \varphi) = \rho + \eta(\vartheta, \varphi)$ with $\rho = b + h$. The amplitude of the water waves is described by the single valued function $\eta(\vartheta, \varphi)$. The dynamical problem we want to consider is that of free surface potential flow of the layer of water under the influence of gravity. For such a flow, since the spherical shell is simply connected, there exists a velocity potential ϕ and the velocity is given by $\vec{u} = \nabla \phi$. The conservation of mass for an incompressible fluid is $\nabla \cdot \vec{u} = 0$. Hence, we should have

$$\Delta \phi = 0 \tag{2.1}$$

in the region occupied by the fluid. On the surface we have

$$\eta_t = \frac{\partial \phi}{\partial r} - \frac{1}{r^2} \frac{\partial \phi}{\partial \vartheta} \frac{\partial \eta}{\partial \vartheta} - \frac{1}{r^2 \sin^2 \vartheta} \frac{\partial \phi}{\partial \varphi} \frac{\partial \eta}{\partial \varphi}, \tag{2.2}$$

and

$$\phi_t = -\frac{1}{2} |\nabla \phi|^2 + \frac{K}{\rho + \eta} - p, \tag{2.3}$$

and at the bottom $r = b$ we have

$$\frac{\partial \phi}{\partial r} = 0. \tag{2.4}$$

The equations of motion (2.1)–(2.4) are obtained from the Euclidean Euler equations (see [L]) by a change of variables: (2.1) is the conservation of mass for an incompressible fluid, (2.4) is the rigid wall boundary condition at the bottom of the sea and (2.2) is the condition, in polar coordinates, that the surface is transported by the flow, or

$$\left[\frac{\partial}{\partial t} + \vec{u} \cdot \nabla\right] F = 0, \tag{2.5}$$

where $F(\vec{r}, t) = r - \rho - \eta(\vartheta, \varphi) = 0$ is the implicit representation of the surface and $\vec{u}$ is the velocity at the surface. The dynamical boundary condition (2.3) follows from Euler's equation at the surface

$$\nabla\left(\frac{\partial \phi}{\partial t} + \frac{1}{2}|\vec{u}|^2 + V(\eta) + p\right) = 0. \tag{2.6}$$

The term $V(\eta) = \frac{K}{\rho+\eta} = \frac{K}{\rho} - g\eta + \cdots$ is the gravitational potential due to the solid sphere of radius b. Physically, $K = GM$ with M the mass of the planet, G the gravitational constant and g the acceleration of gravity at $r = b$. Also we require that the pressure p be constant in space. ($\nabla p \neq 0$ would correspond to the presence of an additional external force.) Note that in (2.3) quantities with no spatial dependence (e.g. p, $\frac{K}{\rho}$) do not play any role and can be set to zero.

The equations of motion suggest that if at any instant t_0 we know the function $\eta(\vartheta, \varphi)$ and the potential ϕ on the surface, i.e., the function $\Phi(\varphi, \vartheta) = \phi(\varphi, \vartheta, \rho + \eta(\vartheta, \varphi))$, we can determine ϕ at t_0 in the whole region occupied by the fluid by solving the boundary value problem: $\Delta\phi = 0$ inside the fluid with $\phi = \Phi$ at the surface $r = \rho + \eta(\vartheta, \varphi)$ and $\frac{\partial \phi}{\partial r} = 0$ at $r = b$. Thus we are essentially interested in the evolution of η and Φ which is given by (2.2) and (2.3). Note that $\frac{\partial \Phi}{\partial t} = \frac{\partial \phi}{\partial t} + \frac{\partial \eta}{\partial t}\frac{\partial \phi}{\partial r}|_{r=\rho+\eta(\vartheta,\varphi)}$. Both equations are nonlocal since they contain the term $\frac{\partial \phi}{\partial r}$. To evaluate $\frac{\partial \phi}{\partial r}$ we need information about the solution of the boundary value problem. A method for writing the equations of motion in terms of η and Φ alone is given in Section 3.

3. Hamiltonian Formulation

We will show that the equations of motion (2.2), (2.3) for η and Φ form a Hamiltonian system. More precisely they can be written as

$$\eta_t = [\eta, H], \qquad \Phi_t = [\Phi, H] \tag{3.1}$$

where $[,]$ is an appropriate Poisson bracket and H is the Hamiltonian of the system.

To express the equations of motion in Hamiltonian form, it is natural to guess that the Hamiltonian should be the physical energy and then try to find the Poisson bracket. In particular, here the Hamiltonian H will be the total energy $K + U$ (kinetic + potential) of the water mass

$$H = \frac{1}{2}\int_{S^2}\int_{r=b}^{r=\rho+\eta(\vartheta,\varphi)}|\nabla\phi|^2 dV + \int_{S^2}\int_{r=b}^{r=\rho+\eta(\vartheta,\varphi)}\left(-\frac{K}{r}\right)dV. \tag{3.2}$$

Integrating by parts the first term,

$$H = \frac{1}{2}\int_{S^2}\Phi R\frac{\partial \phi}{\partial n}\bigg|_{r=\rho+\eta(\vartheta,\varphi)}dA_r + \frac{1}{2}\int_{S^2}(-K)dA_r, \tag{3.3}$$

where $R = [1 + (\frac{1}{r}\frac{\partial \eta}{\partial \vartheta})^2 + (\frac{1}{r \sin \vartheta}\frac{\partial \eta}{\partial \varphi})^2]^{\frac{1}{2}}$ and $\frac{\partial \phi}{\partial n}|_{r=\rho+\eta(\vartheta,\varphi)}$ is the normal derivative at the surface. We also use the notation $dA_r = r^2 \sin \theta d\theta d\varphi$ with $r = \rho + \eta(\vartheta, \varphi)$. The quantity $R dA_r$ is the area element of the water surface.

We introduce the Dirichlet-Neumann ("flux") operator $G(\eta)$ by

$$G(\eta)\Phi = R \left.\frac{\partial \phi}{\partial n}\right|_{r=\rho+\eta(\vartheta,\varphi)} \tag{3.4}$$

and we can write the Hamiltonian as

$$H = \frac{1}{2}\int_{S^2} \Phi G(\eta)\Phi dA_r + \frac{1}{2}\int_{S^2}(-K)dA_r. \tag{3.5}$$

The Poisson bracket $[\,,\,]$ of (3.1) is defined by

$$[F, G] = \int_{S^2}\left(\frac{\rho+\eta}{\rho}\right)^{-2}\left(\frac{\delta F}{\delta \eta}\frac{\delta G}{\delta \Phi} - \frac{\delta G}{\delta \eta}\frac{\delta F}{\delta \Phi}\right)\rho^2 dA_1. \tag{3.6}$$

The variational derivative $\frac{\delta F}{\delta x}$ is defined by $\langle\frac{\delta F}{\delta x}, \delta x\rangle = F'$, where F' is the Frechet derivative of F with respect to x, and $\langle\,,\,\rangle$ is the L^2 inner product on S^2 (with radius ρ). From (3.6) the equations of motion (3.1) become

$$\eta_t = \left(\frac{\rho+\eta}{\rho}\right)^{-2}\frac{\delta H}{\delta \Phi}, \qquad \Phi_t = -\left(\frac{\rho+\eta}{\rho}\right)^{-2}\frac{\delta H}{\delta \eta}. \tag{3.7}$$

It is easy to see that the above bracket indeed satisfies the axioms of Poisson brackets (for the formalism of Poisson brackets for fields see [D], ch. 1). The only axiom that is not immediate is the Jacobi identity, which can be verified by a direct computation.

Proposition 3.1. *Hamilton's equations (3.1) and the original equations of motion (2.2), (2.3) are equivalent.*

Proof. We note that $\frac{\delta U}{\delta \Phi} = 0$ and

$$K(\Phi + \delta\Phi) = \frac{1}{2}\int_{S^2} \Phi G(\eta)\delta\Phi dA_r + \frac{1}{2}\int_{S^2}\delta\Phi G(\eta)\Phi dA_r + K(\Phi) + o(\delta\Phi)$$

so that $\frac{\delta K}{\delta \Phi} = (1 + \frac{\eta}{\rho})^2 G(\eta)\Phi$, and, therefore, Hamilton's equation for η_t is

$$\eta_t = G(\eta)\Phi = R\frac{\partial \phi}{\partial n}|_{r=\rho+\eta(x,y)} = \nabla\phi \cdot \left[1, -\frac{1}{r}\frac{\partial \eta}{\partial \vartheta}, -\frac{1}{r \sin \vartheta}\frac{\partial \eta}{\partial \varphi}\right], \tag{3.8}$$

which is (2.2). For (2.3),

$$\frac{\delta U}{\delta \eta} = -\left(1 + \frac{\eta}{\rho}\right)^2 \frac{K}{\rho+\eta}$$

and

$$K(\eta+\delta\eta) = \frac{1}{2}\int_{S^2}|\nabla\phi|^2\delta\eta dA + \frac{1}{2}\int_{S^2}\int_{r=b}^{r=\rho+\eta(\vartheta,\varphi)}|\nabla\phi(\eta+\delta\eta)|^2 dV + o(\delta\eta). \tag{3.9}$$

Integrating by parts, the second integral is

$$-\int_{S^2}\int_{r=b}^{r=\rho+\eta(\vartheta,\varphi)} \nabla\phi\nabla\left(\frac{\partial\phi}{\partial r}\delta\eta\right) dV + K(\eta) + o(\delta^2\eta)$$

$$= -\int_{S^2} \frac{\partial\phi}{\partial r}\delta\eta\frac{\partial\phi}{\partial n} RdA_r + K(\eta) + o(\delta^2\eta)$$

$$= -\int_{S^2} \frac{\partial\phi}{\partial r}\eta_t\delta\eta dA_r + K(\eta) + o(\delta\eta) \tag{3.10}$$

using Green's identity and the equation for η_t. From (3.6), (3.7), (3.10) we obtain

$$-\left(\frac{\rho+\eta}{\rho}\right)^{-2}\frac{\delta H}{\delta\eta} = -\frac{1}{2}|\nabla\phi|^2 - g\eta + \eta_t\frac{\partial\phi}{\partial r}. \tag{3.11}$$

But we also have $\frac{\partial\Phi}{\partial t} = \frac{\partial\phi}{\partial t_,} + \frac{\partial\eta}{\partial t}\frac{\partial\phi}{\partial r}|_{r=\rho+\eta(\vartheta,\varphi)}$, so that by (3.9), $\frac{\partial\Phi}{\partial t} = -\left(\frac{\rho+\eta}{\rho}\right)^{-2}\frac{\delta H}{\delta\eta}$ is equivalent to (2.3).   □

Remark 3.1. A comparison of the above derivation of the Hamiltonian structure of water wave equations on the sphere with the derivations of the analogous result in Euclidean space ([Z], also [BO]) will reveal that the adoption of the nonstandard bracket was needed because of the form of the surface element in polar coordinates. The departure of our bracket from the standard one is small—in a sense that will be made precise later—but not small enough to be discarded.

It is possible to make a change of variables in such a way that the bracket becomes the standard one. (We will refer to this process as "diagonalising" the bracket.) In particular, choosing a transformation $f(\eta, \Phi) = (\tilde{\eta}, \tilde{\Phi})$ defined by

$$\tilde{\eta} = \eta\left(1 + \frac{\eta}{2\rho}\right), \qquad \tilde{\Phi} = \Phi\left(1 + \frac{\eta}{\rho}\right), \tag{3.12}$$

the bracket is diagonalised: $Df \cdot A \cdot Df^T = J$, where A, J are the respective cosymplectic forms of our bracket and the standard bracket. The transformation of (3.12) is invertible in the range of interest $\sup|\eta| < h$.

4. Applicability of the Model to Physical Problems

The main physical limitation of our model is the potential flow assumption. The absence of vorticity and angular momentum in the rest frame has the consequence that the model can not describe phenomena such as Rossby waves or the bulging of the equator observed in rotating liquids. Thus our model isolates the effects of gravitation. We can also add to our formalism the apparent motions due to the rotation of the observer's reference frame, as well as another gravitational effect, namely the tides. Also, the model can be useful for studying phenomena involving short waves, in particular energy transport by wind-generated ocean waves.

First we consider the water wave equations in a frame rotating with angular velocity Ω around the z-axis ($\vartheta = 0$). We introduce the rotating frame canonical variables $u_\Omega = (\eta_\Omega, \Phi_\Omega)$ related to the rest frame variables $u = (\eta, \Phi)$ by $u_\Omega = A_\Omega^{-1}(t)u$, where $A_\Omega(t)$ acts on functions defined on the sphere by

$$A_\Omega(t)f(\vartheta, \varphi) = f(\vartheta, \varphi + \Omega t).$$

The flow $A_\Omega(t)$ is generated by

$$f_t = [f, \Omega L_z],$$

where $[\ ,\ ]$ is the Poisson bracket introduced previously and L_z, the angular momentum, is

$$L_z = \int_{S^2}\int_{r=b}^{r=\rho+\eta(\vartheta,\varphi)} (\vec{r} \times \nabla\phi) \cdot \hat{z}\,dV = \int_{S^2} \Phi \frac{\partial\eta}{\partial\varphi}\,dA_r. \tag{4.1}$$

Note that $[H, L_z] = 0$ so that if the rest frame variables evolve under $u_t = [u, H]$ the rotating frame observables evolve under

$$\frac{\partial\eta_\Omega}{\partial t} = [\eta_\Omega, H - \Omega L_z], \qquad \frac{\partial\Phi_\Omega}{\partial t} = [\Phi_\Omega, H - \Omega L_z]. \tag{4.2}$$

Also note that the rotating frame velocity u_Ω is given by $u_\Omega = \nabla\phi_\Omega + U_\Omega$ where ϕ_Ω is the solution of the Neumann problem (i.e., (2.1), (2.4) for ϕ_Ω) with Φ_Ω as the boundary value at the surface, and U_Ω is the velocity of rigid rotation with angular velocity Ω. A Hamiltonian formalism for general Euler free boundary flows is given in [LMMR]. There the authors use the unique decomposition of divergenceless velocity field u to $w + \nabla\phi$, with w divergenceless and tangent to the boundary, to write the canonical theory for w, the surface potential and the shape of the boundary. However, the description of the fluid motion by functions of two variables is now lost.

Tides result from the gravitational attraction of another body, in particular, the acceleration of points on the surface of the sphere relative to the acceleration of the center of the sphere. The effect of tides is described by a tidal potential, which can be added to the equation for the evolution of the surface hydrodynamic potential. The tidal potential W at the surface of the fluid layer has, in lowest order in the distance to the attracting body, the form (see [P])

$$W = A P_2(\zeta), \tag{4.3}$$

where $P_2(\zeta)$ is the second Legendre function and A is a physical constant. If we consider a point x on the surface of the layer, the point O at the center of the sphere and the point S at the center of the distant star, ζ is the angle between the lines Ox and OS. The tidal potential can be expressed as a function of the polar and azimuth angles ϑ and φ of the point x and of the polar and azimuth angles δ' and T' of the external body

$$W = A[P_2(\vartheta)P_2(\delta') + \frac{3}{4}\sin 2\varphi \sin 2\delta' \cos(T' + \varphi) + \frac{3}{4}\cos^2\varphi \cos^2\delta' \cos 2(T' + \varphi)]. \tag{4.4}$$

If the sphere rotates with angular velocity Ω, then $T' = \Omega t$, and, similarly, we can consider variations in time of δ'. We can then use the rotating frame formalism of equations (4.2) and the tidal potential to model the effects of the periodic variation of the tidal forces. Notice that in (4.3) we can write $P_2(\zeta) = Y_2^0(\zeta, \varpi)$ where ϖ is the angle around the axis OS. In the second form, (4.4), we are using the terrestrial system of axes and angles. Since the two systems are related by a time-dependent rotation, the tidal potential in the terrestrial system can be written as a time-dependent linear combination of the spherical harmonics Y_l^m with $l = 2, m = -2, -1, \ldots, 2$. Thus the potential tide can be added to the spectral form of the equations of motion in a straightforward way provided that we have the time dependence of the coefficient of each spherical harmonic. Since the harmonics are eigenfunctions of the linearised problem, a simple way to mimic the varying tidal forces is parametric excitation (variation of one of the parameters of the problem) at frequencies that are at resonance with the $l = 2$ harmonics. In fact, the phenomenon of parametric excitation of nonlinear waves (see, for instance, [FS] for water waves) is of independent interest.

One other area of possible application of our model is in the transport of energy by wind-driven sea waves in the ocean. The typical wavelength of these waves is very small compared to the global scales, and the effect of the Coriolis force is minimal. We can include the effect of wind by adding to the equations of motion a pressure (gradient) term computed from given velocity profiles for the motion of the air masses above the sea. Since the approximations used in the following chapters are uniform in spatial wave number of the quantities involved, we expect that the response of short waves will yield rather reliable information on the long-range transport of wind energy by the sea waves.

5. The Flux Operator

We now describe the Dirichlet-Neumann operator $G(\eta)$ as a function of η. It will be assumed that $G(\eta)$ can be expanded around $\eta = 0$ so that we can write $G(\eta) = \sum_{i=0}^{\infty} G_i(\eta)$, the $G_i(\eta)$ being homogeneous of order i in η. Our task will be to calculate the $G_i(\eta)$. Clearly, from (3.5), the expansion of $G(\eta)$ will give us an expansion of the Hamiltonian $H = H_0 + H_1 + \cdots$ with $H_0 = \frac{1}{2} \int \Phi G_0(\eta) \Phi + \frac{1}{2} \int g \eta^2$, $H_1 = \frac{1}{2} \int \Phi G_1(\eta) \Phi$ and so forth. As we will show, it is possible to compute recursively the $G_i(\eta)$ in a way very similar to that in [CG], [CS] and [GAS]. Once the $G_i(\eta)$ are computed explicitly, the Hamiltonian of (3.5) can be systematically expanded in powers of η. If we use as the Hamiltonian some truncation of this expansion, the resulting model can be considered as a small-amplitude approximation of the water waves problem. This formalism, therefore, provides us with a systematic way to produce increasingly more refined small-amplitude expansions that can be computed rather effectively. We remark, however, that the expansion here is formal. We leave the problem of convergence and the choice of appropriate spaces for η and Φ open. For waves on the line the question has been addressed (in a more general context) in [CM].

To calculate the $G_i(\eta)$ we first look for harmonic functions ϕ_γ of the form

$$\phi_\gamma(r, \vartheta, \varphi) = u_\gamma(r) Y_\gamma(\vartheta, \varphi).$$

The functions $Y_\gamma(\vartheta, \varphi)$ are the spherical harmonics, indexed by $\gamma = [l, m]$ with l a positive integer being the total angular momentum number, and $m = -l, -l + 1, \ldots, l$

the azimuthal angular momentum number. The condition $\Delta\phi_\gamma = 0$ implies that

$$r^2 u_\gamma'' + 2r u_\gamma' - l(l+1)u_\gamma = 0. \tag{5.1}$$

The only solution of (5.1) (up to a multiplicative constant) that also satisfies $u_\gamma'(b) = 0$ is

$$u_\gamma(r) = (l+1)\left(\frac{r}{b}\right)^l + l\left(\frac{b}{r}\right)^{l+1}. \tag{5.2}$$

Since the $\phi_\gamma = u_\gamma Y_\gamma$, as above, are harmonic and satisfy $\frac{\partial\phi_\gamma}{\partial r} = 0$ at $r = b$, from the definition of $G(\eta)$, (3.4), we have that for every index γ

$$G(\eta)\phi_\gamma(\vartheta, \varphi, \rho + \eta) = \nabla\phi_\gamma(\vartheta, \varphi, \rho + \eta) \cdot \left[1, -\frac{1}{r}\frac{\partial\eta}{\partial\vartheta}, -\frac{1}{r\sin\vartheta}\frac{\partial\eta}{\partial\varphi}\right] \tag{5.3}$$

or

$$\sum_{i=0}^{\infty} G_i(\eta)u_\gamma(\rho + \eta)Y_\gamma = u_\gamma'(\rho + \eta)Y_\gamma$$

$$-\frac{\partial\eta}{\partial\vartheta}\frac{u_\gamma(\rho + \eta)}{(\rho + \eta)^2}\frac{\partial Y_\gamma}{\partial\vartheta} - \frac{\partial\eta}{\partial\varphi}\frac{u_\gamma(\rho + \eta)}{(\rho + \eta)^2\sin^2\vartheta}\frac{\partial Y_\gamma}{\partial\varphi}. \tag{5.4}$$

Expanding $u_\gamma(r)$ and $\frac{1}{r^2}$ around $\eta = 0$ (i.e., $r = \rho$) and matching powers in η, we obtain, at order 0,

$$u_\gamma'(\rho)Y_\gamma = G_0(\eta)u_\gamma(\rho)Y_\gamma$$

and hence

$$G_0(\eta)Y_\gamma = \frac{1}{u_\gamma(\rho)}u_\gamma'(\rho)Y_\gamma. \tag{5.5}$$

At order 1 we have

$$\eta u_\gamma''(\rho)Y_\gamma - \frac{1}{\rho^2}u_\gamma(\rho)\left[\eta_\vartheta\frac{\partial Y_\gamma}{\partial\vartheta} + \frac{1}{\sin^2\vartheta}\eta_\varphi\frac{\partial Y_\gamma}{\partial\varphi}\right] = G_0(\eta)\eta u_\gamma'(\rho)Y_\gamma + G_1(\eta)u_\gamma(\rho)Y_\gamma$$

and hence

$$G_1(\eta)Y_\gamma = \frac{1}{u_\gamma(\rho)}\left[\eta u_\gamma''(\rho)Y_\gamma - \frac{1}{\rho^2}u_\gamma(\rho)\right.$$

$$\left. \times\left[\eta_\vartheta\frac{\partial Y_\gamma}{\partial\vartheta} + \frac{1}{\sin^2\vartheta}\eta_\varphi\frac{\partial Y_\gamma}{\partial\varphi}\right] - G_0(\eta)\eta u_\gamma'(\rho)Y_\gamma\right]. \tag{5.6}$$

In general, at order k we have

$$\frac{1}{k!}\eta^k u_\gamma^{(k+1)}(\rho)Y_\gamma - \frac{1}{(k-1)!}\eta^{k-1}\left(\frac{u_\gamma}{r^2}\right)^{(k-1)}(\rho)\left[\eta_\vartheta\frac{\partial Y_\gamma}{\partial\vartheta} + \frac{1}{\sin^2\vartheta}\eta_\varphi\frac{\partial Y_\gamma}{\partial\varphi}\right]$$

$$= \left[G_0(\eta)\frac{1}{k!}\eta^k u_\gamma^{(k)}(\rho) + G_1(\eta)\frac{1}{(k-1)!}\eta^{k-1}u_\gamma^{(k-1)}(\rho) + \cdots + G_k(\eta)u_\gamma(\rho)\right]Y_\gamma \tag{5.7}$$

($u_\gamma^{(k)} = k$-th derivative), and we can obtain $G_k(\eta)$ recursively in terms of the $G_i(\eta)$ with $i < k$. The above formulas determine the $G_i(\eta)$ as an operator in a, for the moment, unspecified space of functions in the sphere. Note that the $G_i(\eta)$ are given as Fourier multipliers, and it is clear that they are not local operators.

6. Scales and Dimensionless Variables

It will be advantageous to introduce several dimensionless quantities that take into account the scales of the problem. The relevant dimensionless quantities are (i) the ratio $\epsilon = \frac{A}{h}$ of the typical amplitude A of the waves to the depth h of the fluid layer, (ii) the ratio $\beta = \frac{h}{b}$ of the depth over the radius b of the planet, and in the rotating frame we also have (iii) $R = \frac{\Omega}{\omega_0}$ where Ω is the angular velocity of the frame and ω_0 is a typical frequency of linearised gravity waves in the rest frame. Here we take the reference frequency to be $\omega_0 = \frac{\sqrt{gh}}{b}$, which turns out to be the phase velocity of linearised shallow water waves.

The dimensionless variables η^*, Φ^* and t^* are introduced by

$$\eta^* = \frac{\eta}{A}, \qquad \Phi^* = \frac{\omega_0 \Phi}{gA}, \qquad t^* = \omega_0 t. \tag{6.1}$$

For the Dirichlet-Neumann operator, we observe from the formulas in the previous section that in each $G_k(\eta)$ we can factor out a term $\frac{1}{b^{k+1}}$ arising from the $u_y{}^{(n)}$ so that

$$G_k(\eta)Y_\gamma = \frac{1}{b^{k+1}} G_k^*(\epsilon h \eta^*)Y_\gamma = \frac{1}{b}(\epsilon\beta)^k G_k^*(\eta^*)Y_\gamma. \tag{6.2}$$

The $G_k^*(\eta^*)$ defined by the first equality in (6.2) are dimensionless: they depend only on η^* and β. The Hamiltonian can be written as

$$H = (\epsilon\beta)^2 b^2 \frac{1}{2} \int_{S^2} \left[g\Phi^* \frac{1}{\beta} \sum_k (\epsilon\beta)^k G_k^*(\eta^*)\Phi^* + g(\eta^*)^2 \right] dA_\rho. \tag{6.3}$$

Note that g is the acceleration of gravity at $r = \rho$. Defining $\hat{H}$ by

$$\hat{H} = \frac{1}{2} \int_{S^2} \left[\Phi^* \frac{1}{\beta} \sum_k (\epsilon\beta)^k G_k^*(\eta^*)\Phi^* + (\eta^*)^2 \right] \rho^2 \, dA_1,$$

Hamilton's equations (3.7) become

$$\eta_{t^*}^* = (1 + \epsilon\beta\eta^*)^{-2} \frac{\delta \hat{H}}{\delta \Phi^*}, \tag{6.4}$$

$$\Phi_{t^*}^* = -(1 + \epsilon\beta\eta^*)^{-2} \frac{\delta \hat{H}}{\delta \eta^*}. \tag{6.5}$$

In the rotating frame we can use the time-scale $t^\dagger = Rt^*$ and Hamilton's equations become

$$\eta_{t^\dagger}^* = (1 + \epsilon\beta\eta^*)^{-2} \left[-\frac{\delta \hat{L}}{\delta \Phi^*} + \frac{1}{R} \frac{\delta \hat{H}}{\delta \Phi^*} \right] \tag{6.6}$$

and

$$\Phi_{t^i}^* = -(1 + \epsilon\beta\eta^*)^{-2} \left[-\frac{\delta\hat{L}}{\delta\eta^*} + \frac{1}{R}\frac{\delta\hat{H}}{\delta\eta^*} \right] \tag{6.7}$$

with $\hat{L}_z = \int_{S^2} \Phi^* \frac{\partial\eta^*}{\partial\vartheta} dA_\rho$.

Note that the factor $(1 + \epsilon\beta\eta^*)^{-2}$ which appears in the nonstandard bracket differs from unity in terms which are first order in the dimensionless variables ϵ and β. Hence, the difference between the standard bracket and the nonstandard one has to be considered in expansions in the dimensionless quantities of this order.

The parameter β is the analog of the depth to wavelength ratio in Euclidean space. The small β regime gives us an analog of the "shallow water" regime (e.g., see [Wi]), in which the $G_k^*(\eta^*)$ can be expanded in β. Note that the $G_k^*(\eta^*)$ are already multiplied by β^k in the Hamiltonian so that, for instance, in the $\beta \to 0$ limit (with ϵ nonzero) the evolution becomes linear. It is also possible to prescribe a relation between the parameters ϵ and β. For example, setting $\epsilon = \beta^2 = \mu$ would lead to an analog of the Boussinesq regime in $\mathbf{R}^2$.

In what follows we will be concerned with the small ϵ regime with β arbitrary; in particular, we will use the $G_i(\eta)$ calculated previously to write the equations of motion to second order in ϵ. From the dimensional analysis we have to take into account terms arising from the nonstandard bracket. It is advantageous to present the Hamiltonian in the variables $\tilde{\eta}$, $\tilde{\Phi}$ in which the bracket is diagonal. Note that since ϵ is small the normalizations of the new variables can be taken to be the same as that of the original variables. The $O(\epsilon^2)$ Hamiltonian $H = H_0 + H_1 + H_2$ in $\tilde{\eta}$, $\tilde{\Phi}$ (unnormalized) is

$$H_0 = \frac{1}{2} \int_{S^2} \tilde{\Phi} G_0(\tilde{\eta}) \tilde{\Phi} \rho^2 \, dA_1 + \frac{1}{2} \int_{S^2} (\tilde{\eta})^2 \rho^2 \, dA_1, \tag{6.8}$$

$$H_1 = -\frac{1}{2} \int_{S^2} \tilde{\Phi} G_0(\tilde{\eta}) (\tilde{\Phi}\tilde{\eta}) \rho \, dA_1 + \frac{1}{2} \int_{S^2} \tilde{\Phi} [G_0(\tilde{\eta})\tilde{\Phi}] \tilde{\eta} \rho \, dA_1 + \frac{1}{2} \int_{S^2} \tilde{\Phi} G_1(\tilde{\eta}) \tilde{\Phi} \rho^2 \, dA_1 \tag{6.9}$$

and

$$\begin{aligned}
H_2 &= \frac{3}{4} \int_{S^2} \tilde{\Phi} G_0(\tilde{\eta}) (\tilde{\Phi}\tilde{\eta}^2) \, dA_1 - \frac{1}{2} \int_{S^2} \tilde{\Phi} [G_0(\tilde{\eta})(\tilde{\Phi}\tilde{\eta})] \tilde{\eta} \, dA_1 \\
&\quad - \frac{1}{4} \int_{S^2} \tilde{\Phi} [G_0(\tilde{\eta})\tilde{\Phi}] \tilde{\eta}^2 \, dA_1 - \frac{1}{2} \int_{S^2} \tilde{\Phi} G_1(\tilde{\eta}) (\tilde{\Phi}\tilde{\eta}) \rho \, dA_1 - \frac{1}{4} \int_{S^2} \tilde{\Phi} G_1(\tilde{\eta}^2) \tilde{\Phi} \rho \, dA_1 \\
&\quad + \frac{1}{2} \int_{S^2} \tilde{\Phi} [G_1(\tilde{\eta})\tilde{\Phi}] \tilde{\eta} \rho \, dA_1 + \frac{1}{2} \int_{S^2} \tilde{\Phi} G_2(\tilde{\eta}) \tilde{\Phi} \rho^2 \, dA_1. \tag{6.10}
\end{aligned}$$

If we use the spectral variables η_γ, Φ_γ (we drop the tilde from the notation) defined by $\eta = \sum_\gamma \eta_\gamma Y_\gamma$, $\Phi = \sum_\gamma \Phi_\gamma Y_\gamma$ with $\eta_{[l,m]}^* = \eta_{[l,-m]}$, $\Phi_{[l,m]}^* = \Phi_{[l,-m]}$, the Poisson bracket (which is now diagonalised) becomes

$$[f, g] = \sum_\gamma \left(\frac{\partial f}{\partial \eta_\gamma} \frac{\partial g}{\partial \Phi_\gamma^*} - \frac{\partial f}{\partial \Phi_\gamma} \frac{\partial g}{\partial \eta_\gamma^*} \right) \tag{6.11}$$

and Hamilton's equations are

$$\dot{\eta}_\gamma = \frac{\partial H}{\partial \Phi_\gamma^*}, \qquad \dot{\Phi}_\gamma = -\frac{\partial H}{\partial \eta_\gamma^*}. \tag{6.12}$$

Using the $G_i(\eta)$ from Section 5, the 4-wave Hamiltonian $H = H_0 + H_1 + H_2$ is

$$H_0 = \frac{1}{2}\rho^2 \sum_\gamma \frac{u_\gamma{}'(\rho)}{u_\gamma(\rho)}\Phi_\gamma\Phi_\gamma^* + \frac{1}{2}\rho^2 \sum_\gamma g\eta_\gamma\eta_\gamma^*, \tag{6.13}$$

$$H_1 = \frac{1}{2}\rho^2 \sum_{\gamma_1,\gamma_2,\gamma_3} \left(\frac{u''_{\gamma_2}(\rho)}{u_{\gamma_2}(\rho)} - \frac{u'_{\gamma_1}(\rho)\,u'_{\gamma_2}(\rho)}{u_{\gamma_1}(\rho)\,u_{\gamma_2}(\rho)}\right)\Phi_{\gamma_1}\Phi_{\gamma_2}\eta_{\gamma_3}\int Y_{\gamma_1}Y_{\gamma_2}Y_{\gamma_3}$$

$$+\frac{1}{2}\sum_{\gamma_1,\gamma_2,\gamma_3}\Phi_{\gamma_1}\Phi_{\gamma_2}\eta_{\gamma_3}\int Y_{\gamma_1}\nabla Y_{\gamma_2}\cdot\nabla Y_{\gamma_3}, \tag{6.14}$$

$$\begin{aligned}
H_2 = \frac{1}{2}\sum_{\gamma_1,\gamma_2,\gamma_3,\gamma_4}\Phi_{\gamma_1}\Phi_{\gamma_2}\eta_{\gamma_3}\eta_{\gamma_4}\cdot\Bigg\{&\sum_\gamma\int Y_{\gamma_1}Y_{\gamma_4}Y_\gamma\int Y_{\gamma_2}Y_{\gamma_3}Y_\gamma^* \\
\cdot\Bigg[-\rho\frac{u''_\gamma(\rho)}{u_\gamma(\rho)}&+\frac{\rho\,u''_{\gamma_2}(\rho)}{2\,u_{\gamma_2}(\rho)}+\frac{\rho\,u'_{\gamma_2}(\rho)\,u'_{\gamma_1}(\rho)}{2\,u_{\gamma_2}(\rho)\,u_{\gamma_1}(\rho)}+\frac{\rho^2}{2}\left(\frac{u'''_{\gamma_2}(\rho)}{u_{\gamma_2}(\rho)}-\frac{u''_{\gamma_2}(\rho)\,u'_{\gamma_1}(\rho)}{u_{\gamma_2}(\rho)\,u_{\gamma_1}(\rho)}\right) \\
&-\frac{\rho^2\,u''_{\gamma_2}(\rho)\,u''_\gamma(\rho)}{2\,u_{\gamma_2}(\rho)\,u_\gamma(\rho)}+\frac{\rho^2\,u'_{\gamma_2}(\rho)\,u'_{\gamma_1}(\rho)\,u'_\gamma(\rho)}{2\,u_{\gamma_2}(\rho)\,u_{\gamma_1}(\rho)\,u_\gamma(\rho)}\Bigg] \\
+\left(\frac{1}{\rho}+\frac{u'_{\gamma_1}(\rho)}{u_{\gamma_1}(\rho)}\right)&\sum_\gamma\int Y_{\gamma_1}\nabla Y_{\gamma_4}\cdot\nabla Y_\gamma\int Y_{\gamma_2}Y_{\gamma_3}Y_\gamma^* \\
+\left(\frac{2}{\rho}-\frac{u'_{\gamma_2}(\rho)}{u_{\gamma_2}(\rho)}\right)&\int Y_{\gamma_1}\nabla Y_{\gamma_2}\cdot\nabla Y_{\gamma_3}Y_{\gamma_4}\Bigg\}. \tag{6.15}
\end{aligned}$$

In the rotating frame the quadratic Hamiltonian H_0 is modified to

$$H_0^\Omega = \frac{1}{2R}\rho^2\sum_\gamma\frac{u_\gamma{}'(\rho)}{u_\gamma(\rho)}\Phi_\gamma\Phi_\gamma^* + \frac{1}{2R}\rho^2\sum_\gamma g\eta_\gamma\eta_\gamma^* - i\rho^2\sum_\gamma m_\gamma\Phi_\gamma\eta_\gamma^* \tag{6.16}$$

while the cubic and quartic Hamiltonians are the same as above.

The dispersion relations implied by H_0 and H_0^Ω are

$$\omega^2(\gamma) = \frac{u_\gamma{}'(\rho)}{u_\gamma(\rho)} \tag{6.17}$$

and

$$\omega_\Omega(\gamma) = m_\gamma + \frac{1}{R}\left(\frac{u_\gamma{}'(\rho)}{u_\gamma(\rho)}\right)^{\frac{1}{2}} \tag{6.18}$$

respectively.

From (5.17) we can establish that $\omega(\gamma) = \omega(l)$ viewed as a function of the positive real variable l is smooth, monotonically increasing with $\omega(0) = 0$. In addition, $\omega''(l) < 0$. The function $\omega(l)$ also depends on β, and note that in the limit $\beta \to 0$, $\omega''(l) \to 0$ (but not uniformly in l). For fixed β it is easy to see that for l large $\omega(l) \to \sqrt{l}$, which is the deep water dispersion relation.

We now introduce some extra notation that will be useful in the next section. We define the variables a_γ, a_γ^* by

$$\eta_\gamma = \frac{\sqrt{2}}{2}\sqrt{\frac{\omega_\gamma}{g}}(a_\gamma + a_{\gamma^-}^*), \qquad \Phi_\gamma = -i\frac{\sqrt{2}}{2}\sqrt{\frac{g}{\omega_\gamma}}(a_\gamma - a_{\gamma^-}^*), \tag{6.19}$$

where if $\gamma = [l, m]$ then $\gamma^- = [l, -m]$. Hamilton's equations become

$$\dot{a}_\gamma = -i\frac{\partial H}{\partial a_\gamma^*}.$$

The Hamiltonian in these variables can be readily evaluated using the formulas above: the quadratic Hamiltonian is $H_0 = \sum_\gamma \omega_\gamma a_\gamma a_\gamma^*$ with ω_γ given by the dispersion relation. If we write

$$H_1 = \sum_{\gamma_1,\gamma_2,\gamma_3} \Phi_{\gamma_1}\Phi_{\gamma_2}\eta_{\gamma_3} I_{\gamma_1\gamma_2\gamma_3}, \qquad H_2 = \sum_{\gamma_1,\gamma_2,\gamma_3,\gamma_4} \Phi_{\gamma_1}\Phi_{\gamma_2}\eta_{\gamma_3}\eta_{\gamma_4} I_{\gamma_1\gamma_2\gamma_3\gamma_4}$$

with the $I_{\gamma_1\gamma_2\gamma_3}$, $I_{\gamma_1\gamma_2\gamma_3\gamma_4}$ the coefficients appearing in (6.14) and (6.15), we also write

$$H_1 = \sum_{\gamma_1,\gamma_2,\gamma_3} (A_{\gamma_1\gamma_2\gamma_3} a_{\gamma_1} a_{\gamma_2} a_{\gamma_3} + A_{\gamma_1\gamma_2\gamma_3^-} a_{\gamma_1} a_{\gamma_2} a_{\gamma_3}^*) + c.c.,$$

$$H_2 = \sum_{\gamma_1,\gamma_2,\gamma_3,\gamma_4} (B_{\gamma_1\gamma_2\gamma_3\gamma_4} a_{\gamma_1} a_{\gamma_2} a_{\gamma_3} a_{\gamma_4} + B_{\gamma_1\gamma_2\gamma_3\gamma_4^-} a_{\gamma_1} a_{\gamma_2} a_{\gamma_3} a_{\gamma_4}^*$$
$$+ B_{\gamma_1\gamma_2\gamma_3^-\gamma_4^-} a_{\gamma_1} a_{\gamma_2} a_{\gamma_3}^* a_{\gamma_4}^*) + c.c.$$

with the coefficients given by

$$A_{\gamma_1\gamma_2\gamma_3} = N_{\gamma_1\gamma_2\gamma_3} I_{\gamma_1\gamma_2\gamma_3},$$

$$A_{\gamma_1\gamma_2\gamma_3^-} = N_{\gamma_1\gamma_2\gamma_3}(-I_{\gamma_3^-\gamma_1\gamma_2} - I_{\gamma_1\gamma_3^-\gamma_2} + I_{\gamma_1\gamma_2\gamma_3^-}),$$

$$B_{\gamma_1\gamma_2\gamma_3\gamma_4} = N_{\gamma_1\gamma_2\gamma_3\gamma_4} I_{\gamma_1\gamma_2\gamma_3\gamma_4},$$

$$B_{\gamma_1\gamma_2\gamma_3\gamma_4^-} = N_{\gamma_1\gamma_2\gamma_3\gamma_4}(-I_{\gamma_4^-\gamma_1\gamma_2\gamma_3} - I_{\gamma_1\gamma_4^-\gamma_2\gamma_3} - I_{\gamma_1\gamma_2\gamma_4^-\gamma_3} + I_{\gamma_1\gamma_2\gamma_3\gamma_4^-}),$$

$$B_{\gamma_1\gamma_2\gamma_3^-\gamma_4^-} = N_{\gamma_1\gamma_2\gamma_3\gamma_4}(-I_{\gamma_3^-\gamma_1\gamma_2\gamma_4^-} - I_{\gamma_1\gamma_3^-\gamma_2\gamma_4^-} + I_{\gamma_1\gamma_2\gamma_3^-\gamma_4^-}),$$

$$N_{\gamma_1\gamma_2\gamma_3} = -\frac{\sqrt{2}}{2}\sqrt{\frac{\omega_{\gamma_3}}{\omega_{\gamma_1}\omega_{\gamma_2}}}, \qquad N_{\gamma_1\gamma_2\gamma_3\gamma_4} = -\frac{1}{4}\sqrt{\frac{\omega_{\gamma_3}\omega_{\gamma_4}}{\omega_{\gamma_1}\omega_{\gamma_2}}}.$$

Alternatively, define $z^k = a_1^{k_1} a_2^{k_2} \ldots$, $\bar{z}^{\bar{k}} = (a_1^*)^{\bar{k}_1} (a_2^*)^{\bar{k}_2} \ldots$ where the $k_i, \bar{k}_i$ are nonnegative integers and the subscripts i label (enumerate) the modes γ_i. Also let $|k| = k_1 + k_2 + \cdots$, $|\bar{k}| = \bar{k}_1 + \bar{k}_2 + \cdots$. We can write

$$H_1 = \sum_{\substack{k,\bar{k} \\ |k|+|\bar{k}|=3}} A_{k,\bar{k}} z^k \bar{z}^{\bar{k}}, \qquad H_2 = \sum_{\substack{k,\bar{k} \\ |k|+|\bar{k}|=4}} B_{k,\bar{k}} z^k \bar{z}^{\bar{k}}. \tag{6.20}$$

The coefficients $A_{k,\bar{k}}$ are

$$A_{k_{\gamma_i} k_{\gamma_j} k_{\gamma_l}} = \sum_{\text{perm}.i,j,l} A_{\gamma_i \gamma_j \gamma_l}, \qquad A_{k_{\gamma_i} k_{\gamma_j} \bar{k}_{\gamma_l}} = \sum_{\text{perm}.i,j,l} A_{\gamma_i \gamma_j \gamma_l^-}. \tag{6.21}$$

Similar formulas define the $B_{k,\bar{k}}$.

7. Normal Forms

We will now see that the cubic (3-wave) terms in the Hamiltonian can be eliminated by a formal canonical transformation. We will also try to eliminate the quartic (4-wave) terms, but it will turn out that some of them are resonant and can not be eliminated.

To construct canonical transformations, we will use the "Lie-series" method, which we now briefly review (see, for example [DF], [C]).

Consider a manifold M with a Poisson bracket J defined on $C^\infty(M)$ (we write $J(f, g) = [f, g]$). To every function $g \in C^\infty(M)$ we associate a map $\text{Ad}_g : C^\infty(M) \to C^\infty(M)$ defined by $\text{Ad}_g f = [g, f]$. We also formally define $\exp \text{Ad}_g$ by

$$(\exp \text{Ad}_g) f = f + \sum_k \frac{1}{k!} (\text{Ad}_g)^k = f + [g, f] + \frac{1}{2}[g, [g, f]] + \cdots. \tag{7.1}$$

We can check that for every function g for which the series makes sense, the map $\exp \text{Ad}_g$ preserves the Poisson bracket structure on $C^\infty(M)$, i.e., $(\exp \text{Ad}_g)[f, h] = [(\exp \text{Ad}_g) f, (\exp \text{Ad}_g) h]$ and, therefore, defines a (local) canonical transformation by acting on the components of the coordinate charts of M.

The above computations can be given some meaning beyond formal manipulation, depending on the interpretation of the series of (7.1). For example, the series converges for analytic functions, and then sometimes it may be extended to the whole space of smooth functions. Note also that when this can be done, $\exp(\text{Ad}_g) f$ is the time-1 map of the Hamiltonian vector field of g acting on the function f. Alternatively, if g contains a small parameter, we may consider the series (7.1) as an asymptotic expansion in the small parameter. These considerations pertain to the finite-dimensional case. Additional problems arise in infinite-dimensional systems. As we mentioned before, in this paper we will present only the formal calculations and postpone the analytic discussion to future work.

In the present application, M is the span of the η_γ, Φ_γ, and J is the Poisson bracket given in (6.11). We want a function g such that $\exp \text{Ad}_g$ eliminates the cubic terms from the Hamiltonian. In particular, let $g = \epsilon S_0$, $H = H_0 + \epsilon H_1 + \epsilon^2 H_2$, then

$$(\exp \epsilon \, \mathrm{Ad}_{S_0}) H = H_0 + \epsilon (H_1 + [S_0, H_0]) + \epsilon^2 (H_2 + [S_0, H_1] + \tfrac{1}{2}[S_0, [S_0, H_0]]) + o(\epsilon^2), \tag{7.2}$$

i.e., we want S_0 such that

$$H_1 + [S_0, H_0] = 0. \tag{7.3}$$

If we can solve (7.3) the new Hamiltonian becomes

$$H_{\text{new}} = H_0 + \epsilon^2 \left(H_2 + \frac{1}{2}[S_0, H_1] \right) + o(\epsilon^2). \tag{7.4}$$

To solve (7.3) and calculate $[S_0, H_1]$, we will use the spectral variables a_γ, a_γ^* and the notation developed at the end of the previous section.

Proposition 7.1. *Equation (7.3) can be formally solved or, equivalently, the cubic terms of the Hamiltonian can be eliminated by a formal canonical transformation.*

Proof. The cubic Hamiltonian ϵH_1 is a sum of terms $\epsilon A_{k,\bar{k}} z^k \bar{z}^{\bar{k}}$ with $|k| + |\bar{k}| = 3$. Let $s_0 = \sigma_{k,\bar{k}} z^k \bar{z}^{\bar{k}}$, $|k| + |\bar{k}| = 3$. Using the derivation property of the bracket and induction, we have the formula

$$[z^\mu \bar{z}^{\bar{\mu}}, z^k \bar{z}^{\bar{k}}] = i \sum_j (\bar{\mu}_j k_j - \bar{k}_j \mu_j) \frac{1}{z_j \bar{z}_j} z^{\mu + k} \bar{z}^{\bar{\mu} + \bar{k}} \tag{7.5}$$

for arbitrary $\mu, \bar{\mu}, k, \bar{k}$. In particular, $[s_0, H_0] = -i\sigma_{k,\bar{k}} (\sum_i \omega_i (k_i - \bar{k}_i)) z^k \bar{z}^{\bar{k}}$. If we set

$$\sigma_{k,\bar{k}} = \frac{1}{i} \frac{A_{k,\bar{k}} z^k \bar{z}^{\bar{k}}}{\sum_i \omega_i (k_i - \bar{k}_i)} \tag{7.6}$$

the term $\epsilon A_{k,\bar{k}} z^k \bar{z}^{\bar{k}}$ is eliminated, provided that the "3-wave resonance" condition

$$A_{k,\bar{k}} \neq 0, \qquad \sum_i \omega_i (k_i - \bar{k}_i) = 0, \tag{7.7}$$

is not satisfied.

Letting $S_0 = -i \sum_{k,\bar{k}} \frac{A_{k,\bar{k}}}{\sum_i \omega_i (k_i - \bar{k}_i)} z^k \bar{z}^{\bar{k}}$, all the nonresonant terms are thus eliminated.

Therefore, it is enough to show that (7.7) is never satisfied.

From (6.20) and (6.21) the resonances (7.7) occur if and only if

$$I_{\gamma_1 \gamma_2 \gamma_3} \neq 0, \qquad \omega_{\gamma_1} + \omega_{\gamma_2} + \omega_{\gamma_3} = 0, \tag{7.8}$$

$$I_{\gamma_1 \gamma_2 \bar{\gamma}_3} \neq 0, \qquad \omega_{\gamma_1} + \omega_{\gamma_2} - \omega_{\gamma_3} = 0. \tag{7.9}$$

To show that none of the above equations are satisfied, we examine the dispersion ω_γ and the terms $I_{\gamma_1\gamma_2\gamma_3}$. First, since $\omega_\gamma > 0$, (7.8) cannot be satisfied. Also, $\frac{d\omega}{dl} > 0$ and $\frac{d^2\omega}{dl^2} < 0$, so that

$$\omega(l_3) = \omega(l_1) + \omega(l_2) \Rightarrow l_3 > l_1 + l_2. \tag{7.10}$$

On the other hand, from (6.14)

$$I_{\gamma_1\gamma_2\gamma_3} = b_{\gamma_1\gamma_2\gamma_3} \int Y_{\gamma_1} Y_{\gamma_2} Y_{\gamma_3} + c_{\gamma_1\gamma_2\gamma_3} \int Y_{\gamma_1} \nabla Y_{\gamma_2} \cdot \nabla Y_{\gamma_3} \tag{7.11}$$

with the $b_{\gamma_1\gamma_2\gamma_3}$, $c_{\gamma_1\gamma_2\gamma_3}$ determined by (6.14). Now, $Y_\gamma = Y_l^m(\vartheta, \varphi) = e^{im\varphi} P_l^{|m|}(\mu)$ with $\mu = \cos\vartheta$ and $P_l^m(\mu)$ the associated Legendre functions (for material related to spherical harmonics, we refer to [BL], [McR]). We have

$$\int Y_{\gamma_1} Y_{\gamma_2} Y_{\gamma_3} = \int_0^{2\pi} e^{i\left(\sum_{i=1}^3 m_i\right)\varphi} d\varphi \int_{-1}^1 P_{\gamma_1} P_{\gamma_2} P_{\gamma_3} d\mu$$

and

$$\int_{-1}^1 P_{\gamma_1} P_{\gamma_2} P_{\gamma_3} d\mu \neq 0 \quad \text{only if} \quad |l_1 - l_2| \leq l_3 \leq l_1 + l_2 \quad \text{and} \quad l_1 + l_2 + l_3 = \text{even}.$$

Hence, when the frequency addition rule holds, $\int Y_{\gamma_1} Y_{\gamma_2} Y_{\gamma_3}$ vanishes. For the second integral of (7.11), we have

$$\int Y_{\gamma_1} \nabla Y_{\gamma_2} \cdot \nabla Y_{\gamma_3} = \delta_{123} \int_{-1}^1 \left[(1 - \mu^2) P_{\gamma_1} \frac{dP_{\gamma_2}}{d\mu} \frac{dP_{\gamma_3}}{d\mu} - \frac{m_2 m_3}{1 - \mu^2} P_{\gamma_1} P_{\gamma_2} P_{\gamma_3} \right] d\mu$$

with $\delta_{123} = \int_0^{2\pi} e^{i\left(\sum_{i=1}^3 m_i\right)\varphi} d\varphi$. Using

$$\frac{dP_l^{|m|}}{d\mu} = (1 - \mu^2)^{1/2} P_l^{|m|+1} - |m|\mu(1 - \mu^2)^{1/2} P_l^{|m|}$$

(note that $P_l^{|m|} = 0$ for $|m| > l$), we have

$$\int Y_{\gamma_1} \nabla Y_{\gamma_2} \cdot \nabla Y_{\gamma_3}$$

$$= -\delta_{123} \int_{-1}^1 \frac{\mu}{(1 - \mu^2)^{1/2}} (|m_3| P_{l_1}^{|m_1|} P_{l_2}^{|m_2|+1} P_{l_3}^{|m_3|} - |m_2| P_{l_1}^{|m_1|} P_{l_2}^{|m_2|} P_{l_3}^{|m_3|+1}) d\mu$$

$$+ \delta_{123} \int_{-1}^1 (P_{l_1}^{|m_1|} P_{l_2}^{|m_2|+1} P_{l_3}^{|m_3|+1} - m_2 m_3 P_{l_1}^{|m_1|} P_{l_2}^{|m_2|} P_{l_3}^{|m_3|}) d\mu$$

$$- \delta_{123} \int_{-1}^1 \frac{\mu}{(1 - \mu^2)^{1/2}} (m_2 m_3 P_{l_1}^{|m_1|} P_{l_2}^{|m_2|} P_{l_3}^{|m_3|} - |m_2 m_3| P_{l_1}^{|m_1|} P_{l_2}^{|m_2|} P_{l_3}^{|m_3|}) d\mu.$$

The terms $\frac{\mu}{(1-\mu^2)^{1/2}} P_{l_i}^{|m_{a_i}|+1}$ in the integral can be written as a linear combination of $P_{l_i}^{|m_{a_i}|}$ and $P_{l_i}^{|m_{a_i}|+2}$ (see [McR, p. 115]). This is also the case for the terms of the last integral, unless one of the m_2, m_3 is zero, in which case the integral is zero. Therefore, the triple integral $\int Y_{\gamma_1} \nabla Y_{\gamma_2} \cdot \nabla Y_{\gamma_3}$ can be expressed as a sum of

$$\int_{-1}^{1} P_{l_1}^{m_{a_1}} P_{l_2}^{m_{a_2}} P_{l_3}^{m_{a_3}} \, d\mu$$

and vanishes when the frequency addition rule holds. $\square$

Note that the size of the denominators of S_0 depends on β. In the $\beta \to 0$ limit the linearised system is dispersionless and the denominators of S_0 will vanish. Thus, the proposition concerns more the intermediate (β of order 1) and deep water regimes. We remark that, using the fact that $\omega' > 0$ and $\omega'' < 0$, it is easy to show that all terms $|\omega_{l_1} - \omega_{l_2} - \omega_{l_3}|^{-1}$ appearing in S_0 can be bounded by $|\omega_1 - 2\omega_2|^{-1}$, which can be computed easily for given β.

In the rotating frame we can obtain a similar result for the Hamiltonian $H^\Omega = H_0^\Omega + \frac{1}{R}(\epsilon H_1 + \epsilon^2 H_2 + \cdots)$. Here the quadratic Hamiltonian H_0^Ω involves the dispersion ω_Ω of (6.18).

Corollary 7.2. *In the rotating frame, the cubic terms in the Hamiltonian can be eliminated by a canonical transformation.*

Proof. The normal form calculation is as above, and we have to check that there are no resonances. From (6.18), the resonance condition is now

$$I_{\gamma_1 \gamma_2 \gamma_3} \neq 0, \qquad Rm_{\gamma_1} + \omega_{\gamma_1} + Rm_{\gamma_2} + \omega_{\gamma_2} + Rm_{\gamma_3} + \omega_{\gamma_3} = 0, \qquad (7.12)$$

$$I_{\gamma_1 \gamma_2 \gamma_3^-} \neq 0, \qquad Rm_{\gamma_1} + \omega_{\gamma_1} + Rm_{\gamma_3} + \omega_{\gamma_2} - Rm_{\gamma_3} - \omega_{\gamma_3} = 0. \qquad (7.13)$$

The coefficients are as in (7.11). From the above discussion of the triple integrals of harmonics, $I_{\gamma_1 \gamma_2 \gamma_3} \neq 0 \Rightarrow m_{\gamma_1} + m_{\gamma_1} + m_{\gamma_1} = 0$ and $I_{\gamma_1 \gamma_2 \gamma_3^-} \neq 0 \Rightarrow m_{\gamma_1} + m_{\gamma_1} - m_{\gamma_1} = 0$, and, therefore, the resonance conditions (7.12), (7.13) are equivalent to the rest frame resonance conditions. $\square$

We now consider the problem of eliminating the quartic terms. However, this time there are resonances.

As previously, we try to find a function S_1 such that $(\exp \epsilon^2 \, \mathrm{Ad}_{S_1})(\exp \epsilon \, \mathrm{Ad}_{S_0}) H$ has no ϵ^2 terms, and we are led to the equation

$$\left(H_2 + \frac{1}{2}[S_0, H_1]\right) + [S_1, H_0] = 0. \qquad (7.14)$$

The contribution from $[S_0, H_1]$ can be computed with the aid of the formula (7.5). The calculation is long but straightforward and can be simplified with the use of a diagrammatic method that will appear elsewhere. We write the result as

$$[S_0, H_1] = K + L$$

with

$$K = -i \sum_{\gamma_1, \gamma_2, \gamma_3, \gamma_4, \gamma_q} \left\{ a_{\gamma_1} a_{\gamma_2} a_{\gamma_3} a_{\gamma_4} \left[A'_{\gamma_q \gamma_1 \gamma_2} A_{\gamma_3 \gamma_4 \gamma_q^-} + A'_{\gamma_1 \gamma_q \gamma_2} A_{\gamma_3 \gamma_4 \gamma_q^-} \right. \right.$$

$$+ A'_{\gamma_1 \gamma_2 \gamma_q} A_{\gamma_3 \gamma_4 \gamma_q^-} + A_{\gamma_q \gamma_1 \gamma_2} A'_{\gamma_3 \gamma_4 \gamma_q^-} + A_{\gamma_1 \gamma_q \gamma_2} A'_{\gamma_3 \gamma_4 \gamma_q^-} + A_{\gamma_1 \gamma_2 \gamma_q} A'_{\gamma_3 \gamma_4 \gamma_q^-} \right]$$

$$+ a_{\gamma_1} a_{\gamma_2} a_{\gamma_3} a_{\gamma_4}^* \left[A_{\gamma_3 \gamma_4^- \gamma_q^-} (A'_{\gamma_q \gamma_1 \gamma_2} + A'_{\gamma_1 \gamma_q \gamma_2} + A'_{\gamma_1 \gamma_2 \gamma_q}) \right.$$

$$+ A'_{\gamma_3 \gamma_q^- \gamma_4^-} (A_{\gamma_q \gamma_1 \gamma_2} + A_{\gamma_1 \gamma_q \gamma_2} + A_{\gamma_1 \gamma_2 \gamma_q})$$

$$+ A'_{\gamma_3 \gamma_4^- \gamma_q^-} (A_{\gamma_q \gamma_1 \gamma_2} + A_{\gamma_1 \gamma_q \gamma_2} + A_{\gamma_1 \gamma_2 \gamma_q}) + A_{\gamma_3 \gamma_q^- \gamma_4^-} (A'_{\gamma_q \gamma_1 \gamma_2} + A'_{\gamma_1 \gamma_q \gamma_2} + A'_{\gamma_1 \gamma_2 \gamma_q})$$

$$+ A_{\gamma_3 \gamma_4^- \gamma_q^-} (A_{\gamma_q \gamma_1 \gamma_2^-} - A_{\gamma_1 \gamma_q \gamma_2^-}) + A'_{\gamma_1 \gamma_2 \gamma_q^-} (A_{\gamma_q \gamma_3 \gamma_4^-} - A_{\gamma_3 \gamma_q \gamma_4^-}) \right] \Big\},$$

where $A'_{\gamma_1 \gamma_2 \gamma_5} = A_{\gamma_1 \gamma_2 \gamma_5} (\omega_{\gamma_1} + \omega_{\gamma_2} + \omega_{\gamma_5})^{-1}$, $A'_{\gamma_1 \gamma_2 \gamma_5^-} = A_{\gamma_1 \gamma_2 \gamma_5^-} (\omega_{\gamma_1} + \omega_{\gamma_2} - \omega_{\gamma_5})^{-1}$ and so forth, and

$$L = i \sum_{\gamma_1, \gamma_2, \gamma_3, \gamma_4, \gamma_q} a_{\gamma_1} a_{\gamma_2} a_{\gamma_3}^* a_{\gamma_4}^* C_{\gamma_1 \gamma_2 \gamma_3 \gamma_4},$$

with

$$C_{\gamma_1 \gamma_2 \gamma_3 \gamma_4} = \left(\frac{1}{\omega_{\gamma_1} + \omega_{\gamma_2} + \omega_{\gamma_q}} + \frac{1}{\omega_{\gamma_1} + \omega_{\gamma_2} + \omega_{\gamma_5}} \right)$$

$$\times (A_{\gamma_q \gamma_1 \gamma_2} + A_{\gamma_1 \gamma_q \gamma_2} + A_{\gamma_1 \gamma_2 \gamma_q})(A_{\gamma_q^- \gamma_3^- \gamma_4^-} + A_{\gamma_1^- \gamma_q^- \gamma_4^-} + A_{\gamma_3^- \gamma_4^- \gamma_q^-})$$

$$+ \left(\frac{1}{\omega_{\gamma_1} + \omega_{\gamma_2} - \omega_{\gamma_q}} - \frac{1}{\omega_{\gamma_q} - \omega_{\gamma_3} - \omega_{\gamma_4}} \right) A_{\gamma_1 \gamma_2 \gamma_q} A_{\gamma_q \gamma_3^- \gamma_4^-}$$

$$- \left(\frac{1}{\omega_{\gamma_1} + \omega_{\gamma_q} - \omega_{\gamma_3}} - \frac{1}{\omega_{\gamma_2} - \omega_{\gamma_q} - \omega_{\gamma_4}} \right)$$

$$\times (A_{\gamma_1 \gamma_q \gamma_3^-} + A_{\gamma_q \gamma_1 \gamma_3^-})(A_{\gamma_2 \gamma_q^- \gamma_4^-} + A_{\gamma_2 \gamma_4^- \gamma_q^-}).$$

Writing $H_2 + \frac{1}{2}[S_0, H_1]$ as $\sum_{k, \bar{k}} \tilde{B}_{k, \bar{k}} z^k \bar{z}^{\bar{k}}$ with $|k| + |\bar{k}| = 4$, the resonance condition is now

$$\tilde{B}_{\gamma_1 \gamma_2 \gamma_3 \gamma_4} \neq 0, \qquad \omega_{\gamma_1} + \omega_{\gamma_2} + \omega_{\gamma_3} + \omega_{\gamma_4} = 0, \tag{7.15}$$

$$\tilde{B}_{\gamma_1 \gamma_2 \gamma_3 \gamma_4^-} \neq 0, \qquad \omega_{\gamma_1} + \omega_{\gamma_2} + \omega_{\gamma_3} - \omega_{\gamma_4} = 0, \tag{7.16}$$

$$\tilde{B}_{\gamma_1 \gamma_2 \gamma_3^- \gamma_4^-} \neq 0, \qquad \omega_{\gamma_1} + \omega_{\gamma_2} - \omega_{\gamma_3} - \omega_{\gamma_4} = 0. \tag{7.17}$$

Proposition 7.3. *The resonance conditions (7.15) and (7.16) are never satisfied while condition (7.17) is satisfied for suitable $\gamma_1, \gamma_2, \gamma_3, \gamma_4$.*

Proof. Since $\omega_\gamma > 0$, (7.15) is never satisfied. For (7.16), we have that $\tilde{B}_{\gamma_1 \gamma_2 \gamma_3 \gamma_4^-}$ is of the form

$$\tilde{B}_{\gamma_1 \gamma_2 \gamma_3 \gamma_4^-} = \sum_{Q, R} b_{\gamma_1 \gamma_2 \gamma_3 \gamma_4^-}^{Q, R} \sum_{\gamma} Q_{\gamma_i \gamma_j \gamma} R_{\gamma_n \gamma_m^- \gamma^-}$$

where $Q_{\gamma_i \gamma_j \gamma}$ and $R_{\gamma_i \gamma_j \gamma}$ are $\int Y_{\gamma_i} Y_{\gamma_j} Y_\gamma$ or $\int (\nabla Y_{\gamma_i} \cdot \nabla Y_{\gamma_j}) Y_\gamma^*$. From our previous discussion of the triple integrals of spherical harmonics, a necessary condition for $\tilde{B}_{\gamma_1 \gamma_2 \gamma_3 \gamma_4^-}$ not to vanish is that there exist indices $\gamma = [l, m]$ such that

$$|l_i - l_j| \le l \le |l_i + l_j| \qquad \text{and} \qquad |l_n - l_m| \le l \le |l_n + l_m| \qquad (7.18)$$

(and permutations on the i, j, n, m). However, since $\frac{d\omega}{dl} > 0$ and $\frac{d^2\omega}{dl^2} < 0$, we have that $\omega_{\gamma_i} = \omega_{\gamma_j} + \omega_{\gamma_n} + \omega_{\gamma_m} \Rightarrow l_i > l_j + l_n + l_m$ or $|l_i - l_j| > |l_n + l_m|$ (and permutations) so that (7.18) is not satisfied, and (7.16) cannot be satisfied either. For (7.17), pick $\gamma_1 = \gamma_3$ and $\gamma_2 = \gamma_4$, then the frequency sum rule is satisfied; moreover $\tilde{B}_{\gamma_1 \gamma_2 \gamma_3 \gamma_4}$ is (generically) nonzero (a few of them have been computed). □

Therefore the resonant part of the Hamiltonian is of the form

$$H_{2,res} = \sum_{\gamma_1, \gamma_2, \gamma_3, \gamma_4} R_{\gamma_1 \gamma_2 \gamma_3 \gamma_4} a_{\gamma_1} a_{\gamma_2} a_{\gamma_3}^* a_{\gamma_4}^*$$

with

$$R_{\gamma_1 \gamma_2 \gamma_3 \gamma_4} = B_{\gamma_1 \gamma_2 \gamma_3^- \gamma_4^-} + \frac{1}{2} C_{\gamma_1 \gamma_2 \gamma_3 \gamma_4}. \qquad (7.19)$$

We note that $H_{2,res}$ should include all the "generic" resonances corresponding to $l_3 = l_1, l_4 = l_2$ or $l_3 = l_2, l_4 = l_1$. In addition the resonance condition (7.17) may be satisfied for other integers. A preliminary numerical search has not found any so far, but note that for large l we have $\omega(l) \to \sqrt{l}$, and since $\sqrt{a} + \sqrt{b} = \sqrt{c} + \sqrt{d}$ has other types of integer solutions, we expect that there are quartets of integers that are arbitrarily close to resonance. A more detailed study of this issue and its dynamical implications will be considered in the future.

The equations arising from the above second-order normal form Hamiltonian $H = H_0 + H_{2,res}$ are easily seen to have families of periodic orbits. To simplify the argument we will assume that $H_{2,res}$ contains only the "generic" 4-wave resonances, in which case the subspaces $M_L = a(l, m) = 0$ for $l \ne L$ are invariant under the flow. On these subspaces we can find families of periodic orbits.

Proposition 7.4. *Let ϕ_H be the flow generated by the Hamiltonian vector field of $H = H_0 + H_{2,res}$. Then, (a) the subspaces $T_{(L,M)} = \{a_{l,m} = 0 \text{ for } l \ne L, m \ne M\}$ with $L = 1, 2, 3, \ldots, M = -L, -L+1, \ldots, L$ are invariant under ϕ_H. Moreover, the restriction of the flow ϕ_H to $T_{(L,M)}$ foliates $T_{(L,M)}$ by periodic orbits. (b) The subspaces $S_{(L,\tilde{M})} = \{a_{l,m} = 0 \text{ for } l \ne L, m \ne \pm M\}$ with $L = 1, 2, 3, \ldots, M = -L, -L+1, \ldots, L$ and $3|M| > L$ are invariant under ϕ_H and the restriction of ϕ_H to $S_{(L,\tilde{M})}$ foliates $S_{(L,\tilde{M})}$ by periodic orbits.*

Proof. (a) From the form of $H_{2,res}$ in (7.19) observe that in order for the coefficients of the terms $a_{(l_1,m_1)} a_{(l_2,m_2)} a_{(l_1,m_3)}^* a_{(l_2,m_4)}^*$ to be nonzero, the m_i must satisfy the relation

$$m_1 + m_2 - m_3 - m_4 = 0. \qquad (7.20)$$

To show that the $T_{(L,M)}$ are invariant, first note that $\dot{a}_{(l,m)} = 0$ if $l \ne L$. Also, for $m \ne M$ we have $\dot{a}_{(L,m)} = iC(M, M, m, M) a_{(L,M)} a_{(L,M)} a_{(L,M)}^*$ with the coefficient

$C(M, M, m, M)$ given by (7.19). But the condition $C(M, M, m, M) \neq 0$ and (7.20) require that $m = M$. Thus the $T_{(L,M)}$ are invariant and

$$\dot{a}_{(L,M)} = i\omega_L a_{(L,M)} + iC(M, M, M, M)a_{(L,M)}a_{(L,M)}a^*_{(L,M)}.$$

This equation is a Hamiltonian system generated by

$$H_{L,M} = \omega_L a_{(L,M)} a^*_{(L,M)} + C_M a_{(L,M)} a_{(L,M)} a^*_{(L,M)} a^*_{(L,M)}$$

with $C_M = C(M, M, M, M)$. The Hamiltonian $H_{L,M}$ depends only on the "action" $J_{L,M} = a_{(L,M)} a^*_{(L,M)}$. By a canonical transformation to the variables $J_{L,M}$ and $\theta_{L,M}$ ("action-angle" variables), the equation becomes

$$\dot{J}_{L,M} = 0, \qquad \dot{\theta}_{L,M} = \omega_L + C_M J_{L,M}.$$

The solutions of this equation are all periodic and correspond to traveling waves with amplitude dependent phase velocity.

(b) Similarly, for $a_{(l,m)} \in S_{(L,\tilde{M})}$, $\dot{a}_{(L,\pm M)}$ has terms of the form

$$C(\tilde{M}, \tilde{M}, m, \tilde{M})a_{(L,M)}a_{(L,M)}a^*_{(L,M)}$$

with $\tilde{M}$ either $+M$ or $-M$. From (7.20) the coefficients $C(\tilde{M}, \tilde{M}, m, \tilde{M})$ vanish unless $m = \pm M$ or $m = \pm 3M$. Thus for $3|M| > L$ the subspaces $S_{(L,\tilde{M})}$ are invariant. On $S_{(L,\tilde{M})}$ the equation of motion for the restricted flow is generated by the Hamiltonian

$$H_{L,\tilde{M}} = i\omega_L(a^*_{(L,M)}a^*_{(L,M)} + a^*_{(L,-M)}a^*_{(L,-M)}) + C_{M,M}J_{L,M}J_{L,M}$$

$$+2C_{M,-M}J_{L,M}J_{L,-M} + C_{-M,-M}J_{L,-M}J_{L,-M}$$

with $J_{L,\pm M} = a_{(L,\pm M)}a^*_{(L,\pm M)}$, $C_{A,B} = C(A, B, A, B)$. Using "action-angle" variables we again have

$$\dot{J}_{L,\pm M} = 0, \qquad \dot{\theta}_{L,\pm M} = \omega_L + 2C_{\pm M,\pm M}J_{L,M} + 2C_{M,-M}J_{L,\mp M}.$$

These periodic orbits are superpositions of the previous traveling waves, and, if the amplitudes $J_{L,\pm M}$ are equal, the result is standing waves.   □

Acknowledgments

It is a pleasure to thank W. Craig for getting us interested in this work and supplying many references and correspondence. This work has been supported by NSF and TARP grants. R. L. has also been supported by the AMS Centennial Fellowship and a URI grant of the University of Texas.

References

[B] T. B. Benjamin, P. J. Olver: Hamiltonian structure, symmetries and conservation laws for water waves, *J. Fluid Mech.*, **83**, 137–185 (1982).

[BL] L. C. Biedenharn, J. D. Louck: *Angular momentum in quantum physics—theory and applications*, Addison-Wesley, Reading, Mass. (1981).
[C] J. R. Cary: Lie transform perturbation theory for Hamiltonian systems, *Phys. Reports*, **79**, No. 2, 129–159 (1981).
[CM] R. R. Coifman, Y. Meyer: Non-linear harmonic analysis and analytic dependence, *AMS Proc. Symp. Pure Math.*, **43**, 71–78 (1985).
[CG] W. Craig, M. D. Groves: Hamiltonian long-wave scaling limits of the water wave problem, preprint (1992).
[CS] W. Craig, C. Sulem: Numerical simulation of gravity waves, *J. Comp. Phys.*, **108**, 73–83 (1993).
[DF] A. J. Dragt, J. M. Finn: Lie series and invariant functions for analytic symplectic maps, *J. Math. Phys.*, **17**, 2215–2227 (1976).
[D] B. A. Dubrovin: *Geometry of Hamiltonian evolutionary systems*, Bibliopolis, Napoli (1991).
[FS] Z. C. Feng, P. R. Sethna: Symmetry breaking and bifurcations in resonant surface waves, *J. Fluid Mech.*, **199**, 495–518 (1989).
[GAS] M. Glozman, Y. Agnon, M. Stiassnie: High order formulation of the water wave problem, *Physica D*, **66**, 347–367 (1993).
[LL] L. D. Landau, A. Lifschitz: *Fluid Mechanics*, Pergamon, Oxford (1959).
[L] H. Lamb: *Hydrodynamics*, Dover, New York (1932).
[LMMR] D. Lewis, J. Marsden, R. Montgomery, T. Ratiu: The Hamiltonian structure for dynamic free boundary problems, *Physica D*, **18**, 391–404 (1986).
[M] J. W. Miles: On Hamilton's principle for surface waves, *J. Fluid Mech.*, **83**, 153–158 (1977).
[Mo] J. Moser: Periodic orbits near an equilibrium and a theorem of Alan Weinstein, *Commun. Pure Appl. Math.*, **29**, 727–747 (1976).
[MG] P. M. Morrison, J. M. Greene: Non-canonical Hamiltonian density formulation of hydrodynamics and ideal magneto-hydrodynamics, *Phys. Rev. Let.*, **45**, 790–794 (1980).
[MW] J. Marsden, A. Weinstein: Coadjoint orbits, vortices and Clebsch variables for incompressible fluids, *Physica D*, **7**, 305–323 (1983).
[McR] T. M. MacRoberts: *Spherical harmonics*, Pergamon Press, London, (1967).
[P] G. W. Platzman: Ocean tides and related waves, in *Mathematical Problems in the Geophysical Sciences I*, ed. W. H. Reid, 239–291 (1970).
[We] A. Weinstein: Normal modes for non-linear Hamiltonian systems, *Inv. Math.*, **20**, 47–57 (1973).
[Wi] G. B. Whitham: *Linear and Nonlinear Waves*, Wiley Interscience, New York (1974).
[Z] V. E. Zakharov: Stability of periodic waves of finite amplitude on the surface of a deep fluid, *J. Appl. Tech. Phys.*, **2**, 190–194 (1968).

A Nonlinear Extensible 4-Node Shell Element Based on Continuum Theory and Assumed Strain Interpolations*

P. Betsch and E. Stein

Institut für Baumechanik und Numerische Mechanik, Universität Hannover, Germany

Received October 17, 1995
Communicated by Jerrold Marsden and Stephen Wiggins

This paper is dedicated to the memory of Juan-Carlos Simo

Summary. A quadrilateral continuum-based C^0 shell element is presented, which relies on extensible director kinematics and incorporates unmodified three-dimensional constitutive models. The shell element is developed from the nonlinear enhanced assumed strain (EAS) method advocated by Simo & Armero [1] and formulated in curvilinear coordinates. Here, the EAS-expansion of the material displacement gradient leads to the local interpretation of enhanced covariant base vectors that are superposed on the compatible covariant base vectors. Two expansions of the enhanced covariant base vectors are given: first an extension of the underlying single extensible shell kinematic and second an improvement of the membrane part of the bilinear element. Furthermore, two assumed strain modifications of the compatible covariant strains are introduced such that the element performs well even in the case of very thin shells.

1. Introduction

Many finite shell elements that have been developed in recent years are based on the common kinematic assumption of inextensibility in the thickness direction and the zero normal stress condition; see e.g. [2]–[6]. According to the inextensibility constraint, a rotation tensor is required in order to treat large rotations.

An alternative approach relies on an extensible director kinematic which offers a very attractive numerical formulation since rotational degrees of freedom can be avoided. Thus, the corresponding update procedure for the extensible director is based on simple vector addition. Since the work of Simo, Rifai & Fox [7] on an extensible 4-node shell element, several other extensible shell formulations have been published. The 8-node

---

* In honour of Professor Juan-Carlos Simo who had significant collaboration with our institute and contributed important insights to our research work.

extensible shell element of Büchter & Ramm [8] and Büchter, Ramm & Roehl [9]
incorporates an enhanced assumed strain (EAS) expansion of the covariant thickness
component of the Green-Lagrangian strain tensor. This expansion allows for the implementation of full three-dimensional constitutive models without resorting to the plane
stress assumption. The plane stress condition is imposed in a weak sense; see also Steinmann, Betsch & Stein [10]. Additionally, the membrane part of the 4-node extensible
shell element of Betsch, Gruttmann & Stein [11] has been improved by means of an
expansion of the enhanced Green-Lagrangian strain tensor. The extensible shell formulation of Sansour [12] relies on a higher order shell kinematic that leads to seven nodal
degrees of freedom. The same kinematic assumption has been investigated by Kühhorn
& Schoop [13] within the context of sandwich shells. An alternative extensible shell
element formulation of Parisch [14] employs three nodes in the thickness direction. The
advantages of a six-degrees-of-freedom director shell theory have also been pointed out
by Schoop [15].

Analogous to the works [8], [9] and [11], the present 4-node shell element relies on
an EAS-expansion in order to facilitate the use of three-dimensional constitutive laws
without any shell-specific modifications. However, we enhance the material displacement gradient instead of the Green-Lagrangian strain tensor. This approach coincides
with the nonlinear EAS-method advocated by Simo & Armero [1] where the material
displacement gradient has been used in order to extend the linear EAS-method of Simo
& Rifai [16] to the nonlinear theory.

We recast the nonlinear EAS-method of Simo & Armero in terms of curvilinear
coordinates. Within the curvilinear description the EAS-expansion of the material displacement gradient leads to the local interpretation of enhanced covariant base vectors
which are superposed on the compatible covariant base vectors. Two expansions of the
enhanced covariant base vectors are presented which both extend the underlying single
extensible shell kinematic and improve the membrane part of the bilinear element.

Furthermore, two assumed strain modifications of the compatible covariant strains
are employed in order to eliminate parasitic strain effects of the compatible low-order
C^0 interpolations. The assumed strain interpolation of Dvorkin and Bathe [17] is a
very powerful tool to overcome the shear locking effect of the 4-node displacement
model. This modification of the compatible covariant shear strains has been intensely
investigated by Bathe and Dvorkin [18] Stander, Matzenmiller and Ramm [19] and
Parisch [20] and works well even for initially distorted meshes.

Analogous to the shear locking effect of low order plate and shell elements, the
bilinear interpolation of the *extensible* director field leads to parasitic transverse normal
strains which lead to locking in bending dominated situations, especially in the case
of very thin shells and coarse meshes. To overcome this locking effect Betsch & Stein
[21] have proposed a simple assumed strain interpolation of the compatible transverse
normal strains. It has been shown in [21] that the assumed strain interpolation makes
the extensible shell formulation competitive to more elaborate formulations which rely
on an orthogonal matrix for the description of finite rotations. Thus the pathological
behavior of the simple extensible director interpolation which has also been reported by
Simo, Rifai & Fox [7] is completely eliminated by means of the assumed strain method.

First, we give a comprehensive account of the nonlinear EAS-method advocated
by Simo & Armero [1] in conjunction with a representation in terms of curvilinear

coordinates. After introducing a single extensible director kinematic we present two EAS-expansions of the enhanced covariant base vectors. Then we proceed with the compatible finite element approximation and discuss alternative interpolations of the extensible director field. Finally we present a modification of the compatible constant transverse normal strains based on the assumed strain method [22]. Several numerical examples are used to verify that our newly developed formulation is able to deal with thin-shell structures as well as with large strain problems.

2. Weak Form of the EAS Formulation

Let $X(\theta^i)$ denote the position vector of a material point in the initial configuration $\mathcal{B}_0$ described by a general curvilinear set of coordinates θ^i. The corresponding placement in the actual configuration $\mathcal{B}$ follows from $x(\theta^i) = X(\theta^i) + u(\theta^i)$, where u is the displacement vector. Following Simo & Armero [1], the weak form of the nonlinear enhanced assumed strain method can be written

$$G_{int} + G_{dyn} - G_{ext} = 0, \tag{1}$$

with the internal virtual work contribution

$$G_{int} = \int_{\mathcal{B}_0} \left\{ \mathrm{Grad}[\delta u] : [2F\partial_C W] + \delta\tilde{H} : [2F\partial_C W - P] - \delta P : \tilde{H} \right\} dV. \tag{2}$$

Accordingly, the weak form contains three independent fields: the displacements u, the nominal stress tensor P and the enhanced displacement gradient $\tilde{H}$. W denotes the strain energy density function per unit initial volume and $C = F^T F$. The total deformation gradient is defined by

$$F = \mathrm{Grad}[x] + \tilde{H}. \tag{3}$$

Within a finite element the discontinuous expansion of $\tilde{H}$ is intended to enhance the compatible deformation gradient $\mathrm{Grad}[x]$ which results from the finite element approximation of x due to the pure displacement formulation. In (1) the virtual work of the inertial forces is denoted by

$$G_{dyn} = \int_{\mathcal{B}_0} \delta u \cdot \ddot{x} \rho_0 \, dV \tag{4}$$

and the virtual work of the external loads is represented by

$$G_{ext} = \int_{\mathcal{B}_0} \delta u \cdot b \, dV + \int_{\partial\mathcal{B}_{0t}} \delta u \cdot \bar{t} \, dA. \tag{5}$$

It has been shown by Simo & Armero [1] that the nominal stress field can be eliminated from the weak form when the expansion of the enhanced displacement gradient satisfies the condition

$$\int_{\mathcal{B}_0} \tilde{H} \, dV = 0. \tag{6}$$

This condition ensures that at least constant states of stress can be reproduced by the finite element model. It represents an important stability requirement of the method as has been shown within the infinitesimal theory by Simo & Rifai [16] and Reddy & Simo [23].

Accordingly, a two field formulation in terms of displacements u and the enhanced displacement gradient $\tilde{H}$ remains. The internal virtual work (2) may now be written in the form

$$G_{int} = \int_{B_0} \left\{ \text{Grad}[\delta u]^T F + \delta \tilde{H}^T F \right\} : \widehat{S}\, dV = \int_{B_0} \left\{ \text{grad}[\delta u] + \delta \tilde{h} \right\} : \widehat{\tau}\, dV, \quad (7)$$

where the expressions

$$\text{Grad}[\delta u] = \text{grad}[\delta u] F \qquad \text{and} \qquad \delta \tilde{H} = \delta \tilde{h} F \tag{8}$$

have been used. In order to alleviate the notation, in (7) the 'constitutive' Second Piola Kirchhoff stress tensor

$$\widehat{S} = 2\partial_C W(C) = \hat{S}^{ij} G_i \otimes G_j \tag{9}$$

and the 'constitutive' Kirchhoff stress tensor

$$\widehat{\tau} = 2F\partial_C W(C)F^T = \hat{S}^{ij} g_i \otimes g_j \tag{10}$$

have been defined. Throughout this paper the usual summation convention is used, where Latin indices range from 1 to 3 and Greek indices range from 1 to 2. The covariant base vectors of the reference configuration B_0

$$G_i = \frac{\partial X}{\partial \theta^i} = X_{,i} \tag{11}$$

are directed tangentially along the curvilinear coordinates θ^i. The corresponding covariant base vectors g_i of the deformed configuration B can be obtained with the help of the deformation gradient via $g_i = FG_i$. The additive decomposition of the total deformation gradient (3) directly leads to an additive decomposition of the deformed base vectors into compatible and enhanced parts:

$$\boxed{g_i = x_{,i} + \tilde{g}_i} \tag{12}$$

Hence, the compatible part of the deformation gradient

$$\text{Grad}[x] = x_{,i} \otimes G^i \tag{13}$$

is locally superposed by the enhanced displacement gradient

$$\tilde{H} = \tilde{g}_i \otimes G^i \tag{14}$$

so that the total deformation gradient (3) may be written in the form

$$F = g_i \otimes G^i, \tag{15}$$

where the contravariant base vectors G^i are defined by $G_i \cdot G^j = \delta_i^j$. Variation of the deformation gradient yields

$$\delta F = \delta g_i \otimes G^i \qquad \text{with} \qquad \delta g_i = \delta x_{,i} + \delta \tilde{g}_i. \tag{16}$$

Introducing the expression of the inverse deformation gradient $F^{-1} = G_i \otimes g^i$, one observes that

$$\delta F^T F = \delta g_i \cdot g_k \, G^i \otimes G^k \qquad \text{and} \qquad \delta F F^{-1} = \delta g_i \otimes g^i. \tag{17}$$

Together with $\delta F = \text{Grad}[\delta u] + \delta \tilde{H}$ and (8), the internal virtual work (7) can be written in the form

$$G_{int} = \iiint \delta g_i \cdot g_j \hat{S}^{ij} \sqrt{G} d\theta^1 d\theta^2 d\theta^3 \tag{18}$$

where $\sqrt{G} = G_1 \cdot (G_2 \times G_3)$, so that a volume element of the reference configuration $\mathcal{B}_0$ can be obtained via $dV = \sqrt{G} d\theta^1 d\theta^2 d\theta^3$. Using equations (12) and (16), G_{int} may now be written in the form

$$\boxed{G_{int} = \iiint \left\{ \delta x_{,i} \cdot x_{,j} + \delta x_{,i} \cdot \tilde{g}_j + \delta \tilde{g}_i \cdot x_{,j} + \delta \tilde{g}_i \cdot \tilde{g}_j \right\} \hat{S}^{ij} \sqrt{G} d\theta^1 d\theta^2 d\theta^3} \tag{19}$$

The virtual work contributions (equations (19), (4) and (5)) form the variational foundation of the proposed nonlinear 4-node shell element described in the next section.

3. Extensible Shell Formulation

We start from the essential kinematic assumption of a single director extensible shell kinematic which leads to the compatible finite element interpolations of the mid-surface and the extensible director field. Then we present two discontinuous EAS-expansions of the material displacement gradient which lead to enhanced covariant base vectors $\tilde{g}_i$. The aim of $\tilde{g}_3$ is to enrich the compatible transverse normal strain field and can be motivated by an extension of the underlying extensible shell kinematic. Additionally, the proposed expansion of the enhanced covariant base vectors $\tilde{g}_\alpha$ improves the membrane part of the bilinear shell element, particularly in bending dominated situations and in the incompressible limit.

3.1. Kinematic Shell Assumptions

The formulation of the finite shell element is based on the three-dimensional continuum theory, where the placement of a point in shell space of the deformed configuration $\mathcal{B}$ is restricted by the assumption of a single *extensible* director kinematic

$$x(\theta^i) = \varphi(\theta^\alpha) + \zeta d(\theta^\alpha). \tag{20}$$

Here, $\varphi(\theta^\alpha)$ denotes the position vector of the actual shell mid-surface $\mathcal{M}$ labeled with curvilinear coordinates θ^α. Furthermore, $\theta^3 = \zeta \in [-h_0/2, h_0/2]$ describes the coordinate in the thickness direction, where h_0 is the shell thickness of the reference configuration. The extensible director field $d(\theta^\alpha)$ accounts for deformation dependent thickness stretch $\lambda = \|d\| = h/h_0$ and allows for transverse shearing.

The corresponding quantities of the reference configuration $\mathcal{B}_0$ are denoted by capital letters. Here, the position vector $X(\theta^i)$ of any point $P \in \mathcal{B}_0$ is defined by

$$X(\theta^i) = \Phi(\theta^\alpha) + \zeta D(\theta^\alpha), \tag{21}$$

where $\Phi(\theta^\alpha)$ denotes a configuration of the reference shell mid-surface $\mathcal{M}_0$. A director $D(\theta^\alpha)$ of the initial configuration is defined as a unit vector perpendicular to the shell mid-surface $\mathcal{M}_0$. The kinematic quantities φ and d are approximated by means of compatible finite element interpolations of the corresponding nodal quantities as described in section 3.6. There, it will be shown that the bilinear approximation of the extensible director field requires special attention.

3.2. EAS-Extension of Compatible Shell Kinematics

We propose a discontinuous expansion of the material displacement gradient which can be motivated by higher order shell kinematics. Consider the expression

$$\tilde{H} = \tilde{g}_3 \otimes G^3 \tag{22}$$

of the enhanced material displacement gradient, with the following discontinuous expansion of the enhanced covariant base vector

$$\tilde{g}_3 = \zeta \tilde{\gamma} d. \tag{23}$$

Here, $\tilde{\gamma}(\xi)$ describes the expansion which is based on the natural coordinates $\xi = [\xi^1, \xi^2]$ of the isoparametric domain $\square = [-1, 1] \times [-1, 1]$:

$$\tilde{\gamma} = e^T \Gamma_d \quad \text{with} \quad e^T = [1, \xi^1, \xi^2, \xi^1 \xi^2] \quad \text{and} \quad \Gamma_d = [\Gamma_d^1, \Gamma_d^2, \Gamma_d^3, \Gamma_d^4]^T. \tag{24}$$

Accordingly, Γ_d contains four internal degrees of freedom associated with the interpolation vector $e(\xi)$ which defines the interpolation in the isoparametric domain $\square$. The total covariant base vector g_3 can now be composed of the compatible part $x_{,3}$ which follows from (20) and the enhanced part (23) so that

$$g_3 = x_{,3} + \tilde{g}_3 = (1 + \zeta \tilde{\gamma}) d. \tag{25}$$

The condition (6) requires that

$$\iiint \tilde{g}_3 \otimes G^3 \sqrt{G} d\theta^1 d\theta^2 d\theta^3 = \int_{\square} \left[\int_{-h_0/2}^{h_0/2} \zeta \, d\zeta \right] \tilde{\gamma} d \otimes D \sqrt{A} d\square = 0. \tag{26}$$

This condition is satisfied, since the term in the square bracket is zero. For the sake of simplicity, it has been assumed in (26) that the metric of the initial shell space $\mathcal{B}_0$ coincides with the metric of the initial mid-surface $\mathcal{M}_0$, i.e. $\sqrt{G} = \sqrt{A} = \|\Phi_{,1} \times \Phi_{,2}\|$ which is justified for thin shells.

Remark. The expansion (23) for $\widetilde{g}_3$ can be motivated by means of higher order shell kinematics. According to Naghdi [24] (page 466) a shell theory can be derived from three-dimensional continuum mechanics using the following series for the position vector of the deformed shell configuration

$$x(\theta^i) = \varphi(\theta^\alpha) + \sum_{N=1}^{\infty} \zeta^N d_N(\theta^\alpha). \tag{27}$$

The applied (compatible) single extensible director kinematic (20) follows from (27) by setting $N = 1$ and $d_1 = d$. Thus the higher order terms with $N \geq 2$ are truncated. If we now consider the quadratic term ($N = 2$) with

$$\widetilde{x}(\theta^i) = \zeta^2 d_2(\theta^\alpha) \qquad \text{and} \qquad d_2 = \frac{1}{2}\widetilde{\gamma}d, \tag{28}$$

the corresponding covariant base vector $\widetilde{g}_3$ is given by

$$\widetilde{g}_3 = \widetilde{x}_{,3} = \zeta\widetilde{\gamma}d. \tag{29}$$

3.3. EAS-Improvement of Membrane Part

The performance of the extensible bilinear shell element can be further improved with the help of the EAS method, especially for in-plane bending dominated cases and the incompressible limit. To this end we consider the following expansion of the enhanced material displacement gradient

$$\tilde{H} = \widetilde{\varphi}_\alpha \otimes G^\alpha \tag{30}$$

with discontinuous enhanced covariant base vectors

$$\widetilde{g}_\alpha = \widetilde{\varphi}_\alpha = \sum_{J=1}^{2} a_\alpha^J \Gamma_J^\beta \varphi_{0,\beta} \qquad \text{and} \qquad a_\alpha^J = \frac{\sqrt{A_0}}{\sqrt{A}}\xi^J(G_0^J \cdot G_\alpha). \tag{31}$$

In (31), for all expressions with the subscript $_0$ the corresponding quantity has to be evaluated in the center of the element, i.e. at $\xi = 0$ and $\zeta = 0$. Furthermore, Γ_J^β denote four internal degrees of freedom. The proposed expansion (31) of the enhanced covariant base vectors $\widetilde{g}_\alpha$ can be derived from Wilson's incompatible shape functions (Wilson et al. [25])

$$\tilde{N}^J = \frac{1}{2}\left\{(\xi^J)^2 - 1\right\} \qquad (J = 1, 2). \tag{32}$$

To this end, we use the systematic procedure for the design of interpolations for the enhanced displacement gradient as suggested by Simo, Armero & Taylor [26] which can be applied within the curvilinear formulation as follows.

Define the map $\mathbb{H}$ on the isoparametric domain $\square$ by the interpolation formula

$$\mathbb{H}(\xi^\alpha) = \sum_{J=1}^{2} \Gamma_J \otimes \mathrm{Grad}_\xi[\tilde{N}^J], \tag{33}$$

where $\boldsymbol{\Gamma}_J = [\Gamma_J^1, \Gamma_J^2, 0]$ contains two internal degrees of freedom which correspond to each incompatible shape function (32). Then calculate the enhanced material displacement gradient according to the formula

$$\tilde{H} = \frac{\sqrt{A_0}}{\sqrt{A}} \bar{F}_0 J_0 \mathbb{H}(\xi^\alpha) J_0^{-1}, \tag{34}$$

where $\bar{F}_0$ denotes the compatible deformation gradient evaluated at the center of the element

$$\bar{F}_0 = \text{Grad}_0[x(\xi^i)] = x_{0,i} \otimes G_0^i \tag{35}$$

and J_0 is the constant reference Jacobian, evaluated at the center of the element, so that

$$J_0 = G_{0i} \otimes e_i \quad \text{and} \quad J_0^{-1} = e_i \otimes G_0^i. \tag{36}$$

Introducing equations (33), (35) and (36) into (34) yields

$$\tilde{H} = \sum_{J=1}^2 \frac{\sqrt{A_0}}{\sqrt{A}} \frac{\partial \tilde{N}^J}{\partial \xi^\alpha} \Gamma_J^\beta x_{0,\beta} \otimes G_0^\alpha. \tag{37}$$

After introduction of the appropriate assumption for the bilinear shell element $G_3 \cdot G_0^\alpha = 0$ and realizing that $\tilde{N}_{,\alpha}^J = \xi^J \delta_\alpha^J$, one eventually obtains the proposed expansion (30) for the enhanced material displacement gradient

$$\tilde{H} = \sum_{J=1}^2 a_\alpha^J \Gamma_J^\beta \varphi_{0\beta} \otimes G^\alpha = \tilde{\varphi}_\alpha \otimes G^\alpha \tag{38}$$

with a_α^J from equation $(31)_2$. The condition (6) in combination with (34) leads to

$$\int_{\mathcal{B}_0} \tilde{H} \, dV = \sqrt{A_0} h_0 \bar{F}_0 J_0 \int_\square \mathbb{H}(\xi^\alpha) \, d\square J_0^{-1} = 0 \tag{39}$$

and is automatically satisfied due to the nature of $\mathbb{H}(\xi^\alpha)$, since the restriction

$$\int_\square \mathbb{H}(\xi^\alpha) \, d\square = 0 \tag{40}$$

holds in view of equation (33). The total covariant base vectors g_α can now be obtained from the compatible part $x_{,\alpha}$ which follows from the shell kinematic (20) in conjunction with the enhanced part $(31)_1$, so that

$$g_\alpha = x_{,\alpha} + \tilde{g}_\alpha = \psi_\alpha + \zeta d_{,\alpha}, \tag{41}$$

where

$$\psi_\alpha = \varphi_{,\alpha} + \tilde{\varphi}_\alpha. \tag{42}$$

Finally, from (22) and (30) we obtain the total enhanced material displacement gradient which can be written in the form

$$\tilde{H} = \tilde{\varphi}_\alpha \otimes G^\alpha + \zeta \tilde{\gamma} d \otimes G^3 \tag{43}$$

By using definition (3), i.e. $F = \mathrm{Grad}_X[x] + \tilde{H}$, and introducing the extensible shell kinematic (20), the total deformation gradient of the proposed shell formulation is now given by

$$F = \{\psi_\alpha + \zeta d_{,\alpha}\} \otimes G^\alpha + \{1 + \zeta \tilde{\gamma}\} d \otimes G^3 \tag{44}$$

The corresponding covariant components of the local metric tensor can be calculated via $g_{ij} = G_i \cdot (F^T F) G_j$. Subsequent ordering of the components with respect to the thickness coordinate ζ leads to

$$g_{ij} = g_{ij}^{\{0\}} + \zeta g_{ij}^{\{1\}} + \zeta^2 g_{ij}^{\{2\}} \tag{45}$$

with

$$
\begin{aligned}
& g_{\alpha\beta}^{\{0\}} = \psi_\alpha \cdot \psi_\beta, && g_{\alpha\beta}^{\{1\}} = d_{,\alpha} \cdot \psi_\beta + \psi_\alpha \cdot d_{,\beta}, && g_{\alpha\beta}^{\{2\}} = d_{,\alpha} \cdot d_{,\beta}, \\
& g_{3\alpha}^{\{0\}} = \psi_\alpha \cdot d, && g_{3\alpha}^{\{1\}} = d \cdot d_{,\alpha} + \tilde{\gamma} \psi_\alpha \cdot d, && g_{3\alpha}^{\{2\}} = \tilde{\gamma} d \cdot d_{,\alpha}, \\
& g_{33}^{\{0\}} = d \cdot d, && g_{33}^{\{1\}} = 2 \tilde{\gamma} d \cdot d, && g_{33}^{\{2\}} = \tilde{\gamma}^2 d \cdot d
\end{aligned}
\tag{46}
$$

along with analogous expressions for $G_{ij} = G_i \cdot G_j$ which follow directly from the kinematic assumption (21) for the reference configuration. The state of deformation can now be described using the Green-Lagrangian strain tensor

$$E = \frac{1}{2}(g_{ij} - G_{ij}) G^i \otimes G^j, \tag{47}$$

which may be written analogous to (45) in partitioned form

$$E = E^{\{0\}} + \zeta E^{\{1\}} + \zeta^2 E^{\{2\}}. \tag{48}$$

Accordingly, the covariant thickness strains which are constant across the shell thickness yield

$$E_{33}^{\{0\}} = \frac{1}{2}\{\|d\|^2 - 1\}, \tag{49}$$

whereas the corresponding linear part may be written in the form

$$E_{33}^{\{1\}} = \tilde{\gamma} \|d\|^2. \tag{50}$$

Equation (50) indicates that the EAS-interpolation of the covariant base vector $\tilde{g}_3$ leads to thickness strains which are linear across the shell thickness. Thus, three-dimensional constitutive models can be implemented without any shell-specific modifications.

3.4. Stress-Resultant Weak Form

We derive the stress-resultant counterpart of the EAS-weak form (1) based on the curvilinear representation (19). To this end we define stress-resultants from the three-dimensional theory. The stress across the coordinate surface $\theta^i = $ constant can be represented by a stress vector (see Green & Zerna [27])

$$t^i = \frac{T^i}{\sqrt{gg^{ii}}} \qquad \text{with} \qquad T^i = \sqrt{g}\,\sigma^{ik} g_k = \sqrt{g}\,\sigma g^i, \tag{51}$$

where σ denotes the Cauchy stress tensor, so that $\sqrt{g}\,\sigma = \sqrt{G}\tau$. With the help of (10) we now define

$$\hat{T}^i = \sqrt{G}\,\hat{\tau}g^i = \sqrt{G}\,\hat{S}^{ik} g_k. \tag{52}$$

Introducing (52) into the internal virtual work (18) of the EAS-formulation yields

$$G_{int} = \iiint \left\{ \delta g_\alpha \cdot \hat{T}^\alpha + \delta g_3 \cdot \hat{T}^3 \right\} d\theta^1 d\theta^2 d\theta^3, \tag{53}$$

where the variation of the total covariant base vectors (25) and (41) is given by

$$\begin{aligned}
\delta g_3 &= \delta d + \zeta \delta \tilde{d}, \\
\delta g_\alpha &= \delta \psi_\alpha + \zeta \delta d_{,\alpha},
\end{aligned} \tag{54}$$

and $\tilde{d} = \tilde{\gamma}d$. Further, we define the stress-resultants $\hat{n}^i$ and $\hat{m}^i$ by

$$\hat{n}^i = \frac{1}{\sqrt{a}} \int_{-\frac{h_0}{2}}^{\frac{h_0}{2}} \hat{T}^i \, d\zeta, \qquad \hat{m}^i = \frac{1}{\sqrt{a}} \int_{-\frac{h_0}{2}}^{\frac{h_0}{2}} \zeta \hat{T}^i \, d\zeta. \tag{55}$$

Introducing (54) and (55) into the internal virtual work (53) leads to the corresponding stress-resultant expression

$$G_{int} = \sum_{e=1}^{n_{elem}} \int_\Box \left\{ \hat{n}^\alpha \cdot \delta \psi_\alpha + \hat{m}^\alpha \cdot \delta d_{,\alpha} + \hat{n}^3 \cdot \delta d + \hat{m}^3 \cdot \delta \tilde{d} \right\} \sqrt{a} \, d\Box, \tag{56}$$

where n_{elem} denotes the number of finite shell elements used for the discretization of the continuous problem. If we drop the contributions of the proposed EAS-expansions for $\tilde{\varphi}_\alpha$ and $\tilde{g}_3$, equations (31) and (23), so that $\tilde{d} = 0$ and $\psi_\alpha = \varphi_{,\alpha}$, then the virtual work of the stress-resultants (56) coincides with the corresponding expression used by Simo, Rifai & Fox [7] (equation (2.14a)). The main advantage of the present formulation lies in the appearance of the fourth term in (56) which arises from the extension of the underlying extensible shell kinematic (20) by means of the enhanced covariant base vector $\tilde{g}_3$, equation (23). Because of this additional term, complete three-dimensional constitutive models can be applied without any modifications. In contrast, the bending part of the constitutive model in [7] needs to be modified by the zero normal stress assumption.

Based on the resultant expression (56) one may proceed with the definition of (symmetric) effective stress resultants as e.g. in [7]. Alternatively, the definition of stress resultants can be avoided when the stress components of the continuum theory are employed directly. Consequently, a formulation that is ideally suited for the numerical implementation of various three-dimensional constitutive models can be obtained.

3.5. Weak Form Based on Stress Components

We proceed directly from the three-dimensional curvilinear representation (19) of the EAS-formulation. With the help of the covariant components of the metric tensor which are given by (45), equation (19) may be rewritten for one element

$$G^e_{int} = \frac{1}{2} \int_\Box \left\{ \delta g^{\{0\}}_{ij} \hat{S}^{ij}_{\{0\}} + \delta g^{\{1\}}_{ij} \hat{S}^{ij}_{\{1\}} + \delta g^{\{2\}}_{ij} \hat{S}^{ij}_{\{2\}} \right\} d\Box, \tag{57}$$

where the integration across the shell thickness is confined to the constitutive expressions

$$\hat{S}^{ij}_{\{N\}} = \int_{-\frac{h_0}{2}}^{\frac{h_0}{2}} \zeta^N \hat{S}^{ij} \sqrt{G}\, d\zeta \tag{58}$$

for $N = 0, 1, 2$. Here, $\hat{S}^{ij}$ are the contravariant components of the stress tensor (9) which can be calculated with the help of the strain energy function W. Depending on the three-dimensional constitutive model, the integration across the shell thickness in (58) can be carried out analytically or has to be performed numerically.

For the sake of an efficient numerical implementation, we neglect in (57) $g^{\{2\}}_{33}$ as well as the contributions of the EAS-expansion to the transverse shear strains. Numerical simulations confirm that these terms only play a negligible part. With the help of (46) equation (57) may now be written in the final form

$$\begin{aligned} G^e_{int} = \int_\Box \Big\{ &\delta\boldsymbol{\psi}_\alpha \cdot \boldsymbol{\psi}_\beta \hat{S}^{\alpha\beta}_{\{0\}} + (\delta\boldsymbol{\varphi}_{,\alpha} \cdot \boldsymbol{d} + \delta\boldsymbol{d} \cdot \boldsymbol{\varphi}_{,\alpha}) \hat{S}^{\alpha 3}_{\{0\}} + \delta\boldsymbol{d} \cdot \boldsymbol{d}\, \hat{S}^{33}_{\{0\}} \\ &+ (\delta\boldsymbol{\psi}_\alpha \cdot \boldsymbol{d}_{,\beta} + \delta\boldsymbol{d}_{,\alpha} \cdot \boldsymbol{\psi}_\beta) \hat{S}^{\alpha\beta}_{\{1\}} + (\delta\boldsymbol{d} \cdot \boldsymbol{d}_{,\alpha} + \delta\boldsymbol{d}_{,\alpha} \cdot \boldsymbol{d}) \hat{S}^{\alpha 3}_{\{1\}} \\ &+ (2\tilde{\gamma}\delta\boldsymbol{d} \cdot \boldsymbol{d} + \delta\tilde{\gamma}\|\boldsymbol{d}\|^2) \hat{S}^{33}_{\{1\}} + \delta\boldsymbol{d}_{,\alpha} \cdot \boldsymbol{d}_{,\beta} \hat{S}^{\alpha\beta}_{\{2\}} \Big\}\, d\Box. \end{aligned} \tag{59}$$

Expression (59) leads to an extensible shell formulation which is ideally suited for the implementation of three-dimensional nonlinear constitutive models. Next, we deal with the remaining compatible finite element interpolations of the mid-surface and the extensible director field.

3.6. Compatible Interpolations

According to the isoparametric concept (see e.g. Hughes [28]), the position vector $\varphi(\xi^\alpha)$ of the mid-surface as well as the extensible director $\boldsymbol{d}(\xi^\alpha)$ are interpolated with the help of the standard bilinear shape functions $N^I\colon \Box \to \mathbb{R}$. Within one element the position vector of the mid-surface is given by

$$\varphi(\xi^\alpha) = \sum_{I=1}^{4} N^I(\xi^\alpha)\varphi_I. \tag{60}$$

Interpolation of the director field. The approximation of the extensible director field within the discrete formulation requires special attention in order to obtain a good numerical performance in the thin-shell limit. The simplest kind of parametrization relies on the direct interpolation of the components of the nodal extensible directors d_I

$$d(\xi^\alpha) = \sum_{I=1}^{4} N^I(\xi^\alpha)d_I. \tag{61}$$

The simplicity of this formulation is due to the additive structure of the update $d^{(k+1)} = d^{(k)} + \Delta d$ within an iterative solution procedure. Accordingly, the interpolation (61) does not require any rotational variables defining a rotation tensor. However, as has been shown by Simo, Rifai & Fox [7], the formulation suffers from pathological behavior in the thin-shell limit. As shown below this is due to the appearance of parasitic thickness strains induced by interpolation (61) in the case of curved shells.

Another parametrization of the director field is based on a multiplicative split into an extensible and inextensible part

$$d(\xi^\alpha) = \lambda(\xi^\alpha)\,t(\xi^\alpha). \tag{62}$$

Therefore, the thickness stretch $\lambda = h/h_0$ and the director orientation characterized by the unit vector t are kinematically decoupled. According to the constraint $\|t\| = 1$, a rotation tensor has to be used as in shell formulations based on the zero normal stress condition and inextensibility in the thickness direction. The structure of the multiplicative decomposition (62) is preserved in the discrete model with interpolations

$$\lambda(\xi^\alpha) = \sum_{I=1}^{4} N^I(\xi^\alpha)\lambda_I \qquad \text{and} \qquad \Delta t(\xi^\alpha) = \sum_{I=1}^{4} N^I(\xi^\alpha)\Delta t_I. \tag{63}$$

Application of interpolations (63) leads to a numerical formulation that recovers a plane stress response in the thin-shell limit without any numerical problems, as has been shown in Betsch, Gruttmann & Stein [11].

A third description for the discrete extensible director field relies on the interpolation (61) in combination with the multiplicative decomposition (62) of the nodal director vectors

$$d(\xi^\alpha) = \sum_{I=1}^{4} N^I(\xi^\alpha)\lambda_I t_I. \tag{64}$$

Remarkably, numerical simulations with this kind of director interpolation yield identical results when compared with the less elaborate interpolation formula (61). This indicates that the simple *nodal* director update procedure $d_I^{(k+1)} = d_I^{(k)} + \Delta d_I$ is equivalent to the more elaborate nodal update for the multiplicative decomposition which requires an orthogonal matrix $\Delta\Lambda_I$ in order to achieve $t_I^{(k+1)} = \Delta\Lambda_I t_I^{(k)}$.

In the next section we consider the discrete director field associated with interpolation (61) which causes the pathological behavior in the thin-shell limit.

3.6.1. One-Dimensional Model Problem.

In order to examine the behavior of the interpolation (61) of the director field, it is sufficient to consider the one-dimensional model problem depicted in Fig. 1. We consider a cut through a curved shell structure with corresponding finite element discretization. The shell mid-surface $\varphi(\xi)$ and the

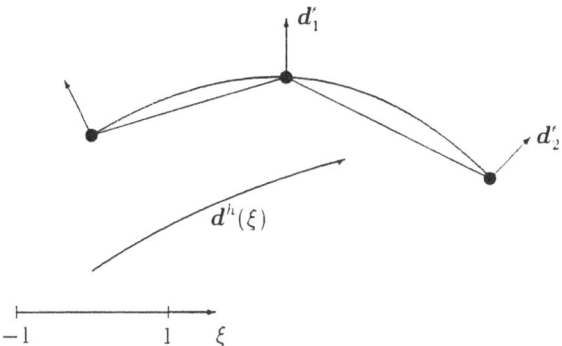

Fig. 1. Cut through a curved shell with discretization.

extensible director field $d(\xi)$ are approximated by piecewise-linear interpolations of the nodal vectors $\varphi_I = \varphi(\xi_I)$ and $d_I = d(\xi_I)$, respectively.

Accordingly, for one representative element one obtains the discrete extensible director field

$$d^h(\xi) = \frac{1}{2}(1 - \xi)d_1^e + \frac{1}{2}(1 + \xi)d_2^e. \tag{65}$$

For the limiting case of a thin shell we consider pure bending deformation, where the inextensibility condition $d \cdot d = 1$ holds along with $E_{33}^{\{0\}} = 0$. Moreover, we assume for the discrete model that the inextensibility condition is fulfilled at the nodal points, i.e. $\|d_1^e\| = 1$ and $\|d_2^e\| = 1$. Application of the interpolation (65) yields

$$d^h \cdot d^h = 1 + \frac{1}{2}(1 - \xi^2)(d_1^e \cdot d_2^e - 1). \tag{66}$$

Thus, for the discrete director field the inextensibility condition holds only if $d_1^e \cdot d_2^e = 1$, i.e. in the case of no curvature. In general the interpolation of the director vector (65) induces thickness change and, therefore, artificial thickness strains which are constant over the shell thickness. Obviously, this defect depends on the curvature and on the mesh density.

Furthermore, for the linearization of the discrete thickness strains one obtains

$$\delta E_{33}^{\{0\}} = d^h \cdot \delta d^h = \frac{1}{4}(1 - \xi^2)(d_1^e \cdot \delta d_2^e + d_2^e \cdot \delta d_1^e) \tag{67}$$

which indicates once more that, even in the case of linear elasticity and initial curvature, artificial straining occurs. Consequently, any bending deformation of the element is accompanied by thickness stretching, thus increasing the bending stiffness. Numerical experiments show that the defect of the interpolation (65) leads to locking response in the case of curved shells. Locking worsens with greater curvature (initial or deformation dependent), especially for coarse meshes.

3.6.2. Assumed Strain Interpolation of $E_{33}^{\{0\}}$. To circumvent the effect of artificial thickness strains, we employ an assumed strain approximation for the covariant component $E_{33}^{\{0\}}$ of the Green-Lagrangian strain tensor. We assume bilinear interpolations of the thickness strain field, where the element nodes serve as sampling points of the compatible thickness strains. Accordingly, we obtain a C^0-continuous strain field of the form

$$\bar{E}_{33}^{\{0\}}(\xi^\alpha) = \sum_{I=1}^{4} N^I(\xi^\alpha) E_{33}^{\{0\}}(\xi_I^\alpha), \tag{68}$$

with the compatible nodal thickness strains (49)

$$E_{33}^{\{0\}}(\xi_I^\alpha) = \frac{1}{2}(d_I \cdot d_I - 1). \tag{69}$$

Variation of equation (68) yields the contribution to the linearized discrete strain-nodal operator matrix

$$\delta\bar{E}_{33}^{\{0\}}(\xi^\alpha) = \sum_{I=1}^{4} N^I(\xi^\alpha) d_I^T \delta d_I. \tag{70}$$

Furthermore, for the contribution to the geometric part of the tangent stiffness matrix, one obtains

$$\Delta\delta\bar{E}_{33}^{\{0\}}(\xi^\alpha) = \sum_{I=1}^{4} \Delta d_I^T N^I(\xi^\alpha) \mathbf{1}\delta d_I. \tag{71}$$

Accordingly the proposed assumed strain interpolation (68) is easy to implement and makes the simple interpolation of the extensible director (61) competitive with more involved formulations that need rotational degrees of freedom.

3.6.3. Assumed Strain Interpolation of $E_{\alpha 3}^{\{0\}}$. The well-known shear locking effect of low order C^0-continuous shell formulations can also be eliminated by means of an assumed strain modification. For the present bilinear shell element we modify the discrete compatible transverse shear strains $E_{\alpha 3}^{\{0\}}$ introducing the assumed strain interpolation of Dvorkin & Bathe [17] (see equation (105) in Appendix A.1), which has been further investigated by Stander, Matzenmiller & Ramm [19] and Parisch [20].

3.7. Implementation Aspects

In order to obtain an efficient numerical implementation, proceeding from expression (59) is suggested, since the explicit integration across the shell thickness is confined to the constitutive terms, i.e. the stress components and the tangent moduli, and can be performed outside the nodal loops. However, for the sake of maximum clearness, we proceed from the implicit version

$$G_{int}^e = \int\int_\square \int_{-\frac{h_0}{2}}^{\frac{h_0}{2}} \Big[\delta g_\alpha \cdot g_\beta \hat{S}^{\alpha\beta} + \{\delta x_{,\alpha} \cdot d + \delta d \cdot x_{,\alpha}\} \hat{S}^{\alpha 3}$$

$$+ \{\delta d \cdot d(1 + \zeta 2\tilde{\gamma}) + \delta\tilde{\gamma}\zeta\|d\|^2\}\hat{S}^{33} \Big] \sqrt{G}\, d\zeta\, d\square \tag{72}$$

which is equivalent to (59), as can be shown by introducing equations (41) and (20) into (72). It is a straightforward way to obtain the corresponding explicit version by ordering with respect to the powers of the thickness coordinate ζ. Introduction of the compatible and enhanced interpolations into (72) leads to the discrete expression

$$G_{int}^e = \delta v_e^T \int\int_\square \int_{-\frac{h_0}{2}}^{\frac{h_0}{2}} B^T \widehat{S} \sqrt{G}\, d\zeta\, d\square + \delta\Gamma_e^T \int\int_\square \int_{-\frac{h_0}{2}}^{\frac{h_0}{2}} \widetilde{B}^T \widehat{S} \sqrt{G}\, d\zeta\, d\square. \tag{73}$$

In (73) $v_e^T = [v_1^T, ..., v_4^T]$ contains the compatible nodal variables $v_I^T = [\varphi_I^T, d_I^T]$, and $\Gamma_e^T = [\Gamma_d^T, \Gamma_1^T, \Gamma_2^T]$ contains the internal degrees of freedom. The symmetric components of the Green-Lagrangian strain tensor are ordered in the convenient column-matrix form

$$E = [E_{11}, E_{22}, 2E_{12}, E_{33}, 2E_{13}, 2E_{23}]^T \tag{74}$$

along with the corresponding ordering of the stress components

$$\widehat{S} = [\widehat{S}^{11}, \widehat{S}^{22}, \widehat{S}^{12}, \widehat{S}^{33}, \widehat{S}^{13}, \widehat{S}^{23}]^T. \tag{75}$$

The linearized strains are given by

$$\delta E = \sum_{I=1}^{4} B_I \delta v_I + \widetilde{B} \delta\Gamma_e, \tag{76}$$

where B_I denotes the compatible nodal operator matrix and $\widetilde{B}$ is the operator matrix of the internal degrees of freedom

$$B_I = \begin{bmatrix} B_{I3\times3}^\varphi & B_{I3\times3}^d \\ 0_{1\times3} & b_{I1\times3}^{33} \\ B_{\varphi I2\times3}^S & B_{d I2\times3}^S \end{bmatrix} \quad \text{and} \quad \widetilde{B} = \begin{bmatrix} 0_{3\times nt} & \widetilde{B}_{I3\times2}^m & \widetilde{B}_{23\times2}^m \\ \widetilde{b}_{1\times nt}^{33} & 0_{1\times2} & 0_{1\times2} \\ 0_{2\times nt} & 0_{2\times2} & 0_{2\times2} \end{bmatrix}. \tag{77}$$

The index nt denotes the number of internal degrees of freedom Γ_d^{nt} in (24). A detailed derivation of the submatrices in B_I and $\widetilde{B}$ is contained in the Appendix A.1. Furthermore, in (73) $B = [B_1, B_2, B_3, B_4]$. Linearization of the discretised static weak form (1)

$$\sum_{e=1}^{n_{elem}} G^e = \sum_{e=1}^{n_{elem}} [G_{int}^e - G_{ext}^e] = 0 \tag{78}$$

yields

$$\sum_{e=1}^{n_{elem}} [G^e + DG^e \cdot (\Delta v_e, \Delta\Gamma_e)] = 0 \tag{79}$$

and leads to an algebraic system of linear equations for the iterative solution procedure by means of Newton's method in combination with an elimination of the internal degrees of freedom on the element level. Accordingly, for a typical Newton iteration (79) may be written in the form

$$\sum_{e=1}^{n_{elem}} [\delta v_e^T, \delta\Gamma_e^T] \left\{ \begin{bmatrix} f_{int}^e - f_{ext}^e \\ \widetilde{f}_{int}^e \end{bmatrix} + \begin{bmatrix} K_{vv} & K_{v\Gamma} \\ K_{\Gamma v} & K_{\Gamma\Gamma} \end{bmatrix} \begin{bmatrix} \Delta v_e \\ \Delta\Gamma_e \end{bmatrix} \right\} = 0. \tag{80}$$

Each submatrix of the Hessian matrix in (80) consists of a material and a geometric part. The derivation of the geometric part is described in the Appendix A.2. Accordingly, we obtain

$$K_{vv} = \iint_{\Box} \int_{-\frac{h_0}{2}}^{\frac{h_0}{2}} B^T C B \sqrt{G}\, d\zeta\, d\Box + K_{vv}^{geo}, \tag{81}$$

$$K_{v\Gamma} = \iint_{\Box} \int_{-\frac{h_0}{2}}^{\frac{h_0}{2}} B^T C \widetilde{B} \sqrt{G}\, d\zeta\, d\Box + K_{v\Gamma}^{geo}, \tag{82}$$

$$K_{\Gamma v} = K_{v\Gamma}^T, \tag{83}$$

$$K_{\Gamma\Gamma} = \iint_{\Box} \int_{-\frac{h_0}{2}}^{\frac{h_0}{2}} \widetilde{B}^T C \widetilde{B} \sqrt{G}\, d\zeta\, d\Box + K_{\Gamma\Gamma}^{geo}, \tag{84}$$

$$f_e^{int} = \iint_{\Box} \int_{-\frac{h_0}{2}}^{\frac{h_0}{2}} B^T \widehat{S} \sqrt{G}\, d\zeta\, d\Box, \tag{85}$$

$$\widetilde{f}_e^{int} = \iint_{\Box} \int_{-\frac{h_0}{2}}^{\frac{h_0}{2}} \widetilde{B}^T \widehat{S} \sqrt{G}\, d\zeta\, d\Box. \tag{86}$$

In (81)–(84), C is a 6×6 matrix which contains the contravariant components of the fourth order constitutive tensor $\mathcal{C} = 4\nabla^2 W(C) = C^{ijkl} G_i \otimes G_j \otimes G_k \otimes G_l$. The components C^{ijkl} are ordered with respect to the column matrix form (74) of the covariant strain components E_{ij}.

Since the internal degrees of freedom Γ_e are only defined on the element level, their increments $\Delta\Gamma_e$ in (80) can be eliminated by static condensation based on the elimination formula

$$\Delta\Gamma_e = -K_{\Gamma\Gamma}^{-1}\left(\widetilde{f}_{int}^e + K_{\Gamma v}\Delta v_e\right). \tag{87}$$

Thus, a generalized displacement problem follows from (80) that is given by

$$\sum_{e=1}^{n_{elem}} \delta v_e \{r_e + K_e \Delta v_e\} = 0 \tag{88}$$

with the condensed element residual vector and tangent matrix

$$r_e = f_{int}^e - f_{ext}^e - K_{v\Gamma} K_{\Gamma\Gamma}^{-1} \widetilde{f}_{int}^e \quad \text{and} \quad K_e = K_{vv} - K_{v\Gamma} K_{\Gamma\Gamma}^{-1} K_{\Gamma v}. \tag{89}$$

After assembly of the element arrays, the corresponding linear algebraic system can be solved for the nodal increments. Then the elimination equation (87) can be used for the update of the internal degrees of freedom, i.e. $\Gamma_e^{(k+1)} = \Gamma_e^{(k)} + \Delta\Gamma_e$, on the element level.

4. Numerical Examples

We present a number of example calculations dealing with thin shell problems as well as large strain applications. The newly developed extensible shell element has been implemented in an enhanced version of the program FEAP documented in [29]. The aim of our numerical simulations is

- to investigate the significance of the proposed EAS-expansion (23) for the enhanced covariant base vector $\widetilde{\boldsymbol{g}}_3$.
- to show the EAS-improvement of the membrane part by means of the proposed expansion (31) for the enhanced covariant base vectors $\widetilde{\boldsymbol{g}}_\alpha$.
- to verify that the proposed assumed strain interpolation (68) for $\bar{E}_{33}^{\{0\}}$ in conjunction with the simple interpolation (61) of the extensible director field improves the performance in the thin-shell limit considerably and makes the developed shell formulation competitive with more elaborate formulations based on rotational degrees of freedom.
- to verify that three-dimensional constitutive models can be implemented without any shell specific modifications. To this end, compressible and quasi-incompressible hyperelastic materials of the St. Venant-Kirchhoff, Neo-Hookean and Mooney-Rivlin type are particularly used.

The following formulations are employed in the numerical simulations:

A) Newly developed extensible shell formulation based on the proposed EAS-expansions of the enhanced covariant base vectors $\widetilde{\boldsymbol{g}}_i$, equations (23) and (31), in conjunction with the assumed strain interpolation (68) of $\bar{E}_{33}^{\{0\}}$.
B) Bilinear inextensible shell formulation in conjunction with the zero normal stress condition and rotational degrees of freedom as suggested by Simo, Fox & Rifai [4].
C) The same as formulation A, but without the assumed strain interpolation (68) for $\bar{E}_{33}^{\{0\}}$.

4.1. Plane Strain Tension Test

This large strain example illustrates the significant improvement of the membrane part of the proposed shell element by means of the expansion (31) for the enhanced covariant base vectors $\widetilde{\boldsymbol{g}}_\alpha$. As in Miehe [30] we consider an initially square block (see Fig. 2) which is stretched within a displacement controlled calculation. Using the present shell formulation A, the plane strain condition is modeled by constraining the thickness extensibility as a boundary condition. We employ the three-dimensional Neo-Hookean material model with the strain energy function

$$W = \tfrac{\mu}{2}(J^{-\frac{2}{3}}I_1 - 3) + U(J), \qquad U(J) = K(\tfrac{1}{2}(J^2 - 1) - \ln J). \qquad (90)$$

Within a penalty formulation, quasi-incompressibility is enforced by the penalty parameter $K = 1.7 \cdot 10^4$ and the material constant $\mu = 33.4$. Due to the symmetry of the problem, only one-quarter of the block has been discretized by $n \times n$ finite element meshes. Fig. 3 depicts the calculated force F for an extension of $u = 40$. The present formulation is compared with the Q1/P0 element described by Simo, Taylor & Pister [31].

We remark further that the pure displacement formulation (obtained in the present formulation by merely dropping the EAS-expansion (31) for $\widetilde{\boldsymbol{g}}_\alpha$) exhibits severe locking in the incompressible limit.

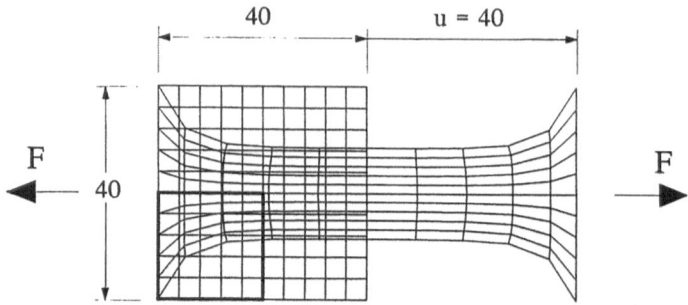

Fig. 2. Plane strain tension test, initial and deformed mesh configuration,
5 × 5 elements (one-quarter).

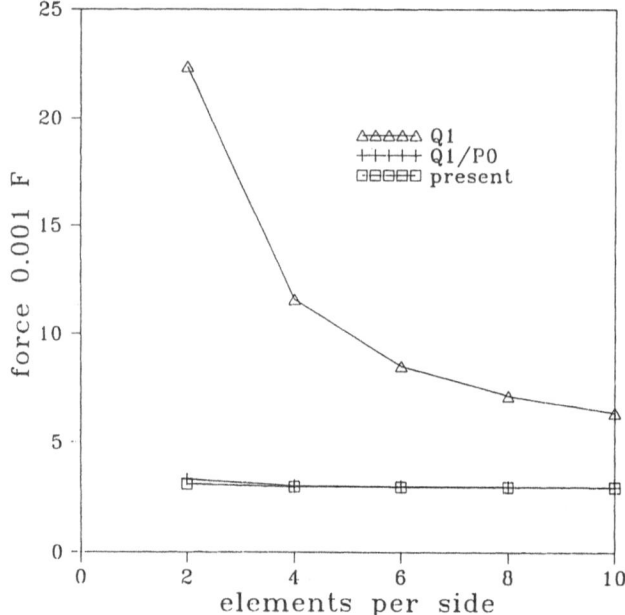

Fig. 3. Plane strain tension test, calculated force for an extension
$u = 40$, $n \times n$ elements (one-quarter).

Finally we consider in Fig. 4 an arbitrarily distorted 4 × 4 mesh (one quarter) of the initial configuration. The corresponding deformed mesh configuration ($u = 40$) of the proposed shell formulation A is plotted with a dash. Even for these irregular element geometries the calculated force deviates from the regular 4 × 4 mesh less than three per cent.

symmetry

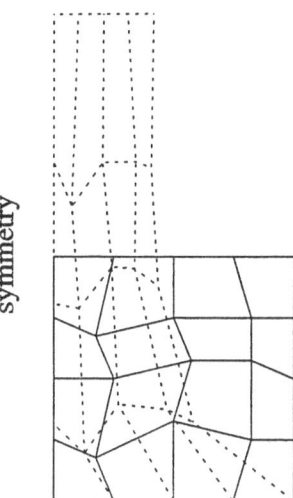

Fig. 4. Plane strain tension test, arbitrarily distorted initial and corresponding deformed 4×4 mesh (one-quarter) for an extension $u = 40$.

4.2. Shearing of a Twisted Ribbon

This linear problem shows the significance of the proposed expansions (31) and (23) for the enhanced covariant base vectors $\widetilde{\boldsymbol{g}}_\alpha$ and $\widetilde{\boldsymbol{g}}_3$, respectively. A shell twisted 90°, with length $l = 12$, width $w = 1.1$, thickness $h_0 = 0.32$, and material properties $E = 2.9 \cdot 10^7$, $\nu = 0.22$ is clamped and subjected to a shear load in thickness direction $F = 1$, see Fig. 5. The displacements in load direction are reported for a sequence of finite element meshes in Table 1. They are normalized with respect to the theoretical solution $u = 1.754 \times 10^{-3}$ [32]. The improved performance of the proposed formulation A due to the enhanced covariant base vectors $\widetilde{\boldsymbol{g}}_\alpha$ and $\widetilde{\boldsymbol{g}}_3$ is obvious.

Table 1. Shearing of a twisted ribbon, normalized displacements in load direction.

Mesh	1×6	2×12	2×24	4×48
With $\widetilde{g}_3$	0.800	0.917	0.970	0.990
With $\widetilde{g}_3$ and $\widetilde{g}_\alpha$	0.947	0.983	0.993	0.996
Without $\widetilde{g}_i$	0.766	0.881	0.934	0.954

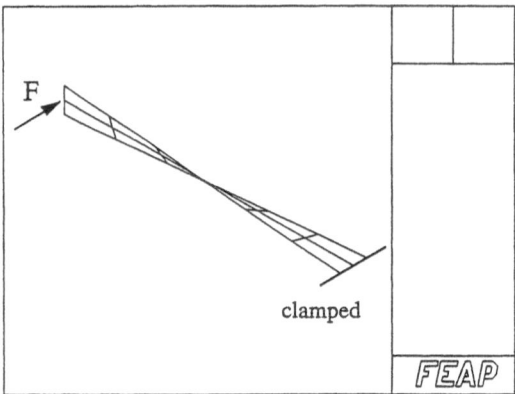

Fig. 5. Shearing of a twisted ribbon. Problem definition
and mesh discretization.

4.3. Roll-Up of a Clamped Beam

An initially flat shell of length $L = 10$, width $w = 1$, thickness $h_0 = 0.01$, and material
properties $E = 1.2 \times 10^7$, $\nu = 0$ is clamped on one end and subjected to a bending
moment on the other end. The exact solution to this problem is a circular curve with
radius $R = EI/M$. An applied bending moment of $M = 2\pi EI/L$ forces the beam to
roll up into a closed circle.

Since formulations A and C contain no rotational degrees of freedom, the bending
moment can not be applied directly. However, as shown in Appendix A.3, a follower load
distribution acting along the edge of the current configuration can be accommodated to
model this load case.

Figures 6 and 7 depict the initial and final deformed mesh configuration for discretiza-
tions with 10 and 20 elements, respectively. Formulation C exhibits severe locking behavior due to parasitic thickness strains. The proposed formulation A yields the same
response as the formulation B (which incorporates rotational degrees of freedom) without any artificial thickness change. Thus, the eminent meaning of the proposed assumed
strain modification (68) of $\bar{E}_{33}^{\{0\}}$ is obvious.

4.4. Pinched Hemisphere with 18° Hole

A hemispheric shell with an 18° hole at the top is loaded at the free edge with two inward
and two outward forces 90° apart. Symmetry conditions are used on this problem and
only one quarter of the shell is modeled. The material properties are $E = 6.825 \times 10^7$
and $\nu = 0.3$, the radius is $R = 10$ and the thickness is $h_0 = 0.04$.

The obtained results of the linear calculation are listed in Table 2 for three mesh configurations. There, the values for the displacements under the unit loads are normalized
with respect to the solution of 0.093, reported in [33]. It can be seen that the formulation
C locks in the case of initial curvature and coarse meshes. On the other hand, formulation

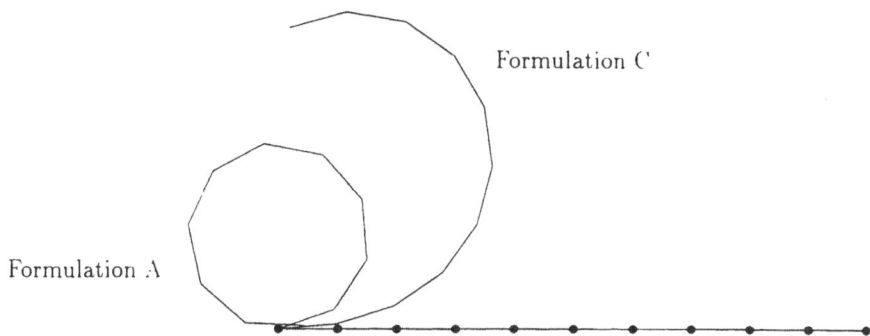

Formulation C

Formulation A

Fig. 6. Roll–up of a clamped beam, 10 elements.

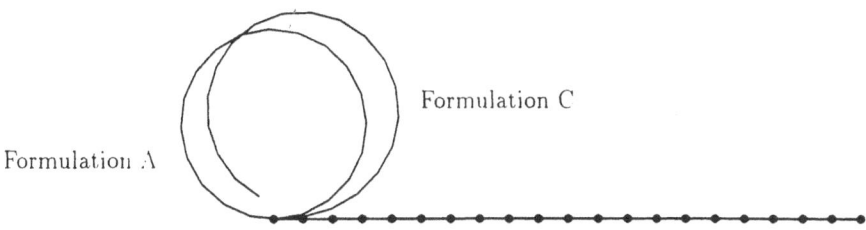

Formulation C

Formulation A

Fig. 7. Roll-up of a clamped beam, 20 elements.

A does not suffer from parasitic transverse normal strains due to the proposed assumed strain modification (68).

For the nonlinear calculation we chose a thickness of $h_0 = 0.01$ in order to yield a radius to thickness ratio of $R/h_0 = 1000$. The same problem has been considered by Simo, Rifai & Fox [7] for the investigation of the thin-shell behavior. A mesh consisting of 16×16 elements has been used to obtain the load-deflection curves plotted in Fig. 8. Formulation C locks progressively with an increasing load due to artificial thickness strains. In contrast to the pathological locking behavior of formulation C, the proposed formulation A yields the same results as the rotational formulation B.

Table 2. Pinched hemisphere, linear solution for the normalized displacements under the load.

Elements	4×4	8×8	16×16
Formulation A	1.026	1.002	0.999
Formulation B	0.985	0.992	0.997
Formulation C	0.083	0.845	0.995

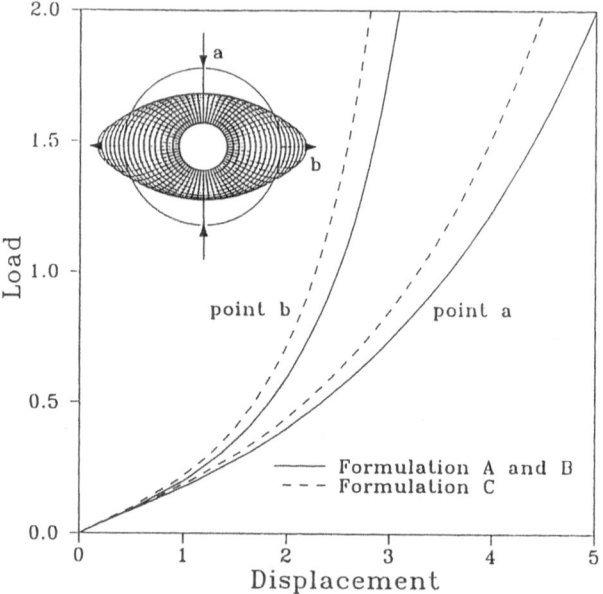

Fig. 8. Pinched hemisphere, $h_0/R = 1 \cdot 10^{-3}$, load-deflection curves of the nonlinear calculation, 16×16 elements.

4.5. Bending and Inflation of a Circular Plate

This large strain problem has been previously investigated by Hughes & Carnoy [34] and Parisch [35]. A circular plate with radius $r = 7.5$ and thickness $h_0 = 0.5$ is simply supported at the circumference and subjected to uniform pressure acting perpendicular to the current configuration. The deformation dependent pressure p leads to an additional (nonsymmetric) contribution to the tangential stiffness matrix $(81)_1$, see e.g. Schweizerhof & Ramm [36]. Because of the symmetry of the problem, only one quarter of the plate has been discretized.

Infinitesimal theory. First, we demonstrate the significance of the proposed EAS-expansion (23) for the enhanced covariant base vector $\widetilde{\boldsymbol{g}}_3$ of formulation A by means of the linear calculations. The material parameters of the three-dimensional St. Venant-Kirchoff model are $E = 600$ and $v = 0.4999$ so that a quasi-incompressible response is enforced. For $p = 0.01$ and a discretization with 108 elements, the plane stress formulation B yields a deflection $w_R = 0.2190$ of the plate center. Table 3 reports the normalized results w/w_R for different meshes. There, nt denotes the employed number of internal degrees of freedom for $\widetilde{\boldsymbol{g}}_3$ in (23). The significance of the proposed EAS-expansion in the bending dominated case is evident.

Table 3. Bending and inflation of a circular plate, normalized deflection w/w_R of the plate center of the linear calculations. nt: number of internal degrees of freedom.

Number of elements		3	12	27	48	108
Inextensible theory, plane stress		0.915	0.986	0.995	0.998	1.000
Extensible theory	$nt = 4$	0.913	0.984	0.994	0.997	0.999
Formulation A	$nt = 1$	0.673	0.784	0.818	0.839	0.873
	$nt = 0$	0.006	0.007	0.007	0.007	0.007

Nonlinear theory. In the nonlinear regime the plate is made of an incompressible hyperelastic material described by the Mooney-Rivlin model with strain energy function

$$W = c_1(I_1 - 3) + c_2(I_2 - 3) + U(J),$$

$$U(J) = \frac{\lambda}{2}(\ln J)^2 - 2(c_1 + 2c_2) \ln J. \tag{91}$$

In (91), $U(J)$ denotes the extension to the compressible range of the constitutive model for the incompressible material. The material constants are $c_1 = 80$ and $c_2 = 20$. We enforce the incompressibility constraint by means of a penalty-parameter $\lambda = 50000$. In Fig. 9 the applied value of the pressure is plotted versus the deflection of the plate center. The computed results of the proposed formulation A, obtained with three Gauss integration points through the thickness and $nt = 4$, are compared to calculations using the membrane element presented by Gruttmann & Taylor [37] and to the shell element of Parisch [35], respectively. Both formulations satisfy incompressibility in an exact manner using the plane stress condition. It can be seen that the results fit very well. Fig. 10 depicts a sequence of deformed configurations for increasing pressure loads.

5. Conclusions

A nonlinear continuum-based 4-node shell element is presented which relies on extensible director kinematics. Within the proposed curvilinear representation of the nonlinear EAS method of Simo & Armero [1], it is shown that the local expansion of the enhanced material displacement gradient leads to enhanced covariant base vectors which are superposed on the compatible covariant base vectors. Two expansions of the enhanced covariant base vectors are developed: First, the underlying extensible shell kinematic is extended and, as a consequence, full three-dimensional constitutive models can be implemented without resorting to the plane stress assumption. Second, the membrane part of the bilinear compatible finite element interpolation is improved. Furthermore, parasitic strain effects of the compatible interpolation are eliminated by means of two assumed strain modifications.

Numerical examples confirm that the element performance is competitive with more elaborate formulations which require an orthogonal matrix in order to describe finite rotations, even in the case of very thin shells. Furthermore, it is shown that the developed extensible shell element is able to deal with large strain problems.

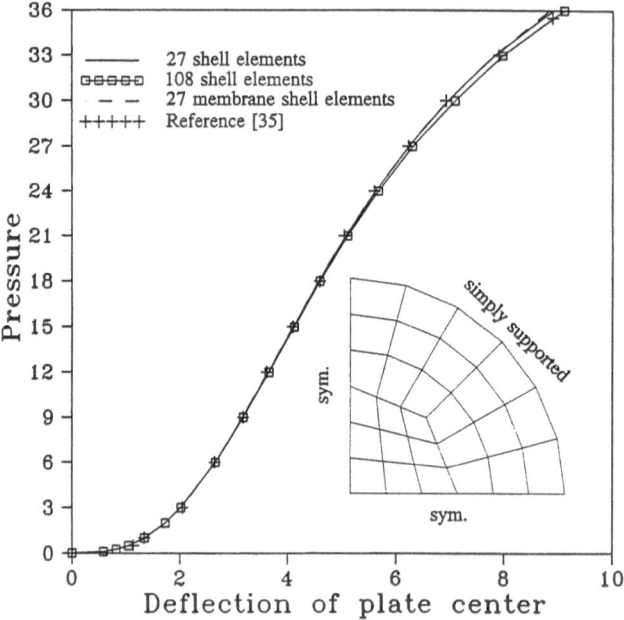

Fig. 9. Bending and inflation of a circular plate. Discretized quarter of the plate (27 elements) and plot of pressure versus deflection at plate center.

Appendix A

A.1. Discrete Operator Matrices

In this section we derive the submatrices of the discrete operator matrices B_I and $\widetilde{B}$ in (77).

Contribution of the tangential strains. Using (42), i.e. $\psi = \varphi_{,\alpha} + \widetilde{\varphi}_\alpha$, together with the EAS-expansion (31) for $\widetilde{\varphi}_\alpha$ leads to

$$\psi_\alpha = \varphi_{,\alpha} + \sum_{J=1}^{2} a_\alpha^J \Gamma_J^\beta \varphi_{0,\beta} \tag{92}$$

$$\psi_\alpha = \sum_{I=1}^{4} M_\alpha^I \varphi_I \quad \text{with} \quad M_\alpha^I = N_{,\alpha}^I + \sum_{J=1}^{2} a_\alpha^J \Gamma_J^\beta N_{0,\beta}^I, \tag{93}$$

where the compatible interpolation (60) of the mid-surface has been introduced. Next, the discrete form of the linearized covariant base vectors (41), i.e. $\delta g_\alpha = \delta \psi_\alpha + \zeta \delta d_{,\alpha}$, yields

$$\delta g_\alpha = \sum_{I=1}^{4} [M_\alpha^I \delta \varphi_I + \zeta N_{,\alpha}^I \delta d_I] + \sum_{J=1}^{2} a_\alpha^J \varphi_{0,\gamma} \delta \Gamma_J^\gamma. \tag{94}$$

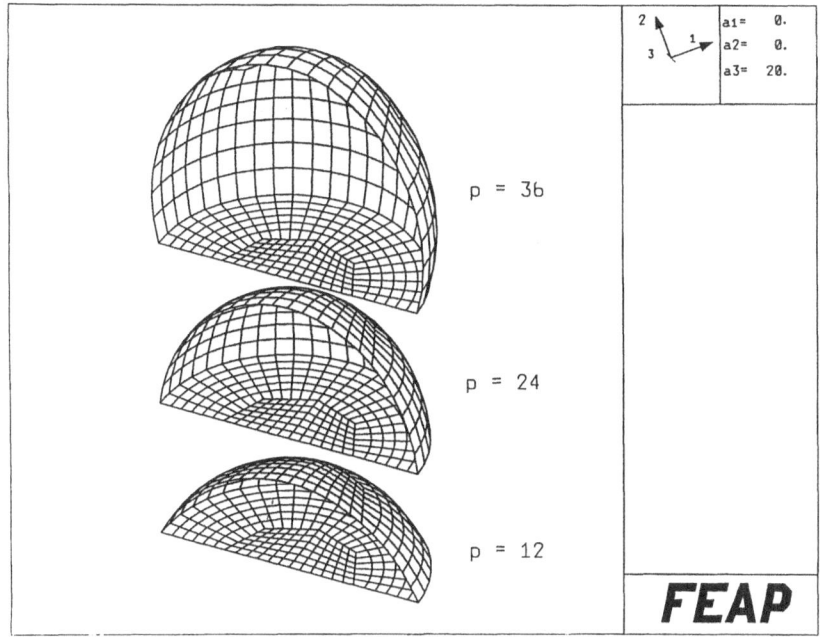

Fig. 10. Bending and inflation of a circular plate. Sequence of deformed configurations (mesh consists of 108 elements per quarter).

Therefore, the variation of the discrete tangential strains can be calculated via

$$\delta E_{\alpha\beta} = \frac{1}{2} \left[\delta \mathbf{g}_\alpha \cdot \mathbf{g}_\beta + \mathbf{g}_\alpha \cdot \delta \mathbf{g}_\beta \right] \tag{95}$$

so that one obtains

$$\begin{bmatrix} \delta E_{11} \\ \delta E_{22} \\ 2\delta E_{12} \end{bmatrix} = \sum_{I=1}^{4} [\mathbf{B}_I^\varphi \delta\boldsymbol{\varphi}_I + \mathbf{B}_I^d \delta \mathbf{d}_I] + \sum_{J=1}^{2} \widetilde{\mathbf{B}}_J^m \delta\Gamma_J, \tag{96}$$

where

$$\mathbf{B}_I^\varphi = \begin{bmatrix} M_1^I \mathbf{g}_1^T \\ M_2^I \mathbf{g}_2^T \\ M_1^I \mathbf{g}_2^T + M_2^I \mathbf{g}_1^T \end{bmatrix} \quad \text{and} \quad \mathbf{B}_I^d = \zeta \begin{bmatrix} N_1^I \mathbf{g}_1^T \\ N_2^I \mathbf{g}_2^T \\ N_1^I \mathbf{g}_2^T + N_2^I \mathbf{g}_1^T \end{bmatrix} \tag{97}$$

and

$$\widetilde{\mathbf{B}}_J^m = \begin{bmatrix} a_1^J b_{11} & a_1^J b_{21} \\ a_2^J b_{12} & a_2^J b_{22} \\ a_1^J b_{12} + a_2^J b_{11} & a_1^J b_{22} + a_2^J b_{21} \end{bmatrix} \quad \text{with} \quad b_{\alpha\beta} = \boldsymbol{\varphi}_{0,\alpha} \cdot \mathbf{g}_\beta. \tag{98}$$

Contribution of the thickness strains. The variation of the discrete thickness strains is given by

$$\delta E_{33} = \delta \bar{E}_{33}^{\{0\}} + \zeta \delta E_{33}^{\{1\}}, \tag{99}$$

where the first term has been derived in (70) from the proposed assumed strain interpolation (68) so that

$$\delta \bar{E}_{33}^{\{0\}} = \sum_{I=1}^{4} N^I d_I^T \delta d_I. \tag{100}$$

The second term in (99) follows from the proposed EAS-extension in section 3.2 and can be written in the form

$$\delta E_{33}^{\{1\}} = 2 \tilde{\gamma} d \cdot \delta d + \|d\|^2 \delta \tilde{\gamma} \tag{101}$$

$$\delta E_{33}^{\{1\}} = \sum_{I=1}^{4} [2 \tilde{\gamma} N^I d^T] d_I + \|d\|^2 e^T \delta \Gamma_d. \tag{102}$$

As a result we obtain

$$\delta E_{33} = \sum_{I=1}^{4} b_I^{33} \delta d_I + \tilde{b}^{33} \delta \Gamma_d \tag{103}$$

with

$$b_I^{33} = N^I d_I^T + 2 \zeta \tilde{\gamma} N^I d^T \qquad \text{and} \qquad \tilde{b}^{33} = \zeta \|d\|^2 e^T. \tag{104}$$

Contribution of the transverse shear strains. Following Dvorkin & Bathe [17], the assumed strain interpolation of the constant transverse shear strains $\bar{E}_{\alpha 3}^{\{0\}}$ is based on the compatible transverse shear strains $E_{\alpha 3}^{\{0\}}$ collocated at the midpoints $Q = A, B, C, D$ of the element boundaries:

$$\begin{bmatrix} 2 \bar{E}_{13}^{\{0\}} \\ 2 \bar{E}_{23}^{\{0\}} \end{bmatrix} = \begin{bmatrix} (1 - \xi^2) E_{13B}^{\{0\}} + (1 + \xi^2) E_{13D}^{\{0\}} \\ (1 - \xi^1) E_{23A}^{\{0\}} + (1 + \xi^1) E_{23C}^{\{0\}} \end{bmatrix}, \tag{105}$$

where the compatible covariant transverse shear strains are given by

$$E_{\alpha 3}^{\{0\}} = \frac{1}{2} [\varphi_{,\alpha} \cdot d - \Phi_{,\alpha} \cdot D]. \tag{106}$$

The total transverse shear strains can be written in the form

$$E_{\alpha 3} = \bar{E}_{\alpha 3}^{\{0\}} + \zeta E_{\alpha 3}^{\{1\}} \qquad \text{with} \qquad E_{\alpha 3}^{\{1\}} = \frac{1}{2} [d \cdot d_{,\alpha} - D \cdot D_{,\alpha}]. \tag{107}$$

Variation of (107) yields

$$\begin{bmatrix} 2 \delta E_{13} \\ 2 \delta E_{23} \end{bmatrix} = \bar{B}_{\varphi}^S \delta \varphi_e + B_d^S \delta d_e, \tag{108}$$

with

$$\bar{B}_\varphi^S = \frac{1}{4}\left[\begin{array}{cccc} -(1-\xi^2)d_B^T & (1-\xi^2)d_B^T & (1+\xi^2)d_D^T & -(1+\xi^2)d_D^T \\ -(1-\xi^1)d_A^T & -(1+\xi^1)d_C^T & (1+\xi^1)d_C^T & (1-\xi^1)d_A^T \end{array}\right],$$

$$B_d^S = \bar{B}_d^S + \zeta B_{(\alpha3)}^{\{1\}}$$

$$\bar{B}_d^S = \frac{1}{4}\left[\begin{array}{cccc} (1-\xi^2)\varphi_{,1}^{B^T} & (1-\xi^2)\varphi_{,1}^{B^T} & (1+\xi^2)\varphi_{,1}^{D^T} & (1+\xi^2)\varphi_{,1}^{D^T} \\ (1-\xi^1)\varphi_{,2}^{A^T} & (1+\xi^1)\varphi_{,2}^{C^T} & (1+\xi^1)\varphi_{,2}^{C^T} & (1-\xi^1)\varphi_{,2}^{A^T} \end{array}\right],$$

$$B_{(\alpha3)I}^{\{1\}} = \left[\begin{array}{c} N^I d_{,1}^T + N_{,1}^I d^T \\ N^I d_{,2}^T + N_{,2}^I d^T \end{array}\right].$$

A.2. The Geometric Part of the Tangent Matrix

In this section we derive the geometric part of the tangent submatrices in (81)–(84). The contribution of the nodes I and K to the tangent submatrix K_{vv} in (81) can be written in the form

$$(K_{vv}^{geo})_{IK} = \left[\begin{array}{cc} K_{\varphi\varphi}^{geo} & K_{\varphi d}^{geo} \\ K_{d\varphi}^{geo} & K_{dd}^{geo} \end{array}\right]_{IK} \tag{109}$$

with

$$(K_{\varphi\varphi}^{geo})_{IK} = (K_{\varphi\varphi}^{\alpha\beta})_{IK}, \tag{110}$$

$$(K_{\varphi d}^{geo})_{IK} = (K_{\varphi d}^{\alpha\beta})_{IK} + (K_{\varphi d}^{\alpha3})_{IK}, \tag{111}$$

$$(K_{d\varphi}^{geo})_{IK} = (K_{d\varphi}^{\alpha\beta})_{IK} + (K_{d\varphi}^{\alpha3})_{IK}, \tag{112}$$

$$(K_{dd}^{geo})_{IK} = (K_{dd}^{\alpha\beta})_{IK} + (K_{dd}^{33})_{IK} + (K_{dd}^{\alpha3})_{IK}, \tag{113}$$

where the superscript $\alpha\beta$ denotes the contribution of the tangential strains. Corresponding to this, the superscripts $\alpha3$ and 33 denote the contribution of the transverse shear strains and the transverse normal strains, respectively.

Analogously, the tangent submatrix $K_{v\Gamma}$ in (82) is given by

$$(K_{v\Gamma}^{geo})_I = \left[\begin{array}{c} K_{\varphi\Gamma}^{geo} \\ K_{d\Gamma}^{geo} \end{array}\right]_I = \left[\begin{array}{ccc} 0_{3\times nt} & K_{\varphi\Gamma_1}^{\alpha\beta} & K_{\varphi\Gamma_2}^{\alpha\beta} \\ K_{d\Gamma_d}^{33} & K_{d\Gamma_1}^{\alpha\beta} & K_{d\Gamma_2}^{\alpha\beta} \end{array}\right]_I, \tag{114}$$

and the tangent submatrix $K_{\Gamma\Gamma}$ in (84) can be written in the form

$$K_{\Gamma\Gamma}^{geo} = \left[\begin{array}{ccc} 0_{nt\times nt} & 0_{nt\times 2} & 0_{nt\times 2} \\ 0_{2\times nt} & K_{\Gamma_1\Gamma_1}^{\alpha\beta} & K_{\Gamma_1\Gamma_2}^{\alpha\beta} \\ 0_{2\times nt} & K_{\Gamma_2\Gamma_1}^{\alpha\beta} & K_{\Gamma_2\Gamma_2}^{\alpha\beta} \end{array}\right]. \tag{115}$$

Next, we derive the submatrices in (110)–(115).

Contribution of the tangential strains. The contribution of the tangential strains to the geometric part of the tangent matrix is given by

$$\iint_\Box \int_{-\frac{h_0}{2}}^{\frac{h_0}{2}} \Delta\{\delta \mathbf{g}_\alpha \cdot \mathbf{g}_\beta\} \hat{S}^{\alpha\beta} \sqrt{G} \, d\zeta \, d\Box$$

$$= \iint_\Box \int_{-\frac{h_0}{2}}^{\frac{h_0}{2}} \{\delta \mathbf{g}_\alpha \cdot \Delta \mathbf{g}_\beta + \Delta\delta \mathbf{g}_\alpha \cdot \mathbf{g}_\beta\} \hat{S}^{\alpha\beta} \sqrt{G} \, d\zeta \, d\Box.$$

The second variation of the discrete covariant base vectors (94) is given by

$$\Delta\delta \mathbf{g}_\alpha = \sum_{I=1}^{4}\sum_{J=1}^{2} \delta\varphi_I N_{0,\gamma}^I a_\alpha^J \Delta\Gamma_J^\gamma + \sum_{M=1}^{2}\sum_{K=1}^{4} \delta\Gamma_M^\gamma a_\alpha^M N_{0,\gamma}^K \Delta\varphi_K. \tag{116}$$

Thus, we obtain the following contributions to the equations (110–113), (114) and (115):

$$(K_{vv}^{\alpha\beta})_{IK} = \iint_\Box \int_{-\frac{h_0}{2}}^{\frac{h_0}{2}} \begin{bmatrix} M_\alpha^I M_\beta^K \hat{S}^{\alpha\beta} \mathbf{1}_{3\times3} & \zeta M_\alpha^I N_{,\beta}^K \hat{S}^{\alpha\beta} \mathbf{1}_{3\times3} \\ \zeta N_{,\alpha}^I M_\beta^K \hat{S}^{\alpha\beta} \mathbf{1}_{3\times3} & \zeta^2 N_{,\alpha}^I N_{,\beta}^K \hat{S}^{\alpha\beta} \mathbf{1}_{3\times3} \end{bmatrix} \sqrt{G} \, d\zeta \, d\Box,$$

$$(K_{\varphi\Gamma_J}^{\alpha\beta})_I = \iint_\Box \int_{-\frac{h_0}{2}}^{\frac{h_0}{2}} (M_\alpha^I a_\beta^J \hat{S}^{\alpha\beta}) \sqrt{G} \, d\zeta \, d\Box \begin{bmatrix} \varphi_{0,1} & \varphi_{0,2} \end{bmatrix}$$

$$+ \iint_\Box \int_{-\frac{h_0}{2}}^{\frac{h_0}{2}} \begin{bmatrix} (a_\alpha^J \mathbf{g}_\beta \hat{S}^{\alpha\beta}) N_{0,1}^I & (a_\alpha^J \mathbf{g}_\beta \hat{S}^{\alpha\beta}) N_{0,2}^I \end{bmatrix} \sqrt{G} \, d\zeta \, d\Box,$$

$$(K_{d\Gamma_J}^{\alpha\beta})_I = \iint_\Box \int_{-\frac{h_0}{2}}^{\frac{h_0}{2}} \zeta (N_{,\alpha}^I a_\beta^J \hat{S}^{\alpha\beta}) \sqrt{G} \, d\zeta \, d\Box \begin{bmatrix} \varphi_{0,1} & \varphi_{0,2} \end{bmatrix},$$

$$K_{\Gamma_J\Gamma_M}^{\alpha\beta} = \iint_\Box \int_{-\frac{h_0}{2}}^{\frac{h_0}{2}} (a_\alpha^J a_\beta^M \hat{S}^{\alpha\beta}) \sqrt{G} \, d\zeta \, d\Box \begin{bmatrix} \varphi_{0,1}\cdot\varphi_{0,1} & \varphi_{0,1}\cdot\varphi_{0,2} \\ \varphi_{0,2}\cdot\varphi_{0,1} & \varphi_{0,2}\cdot\varphi_{0,2} \end{bmatrix}.$$

Contribution of the thickness strains. The contribution of the thickness strains to the geometric part of the tangent matrix is given by

$$\iint_\Box \int_{-\frac{h_0}{2}}^{\frac{h_0}{2}} \Delta\delta E_{33}\hat{S}^{33} \sqrt{G} \, d\zeta \, d\Box. \tag{117}$$

From (103) we obtain the following contributions to (113) and (114):

$$(K_{dd}^{33})_{IK} = \iint_\Box \int_{-\frac{h_0}{2}}^{\frac{h_0}{2}} N^I (\delta_I^K + 2\zeta\tilde{\gamma}N^K)\hat{S}^{33} \sqrt{G} \, d\zeta \, d\Box \mathbf{1}_{3\times3},$$

$$(K_{d\Gamma_d}^{33})_I = \iint_\Box \int_{-\frac{h_0}{2}}^{\frac{h_0}{2}} (2\zeta N^I \hat{S}^{33} d \otimes e)\sqrt{G} \, d\zeta \, d\Box.$$

Contribution of the Transverse Shear Strains. The contribution of the transverse shear strains to the geometric part of the tangent matrix is given by

$$\iint_\square \int_{-\frac{h_0}{2}}^{\frac{h_0}{2}} \Delta \delta E_{\alpha 3} \hat{S}^{\alpha 3} \sqrt{G} \, d\zeta \, d\square. \tag{118}$$

From (108) we obtain the following contributions to the equations (111–113):

$$\mathbf{K}_{\varphi d}^{\alpha 3} = \iint_\square \int_{-\frac{h_0}{2}}^{\frac{h_0}{2}} \begin{bmatrix} -(A+B)\mathbf{1} & -B\mathbf{1} & \mathbf{0} & -A\mathbf{1} \\ B\mathbf{1} & (B-C)\mathbf{1} & -C\mathbf{1} & \mathbf{0} \\ \mathbf{0} & C\mathbf{1} & (C+D)\mathbf{1} & D\mathbf{1} \\ A\mathbf{1} & \mathbf{0} & -D\mathbf{1} & (A-D)\mathbf{1} \end{bmatrix} \sqrt{G} \, d\zeta \, d\square,$$

$$A = \tfrac{1}{4}(1-\xi^1)\hat{S}^{23} \quad B = \tfrac{1}{4}(1-\xi^2)\hat{S}^{13} \quad C = \tfrac{1}{4}(1+\xi^1)\hat{S}^{23} \quad D = \tfrac{1}{4}(1+\xi^2)\hat{S}^{13},$$

$$\mathbf{K}_{d\varphi}^{\alpha 3} = \mathbf{K}_{\varphi d}^{\alpha 3^T},$$

$$(\mathbf{K}_{dd}^{\alpha 3})_{IK} = \iint_\square \int_{-\frac{h_0}{2}}^{\frac{h_0}{2}} \zeta [N^I N_{,\alpha}^K + N_{,\alpha}^I N^K] \hat{S}^{\alpha 3} \sqrt{G} \, d\zeta \, d\square \mathbf{1}_{3\times 3}.$$

A.3. Load Case Bending Moment

The numerical example 4.3 'roll-up of a clamped beam' requires the application of a bending moment at the end of the beam. Since formulations A and C do not incorporate any rotational degrees of freedom, we subject the shell to follower end loads to model this load case. In this Appendix g_i and g^i denote the compatible covariant and contravariant base vectors, respectively. We consider a stress distribution, linear over the shell thickness, acting along the edge $\xi^1 = $ constant of the current configuration. Accordingly, the first Piola-Kirchhoff stress tensor

$$\mathbf{P} = \frac{M}{I} \zeta \frac{g^1}{\|g^1\|} \otimes \mathbf{N} \quad \text{with} \quad g^1 = \frac{1}{\sqrt{8}} g_2 \times d \quad \text{and} \quad I = \frac{w\, h_0^3}{12} \tag{119}$$

leads to the external loading contribution to the weak form

$$G_{ext} = \int_{\partial B_{0t}} \delta x \cdot \mathbf{P} \mathbf{N} \, dA. \tag{120}$$

For the present example, the center-line of the initial configuration coincides with the $x-$ axis, i.e. $\mathbf{N} = e_1$, and the axis of rotation coincides with the $y-$ axis, i.e. $g_2/\|g_2\| = e_2$. Therefore

$$\frac{g^1}{\|g^1\|} = e_2 \times \frac{d}{\|d\|} \quad \text{and} \quad dA = dy\, d\zeta. \tag{121}$$

Applying shell kinematics leads to the contribution to the weak form of momentum balance

$$G_{ext} = \delta d^T M \left(e_2 \times \frac{d}{\|d\|} \right). \tag{122}$$

The linearization of equation (122) has to be taken into account in order to reach asymptotically quadratic rates of convergence for the Newton iteration solution procedure. One obtains the contribution to the tangent stiffness matrix

$$DG_{ext} \cdot \Delta d = \delta d^T M \frac{1}{\|d\|} \left(\widehat{e}_2 + \left[\frac{d}{\|d\|} \times e_2 \right] \otimes \frac{d}{\|d\|} \right) \Delta d, \tag{123}$$

where $\widehat{e}_2$ denotes the skew–symmetric matrix associated with the axial vector e_2, defined by the relation $\widehat{e}_2 a = e_2 \times a$ for any $a \in \mathbb{R}^3$.

References

[1] J.-C. Simo and F. Armero, Geometrically nonlinear enhanced strain mixed methods and the method of incompatible modes, *Int. J. Num. Meth. Eng.* **33** (1992) 1413–1449.

[2] E. Ramm, Geometrisch nichtlineare Elastostatik und finite Elemente, Habilitation, Bericht Nr. 76-2, Institut für Baustatik, Universität Stuttgart, (1976).

[3] T. J. R. Hughes and W. K. Liu, Nonlinear finite element analysis of shells: Part I. Three–dimensional shells, *Comp. Meth. Appl. Mech. Eng.* **26** (1981) 331–362.

[4] J.-C. Simo, D. D. Fox and M. S. Rifai, On a stress resultant geometrically exact shell model. Part III: Computational aspects of the nonlinear theory, *Comp. Meth. Appl. Mech. Eng.* **79** (1990) 21–70.

[5] F. Gruttmann, W. Wagner, P. Wriggers, A nonlinear quadrilateral shell element with drilling degrees of freedom, *Archive Appl. Mech.* **62** (1992) 474–486.

[6] Y. Basar, Y. Ding and W. B. Krätzig, Finite-rotatation shell elements via mixed formulation, *Comp. Mech.* **10** (1992) 289–306.

[7] J.-C. Simo, M. S. Rifai and D. D. Fox, On a stress resultant geometrically exact shell model. Part IV: Variable thickness shells with through-the-thickness stretching, *Comp. Meth. Appl. Mech. Eng.* **81** (1990) 91–126.

[8] N. Büchter and E. Ramm, 3d–extension of nonlinear shell equations based on the enhanced assumed strain concept, *Comp. Meth. Appl. Sci.* (1992) 55–62.

[9] N. Büchter, E. Ramm and D. Roehl, Three-dimensional extension of nonlinear shell formulations based on the enhanced assumed strain concept, *Int. J. Num. Meth. Eng.* **37** (1994) 2551–2568.

[10] P. Steinmann, P. Betsch and E. Stein, FE plane stress analysis incorporating arbitrary 3d large strain constitutive models, submitted to *Eng. Comp.* (1994).

[11] P. Betsch, F. Gruttmann and E. Stein, A 4–node finite shell element for the implementation of general hyperelastic 3D–elasticity at finite strains, to appear in *Comp. Meth. Appl. Mech. Eng.* (1994).

[12] C. Sansour, A theory and finite element formulation of shells at finite deformations including thickness change: Circumventing the use of a rotation tensor, *Arch. Appl. Mech.* **65** (1995) 194–216.

[13] A. Kühhorn and H. Schoop, A nonlinear theory for sandwich shells including the wrinkling phenomenon, *Arch. Appl. Mech.* **62** (1992) 413–427.

[14] H. Parisch, A continuum-based shell theory for non-linear applications, *Int. J. Num. Meth. Eng.* **38** (1995) 1855–1883.

[15] H. Schoop, A simple nonlinear flat element for large displacement structures, *Computers & Structures* **32** (1989) 379–385.

[16] J.-C. Simo and M. S. Rifai, A class of mixed assumed strain methods and the method of incompatible modes, *Int. J. Num. Meth. Eng.* **29** (1990) 1595–1638.

[17] E. N. Dvorkin and K.-J. Bathe, A continuum mechanics based four-node shell element for general nonlinear analysis, *Eng. Comput.* **1** (1984) 77–88.

[18] K. J. Bathe and E. N. Dvorkin, A four-node plate bending element based on Mindlin/Reissner plate theory and a mixed interpolation, *Int. J. Num. Meth. Eng.* **21** (1985) 367–383.

[19] N. Stander, A. Matzenmiller and E. Ramm, An assessment of assumed strain methods in finite rotation shell analysis, *Eng. Comput.* **6** (1989) 58–66.

[20] H. Parisch, An investigation of a finite rotation four-node assumed strain shell element, *Int. J. Num. Meth. Eng.* **31** (1991) 127–150.

[21] P. Betsch and E. Stein, An assumed strain approach avoiding artificial thickness straining for a nonlinear 4-node shell element, *Comm. Num. Meth. Eng.* **11** (1995) 899–909.

[22] J.-C. Simo and T. J. R. Hughes, On the variational foundations of assumed strain methods, *J. Appl. Mech.* **53** (1986) 51–54.

[23] B. D. Reddy and J.-C. Simo, Stability and convergence of a class of enhanced strain methods, preprint (1994).

[24] P. M. Naghdi, The theory of plates and shells, in *Handbuch der Physik*, Band VIa/2 (Springer, Berlin, 1972) 425–640.

[25] E. L. Wilson, R. L. Taylor, W. P. Doherty and J. Ghaboussi, Incompatible Displacement Models, Numerical and Computer Models in Structural Mechanics, eds. S. J. Fenves, N. Perrone, A. R. Robinson and W. C. Schnobrich. New York: Academic Press (1973) 43–57.

[26] J.-C. Simo, F. Armero and R. L. Taylor, Improved versions of assumed enhanced strain tri–linear elements for 3D finite deformation problems, *Comp. Meth. Appl. Mech. Eng.* **110** (1993) 359–386.

[27] A. E. Green and W. Zerna, *Theoretical Elasticity* (Oxford University Press, London, 1954).

[28] T. J. R. Hughes, *The Finite Element Method* (Prentice Hall, Englewood Cliffs, NJ, 1987).

[29] O. C. Zienkiewicz and R. L. Taylor, *The Finite Element Method, Volume 1, Basic Formulation and Linear Problems* (McGraw-Hill, London, 1989).

[30] C. Miehe, Aspects of the formulation and finite element implementation of large strain isotropic elasticity, *Int. J. Num. Meth. Eng.* **37** (1994) 1981–2004.

[31] J.-C. Simo, R. L. Taylor and K. S. Pister, Variational and projection methods for the volume constraint in finite deformation elasto-plasticity, *Comp. Meth. Appl. Mech. Eng.* **51** (1985) 177–208.

[32] R. H. MacNeal and R. L. Harder, A proposed standard set of problems to test finite element accuracy, *Finite Elements in Analysis and Design* **1** (1985) 3–20.

[33] J.-C. Simo, D. D. Fox and M. S. Rifai, On a stress resultant geometrically exact shell model. Part II: The linear theory; Computational aspects, *Comp. Meth. Appl. Mech. Eng.* **73** (1989) 53–92

[34] T. J. R. Hughes and E. Carnoy, Nonlinear finite element shell formulation accounting for large membrane strains, *Comp. Meth. Appl. Mech. Eng.* **39** (1983) 69–82.

[35] H. Parisch, Efficient non-linear finite element shell formulation involving large strains, *Eng. Comput.* **3** (1986) 121–128.

[36] K. Schweizerhof and E. Ramm, Displacement dependent pressure loads in nonlinear finite element analyses, *Comput. Struct.* **18** (1984) 1099–1114.

[37] F. Gruttmann and R. L. Taylor, Theory and finite element formulation of rubberlike membrane shells using principal stretches, *Int. J. Num. Meth. Eng.* **35** (1992) 1111–1126.

Multilayer Beams: A Geometrically Exact Formulation

L. Vu-Quoc, H. Deng, and I.K. Ebcioğlu

Aerospace Engineering, Mechanics & Engineering Science, University of Florida, Gainesville, FL 32611, USA

Received September 25, 1995; revised version accepted February 16, 1996
Communicated by Jerrold Marsden and Stephen Wiggins

Dedicated to the memory of Professor Juan-Carlos Simo,
whose early demise is a great loss for the
applied and computational mechanics community

Summary. We review and extend our recent work on a new theory of multilayer structures, with particular emphasis on sandwich beams/1-D plates. Both the formulation of the equations of motion in the general dynamic case and the computational formulation of the resulting nonlinear equations of equilibrium in the static case based on a Galerkin projection are presented. Finite rotations of the layer cross sections are allowed, with shear deformation accounted for in each layer. There is no restriction on the layer thickness; the number of layers can vary between one and three. The deformed profile of a beam cross section is continuous, piecewise linear, with a motion in 2-D space identical to that of a planar multibody system that consists of three rigid links connected by hinges. With the dynamics of this multi (rigid/flexible) body being referred directly to an inertial frame, the equations of motion are derived via the balance of (1) the rate of kinetic energy and the power of resultant contact (internal) forces/couples, and (2) the power of assigned (external) forces/couples. The present formulation offers a general method for analyzing the dynamic response of flexible multilayer structures undergoing large deformation and large overall motion. With the layers *not* required to have equal length, the formulation permits the analysis of an important class of multilayer structures with ply drop-off. For sandwich structures, an approximated theory with infinitesimal relative outer-layer rotations superimposed onto finite core-layer rotation is deduced from the general nonlinear equations in a consistent manner. The classical linear theory of sandwich beams/1-D plates is recovered upon a consistent linearization. Using finite element basis functions in the Galerkin projection, we provide extensive numerical examples to verify the theoretical formulation and to illustrate its versatility.

1. Introduction

We review here recent developments on the theory of geometrically exact multilayer beams, also applicable to 1-D plates, accounting for bending stiffness and shear deformation in all layers (see Vu-Quoc & Ebcioğlu [1995a]). Fully nonlinear equations of motion are presented first, followed by approximated equations of motion for two cases: (i) the case where the difference between the outer-layer rotations and the core-layer rotation are infinitesimal, (ii) the case where all rotations are infinitesimal. A Galerkin projection of the fully nonlinear governing equations of equilibrium of the new theory is also presented. Some additional results are provided.

In Vu-Quoc & Ebcioğlu [1995a], we assume that the cross section in the deformed beam is continuous and piecewise linear, admitting finite rotations in the layer cross section, with the continuity of the displacement field at the layer interfaces exactly enforced. Thus the motion of a typical cross section of a geometrically exact sandwich beam can be thought of as the motion of a planar *multibody system* consisting of three rigid links, connected to each other by hinges, and moving in a plane *without* any restriction on the magnitude of motion. The reader is referred to Kamman & Huston [1984] for details on the dynamic formulation of a chain of rigid links. To account for shear deformation, the cross section (a rigid link) in each layer remains straight, but not orthogonal to the deformed centroidal line of that layer (i.e., first-order shear deformation theory). The overall deformation of the sandwich beam can be described by the deformation of a reference layer (which is the core layer in the present finite element formulation), together with two outer-layer rotations.[1] Thus the principal unknowns in the formulation are two linear displacement components of the deformed centroidal line of the reference layer, and the three finite rotations of each layer.

The starting point for the development of the Galerkin projection of the nonlinear governing equations of motion is the weak form based on the balance of stress power of the sandwich beam, from which expressions of fully nonlinear strain measures are obtained (see Vu-Quoc & Ebcioğlu [1995a]). A linearization of this nonlinear weak form is performed for use in the solution for the kinematic quantities via Newton's method. The result of this linearization is an unsymmetric tangent operator, which is composed of a tangent geometric stiffness operator, a tangent material stiffness operator and a tangent inertia operator. The tangent geometric and material stiffness operators are symmetric as given in Vu-Quoc & Deng [1995c], whereas the tangent inertia operator is unsymmetric,[2] thus making the total dynamic tangent stiffness of the sandwich beam unsymmetric. However, focusing on the static case, we will discuss here the linearization of the equations of equilibrium only, i.e., the symmetric tangent geometric and material stiffness operators. Linear and quadratic finite element basis functions are chosen to form a basis for the Galerkin projection of the linearized equilibrium equations into a finite-dimensional space of trial solutions. The symmetric tangent stiffness

---

[1] In Vu-Quoc & Ebcioğlu [1996], the number of layers is arbitrary, and the reference layer with respect to which the deformation is described can be chosen to be any layer.

[2] The reader is referred to Vu-Quoc & Deng [1997b] and Vu-Quoc & Deng [1997a] for the expression for the unsymmetric tangent inertia operator.

matrix is evaluated using uniformly reduced integration in all layers to avoid shear locking.

Several numerical examples, including the roll-up maneuver of a sandwich beam with three identical layers as compared with the equivalent single-layer beams, are given to verify the correctness of the present formulation and to illustrate its salient features. In particular, we present the important case of multilayer beams with ply drop-off, as compared with the results obtained with single-layer beams with symmetric ply drop-off and the results obtained in Zinno & Barbero [1994]. Notice that there is no a priori assumption made on the distribution of mass over the layer cross section, and on material constants in each layer; we can then model an important class of ideal sandwich beams, which have a thick and soft core layer, and thin and hard facings, i.e., outer layers. These results are good compared with those obtained in Sayir & Koller [1986].

2. Kinematics of Deformation

Shown in Figure 1 is the profile of a sandwich structure with three layers: the bottom layer (1), the core layer (2), and the top layer (3). Each layer may have a different thickness. The thickness of layer (ℓ)—for $\ell = 1, 2, 3$—is denoted by $_{(\ell)}H := 2\,_{(\ell)}h$.

The material configuration is defined by means of the material basis vectors $\{\mathbf{E}_1, \mathbf{E}_2\}$, with the associated coordinates (X^1, X^2). The undeformed beam lies along the X^1 axis. The spatial configuration is defined by the basis vectors $\{\mathbf{e}_1, \mathbf{e}_2\}$, with $\mathbf{e}_3$ normal to the previous two vectors according to the right-hand rule. For convenience, we often choose to have $\mathbf{E}_\alpha \equiv \mathbf{e}_\alpha$ for $\alpha = 1, 2$. The time parameter is denoted by $t \in \mathbb{R}^+$, with $\mathbb{R}^+ := [0, +\infty[$. Let $\mathcal{L} =]0, L[$ denote the interval, with the axial coordinate X^1, covering the length of the undeformed beam, and $\widehat{\mathfrak{U}}$ the interval, with the transverse coordinate X^2, representing the cross section of the multilayer structure. Let $\widehat{\Omega} = \mathcal{L} \times \widehat{\mathfrak{U}} \subset \mathbb{R}^2$ denote the domain of the undeformed beam. We define $_{(\ell)}\mathfrak{U} :=]\,(_{(\ell)}Z - _{(\ell)}h),\ (_{(\ell)}Z + _{(\ell)}h)[$ as the cross section of layer (ℓ), where $_{(\ell)}Z$ designates the X^2-coordinate of the midpoint of layer (ℓ).[3] Setting the origin of coordinate X^2 to coincide with the midpoint of core layer (2) (see Figure 1), we have $_{(1)}Z = -\big(_{(1)}h + _{(2)}h\big)$, $_{(2)}Z = 0$, and $_{(3)}Z = _{(2)}h + _{(3)}h$; also $\widehat{\mathfrak{U}} = \bigcup_\ell \,_{(\ell)}\mathfrak{U} = \,]-\big(_{(1)}H + _{(2)}h\big),\ \big(_{(2)}h + _{(3)}H\big)[$. With the domain of layer (ℓ) denoted by $_{(\ell)}\Omega := \mathcal{L} \times _{(\ell)}\mathfrak{U}$, we have $\widehat{\Omega} = \bigcup_\ell \,_{(\ell)}\Omega$.

The deformation map for layer (ℓ), denoted by $_{(\ell)}\Phi\colon\ _{(\ell)}\Omega \times \mathbb{R}^+ \to \mathbb{R}^2$ is described as follows: A material point[4] $X^\alpha \mathbf{E}_\alpha$ in layer (ℓ) is mapped to a spatial point $_{(\ell)}\Phi(X^1, X^2, t) \in \mathbb{R}^2$ defined by

$$_{(\ell)}\Phi(X^1, X^2, t) := \,_{(\ell)}\Phi_0(X^1, t) + \big(X^2 - _{(\ell)}Z\big)\,_{(\ell)}\mathbf{t}_2(X^1, t), \qquad (2.1)$$

for $X^1 \in \mathcal{L}$, and $X^2 \in \,_{(\ell)}\mathfrak{U}$, and for $\ell = 1, 2, 3$. In (2.1), $_{(\ell)}\Phi_0\colon \mathcal{L} \times \mathbb{R}^+ \to \mathbb{R}^2$ is the deformation map of the centroidal line of layer (ℓ). Describing the orientation of the

---

[3] The midpoint of a layer cross section coincides with its centroid when the mass distribution is symmetric within the layer cross section.

[4] Summation convention on repeated indices applies, with Greek indices taking values in $\{1, 2\}$.

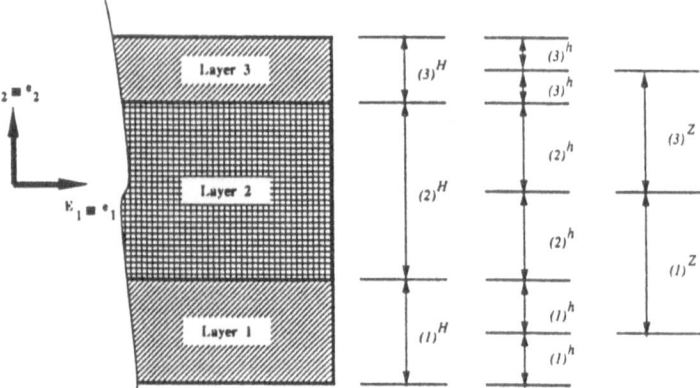

Fig. 1. Sandwich beam/1-D plate: Profile of a sandwich structure. The layer thicknesses and the position of the layer midpoints are indicated.

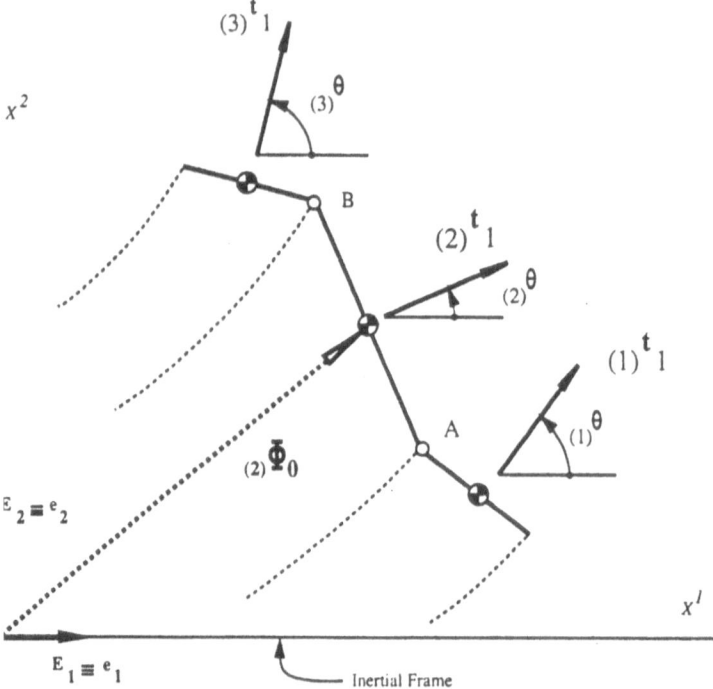

Fig. 2. Multibody dynamics. Motion of cross section as a chain of rigid links. For clarity, the basis vectors $_{(\ell)}\mathbf{t}_2$ for layers $\ell = 1, 2, 3$ are not drawn.

cross section of layer (ℓ) is a set of orthonormal basis vectors $\{ \ _{(\ell)}\mathbf{t}_1, \ _{(\ell)}\mathbf{t}_2 \}$ attached to that cross section, such that $_{(\ell)}\mathbf{t}_\alpha(X^1, 0) \equiv \mathbf{e}_\alpha$, for $\alpha = 1, 2$, at time $t = 0$.

The motion of the cross section is that of a multibody system consisting of a chain of three rigid links connected by hinges as shown in Figure 2, with $_{(\ell)}\Phi_0$ being the position of the centroid of link (layer) (ℓ). The following kinematic constraint equations express $_{(1)}\Phi_0$ and $_{(3)}\Phi_0$ in terms of $_{(2)}\Phi_0$, which is chosen as a principal unknown function

$$_{(1)}\Phi_0 = {}_{(2)}\Phi_0 - {}_{(1)}h \,{}_{(1)}\mathbf{t}_2 - {}_{(2)}h \,{}_{(2)}\mathbf{t}_2, \qquad (2.2)$$

$$_{(3)}\Phi_0 = {}_{(2)}\Phi_0 + {}_{(2)}h \,{}_{(2)}\mathbf{t}_2 + {}_{(3)}h \,{}_{(3)}\mathbf{t}_2. \qquad (2.3)$$

The continuity conditions for displacements are exactly satisfied through the use of (2.1)–(2.3).

Let $S \equiv X^1$, and $\mathbf{u}(S,t) = u^\alpha(S,t)\mathbf{e}_\alpha$ be the displacement vector of the material point $S \in \mathcal{L}$. Then the deformation map of the centroidal line of the core layer (2) takes the form

$$_{(2)}\Phi_0 := \left[S + u^1(S,t)\right]\mathbf{e}_1 + u^2(S,t)\mathbf{e}_2. \qquad (2.4)$$

Let $_{(\ell)}\theta(S,t)$ be the rotation angle of the cross section of layer (ℓ), then

$$_{(\ell)}\mathbf{t}_\alpha(S,t) = {}_{(\ell)}\Lambda(S,t)\,\mathbf{E}_\alpha, \qquad (2.5)$$

$$_{(\ell)}\Lambda = {}_{(\ell)}\Lambda^\alpha_\beta\,\mathbf{e}_\alpha \otimes \mathbf{E}^\beta, \quad \left[{}_{(\ell)}\Lambda^\alpha_\beta\right] := \begin{bmatrix} \cos\,_{(\ell)}\theta & -\sin\,_{(\ell)}\theta \\ \sin\,_{(\ell)}\theta & \cos\,_{(\ell)}\theta \end{bmatrix}, \qquad (2.6)$$

where $_{(\ell)}\Lambda$ is an orthogonal two-point tensor, whose matrix of components with respect to the basis $\mathbf{e}_\alpha \otimes \mathbf{E}^\beta$ is denoted by $\left[{}_{(\ell)}\Lambda^\alpha_\beta\right]$, where in $_{(\ell)}\Lambda^\alpha_\beta$, the superscript α denotes the row index and the subscript β the column index. It follows that $\mathbf{E}^\alpha \equiv \mathbf{E}_\alpha$ for Cartesian coordinates. Thus from (2.5), (2.6) we obtain $_{(\ell)}\mathbf{t}_\alpha = {}_{(\ell)}\Lambda^\beta_\alpha\mathbf{e}_\beta$ (see, e.g., Marsden & Hughes [1983], Vu-Quoc [1986]).

The five principal unknown functions of (S,t) for a three-layer sandwich structure are the following kinematic quantities: u^1, u^2, $_{(1)}\theta$, $_{(2)}\theta$, and $_{(3)}\theta$. Using (2.2), (2.3), and (2.4)–(2.6), we can relate the deformation maps $_{(\ell)}\Phi$ of the three layers $(\ell = 1, 2, 3)$ to the above five principal unknown functions.

3. Equations of Motion

3.1. *Power of Contact and Assigned Forces/Couples*

The mechanical power of contact forces/couples in a three-layer sandwich structure is given by

$$\mathcal{P}_c := \sum_{\ell=1}^{3} \int_{\mathcal{L}} \left[{}_{(\ell)}\mathbf{n}\cdot{}_{(\ell)}\dot{\gamma}^{[\ell]} + {}_{(\ell)}m\,{}_{(\ell)}\dot{\theta},_s\right]dS, \qquad (3.1)$$

where $_{(\ell)}\mathbf{n} = {}_{(\ell)}n^\alpha\mathbf{e}_\alpha$ is the resultant contact (internal) force for layer (ℓ), $_{(\ell)}m$ the resultant contact couple with respect to the centroid of the layer (ℓ) cross section, whereas $_{(\ell)}\gamma$ and $_{(\ell)}\theta,_s$ are the strain measures conjugated to $_{(\ell)}\mathbf{n}$ and $_{(\ell)}m$, respectively (see Simo & Vu-Quoc [1986a] for the single-layer case). In (3.1), the overhead dot "˙" designates time differentiation, whereas the overhead symbol "$^{[\ell]}$" designates the time differentiation of a vector quantity, say, $\mathbf{v}$, expressed in the cross-section basis $\{ _{(\ell)}\mathbf{t}_1, \ _{(\ell)}\mathbf{t}_2 \}$ of layer (ℓ) while keeping *this* basis *fixed* in the following sense:

$$_{(\ell)}\mathbf{t}_\alpha^{[\ell]} \equiv 0 \qquad \text{and} \qquad \mathbf{v} = v^\alpha \ _{(\ell)}\mathbf{t}_\alpha \Longrightarrow \mathbf{v}^{[\ell]} := \dot{v}^\alpha \ _{(\ell)}\mathbf{t}_\alpha. \qquad (3.2)$$

Using (3.2), the objective rates of the spatial strains for each layer $\ell = 1, 2, 3$ are given by

$$_{(\ell)}\gamma := \frac{\partial \ _{(\ell)}\Phi_0}{\partial S} - \ _{(\ell)}\mathbf{t}_1 \quad \Longrightarrow \quad _{(\ell)}\gamma^{[\ell]} = \left(\frac{\partial \ _{(\ell)}\Phi_0}{\partial S} \right)^{[\ell]} =: \ _{(\ell)}\Phi_{0,S}^{[\ell]}. \qquad (3.3)$$

We refer to Vu-Quoc & Ebcioğlu [1995a] for the detailed expressions for $_{(\ell)}\gamma^{[\ell]}$, for $\ell = 1, 2, 3$, and for expressions of the power of contact forces and contact couples $\mathcal{P}_c$.

Let $\widetilde{\overline{\mathbf{n}}}(S, t)$ be the resultant assigned forces for all layers, and $_{(\ell)}\mathcal{M}(S, t)$ the resultant assigned couple associated with layer (ℓ). The power due to assigned forces/couples for a three-layer sandwich structure is given by

$$\mathcal{P}_a := \int_{\mathfrak{L}} \left\{ \widetilde{\overline{\mathbf{n}}} \cdot \dot{\mathbf{u}} + \sum_{\ell=1}^{3} {}_{(\ell)}\overline{\mathcal{M}} \ _{(\ell)}\dot{\theta} \right\} dS + \left[\widetilde{\overline{\overline{n}}} \cdot \dot{\mathbf{u}} + \sum_{\ell=1}^{3} {}_{(\ell)}\overline{\overline{\mathcal{M}}} \ _{(\ell)}\dot{\theta} \right]_{S=0}^{S=L}. \qquad (3.4)$$

3.2. Rate of Kinetic Energy

Let $_{(\ell)}\rho$ denote the mass per unit volume in layer (ℓ).[5] Then the kinetic energy $\mathcal{K}$ and its rate are

$$\mathcal{K} := \frac{1}{2} \sum_{\ell=1}^{3} \int_{_{(\ell)}\Omega} {}_{(\ell)}\rho \ _{(\ell)}\dot{\Phi} \cdot \ _{(\ell)}\dot{\Phi} \, d\Omega \quad \Longrightarrow \quad \frac{d}{dt}\mathcal{K} = \sum_{\ell=1}^{3} \int_{_{(\ell)}\Omega} {}_{(\ell)}\rho \ _{(\ell)}\ddot{\Phi} \cdot \ _{(\ell)}\dot{\Phi} \, d\Omega.$$
$$(3.5)$$

Using (2.1) and the kinematic constraint equations (2.2)–(2.3), we can express the deformation maps $_{(\ell)}\Phi$ (for $\ell = 1, 2, 3$) in terms of the principal unknown functions $_{(2)}\Phi_0$, $_{(1)}\theta$, $_{(2)}\theta$, $_{(3)}\theta$. One can then derive the expression for the rate of the kinetic energy in terms of the rates of the five principal unknown functions as follows (Vu-Quoc & Ebcioğlu [1995a]):

$$\frac{d}{dt}\mathcal{K} = \int_{\mathfrak{L}} \left[\mathbf{f} \cdot \dot{\mathbf{u}} + \sum_{\ell=1}^{3} {}_{(\ell)}C \ _{(\ell)}\dot{\theta} \right] dS, \qquad (3.6)$$

where the inertia force and the inertia couples $_{(\ell)}C$, for $\ell = 1, 2, 3$, can be expressed in terms of the five principal unknown functions u^1, u^2, $_{(1)}\theta$, $_{(2)}\theta$ and $_{(3)}\theta$ (see Vu-Quoc & Ebcioğlu [1995a]).

---

[5] The mass density $_{(\ell)}\rho$ is assumed constant in time.

3.3. Balance of Power: Equations of Motion

To derive the equations of motion, we employ the balance of (i) the power of assigned forces/couples and (ii) the combined rate of kinetic energy and power of contact forces/couples

$$\frac{d}{dt}\mathcal{K} + \mathcal{P}_c = \mathcal{P}_a, \tag{3.7}$$

for all admissible rates $\dot{\mathbf{u}}$, $_{(1)}\dot{\theta}$, $_{(2)}\dot{\theta}$, $_{(3)}\dot{\theta}$, to obtain (Vu-Quoc & Ebcioğlu [1995a])

$$\widehat{\mathbf{n}}_{,S} + \overline{\widehat{\mathbf{n}}} = \mathbf{f}, \tag{3.8}$$

$$_{(1)}m_{,S} + \big[\,_{(2)}\Phi_{0,S} \times {}_{(1)}\mathbf{n}\big] \cdot \mathbf{e}_3 + {}_{(1)}h\big(\,_{(1)}\mathbf{n} \cdot {}_{(1)}\mathbf{t}_1\big)_{,S}$$
$$+ {}_{(2)}h \,_{(2)}\theta_{,S}\big(\,_{(1)}\mathbf{n} \cdot {}_{(2)}\mathbf{t}_2\big) + {}_{(1)}\overline{\mathcal{M}} = {}_{(1)}C, \tag{3.9}$$

$$_{(2)}m_{,S} + \big[\,_{(2)}\Phi_{0,S} \times {}_{(2)}\mathbf{n}\big] \cdot \mathbf{e}_3 + {}_{(2)}h\big(\,_{(1)}\mathbf{n}_{,S} - {}_{(3)}\mathbf{n}_{,S}\big) \cdot {}_{(2)}\mathbf{t}_1 + {}_{(2)}\overline{\mathcal{M}} = {}_{(2)}C, \tag{3.10}$$

$$_{(3)}m_{,S} + \big[\,_{(2)}\Phi_{0,S} \times {}_{(3)}\mathbf{n}\big] \cdot \mathbf{e}_3 - {}_{(2)}h \,_{(2)}\theta_{,S}\big(\,_{(3)}\mathbf{n} \cdot {}_{(2)}\mathbf{t}_2\big)$$
$$- {}_{(3)}h\big(\,_{(3)}\mathbf{n} \cdot {}_{(3)}\mathbf{t}_1\big)_{,S} + {}_{(3)}\overline{\mathcal{M}} = {}_{(3)}C, \tag{3.11}$$

and the boundary conditions at $S = 0$ and $S = L$ are

$$\widehat{\mathbf{n}} = \overline{\overline{\widehat{n}}}, \tag{3.12}$$

$$_{(1)}m + {}_{(1)}h\big(\,_{(1)}\mathbf{n} \cdot {}_{(1)}\mathbf{t}_1\big) = {}_{(1)}\overline{\overline{\mathcal{M}}}, \tag{3.13}$$

$$_{(2)}m + {}_{(2)}h\big(\,_{(1)}\mathbf{n} - {}_{(3)}\mathbf{n}\big) \cdot {}_{(2)}\mathbf{t}_1 = {}_{(2)}\overline{\overline{\mathcal{M}}}, \tag{3.14}$$

$$_{(3)}m - {}_{(3)}h\big(\,_{(3)}\mathbf{n} \cdot {}_{(3)}\mathbf{t}_1\big) = {}_{(3)}\overline{\overline{\mathcal{M}}}. \tag{3.15}$$

We refer the reader to Vu-Quoc & Ebcioğlu [1995a] for a geometric interpretation of $_{(\ell)}\overline{\overline{\mathcal{M}}}$, for $\ell = 1, 2, 3$.

Remark 3.1. Global balance of angular momentum equation. The third term in (3.11) can be shown to have an equivalent form (Vu-Quoc & Ebcioğlu [1995a])

$$\big(\,_{(1)}\mathbf{n}_{,S} - {}_{(3)}\mathbf{n}_{,S}\big) \cdot {}_{(2)}\mathbf{t}_1 = \big[\big(\,_{(1)}\mathbf{n} - {}_{(3)}\mathbf{n}\big) \cdot {}_{(2)}\mathbf{t}_1\big]_{,S}$$
$$- {}_{(2)}\theta_{,S}\big(\,_{(1)}\mathbf{n} - {}_{(3)}\mathbf{n}\big) \cdot {}_{(2)}\mathbf{t}_2. \tag{3.16}$$

Equations (3.13)–(3.16) then suggest the following definition of moments, for $S \in \mathcal{L}$,

$$_{(1)}\mathcal{M} := {}_{(1)}m + {}_{(1)}h\big(\,_{(1)}\mathbf{n} \cdot {}_{(1)}\mathbf{t}_1\big), \tag{3.17}$$

$$_{(2)}\mathcal{M} := {}_{(2)}m + {}_{(2)}h\big(\,_{(1)}\mathbf{n} - {}_{(3)}\mathbf{n}\big) \cdot {}_{(2)}\mathbf{t}_1, \tag{3.18}$$

$$_{(3)}\mathcal{M} := {}_{(3)}m - {}_{(3)}h\big(\,_{(3)}\mathbf{n} \cdot {}_{(3)}\mathbf{t}_1\big), \tag{3.19}$$

which in turn lead to an alternative form of the moment equations (3.10), (3.11):

$$_{(1)}\mathcal{M}_{,S} + \big[\,_{(2)}\Phi_{0,S} \times {}_{(1)}\mathbf{n}\big] \cdot \mathbf{e}_3 + {}_{(2)}h \,_{(2)}\theta_{,S}\big(\,_{(1)}\mathbf{n} \cdot {}_{(2)}\mathbf{t}_2\big) + {}_{(1)}\overline{\mathcal{M}} = {}_{(1)}C, \tag{3.20}$$

$$_{(2)}\mathcal{M}_{,S} + \big[\,_{(2)}\Phi_{0,S} \times {}_{(2)}\mathbf{n}\big] \cdot \mathbf{e}_3 - {}_{(2)}h \,_{(2)}\theta_{,S}\big(\,_{(1)}\mathbf{n} - {}_{(3)}\mathbf{n}\big) \cdot {}_{(2)}\mathbf{t}_2 + {}_{(2)}\overline{\mathcal{M}} = {}_{(2)}C, \tag{3.21}$$

$$_{(3)}\mathcal{M},_S + \left[{}_{(2)}\Phi_0,_S \times {}_{(3)}\mathbf{n}\right] \cdot \mathbf{e}_3 - {}_{(2)}h\ {}_{(2)}\theta,_S \left({}_{(3)}\mathbf{n}\cdot {}_{(2)}\mathbf{t}_2\right) + {}_{(3)}\overline{\mathcal{M}} = {}_{(3)}C, \quad (3.22)$$

where the assigned moments ${}_{(\ell)}\overline{\mathcal{M}}$ had been introduced in (3.4), and boundary conditions (3.13)–(3.15) at $S = 0$ and $S = L$

$$_{(\ell)}\mathcal{M} = {}_{(\ell)}\overline{\overline{\mathcal{M}}}, \qquad \text{for } \ell = 1, 2, 3. \tag{3.23}$$

Finally, we note that by introducing the definition of the *total* resultant couple over the whole beam cross section

$$\widehat{m} := \sum_{\ell=1}^{3} {}_{(\ell)}\mathcal{M}, \qquad \widehat{\overline{m}} := \sum_{\ell=1}^{3} {}_{(\ell)}\overline{\mathcal{M}}, \tag{3.24}$$

and then by summing (3.20)–(3.22), we obtain the global balance of angular momentum equation

$$\widehat{m},_S + \left[{}_{(2)}\Phi_0,_S \times \widehat{\mathbf{n}}\right] \cdot \mathbf{e}_3 + \widehat{\overline{m}} = \widehat{C}, \tag{3.25}$$

where

$$\widehat{C} := \sum_{\ell=1}^{3} {}_{(\ell)}C. \tag{3.26}$$

In (3.25), we find the same mathematical structure as in the balance of angular momentum equation for a single-layer beam (see Simo & Vu-Quoc [1986a]). We also refer the reader to Vu-Quoc & Ebcioğlu [1995a] for an alternative derivation of (3.25).

Remark 3.2. It should be emphasized that the present theory remains valid for multilayer structures with any number of layers between one and three. In addition, the present theory can accommodate the important case of multilayer structures with ply drop-off. These points are clearly discussed in Vu-Quoc & Deng [1995c], where we present the computational formulation and the numerical results for the present theory.

4. Constitutive Laws

We define the material contact force ${}_{(\ell)}\mathbf{N}$ and the material strain measure ${}_{(\ell)}\Gamma$ pertaining to layer (ℓ) as follows:

$$_{(\ell)}\mathbf{N} = {}_{(\ell)}N^\alpha \mathbf{E}_\alpha, \qquad {}_{(\ell)}\mathbf{N} := {}_{(\ell)}\Lambda^T\ {}_{(\ell)}\mathbf{n}, \tag{4.1}$$

$$_{(\ell)}\Gamma = {}_{(\ell)}\Gamma^\alpha \mathbf{E}_\alpha, \qquad {}_{(\ell)}\Gamma := {}_{(\ell)}\Lambda^T\ {}_{(\ell)}\gamma, \tag{4.2}$$

where the orthogonal two-point tensor ${}_{(\ell)}\Lambda$ had been defined in (2.5), (2.6), and ${}_{(\ell)}\mathbf{n}$ and ${}_{(\ell)}\gamma$ are the spatial contact force and spatial strain measure, respectively. For layer (ℓ), we employ the constitutive relations

$$_{(\ell)}\mathbf{N} = {}_{(\ell)}\mathbf{D}\ {}_{(\ell)}\Gamma, \qquad {}_{(\ell)}\mathbf{D} = {}_{(\ell)}D^\alpha_\beta \mathbf{E}_\alpha \otimes \mathbf{E}^\beta, \qquad {}_{(\ell)}m = {}_{(\ell)}D^3_3\ {}_{(\ell)}\theta,_S, \tag{4.3}$$

$$\left[_{(\ell)}D_j^i \right] := \begin{bmatrix} _{(\ell)}EA & 0 & 0 \\ 0 & _{(\ell)}GA_s & 0 \\ 0 & 0 & _{(\ell)}EI \end{bmatrix} \in \mathbb{R}^{3\times3}, \qquad (4.4)$$

or

$$\left[_{(\ell)}D_j^i \right] := \begin{bmatrix} \dfrac{2\,_{(\ell)}E\,_{(\ell)}h}{1 - _{(\ell)}v^2} & 0 & 0 \\ 0 & _{(\ell)}GA_s & 0 \\ 0 & 0 & \dfrac{2\,_{(\ell)}E\,_{(\ell)}h^3}{3(1 - _{(\ell)}v^2)} \end{bmatrix} \in \mathbb{R}^{3\times3}, \qquad (4.5)$$

where, for beams, $_{(\ell)}EA$, $_{(\ell)}GA_s$, and $_{(\ell)}EI$ denote the extensional stiffness, shear stiffness, and bending stiffness of layer (ℓ), respectively, whereas, for plates, $_{(\ell)}E$ and $_{(\ell)}v$ are Young's modulus and Poisson's ratio of layer ℓ.[6] The effects of stretching, shearing, and bending are uncoupled in the above constitutive relations (4.3)–(4.5) as a result of choosing $_{(\ell)}\Phi_0$ to be the deformation map of the centroidal line of layer (ℓ).[7] Of particular usefulness is the relation between the spatial contact force $_{(\ell)}\mathbf{n}$ and the five principal unknown functions u^1, u^2, $_{(1)}\theta$, $_{(2)}\theta$, and $_{(3)}\theta$ given by

$$_{(\ell)}\mathbf{n} = \,_{(\ell)}\mathbf{\Lambda}\,_{(\ell)}\mathbf{D}\,_{(\ell)}\mathbf{\Lambda}^T\,_{(\ell)}\boldsymbol{\gamma}, \qquad (4.6)$$

which is obtained from (4.1)–(4.3). The expressions for the spatial strain measures $_{(\ell)}\boldsymbol{\gamma}$, for $\ell = 1, 2, 3$, in terms of the five principal unknown functions can be obtained easily, and are given in Vu-Quoc & Ebcioğlu [1995a].

5. Approximated Equations of Motion

5.1. Infinitesimal Relative Rotations in Outer Layers

In this section, the rotation $_{(2)}\theta$ of the core layer (2) remains *finite*, but the rotations of the outer layers (1) and (3) are assumed to be small relative to the core rotation $_{(2)}\theta$, i.e.,

$$| \psi_{2\ell} | = | _{(2)}\theta - _{(\ell)}\theta | \ll 1, \qquad \cos\psi_{2\ell} \approx 1, \qquad \sin\psi_{2\ell} \approx \psi_{2\ell}. \qquad (5.1)$$

Also we shall assume that

$$_{(\ell)}\theta_{,S} \approx \,_{(2)}\theta_{,S}, \qquad _{(\ell)}\dot\theta \approx \,_{(2)}\dot\theta, \qquad _{(\ell)}\ddot\theta \approx \,_{(2)}\ddot\theta. \qquad (5.2)$$

The five principal unknown functions are now u^1, u^2, $_{(2)}\theta$, ψ_{21} and ψ_{23}. The approximated constitutive laws (4.6) and (4.3)–(4.5) now take the following form (see Vu-Quoc & Ebcioğlu [1995a]):

$$_{(\ell)}\mathbf{n} \approx \left[\mathfrak{I} - \check\psi_{2\ell} \right] _{(2)}\mathbf{\Lambda}\,_{(\ell)}\mathbf{D}\,_{(\ell)}\tilde{\mathbf{\Gamma}}, \qquad (5.3)$$

---

[6] See, e.g., Chadwick [1976] or Marsden & Hughes [1983] for the tensor notation and rules of calculation. The matrix $\left[_{(\ell)}D_j^i \right] \in \mathbb{R}^{3\times3}$ in (4.4), (4.5) has the representative element $_{(\ell)}D_j^i$ with the upper index i designating the row index, and the lower index j the column index.

[7] That is, we are restricting our discussion to the case where the centroid of a layer cross section coincides with the midpoint of that cross section.

$$_{(\ell)}m \approx {}_{(\ell)}EI \; _{(2)}\theta,_S , \tag{5.4}$$

where $\mathfrak{I}$ designates the identity two-tensor $\mathfrak{I} := \delta_\beta^\alpha \mathbf{e}_\alpha \otimes \mathbf{e}^\beta$, with δ_β^α being the Kronecker delta, and

$$\check{\psi}_{2\ell} = \left\{ \check{\psi}_{2\ell} \right\}_\beta^\alpha \mathbf{e}_\alpha \otimes \mathbf{e}^\beta, \qquad \left[\left\{ \check{\psi}_{2\ell} \right\}_\beta^\alpha \right] := \begin{bmatrix} 0 & -\psi_{2\ell} \\ \psi_{2\ell} & 0 \end{bmatrix}. \tag{5.5}$$

Relations (5.3), (5.4) are intended for the outer layers (1) and (3). If one introduces, however, the identity $_{(2)}\widetilde{\Gamma} \equiv {}_{(2)}\Gamma$, then expressions (5.3), (5.4) are also valid for the core layer (2) since $\psi_{22} = 0$; hence, the approximation in (5.3), (5.4) becomes an exact equality for the core layer (2). The approximated stiffness operators (left-hand sides) of the equations of motion (3.9)–(3.11) can be obtained in a consistent[8] manner using the above approximations for the layer resultant forces/couples.

The inertia force $\mathbf{f}$ and the inertia couples in (3.6) can be expressed completely in terms of the five unknown functions $u^1, u^2, {}_{(2)}\theta, \psi_{21}$ and ψ_{23} (see Vu-Quoc & Ebcioğlu [1995a]).

5.2. All Infinitesimal Rotations: Linearized Equations

In this section, all rotation angles are assumed small, such that $| \; _{(\ell)}\theta \; | \ll 1, \cos {}_{(\ell)}\theta \approx 1$, and $\sin {}_{(\ell)}\theta \approx {}_{(\ell)}\theta$, for $\ell = 1, 2, 3$. The five principal unknown functions are the usual $u^1, u^2, {}_{(1)}\theta, {}_{(2)}\theta$ and $_{(3)}\theta$. We will neglect all nonlinear terms and retain only the linear terms in a consistent manner. It follows from (4.6) that

$$_{(\ell)}\mathbf{n} \approx {}_{(\ell)}\mathbf{D} \; _{(\ell)}\widetilde{\widetilde{\Gamma}}, \tag{5.6}$$

where

$$_{(1)}\widetilde{\widetilde{\Gamma}} := \left[u^1,_S + {}_{(1)}h \; _{(1)}\theta,_S + {}_{(2)}h \; _{(2)}\theta,_S \right] \mathbf{E}_1 + \left[- {}_{(1)}\theta + u^2,_S \right] \mathbf{E}_2, \tag{5.7}$$

$$_{(2)}\widetilde{\widetilde{\Gamma}} := u^1,_S \mathbf{E}_1 + \left[- {}_{(2)}\theta + u^2,_S \right] \mathbf{E}_2, \tag{5.8}$$

$$_{(3)}\widetilde{\widetilde{\Gamma}} := \left[u^1,_S - {}_{(2)}h \; _{(2)}\theta,_S - {}_{(3)}h \; _{(3)}\theta,_S \right] \mathbf{E}_1 + \left[- {}_{(3)}\theta + u^2,_S \right] \mathbf{E}_2. \tag{5.9}$$

The constitutive law for the moment $_{(\ell)}m$ remains identical to (4.3)$_3$.

The linearization of the inertia force $\mathbf{f}$, which we omit here, can be obtained easily. The linearized equation of balance of linear momentum has a form identical to that in (3.9).

Next, consider the equations for balance of angular momentum (3.10), (3.11). Using the following approximations,

$$\left[_{(2)}\Phi_0,_S \times {}_{(\ell)}\mathbf{n} \right] \approx {}_{(\ell)}n^2, \tag{5.10}$$

$$\left(_{(\ell)}\mathbf{n} \cdot {}_{(\ell)}\mathbf{t}_1 \right),_S \approx {}_{(\ell)}n^1,_S , \tag{5.11}$$

$$\left(_{(1)}\mathbf{n},_S - {}_{(3)}\mathbf{n},_S \right) \cdot {}_{(2)}\mathbf{t}_1 \approx {}_{(1)}n^1,_S - {}_{(3)}n^1,_S , \tag{5.12}$$

---

[8] We refer to Vu-Quoc & Ebcioğlu [1995a] for a discussion on the consistent linearization of the nonlinear quantities involved.

and neglecting all nonlinear terms in the stiffness operators, we obtain the linearized equations of balance of angular momentum

$$_{(1)}m_{,s} + {}_{(1)}n^2 + {}_{(1)}h \, {}_{(1)}n^1_{,s} + {}_{(1)}\overline{\mathcal{M}} = {}_{(1)}\widetilde{\widetilde{C}}, \tag{5.13}$$

$$_{(2)}m_{,s} + {}_{(2)}n^2 + {}_{(2)}h \left({}_{(1)}n^1_{,s} - {}_{(3)}n^1_{,s} \right) + {}_{(2)}\overline{\mathcal{M}} = {}_{(2)}\widetilde{\widetilde{C}}, \tag{5.14}$$

$$_{(3)}m_{,s} + {}_{(3)}n^2 - {}_{(3)}h \, {}_{(3)}n^1_{,s} + {}_{(3)}\overline{\mathcal{M}} = {}_{(3)}\widetilde{\widetilde{C}}. \tag{5.15}$$

The reader is referred to Vu-Quoc & Ebcioğlu [1995a] for the expressions of the linearized inertia couples ${}_{(\ell)}\widetilde{\widetilde{C}}$.

6. Nonlinear Weak Form of Equilibrium Equations

6.1. Strong Form

In the present paper, we assume that the material basis vectors $\{\mathbf{E}_\alpha\}$ coincide with the spatial basis vectors $\{\mathbf{e}_\alpha\}$; i.e., $\mathbf{E}_\alpha \equiv \mathbf{e}_\alpha$, for $\alpha = 1, 2$. The reader is referred to Vu-Quoc & Deng [1995c] and to Vu-Quoc [1995] for the more general formulation where the above restrictive assumption is not made. We recall below the equation of equilibrium of geometrically exact sandwich beams as obtained from (3.9)–(3.11):

$$\widehat{\mathbf{n}}_{,s} + \widetilde{\overline{\mathbf{n}}} = 0,$$

$$_{(1)}m_{,s} + \left[{}_{(2)}\Phi_{0,s} \times {}_{(1)}\mathbf{n} \right] \cdot \mathbf{e}_3 + {}_{(1)}h \left({}_{(1)}\mathbf{n} \cdot {}_{(1)}\mathbf{t}_1 \right)_{,s}$$
$$+ {}_{(2)}h \, {}_{(2)}\theta_{,s} \left({}_{(1)}\mathbf{n} \cdot {}_{(2)}\mathbf{t}_2 \right) + {}_{(1)}\overline{\mathcal{M}} = 0,$$

$$_{(2)}m_{,s} + \left[{}_{(2)}\Phi_{0,s} \times {}_{(2)}\mathbf{n} \right] \cdot \mathbf{e}_3 + {}_{(2)}h \left({}_{(1)}\mathbf{n}_{,s} - {}_{(3)}\mathbf{n}_{,s} \right) \cdot {}_{(2)}\mathbf{t}_1 + {}_{(2)}\overline{\mathcal{M}} = 0,$$

$$_{(3)}m_{,s} + \left[{}_{(2)}\Phi_{0,s} \times {}_{(3)}\mathbf{n} \right] \cdot \mathbf{e}_3 - {}_{(3)}h \left({}_{(3)}\mathbf{n} \cdot {}_{(3)}\mathbf{t}_1 \right)_{,s}$$
$$- {}_{(2)}h \, {}_{(2)}\theta_{,s} \left({}_{(3)}\mathbf{n} \cdot {}_{(2)}\mathbf{t}_2 \right) + {}_{(3)}\overline{\mathcal{M}} = 0.$$

$$\tag{6.1}$$

Let $\{\overline{\mathbf{u}}, {}_{(1)}\overline{\theta}, {}_{(2)}\overline{\theta}, {}_{(3)}\overline{\theta}\}$ represent the assigned displacement and the assigned rotations, respectively. The boundary conditions at $S = 0$ and $S = L$ are then stated as follows:

$$
\begin{array}{llllll}
\text{Either} & \mathbf{u} &= \overline{\mathbf{u}} & \text{or} & \widehat{\mathbf{n}} &= \widetilde{\overline{\mathbf{n}}}, \\
\text{either} & {}_{(1)}\theta &= {}_{(1)}\overline{\theta} & \text{or} & {}_{(1)}m + {}_{(1)}h \left({}_{(1)}\mathbf{n} \cdot {}_{(1)}\mathbf{t}_1 \right) &= {}_{(1)}\overline{\mathcal{M}}, \\
\text{either} & {}_{(2)}\theta &= {}_{(2)}\overline{\theta} & \text{or} & {}_{(2)}m + {}_{(2)}h \left({}_{(1)}\mathbf{n} - {}_{(3)}\mathbf{n} \right) \cdot {}_{(2)}\mathbf{t}_1 &= {}_{(2)}\overline{\mathcal{M}}, \\
\text{either} & {}_{(3)}\theta &= {}_{(3)}\overline{\theta} & \text{or} & {}_{(3)}m - {}_{(3)}h \left({}_{(3)}\mathbf{n} \cdot {}_{(3)}\mathbf{t}_1 \right) &= {}_{(3)}\overline{\mathcal{M}}.
\end{array}
\tag{6.2}
$$

6.2. Weak Form

We now develop the operator expressions for the weak form of the equilibrium equations based on the balance of power in Section 3.1. These expressions will be linearized and subsequently used in a Galerkin projection for the solution process.

For a sandwich beam, let the five unknown kinematic quantities be gathered in $\Phi \colon \mathcal{L} \to \mathbb{R}^5$ such that

$$\Phi(S) := \left\{ u^1, \ u^2, \ _{(1)}\theta, \ _{(2)}\theta, \ _{(3)}\theta \right\}^T \Big|_{(S)}, \tag{6.3}$$

and let $\mathbf{w} \colon \mathcal{L} \to \mathbb{R}^5$ be the weighting functions corresponding to the five kinematic quantities in (6.3)

$$\mathbf{w} := \left\{ v^1, \ v^2, \ _{(1)}\varphi, \ _{(2)}\varphi, \ _{(3)}\varphi \right\}^T, \tag{6.4}$$

instead of velocities in (3.1). Gathered in $_{(\ell)}\mathbf{f}_c \colon \mathcal{L} \times \mathbb{R}_+ \to \mathbb{R}^3$ such that

$$_{(\ell)}\mathbf{f}_c := \left\{ _{(\ell)}n_1, \ _{(\ell)}n_2, \ _{(\ell)}m \right\}^T, \tag{6.5}$$

are the contact forces/couple for layer (ℓ). Since the core layer (2) is the reference layer based on which the deformation of all layers is referred to, we first focus our attention on layer (2). Based on the expression for $_{(\ell)}\mathcal{P}_c$ in (3.1)$_2$, the weak form for the contact forces/couple in layer (2) can be written as

$$_{(2)}G_c(\Phi, \mathbf{w}) := \int_{\mathfrak{L}} \left[_{(2)}\mathbf{n} \cdot {}_{(2)}\gamma^{[2]} + {}_{(2)}m \ _{(2)}\varphi,_s \right] dS, \tag{6.6}$$

where, in terms of the weighting functions (6.4),

$$_{(2)}\gamma^{[2]} = \left[v^1,_s + u^2,_s \ _{(2)}\varphi \right] \mathbf{e}_1 + \left[v^2,_s - \left(1 + u^1,_s \right) {}_{(2)}\varphi \right] \mathbf{e}_2. \tag{6.7}$$

The integrand in (6.6) can be rearranged in matrix form as follows (Vu-Quoc & Deng [1995c]):

$$\boxed{_{(2)}G_c(\Phi, \mathbf{w}) = \int_{\mathfrak{L}} \left[_{(2)}\Xi(\Phi)\mathbf{w} \right] \cdot {}_{(2)}\mathbf{f}_c(\Phi) \ dS,} \tag{6.8}$$

where

$$_{(2)}\Xi(\Phi) := \begin{bmatrix} \dfrac{d}{dS} & 0 & 0 & _{(2)}\Xi_1 & 0 \\[2mm] 0 & \dfrac{d}{dS} & 0 & _{(2)}\Xi_2 & 0 \\[2mm] 0 & 0 & 0 & \dfrac{d}{dS} & 0 \end{bmatrix}_{3 \times 5}, \tag{6.9}$$

and

$$_{(2)}\Xi_1 := u^2,_S, \qquad _{(2)}\Xi_2 := -\left[1 + u^1,_S \right]. \tag{6.10}$$

The stress power $_{(\ell)}\mathcal{P}_c$ for layer (1) and layer (3) as given in (3.1) can be rewritten as a weak form

$$_{(\ell)}G_c(\Phi, \mathbf{w}) := \int_{\mathfrak{L}} \left[_{(1)}\mathbf{n} \cdot {_{(1)}\gamma^{[1]}} + {_{(1)}m} \, _{(1)}\varphi_{,s} \right] dS, \tag{6.11}$$

where, in terms of the weighting functions (6.4),

$$_{(1)}\gamma^{[1]} = \left[v^1_{,s} + u^2_{,s} \, _{(1)}\varphi \right] \mathbf{e}_1 + \left[v^2_{,s} - \left(1 + u^1_{,s} \right) {_{(1)}\varphi} \right] \mathbf{e}_2$$
$$+ {_{(1)}h} \, _{(1)}\varphi_{,s} \, _{(1)}\mathbf{t}_1 + {_{(2)}h} \, _{(2)}\varphi_{,s} \, _{(2)}\mathbf{t}_1 + {_{(2)}h} \, _{(2)}\theta_{,s} \left[_{(2)}\varphi - {_{(1)}\varphi} \right] _{(2)}\mathbf{t}_2, \tag{6.12}$$

$$_{(3)}\gamma^{[3]} = \left[v^1 + u^2_{,s} \, _{(3)}\varphi \right] \mathbf{e}_1 + \left[v^2 - \left(1 + u^1_{,s} \right) {_{(3)}\varphi} \right] \mathbf{e}_2$$
$$- {_{(3)}h} \, _{(3)}\varphi_{,s} \, _{(3)}\mathbf{t}_1 - {_{(2)}h} \, _{(2)}\varphi_{,s} \, _{(2)}\mathbf{t}_1 - {_{(2)}h} \, _{(2)}\theta_{,s} \left[_{(2)}\varphi - {_{(3)}\varphi} \right] _{(2)}\mathbf{t}_2 \tag{6.13}$$

are the expressions for the variation of the strain $_{(1)}\gamma$ and $_{(3)}\gamma$ by virtue of (3.2), (3.3), (2.2), (2.3) and (2.6). Rearranging (6.11) in matrix form with the aid of (6.5), (6.12) and (6.13), we obtain

$$_{(\ell)}G_c(\Phi, \mathbf{w}) = \int_{\mathfrak{L}} \left[_{(\ell)}\Xi(\Phi)\mathbf{w} \right] \cdot {_{(\ell)}\mathbf{f}_c}(\Phi) \, dS, \tag{6.14}$$

where

$$_{(1)}\Xi(\Phi) := \begin{bmatrix} \dfrac{d}{dS} & 0 & _{(1)}\Xi_1 & _{(1)}\Xi_3 & 0 \\[2mm] 0 & \dfrac{d}{dS} & _{(1)}\Xi_2 & _{(1)}\Xi_4 & 0 \\[2mm] 0 & 0 & \dfrac{d}{dS} & 0 & 0 \end{bmatrix}_{3\times 5}, \tag{6.15}$$

$$_{(3)}\Xi(\Phi) := \begin{bmatrix} \dfrac{d}{dS} & 0 & 0 & _{(3)}\Xi_3 & _{(3)}\Xi_1 \\[2mm] 0 & \dfrac{d}{dS} & 0 & _{(3)}\Xi_4 & _{(3)}\Xi_2 \\[2mm] 0 & 0 & 0 & 0 & \dfrac{d}{dS} \end{bmatrix}_{3\times 5}, \tag{6.16}$$

in which

$$_{(1)}\Xi_1 := \underbrace{+u^2_{,s}} + {_{(1)}h} \cos {_{(1)}\theta} \frac{d}{dS} + {_{(2)}h} \, _{(2)}\theta_{,s} \sin {_{(2)}\theta}, \tag{6.17}$$

$$_{(1)}\Xi_2 := \underbrace{-\left[1 + u^1_{,s} \right]} + {_{(1)}h} \sin {_{(1)}\theta} \frac{d}{dS} - {_{(2)}h} \, _{(2)}\theta_{,s} \cos {_{(2)}\theta}, \tag{6.18}$$

$$_{(1)}\Xi_3 := {_{(2)}h} \cos {_{(2)}\theta} \frac{d}{dS} - {_{(2)}h} \, _{(2)}\theta_{,s} \sin {_{(2)}\theta}, \tag{6.19}$$

$$_{(1)}\Xi_4 := {_{(2)}h} \sin {_{(2)}\theta} \frac{d}{dS} + {_{(2)}h} \, _{(2)}\theta_{,s} \cos {_{(2)}\theta}, \tag{6.20}$$

$$_{(3)}\Xi_1 := \underbrace{+u^2,_S} - _{(2)}h \, _{(2)}\theta,_S \sin \, _{(2)}\theta - _{(3)}h \cos \, _{(3)}\theta \frac{d}{dS}, \tag{6.21}$$

$$_{(3)}\Xi_2 := \underbrace{-\left[1 + u^1,_S\right]} + _{(2)}h \, _{(2)}\theta,_S \cos \, _{(2)}\theta - _{(3)}h \sin \, _{(3)}\theta \frac{d}{dS}, \tag{6.22}$$

$$_{(3)}\Xi_3 := \quad _{(2)}h \, _{(2)}\theta,_S \sin \, _{(2)}\theta - _{(2)}h \cos \, _{(2)}\theta \frac{d}{dS}, \tag{6.23}$$

$$_{(3)}\Xi_4 := - _{(2)}h \, _{(2)}\theta,_S \cos \, _{(2)}\theta - _{(2)}h \sin \, _{(2)}\theta \frac{d}{dS}. \tag{6.24}$$

The total weak form of contact forces/couples for a geometrically exact sandwich beam is then given by

$$G_c(\Phi, \mathbf{w}) = \sum_{\ell=1}^{3} {}_{(\ell)}G_c(\Phi, \mathbf{w}). \tag{6.25}$$

We refer the reader to Vu-Quoc & Deng [1995c] for more general expressions for $_{(\ell)}G_c(\Phi, \mathbf{w})$ in the case where $\mathbf{E}_\alpha \neq \mathbf{e}_\alpha$. Gathered in $\overline{\mathcal{F}}: \mathcal{L} \to \mathbb{R}^5$ are the assigned forces/couples distributed in $\mathcal{L}$

$$\overline{\mathcal{F}} := \left\{ \overline{\vec{n}}^1, \, \overline{\vec{n}}^2, \, _{(1)}\overline{\mathcal{M}}, \, _{(2)}\overline{\mathcal{M}}, \, _{(3)}\overline{\mathcal{M}} \right\}^T, \tag{6.26}$$

and in $\overline{\overline{\mathcal{F}}} \in \mathbb{R}^5$ the assigned forces/couples at the boundaries

$$\overline{\overline{\mathcal{F}}} := \left\{ \overline{\overline{\vec{n}}}^1, \, \overline{\overline{\vec{n}}}^2, \, _{(1)}\overline{\overline{\mathcal{M}}}, \, _{(2)}\overline{\overline{\mathcal{M}}}, \, _{(3)}\overline{\overline{\mathcal{M}}} \right\}^T. \tag{6.27}$$

The weak form for assigned forces/couples is as follows:

$$G_a(\mathbf{w}) = \int_{\mathfrak{L}} \overline{\mathcal{F}} \cdot \mathbf{w} dS + \left[\overline{\overline{\mathcal{F}}} \cdot \mathbf{w} \right]_{S=0}^{S=L}. \tag{6.28}$$

Let $\mathcal{V}$ designate the set of admissible weighting functions (see, e.g., Hughes [1987]). The weak form of the geometrically exact multilayer beam problem is to

> Find Φ such that
> $$G_c(\Phi, \mathbf{w}) = G_a(\mathbf{w}), \qquad \forall \mathbf{w} \in \mathcal{V}. \tag{6.29}$$

To solve the above nonlinear problem, one can use Newton's iterative method. To this end, a consistent linearization of (6.29) has to be carried out. We then discretize the spatial coordinate of the linearized weak form via a Galerkin projection employing finite element basis functions. All these issues are addressed in detail in Vu-Quoc & Deng [1995c], to which the reader is referred.

7. Numerical Examples

The finite-element formulation for geometrically exact sandwich beams presented in Section 6 has been implemented in the Finite Element Analysis Program (FEAP), developed by R. L. Taylor (see Zienkiewicz & Taylor [1989]) and run on a DEC 5000 with the Ultrix 4.2 operating system. The finite-element basis functions used in the examples in this section are linear or quadratic. To avoid shear locking, uniform reduced integration is used to evaluate the tangent stiffness matrix and the residual matrix.

7.1. Verification of FE Implementation with Single-Layer Beam

To check the correctness of the formulation and the implementation, we consider the case where the outer layers of the sandwich-beam element do not exist, and compare the results to those from single-layer beam theory: The results must be identical for the implementation to be correct. To do so, we simply set the material constants of the outer layers of sandwich-beam elements to zero. We select the following material constants:

$$_{(\ell)}EA = 0, \quad _{(\ell)}GA_s = 0, \quad _{(\ell)}EI = 0, \text{ for } \ell = 1, 3, \tag{7.1}$$

$$_{(2)}EA = 2, \quad _{(2)}GA_s = 2, \quad _{(2)}EI = 2. \tag{7.2}$$

Consider a cantilever beam of length $L = 1$. Two loading tests are performed, both with 5 linear sandwich-beam elements. In the first test, the beam is subjected to a transverse tip load of $F^2 = 0.1$. A transverse tip displacement of 0.067 is obtained, which corresponds exactly to the theoretical results using the linear Timoshenko beam theory. In the second loading test, we subject the cantilever beam to a pure bending by applying a moment of $M = 12.566 \approx 4\pi$ at the beam tip to force the beam to roll up into an exact circle (see Vu-Quoc [1986]). In this case, the beam tip is displaced to coincide exactly with the clamped end, with a displacement having a component of -1 along the beam length, a zero transverse component, and a rotation of 2π rad; the numerical results are very close to the exact results[9]:

u^1	u^2	$_{(2)}\theta$
-1.00003	$2.92110E{-}09$	6.28300

Figure 3 depicts the deformed shape as a pentagon because five linear elements are used; the nodes lie exactly on the circle circumscribing the pentagon. Convergence to the deformed shape shown is obtained in two iterations; the same rate of convergence was obtained for single-layer beams in Simo & Vu-Quoc [1986b] and Vu-Quoc [1986]. It should be noted, however, that such a rapid rate of convergence is obtained only when we set $_{(\ell)}EA = {}_{(\ell)}GA_s = {}_{(\ell)}EI$, as in (7.2). In other words, with $EA \neq GA_s \neq EI$, a higher number of iterations to convergence is expected. Indeed, with the material properties $EA = 5$, $GA_s = 1$, $EI = 2$, we need eight iterations to converge, but the converged result is identical to the case where $_{(\ell)}EA = {}_{(\ell)}GA_s = {}_{(\ell)}EI$, as in (7.2)

---

[9] All rotations are given in radians.

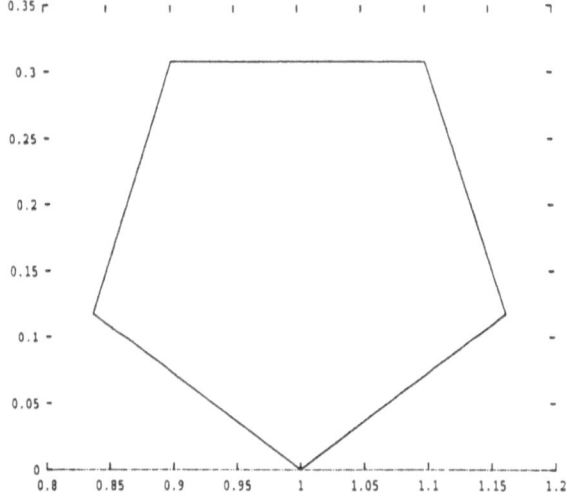

Fig. 3. Verification of FE implementation. "Sandwich" beam having only core layer (2). Roll-up maneuver. Comparison with single-layer beam.

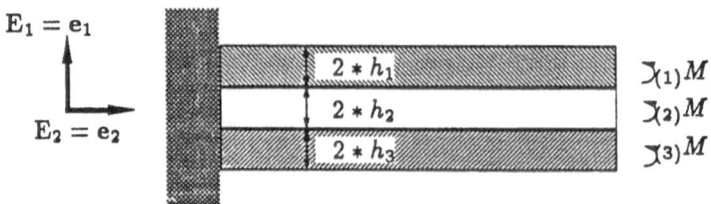

Fig. 4. Sandwich beam with identical layers. Assigned layer moments at beam tip.

above. We have thus demonstrated that the present sandwich-beam element contains the single-layer beam element as a particular case.

7.2. Sandwich Beam with Identical Layers

Consider a cantilever sandwich beam with three identical layers subjected to assigned tip moments as shown in Figure 4. Thus, let $_{(\ell)}EA = EA$, $_{(\ell)}GA_s = GA_s$, $_{(\ell)}EI = EI$, for $\ell = 1, 2, 3$. The cantilever boundary conditions at $S = 0$ are

$$u^1(0) = u^2(0) = \ _{(1)}\theta(0) = \ _{(2)}\theta(0) = \ _{(3)}\theta(0) = 0. \qquad (7.3)$$

7.2.1. Normal Moment Distribution. A sandwich beam will behave as a single-layer beam when the rotations in the three layers are constrained to be the same so that the overall beam cross section remains plane (no hinge formation) in the deformed configuration, i.e.,

$$_{(1)}\theta = {}_{(2)}\theta = {}_{(3)}\theta = \theta. \tag{7.4}$$

In a force-driven problem, the above constraint is not applied explicitly, but an appropriate distribution of assigned layer moments must be applied to keep the layer rotations the same as in (7.4). To find such moment distribution, we begin by computing the tangent material stiffness matrix of a cantilever sandwich beam with one *linear* element according to the following equation (Vu-Quoc & Deng [1995c]):

$$D^m {}_{(\ell)}G_c(\Phi^{(i)}, \mathbf{w}) \cdot \Delta\Phi^{(i+1)}$$
$$= \int_{\mathcal{L}} \left[{}_{(\ell)}\Xi(\Phi^{(i)})\mathbf{w} \right] \cdot {}_{(\ell)}\Lambda^{(i)} {}_{(\ell)}C {}_{(\ell)}\Lambda^{(i)T} {}_{(\ell)}\Xi(\Phi^{(i)})\Delta\Phi^{(i)}dS. \tag{7.5}$$

There are thus five degrees of freedom at the beam tip. Let $\theta_o = 0$ so that ${}_{(\ell)}\Lambda = \mathfrak{J}_3$, for $\ell = 1, 2, 3$.[10] Thus,

$$\Delta\Phi(S) = \mathbf{d}N(S) \in \mathbb{R}^{5\times 1}, \qquad \mathbf{w}(S) = \mathbf{c}N(S) \in \mathbb{R}^{5\times 1}, \tag{7.6}$$

where $\mathbf{d} \in \mathbb{R}^{5\times 1}$ contains the two displacement components and three layer rotations at the beam tip and $N(\cdot)$ is a *linear* basis function. The tangent material stiffness matrix (7.5) becomes

$$\mathbf{K}^m = \sum_{\ell=1}^{3} \int_{\mathcal{L}} \left\{ {}_{(\ell)}\Xi(\Phi)\left[N\mathfrak{J}_5\right] \right\}^T {}_{(\ell)}C {}_{(\ell)}\Xi(\Phi)N\mathfrak{J}_5 \, dS$$

$$= \int_{\mathcal{L}} \begin{bmatrix} 3EA\,(N,_S)^2 & 0 & EAh\,(N,_S)^2 & 0 & -EAh\,(N,_S)^2 \\ 0 & 3GA_s\,(N,_S)^2 & -GA_sNN,_S & -GA_sNN,_S & -GA_sNN,_S \\ EAh\,(N,_S)^2 & -GA_sNN,_S & K_{33} & EAh^2\,(N,_S)^2 & 0 \\ 0 & -GA_sNN,_S & EAh^2\,(N,_S)^2 & K_{44} & EAh^2\,(N,_S)^2 \\ -EAh\,(N,_S)^2 & -GA_sNN,_S & 0 & EAh^2\,(N,_S)^2 & K_{55} \end{bmatrix}_{5\times5} dS. \tag{7.7}$$

where

$$K_{33} = K_{55} := EAh^2\,(N,_S)^2 + GA_s N^2 + EI\,(N,_S)^2, \tag{7.8}$$
$$K_{44} := 2EAh^2\,(N,_S)^2 + GA_s N^2 + EI\,(N,_S)^2. \tag{7.9}$$

Let the tip forces and moments be gathered in

$$\mathbf{F} = \left\{ F^1, \ F^2, \ {}_{(1)}M, \ {}_{(2)}M, \ {}_{(3)}M \right\}^T = \mathbf{K}^m\mathbf{d}. \tag{7.10}$$

For pure bending problems, such as the roll-up maneuver presented in Figure 3, shear deformation is absent. Thus to find the distribution of moments that replicates the conditions of pure bending in single-layer beams, we set tip displacements to zero (as if the

---

[10] $\mathfrak{J}_3$ is a 3×3 unit matrix.

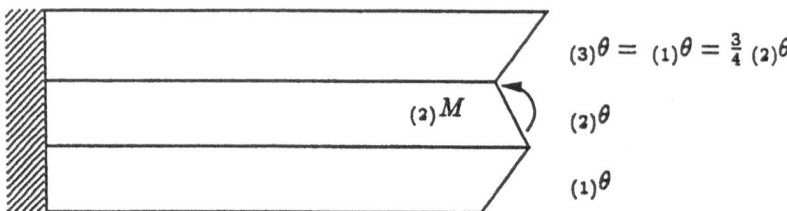

Fig. 5. Sandwich beam with identical layers. Moment assigned to core layer only, and distribution of layer rotations in the *linear* case.

tip is simply supported) and the layer rotations at the beam tip to be the same as in (7.4), i.e.,

$$\mathbf{d} = \{0,\ 0,\ \theta,\ \theta,\ \theta\}^T. \tag{7.11}$$

Consider the beam as having *unit width*. Then from (7.7) and (7.10) with $GA_S = 0$, we obtain the following moment distribution that makes the sandwich beam behave as a single-layer beam

$$_{(1)}M =\ _{(3)}M = \frac{14}{3}Eh^3\theta, \qquad _{(2)}M = \frac{26}{3}Eh^3\theta. \tag{7.12}$$

Such moment distribution is referred to as the *normal* moment distribution *for a sandwich beam with three identical layers.*[11] It can be verified that

$$_{(1)}M +\ _{(2)}M +\ _{(3)}M = E\frac{(6h)^3}{12}\theta = EI_{\text{total}}\theta; \tag{7.13}$$

i.e., the sandwich beam with the moment distribution (7.12) has the same bending stiffness as the single-layer beam. The normal moment distribution (7.12) will be used in the roll-up of cantilever sandwich beams presented below.

7.2.2. Assigned Moment on Core Layer Only.

For the same linear sandwich beam with one end clamped and one end simply supported as in Section 7.2.1, consider now the case where only the moment assigned on the core layer (2) exists, i.e.,

$$_{(1)}M = 0, \qquad _{(2)}M \neq 0, \qquad _{(3)}M = 0. \tag{7.14}$$

Then from (7.10), we obtain the following layer rotations

$$_{(1)}\theta =\ _{(3)}\theta = -\frac{9}{20}\frac{_{(2)}M}{Eh^3}, \qquad _{(2)}\theta = \frac{3}{5}\frac{_{(2)}M}{Eh^3}, \tag{7.15}$$

as shown in Figure 5. The computer implementation has been verified against the above analytical results.

---

[11] In fact, the ratio (7.29), which comes from (7.12), is independent of both the shape function used and the beam length, as shown in Vu-Quoc & Deng [1997b].

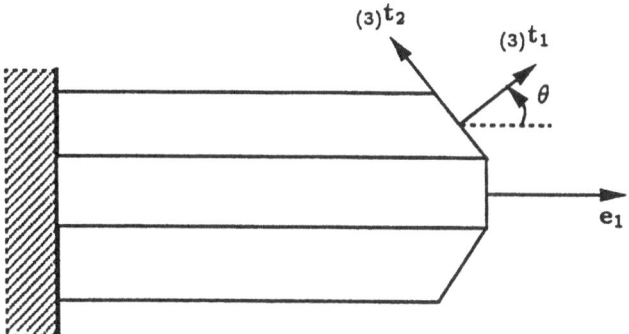

Fig. 6. Sandwich beam with identical layers. Applied axial force at core layer. Deformed shape.

7.2.3. Assigned Axial Force on Core Layer. Consider now a cantilever sandwich beam having identical layers with half-layer thickness $h = 0.001$ and with equal length $L = 1$. The beam is subjected to an assigned axial force $F^1 = 3$ at the tip (all other components of $\mathbf{F}$ in (7.10) are zero); the force acts on the centroidal line of the core layer by virtue of the formulation of the present theory. The material properties are

$$_{(\ell)}EA = 2, \qquad _{(\ell)}GA_s = 2, \qquad _{(\ell)}EI = 2, \quad \text{for } \ell = 1, 2, 3. \qquad (7.16)$$

The beam is discretized using five linear elements. The displacements and rotations at the beam tip are given below.

$u^1(1)$	$u^2(1)$	$_{(1)}\theta(1)$	$_{(2)}\theta(1)$	$_{(3)}\theta(1)$
5.00000E−01	−7.91110E−21	−3.44091E−04	−6.15256E−21	3.44091E−04

It is observed from the numerical results that the axial displacement is the same as that in an equivalent single-layer beam with axial stiffness $3EA = 6$ (i.e., $F^1/3EA = 0.5$), while the two outer layers have non-zero, symmetric rotations. This observation can be explained by looking at the first equation of (7.10), i.e., at $S = 1$

$$F^1 = 3EA \,(N_{,s})^2 \, u^1(1) + EAh \,(N_{,s})^2 \left[_{(1)}\theta(1) - _{(3)}\theta(1) \right], \qquad (7.17)$$

obtained with the aid of (7.7). In (7.17), it can be seen that the contribution of the outer rotations $_{(1)}\theta$ and $_{(3)}\theta$ (values from the table above) to the axial force $F^1 = 3$ is of order

$$EAh \,(N_{,s})^2 \,_{(1)}\theta \approx (2)(0.002)(1)(3 \times 10^{-4}) = \mathcal{O}(10^{-6}), \qquad (7.18)$$

and is thus negligible in front of the contribution of the axial displacement u^1 (value from the table above) of order

$$3EA \,(N_{,s})^2 \, u^1(1) \approx (3)(2)(1)^2(0.5) = 3. \qquad (7.19)$$

Thus, when the outer layer rotations are not small, or when the beam is thick, the axial displacement will not be the same as that in an equivalent single-layer beam, as shown below. Let the beam now have a thickness parameter of $h = 0.05$ (i.e., 50 times thicker than in the above example) and the same material properties as in (7.16). The displacements and rotations at the beam tip are now

$u^1(1)$	$u^2(1)$	$_{(1)}\theta(1)$	$_{(2)}\theta(1)$	$_{(3)}\theta(1)$
5.00545E−01	7.80458E−20	−1.71837E−02	1.73529E−19	1.71837E−02

Furthermore, if we increase the depth of the beam by another 10 times, i.e., $h = 0.5$, we obtain the following results:

$u^1(1)$	$u^2(1)$	$_{(1)}\theta(1)$	$_{(2)}\theta(1)$	$_{(3)}\theta(1)$
5.48942E−01	2.03216E−18	−1.54095E−01	1.09290E−18	1.54095E−01

It can be seen clearly that the axial tip displacement $u^1(1)$ of the sandwich beam is not the same as that of a single-layer beam with an equivalent axial stiffness, due to the coupling between $u^1(1)$ and the outer-layer rotations $_{(\ell)}\theta$, for $\ell = 1, 3$ (see Figure 6). This coupling becomes important as the beam thickness increases, as shown in the numerical results above. The deformed shape is depicted in Figure 6.

7.2.4. Reduced Equations of Equilibrium.

We now develop the reduced nonlinear equations of equilibrium for a sandwich beam under tip axial loading. By symmetry of the problem, we have only *two* unknown kinematic quantities, the axial displacement u^1 and the outer-layer rotation $_{(1)}\theta = -\, _{(3)}\theta$, since $u^2 = \,_{(2)}\theta = 0$. Hence the five equilibrium equations are reduced to two equations for u^1 and $_{(3)}\theta$.

The reduced equations of equilibrium for this particular case are as follows: From the first equation in (2.4), and since $\mathbf{E}_1 = \mathbf{e}_1$ and $u^2 = 0$, we obtain

$$_{(2)}\Phi_0(S) = (S + u^1)\mathbf{e}_1. \tag{7.20}$$

Also, since $_{(2)}\mathbf{t}_1 = \mathbf{e}_1$ and $_{(2)}\mathbf{t}_2 = \mathbf{e}_2$, using (2.6) and (2.3), we obtain

$$_{(3)}\Phi_0 = \left(S + u^1,_S - h\sin\,_{(3)}\theta\right)\mathbf{e}_1 + h\left(1 + \cos\,_{(3)}\theta\right)\mathbf{e}_2. \tag{7.21}$$

Now from (3.3), it follows that

$$_{(3)}\gamma = \left(1 + u,_S - h\,_{(3)}\theta,_S\cos\,_{(3)}\theta - \cos\,_{(3)}\theta\right)\mathbf{e}_1 - \sin\,_{(3)}\theta\left(1 + h\,_{(3)}\theta,_S\right)\mathbf{e}_2, \tag{7.22}$$

which then yields the expression

$$_{(3)}\mathbf{n} = 2\left(1 + u,_S - h\,_{(3)}\theta,_S\cos\,_{(3)}\theta - \cos\,_{(3)}\theta\right)\mathbf{e}_1 - 2\sin\,_{(3)}\theta\left(1 + h\,_{(3)}\theta,_S\right)\mathbf{e}_2, \tag{7.23}$$

by virtue of relation (4.6) and the material properties (7.16). Similarly, the resultant contact force for the layer (1) takes the form

$$_{(1)}\mathbf{n} = 2\left(1 + u,_S - h\,_{(3)}\theta,_S\cos\,_{(3)}\theta - \cos\,_{(3)}\theta\right)\mathbf{e}_1 + 2\sin\,_{(3)}\theta\left(1 + h\,_{(3)}\theta,_S\right)\mathbf{e}_2. \tag{7.24}$$

It is easy to see from (7.20) that

$$_{(2)}\mathbf{n} = 2u_{,S}\,\mathbf{e}_1. \tag{7.25}$$

Notice that both $\mathbf{f}$ and $_{(3)}C$ are zero; we obtain from (3.9) the equilibrium equation for the axial force

$$6u_{,S} + 4\left(1 - \cos{_{(3)}\theta}\right) - 4h\,_{(3)}\theta_{,S}\cos{_{(3)}\theta} = 3, \tag{7.26}$$

and (3.11) yields the equation of equilibrium for the outer-layer moment

$$\left(1 + h^2\right)\,_{(3)}\theta_{,SS} - hu_{,SS}\cos{_{(3)}\theta} - 2\left(1 + u_{,S}\right)\sin{_{(3)}\theta} = 0. \tag{7.27}$$

The boundary conditions are given as follows:

At $S = 0$, $\quad u^1(0) = 0$, $\quad\quad\quad\quad\quad _{(3)}\theta(0) = 0$,

At $S = 1$, $\quad u^1_{,S}(1) = \dfrac{F^1}{EA} = 0.5$, $\quad _{(3)}\theta_{,S}(1) = 0$. $\tag{7.28}$

An analytical solution to problem (7.26)–(7.28) can be developed using an iterative procedure as follows. First assume that u^1 is the same as the axial displacement u_0^1 of a linear single-layer beam, and replace this solution in (7.27) to obtain an equation for $_{(3)}\theta_0$. Then solve for $_{(3)}\theta_0$, which is then replaced into (7.26) to solve for a more accurate solution for u_1^1. And so on and so forth.

7.2.5. Roll-up Maneuver with Normal Moment Distribution.

We now apply the *normal* moment distribution for a cantilever sandwich beam with the ratio

$$_{(1)}M:\,_{(2)}M:\,_{(3)}M = 7:13:7, \tag{7.29}$$

as derived in Section 7.2.1 (see (7.12)).

The layer half-thickness is chosen to be

$$h = 0.01732 \approx 0.01\sqrt{3}. \tag{7.30}$$

The material properties are as follows:

$$_{(\ell)}EA = 2,000,000, \quad _{(\ell)}GA_s = 2,000,000, \quad _{(\ell)}EI = 200, \text{ for } \ell = 1, 2, 3, \tag{7.31}$$

thus making the total bending stiffness equivalent to a single-layer beam

$$EI_{\text{total}} = 27EI = 5,400. \tag{7.32}$$

The beam is discretized using five linear sandwich elements. A cantilever beam subjected to assigned moment M at the free end will undergo pure bending; the deformed shape will be that of an arc of a circle of radius

$$R = \frac{EI_{\text{total}}}{M}, \tag{7.33}$$

depending on the magnitude of the assigned moment. Thus to have the beam of length $L = 1$ rolled up into one full circle, the radius will be

$$R = \frac{L}{2\pi},$$ $\bullet$ (7.34)

and the desired total moment will be

$$M = E I_{\text{total}} \frac{2\pi}{L}.$$ (7.35)

Thus using the ratio (7.29), the moment on each layer of the sandwich beam is

$$_{(1)}M = {}_{(3)}M = \frac{7}{27}M = 8796.46, \qquad _{(2)}M = \frac{13}{27}M = 16336.3. \qquad (7.36)$$

It takes two iterations to converge to the final deformed shape (i.e., the full circle). The tip displacements of the sandwich beam are

u^1	u^2	$_{(1)}\theta$	$_{(2)}\theta$	$_{(3)}\theta$
-0.999999	$-2.15292\text{E}{-}09$	6.28319	6.28319	6.28319

Note that the beam tip moved back to coincide with the cantilever end, with a horizontal displacement of 1 and a vertical displacement zero,[12] and the cross section rotates by $6.28319 \approx 2\pi$ rad.

For an equivalent single-layer beam with the same bending stiffness as in (7.32), and the same total assigned moment as in (7.35), the converged tip displacements and rotation, obtained with two iterations as in Simo & Vu-Quoc [1986b], are given below.

u^1	u^2	θ
-1.00000	$1.25011\text{E}{-}15$	6.28319

The deformed shape (a pentagon), given in Figure 3, is exactly the same as that of the single-layer beam. Despite the presence of three layers in the sandwich beam and the more complex system of equations, the rate of convergence is the same as that of the single-layer beam.

7.3. Two-Layer Beams

An example of the application of the present formulation to the important class of multilayer structures with ply drop-off is now presented. A two-layer beam used in this section is depicted in Figure 7, where the top layer is shorter than the bottom layer by half, creating a ply drop-off. There is thus a sharp discontinuity in the cross section in the middle of the beam. The present example illustrates both the variability of the number of layers in the present multilayer beam element (since elements with two layers and elements with one layer are used to model the beam), and the ability to accommodate

---

[12] There are round-off errors.

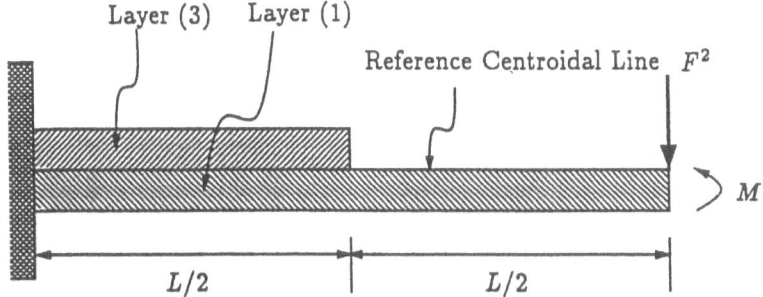

Fig. 7. Two-layer beam with ply drop-off. Long layer (1), nonexistent layer (2), short layer (3). Reference centroidal line is top fiber of layer (1).

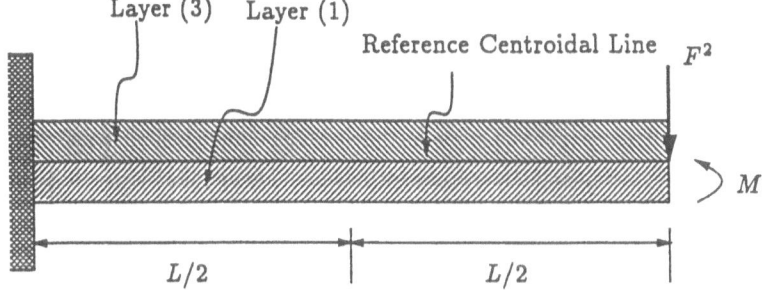

Fig. 8. Two-layer beam without ply drop-off. Reference centroidal line is top fiber of layer (1).

ply drop-off naturally. The case with small deformation is analyzed in Zinno & Barbero [1994]. The following material properties are selected[13]:

$$E = 29,000, \ \nu = 0.294, \ \chi = \frac{5}{6}, \tag{7.37}$$

i.e., $_{(\ell)}EA = 14,500, \ _{(\ell)}GA_s = 5,602.8, \ _{(\ell)}EI = 302.1, \qquad \text{for } \ell = 1, 3. \tag{7.38}$

The beam has a length of $L = 5$, and a half-layer thickness $h = 0.25$.

7.3.1. Case without Ply Drop-off. To verify the modeling, we first consider the case without ply drop-off for which an exact solution for small deformation exists. Thus, for now, the two layers have equal length $L = 5$. A transverse force $F^2 = 1$ is applied at the beam tip. Using the linear Timoshenko beam theory for an equivalent *single*-layer beam, we obtain a tip displacement of

$$u^2(L) = \frac{F^2 L^3}{3EI} + \frac{F^2 L}{GA_s} = 0.017775, \tag{7.39}$$

---

[13] Young's modulus E, Poisson's ratio ν, shear coefficient χ such that $A_s = \chi A$.

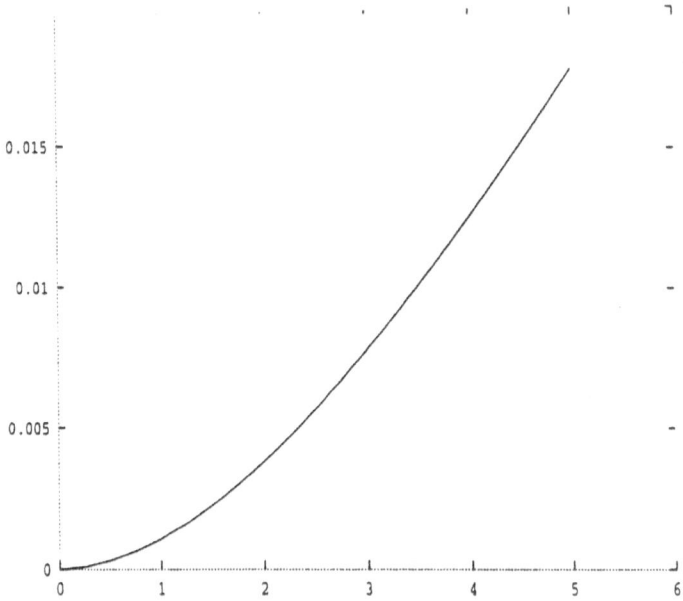

Fig. 9. Two-layer beam without ply drop-off. Deformed shape.

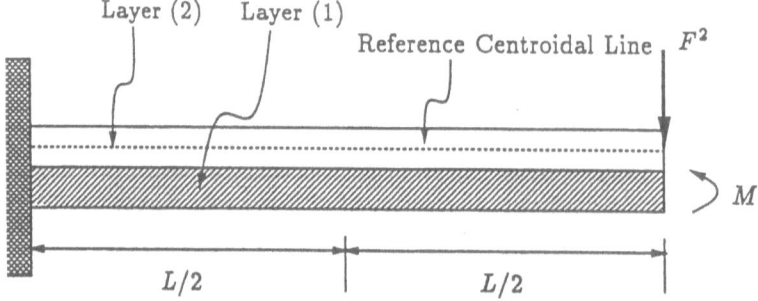

Fig. 10. Two-layer beam without ply drop-off. There is no layer (3). The reference centroidal line is centroidal line of layer (2).

as compared to a value of 0.01750 obtained in Zinno & Barbero [1994]. The two-layer beam is modeled using 20 linear sandwich-beam elements with core layers having zero height ($h = 0$) and zero material properties. In this case, with the core layer (2) having zero height, the top fiber of layer (1) coincides with the bottom fiber of layer (3); the deformation of the two-layer beam thus modeled is described by the top fiber of layer (1), together with the two layer rotations. The computed transverse tip displacement is 0.017766, i.e., a relative error of 0.05% compared to the equivalent single-layer Timoshenko beam, and 1.52% compared to Zinno & Barbero [1994]. The deformed shape is shown in Figure 9.

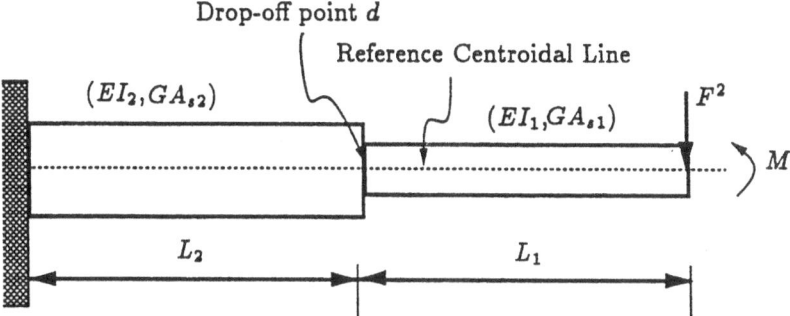

Fig. 11. Linear Timoshenko beam with symmetric discontinuous variation of cross section. Model used to calculate theoretical results for comparison.

Remark 7.1. That there are three ways to model a two-layer beam using the proposed sandwich-beam element is yet another aspect of the versatility of the present formulation. Setting the height and the material properties of the core layer (2) to zero as done above is only one way: see Figure 8. One could conceivably do the same to layer (1) or to layer (3), i.e., setting the height and material properties to zero (see Figure 10). But when we consider the assigned force, these three ways of modeling actually correspond to three different loading conditions: Recall that, by expressing the deformed centroidal lines of all layers relative to that of the core layer (2), the assigned force always acts on the centroidal line of the core layer. Hence, as noted above, when layer (2) is nonexistent, the assigned force is applied at the top fiber of layer (1), which coincides with the bottom fiber of layer (3). When layer (1) or layer (3) is nonexistent, the assigned force is applied at the centroidal line of layer (2): See Figure 10 for the case where the drop-off occurs in layer (2).

7.3.2. Case with Ply Drop-off. Consider now the beam with ply drop-off as shown in Figure 7. First, a transverse force $F^2 = 1$ is applied at the tip. The displacement obtained from the use of the linear Timoshenko beam with *symmetric* discontinuous variation of cross section as shown in Figure 11 will serve for comparison. Unlike the multilayer theory, all sections in the linear Timoshenko beam, including the one at the symmetric drop-off, remain plane after deformation. The transverse tip displacement W_t of the linear Timoshenko beam is then given by

$$W_t = W_d + \theta_d \cdot L_2 + W_r = 0.03393, \qquad (7.40)$$

where W_d and θ_d are respectively the transverse displacement and section rotation at the symmetric drop-off point, W_r is the transverse tip displacement relative to the drop-off point, and all other quantities are as explained in Figure 11. The finite element formulation for small-deformation multilayer structures by Zinno & Barbero [1994] yields a computed transverse displacement of 0.034846 for the two-layer beam with ply drop-off at midspan as shown in Figure 7. Using our formulation, we now model the beam in Figure 7 using 20 linear sandwich-beam elements with the layer (2) nonexistent (as explained above), and with layer (3) removed from midspan ($S = 2.5$) to the beam tip ($S = 5$). The computed tip displacement, for our formulation and with the mentioned modeling of ply drop-off, is 0.0354247. Figure 12 shows the comparison of the deformed shape of the beam with ply drop-off and the deformed shape of the beam without ply drop-off.

Remark 7.2. Unlike Remark 7.3.1 for beams without ply drop-off, there are only *two* ways to model two-layer beams *with* ply drop-off using the present sandwich-beam elements (see Figures 7 and 13). If the two-layer beam with ply drop-off were modeled using layers (2) and (3), with layer (1) nonexistent, and with layer (3) shorter than layer (2) by half (see Figure 13), then the transverse force F^2 at the beam tip would be applied on the centroidal line of layer (2). Note that the point of application of the transverse force F^2 in Figure 7 is different from that in Figure 13. The eccentricity in Figure 7 will result in an assigned moment to the centroid of layer (1) as the beam deforms, in addition to the assigned force F^2, and therefore a larger transverse displacement. Indeed, using the modeling of the beam as described in Figure 13 (without eccentricity), we obtain a computed tip displacement of 0.0353380, which is smaller than the tip displacement of 0.0354247 obtained above.

Now if the two-layer beam with ply drop-off were modeled using layers (1) and (2), with layer (3) nonexistent, and with layer (2) shorter than layer (1) by half[14] (see Figure 14), then the transverse force F^2 at the beam tip is applied, not anywhere within layer (1), but on the extended centroidal line of layer (2). Such modeling leaves $(6.1)_3$ meaningless, i.e.,

$$_{(2)}h \; _{(1)}\mathbf{n}, s \cdot _{(2)}\mathbf{t}_1 = 0, \tag{7.41}$$

since for the ply drop-off part of layer (2) ($S \in [L/2, L]$) the applied moment $_{(2)}\overline{\mathcal{M}}$, the resultant force $_{(2)}\mathbf{n}$, and the moment $_{(2)}m, s$ for layer (2), and the resultant force[15] $_{(3)}\mathbf{n}$ are all zero. Hence $_{(2)}h \neq 0$ is impossible for (7.41) to be satisfied. A way to overcome this problem is to set $_{(2)}h = 0$ for $S \in [L/2, L]$, i.e., after the drop-off point, thus making the top fiber of layer (1) a referential line within this portion. In this case, the overall referential line is no longer a straight line, but a discontinuous piecewise-constant line, because the referential line for $S \in [0, L/2]$ is the centroidal line of layer (2), while the referential line for $S \in [L/2, L]$ is the top fiber of layer (1).

In summary, we now compare our results to those obtained using (i) the linear Timoshenko beam theory with symmetric discontinuous variation of cross section as depicted in Figure (11), and (ii) the Zinno & Barbero [1994] formulation, and this for each geometry of the beam. For the two-layer beam *without* ply drop-off, our results differ from the linear Timoshenko theory by 0.05%, and from Zinno & Barbero [1994] by 1.52%. For the two-layer beam *with* ply drop-off, our results differ from the linear Timoshenko theory by 4.22%, and from Zinno & Barbero [1994] by 1.63%. The above comparison suggests that the Zinno & Barbero [1994] formulation is not accurate for the beam without ply drop-off. On the other hand, for the beam with ply drop-off, the results using

---

[14] By setting the material properties of layer (2) to zero, but leaving the thickness unchanged.

[15] Because layer (3) is nonexistent.

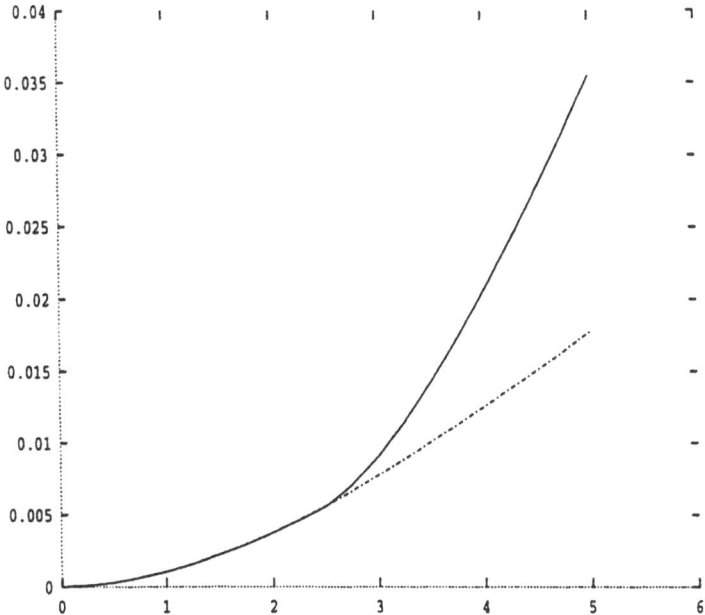

Fig. 12. Two-layer beam. Deformed shape. Solid line: With ply drop-off at midspan. Dotted line: Without ply drop-off.

the Timoshenko theory (Figure (11)) are clearly erroneous since the nonsymmetry in the ply drop-off cannot be modeled, and also since the cross section at the drop-off point remains plane (i.e., cannot develop a hinge). As a consequence, the beam modeled using the linear Timoshenko theory is stiffer than our beam model. The same can be said for the Zinno & Barbero [1994] formulation, since the error between our results and those by Zinno & Barbero [1994] is only 1.63%.

Thus far, a unit transverse force $F^2 = 1$ is applied at the beam tip. To produce large deformation in the beam, the assigned force is increased to $F^2 = 100$. Shown in the table below are the computed results using the present sandwich-beam elements, compared to those obtained by using the single-layer geometrically exact beam (i.e., nonlinear Timoshenko theory) to model beams with symmetric discontinuous variation of cross section[16] (see Figure 11).

Beam	u^1	u^2	$_{(1)}\theta$
Sandwich	−0.798692	−2.45561	−1.14239
Single-layer	−0.769361	−2.16970	−0.979568

As explained in Remark 7.3.1, the displacements u^1 and u^2 in the above table are those of the top fiber of layer (1), and the rotation $_{(1)}\theta$ is that of layer (1) cross section. Similar to

---

[16] As in the linear Timoshenko beam, the cross section in the nonlinear Timoshenko beam at the drop-off point remains plane after deformation, and cannot develop a hinge.

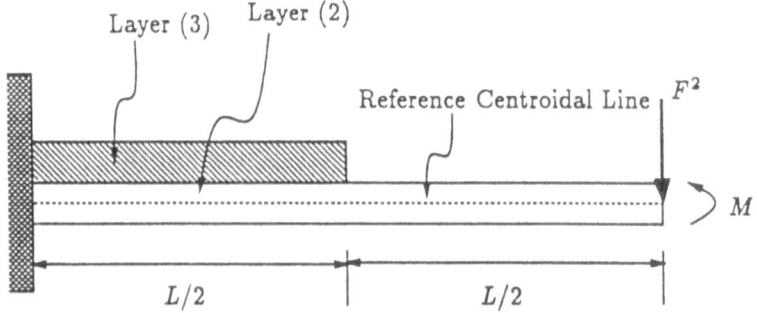

Fig. 13. Two-layer beam with ply drop-off. Nonexistent layer (1), long layer (2), short layer (3). Reference centroidal line is centroidal line of layer (2).

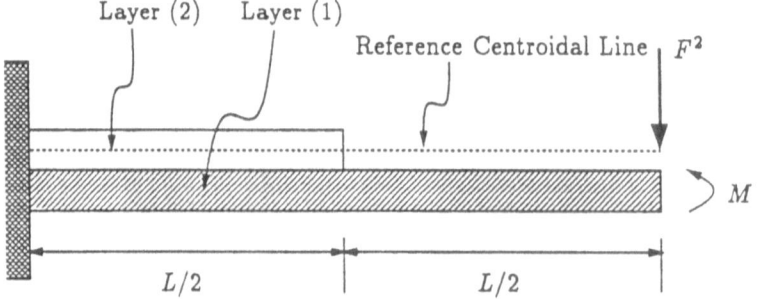

Fig. 14. Two-layer beam with ply drop-off. Long layer (1), short layer (2), nonexistent layer (3). Reference centroidal line is centroidal line of layer (2).

the small deformation case, the (nonlinear) Timoshenko theory produces a stiffer model yielding smaller displacements.

7.4. Ideal Sandwich Beams

We now consider what is sometimes called the *ideal* sandwich beam, with symmetrical cross section having a soft core layer, together with very thin, but axially very stiff, outer layers (see Sayir & Koller [1986]). Such thin and stiff outer layers are expected to allow primarily shear deformation in the sandwich beam, since they prevent bending by means of membrane deformation almost completely. Consider a simply supported ideal sandwich beam, with overhangs, shown in Figure 15. Let $L = 10$ and let a concentrated force $F^2 = -30$ be applied at midspan, i.e., at $S = L$. Also consider the cross section of the beam to be of unit width. The material properties used are as follows:

$$\chi = \frac{_{(s)}A}{A} = 1, \qquad \nu = 0.25, \tag{7.42}$$

$$_{(1)}E = {}_{(3)}E = 4 * 10^5, \qquad _{(1)}h = {}_{(3)}h = 0.025,$$

$$_{(2)}E = 10^3, \qquad _{(2)}h = 0.5, \tag{7.43}$$

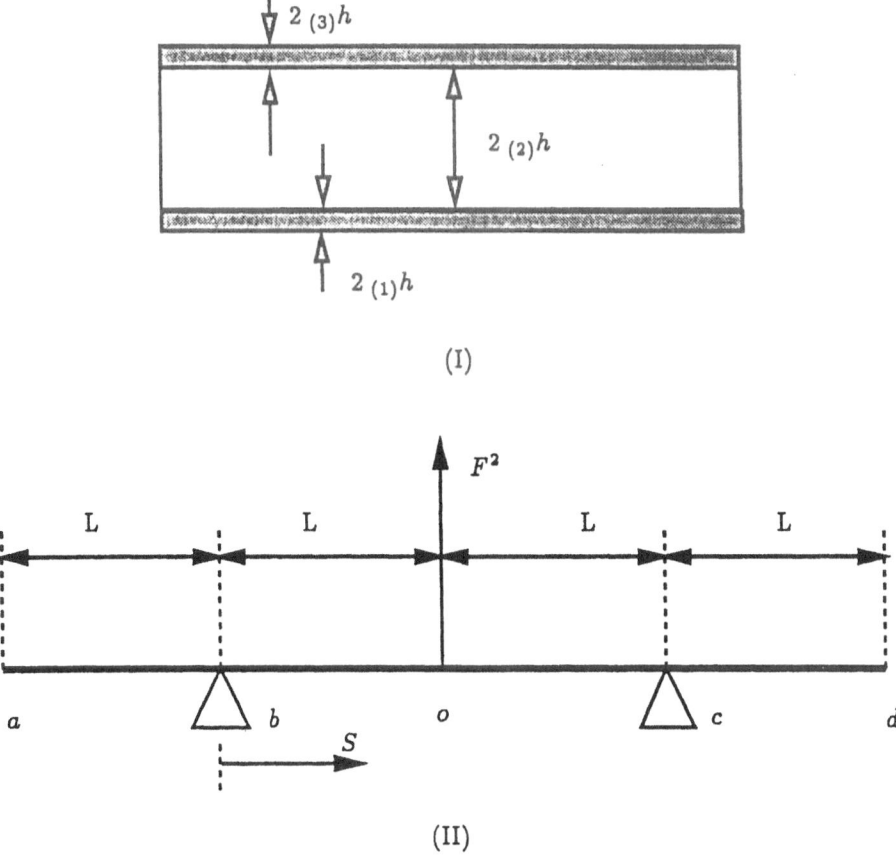

Fig. 15. Ideal sandwich beam. (I) Profile: Soft core layer, thin and axially stiff outer layers. (II) Simply supported beam with overhangs: Loading at midspan; origin of coordinate S at support b.

which imply the following elastic moduli for the three layers:

$$_{(\ell)}EA = 20,000, \quad _{(\ell)}EI = 4.1667, \quad _{(\ell)}GA_s = 8,000, \quad \text{for } \ell = 1, 3. \quad (7.44)$$

$$_{(2)}EA = 1,000, \quad _{(2)}EI = 83.33, \quad _{(2)}GA_s = 400. \quad (7.45)$$

A uniform discretization of 20 linear sandwich beam elements yields a transverse deflection at midspan of $u^2(L) = -0.773186$.

7.4.1. Heuer's Theory for Sandwich Beams. Next, we calculate the deflection of the sandwich beam using the theory developed in Heuer [1992], which essentially yields a linear Timoshenko beam equivalent to the sandwich beam. Heuer [1992] derives the following equation of equilibrium:

$$\mathcal{E}\mathcal{I}_{\text{total}} u^2,_{SSSS} = \widetilde{\overline{n}}^2 - s\mathcal{E}\mathcal{I}_{\text{total}}\widetilde{\overline{n}}^2,_{SS}, \tag{7.46}$$

where $\mathcal{E}\mathcal{I}_{\text{total}}$ is the equivalent bending stiffness, u^2 the transverse deflection, s the "shear tracer" coefficient, and $\widetilde{\overline{n}}^2$ the distributed transverse load. Following Heuer [1992], the equivalent bending stiffness $\mathcal{E}\mathcal{I}_{\text{total}}$ is given by

$$\mathcal{E}\mathcal{I}_{\text{total}} = 11,116.7. \tag{7.47}$$

Remark 7.3. Equivalent Euler-Bernoulli beam. To verify the bending stiffness $\mathcal{E}\mathcal{I}_{\text{total}}$ obtained in (7.47), we compute the equivalent total bending stiffness using the Euler-Bernoulli theory:

$$EI_{\text{total}} = 2 \;_{(1)}E \int_{0.5}^{0.55} x^2\, dx + \;_{(2)}E \int_{-0.5}^{0.5} x^2\, dx = 11,116.66. \tag{7.48}$$

Comparing (7.48) and (7.47), it can be seen that $\mathcal{E}\mathcal{I}_{\text{total}} \equiv EI_{\text{total}}$. The transverse tip deflection according to the Euler-Bernoulli theory is then

$$u^2(L) = \frac{FL^3}{6EI_{\text{total}}} = 0.44978. \tag{7.49}$$

For the present problem, $\widetilde{\overline{n}}^2 = F^2\delta(S - L)$. Thus with $\chi = 1$ as stated in (7.42)$_1$, it follows from Heuer [1992] that $s = 0.0021678$, and the midspan transverse deflection is computed from (7.46) to be $u^2(L) = -0.77497$. So the computed midspan deflection using our sandwich beam elements (i.e., $u^2(L) = -0.773186$) is smaller than that from Heuer's theory by 0.23%.

Remark 7.4. If one uses the geometrically exact single-layer beam theory (nonlinear Timoshenko beam) with the equivalent material properties EA and GA_s obtained by simply summing the material properties of each layer, i.e.,[17]

$$EA = \sum_{\ell=1}^{3} {}_{(\ell)}EA = 41,000,$$

$$GA_s = \sum_{\ell=1}^{3} {}_{(\ell)}GA_s = 16,400, \tag{7.50}$$

$$EI = EI_{\text{total}} = 11,116.67,$$

then a midspan deflection $u^2(L) = -0.453498$ is obtained. Such a small deflection, as compared with the values obtained by the present sandwich beam theory and by Heuer's theory, demonstrates the result of a complex coupling between axial, shear and bending

---

[17] EI is as computed in (7.48) or in (7.47).

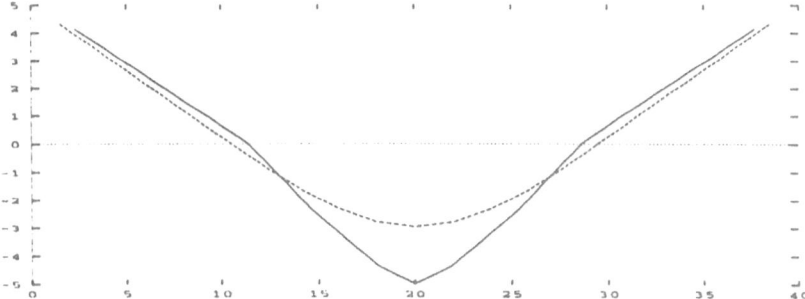

Fig. 16. Ideal sandwich beam. Simply supported beam with overhangs. Deformed shape. Comparison with geometrically exact single-layer beam (dotted line).

deformations in sandwich beams. In a sense, Heuer's theory accounts for this coupling through the shear tracer coefficient s.

Remark 7.5. To show clearly the difference in behavior between sandwich beams and single-layer beams, as demonstrated in experiments (Sayir & Koller [1986]), we increase the magnitude of the assigned force to $F^2 = -210$. The same material properties as in (7.50) are used. The deformed shapes are shown in Figure 16, where it can be seen that the overhangs of the sandwich beam do not rise as high as those of the equivalent single-layer beam. In addition, there is a clear break in the beam slope at the supports when using the proposed sandwich beam formulation. Such behavior of sandwich beams is corroborated by experiments (Sayir & Koller [1986]).

8. Conclusion

A new theory of geometrically exact multilayer beams/1-D plates has been presented. In addition to the general fully nonlinear theory, we derive several particular cases, with restrictive assumptions, but with important applications. The case of infinitesimal outer layer rotations superposed onto the finite core layer rotation is of practical importance. The classical linear theory of Yu [1959] is shown to be yet another more particular case. The theory can be applied to a wide class of problems involving multilayer structures undergoing large deformation and/or large overall motion. We have successfully formulated a Galerkin projection of the equations developed herein; this formulation and the numerical results are presented in Vu-Quoc & Deng [1995c]. As mentioned in the introduction, the present formulation is not restricted to multilayer structures having three layers, but the number of layers can vary between one and three. Furthermore, the lengths of the layers need not be equal as in multilayer structures with ply drop-off. The analysis of the important class of structures with a variable number of layers and with ply drop-off is discussed in detail in Vu-Quoc & Deng [1995c]. For shell structures, a formulation for geometrically exact multilayer shells using the methodology employed above can be found in Vu-Quoc & Ebcioğlu [1998].

Acknowledgments

Our research is supported by a grant from the National Science Foundation and from the SRAP program of the Florida Space Grant Consortium. This support is appreciated.

References

[1] P. Chadwick [1976], *Continuum Mechanics*, John Wiley & Sons, New York.
[2] R. Heuer [1992], Static and dynamic analysis of transversely isotropic, moderately thick sandwich beams by analogy, *Acta Mechanica* **91**, 1–9.
[3] T. J. R. Hughes [1987], *The Finite Element Method*, Prentice-Hall, Englewood Cliffs, NJ.
[4] J. W. Kamman & Huston, R. L. [1984], Dynamics of constrained multibody systems, *ASME Journal of Applied Mechanics* **51**, 889–903.
[5] J. E. Marsden & Hughes, T. J. R. [1983], *Mathematical Foundations of Elasticity*, Prentice-Hall, Englewood Cliffs, NJ.
[6] M. Sayir & Koller, M. G. [1986], Dynamic behaviour of sandwich plates, *Zeitschrift für Angewandte Mathematik und Physik (ZAMP Journal of Applied Mathematics and Physics)* **37**(1), 78–103.
[7] J.-C. Simo & Vu-Quoc, L. [1986*a*], On the dynamics of flexible beams under large overall motions—The plane case: Part I, *ASME Journal of Applied Mechanics* **53**(4), 849–854.
[8] J.-C. Simo & Vu-Quoc, L. [1986*b*], On the dynamics of flexible beams under large overall motions—The plane case: Part II, *ASME Journal of Applied Mechanics* **53**(4), 855–863.
[9] L. Vu-Quoc [1986], *Dynamics of Flexible Structures Performing Large Overall Motions: A Geometrically-Nonlinear Approach*, PhD thesis, University of California at Berkeley, Berkeley, CA 94720. Electronics Research Laboratory, Memorandum No. UCB/ERL M86/36.
[10] L. Vu-Quoc [1995], A formulation of geometrically-exact structural finite elements for efficient implementation and computation, *International Journal for Numerical Methods in Engineering*. In preparation.
[11] L. Vu-Quoc & Deng, H. [1997*a*], Dynamics of geometrically-exact sandwich structures, *International Journal of Mechanical Sciences* **40**(5), 421–441.
[12] L. Vu-Quoc & Deng, H. [1997*b*], Dynamics of geometrically-exact multilayer beams: Computational aspects, *Computer Methods in Applied Mechanics and Engineering* **146**, 135–172.
[13] L. Vu-Quoc & Deng, H. [1995*c*], Galerkin projection for geometrically-exact multilayer beams allowing for ply drop-off, *ASME Journal of Applied Mechanics* **62**, 479–488.
[14] L. Vu-Quoc & Ebcioğlu, I. K. [1995*a*], Dynamic formulation for geometrically-exact sandwich beams and 1-D plates, *ASME Journal of Applied Mechanics* **62**, 756–763.
[15] L. Vu-Quoc & Ebcioğlu, I. K. [1998], Formulation of equations of motion for multilayered geometrically-exact shells accommodating large deformation and large overall motions. Submitted.
[16] L. Vu-Quoc & Ebcioğlu, I. K. [1996], General multilayer geometrically-exact beams/1-D plates with piecewise linear section deformation, *Zeitschrift für Angewandte Mathematik und Mechanik (ZAMM Journal of Applied Mathematics and Mechanics)* **76**(7), 391–409.
[17] Y. Y. Yu [1959], A new theory of sandwich plates—General case, Technical Report Contract AF49(638)–455, Air Force Office of Scientific Research.
[18] O. C. Zienkiewicz & Taylor, R. L. [1989], *The Finite Element Method. Volume 1: Basic Formulation and Linear Problems*, 4th ed., McGraw-Hill, New York.
[19] R. Zinno & Barbero, E. J. [1994], A three-dimensional layer-wise constant shear element for general anisotropic shell-type structures, *International Journal for Numerical Methods in Engineering* **37**(14), 2445.

Obstructions to Quantization

M. J. Gotay[1]

[1] Department of Mathematics, University of Hawaii, 2565 The Mall, Honolulu, HI 96822, USA

Received July 24, 1998; revised manuscript received October 18, 1999

This paper is dedicated to the memory of Juan-Carlos Simo

Summary. Quantization is not a straightforward proposition, as demonstrated by Groenewold's and Van Hove's discovery, more than fifty years ago, of an "obstruction" to quantization. Their "no-go theorems" assert that it is in principle impossible to consistently quantize every classical polynomial observable on the phase space $\mathbf{R}^{2n}$ in a physically meaningful way. Similar obstructions have been recently found for S^2 and T^*S^1, buttressing the common belief that no-go theorems should hold in some generality. Surprisingly, this is not so—it has just been proven that there are no obstructions to quantizing either T^2 or $T^*\mathbf{R}_+$.

In this paper we work towards delineating the circumstances under which such obstructions will appear, and understanding the mechanisms which produce them. Our objectives are to conjecture—and in some cases prove—generalized Groenewold-Van Hove theorems, and to determine the maximal Lie subalgebras of observables which can be consistently quantized. This requires a study of the structure of Poisson algebras of symplectic manifolds and their representations. To these ends we include an exposition of both prequantization (in an extended sense) and quantization theory, here formulated in terms of "basic algebras of observables." We then review in detail the known results for $\mathbf{R}^{2n}$, S^2, T^*S^1, T^2, and $T^*\mathbf{R}_+$, as well as recent theoretical work. Our discussion is independent of any particular method of quantization; we concentrate on the structural aspects of quantization theory which are common to all Hilbert space-based quantization techniques.

1. Introduction

Quantization—the problem of constructing the quantum formulation of a system from its classical description—has always been one of the great mysteries of mathematical physics. It is generally acknowledged that quantization is an ill-defined procedure which cannot be consistently applied to all classical systems. While there is certainly no extant quantization procedure which works well in all circumstances, this assertion nonetheless bears closer scrutiny.

Already from first principles one encounters difficulties. Given that the classical description of a system is an approximation to its quantum description, obtained in a macroscopic limit (when $\hbar \rightarrow 0$), one expects that some information is lost in the limit. So quantization should somehow have to compensate for this. But how can a given quantization procedure select, from amongst the myriad of quantum theories all of which have the same classical limit, the physically correct one?

In view of this ambiguity it is not surprising that the many quantization schemes which have been developed over the years—such as the physicists' original "canonical quantization" [Di] (and its modern formulations, such as geometric quantization [Ki, So, Wo]), Weyl quantization [Fo] (and its successor deformation quantization [BFFLS, Ri2, Ri3, Ri4]), path integral quantization [GJ], and the group theoretic approach to quantization [AA, Is], to cite just some—have shortcomings. Rather, is it amazing that they work as well as they do!

But there are deeper, subtler problems, involving the Poisson algebras of classical systems and their representations. In this context the conventional wisdom is that it is impossible to "fully" quantize any given classical system—regardless of the particular method employed—in a way which is consistent with the physicists' Schrödinger quantization of $\mathbf{R}^{2n}$. (We will make this somewhat nebulous statement precise later.) In other words, the assertion is that there exists a universal "obstruction" which forces one to settle for something less than a complete and consistent quantization of *any* system. Each quantization procedure listed above evinces this defect in various examples.

That there are problems in quantizing even simple systems was observed very early on. One difficulty was to identify the analogue of the multiplicative structure of the classical observables in the quantum formalism. For instance, consider the quantization of $\mathbf{R}^{2n}$ with canonical coordinates $\{q^i, p_i \mid i = 1, \ldots, n\}$, representing the phase space of a particle moving in $\mathbf{R}^n$. For simple observables the "product $\rightarrow$ anti-commutator" rule worked well. But for more complicated observables (say, ones which are quartic polynomials in the positions and momenta), this rule leads to inconsistencies. (See [AB, §4], [Fo, §1.1] and §§5.1 and 6.5 for discussions of these factor-ordering ambiguities.) Of course this, in and by itself, might only indicate the necessity of coming up with some subtler symmetrization rule. But attempts to construct a quantization map also conflicted with Dirac's "Poisson bracket $\rightarrow$ commutator" rule. This was implicitly acknowledged by Dirac [Di, p. 87], where he made the now famous hedge:

> *"The strong analogy between the quantum P.B. . . . and the classical P.B. . . . leads us to make the assumption that the quantum P.B.s, or at any rate the simpler ones of them, have the same values as the corresponding classical P.B.s."*

In any case, as a practical matter, one was forced to limit the quantization to relatively "small" Lie subalgebras of classical observables which could be handled without ambiguity (e.g., polynomials which are at most quadratic in the p_i and the q^i, or observables which are affine functions of the positions or of the momenta).

Then, in 1946, Groenewold [Gro] showed that the search for an "acceptable" quantization map was futile. His "no-go" theorem states that one cannot consistently quantize the Poisson algebra of all polynomials in the q^i and p_i on $\mathbf{R}^{2n}$ as symmetric operators on some Hilbert space, subject to the requirement that the q^i and p_i be irreducibly represented. Van Hove subsequently refined Groenewold's result [VH1]. Thus it is *in principle* impossible to quantize—by *any* means—every classical observable on $\mathbf{R}^{2n}$, or even every polynomial observable, in a way consistent with Schrödinger quantization (which, according to the Stone-Von Neumann theorem, is the import of the irreducibility requirement on the p_i and q^i). At most, one can consistently quantize certain Lie subalgebras of observables, for instance the ones mentioned in the preceding paragraph.

Of course, Groenewold's remarkable result is valid only for the classical phase space $\mathbf{R}^{2n}$. The immediate problem is to determine whether similar obstructions appear when trying to quantize other symplectic manifolds. Little was known in this regard, and only in the mid 1990s have other examples come to light. A few years ago an obstruction was found for S^2, representing the (internal) phase space of a massive spinning particle [GGH]. It was shown that one cannot consistently quantize the Poisson algebra of spherical harmonics (thought of as polynomials in the components S_i of the spin angular momentum vector $\mathbf{S}$), subject to the requirement that the S_i be irreducibly represented. This is a direct analogue for S^2 of Groenewold's theorem. Moreover, just recently it was shown that the symplectic cylinder T^*S^1, which plays a role in geometric optics, exhibits a similar obstruction [GGru1]. Combined with the observations that S^2 is in a sense at the opposite extreme from $\mathbf{R}^{2n}$ insofar as symplectic manifolds go, and that T^*S^1 lies somewhere in between, these results indicate that no-go theorems can be expected to hold in some generality. But, interestingly enough, they are *not* universal: It is possible to explicitly construct a quantization of the full Poisson algebra of the torus T^2 in which a suitable irreducibility requirement is imposed [Go3]. It is also possible to quantize certain noncompact phase spaces, e.g. $T^*\mathbf{R}_+$ [GGra1]. An important point, therefore, is to understand the mechanisms which are responsible for these divergent outcomes.

Our goal here is to study obstructions to the quantization of the Poisson algebra of a symplectic manifold. We will review the known examples in some detail, and give a careful presentation of prequantization (in an extended sense) and quantization, with a view to conjecturing a generalized Groenewold-Van Hove theorem and in particular delineating the circumstances under which it can be expected to hold. Already some results have been established along these lines, to the effect that under certain circumstances there are obstructions to quantizing both compact and noncompact symplectic manifolds [GGG, GGra1, GGru2, GM]. Despite these recent advances, many interesting and difficult problems remain. Our discussion will be independent of any particular method of quantization; we concentrate on the structural aspects of quantization theory which are common to all Hilbert space-based quantization techniques.

The present paper is a revised and updated version of the review article "Obstruction Results in Quantization Theory," which was published in 1996 in the *Journal of Nonlinear Science* [GGT]. Since a number of new results and examples have been obtained since that article appeared, we thought it useful to provide a more current summary of the field. As well, a number of the concepts and constructions of that paper have evolved over time, and we have amended the paper accordingly. We have also taken this opportunity to correct a number of misprints and minor errors.

We express our appreciation to V. Aldaya, C. Atkin, P. Chernoff, G. Folland, V. Ginzburg, P. Jorgensen, J. Velhinho, and N. Wildberger, for many helpful conversations and for sharing insights, and to J. Grabowski, G. Tuynman, and especially H. Grundling, for fruitful collaborations. This research was supported in part by NSF grant DMS 96-23083.

2. Prequantization

Let (M, ω) be a fixed $2n$-dimensional connected symplectic manifold with associated Poisson algebra $(C^\infty(M), \{\cdot, \cdot\})$, where $\{\cdot, \cdot\}$ is the Poisson bracket.

To start the discussion, we state what it means to "prequantize" a Lie algebra of observables.

Definition 1. Let $\mathcal{O}$ be a Lie subalgebra of $C^\infty(M)$. A *prequantization* of $\mathcal{O}$ is a linear map $\mathcal{Q}$ from $\mathcal{O}$ to the linear space $\mathrm{Op}(D)$ of symmetric operators which preserve a fixed dense domain D in some separable Hilbert space $\mathcal{H}$, such that for all $f, g \in \mathcal{O}$,

(Q1) $\mathcal{Q}(\{f, g\}) = \frac{i}{\hbar}[\mathcal{Q}(f), \mathcal{Q}(g)]$,

(Q2) if the constant function $1 \in \mathcal{O}$, then $\mathcal{Q}(1) = I$, and

(Q3) if the Hamiltonian vector field X_f of f is complete, then $\mathcal{Q}(f)$ is essentially self-adjoint on D.

If $\mathcal{O} = C^\infty(M)$, the prequantization is said to be *full*. A prequantization $\mathcal{Q}$ is *nontrivial* provided codim ker $\mathcal{Q} > 1$; otherwise $\mathcal{Q}$ factors through a representation of $\mathcal{O}/\ker \mathcal{Q}$ with $\dim(\mathcal{O}/\ker \mathcal{Q}) \leq 1$.

Remarks. 1. By virtue of (Q1) a prequantization $\mathcal{Q}$ of $\mathcal{O}$ is essentially a Lie representation of $\mathcal{O}$ by symmetric operators. (More precisely: If we set $\pi(f) = \frac{i}{\hbar} \mathcal{Q}(f)$, then π is a true Lie representation by skew-symmetric operators on D equipped with the commutator bracket. We will blur the distinction between π and $\mathcal{Q}$.) In this context there are several additional requirements we could place upon $\mathcal{Q}$, such as irreducibility and integrability. However, we do not want to be too selective at this point, so we do not insist on these; they will be discussed as the occasion warrants.

2. Condition (Q2) reflects the fact that if an observable f is a constant c, then the probability of measuring $f = c$ is one regardless of which quantum state the system is in. It also serves to eliminate some "trivial" possibilities, such as the regular representation $f \mapsto -i\hbar X_f$ on $L^2(M, \omega^n)$.

3. Regarding (Q3), we remark that in contradistinction with Van Hove [VH1], we do not confine our considerations to only those classical observables whose Hamiltonian vector fields are complete. Rather than taking the point of view that "incomplete" classical observables cannot be quantized, we simply do not demand that the corresponding quantum operators be essentially self-adjoint ("e.s.a."). We do not imply by this that

symmetric operators which are not e.s.a. are acceptable as physical observables; as is well known, this is a controversial point.

4. Notice that no assumptions are made at this stage regarding the multiplicative structure on $C^\infty(M)$ vis-à-vis $\mathcal{Q}$. This is partly for historical reasons: In classical mechanics the Lie algebra structure has played a more dominant role than the associative algebra structure, so it is natural to concentrate on the former. This is also the approach favored by Dirac [Di] and the geometric quantization theorists [So, Wo]. For more algebraic treatments, see [As, Em, VN]. The associative algebra structure is emphasized to a much greater degree in deformation quantization theory [BFFLS, Ri2, Ri3]. We shall make some comments on this as we go along; see especially §§5.1 and 6.5.

Prequantizations in this broad sense (even full ones) are usually easy to construct, cf. [Ch3, Ur, Wo]. Van Hove was the first to construct a full prequantization of $C^\infty(\mathbf{R}^{2n})$ [VH1]. It goes as follows: The Hilbert space $\mathcal{H}$ is $L^2(\mathbf{R}^{2n})$, the domain D is the Schwartz space $\mathcal{S}(\mathbf{R}^{2n},\mathbf{C})$ of rapidly decreasing smooth complex-valued functions (for instance), and for $f \in C^\infty(\mathbf{R}^{2n})$,

$$\mathcal{Q}(f) = -i\hbar \sum_{k=1}^{n} \left[\frac{\partial f}{\partial p_k} \left(\frac{\partial}{\partial q^k} - \frac{i}{\hbar} p_k \right) - \frac{\partial f}{\partial q^k} \frac{\partial}{\partial p_k} \right] + f. \tag{1}$$

As luck would have it, however, prequantization representations tend to be flawed. For example, the Van Hove prequantization of $C^\infty(\mathbf{R}^{2n})$, when restricted to the Heisenberg subalgebra $\mathrm{h}(2n) \cong \mathrm{span}\{1, p_i, q^i \mid i = 1, \ldots, n\}$, is not unitarily equivalent to the Schrödinger representation (which it ought to be, in the context of a particle moving in $\mathbf{R}^n$ with no superselection rules) [Bl1, Ch1]. (Recall that the *Schrödinger representation* of $\mathrm{h}(2n)$ is defined to be[1]

$$q^i \mapsto q^i, \quad p_j \mapsto -i\hbar\, \partial/\partial q^j, \quad \text{and} \quad 1 \mapsto I \tag{2}$$

on the domain $\mathcal{S}(\mathbf{R}^n,\mathbf{C}) \subset L^2(\mathbf{R}^n)$. It is irreducible in the sense given in §4.) There are various ways to see this; we give Van Hove's original proof [VH1, §17] as it will be useful later. Take $n = 1$ for simplicity. First, define a unitary operator F on $L^2(\mathbf{R}^2)$ by

$$(F\psi)(p,q) = \frac{1}{\sqrt{h}} \int_{-\infty}^{\infty} e^{ipv/\hbar} \psi(v, q-v)\, dv.$$

Then for each fixed $j = 0, 1, \ldots$ take $\mathcal{H}_j$ to be the closure in $L^2(\mathbf{R}^2)$ of the linear span of elements of the form Fh_{jk}, where $h_{jk}(p,q) = h_j(p)h_k(q)$ and

$$h_k(q) = e^{q^2/2} \frac{d^k}{dq^k} e^{-q^2} \tag{3}$$

is the Hermite function of degree k. Now from (1),

$$\mathcal{Q}(q) = i\hbar \frac{\partial}{\partial p} + q, \quad \mathcal{Q}(p) = -i\hbar \frac{\partial}{\partial q}.$$

---

[1] We denote multiplication operators as functions.

These operators are e.s.a. on $\mathcal{S}(\mathbf{R}^2,\mathbf{C})$, and one may verify that they strongly commute with the orthogonal projectors onto the closed subspaces $\mathcal{H}_j$.[2] Thus the Van Hove prequantization of $C^\infty(\mathbf{R}^2)$ is reducible when restricted to the Heisenberg subalgebra and hence does not produce the Schrödinger representation. Moreover the association $F h_{jk}(p, q) \mapsto c_j h_k(q)$, where the c_j are normalization constants, provides a unitary equivalence of each subrepresentation of h(2) on $\mathcal{H}_j$ with the Schrödinger representation on $L^2(\mathbf{R})$, from which we see that the multiplicity of the latter is infinite in the Van Hove representation. The Van Hove representation suffers from other defects as well [Zi, §4.5.B].

Likewise, the Kostant-Souriau prequantizations of S^2 do not reproduce the familiar spin representations of the special unitary algebra su(2). We realize S^2 as a coadjoint orbit in su(2)* according to $\mathbf{S} \cdot \mathbf{S} = s^2$, where $\mathbf{S} = (S_1, S_2, S_3)$ is the spin vector and $s > 0$ is the classical spin. It comes equipped with the symplectic form

$$\omega = \frac{1}{2s^2} \sum_{i,j,k=1}^{3} \epsilon_{ijk} \, S_i \, dS_j \wedge dS_k. \tag{4}$$

Now the de Rham class $[\omega/h]$ is integral iff $s = \frac{n}{2}\hbar$, where n is a positive integer, and the corresponding Kostant-Souriau prequantum line bundles can be shown to be $L^{\otimes n}$ where L is the dual of the universal line bundle over S^2 [Ki]. The corresponding prequantum Hilbert spaces $\mathcal{H}_n$ can thus be identified with spaces of square integrable sections ψ of these bundles w.r.t. the inner product

$$\langle \psi, \phi \rangle = \frac{i}{2\pi} \int_{\mathbf{C}} \frac{\overline{\psi(z)}\phi(z) \, dz \wedge d\bar{z}}{(1 + z\bar{z})^{n+2}}$$

where $z = (S_1 + i S_2)/(s - S_3)$, cf. [Wo]. But these $\mathcal{H}_n$ are infinite-dimensional, whereas the standard representation spaces for quantum spin $s = \frac{n}{2}\hbar$ have dimension $n + 1$.

In both examples the prequantization Hilbert spaces are "too big." The main problem is how to remedy this, in other words, how to modify the notion of a prequantization so as to yield a genuine *quantization*.

It is here that the ideas start to diverge, because there is less agreement in the literature as to what constitutes a quantization map. Some versions define it as a prequantization, not necessarily defined on the whole of $C^\infty(M)$, which is irreducible on a "basic set" $\mathfrak{b} \subset C^\infty(M)$ [Ki]. This is in line with the group theoretical approach to quantization [Is], in which context $\mathfrak{b}$ is realized as the Lie algebra of a symmetry group;[3] quantization should then yield an irreducible representation of this algebra. For example, when $M = \mathbf{R}^{2n}$, one usually takes $\mathfrak{b}$ to be the Heisenberg algebra h(2n) $\cong$ span$\{1, p_i, q^i \mid i = 1, \ldots, n\}$ of polynomials of degree at most one. Similarly, when $M = S^2$, one takes for $\mathfrak{b}$ the special unitary algebra su(2) $\cong$ span$\{S_1, S_2, S_3\}$ of spherical harmonics of degree one. We will plumb in detail the rationale behind these choices of $\mathfrak{b}$ in the next section.

---

[2] Recall that two e.s.a. (or, more generally, normal) operators *strongly commute* iff their spectral resolutions commute, cf. [ReSi, §VIII.5]. Two operators A, B *weakly commute* on a domain D if they commute in the ordinary sense, i.e., $[A, B]$ is defined on D and vanishes.

[3] We typically identify an abstract Lie algebra with its isomorph in $C^\infty(M)$.

A different approach to quantization is to require a prequantization $\mathcal{Q}$ to satisfy some "Von Neumann rule," that is, some given relation between the classical multiplicative structure of $C^\infty(M)$ and operator multiplication on $\mathcal{H}$. (Note that thus far in our discussion the multiplication on $C^\infty(M)$ has been ignored, and it is reasonable to require that quantization preserve at least some of the associative algebra structure of $C^\infty(M)$, given that the Leibniz rule intertwines pointwise multiplication with the Poisson bracket.) There are many different types of such rules [Co, Fo, KLZ, KS, Ku, MC, VN], the simplest being of the form

$$\mathcal{Q}(\varphi \circ f) = \varphi\big(\mathcal{Q}(f)\big) \qquad (5)$$

for some distinguished observables $f \in C^\infty(M)$, and certain smooth functions $\varphi \in C^\infty(\mathbf{R})$. (Technically, if φ is not a polynomial, then $\mathcal{Q}(f)$ must be e.s.a. for $\varphi\big(\mathcal{Q}(f)\big)$ to be defined.) In the case of $M = \mathbf{R}^{2n}$, Von Neumann states that the physical interpretation of the quantum theory requires (5) to hold for *all* $f \in C^\infty(M)$ and $\varphi \in C^\infty(\mathbf{R})$ [VN]. However, it is easy to see that this is impossible (simple demonstrations are given in [AB, Fo] as well as §5.1 following); hence the qualifiers in the definition above. In this example, one typically ends up imposing the squaring Von Neumann rule $\varphi(x) = x^2$ on elements of h($2n$). The relevant rule for the sphere turns out to be less intuitive; it takes the form $\mathcal{Q}(S_i{}^2) = a\mathcal{Q}(S_i)^2 + cI$ for $i = 1, 2, 3$, where a, c are undetermined (representation-dependent) constants subject only to the constraint that $a^2 + c^2 \neq 0$. Derivations of these rules in these two examples are given in §5 and [GGH].

Another type of quantization is obtained by "polarizing" a prequantization representation [Wo]. Following Blattner [Bl1], we paraphrase it algebraically as follows. Start with a *polarization*, i.e., a maximally commuting Lie subalgebra $\mathcal{A}$ of $C^\infty(M)$. Then require for the quantization map $\mathcal{Q}$ that the image $\mathcal{Q}(\mathcal{A})$ be "maximally commuting" as operators. (If $\mathcal{Q}(\mathcal{A})$ consists of bounded operators, this means that the weak operator closure of the *-algebra generated by $\mathcal{Q}(\mathcal{A})$ ($= \mathcal{Q}(\mathcal{A})''$) is maximally commuting in $B(\mathcal{H})$. If $\mathcal{Q}(\mathcal{A})$ contains unbounded operators, one should look for a generating set of normal operators in $\mathcal{Q}(\mathcal{A})$, and require that the Von Neumann algebra generated by their spectral projections is maximally commuting.) One can then realize the Hilbert space $\mathcal{H}$ as an L^2-space over the spectrum of this Von Neumann algebra on which this algebra acts as multiplication operators. There will also be a cyclic and separating vector for such an algebra, which provides a suitable candidate for a vacuum vector. Thus another motivation for polarizations is that a maximally commuting set of observables provides a set of compatible measurements, which can determine the state of a system. When $M = \mathbf{R}^{2n}$, one often takes the "vertical" polarization $\mathcal{A} = \{f(q^1, \ldots, q^n)\}$, in which case one recovers the standard position or coordinate representation. However, in some instances, such as S^2, it is useful to broaden the notion of polarization to that of a maximally commuting subalgebra of the complexified Poisson algebra $C^\infty(M, \mathbf{C})$. Then, thinking of S^2 as $\mathbf{C}P^1$, we may take the "antiholomorphic" polarization $\mathcal{A} = \{f(z)\}$, which leads to the usual representations for spin. For treatments of polarizations in the context of deformation quantization, see [Fr, He].

Thus, informally, a "quantization" could be defined as a prequantization which incorporates one (or more) of the three additional requirements above (or possibly even others). Before proceeding, however, there are two points we would like to make.

The first is that it is of course not enough to simply state the requirements that a quantization map should satisfy; one must also devise methods for implementing them in practice. Thus geometric quantization theory, for example, provides a specific technique for polarizing certain (Kostant-Souriau) prequantization representations [Bl1, Ki, So, Wo]. However, as we are interested here in the structural aspects of quantization theory, and not in specific quantization schemes, we do not attempt to find such implementations.

Second, these three approaches to a quantization map are not independent; there exist subtle connections between them which are not well understood. For instance, demanding that a prequantization be irreducible on some basic algebra typically leads to the appearance of Von Neumann rules; this is how the Von Neumann rules for $\mathbf{R}^{2n}$ and S^2 mentioned above arise. We will delineate these connections in specific cases in §5, and more generally in §7.

At the core of each of the approaches above is the imposition—in some guise—of an irreducibility requirement, which is used to "cut down" a prequantization representation. Since this is most apparent in the first approach, we will henceforth concentrate on it. We will tie in the two remaining approaches as we go along.

So let $\mathcal{O}$ be a Lie subalgebra of $C^\infty(M)$, and suppose that $\mathfrak{b} \subset \mathcal{O}$ is a "basic algebra" of observables. Provisionally, we take a *quantization* of the pair $(\mathcal{O}, \mathfrak{b})$ to mean a prequantization $\mathcal{Q}$ of $\mathcal{O}$ which (among other things) irreducibly represents $\mathfrak{b}$. In the next two sections we will make this more precise, as well as examine in detail the criteria that $\mathfrak{b}$ should satisfy.

Natural issues to address for quantizations are existence, uniqueness and classification, and functoriality. For *pre*quantizations these questions already have partial answers from geometric quantization theory. So in particular we know that if (M, ω) satisfies the integrality condition $[\omega/h] \in H^2(M, \mathbf{Z})$, then full prequantizations of the Poisson algebra $C^\infty(M)$ exist, and that certain types of these—the Kostant-Souriau prequantizations— can be classified cohomologically [Ur, Wo]. For some limited types of manifolds the functorial properties of these prequantizations were considered by Blattner [Bl1]. However, as there are prequantizations not of the Kostant-Souriau type [Av3, Ch3], these questions are still open in general (especially for manifolds which violate the integrality condition [We]).

For quantization maps these questions are far more problematic. Our main focus will be on the existence of both *full quantizations*, by which we mean a quantization of $(C^\infty(M), \mathfrak{b})$ for some appropriately chosen basic algebra $\mathfrak{b}$, and *polynomial quantizations*, by which we mean a quantization of $(P(\mathfrak{b}), \mathfrak{b})$, where $P(\mathfrak{b})$ is the Poisson algebra of polynomials generated by $\mathfrak{b}$. As indicated earlier, these are not completely understood in general, although substantial progress has been made in the past several years. In our terminology, the classical result of Groenewold states that there is no quantization of $\big(P(\mathrm{h}(2n)), \mathrm{h}(2n)\big)$ on $\mathbf{R}^{2n}$, while the more recent results of [GGH] and [GGru1] imply the same for $\big(P(\mathrm{su}(2)), \mathrm{su}(2)\big)$ on S^2 and $\big(P(\mathrm{e}(2)), \mathrm{e}(2)\big)$ on T^*S^1, respectively, where e(2) is the Euclidean algebra (cf. §5.3). On the other hand, nontrivial polynomial quantizations do exist: One can construct such a quantization of $T^*\mathbf{R}_+$ with the affine algebra a(1) [GGra1]. In fact, full quantizations exist as well; there is one of T^2 with $\mathfrak{b}$ the Lie algebra of trigonometric polynomials of mean zero [Go3]. However, it does seem that nonexistence results are the rule. In the absence of a full (resp. a polynomial) quantization, then, it is important to determine the largest Lie subalgebras $\mathcal{O}$ of $C^\infty(M)$ (resp.

$P(\mathfrak{b})$) for which $(\mathcal{O}, \mathfrak{b})$ can be quantized. This we will investigate for $\mathbf{R}^{2n}$, S^2, and T^*S^1 in §5. At present, questions of uniqueness and classification can only be answered in specific examples.

3. Basic Algebras of Observables

Our first goal here is to make clear what we mean by a basic algebra of observables $\mathfrak{b} \subset C^\infty(M)$. Such algebras, in one way or another, play an important role in many quantization methods, such as geometric quantization [Ki], deformation quantization [BFFLS, Fr] and also the group theoretic approach [Is].

We start with the most straightforward case, that of an "elementary system" in the terminology of Souriau [So, Wo]. This means that M is a homogeneous space for a Hamiltonian action of a finite-dimensional Lie group G. The appeal of an elementary system is that it is a classical version of an irreducible representation: Using the transitive action of G, one can obtain any classical state from any other one, in direct analogy with the fact that every nonzero vector in a Hilbert space $\mathcal{H}$ is cyclic for an irreducible unitary representation ("IUR") of G on $\mathcal{H}$ [BaRa, §5.4]. Now notice that the span $\mathfrak{j}$ of the components of the associated (equivariant) momentum map satisfies:
    (J1) $\mathfrak{j}$ is a finite-dimensional Lie subalgebra of $C^\infty(M)$,

    (J2) the Hamiltonian vector fields X_f, $f \in \mathfrak{j}$, are complete, and

    (J3) $\{X_f \mid f \in \mathfrak{j}\}$ spans TM.

For both $M = \mathbf{R}^{2n}$ and S^2, the basic algebras are precisely of this type: From the elementary systems of the Heisenberg group $H(2n)$ acting on $\mathbf{R}^{2n}$, and the special unitary group $SU(2)$ acting on S^2, we have for $\mathfrak{j}$ the spaces $\operatorname{span}\{1, p_i, q^i \mid i = 1, \ldots, n\}$ and $\operatorname{span}\{S_1, S_2, S_3\}$, respectively. The same is true for $M = T^*S^1$ and $T^*\mathbf{R}_+$, as explained in §5.

Property (J3) is just an infinitesimal restatement of transitivity, and so we call a subset of $C^\infty(M)$ *transitive* if it satisfies this condition. Kirillov [Ki] uses the terminology "complete," motivated by the fact that such a set of observables locally separates classical states. (If a set of observables *globally* separates classical states, we call it *separating*.) In this regard, the finite-dimensionality criterion in (J1) plays an important role operationally: It guarantees that a finite number of measurements using this collection of observables will suffice to distinguish any two nearby states.

A Lie subalgebra $\mathfrak{b} \subset C^\infty(M)$ satisfying (J1)–(J3) is a prototypic basic algebra. However, there need not exist basic algebras in this sense for arbitrary M. For instance, if $M = T^2$, the self-action of the torus is not Hamiltonian, so there is no momentum map. Thinking of T^2 as $\mathbf{R}^2/\mathbf{Z}^2$, a natural choice of basic algebra is then the Lie algebra $\mathfrak{t}$ generated by

$$\mathcal{T} = \{\sin 2\pi x, \cos 2\pi x, \sin 2\pi y, \cos 2\pi y\}.$$

This Lie algebra—viz. the trigonometric polynomials of mean zero—is infinite-dimensional. While perhaps unpleasant, this is in fact unavoidable: It follows from Proposition 2 below that there is no finite-dimensional basic algebra on T^2. However, in keeping with the discussion above, note that t is finitely generated, and one can use this generating set to separate states.

We will therefore dispense with the finite-dimensionality assumption, and instead merely require that b be finitely generated. One then still has a finite number of "basic observables" with which to distinguish states. Thus we make:

Definition 2. A *basic algebra of observables* b is a Lie subalgebra of $C^\infty(M)$ such that:[4]
(B1) b is finitely generated,

(B2) the Hamiltonian vector fields X_f, $f \in b$, are complete,

(B3) b is transitive and separating, and

(B4) b is a minimal Lie algebra satisfying these requirements.

We spend some time elaborating on this definition. First, the completeness condition (B2) guarantees that a basic observable generates a one-parameter group of canonical transformations. In view of (Q3), it is the classical analogue of the requirement that an operator representing a physically observable quantity should be e.s.a., whence it generates a one-parameter group of unitary transformations.

Next consider the transitivity requirement in (B3). When b is finite-dimensional, it together with (B2) enables us to integrate b to a transitive group action on M. Indeed, the map $f \mapsto X_f$ can be thought of as an action of b on M. By (B2) the vector fields X_f are complete and so by a theorem of Palais [Va, Thm. 2.16.13] this action of b can be integrated to an action of the corresponding simply connected Lie group B. Condition (B3) implies that this action is locally transitive and thus globally transitive as M is connected.

As part of (B3) we also require that b globally separate classical states. This ensures that b accurately reflects the topology of M [Ve]. Without it, e.g., the Lie algebra t defined above could equally well live on either $\mathbf{R}^2$ or T^2 (or even "halfway between," on T^*S^1); measurements using t could not distinguish amongst these phase spaces.

Although a transitive set of observables is locally separating, it need not be (globally) separating. Conversely, a separating set of observables need not be everywhere transitive. So these two conditions are distinct.

While (B3) is geometrically natural, there are other conditions one might use in place of it. By way of motivation, consider a unitary representation U of a Lie group G on a Hilbert space $\mathcal{H}$. The representation U is irreducible iff the *-algebra $\mathcal{U}$ of bounded operators generated by $\{U(g) \mid g \in G\}$ is irreducible, in which case we have the following equivalent characterizations of irreducibility:
(I1) The commutant $\mathcal{U}' = \mathbf{C}I$, and

(I2) the weak operator closure of $\mathcal{U}$ is the algebra of all bounded operators: $\overline{\mathcal{U}}^w = B(\mathcal{H})$ $(= \mathcal{U}'')$.

---

[4] This definition differs from that given in [GGT] in three regards: It is phrased in terms of basic *algebras* as opposed to basic *sets*, we no longer insist that $1 \in b$ (this is superfluous), and we have strengthened (B3) by requiring that b be separating.

That (I1) is equivalent to irreducibility is the content of Schur's Lemma. Property (I2) means that all bounded operators can be built from those in $\mathcal{U}$ by weak operator limits. It follows from (I1), the Von Neumann density theorem [BrRo, Cor. 2.4.15], and the fact that $\mathcal{U}' = \left(\overline{\mathcal{U}}^w\right)'$. Clearly (I2) implies (I1).

These restatements of irreducibility have the following classical analogues for a set $\mathcal{F}$ of observables:

(C1) $\{f, g\} = 0$ for all $f \in \mathcal{F}$ implies g is constant, and

(C2) the Poisson algebra of polynomials generated by $\mathcal{F}$ forms a dense subspace in $C^\infty(M)$.

For (C2) a topology on $C^\infty(M)$ must be decided on, and we will use the topology of uniform convergence on compacta of a function as well as its derivatives.

Because the algebraic structures of classical and quantum mechanics are different, (C1) and (C2) lead to inequivalent notions of "classical irreducibility." It is not difficult to verify that (C1) $\Leftarrow$ (B3) $\Leftarrow$ (C2) strictly. In principle either of (C1) or (C2) could serve in place of (B3). Indeed, since on $C^\infty(M)$ one has two algebraic operations, it is natural to consider irreducibility in either context: in terms of the multiplicative structure (C2), or the Poisson bracket (C1). However, it turns out that (C1) is too weak for our purposes, while (C2) is too strong.

The nondegeneracy condition (C1) is equivalent to the statement that observables in $\mathfrak{b}$ locally separate states almost everywhere [Ki]. It is also equivalent to the statement that the Hamiltonian vector fields of elements of $\mathfrak{b}$ span the tangent spaces to M almost everywhere. Consequently, it will not do to replace (B3) by (C1) in the definition of basic algebra, for then as shown below the Lie algebra $\mathfrak{t}$ on T^2 would no longer be minimal, which seems both awkward and unreasonable. Furthermore, unlike (B3), (C1) has the defect that the simply connected covering group of $\mathfrak{b}$ need not act transitively on M. This happens for the symplectic algebra $\mathrm{sp}(2, \mathbf{R}) \cong \mathrm{span}\{p^2, pq, q^2\}$ on $\mathbf{R}^2$. Condition (C2) fails for the affine algebra $\mathfrak{a}(1) \cong \mathrm{span}\{pq, q^2\}$ on $T^*\mathbf{R}_+$ since, e.g., $C^\infty(T^*\mathbf{R}_+)$ contains functions which blow up as $q \to 0$ along with all their q-derivatives, and such functions cannot be approximated by polynomials in the elements of $\mathfrak{a}(1)$. On the other hand, all these examples satisfy (B3), which shows that this is a reasonable condition to impose.

Finally, the minimality condition (B4) is crucial. From a physical or operational point of view, it is not obvious that it is necessary, as long as $\mathfrak{b}$ is finitely generated. But the quantization of a pair $(\mathcal{O}, \mathfrak{b})$ with $\mathfrak{b}$ nonminimal in this sense can lead to physically incorrect results.

Here is an illustration. First observe that the extended symplectic group $\mathrm{HSp}(2n,\mathbf{R})$ (which is the semidirect product of the symplectic group $\mathrm{Sp}(2n,\mathbf{R})$ with the Heisenberg group $\mathrm{H}(2n)$) acts transitively on $\mathbf{R}^{2n}$. This action has a momentum map whose components consist of all inhomogeneous quadratic polynomials in the q^i and p_i. The corresponding Lie subalgebra $\mathfrak{j} \cong \mathrm{hsp}(2n, \mathbf{R})$ satisfies all the requirements for a basic algebra save minimality, since $\mathrm{h}(2n)$ is a separating transitive subalgebra of $\mathrm{hsp}(2n,\mathbf{R})$. Now consider again the Van Hove prequantization $\mathcal{Q}$ of $C^\infty(\mathbf{R}^{2n})$ for $n = 1$. In [VH1, §17] it is shown that $\mathcal{Q}$ is completely reducible when restricted to $\mathfrak{j}$. In fact, there exist exactly two nontrivial HSp(2,**R**)-invariant closed subspaces $\mathcal{H}_\pm$ in $L^2(\mathbf{R}^2)$, namely

$$\mathcal{H}_+ = \bigoplus_{j \text{ even}} \mathcal{H}_j \quad \text{and} \quad \mathcal{H}_- = \bigoplus_{j \text{ odd}} \mathcal{H}_j,$$

cf. §2. If we denote the corresponding subrepresentations of j on $\mathcal{S}(\mathbf{R}^2, \mathbf{C}) \cap \mathcal{H}_\pm$ by $\mathcal{Q}_\pm$, then it follows that $\mathcal{Q}_\pm$ are quantizations of the pair (j, j). But these quantizations are physically unacceptable, since—just like the full prequantization $\mathcal{Q}$—they are reducible when further restricted to h(2). On the one hand, asking for a quantization of (j, j) in this context is clearly the wrong thing to do, since compatibility with Schrödinger quantization devolves upon the irreducibility of an h(2) algebra, not an hsp(2, **R**) one. But on the other hand, this example does make our point.

To illustrate the appropriateness of Definition 2, consider again the torus and let t_k be the Lie algebras generated by the sets

$$\mathcal{T}_k = \{\sin 2\pi kx, \ \cos 2\pi kx, \ \sin 2\pi ky, \ \cos 2\pi ky\}$$

for $k = 1, 2, \ldots$ Each t_k is transitive. But without the separation axiom in (B3), none of the t_k would be minimal, since each contains the infinite descending series $t_k \supset t_{2k} \supset \cdots$. However, only $t = t_1$ is separating, and in fact it is a minimal *separating* transitive subalgebra.

Other properties that basic algebras might be required to satisfy are discussed in [Is]. For our purposes, (B1)–(B4) will suffice.

It is difficult to characterize basic algebras on general symplectic manifolds. In the compact case, however, we can be quite precise.

Proposition 1. *Let* b *be a finite-dimensional basic algebra on a compact symplectic manifold. Then* b *is compact and semisimple. In particular, its center must be zero.*

Proof. Define an inner product on b according to

$$\langle f, g \rangle = \int_M fg \ \omega^n.$$

Using the identity

$$\{f, g\} \omega^n = n \, d(f \, dg \wedge \omega^{n-1}) \, . \tag{6}$$

together with Stokes' Theorem, we immediately verify that

$$\langle \{f, g\}, h \rangle + \langle g, \{f, h\} \rangle = 0$$

whence b is compact [On, §1.2.6]. As a consequence, b splits as the Lie algebra direct sum $\mathfrak{z} \oplus \mathfrak{s}$, where $\mathfrak{z}$ is the center of b and $\mathfrak{s}$ is semisimple [On, Prop. 1.2.8].

Now transitivity implies that any function which Poisson commutes with every element of b must be a constant, so that $\mathfrak{z} \subseteq \mathbf{R}$. But if equality holds then $\mathfrak{s}$ would be a separating transitive subalgebra, thereby violating (B4). Thus $\mathfrak{z} = \{0\}$ and b is semisimple. $\square$

In particular, the proof shows that any reductive (and consequently any compact) basic algebra must be semisimple.

There is no guarantee that a given symplectic manifold will carry a basic algebra. Indeed, the next proposition shows that those phase spaces which admit basic algebras form a quite restricted class.

Proposition 2. *If a connected symplectic manifold M admits a finite-dimensional basic algebra* $\mathfrak{b}$, *then M is a coadjoint orbit in* $\mathfrak{b}^*$. *In particular, when M is compact it must be simply connected.*

Proof. For if $\mathfrak{b}$ is a finite-dimensional basic algebra on M then, by our considerations above, M must be a homogeneous Hamiltonian B-space, where B is the simply connected covering group of $\mathfrak{b}$ and the momentum map J is given by $\langle J(m), b \rangle = b(m)$ for $b \in \mathfrak{b}$ and $m \in M$. The Kirillov-Kostant-Souriau Coadjoint Orbit Covering Theorem [MR, Thm. 14.6.5] then implies that $J : M \to \mathfrak{b}^*$ is a symplectic local diffeomorphism of M onto a coadjoint orbit $O \subset \mathfrak{b}^*$. Since by (B3) $\mathfrak{b}$ is separating, it follows that J is injective (for otherwise elements of $\mathfrak{b}$ cannot separate points in $J^{-1}(\mu)$ for $\mu \in O$.) Thus M is symplectomorphic to O.

If M is compact, then by Proposition 1 $\mathfrak{b}$ is compact and semisimple. We conclude that B is compact [On, p. 29]. But the coadjoint orbits of a compact connected Lie group are simply connected [Fi, Thm. 2.3.7]. ☐

As M is a homogeneous space for B, the last paragraph of this proof shows that M is compact iff $\mathfrak{b}$ is compact iff B is compact.

Thus the symplectic algebra sp(2,$\mathbf{R}$) $\cong$ span$\{p^2, pq, q^2\}$ is not a basic algebra on $\mathbf{R}^2 \setminus \{0\}$, since the latter is not a coadjoint orbit (but rather a double covering of one). (Note that sp(2,$\mathbf{R}$) satisfies all the criteria for a basic algebra save the separation axiom.) Even if $M \subset \mathfrak{g}^*$ is a coadjoint orbit, $\mathfrak{g}$ need *not* form a basic algebra on M. An example is provided by S^2, which is a coadjoint orbit in su(2)$^* \oplus \{0\} \subset$ su(2)$^* \oplus$ su(2)$^* \cong$ o(4)* with basic algebra su(2), not o(4). In the compact case we can be more explicit:

Proposition 3. *Let M be a coadjoint orbit in* $\mathfrak{g}^*$, *where* $\mathfrak{g}$ *is a compact semisimple Lie algebra. If either M is principal or* $\mathfrak{g}$ *is simple, then M admits* $\mathfrak{g}$ *as a basic algebra.*

The proof is given in [GGG]. This result is not true when $\mathfrak{g}$ is noncompact, cf. §6.5.

Despite all this, M may still carry *infinite*-dimensional basic algebras, as happens for T^2. Not much is known regarding these, and we refer the reader to [Is] for further discussion (cf. especially §4.8.4).

We denote by $P(\mathfrak{b})$ the polynomial algebra generated by $\mathfrak{b}$. Since $\mathfrak{b}$ is a Lie algebra, $P(\mathfrak{b})$ is a Poisson algebra. Note that (*i*) $P(\mathfrak{b})$ is not necessarily freely generated by $\mathfrak{b}$ as an associative algebra (cf. the examples in §5), and (*ii*) by definition $\mathbf{R} \subset P(\mathfrak{b})$. When $P(\mathfrak{b})$ is freely generated by $\mathfrak{b}$, it can be identified with the symmetric algebra $S(\mathfrak{b})$ over $\mathfrak{b}$, but otherwise $P(\mathfrak{b})$ is realized as the quotient of $S(\mathfrak{b})$ by the associative ideal generated by elements of the form $C - c$, where C is a "Casimir" and c is some constant (depending upon M).[5,6] Note that $S(\mathfrak{b})$ is itself a unital Poisson algebra, and that the canonical projection is a Poisson algebra homomorphism. In general we will not distinguish between $P(\mathfrak{b})$ and $S(\mathfrak{b})$, and in examples where Casimirs are present we will often work with representatives, i.e., on $S(\mathfrak{b})$, without explicitly stating so. Let $P^k(\mathfrak{b})$ denote the subspace of polynomials of minimal degree at most k. (Since $P(\mathfrak{b})$ is not necessarily freely generated by $\mathfrak{b}$, the notion of "degree" may not be well-defined, but that of "minimal degree" is.) In the cases when degree does make sense, we let $P_k(\mathfrak{b})$ denote the subspace of homogeneous polynomials of degree k, so that $P^k(\mathfrak{b}) = \oplus_{l=0}^{k} P_l(\mathfrak{b})$ (vector space direct sum). We then also introduce $P_{(k)}(\mathfrak{b}) = +_{l \geq k} P_l(\mathfrak{b})$. Notice that $P^1(\mathfrak{b}) = \mathfrak{b}$ or $\mathbf{R} \oplus \mathfrak{b}$, depending upon whether $1 \in \mathfrak{b}$ or not. When $\mathfrak{b}$ is fixed in context, we simply write $P = P(\mathfrak{b})$, etc.

4. Quantization

We are now ready to discuss what we mean by a "quantization." Let $\mathcal{O}$ be a Lie subalgebra of $C^\infty(M)$, and suppose that $\mathfrak{b} \subset \mathcal{O}$ is a basic algebra of observables. Two eminently reasonable requirements to place upon a quantization are irreducibility and integrability [BaRa, Fl, Is, Ki].

Irreducibility is of course one of the pillars of the quantum theory, and we have already seen the necessity of requiring that quantization represent $\mathfrak{b}$ irreducibly. We must however be careful to give a precise definition since the operators $\mathcal{Q}(f)$ are in general unbounded (although, according to (B2) and (Q3), all elements of $\mathcal{Q}(\mathfrak{b})$ are e.s.a.). So let $\mathcal{X}$ be a set of e.s.a. operators defined on a common invariant dense domain D in a Hilbert space $\mathcal{H}$. Then $\mathcal{X}$ is *irreducible* provided the only bounded self-adjoint operators which strongly commute with all $X \in \mathcal{X}$ are multiples of the identity. While this definition is fairly standard, and well suited to our needs, we note that other notions of irreducibility can be found in the literature [BaRa, MMSV, Tu]; cf. also the discussion at the end of this section.

Given such a set $\mathcal{X}$ of operators, let $\mathcal{U}(\mathcal{X})$ be the *-algebra generated by the unitary operators $\{ \exp(it\overline{X}) \mid t \in \mathbf{R}, \ X \in \mathcal{X} \}$, where $\overline{X}$ is the closure of X. Then by Schur's Lemma $\mathcal{X}$ is irreducible iff the only closed subspaces of $\mathcal{H}$ which are invariant under $\mathcal{U}(\mathcal{X})$ are $\{0\}$ and $\mathcal{H}$.

Turning now to integrability, we first consider the case when basic algebra is finite-dimensional. Then it is natural to demand that the Lie algebra representation $\mathcal{Q}(\mathfrak{b})$ on $D \subset \mathcal{H}$ be *integrable* in the following sense: There exists a unitary representation Π of some Lie group with Lie algebra $\mathfrak{b}$ on $\mathcal{H}$ such that $\mathcal{Q}(f) = -i\hbar \, d\Pi(f) \restriction D$ for all $f \in \mathfrak{b}$, where $d\Pi$ is the derived representation of Π. For this it is *not* sufficient that elements of $\mathfrak{b}$ quantize to e.s.a. operators on D [ReSi, §VIII.5]. But integrability will follow from the following result of Flato et al., cf. [Fl] and [BaRa, Ch. 11].

Proposition 4. *Let $\mathfrak{g}$ be a real finite-dimensional Lie algebra, and let π be a rep-*

---

[5] A *Casimir* is an element of the Lie center of $S(\mathfrak{b})$ which has no constant term.

[6] Should $1 \in \mathfrak{b}$, we identify it with the units in both $P(\mathfrak{b})$ and $S(\mathfrak{b})$.

resentation of $\mathfrak{g}$ *by skew-symmetric operators on a common dense invariant domain* D *in a Hilbert space* $\mathcal{H}$. *Suppose that* $\{\xi_1, \ldots, \xi_k\}$ *generates* $\mathfrak{g}$ *by linear combinations and repeated brackets. If* D *contains a dense set of separately analytic vectors for* $\{\pi(\xi_1), \ldots, \pi(\xi_k)\}$, *then there exists a unique unitary representation* Π *of the connected simply connected Lie group with Lie algebra* $\mathfrak{g}$ *on* $\mathcal{H}$ *such that* $d\Pi(\xi) \restriction D = \pi(\xi)$ *for all* $\xi \in \mathfrak{g}$.

We recall that a vector ψ is *analytic* for an operator X on $\mathcal{H}$ provided the series

$$\sum_{k=0}^{\infty} \frac{\|X^k \psi\|}{k!} t^k$$

is defined and converges for some $t > 0$. If $\{X_1, \ldots, X_k\}$ is a set of operators defined on a common invariant dense domain D, a vector $\psi \in D$ is *separately analytic* for $\{X_1, \ldots, X_k\}$ if ψ is analytic for each X_j. By a slight abuse of terminology, we will say that a vector is separately analytic for a Lie algebra of operators $\mathcal{X}$ if it is separately analytic for some Lie generating set $\{X_1, \ldots, X_k\}$ of $\mathcal{X}$.

If it happens that $\mathfrak{b}$ is infinite-dimensional, then there need not exist a Lie group having $\mathfrak{b}$ as its Lie algebra. Even if such a Lie group existed, integrability is far from automatic, and technical difficulties abound. Thus we will not insist that a quantization be integrable in general. On the other hand, the analyticity requirement in Proposition 4 makes sense under all circumstances,[7] and does guarantee integrability when $\mathfrak{b}$ is finite-dimensional, so we will adopt it in lieu of integrability.

Finally, we will require that a quantization $\mathcal{Q}$ be faithful on $\mathfrak{b}$. While faithfulness is not usually assumed in the definition of a quantization, it seems to us a reasonable requirement in that a classical observable can hardly be regarded as "basic" in a physical sense if it is in the kernel of a quantization map. In this case, it cannot be obtained in any classical limit from a quantum theory.

Therefore we have at last:

Definition 3. A *quantization* of the pair $(\mathcal{O}, \mathfrak{b})$ is a prequantization $\mathcal{Q}$ of $\mathcal{O}$ on $\mathrm{Op}(D)$ satisfying

(Q4) $\mathcal{Q} \restriction \mathfrak{b}$ is irreducible,

(Q5) D contains a dense set of separately analytic vectors for $\mathcal{Q}(\mathfrak{b})$, and

(Q6) $\mathcal{Q} \restriction \mathfrak{b}$ is faithful.

Remarks. 5. There are a number of analyticity assumptions similar to (Q5) that one could make [Fl]; we have chosen the weakest possible one.

6. (Q5) is not a severe restriction: When $\mathfrak{b}$ is finite-dimensional, it is always possible to find representations of it on domains D which satisfy this property [Fl]. On the other hand, nonintegrable representations do exist in general [Fl, p. 247].

---

[7] As long as $\mathfrak{b}$ is finitely generated, which is assured by (B1).

7. Proposition 4 requires that a specific generating set for $\mathcal{Q}(\mathfrak{b})$ be singled out. This also is not a serious restriction: In examples, $\mathfrak{b}$ is often specified in this manner. It is possible that (Q5) could be satisfied for one such set but not another, but Remark 6 shows that the domain D can be chosen in such a way that this cannot happen if $\mathfrak{b}$ is finite-dimensional.

8. It is important to realize that irreducibility does not imply integrability. For instance, there is an irreducible representation of h(2) which is not integrable [ReSi, p. 275].

We briefly comment on the domains D appearing in Definition 3. For a representation π of a Lie algebra $\mathfrak{g}$ on a Hilbert space $\mathcal{H}$, there is typically a multitude of common, invariant dense domains that one can use as carriers of the representation. (See [BaRa, §11.2] for a discussion of some of the possibilities.) But what is ultimately important for our purposes are the closures $\overline{\pi(\xi)}$ for $\xi \in \mathfrak{g}$, and not the $\pi(\xi)$ themselves. So we do not want to distinguish between two representations π on Op(D) and π' on Op(D') whenever $\overline{\pi(\xi)} = \overline{\pi'(\xi)}$, in which case we say that π and π' are *coextensive*. In particular, it may happen that the given domain D for a representation π does not satisfy (Q5), but there is an extension to a coextensive representation π' on a domain D' that does.[8] In such cases we will suppose that the representation has been so extended.

We end this section with a refinement of the irreducibility condition (Q4). There is another, simpler notion of irreducibility which is very useful for our purposes: We say that $\mathcal{Q}(\mathfrak{b})$ is *algebraically irreducible* provided the only operators in Op(D) which (weakly) commute with all elements of $\mathcal{Q}(\mathfrak{b})$ are scalar multiples of the identity. It turns out that a quantization is automatically algebraically irreducible.

Proposition 5. *Let $\mathcal{Q}$ be a representation of a finite-dimensional Lie algebra $\mathfrak{b}$ by symmetric operators on an invariant dense domain D in a separable Hilbert space $\mathcal{H}$. If $\mathcal{Q}$ satisfies (Q4) and (Q5), then $\mathcal{Q}(\mathfrak{b})$ is algebraically irreducible.*

The proof, which hinges on an unbounded version of Schur's lemma [Ro], is given in [GGra1, Prop. 3]; cf. also [Go4].

5. Examples

In this section we present the gist of the arguments—more or less as they originally appeared—that there are no nontrivial polynomial quantizations of either $\mathbf{R}^{2n}$, S^2, or T^*S^1, with the basic algebras h($2n$), su(2), and e(2), respectively. The complete proofs can be found in [AM, Ch1, Fo, Go1, Go4, Gro, GS, VH1, VH2] for $\mathbf{R}^{2n}$, [GGH] for S^2, and [GGru1] for T^*S^1. The proofs in all three examples require a detailed knowledge of the structure of the Poisson algebras involved and their representations. Finally, we show following [GGra1] that there is a polynomial quantization of $T^*\mathbf{R}_+$ with the basic algebra a(1), and following [Go3] that there is a full quantization of T^2 with the basic

---

[8] A simple illustration is provided by the Schrödinger representation (2) with $D = C_c^\infty(\mathbf{R}^n, \mathbf{C})$ and $D' = \mathcal{S}(\mathbf{R}^n, \mathbf{C})$.

algebra t. We also take this opportunity to repair a defect in the standard presentations of the Groenewold-Van Hove theorem for $\mathbf{R}^{2n}$.

As an aside, we point out that many of the calculations in §§5.2 and 5.3 were done using the **Mathematica** package *NCAlgebra* [HM].

5.1. $\mathbf{R}^{2n}$

Before proceeding with the no-go theorem for $\mathbf{R}^{2n}$, we remark that already at a purely mathematical level one can observe a suggestive structural mismatch between the classical and the quantum formalisms. Since a prequantization is essentially a Lie algebra homomorphism, it "compares" the Lie algebra structure of $C^\infty(\mathbf{R}^{2n})$ with the Lie algebra of (skew-) symmetric operators (preserving a dense domain D) equipped with the commutator bracket. But if we take $P \subset C^\infty(\mathbf{R}^{2n})$ to be the subalgebra of polynomials, Joseph [Jo] has shown that P has outer derivations, but the enveloping algebra of the Heisenberg algebra h($2n$)—and hence that of the Schrödinger representation thereof on $L^2(\mathbf{R}^n)$—has none. In the next section we generalize this line of reasoning, and present another such "algebraic" no-go theorem to the effect that a unital Poisson algebra can never be realized as an associative algebra with the commutator bracket.

In particular, one can see at the outset that it is impossible for a prequantization to satisfy the "product $\rightarrow$ anti-commutator" rule. Taking $n = 1$ for simplicity, suppose $\mathcal{Q}$ were a prequantization of the polynomial algebra $P = \mathbf{R}[q, p]$ for which

$$\mathcal{Q}(fg) = \tfrac{1}{2}\big(\mathcal{Q}(f)\mathcal{Q}(g) + \mathcal{Q}(g)\mathcal{Q}(f)\big)$$

for all $f, g \in P$. Take $f(p, q) = p$ and $g(p, q) = q$. Then

$$\tfrac{1}{4}\big(\mathcal{Q}(p)\mathcal{Q}(q) + \mathcal{Q}(q)\mathcal{Q}(p)\big)^2 = \mathcal{Q}(pq)^2$$

$$= \mathcal{Q}(p^2 q^2) = \tfrac{1}{2}\big(\mathcal{Q}(p)^2\mathcal{Q}(q)^2 + \mathcal{Q}(q)^2\mathcal{Q}(p)^2\big).$$

Now by (Q1) we have $[\mathcal{Q}(p), \mathcal{Q}(q)] = -i\hbar I$, so that the L.H.S. reduces to

$$\mathcal{Q}(q)^2\mathcal{Q}(p)^2 - 2i\hbar\mathcal{Q}(q)\mathcal{Q}(p) - \tfrac{1}{4}\hbar^2 I$$

while the R.H.S. becomes

$$\mathcal{Q}(q)^2\mathcal{Q}(p)^2 - 2i\hbar\mathcal{Q}(q)\mathcal{Q}(p) - \hbar^2 I.$$

As the product $\rightarrow$ anti-commutator rule is equivalent to the squaring Von Neumann rule $\mathcal{Q}(f^2) = \mathcal{Q}(f)^2$, this contradiction also shows that the latter is inconsistent with prequantization. Note that the contradiction is obtained on quartic polynomials; there is no problem if consideration is limited to observables which are at most cubic.

This argument only used axiom (Q1) in the specific instance $[\mathcal{Q}(p), \mathcal{Q}(q)] = -i\hbar I$. Consequently, one still obtains a contradiction if one drops (Q1) and instead insists that $\mathcal{Q}$ be consistent with Schrödinger quantization (in which context this one commutation relation remains valid, cf. (2)). This *manifest* impossibility of satisfying the product $\rightarrow$ anti-commutator rule while being consistent with Schrödinger quantization is one reason we have decided to focus on the Lie structure as opposed to the associative structure of $C^\infty(M)$. See [AB] for further results in this direction.

We now turn to the no-go theorem for $\mathbf{R}^{2n}$. We shall state the main results for $\mathbf{R}^{2n}$ but, for convenience, usually prove them only for $n = 1$. The proofs for higher dimensions are immediate generalizations of these. In what follows $P = \mathbf{R}[q^1, \ldots, q^n, p_1, \ldots, p_n]$; note that $P^1 \cong h(2n)$, $P_2 \cong sp(2n, \mathbf{R})$, and $P^2 \cong hsp(2n, \mathbf{R})$.

Groenewold's celebrated result is:

Theorem 6. *There is no quantization of* (P, P^1).

Proof. Set $n = 1$ and let $\mathcal{Q}$ be a quantization of (P, P^1). We will show that a contradiction arises when cubic polynomials are considered.

We begin by determining $\mathcal{Q}(q^2)$. Set $\Delta = \mathcal{Q}(q^2) - \mathcal{Q}(q)^2$. Using (Q1) we readily verify that $[\Delta, \mathcal{Q}(q)] = 0$ and $[\Delta, \mathcal{Q}(p)] = 0$. But now algebraic irreducibility (cf. Proposition 5) implies that $\Delta = EI$ for some real constant E. Thus $\mathcal{Q}(q^2) = \mathcal{Q}(q)^2 + EI$.

An identical argument yields $\mathcal{Q}(p^2) = \mathcal{Q}(p)^2 + FI$. Quantizing the relation $4pq = \{p^2, q^2\}$ and using these formulæ then gives

$$\mathcal{Q}(pq) = \tfrac{1}{2}\big(\mathcal{Q}(p)\mathcal{Q}(q) + \mathcal{Q}(q)\mathcal{Q}(p)\big).$$

But upon quantizing $2q^2 = \{pq, q^2\}$ we find that $E = 0$. Similarly $F = 0$. Thus we have the Von Neumann rules

$$\mathcal{Q}(q^2) = \mathcal{Q}(q)^2, \quad \mathcal{Q}(p^2) = \mathcal{Q}(p)^2 \tag{7}$$

and

$$\mathcal{Q}(qp) = \tfrac{1}{2}\big(\mathcal{Q}(q)\mathcal{Q}(p) + \mathcal{Q}(p)\mathcal{Q}(q)\big). \tag{8}$$

These in turn lead to higher degree Von Neumann rules.

Lemma 1. *For all real-valued polynomials r,*

$$\mathcal{Q}\big(r(q)\big) = r\big(\mathcal{Q}(q)\big), \quad \mathcal{Q}\big(r(p)\big) = r\big(\mathcal{Q}(p)\big),$$

$$\mathcal{Q}\big(r(q)p\big) = \tfrac{1}{2}\big[r\big(\mathcal{Q}(q)\big)\mathcal{Q}(p) + \mathcal{Q}(p)r\big(\mathcal{Q}(q)\big)\big],$$

and

$$\mathcal{Q}\big(qr(p)\big) = \tfrac{1}{2}\big[\mathcal{Q}(q)r\big(\mathcal{Q}(p)\big) + r\big(\mathcal{Q}(p)\big)\mathcal{Q}(q)\big].$$

Proof. We illustrate this for $r(q) = q^3$. The other rules follow similarly using induction. Now $\{q^3, q\} = 0$ whence by (Q1) we have $[\mathcal{Q}(q^3), \mathcal{Q}(q)] = 0$. Since also $[\mathcal{Q}(q)^3, \mathcal{Q}(q)] = 0$, we may write $\mathcal{Q}(q^3) = \mathcal{Q}(q)^3 + T$ for some operator T which (weakly) commutes with $\mathcal{Q}(q)$. We likewise have using (7)

$$[\mathcal{Q}(q^3), \mathcal{Q}(p)] = -i\hbar\, \mathcal{Q}(\{q^3, p\}) = 3i\hbar\, \mathcal{Q}(q^2) = 3i\hbar\, \mathcal{Q}(q)^2 = [\mathcal{Q}(q)^3, \mathcal{Q}(p)]$$

from which we see that T commutes with $\mathcal{Q}(p)$ as well. Consequently, T also commutes with $\mathcal{Q}(q)\mathcal{Q}(p) + \mathcal{Q}(p)\mathcal{Q}(q)$. But then from (8),

$$
\begin{aligned}
\mathcal{Q}(q^3) &= \tfrac{1}{3}\mathcal{Q}(\{pq, q^3\}) = \tfrac{i}{3\hbar}[\mathcal{Q}(pq), \mathcal{Q}(q^3)] \\
&= \tfrac{i}{3\hbar}\left[\tfrac{1}{2}(\mathcal{Q}(q)\mathcal{Q}(p) + \mathcal{Q}(p)\mathcal{Q}(q)), \mathcal{Q}(q)^3 + T\right] \\
&= \tfrac{i}{6\hbar}[\mathcal{Q}(q)\mathcal{Q}(p) + \mathcal{Q}(p)\mathcal{Q}(q), \mathcal{Q}(q)^3] = \mathcal{Q}(q)^3. \qquad \triangledown
\end{aligned}
$$

With this lemma in hand, it is now a simple matter to prove the no-go theorem. Consider the classical equality

$$
\tfrac{1}{9}\{q^3, p^3\} = \tfrac{1}{3}\{q^2 p, p^2 q\}.
$$

Quantizing and then simplifying this, the formulæ in Lemma 1 give

$$
\mathcal{Q}(q)^2\mathcal{Q}(p)^2 - 2i\hbar\,\mathcal{Q}(q)\mathcal{Q}(p) - \tfrac{2}{3}\hbar^2 I
$$

for the L.H.S., and

$$
\mathcal{Q}(q)^2\mathcal{Q}(p)^2 - 2i\hbar\,\mathcal{Q}(q)\mathcal{Q}(p) - \tfrac{1}{3}\hbar^2 I
$$

for the R.H.S., which is a contradiction. $\square$

Remarks. 9. Note that this proof of the no-go theorem does not use the Stone-Von Neumann theorem.

10. There is a small gap in Groenewold's original proof of Theorem 6 [Gro] in that the Von Neumann rules (7) and (8) were not rigorously derived. In effect, Groenewold took the weak commutativity of Δ with $\mathcal{Q}(q)$ to mean strong commutativity.[9] Van Hove supplied an extra assumption which remedied the situation, and which in particular implies: If the Hamiltonian vector fields of f, g are complete and $\{f, g\} = 0$, then $\mathcal{Q}(f)$ and $\mathcal{Q}(g)$ *strongly* commute [VH1]. This assumption along with the Stone-Von Neumann theorem is used to derive the Von Neumann rules (7) and (8) in [AM, ?]. However, as our argument based on Proposition 5 shows, it is possible obtain the desired result directly from the quantization axioms, *without* introducing additional assumptions. See [Go4] for further discussion.

11. Van Hove [VH1] gave a slightly different analysis using only those observables $f \in C^\infty(\mathbf{R}^{2n})$ with complete Hamiltonian vector fields, and still obtained an obstruction (but now to quantizing all of $C^\infty(\mathbf{R}^{2n})$). Yet another variant of Groenewold's theorem will be presented in §6.5.

Even though all of P cannot be quantized, there does exist a quantization $d\varpi$ of the pair (P^2, P^1), given by the familiar formulæ

$$
d\varpi(q^i) = q^i, \quad d\varpi(1) = I, \quad d\varpi(p_j) = -i\hbar\frac{\partial}{\partial q^j},
$$

---

[9] This is quite understandable: Groenewold presented his proof in 1946, but the distinction between weak and strong commutativity was apparently not fully appreciated until the late 1950s with the work of Nelson, cf. [ReSi, §VIII.5].

$$d\varpi(q^i q^j) = q^i q^j, \quad d\varpi(p_i p_j) = -\hbar^2 \frac{\partial^2}{\partial q^i \partial q^j}, \tag{9}$$

$$d\varpi(q^i p_j) = -i\hbar\left(q^i \frac{\partial}{\partial q^j} + \frac{1}{2}\delta^i_j\right),$$

on the domain $S(\mathbf{R}^n, \mathbf{C}) \subset L^2(\mathbf{R}^n)$. Properties (Q1)–(Q3) and (Q6) are readily verified. (Q4) follows automatically since the restriction of $d\varpi$ to P^1 is just the Schrödinger representation. For (Q5), we recall the well-known fact that the Hermite functions (3) form a dense set of separately analytic vectors for $d\varpi(P^1)$. We call $d\varpi$ the "extended metaplectic quantization"; detailed discussions of it may be found in [Fo, GS].

In fact, $d\varpi$ is the "only" quantization of (P^2, P^1):

Proposition 7. *Up to restriction of representations, any quantization of (P^2, P^1) is unitarily equivalent to the extended metaplectic quantization.*

Proof. Suppose $\mathcal{Q}$ is a quantization of (P^2, P^1) on a domain D in a Hilbert space $\mathcal{H}$. Then by (Q4) and (Q5) $\mathcal{Q}(P^1)$ can be integrated to an irreducible unitary representation τ of H($2n$). The Stone-Von Neumann theorem then states that τ is unitarily equivalent to the Schrödinger representation Π, and hence $\tau = U\Pi U^{-1}$ for some unitary map $U : L^2(\mathbf{R}^n) \to \mathcal{H}$. Consequently, $\mathcal{Q}(f) = U\overline{d\Pi(f)}U^{-1}\upharpoonright D$ for all $f \in P^1$. It now follows from (2), the invariance of the domain D, and Sobolev's lemma that $U^{-1}D \subset S(\mathbf{R}^n)$, so that $U^{-1}(\mathcal{Q}\upharpoonright P^1)U$ is in fact the *restriction* of $d\Pi$ to $U^{-1}D \subset S(\mathbf{R}^n)$. Thus without loss of generality we may assume that $\mathcal{Q}(P^1)$ is the Schrödinger representation (2) and that the domain $D \subset S(\mathbf{R}^n)$.

Then the first part of the proof of Theorem 6 yields the Von Neumann rules (7) and (8), which in view of (2) imply that $\mathcal{Q}(P^2)$ is given by (9). Thus, up to unitary equivalence, $\mathcal{Q}$ must be either $d\varpi$ or a restriction thereof. $\square$

Since P^2 is a maximal Lie subalgebra of P [GS, §16], (Q1) implies that any quantization which extends $d\varpi$ must be defined on all of P. Theorem 6 then implies

Corollary 8. *The extended metaplectic quantization of (P^2, P^1) cannot be extended beyond P^2 in P.*

We hasten to add that there are subalgebras of P other than P^2 which can be quantized. For example, let

$$C = \left\{\sum_{i=1}^{n} f^i(q)p_i + g(q)\right\},$$

where f^i and g are polynomials. Then it is straightforward to verify that for each $\eta \in \mathbf{R}$, the map $\mathcal{Q}_\eta : C \to \mathrm{Op}(S(\mathbf{R}^n, \mathbf{C}))$ given by

$$\mathcal{Q}_\eta\left(\sum_{i=1}^{n} f^i(q)p_i + g(q)\right) = -i\hbar\sum_{i=1}^{n}\left(f^i(q)\frac{\partial}{\partial q^i} + \left[\frac{1}{2} + i\eta\right]\frac{\partial f^i}{\partial q^i}\right) + g(q)$$

is a quantization of (C, P^1). $\mathcal{Q}_0$ is the familiar "position" or "coordinate representation." The significance of the parameter η is explained in [ADT]. Since C is also a maximal subalgebra of P, Theorem 6 yields

Corollary 9. *The quantizations $\mathcal{Q}_\eta$ of (C, P^1) cannot be extended beyond C in P.*

We furthermore point out that Proposition 7 in [Go4] yields "uniqueness": Any quantization of (C, P^1) must be unitarily equivalent to (a restriction of) $\mathcal{Q}_\eta$ for some $\eta \in \mathbf{R}$.

A similar analysis applies to the "Fourier transform" of the subalgebra C, i.e. the "momentum subalgebra" of all polynomials which are at most affine in the coordinates q^i. In fact [Go4], it turns out for $n = 1$ that P^2 and C exhaust the list of isomorphism classes of maximal Lie subalgebras of P which contain P^1. This is not true in higher dimensions, however: The subalgebra

$$\{f(q^1)p_1 + g(q^1, q^2, p_2)\}$$

on $\mathbf{R}^4$ is maximal, but not isomorphic to either C or P^2. Furthermore, the subalgebra thereof for which g is at most quadratic in q^2, p_2 is maximal quantizable, but also not isomorphic to either C or P^2. It remains an open problem to determine the largest quantizable Lie subalgebras of P for $n > 1$.

5.2. S^2

Now we turn our attention to the sphere. Since S^2 is compact, all classical observables are complete. Moreover, $su(2) \cong \mathrm{span}\{S_1, S_2, S_3\}$ is a compact simple Lie algebra. Consequently the functional analytic subtleties present in the case of $\mathbf{R}^{2n}$ disappear. But the actual computations, which were fairly routine for $\mathbf{R}^{2n}$, turn out to be much more complicated for S^2.

The Poisson bracket on $C^\infty(S^2)$ corresponding to (4) is

$$\{f, g\} = - \sum_{i,j,k=1}^{3} \epsilon_{ijk} S_i \frac{\partial f}{\partial S_j} \frac{\partial g}{\partial S_k}.$$

In particular, we have the relations $\{S_j, S_k\} = - \sum_{l=1}^{3} \epsilon_{jkl} S_l$. In this example P is the polynomial algebra in the components of the spin vector, subject to the relation

$$S_1{}^2 + S_2{}^2 + S_3{}^2 = s^2. \tag{10}$$

We may identify P with the space of spherical harmonics. We have $P_1 \cong su(2)$ and $P^1 \cong u(2)$.

Let $\mathcal{Q}$ be a quantization of (P, P_1) on a Hilbert space $\mathcal{H}$, whence

$$[\mathcal{Q}(S_j), \mathcal{Q}(S_k)] = i\hbar \sum_{l=1}^{3} \epsilon_{jkl} \mathcal{Q}(S_l) \tag{11}$$

and

$$\mathcal{Q}(\mathbf{S}^2) = s^2 I. \tag{12}$$

By (Q5) and Proposition 4, $\mathcal{Q}(su(2))$ can be exponentiated to a unitary representation of SU(2) which, according to (Q4), is irreducible. Therefore $\mathcal{H}$ must be finite-dimensional, and $\mathcal{Q}(su(2))$ must be one of the usual spin angular momentum representations, labeled by $j = 0, \frac{1}{2}, 1, \ldots$ For a fixed value of j, $\dim \mathcal{H} = 2j + 1$ and

$$\sum_{i=1}^{3} \mathcal{Q}(S_i)^2 = \hbar^2 j(j+1)I. \tag{13}$$

Our goal is show that no such (nontrivial) quantization exists. Patterning our analysis after that for $\mathbf{R}^{2n}$, we use irreducibility to derive some Von Neumann rules.

Lemma 2. *For $i = 1, 2, 3$ we have*

$$\mathcal{Q}(S_i^2) = a\mathcal{Q}(S_i)^2 + cI \tag{14}$$

where a, c are representation dependent real constants with $a^2 + c^2 \neq 0$.

The proof is in [GGH]. From this we also derive

$$\mathcal{Q}(S_i S_k) = \frac{a}{2}\left(\mathcal{Q}(S_i)\mathcal{Q}(S_k) + \mathcal{Q}(S_k)\mathcal{Q}(S_i)\right) \tag{15}$$

for $i \neq k$. (As an aside, these formulæ show that a quantization, if it exists, may be badly behaved with respect to the multiplicative structure on $C^\infty(S^2)$; in particular, the product $\to$ anti-commutator rule need not hold. Remarkably, this is as it should be: For *if* this rule were valid, then – subject to a few mild assumptions on $\mathcal{Q}$—the classical spectrum of S_3, say, would have to coincide with that of $\mathcal{Q}(S_3)$ which is contrary to experiment [GGH].) With these tools, we can now prove the main result:

Theorem 10. *There is no nontrivial quantization of (P, P_1).*

Proof. Fix $j > 0$, as $j = 0$ produces a trivial quantization. Assuming that $\mathcal{Q}$ is a quantization of (P, P_1), we can use (11)-(15) to quantize the classical relation

$$s^2 S_3 = \left\{S_1^2 - S_2^2, \ S_1 S_2\right\} - \left\{S_2 S_3, \ S_3 S_1\right\}, \tag{16}$$

thereby obtaining

$$s^2 = a^2 \hbar^2 \left(j(j+1) - \tfrac{3}{4}\right) \tag{17}$$

which contradicts $s > 0$ for $j = \frac{1}{2}$. Now assume $j > \frac{1}{2}$, and quantize

$$2s^2 S_2 S_3 = \left\{S_2^2, \{S_1 S_2, S_1 S_3\}\right\} - \tfrac{3}{4}\left\{S_1^2, \{S_1^2, S_2 S_3\}\right\}, \tag{18}$$

similarly obtaining

$$s^2 = a^2 \hbar^2 \left(j(j+1) - \tfrac{9}{4}\right)$$

which contradicts (17). Thus we have derived contradictions for all $j > 0$, and the theorem is proven. $\square$

In view of the impossibility of quantizing (P, P_1), one can ask what the maximal Lie subalgebras in P are to which we can extend an irreducible representation of P_1. The following chain of results, which we quote without proof (cf. [GGH]), provides the answer.

Proposition 11. *P^1 is a maximal Lie subalgebra of $\mathbf{R} \oplus O \subset P$, where O is the Poisson algebra consisting of polynomials containing only terms of odd degree.*

Next we establish a no-go theorem for $(\mathbf{R} \oplus O, P_1)$. However, the Von Neumann rules listed in Lemma 2 involve only even degree polynomials, so these are not applicable in O. Fortunately, we have another set of Von Neumann rules, also implied by the irreducibility of $\mathcal{Q}(P_1)$, involving only terms of odd degree.

Lemma 3. *If $\mathcal{Q}$ is a quantization of $(\mathbf{R} \oplus O, P_1)$, then for $i = 1, 2, 3$,*

$$\mathcal{Q}(S_i{}^3) = b\mathcal{Q}(S_i)^3 + e\mathcal{Q}(S_i)$$

where $b, e \in \mathbf{R}$.

From this we prove (with far greater effort):

Theorem 12. *There is no nontrivial quantization of $(\mathbf{R} \oplus O, P_1)$.*

Now $\mathbf{R} \oplus O$ is itself a maximal Lie subalgebra of P, and in fact the only Lie subalgebras of P strictly containing P_1 are P^1, $\mathbf{R} \oplus O$, and P itself. On the other hand, $P^1 = \mathbf{R} \oplus P_1$ is obviously quantizable. Thus Theorem 12 and Proposition 11 combine to yield our sharpest result for the sphere:

Corollary 13. *No nontrivial quantization of (P^1, P_1) can be extended beyond P^1 in P.*

Thus within the algebra of polynomials, $\big(u(2), su(2)\big)$ is the most one can quantize.

There is a crucial structural difference between the Groenewold-Van Hove analysis of $\mathbf{R}^2$ and the current analysis of S^2. Within $P = \mathbf{R}[q, p]$ the Heisenberg algebra has as its Lie normalizer the algebra of polynomials of degree at most 2, and there is no obstruction to quantization in this algebra: The obstruction comes from the cubic polynomials. On the other hand, for the sphere, the special unitary algebra has as its normalizer the algebra of polynomials of degree at most one; we obtain an obstruction in the quadratic polynomials, and find that there is no extension possible for a quantization of P^1.

5.3. T^*S^1

Our final example of an obstruction is provided by the symplectic cylinder, which appears in geometric optics [GS, §17]. Endow T^*S^1 with the canonical Poisson bracket

$$\{f, g\} = \frac{\partial f}{\partial \ell}\frac{\partial g}{\partial \theta} - \frac{\partial f}{\partial \theta}\frac{\partial g}{\partial \ell},$$

where ℓ is the angular momentum conjugate to θ. While the symplectic self-action of T^*S^1 is not Hamiltonian (thinking of T^*S^1 as $S^1 \times \mathbf{R}$), the cylinder can nonetheless be realized as a coadjoint orbit of the special Euclidean group SE(2) [MR, §14.8]. The corresponding momentum map $T^*S^1 \to \mathrm{e}(2)^*$ has components $\{\sin\theta, \cos\theta, \ell\}$; therefore we take as a basic algebra

$$\mathrm{e}(2) \cong \mathrm{span}\{\sin\theta, \cos\theta, \ell\}.$$

The polynomial algebra P generated by this basic algebra consists of sums of multiples of terms of the form $\ell^r \sin^m\theta \cos^n\theta$ of total degree $r + m + n$ with r, m, n nonnegative integers. Then $P_1 \cong \mathrm{e}(2)$.

Our first task is to determine all possible quantizations of the basic algebra $\mathrm{e}(2)$. By virtue of (Q3) and (Q4), for this it suffices to compute the derived representations corresponding to the IURs of the universal covering group of SE(2), which is the semidirect product $\mathbf{R} \ltimes \mathbf{R}^2$ with the composition law

$$(t, x, y) \cdot (t', x', y') = (t + t', x' \cos t + y' \sin t + x, y' \cos t - x' \sin t + y).$$

From the theory of induced representations of semidirect products [Ma] (see also [Is, §5.8]), we find that the only nontrivial IURs are infinite-dimensional; up to unitary equivalence they take the form

$$\left(U_\nu^\lambda(t, x, y)\psi\right)(\theta) = e^{i\lambda(x\cos\theta + y\sin\theta)} e^{i\nu t} \psi(\theta + t)$$

on $L^2(S^1)$. Here λ, ν are real parameters satisfying $\lambda > 0$ and $0 \le \nu < 1$. We identify λ with the reciprocal of $\hbar$, cf. [GGru1, Is]. After rescaling appropriately, the corresponding derived representations become

$$dU_\nu(\ell) = -i\hbar\left(\frac{d}{d\theta} + i\nu I\right), \quad dU_\nu(\sin\theta) = \sin\theta, \quad dU_\nu(\cos\theta) = \cos\theta \qquad (19)$$

on the domain $C^\infty(S^1, \mathbf{C})$. Each representation dU_ν is a quantization of P_1; in particular, $\{e^{in\theta} \mid n \in \mathbf{Z}\} \subset C^\infty(S^1, \mathbf{C})$ is a dense set of separately analytic vectors for the above basis of $dU_\nu(P_1)$.

Now suppose that $\mathcal{Q}$ is a quantization of (P, P_1) on a domain D. Arguing as in the proof of Proposition 7, we find that up to unitary equivalence $\mathcal{Q} \upharpoonright P_1$ coincides with dU_ν for some ν and that $D \subset C^\infty(S^1, \mathbf{C})$. Up to coextension of representations (cf. the discussion at the end of §4), then, we may as well assume that $D = C^\infty(S^1, \mathbf{C})$.

Just as with our previous examples, we use irreducibility to obtain Von Neumann rules. In [GGru1] we compute

$$\mathcal{Q}(\ell^2) = \mathcal{Q}(\ell)^2 + cI,$$

where $c \in \mathbf{R}$ is arbitrary. From this and (19) we eventually derive

$$\mathcal{Q}(\ell^2 \sin\theta) = \mathcal{Q}(\sin\theta)\mathcal{Q}(\ell)^2 - i\hbar\mathcal{Q}(\cos\theta)\mathcal{Q}(\ell) + \frac{\hbar^2}{4}\mathcal{Q}(\sin\theta) \qquad (20)$$

and

$$\mathcal{Q}(\ell^2 \cos\theta) = \mathcal{Q}(\cos\theta)\mathcal{Q}(\ell)^2 + i\hbar\mathcal{Q}(\sin\theta)\mathcal{Q}(\ell) + \frac{\hbar^2}{4}\mathcal{Q}(\cos\theta). \qquad (21)$$

Theorem 14. *There is no nontrivial quantization of* (P, P_1).

Proof. We merely use (19)–(21) to quantize the bracket relation

$$2\{\{\ell^2 \sin\theta, \ell^2 \cos\theta\}, \cos\theta\} = 12\ell^2 \sin\theta. \tag{22}$$

After simplifying, the left hand side reduces to

$$12\mathcal{Q}(\sin\theta)\mathcal{Q}(\ell)^2 - 12i\hbar\mathcal{Q}(\cos\theta)\mathcal{Q}(\ell) + 5\hbar^2\mathcal{Q}(\sin\theta),$$

whereas the right hand side is

$$12\mathcal{Q}(\sin\theta)\mathcal{Q}(\ell)^2 - 12i\hbar\mathcal{Q}(\cos\theta)\mathcal{Q}(\ell) + 3\hbar^2\mathcal{Q}(\sin\theta),$$

and the required contradiction is evident.   □

We next determine the maximal Lie subalgebras of P to which we can extend an irreducible representation of P_1. Such subalgebras certainly exist: For instance, there is a two-parameter family of quantizations of the pair (L^1, P_1), where L^1 denotes the Lie subalgebra of polynomials which are at most affine in ℓ. They are the "position representations" on $C^\infty(S^1, \mathbf{C}) \subset L^2(S^1)$ given by

$$\mathcal{Q}_{v,\eta}\big(f(\theta)\ell + g(\theta)\big) = -i\hbar\left(f(\theta)\frac{d}{d\theta} + \left[\frac{1}{2} + i\eta\right]f'(\theta) + i\nu f(\theta)\right) + g(\theta), \tag{23}$$

where ν labels the IURs of the universal cover of SE(2) and η is real.

To this end we classify the maximal Lie subalgebras of P containing P_1. For each $\alpha \in \mathbf{R}$ let V_α be the Lie subalgebra generated by

$$\{1, \ \sin\theta, \ \cos\theta, \ \ell, \ \ell(\ell + \alpha)\cos(2N + 1)\theta, \ \ell(\ell + \alpha)\sin(2N + 1)\theta \mid N \in \mathbf{N}\}.$$

Although far from obvious, it turns out that [GGru1]

Proposition 15. L^1 *and* $V_\alpha, \alpha \in \mathbf{R}$, *are the only proper maximal Lie subalgebras of* P *strictly containing* P_1.

In contrast to L^1, it is possible to show that there is *no* nontrivial quantization of any V_α which represents P_1 irreducibly. (While the method of proof is the same as that of the no-go theorem for P presented above, we must make sure that all constructions take place in V_α. The details may be found in [GGru1].) Since L^1 is maximal, Theorem 14 implies that none of the quantizations $\mathcal{Q}_{v,\eta}$ can be extended beyond L^1 in P. Furthermore [GGru1], the quantizations (23) of L^1 are the only possible ones:

Theorem 16. *If* $\mathcal{Q}$ *is a nontrivial quantization of* (L^1, P_1), *then* $\mathcal{Q} = \mathcal{Q}_{v,\eta}$ *for some* $\nu \in [0, 1)$ *and* $\eta \in \mathbf{R}$.

Taken together, these results completely characterize the polynomial quantizations for the basic algebra e(2).

Since T^*S^1 is covered by $\mathbf{R}^2$, and as e(2) is the natural analogue for the cylinder of h(2) for the plane, the quantization of the former might be expected to share some of the features of that of the latter, and we see from the above that in most respects this is so. In both examples there is an obstruction, and a maximal Lie subalgebra of polynomial observables that can be consistently quantized consists of those polynomials which are affine in the momentum.

There are some differences, however, which reflect the non-simple connectivity of T^*S^1. For instance, on $\mathbf{R}^2$, there are exactly two isomorphism classes of maximal polynomial Lie subalgebras containing the basic algebra span$\{1, q, p\}$ which can be consistently quantized, whereas according to Proposition 15 there is only one such containing span$\{\sin\theta, \cos\theta, \ell\}$ for the cylinder. (Since P^2 is not a Lie subalgebra of P, there is no analogue of the metaplectic representation for T^*S^1 and, since θ is an angular variable, there is also no cylindrical counterpart of the momentum representation.) Thus the possible polynomial quantizations of T^*S^1 are more limited than those of $\mathbf{R}^2$.

One topic for future exploration would be to consider the higher-dimensional analogues of the cylinder, viz. T^*S^n with basic algebra e(n).

5.4. $T^*\mathbf{R}_+$

We have encountered obstructions to quantization in the three examples presented so far, despite the fact that $\mathbf{R}^2$, T^*S^1, and S^2 are quite different structurally. Topologically these phase spaces range from contractible to compact, and algebraically the basic algebras h(2), e(2), and su(2) are nilpotent, solvable, and simple, respectively. Moreover, the representations of these algebras were in some instances unique and in others not, and they were finite- as well as infinite-dimensional. This wide array of behaviors strongly suggests that such obstructions should be ubiquitous. Therefore it comes as a surprise that this is *not* so [GGra1]: there is no obstruction to polynomially quantizing $T^*\mathbf{R}_+ = \{(q, p) \in \mathbf{R}^2 \mid q > 0\}$ with the "affine" basic algebra

$$a(1) \cong \text{span}\{pq, q^2\}.$$

The simply connected covering group of a(1) is isomorphic to the group $A_+(1) = \mathbf{R} \rtimes \mathbf{R}_+$ of orientation-preserving affine transformations of the line (hence the terminology). It is straightforward to check that $T^*\mathbf{R}_+$ with the canonical Poisson bracket can be realized as a coadjoint orbit in a(1)* [MR, §14.1(b)].

Upon writing

$$x = pq, \quad y = q^2$$

the bracket relation becomes $\{x, y\} = 2y$. The corresponding polynomial algebra $P = \mathbf{R}[x, y]$ is freely generated, and has the crucial feature that for each $k \geq 0$, the subspaces P_k are *ad*-invariant, i.e.,

$$\{P_1, P_k\} \subset P_k. \tag{24}$$

(Note that $P_1 \cong a(1)$). Because of this $\{P_k, P_l\} \subset P_{k+l}$, whence each $P_{(k)}$ is a Lie ideal. We thus have the semidirect sum decomposition

$$P = P^1 \ltimes P_{(2)}. \tag{25}$$

Now on to quantization. Since $P_{(2)}$ is a Lie ideal, we can obtain a quantization $\mathcal{Q}$ of *all* of P simply by finding an appropriate representation of $P^1 = \mathbf{R} \oplus P_1$ and setting $\mathcal{Q}(P_{(2)}) = \{0\}$!

Since $A_+(1)$ is a semidirect product we can generate the required representation of P_1 by induction. Following the recipe in [BaRa, §17.1] we obtain the two one-parameter families of unitary representations $U^\mu_\pm$ of $A_+(1)$ on $L^2(\mathbf{R}_+, dq/q)$ given by

$$\left(U^\mu_\pm(v, \lambda)\psi\right)(q) = e^{\pm i\mu v q^2}\psi(\lambda q)$$

with $\mu > 0$. As in the previous subsection, we identify the parameter μ with $\hbar^{-1}$. According to Theorems 4 and 5 in [BaRa, §17.1] the remaining two representations (one for each choice of sign) are irreducible and inequivalent; moreover, these are the *only* irreducible infinite-dimensional unitary ones.

Let $D \subset L^2(\mathbf{R}_+, dq/q)$ be the linear span of the functions $\sqrt{q}\, h_k(q)$, where the h_k are the Hermite functions. Writing $\varrho_\pm = -i\hbar\, dU_\pm$ we get the representations of $a(1)$ on $L^2(\mathbf{R}_+, dq/q)$:

$$\varrho_\pm(pq) = -i\hbar q\frac{d}{dq}, \quad \varrho_\pm(q^2) = \pm q^2.$$

Extend these to P^1 by demanding that $\varrho_\pm(1) = I$ and set $\mathcal{Q}_\pm = \varrho_\pm \oplus 0$, cf. (25). This is clearly a prequantization of P, by construction (Q4) is satisfied, and $\mathcal{Q}_\pm \lceil a(1) = \varrho_\pm$ is faithful. Finally, it is straightforward to verify that D consists of analytic vectors for both $\varrho_\pm(pq)$ and $\varrho_\pm(q^2)$. Thus $\mathcal{Q}_\pm$ are the required quantization(s) of (P, P_1).

Remarks. 12. The $+$ quantization of $a(1)$ is exactly what one obtains by geometrically quantizing M in the vertical polarization. Carrying this out, one gets $H = L^2(\mathbf{R}_+, dq)$ and

$$pq \mapsto -i\hbar\left(q\frac{d}{dq} + \frac{1}{2}\right), \quad q^2 \mapsto q^2.$$

The $+$ quantization is equivalent to this via the unitary transformation $L^2(\mathbf{R}_+, dq/q) \to L^2(\mathbf{R}_+, dq)$ which takes $f(q) \mapsto f(q)/\sqrt{q}$.

13. Note that $a(1) \subset sp(2, \mathbf{R})$. In fact, the $+$ quantization is equivalent to the restrictions to $a(1)$ of the metaplectic representations of $sp(2, \mathbf{R})$ on both $L^2_{\mathrm{even}}(\mathbf{R}, dq)$ and $L^2_{\mathrm{odd}}(\mathbf{R}, dq)$ (cf. §5.1 and Remark 12.

14. Since $\mathcal{Q}(P_{(2)}) = 0$, the quantization is somewhat 'trivial.' However, there are quantizations which are nonzero on $P_{(2)}$: for instance, set $\mathcal{Q}(x^k) = k\mathcal{Q}(x)$ for $k > 0$, $\mathcal{Q}(x^l y) = \mathcal{Q}(y)$, and $\mathcal{Q}(x^l y^m) = 0$ for $m > 1$.

15. This quantization of $T^*\mathbf{R}_+$ should be contrasted with that given in [Is, §4.5]. Also, we observe that this example is symplectomorphic to $\mathbf{R}^2$ with the basic algebra span$\{p, e^{2q}\}$.

What makes this example work? After comparing it with our other examples, it is clear that this polynomial quantization exists because we can never decrease degree in P by taking Poisson brackets. Due to this we have (24) as opposed to merely

$$\{P_1, P_k\} \subset P^k.$$

We shall pursue this line of investigation in a more general setting in §7.

5.5. T^2

We have just exhibited a polynomial quantization of $T^*\mathbf{R}_+$. But we can do even more: Here we exhibit a quantization of the *full* Poisson algebra of the torus.

Consider the torus T^2 thought of as $\mathbf{R}^2/\mathbf{Z}^2$, with symplectic form

$$\omega = B\, dx \wedge dy.$$

We study the basic algebra t generated by the set

$$\mathcal{T} = \{\sin 2\pi x, \cos 2\pi x, \sin 2\pi y, \cos 2\pi y\}.$$

We already know from Proposition 2 that there are no finite-dimensional basic algebras on the torus; thus t is the most natural choice.

Now (T^2, ω) is (geometrically) quantizable provided $B = Nh$ for some nonzero integer N. Fix $N = 1$ and let L be the corresponding Kostant-Souriau prequantization line bundle over T^2 [Ki]. Then the space of smooth sections $\Gamma(L)$ can be identified with the space of "quasi-periodic" functions $\varphi \in C^\infty(\mathbf{R}^2, \mathbf{C})$ satisfying

$$\varphi(x + m,\ y + n) = e^{2\pi imy}\varphi(x,\ y), \quad n,\ m \in \mathbf{Z},$$

and the prequantization Hilbert space $\mathcal{H}$ with the (completion of) the set of those quasi-periodic φ which are L^2 on $[0, 1) \times [0, 1)$. The associated prequantization map $\mathcal{Q}:$ $C^\infty(M) \to \mathrm{Op}\big(\Gamma(L)\big)$ (for a specific choice of connection on L) is defined by

$$\mathcal{Q}(f) = -i\hbar\left[\frac{\partial f}{\partial x}\left(\frac{\partial}{\partial y} - \frac{i}{\hbar}x\right) - \frac{\partial f}{\partial y}\frac{\partial}{\partial x}\right] + f. \tag{26}$$

As the torus is compact, these operators are essentially self-adjoint on $\Gamma(L) \subset \mathcal{H}$.

Theorem 17. $\mathcal{Q}$ is a quantization of $\big(C^\infty(T^2), \mathfrak{t}\big)$.

Proof. Since $\mathcal{Q}$ is a prequantization, we need only verify (Q4) and (Q5), (Q6) being obvious from (26). To this end it is convenient to use complex notation and view

$$\mathcal{T}_\mathbf{C} = \big\{e^{\pm 2\pi ix}, e^{\pm 2\pi iy}\big\}.$$

The analysis is simplified by applying the Weil-Brezin-Zak transform Z [Fo, §1.10] to the above data. Define a unitary map $Z : \mathcal{H} \to L^2(\mathbf{R})$ by

$$(Z\phi)(x) = \int_0^1 \phi(x, y)\, dy$$

with inverse

$$(Z^{-1}\psi)(x, y) = \sum_{m\in\mathbf{Z}} \psi(x + m)e^{-2\pi imy}.$$

Under Z the domain $\Gamma(L)$ maps onto the Schwartz space $\mathcal{S}(\mathbf{R},\mathbf{C})$ [Ki]. Setting $A_\pm :=$ $ZQ(e^{\pm 2\pi i x})Z^{-1}$ and $B_\pm := ZQ(e^{\pm 2\pi i y})Z^{-1}$ we compute, as operators on $\mathcal{S}(\mathbf{R},\mathbf{C})$,

$$(A_\pm \psi)(x) = e^{\pm 2\pi i x}(1 \mp 2\pi i x)\psi(x)$$

$$(B_\pm \psi)(x) = \left(1 \mp 2\pi\hbar\frac{d}{dx}\right)\psi(x \pm 1).$$

Then $A_\pm{}^* = \overline{A_\mp}$ on the domain $\{\psi \mid x\psi \in L^2(\mathbf{R})\}$, and likewise $B_\pm{}^* = \overline{B_\mp}$ on $\{\psi \mid d\psi/dx \in L^2(\mathbf{R})\}$.[10] In fact $\overline{A_\pm}$ and $\overline{B_\pm}$ are normal operators.

To show that $\mathcal{Q}(t)$ is an irreducible set, let us suppose that T is a bounded s.a. operator on $L^2(\mathbf{R})$ which strongly commutes with $\overline{A_\pm}$ and $\overline{B_\pm}$. Then T must commute (in the weak sense) with these operators on their respective domains.[11] Consequently T commutes with both

$$\overline{A_- A_+} = I + 4\pi^2 x^2 \tag{27}$$

on the domain $\{\psi \mid x^2\psi \in L^2(\mathbf{R})\}$, and

$$\overline{B_- B_+} = I - 4\pi^2\hbar^2\frac{d^2}{dx^2}$$

on $\{\psi \mid d^2\psi/dx^2 \in L^2(\mathbf{R})\}$. From these equations we see that T commutes, and therefore strongly commutes, with the closures of two of the three generators of the metaplectic representation (9) of $\mathfrak{sp}(2,\mathbf{R})$ on $\mathcal{S}(\mathbf{R},\mathbf{C})$.

Suppose that μ denotes the metaplectic representation of the metaplectic group $\mathrm{Mp}(2,\mathbf{R})$ on $L^2(\mathbf{R})$. We have in effect just established that T commutes with the one-parameter groups $\exp(is\,x^2)$ and $\exp(-it\hbar^2\,d^2/dx^2)$. Now classically the exponentials $\exp(sx^2)$ and $\exp(ty^2)$ generate $\mathrm{Sp}(2,\mathbf{R})$ [GS, §4]. As $\mathrm{Mp}(2,\mathbf{R}) \to \mathrm{Sp}(2,\mathbf{R})$ is a double covering, the corresponding exponentials in $\mathrm{Mp}(2,\mathbf{R})$ generate a neighborhood of the identity in the metaplectic group. Since $\mu\big[\exp(sx^2)\big] = \exp(is\,x^2)$ and $\mu\big[\exp(ty^2)\big] = \exp(-it\hbar^2\,d^2/dx^2)$, it follows that T commutes with $\mu(\mathcal{M})$ for all $\mathcal{M}$ in a neighborhood of the identity in $\mathrm{Mp}(2,\mathbf{R})$ and hence, as this group is connected, for all $\mathcal{M} \in \mathrm{Mp}(2,\mathbf{R})$.

Although the metaplectic representation μ is reducible, the subrepresentations μ_e and μ_o on each invariant summand of $L^2(\mathbf{R}) = L_e^2(\mathbf{R}) \oplus L_o^2(\mathbf{R})$ of even and odd functions are irreducible [Fo, §4.4]. Writing $T = P_eT + P_oT$, where P_e and P_o are the even and odd projectors, one has

$$[P_eT, \mu(\mathcal{M})] = 0 \tag{28}$$

for any $\mathcal{M} \in \mathrm{Mp}(2,\mathbf{R})$. It then follows from the irreducibility of the subrepresentation μ_e that $P_eT = k_e P_e + R P_o$ for some constant k_e and some operator $R : L_o^2(\mathbf{R}) \to L_e^2(\mathbf{R})$. Substituting this expression into (28) yields $[R P_o, \mu(\mathcal{M})] = 0$, and Schur's Lemma then implies that R is either an isomorphism or is zero. But R cannot be an isomorphism as

---

[10] $d\psi/dx$ is to be understood in the sense of tempered distributions.

[11] Here and in what follows we use the fact that a bounded operator weakly commutes with an (unbounded) normal operator iff they strongly commute.

the representations μ_e and μ_o are inequivalent [Fo, Thm. 4.56]. (Recall that two unitary representations are similar iff they are unitarily equivalent.) Thus $P_e T = k_e P_e$. Similarly $P_o T = k_o P_o$, whence $T = k_e P_e + k_o P_o$.

But now a short calculation shows that T commutes with

$$\overline{A_+} - \overline{A_-} = 2i(\sin 2\pi x - 2\pi x \cos 2\pi x)$$

only if $k_e = k_o$. Thus T is a multiple of the identity, and so $\{A_\pm, B_\pm\}$ is an irreducible set, as was to be shown. Thus in particular (Q4) is satisfied.

For (Q5), we claim that the linear span of the Hermite functions form a dense set of separately analytic vectors for the e.s.a. components of $\{A_\pm, B_\pm\}$. From the expression above for $A_\pm$, it is clear that a vector will be analytic for the e.s.a. components of $A_\pm$ iff it is analytic for multiplication by x. But it is well known that the Hermite functions are analytic for this latter operator. The result for $B_\pm$ is obtained directly from this by means of the Fourier transform. $\square$

Remark. 16. The proof also works for $N = -1$ but breaks down when $|N| \neq 1$ [Go3]. It is not known to what extent this theorem will remain valid in general (but see §7). As a consequence the classical limit is unclear; to compute it, one needs to study how the torus quantization behaves for large values of the quantum number N. But for $N > 1$, the *prequantizations* with Chern class N may not be actual quantizations. If they are not, then one must construct a series of quantizations $Q_1, \ldots, Q_N, \ldots$ with $Q_1 = Q$ and see what happens to Q_N as N grows. Without these "interpolating quantizations," the classical limit of Q cannot be determined.

This full quantization has several remarkable features. (See [Go3, Ve] for detailed discussions). First, in previous examples the irreducibility requirement typically led to Von Neumann rules. But for T^2 both $Q(f^2)$ and $Q(f)^2$ are completely determined for any observable f by the simple fact that Q is a Kostant-Souriau prequantization; irreducibility is irrelevant. Moreover, one sees from (26) that $Q(f^2)$ is a first order differential operator whereas $Q(f)^2$ is of second order, indicating that this quantization will not respect the classical multiplicative structure at all.

This is particularly evident when one considers the classical identity $\cos^2 2\pi x + \sin^2 2\pi x = 1$, as emphasized by [Ve]. In view of (27)

$$[Q(\cos 2\pi x)]^2 + [Q(\sin 2\pi x)]^2 = I + 4\pi^2 x^2, \tag{29}$$

which bears scant resemblance to

$$Q(\cos^2 2\pi x) + Q(\sin^2 2\pi x) = I.$$

So the torus quantization dramatically violates Souriau's requirement that "the quantum spectrum of commuting observables should be concentrated on their classical range" [Zi]. As reflected by (29), the bounded observables $\cos 2\pi x$ and $\sin 2\pi x$ quantize to unbounded operators. While this may be seen as a flaw of the quantization, it cannot be helped: A theorem of Avez states that when the phase space M is compact, the only possible prequantization of $C^\infty(M)$ by *bounded* operators is $f \mapsto \bar{f} I$, where $\bar{f}$ is the mean value f [Av1]. If the torus is to be fully (and nontrivially) quantized, the representation space must thus be infinite-dimensional, whence a certain "amount" of unboundedness must ensue. So in this regard, the torus is not really behaving badly; there is a trade-off involved here.

Finally, the salient feature of this example is that the basic algebra $\mathfrak{t}$ is infinite-dimensional. This also did not happen in any of our other examples. As a consequence the irreducibility requirement on T^2 is substantially weaker than the corresponding requirements on either $\mathbf{R}^{2n}$, S^2, or T^*S^1, and is likely the underlying reason why $\mathcal{Q}$ provides a full quantization of $(C^\infty(T^2), \mathfrak{t})$.

6. No-Go Theorems

Our treatment of the examples in §5 relied heavily on an intimate knowledge of the representations of the relevant basic algebras, and involved detailed calculations. Here we present some general results on the occurrence of obstructions. To accomplish this, we focus on the Lie and Poisson structures of basic algebras and the polynomial algebras they generate; necessarily, the representations of these objects now play a more subdued role. Background on Poisson algebras is given in [At, Gra1, Gra2].

The first key result appeared in 1974 and is due to Avez [Av1, Av2]. Recall that the mean value of $f \in C^\infty(M)$ is

$$\bar{f} = \frac{1}{\text{vol}(M)} \int_M f \, \omega^n.$$

Theorem 18. *The only full prequantization of a compact symplectic manifold by bounded operators is given by $f \mapsto \bar{f}I$.*

Thus there can be no nontrivial finite-dimensional full prequantizations of a compact phase space. In the noncompact case, there is the following complementary result due to Doebner and Melsheimer [DM].

Proposition 19. *A nonzero infinite-dimensional representation of a noncompact finite-dimensional Lie algebra by skew-symmetric operators contains at least one unbounded operator.*

Combining these two results, we see that an infinite-dimensional quantization will necessarily involve unbounded operators. Whereas Theorem 18 uses the Poisson structure on $C^\infty(M)$, Proposition 19 is purely representation theoretic. We shall encounter this dichotomy again in §6.3.

The next advance was made by Ginzburg and Montgomery [GM], who generalized Avez's theorem to noncompact M. Let $C_c^\infty(M)$ denote the Poisson algebra of compactly supported smooth functions on M.

Theorem 20. *There is no nontrivial finite-dimensional Lie representation of $C_c^\infty(M)$.*

We do not give the proof, as it is similar to that of Theorem 22 following. Since a prequantization is simply a special type of Lie representation, Theorems 18 and 20 yield the no-go result:

Corollary 21. *There exists no nontrivial finite-dimensional full prequantization of any symplectic manifold M.*

Inspired by this work, we generalize both Theorem 20 and Corollary 21 to polynomial quantizations. Let $\mathfrak{b}$ be a basic algebra of observables and $P(\mathfrak{b})$ the Poisson algebra of polynomials generated by $\mathfrak{b}$. *Throughout this section we assume that $\mathfrak{b}$ is finite-dimensional.* We break the analysis up into four cases, depending upon whether $\mathfrak{b}$, or equivalently M, is compact and its representations are finite-dimensional. It turns out that we are able to obtain obstructions to quantizing $(P(\mathfrak{b}), \mathfrak{b})$ in three of these cases. And in the remaining case (viz. when $\mathfrak{b}$ is noncompact and the representation space is infinite-dimensional), there is no universal obstruction. In this gross sense, then, we have solved the Groenewold-Van Hove problem for polynomial quantizations.

6.1. M Compact, Finite-dimensional Representations

The main result is:

Theorem 22. *Let $\mathfrak{b}$ be a finite-dimensional basic algebra on a compact symplectic manifold M. There exists no nontrivial finite-dimensional Lie representation of $P(\mathfrak{b})$.*

We begin with a purely algebraic lemma, whose proof is given in [GGG].

Lemma 4. *If L is a finite-codimensional Lie ideal of an infinite-dimensional Poisson algebra $\mathcal{P}$ with identity, then either L contains the commutator ideal $\{\mathcal{P}, \mathcal{P}\}$ or there is a maximal finite-codimensional associative ideal J of $\mathcal{P}$ such that $\{\mathcal{P}, \mathcal{P}\} \subseteq J$.*

Proof of Theorem 22. Suppose that $\mathcal{Q}$ were a Lie representation of $P(\mathfrak{b})$ on some finite-dimensional vector space. Then $L = \ker \mathcal{Q}$ is a finite-codimensional Lie ideal of $P(\mathfrak{b})$. We will show that L has codimension at most 1, whence the representation is trivial. We accomplish this in two steps, by showing that:

(a) The derived ideal $\{P(\mathfrak{b}), P(\mathfrak{b})\}$ has codimension 1 in $P(\mathfrak{b})$, and

(b) $L \supseteq \{P(\mathfrak{b}), P(\mathfrak{b})\}$.

Let $A(\mathfrak{b})$ denote the Lie ideal of polynomials of zero mean. The decomposition $f \mapsto \bar{f} + (f - \bar{f})$ gives $P(\mathfrak{b}) = \mathbf{R} \oplus A(\mathfrak{b})$. Thus, if we prove that $\{P(\mathfrak{b}), P(\mathfrak{b})\} = A(\mathfrak{b})$, (a) will follow.

Using (6) along with Stokes' Theorem, we immediately have that $\{P(\mathfrak{b}), P(\mathfrak{b})\} \subseteq A(\mathfrak{b})$. To show the reverse inclusion, let $\{b_1, \ldots, b_N\}$ be a basis for $\mathfrak{b}$, so that

$$\{b_i, b_j\} = \sum_{k=1}^{N} c_{ij}^k b_k$$

for some constants c_{ij}^k. Following Avez [Av2], define the "symplectic Laplacian"

$$mathnormal\, \Delta f = -\sum_{i=1}^{N} \{b_i, \{b_i, f\}\}.$$

It is clear from these two expressions and the Leibniz rule that the linear operator Δ maps $P^k(\mathfrak{b})$ into $A^k(\mathfrak{b})$. Furthermore, taking into account the transitivity of $\mathfrak{b}$, we can apply [Av2, Prop. 1(4)] to conclude that $\Delta f = 0$ only if f is constant. Thus for each $k \geq 0$, the decomposition $P^k(\mathfrak{b}) = \mathbf{R} \oplus A^k(\mathfrak{b})$ implies $\Delta(P^k(\mathfrak{b})) = A^k(\mathfrak{b})$. It follows that $A(\mathfrak{b}) \subseteq \{P(\mathfrak{b}), P(\mathfrak{b})\}$.

If (b) does not hold, then by Lemma 4 there must be a proper associative ideal J in $P(\mathfrak{b})$ with $\{P(\mathfrak{b}), P(\mathfrak{b})\} \subseteq J$. Since $\{P(\mathfrak{b}), P(\mathfrak{b})\} = A(\mathfrak{b})$ has codimension 1, $A(\mathfrak{b}) = J$. This is, however, impossible, since f^2 has zero mean only if $f = 0$.   □

Corollary 23. *Let $\mathfrak{b}$ be a finite-dimensional basic algebra on a compact symplectic manifold M. There exists no nontrivial finite-dimensional prequantization of $P(\mathfrak{b})$. In particular, there exists no nontrivial finite-dimensional quantization of $(P(\mathfrak{b}), \mathfrak{b})$.*

Although not surprising on mathematical grounds, since $P(\mathfrak{b})$ is "large," these corollaries do have physical import, as one expects the quantization of a compact phase space to yield a *finite*-dimensional Hilbert space.

6.2. M Compact, Infinite-dimensional Representations

We reduce this to the previous case as follows. Suppose that Q were a quantization of $(P(\mathfrak{b}), \mathfrak{b})$ on a Hilbert space. By conditions (Q3) and (Q5), $Q(\mathfrak{b})$ can be exponentiated to a unitary representation of the simply connected Lie group B with Lie algebra $\mathfrak{b}$ (recall that $\mathfrak{b}$ is assumed finite-dimensional) which, according to (Q4), is irreducible. Since M is compact, B is compact. The representation space must thus be finite-dimensional, and so Corollary 23 applies. This proves

Theorem 24. *Let $\mathfrak{b}$ be a finite-dimensional basic algebra on a compact symplectic manifold M. There exists no nontrivial quantization of $(P(\mathfrak{b}), \mathfrak{b})$.*

Thus, there is an obstruction to polynomially quantizing a compact symplectic manifold *regardless* of the dimensionality of the representation.

6.3. M Noncompact, Finite-dimensional Representations

Now suppose that M is noncompact. On physical grounds one expects a quantization of M, if it exists, to be infinite-dimensional. This is what we rigorously prove here, following [GGru2].

Already on the basis of representation theory, one can see that it will be difficult to obtain finite-dimensional quantizations of noncompact basic algebras. For instance, it is known that a Lie algebra admits a nontrivial finite-dimensional irreducible representation by skew-symmetric operators iff its Levi factor contains a nontrivial compact ideal [BaRa, Prop. 8.7.3]. Thus in particular a solvable algebra has no nontrivial finite-dimensional irreducible representations. We now prove that a *basic* algebra cannot admit any faithful finite-dimensional representations at all, irreducible or not.

Theorem 25. *Let* $\mathfrak{b}$ *be a finite-dimensional basic algebra on a noncompact symplectic manifold. Then* $\mathfrak{b}$ *has no faithful finite-dimensional representations by symmetric operators.*

Proof. We argue by contradiction. Suppose there exists a representation ϱ of $\mathfrak{b}$ on some $\mathbf{C}^k$. As $\varrho(\mathfrak{b})$ consists of hermitian matrices, ϱ is completely reducible. Since by assumption ϱ is faithful, one deduces from [Va, Theorem 3.16.3] that $\mathfrak{b}$ is reductive. By the comment following the the proof of Proposition 1, $\mathfrak{b}$ must then be semisimple.

Since M is noncompact, so is the simply connected covering group B of the semi-simple algebra $\mathfrak{b}$. Now consider a unitary representation U of B on $\mathbf{C}^k$. Decompose B into a product $B_1 \times \cdots \times B_K$ of simple groups. Then (at least) one of these, say B_1, must be noncompact. But it is well-known that a connected, simple, noncompact Lie group has no nontrivial finite-dimensional unitary representations [BaRa, Theorem. 8.1.2]. Thus $U(b) = I$ for all $b \in B_1$. Since every finite-dimensional representation ϱ of $\mathfrak{b}$ by symmetric operators is a derived representation of some unitary representation U of B, it follows that $\varrho \upharpoonright \mathfrak{b}_1 = 0$, and so ϱ cannot be faithful. $\square$

Since every quantization of $(\mathcal{O}, \mathfrak{b})$ must be faithful on $\mathfrak{b}$, we conclude that *there is no nontrivial finite-dimensional quantization of* $(\mathcal{O}, \mathfrak{b})$ *on a noncompact symplectic manifold,* where $\mathcal{O}$ is *any* Lie algebra containing $\mathfrak{b}$. Combining this with Corollary 23 we can now assert—roughly speaking—that no symplectic manifold with a (finite-dimensional) basic algebra has a finite-dimensional quantization.

6.4. M Noncompact, Infinite-dimensional Representations

So far we have encountered obstructions in every instance. The present case is the exception: We know from §5.4 that there exists a polynomial quantization of $T^*\mathbf{R}_+$ with the basic algebra a(1).

The behavior exhibited by this example is not characteristic of solvable algebras such as a(1), since e(2) for the cylinder is also solvable yet exhibits an obstruction. Likewise, the Heisenberg algebra is nilpotent and is obstructed as well.

6.5. Discussion and Further Results

Theorem 24 asserts that the polynomial algebra $P(\mathfrak{b})$ generated by any finite-dimensional basic algebra $\mathfrak{b}$ on a compact symplectic manifold cannot be consistently quantized. As the torus illustrates, this need not be true if $\mathfrak{b}$ is allowed to be infinite-dimensional. Similarly Theorem 22 and Corollary 23 can fail when the representation space is allowed to be infinite-dimensional: as is well-known, full prequantizations exist provided ω/h is integral. Thus Corollary 23 and Theorem 24 are the optimal no-go results for compact phase spaces.

Proposition 2 enables us to identify $P(\mathfrak{b})$ with the Poisson algebra of polynomials on $\mathfrak{b}^*$ restricted to the coadjoint orbit M. In particular, we can take $M = S^2 \subset \mathrm{su}(2)^*$, $\mathfrak{b}$ the space of spherical harmonics of degree one ($\mathfrak{b} \cong \mathrm{su}(2)$), and $P(\mathfrak{b})$ the space of all spherical harmonics. Thus Theorem 10 follows immediately from Theorem 24. A similar analysis applies to $\mathbf{C}P^n \subset \mathrm{su}(n+1)^*$.

Our results in the compact case lean heavily on the algebraic structure of $P(\mathfrak{b})$, and in particular on the property that $\{P(\mathfrak{b}), P(\mathfrak{b})\}$ has codimension 1 in $P(\mathfrak{b})$. When M is noncompact, codim $\{P(\mathfrak{b}), P(\mathfrak{b})\}$ is not fixed; it takes on the values 0, 1, and even ∞ in examples. Thus the Poisson theoretic techniques that worked for compact phase spaces will not apply to noncompact ones. This partly explains why Theorem 25 is a representation theoretic result. Furthermore, this theorem hinges on the fact that $\mathfrak{b}$, being noncompact and semisimple, cannot have faithful finite-dimensional representations by Hermitian matrices. But when M is compact, $\mathfrak{b}$ is compact semisimple, and these algebras *do* have such representations. Thus the compact and noncompact cases require entirely different approaches.

It is useful to keep track of which hypotheses the five theorems in this section require. They all use (Q1), and Theorems 20 and 22 require only this. Theorem 18 needs (Q2) as well. Theorem 24 uses also (Q3)–(Q5), and lastly Theorem 25 assumes in addition only (Q6). We do not know if a no-go theorem can be proven in the noncompact, finite-dimensional case without the faithfulness assumption (Q6). Irreducibility was only used in the proof of Theorem 24; in the other cases the finite-dimensionality assumption forced the representation to be "small."

We are thus left with trying to understand the noncompact, infinite-dimensional case, which is naturally the most difficult one. Here one has little control over either the types of basic algebras that can appear (in examples they range from solvable to simple; compare Proposition 1), the structure of the polynomial algebras they generate (cf. the above), or their representations. Thus one should try a different tack. Following the lead of Joseph [Jo] (cf. §5.1), let us try to compare the algebraic structures of Poisson algebras on the one hand with associative algebras of operators with the commutator bracket on the other. Grabowski has adopted this approach, and has produced the following "algebraic" no-go theorem, which is proved in [GGra1].

Theorem 26. *Let $\mathcal{P}$ be a unital Poisson subalgebra of $C^\infty(M, \mathbf{C})$. If as a Lie algebra $\mathcal{P}$ is not commutative, it cannot be realized as an associative algebra with the commutator bracket.*

Apply this result to polynomial quantizations. Suppose that $\mathcal{Q} : P(\mathfrak{b}) \to \mathrm{Op}(D)$ were a quantization of $(P(\mathfrak{b}), \mathfrak{b})$ on some dense invariant domain D in a Hilbert space. By requiring $\mathcal{Q}$ to be complex linear, we may view it as a quantization of the complexified polynomial algebra $\mathcal{P} = P(\mathfrak{b})_{\mathbf{C}}$. Take $\mathcal{A} \subset \mathrm{Op}(\mathcal{D})$ to be the associative algebra generated over $\mathbf{C}$ by $\{\mathcal{Q}(f) \mid f \in \mathfrak{b}\}$ together with I (if $1 \notin \mathfrak{b}$). If it can be shown that $\mathcal{Q}$ must be a Lie algebra isomorphism of $\mathcal{P}$ onto $\mathcal{A}$, then the algebraic no-go theorem will yield a contradiction.

To see how this works in practice, let us once again look at the Heisenberg algebra on $\mathbf{R}^2$. We shall prove inductively that

$$\mathcal{Q}(q^k p^l) = X^k Y^l + \sum_{k'+l'<k+l} a_{k'l'}^{kl} X^{k'} Y^{l'} \tag{30}$$

for some constants $a_{k'l'}^{kl}$, where $X = \mathcal{Q}(q)$, $Y = \mathcal{Q}(p)$. Indeed,

$$[\mathcal{Q}(q^k p^l), Y] = -i\hbar \mathcal{Q}(\{q^k p^l, p\}) = i\hbar k \mathcal{Q}(q^{k-1} p^l)$$
$$= i\hbar k X^{k-1} Y^l + \textit{lower degree terms},$$

where we have used the inductive assumption. Similarly

$$[\mathcal{Q}(q^k p^l), X] = -i\hbar l X^k Y^{l-1} + \textit{lower degree terms}$$

and, due to $ad_X \circ ad_Y = ad_Y \circ ad_X$, we can find $F^{kl} = X^k Y^l + \textit{lower degree terms}$, which has the same commutators with X and Y as $\mathcal{Q}(q^k p^l)$. Since by Proposition 5 $\mathcal{Q}$ is algebraically irreducible in the sense that the the only elements of $\mathcal{A}$ which commute with $\mathcal{Q}(h(2))$ are multiples of the identity, $\mathcal{Q}(q^k p^l)$ differs from F^{kl} by a constant, and that proves the inductive step. It now follows from (30) that $\mathcal{Q}$ is valued in $\mathcal{A}$, and that it is surjective.

It is easy to see that every nontrivial Lie ideal of $P = \mathbf{R}[q, p]$ intersects P^1. In particular if $\ker \mathcal{Q} \neq \{0\}$, then we contradict (Q6). Thus $\mathcal{Q}$ must be injective, and so we have an algebraic obstruction to quantizing (P, P^1).

The main difficulty in correlating the algebraic approach with our previous consider-ations is that there is no *a priori* reason why $\mathcal{Q}(\mathcal{P}) \subset \mathcal{A}$. This requirement is reminiscent of a Von Neumann rule, so one might expect that irreducibility can be used to establish this inclusion as in the example above; this is actually the case for nilpotent basic algebras since then $\mathcal{A}$ must be isomorphic to a Weyl algebra, cf. [GGra1, Prop. 8] for the details on how this works.

In fact, the entire argument for h(2n) can be extended to *any* nilpotent basic algebra [GGra1]. First we observe that if $\mathfrak{b}$ is a finite-dimensional nilpotent basic algebra on M, then M must be symplectomorphic to some $\mathbf{R}^{2n}$. (This follows from Proposition 2 and the well-known fact that a coadjoint orbit of a nilpotent Lie algebra must be symplec-tomorphic to some $\mathbf{R}^{2n}$ [Wi].) The canonical example of a nilpotent basic algebra on $\mathbf{R}^{2n}$ is of course h(2n). It is not difficult to see that, up to isomorphism, h(2) is the only nilpotent *basic* algebra on $\mathbf{R}^2$. This is not true in higher dimensions, however:

$$\mathrm{span}\{1, q_1, p_2, q_1 p_2 + q_2, p_1\}$$

is a nilpotent basic algebra on $\mathbf{R}^4$ which is not isomorphic to h(4).

Theorem 27. *Let $\mathfrak{b}$ be a finite-dimensional nilpotent basic algebra on a connected symplectic manifold. Then there is no quantization of $(P(\mathfrak{b}), \mathfrak{b})$.*

Theorem 27 does not carry over to more general basic algebras. Indeed, it is not true for a(1) on $T^*\mathbf{R}_+$ as we have already seen; note that a(1) is the simplest example of a solvable algebra which is not nilpotent.

Regardless, it appears that this algebraic approach holds promise; at least it enables us to partially suppress the representational aspects over which we have little control.

We turn now to the other extreme case, viz. when $\mathfrak{b}$ is semisimple. Identifying coadjoint and adjoint orbits by means of the Killing form, we know that M must be an adjoint orbit in $\mathfrak{b}$. Unlike the compact case, however, it is difficult to say which adjoint orbits are "basic," i.e. admit $\mathfrak{b}$ as a basic algebra. For example, in $sl(2, \mathbf{R})$ the nonzero adjoint orbits of which are either open half-cones, hyperboloids of one sheet, or components of hyperboloids of two sheets. One can verify that the open half-cones as well as the hyperboloids of one sheet are basic for $sl(2, \mathbf{R})$, but that the components of the hyperboloids of two sheets are not. (Instead they are basic for the subalgebra of upper triangular matrices.) Thus there is no apparent analogue of Proposition 3 in this context. Nonetheless, we are able to prove [GGra2]

Theorem 28. *Let M be a basic nilpotent adjoint orbit in $\mathfrak{b}$, where $\mathfrak{b}$ is a finite-dimensional semisimple Lie algebra. Then there is a nontrivial quantization of $(P(\mathfrak{b}), \mathfrak{b})$.*

This result is actually a corollary of Theorem 30 in §7.

Other than Theorem 28, nothing firm is known regarding obstructions to obtaining infinite-dimensional polynomial quantizations of noncompact semisimple basic algebras (but see Conjecture 1 in §7).

7. Speculations

In view of the theorems in the previous section, obstructions to quantization are guaranteed to exist except when the phase space is noncompact and the representations under consideration are infinite-dimensional. Three of our examples fall into this category: $\mathbf{R}^{2n}$, T^*S^1, and $T^*\mathbf{R}_+$. The first two exhibit obstructions, while the last does not. Comparing the behavior of these examples, as well as that of S^2, which is also obstructed, we attempt to extract the key features which govern the appearance of obstructions to a polynomial quantization.

Of course, any conclusions that we can draw at this point are necessarily tentative, due to the paucity of examples against which to test them. There are also various aspects of these examples that still are not completely understood. Nonetheless, some interesting observations can be made, which may prove helpful in subsequent investigations.

A detailed look at the derivations of the Von Neumann rules for $\mathbf{R}^{2n}$, T^*S^1, and S^2, and how they engender obstructions, shows that the controlling factor is apparently that one can decrease degree *in* $P(\mathfrak{b})$ by taking Poisson brackets. This is particularly evident in the classical Poisson bracket relations (16) and (18), and (22), which led to the contradictions for S^2 and T^*S^1, respectively. The situation for $\mathbf{R}^2$ is subtler, but one can spot this phenomenon in the proof of Lemma 1. The analysis in §5.4 shows that it is *not* possible to decrease degree in $P(\mathfrak{b})$ by taking Poisson brackets on $T^*\mathbf{R}_+$.

There are two—and only two—circumstances under which taking Poisson brackets in $P(\mathfrak{b})$ can decrease degree:[12]

---

[12] *A priori,* a third circumstance would be if $1 \in \mathfrak{b}$. Using the minimality condition (B4), it is not difficult to

(D1) $1 \in \{P(\mathfrak{b}), P(\mathfrak{b})\}$, and

(D2) There exist *nonzero* Casimirs in the symmetric algebra $S(\mathfrak{b})$ of $\mathfrak{b}$.[13]

According to the discussion at the end of §3, (D2) implies that $P(\mathfrak{b})$ is not freely generated by $\mathfrak{b}$ as an associative algebra. Specifically, $S(\mathfrak{b})$ will have nonzero Casimirs whenever $\mathfrak{b}$ is semisimple and has a nonzero compact ideal, and in particular when it is compact (cf. the comment following the proof of Proposition 1). At the other extreme, when $\mathfrak{b}$ is nilpotent, (D1) holds. Indeed, a nilpotent algebra has a center, and (B3) implies that this center consists of constants. An examination of the descending central series for $\mathfrak{b}$ then shows that $1 \in \{\mathfrak{b}, \mathfrak{b}\}$. In the examples, $\mathbf{R}^{2n}$ satisfies (D1) but not (D2), S^2 satisfies (D2) by virtue of (10) but not (D1), and T^*S^1 satisfies both because of

$$1 = \cos^2\theta + \sin^2\theta = \tfrac{1}{2}\{\{\ell^2, \sin\theta\}, \sin\theta\} + \tfrac{1}{2}\{\{\ell^2, \cos\theta\}, \cos\theta\}.$$

On the other hand, $T^*\mathbf{R}_+$ satisfies neither condition.

On the basis of this "anecdotal" evidence, we propose that a general Groenewold-Van Hove theorem takes the form:

Conjecture 1. *Let M be a symplectic manifold with a finite-dimensional basic algebra $\mathfrak{b}$. Suppose that the polynomial algebra $P(\mathfrak{b})$ satisfies either (D1) or (D2). Then there is no nontrivial quantization of $(P(\mathfrak{b}), \mathfrak{b})$.*

Indeed, is possible to directly verify this conjecture under certain circumstances.

Theorem 29. *Conjecture 1 is valid when either M is compact or the representation space is finite-dimensional.*

Proof. According to Proposition 1, when M is compact $\mathfrak{b}$ is compact. Just as in §6.2 we may then use (Q3)–(Q5) to reduce the case of infinite-dimensional representations to that of finite-dimensional ones. Thus it suffices to prove the theorem for the case when $\mathcal{Q}$ is a quantization of $P(\mathfrak{b})$ on a finite-dimensional Hilbert space, whence $L = \ker \mathcal{Q}$ has finite codimension in $P(\mathfrak{b})$.

Arguing as in the proof of Theorem 25, we have from (Q6) that $\mathfrak{b}$ is semisimple.

We apply Lemma 4 to L. First suppose that $\{P(\mathfrak{b}), P(\mathfrak{b})\} \subseteq L$. Then semisimplicity gives $\mathfrak{b} = \{\mathfrak{b}, \mathfrak{b}\} \subset L$, and so $\mathcal{Q}\!\upharpoonright\!\mathfrak{b} = 0$, which contradicts (Q6).

Thus there must exist a maximal finite-codimensional associative ideal J in $P(\mathfrak{b})$ with $\{P(\mathfrak{b}), P(\mathfrak{b})\} \subseteq J$. If (D1) holds, then $1 \in J$, which cannot be as J is proper. Now suppose (D2) holds, so that there is a nonzero Casimir $C \in S(\mathfrak{b})$. If ρ is the projection $S(\mathfrak{b}) \to P(\mathfrak{b})$, then $K = \rho^{-1}(J)$ is a maximal finite-codimensional associative ideal in $S(\mathfrak{b})$ with $\{S(\mathfrak{b}), S(\mathfrak{b})\} \subseteq K$. Since $\mathfrak{b} = \{\mathfrak{b}, \mathfrak{b}\} \subset \{S(\mathfrak{b}), S(\mathfrak{b})\} \subseteq K$, and since $1 \notin K$ (as K is proper), it follows that K is the associative ideal generated by $\mathfrak{b}$. (Actually, this shows that $S(\mathfrak{b}) = \mathbf{R} \oplus K$.)

---

prove that then $1 \in \{\mathfrak{b}, \mathfrak{b}\}$, so this is actually a subcase of (D1).

[13] By this we mean that if C is a Casimir, then its projection to $P(\mathfrak{b})$ is nonvanishing; in other words, when viewed as a function on M, C is nonzero.

Since C is nonzero, transitivity implies that $\rho(C) = c$ for some constant $c \neq 0$. By the definition of a Casimir and the above remarks $C \in K$. But then $C - c \notin K$, which is a contradiction since $C - c \in \ker \rho \subset K$. $\qquad\square$

While similar to the proof of Theorem 22, this proof has a key advantage: It does not require us to know the detailed structure of the commutator ideal (which we do not, when $\mathfrak{b}$ is noncompact).

Thus Conjecture 1 is consistent with the results of §6. Furthermore, the hypotheses of Conjecture 1 are certainly *necessary*.

Theorem 30. *Suppose that the polynomial algebra $P(\mathfrak{b})$ satisfies neither condition* (D1) *nor* (D2). *Then any nontrivial quantization of $\mathfrak{b}$ extends to a quantization of $(P(\mathfrak{b}), \mathfrak{b})$.*

Proof. For if $P(\mathfrak{b})$ satisfies neither of these conditions, then the notion of homogeneous polynomial is well-defined and it is not possible to lower degree in $P(\mathfrak{b})$ by taking Poisson brackets. Just as in §5.4, $P_{(2)}(\mathfrak{b})$ is then an ideal in $P(\mathfrak{b})$, and $P(\mathfrak{b}) = P^1(\mathfrak{b}) \ltimes P_{(2)}(\mathfrak{b})$. Let ϱ be the assumed representation of $\mathfrak{b}$; this extends to a representation of $P^1(\mathfrak{b})$. Then $\mathcal{Q} = \varrho \oplus 0$ is the required quantization of $(P(\mathfrak{b}), \mathfrak{b})$. $\qquad\square$

As a specific application of this result [GGra2], suppose that $\mathfrak{b}$ is semisimple and M is a basic nilpotent adjoint orbit in $\mathfrak{b}$. (For instance, when $\mathfrak{b} = \mathrm{sl}(2, \mathbf{R})$ one may take M to be either of the open half-cones.) Now a nilpotent orbit is invariant under the scaling action of $\mathbf{R}_+$ on $\mathfrak{b}$. But by Chevalley's theorem, the ideal of Casimirs of $S(\mathfrak{b})$ is generated by a finite collection of homogeneous polynomials of degree two or higher. Since Casimirs are constant on adjoint orbits, they must therefore vanish on conical ones. Thus (D2) cannot be satisfied. Furthermore, (D1) cannot hold either: If $1 \in \{P(\mathfrak{b}), P(\mathfrak{b})\}$, then $1 = \sum_{i=1}^k \{f_i, g_i\}$ for some polynomials f_i, g_i, whence $\sum_{i=1}^k \{f_i, g_i\}$ is a nonzero Casimir. So such orbits are polynomially quantizable.

Lastly, we observe that the finite-dimensionality assumption on $\mathfrak{b}$ in Conjecture 1 is necessary as well: The symmetric algebra $S(\mathfrak{t})$ on T^2 certainly contains Casimirs, but violates the conjecture.

Of our five examples, the torus is clearly much different than the others. It is not a Hamiltonian homogeneous space, and the basic algebra $\mathfrak{t}$ is infinite-dimensional. Because of this, the irreducibility requirement (Q4) loses much of its force – so much so that it precludes the existence of an obstruction. So it seems equally reasonable to propose

Conjecture 2. *Let M be a symplectic manifold and $\mathfrak{b}$ a basic algebra with $P^1(\mathfrak{b})$ dense in $C^\infty(M)$.*[14] *Then there exists a nontrivial quantization of $(C^\infty(M), \mathfrak{b})$.*

A necessary condition for $\mathcal{Q}$ to be a full quantization of $(C^\infty(M), \mathfrak{b})$ is that $\mathcal{Q}$ represent $C^\infty(M)$ itself irreducibly. It turns out [Ch2, Tu] that this is so for all Kostant-Souriau

---

[14] We use $P^1(\mathfrak{b})$ here to ensure that 1 is present: On the torus, $\mathfrak{b}$ consists only of trigonometric polynomials of mean zero, whereas $P^1(\mathfrak{b})$ comprises all trigonometric polynomials.

prequantizations[15]; thus it is natural to consider the case when M is prequantizable in this sense. In fact, in this context [Tu] gives even more:

Proposition 31. *Let M be an integral symplectic manifold, L a Kostant-Souriau prequantization line bundle over M and $\mathcal{Q}_L$ the corresponding prequantization map. Let $\mathfrak{b}$ be a basic algebra with $P^1(\mathfrak{b})$ dense in $C^\infty(M)$. Then $\mathcal{Q}_L$ represents $\mathfrak{b}$ irreducibly on the domain which consists of compactly supported sections of L.*

Set $D_c = \Gamma(L)_c$, the compactly supported sections of L. By construction $\mathcal{Q}_L :$ $C^\infty(M) \to \mathrm{Op}(D_c)$ satisfies (Q1)–(Q3) and (Q6). This proposition states that $\mathcal{Q}_L$ satisfies (Q4) as well. Thus to obtain a full quantization it remains to verify (Q5)—perhaps on some appropriately chosen coextensive domain D; unfortunately, it does not seem possible to do this except in specific instances. A first test would be to understand what happens for $\left(C^\infty(T^2), \mathfrak{t}\right)$ with $|N| \neq 1$. In any event, Proposition 31 does provide a certain amount of support for Conjecture 2.

The "gray area" between these two conjectures consists of symplectic manifolds with basic algebras $\mathfrak{b}$ for which $P^1(\mathfrak{b})$ is infinite-dimensional, yet not dense in $C^\infty(M)$. Maybe the infinite-dimensionality of $\mathfrak{b}$ alone is enough to guarantee the existence of a full quantization?

Completing the proof of Conjecture 1—that is, when M is noncompact and the quantizations are infinite-dimensional—seems to be a difficult problem. Perhaps the "algebraic approach" sketched at the end of §6 will continue to prove useful. It will likely be necessary to work through a few more examples of Groenewold-Van Hove obstructions before one is able to gain sufficient insight into this problem. One example worth studying are the various coadjoint orbits for sp($2n$, **R**). As well, it would be useful to consider basic algebras of a more general type than the ones we have encountered thus far (which were all either solvable or semisimple). We have also restricted consideration to polynomial subalgebras to a large extent, but there are other subalgebras $\mathcal{O}$ which are of interest (e.g., on **R**^{2n}, those functions which are constant outside some compact set [Ch3]).

A negative answer to the conjecture might indicate that one should strengthen the conditions defining a basic algebra by, e.g., replacing (B3) by (C2) as discussed in §3 (although this specific change would eliminate a(1) on T^***R**$_+$ from the ranks of basic algebras.) One could also modify the axioms for a quantization, for instance by adopting Souriau's requirement that classical observables with bounded spectra should quantize to operators with bounded spectra. Or, if the conjecture still seems undecidable, perhaps one should abandon the definition of a quantization map solely in terms of basic algebras and consider an alternative. However, the two other ways to define a quantization map listed previously suffer from serious flaws. If one imposes Von Neumann rules at the outset, then one tends to run into difficulties rather quickly—especially if one tries to enforce the rules on all of $C^\infty(M)$ and not some basic algebra thereof—as was shown in §5.1. Furthermore, it is unclear what form Von Neumann rules should take in general,

---

[15] However, there are other prequantizations which do not represent $C^\infty(M)$ irreducibly; for instance, the prequantization of Avez [Av3, Ch3].

as is illustrated by the unintuitive rules (14) for the sphere. For instance, mimicking the situation for $\mathbf{R}^{2n}$, one might simply postulate that $\mathcal{Q}(f^2) = \mathcal{Q}(f)^2$ for $f \in$ su(2). While the squaring rule for angular momentum is compatible with (14), one would still "miss" various possibilities (corresponding to the freedom in the choice of parameters a, c), which do occur in specific representations.[16] And in the case of the torus, Von Neumann rules are effectively moot, since the explicit prequantization map $\mathcal{Q}$ itself determines the quantization of every observable. Von Neumann rules are also irrelevant in the $T^*\mathbf{R}_+$ example, because of the peculiar structure (25) of $P(a(1))$. All in all, it appears as if the Von Neumann rules play a secondary role; the basic algebra $\mathfrak{b}$ is the primary object. It is also more compelling physically and pleasing æsthetically to require $\mathcal{Q}$ to satisfy an irreducibility requirement than a Von Neumann rule. Still, one can argue that such rules serve an important purpose [As, Ve].

There are problems with the polarization approach as well. For one thing, symplectic manifolds need not be polarizable [Go2]. This rare occurrence notwithstanding, there are quantizations which cannot be obtained by polarizing a prequantization: A well-known example is the extended metaplectic quantization of $\big($hsp$(2n,\mathbf{R})$,h$(2n)\big)$ [Bl2]. As we shall see presently, the specific predictions of geometric quantization theory are also off the mark in a number of instances.

Finally, it should be emphasized that these three approaches to quantization typically lead to obstructions in one way or another. We have already seen in §5 that Von Neumann rules play a crucial role in deriving the Groenewold-Van Hove obstructions for $\mathbf{R}^{2n}$, S^2 and T^*S^1. In the context of polarizations, the only observables which are consistently quantizable *ab initio* are those whose Hamiltonian vector fields preserve a given polarization [Bl1, Wo]. While this does not preclude the possibility of quantizing more general observables, attempts to quantize observables outside this class in specific examples usually result in inconsistencies. In *all* instances, the set of *a priori* quantizable observables relative to a given polarization forms a proper Lie subalgebra of the Poisson algebra of the given symplectic manifold. This observation provides further corroboration that Groenewold-Van Hove obstructions to quantization should be the rule rather than the exception.

Setting aside the question of the existence of obstructions, let us now suppose that there is an obstruction to, say, a polynomial quantization, so that it is impossible to consistently quantize all of $P(\mathfrak{b})$. The question is: What are the largest Lie subalgebras $\mathcal{O} \subset P(\mathfrak{b})$ containing the given basic algebra $\mathfrak{b}$ such that $(\mathcal{O}, \mathfrak{b})$ can be quantized? Modulo technical issues, given a representation $\mathcal{Q}$ of $\mathfrak{b}$ on a Hilbert space $\mathcal{H}$, one ought to be able to induce a representation of its Lie normalizer $\mathfrak{n}(\mathfrak{b})$ in $P(\mathfrak{b})$ on $\mathcal{H}$. (Indeed, the structure $(\mathfrak{n}(\mathfrak{b}), \mathfrak{b})$ brings to mind an infinitesimal version of a Mackey system of imprimitivity [BaRa].) Thus it seems reasonable to assert:

Conjecture 3. *Let $\mathfrak{b}$ be a finite-dimensional basic algebra. Then every quantization of $\mathfrak{b}$ can be extended to a quantization of $(\mathfrak{n}(\mathfrak{b}), \mathfrak{b})$.*[17]

---

[16] Because of this, [KLZ] would refer to (14) as "*non*-Neumann rules"!

[17] In [GGT] quantizations which satisfy this condition are termed "strong."

This is in exact agreement with the examples. In particular, for $\mathbf{R}^{2n}$ one has $\mathfrak{n}\big(\mathfrak{h}(2n)\big) = \mathfrak{hsp}(2n, \mathbf{R})$, and for S^2 one computes $\mathfrak{n}\big(\mathfrak{su}(2)\big) = \mathfrak{u}(2)$. In both cases, we have shown that these normalizers are in fact the maximal polynomial subalgebras that can be consistently quantized. It is therefore tempting to conjecture that:

No nontrivial quantization of $(\mathfrak{n}(\mathfrak{b}), \mathfrak{b})$ *can be extended beyond* $\mathfrak{n}(\mathfrak{b})$.

If true, this would point where to look for a Groenewold-Van Hove contradiction, viz. just outside the normalizer. Alas, this is *false*: For the cylinder $\mathfrak{n}(\mathfrak{e}(2)) = \mathbf{R} \oplus \mathfrak{e}(2)$. But from §5.3, we know that the representation (19) can be extended, in infinitely many ways, to the quantizations (23) of (L^1, P_1), where L^1 is the Lie subalgebra of observables which are affine in the angular momentum ℓ. It is not clear how one could "discover" this subalgebra given just the basic algebra $\mathfrak{e}(2)$ (but see below). The situation for $T^*\mathbf{R}_+$ is of course even worse than for T^*S^1. An outstanding problem is therefore to determine the maximal Lie subalgebras of quantizable observables.

This is reminiscent of the situation in geometric quantization with respect to polarizations. Suppose that $\mathcal{A}$ is a polarization of $C^\infty(M, \mathbf{C})$. Then one knows that one can consistently quantize those observables which preserve $\mathcal{A}$, i.e., which belong to the real part of $\mathfrak{n}(\mathcal{A})$ [Bl1, Wo]. In this way one obtains a "lower bound" on the set of quantizable functions for a given polarization. If one takes the antiholomorphic polarization on S^2, then it turns out that the set of *a priori* quantizable functions obtained in this manner is precisely the $\mathfrak{u}(2)$ subalgebra $\mathrm{span}\{1, S_1, S_2, S_3\}$. But it may happen that the real part of $\mathfrak{n}(\mathcal{A})$ is too small, as for $\mathbf{R}^{2n}$ with the antiholomorphic polarization. In this case the real part of $\mathfrak{n}(\mathcal{A})$ is only a proper subalgebra of P^2, and in particular is not maximal. This illustrates the fact, alluded to previously, that the extended metaplectic representation cannot be derived via geometric quantization. Furthermore, in the case of the torus, introducing a polarization will drastically cut down the set of *a priori* quantizable functions, which is at odds with the existence of a full quantization of this space. So geometric quantization is not a reliable guide insofar as computing maximally quantizable Lie subalgebras of observables. On the other hand, the position subalgebra $S = \{f(q)p + g(q)\}$ (resp. L^1) is just the normalizer of the vertical polarization $\mathcal{A} = \{h(q)\}$ on $\mathbf{R}^2$ (resp. $\{h(\theta)\}$ on T^*S^1), so these subalgebras find natural interpretations in the context of polarizations.

Clearly, there must be some connection between polarizations and basic algebras that awaits elucidation. It would be interesting to determine if there is a way to recast the Groenewold-Van Hove results in terms of polarizations. It would also be worthwhile, assuming that it is somehow possible to predict the maximal set(s) of quantizable observables *a priori*, to see whether one can use this knowledge to refine geometric quantization theory, or to develop a new quantization procedure, which is adapted to the Groenewold-Van Hove obstruction in that it will automatically be able to quantize this maximal set.

Here we have focused on the quantization of symplectic manifolds. It is natural to wonder to what extent these results will carry over to Poisson manifolds, or even to abstract Poisson algebras.

One of our goals in this paper was to obtain results which are independent of the particular quantization scheme employed, as long as it is Hilbert-space based. Therefore it is interesting that some of the go and no-go results described in this proposal have direct analogues in deformation quantization theory, since this theory was developed, at least in part, to avoid the use of Hilbert spaces altogether [BFFLS]. So for example, the no-go result for S^2 is mirrored by the fact that there are no strict SU(2)-invariant deformation quantizations of $C^\infty(S^2)$ [Ri1], while the go theorem for T^2 has as a counterpart the result that there do exist strict deformation quantizations of the torus [Ri1]. It is generally believed that the existence of Groenewold-Van Hove obstructions necessitates a weakening of the Poisson bracket $\rightarrow$ commutator rule (by insisting that it hold only to order $\hbar$), but these observations indicate that this may not suffice to remove the obstructions. There are undoubtedly important things to be learned by getting to the heart of this analogy.

References

[AM] Abraham, R. & Marsden, J.E. [1978] *Foundations of Mechanics*. Second Ed. (Benjamin-Cummings, Reading, MA).

[AA] Aldaya, V. & Azcárraga, J.A. [1982] Quantization as a consequence of the symmetry group: An approach to geometric quantization. *J. Math. Phys.* **23**, 1297–1305.

[ADT] Angermann, B., Doebner, H.-D. & Tolar, J. [1983] Quantum kinematics on smooth manifolds. In: *Nonlinear Partial Differential Operators and Quantization Procedures*. Andersson, S.I. & Doebner, H.-D., Eds. *Lecture Notes in Math.* **1087**, 171–208.

[AB] Arens, R. & Babbit, D. [1965] Algebraic difficulties of preserving dynamical relations when forming quantum-mechanical operators *J. Math. Phys.* **6**, 1071–1075.

[As] Ashtekar, A. [1980] On the relation between classical and quantum variables. *Commun. Math. Phys.* **71**, 59–64.

[At] Atkin, C.J. [1984] A note on the algebra of Poisson brackets. *Math. Proc. Camb. Phil. Soc.* **96**, 45–60.

[Av1] Avez, A. [1974] Représentation de l'algèbre de Lie des symplectomorphismes par des opérateurs bornés. *C.R. Acad. Sc. Paris Sér. A,* **279**, 785–787.

[Av2] Avez, A. [1974–1975] Remarques sur les automorphismes infinitésimaux des variétés symplectiques compactes. *Rend. Sem. Mat. Univers. Politecn. Torino*, **33**, 5–12.

[Av3] Avez, A. [1980] Symplectic group, quantum mechanics and Anosov's systems. In: *Dynamical Systems and Microphysics*. Blaquiere, A. et al., Eds. (Springer, New York) 301–324.

[BaRa] Barut, A.O. & Rączka, R. [1986] *Theory of Group Representations and Applications*. Second Ed. (World Scientific, Singapore).

[BFFLS] Bayen, F., Flato, M., Fronsdal, C., Lichnerowicz, A., & Sternheimer, D. [1978] Deformation theory and quantization I, II. *Ann. Phys.* **110**, 61–110, 111–151.

[Bl1] Blattner, R.J. [1983] On geometric quantization. In: *Non-Linear Partial Differential Operators and Quantization Procedures*. Andersson, S.I. & Doebner, H.-D., Eds. *Lecture Notes in Math.* **1087**, 209–241.

[Bl2] Blattner, R.J. [1991] Some remarks on quantization. In: *Symplectic Geometry and Mathematical Physics*. Donato, P. et al., Eds. *Progress in Math.* **99** (Birkhäuser, Boston) 37–47.

[BrRo] Bratteli, O. & Robinson, D.W. [1979] *Operator Algebras and Quantum Statistical Mechanics I*. (Springer, New York).

[Ch1] Chernoff, P.R. [1981] Mathematical obstructions to quantization. *Hadronic J.* **4**, 879–898.

[Ch2] Chernoff, P.R. [1988] Seminar on representations of diffeomorphism groups. Unpublished notes.

[Ch3] Chernoff, P.R. [1995] Irreducible representations of infinite dimensional transformation groups and Lie algebras I. *J. Funct. Anal.* **130**, 255–282.

[Co] Cohen, L. [1966] Generalized phase-space distribution functions. *J. Math. Phys.* **7**, 781–786.

[Di] Dirac, P.A.M. [1967] *The Principles of Quantum Mechanics*. Revised Fourth Ed. (Oxford Univ. Press, Oxford).

[DM] Doebner, H.D. and Melsheimer, O. [1968] Limitable dynamical groups in quantum mechanics I. *J. Math. Phys.* **9**, 1638–1656.

[Em] Emch, G.G. [1972] *Algebraic Methods in Statistical Mechanics and Quantum Field Theory*. (Wiley, New York).

[Fi] Filippini, R.J. [1995] The symplectic geometry of the theorems of Borel-Weil and Peter-Weyl. Thesis, University of California at Berkeley.

[Fl] Flato, M. [1976] Theory of analytic vectors and applications. In: *Mathematical Physics and Physical Mathematics*. Maurin, K. & Rączka, R., Eds. (Reidel, Dordrecht) 231–250.

[Fo] Folland, G.B. [1989] *Harmonic Analysis in Phase Space*. Ann. Math. Ser. **122** (Princeton University Press, Princeton).

[Fr] Fronsdal, C. [1978] Some ideas about quantization. *Rep. Math. Phys.* **15**, 111–145.

[GM] Ginzburg, V.L. & Montgomery, R. [1997] Geometric quantization and no-go theorems. Preprint dg-ga/9703010.

[GJ] Glimm, J. & Jaffe, A. [1981] *Quantum Physics. A Functional Integral Point of View*. (Springer Verlag, New York).

[Go1] Gotay, M.J. [1980] Functorial geometric quantization and Van Hove's theorem. *Int. J. Theor. Phys.* **19**, 139–161.

[Go2] Gotay, M.J. [1987] A class of non-polarizable symplectic manifolds. *Mh. Math.* **103**, 27–30.

[Go3] Gotay, M.J. [1995] On a full quantization of the torus. In: *Quantization, Coherent States and Complex Structures*, Antoine, J.-P. et al., Eds. (Plenum, New York) 55–62.

[Go4] Gotay, M.J. [1999] On the Groenewold-Van Hove problem for $\mathbf{R}^{2n}$. *J. Math. Phys.* **40**, 2107–2116.

[GGra1] Gotay, M.J. & Grabowski, J. [1999] On quantizing nilpotent and solvable basic algebras. Preprint math-ph/9902012.

[GGra2] Gotay, M.J. & Grabowski, J. [2000] On quantizing semisimple basic algebras. In preparation.

[GGG] Gotay, M.J., Grabowski, J., & Grundling, H.B. [2000] An obstruction to quantizing compact symplectic manifolds. *Proc. Amer. Math. Soc.* **128**, 237–243.

[GGru1] Gotay, M.J. & Grundling, H.B. [1997] On quantizing T^*S^1. *Rep. Math. Phys.* **40**, 107–123.

[GGru2] Gotay, M.J. & Grundling, H. [1999] Nonexistence of finite-dimensional quantizations of a noncompact symplectic manifold. In: *Differential Geometry and Applications*, Kolář, I. et al., Eds. (Masaryk Univ., Brno) 593–596.

[GGH] Gotay, M.J., Grundling, H., & Hurst, C.A. [1996] A Groenewold-Van Hove theorem for S^2. *Trans. Amer. Math. Soc.* **348**, 1579–1597.

[GGT] Gotay, M.J., Grundling, H., & Tuynman, G.T. [1996] Obstruction results in quantization theory. *J. Nonlinear Sci.* **6**, 469–498.

[Gra1] Grabowski, J. [1978] Isomorphisms and ideals of the Lie algebras of vector fields. *Invent. Math.* **50**, 13–33.

[Gra2] Grabowski, J. [1985] The Lie structure of C^* and Poisson algebras. *Studia Math.* **81**, 259–270.

[Gro] Groenewold, H.J. [1946] On the principles of elementary quantum mechanics. *Physica* **12**, 405–460.

[GS] Guillemin, V. & Sternberg, S. [1984] *Symplectic Techniques in Physics*. (Cambridge Univ. Press, Cambridge).

[HM] Helton, J.W. & Miller, R.L. [1994] NC Algebra: A Mathematica Package for Doing Non Commuting Algebra. v0.2, ncalg@ucsd.edu. (USCD, La Jolla).

[He] Hennings, M.A. [1986] Fronsdal *-quantization and Fell inducing. *Math. Proc. Camb. Phil. Soc.* **99**, 179–188.

[Is] Isham, C.J. [1984] Topological and global aspects of quantum theory. In: *Relativity, Groups, and Topology II.* DeWitt, B.S. & Stora, R., Eds. (North-Holland, Amsterdam) 1059–1290.

[Jo] Joseph, A. [1970] Derivations of Lie brackets and canonical quantization. *Commun. Math. Phys.* **17**, 210–232.

[KS] Kerner, E.H. & Sutcliffe, W.G. [1970] Unique Hamiltonian operators via Feynman path integrals. *J. Math. Phys.* **11**, 391–393.

[Ki] Kirillov, A.A. [1990] Geometric quantization. In: *Dynamical Systems IV: Symplectic Geometry and Its Applications.* Arnol'd, V.I. and Novikov, S.P., Eds. *Encyclopædia Math. Sci.* **IV**. (Springer, New York) 137-172.

[Ku] Kuryshkin, V.V. [1972] La mécanique quantique avec une fonction non-négative de distribution dans l'espace des phases. *Ann. Inst. H. Poincaré* **17**, 81–95.

[KLZ] Kuryshkin, V.V., Lyabis, I.A., & Zaparovanny, Y.I. [1978] Sur le problème de la regle de correspondence en théorie quantique. *Ann. Fond. L. de Broglie.* **3**, 45–61.

[Ma] Mackey, G.W. [1976] *The Theory of Unitary Group Representations* (University of Chicago Press, Chicago).

[MC] Margenau, H. & Cohen, L. [1967] Probabilities in quantum mechanics. In: *Quantum Theory and Reality.* Bunge, M., Ed. (Springer-Verlag, New York), 71–89.

[MR] Marsden, J.E. & Ratiu, T.S. [1994] *Introduction to Mechanics and Symmetry.* (Springer-Verlag, New York).

[MMSV] Mnatsakanova, M., Morchio, G., Strocchi, F., & Vernov, Yu. [1998] Irreducible representations of the Heisenberg algebra in Krein spaces. *J. Math. Phys.* **39**, 2969–2982.

[On] Onishchik, A.L. [1994] *Topology of Transitive Transformation Groups.* (Johann Ambrosius Barth, Leipzig).

[ReSi] Reed, M. & Simon, B. [1972] *Functional Analysis I.* (Academic Press, New York).

[Ri1] Rieffel, M.A. [1989] Deformation quantization of Heisenberg manifolds. *Commun. Math. Phys.* **122**, 531–562.

[Ri2] Rieffel, M.A. [1990] Deformation quantization and operator algebras. *Proc. Sym. Pure Math.* **45**, 411–423.

[Ri3] Rieffel, M.A. [1993] Quantization and C^*-algebras. In: *C^*-Algebras: 1943-1993, A Fifty Year Celebration.* Doran, R.S., Ed. *Contemp. Math.* **167**, 67–97.

[Ri4] Rieffel, M.A. [1998] Questions on quantization. In: *Operator Algebras and Operator Theory. Contemp. Math.* **228**, 315–326.

[Ro] Robert, A. [1983] *Introduction to the Representation Theory of Compact and Locally Compact Groups.* London Math. Soc. Lect. Note Ser. **80** (Cambridge U. P., Cambridge).

[So] Souriau, J.-M. [1997] *Structure of Dynamical Systems.* (Birkhäuser, Boston).

[Tu] Tuynman, G.M. [1998] Prequantization is irreducible. *Indag. Mathem.* **9**, 607–618.

[Ur] Urwin, R.W. [1983] The prequantization representations of the Poisson Lie algebra. *Adv. Math.* **50**, 126–154.

[VH1] van Hove, L. [1951] Sur certaines représentations unitaires d'un groupe infini de transformations. *Proc. Roy. Acad. Sci. Belgium* **26**, 1–102.

[VH2] van Hove, L. [1951] Sur le problème des relations entre les transformations unitaires de la mécanique quantique et les transformations canoniques de la mécanique classique. *Acad. Roy. Belgique Bull. Cl. Sci.* (5) **37**, 610–620.

[Va] Varadarajan, V.S. [1984] *Lie Groups, Lie Algebras and Their Representations.* (Springer-Verlag, New York).

[Ve] Velhinho, J. [1998] Some remarks on a full quantization of the torus. *Int. J. Mod. Phys.* **A13**, 3905–3914.

[VN] von Neumann, J. [1955] *Mathematical Foundations of Quantum Mechanics.* (Princeton. Univ. Press, Princeton).

[We] Weinstein, A. [1989] Cohomology of symplectomorphism groups and critical values of hamiltonians. *Math. Z.* **201**, 75–82.

[Wi] Wildberger, N. [1983] Quantization and harmonic analysis on Lie groups. Dissertation, Yale University.

[Wo] Woodhouse, N.M.J. [1992] *Geometric quantization*. Second Ed. (Clarendon Press, Oxford).

[Zi] Ziegler, F. [1996] Quantum representations and the orbit method. Thesis, Université de Provence.

An Impetus-Striction Simulation of the Dynamics of an Elastica

D.J. Dichmann* and J.H. Maddocks[†]

[1] The Aerospace Corporation, 2350 E. El Segundo Blvd., El Segundo, CA 90245

[2] Department of Mathematics, Swiss Federal Institute of Technology-Lausanne, CH-1015 Lausanne

Received October 18, 1995; revised version accepted February 16, 1996
Communicated by Jerrold Marsden and Stephen Wiggins**

Summary. This article concerns the three-dimensional, large deformation dynamics of an inextensible, unshearable rod. To enforce the conditions of inextensibility and unshearability, a technique we call the impetus-striction method is exploited to reformulate the constrained Lagrangian dynamics as an unconstrained Hamiltonian system in which the constraints appear as integrals of the evolution. We show here that this impetus-striction formulation naturally leads to a numerical scheme which respects the constraints and conservation laws of the continuous system. We present simulations of the dynamics of a rod that is fixed at one end and free at the other.

Dedication: Juan-Carlos Simo and I shared many common interests in Hamiltonian systems, stability analyses, and the theory of rods. We rarely agreed on the best way of viewing problems, but we both always enjoyed debating the issues. He would undoubtedly have held strong opinions about this article, which is dedicated to him. He is sorely missed.

—JHM

1. Introduction

An elastica is an inextensible, unshearable elastic rod. The inextensibility and unshearability constraints imply that the net force n acting across a cross-section is a basic unknown in the Lagrangian form of the dynamics. This net force n can be viewed as a

---

* Research supported by the NSF, NASA GSFC and Computer Sciences Corporation.

[†] Research supported by AFOSR and ONR.

vector of three Lagrange multipliers enforcing the constraints. A difficulty in evolving such constrained Lagrangian systems is that no time derivative of the multipliers appears in the evolution equations. The impetus-striction approach, developed in [6], [8], [19], is a method for reformulating such constrained Lagrangian systems as unconstrained Hamiltonian systems in which the constraints are manifested as integrals of motion. As is more fully explained in Section 2, the impetus-striction technique is a variant of 'vakonomic' mechanics as described in [2], but with modifications tailored to a change in primary focus from ordinary differential equations subject to nonholonomic constraints, to partial differential equations subject to holonomic constraints. The impetus-striction formulation of the three-dimensional motion of an elastica was first introduced in [6], [7]. In [6], [8], this new formulation was exploited to obtain analytical stability results for solitary waves. Our purpose here is to show that the impetus-striction formulation also leads naturally to accurate numerical schemes for the computation of constrained dynamics.

Our Hamiltonian formulation of rod dynamics comprises a set of fourteen nonlinear, partial differential equations in one space dimension and one time dimension. The Hamiltonian also involves a set of auxiliary variables, or multipliers $\Lambda(s, t)$ that we call strictions, which must be determined through a minimization procedure whenever the Hamiltonian is evaluated. In the context of unshearable, inextensible rods, the determination of the strictions involves the solution of elliptic boundary value problems that correspond, in some sense, to the limiting case of infinite wave speeds associated with compression and shear motions. Moreover, the force $n(s, t)$ and the strictions $\Lambda(s, t)$ are related through $\Lambda_t = n$. As a consequence of this identity, we demonstrate that pointwise evaluation of the strictions Λ arises in an invariant of the motion for rod dynamics related to linear momentum. Thus, although striction is an unfamiliar quantity, it does have a natural physical interpretation in the context of rod dynamics.

To numerically approximate the rod dynamics, we adopt a method-of-lines approach in which we first discretize the constrained Lagrangian with respect to arc-length to obtain a differential-algebraic system involving second-order ordinary differential equations. We then apply the impetus-striction method to obtain an equivalent unconstrained first-order Hamiltonian system of ordinary differential equations in which the spatially discretized constraints are integrals. The resulting equations can then be recognized as a spatial discretization of the continuous impetus-striction Hamiltonian system, but with an unusual differencing scheme that retains as many symmetries and integrals of the continuous system as possible, and which consequently has discrete versions of all of the continuous constraints. Moreover, for our problem the constraints are all quadratic in the configuration variables. The midpoint rule, which is the simplest symplectic Gaussian Runge-Kutta algorithm, is then employed in the time integration, and all of the quadratic invariants of the discretized system are thereby conserved in the discrete time evolution. Because the constraint functions are integrals of motion on the entire phase space of our Hamiltonian formulation, the choice of the midpoint rule for time integration reduces drift in the constraints. We discuss numerical simulations of the particular boundary value problem modelling the motion of a strut, that is, a rod which is fixed at one end and free at the other.

Numerical treatments of Hamiltonian formulations of unconstrained, i.e., shearable, extensible rod dynamics are presented by Simo et al. in [21], [22], [23]. Gonzalez &

Simo [12] have investigated conservative time integration schemes for more general problems in nonlinear elasticity. Tabor & Klapper [24] and Klapper [14] describe another approach to constrained rod dynamics, but their numerical implementations are for cases where comparatively large dissipation is incorporated. In this article, our focus is on the conservative, constrained dynamics of inextensible and unshearable rods. As previously remarked, after the Lagrangian version of the partial differential equations in our model are discretized in space, a system of differential-algebraic equations is obtained. The numerical solution of differential-algebraic equations is itself an active area of research. For example, numerical treatments of constrained Lagrangian and Hamiltonian dynamics are discussed in [4], [11], [16], [17], and the impetus-striction formulation can be regarded as an alternative to these methods.

In addition to the novelty of the impetus-striction approach, a second unusual feature of our treatment is the use of Euler parameters to represent the three-dimensional orientation of each cross-section of the rod. Euler parameters are now widely realized to be valuable in the modelling and analysis of rigid body dynamics (cf., e.g., [3], [13], [15]), but their use in continuum mechanics is not typical. In elasto-dynamics, Simo was among the first to exploit Euler parameters to model three-dimensional motions with finite strain (cf. [21], [22], [23]). Because they involve only ratios of quadratic functions, Euler parameters have significant computational advantages over Euler angles, which require the evaluation of ubiquitous trigonometric functions. A set of four Euler parameters is constrained to have unit norm, a condition which can lead to some awkwardness in numerics, but which combines naturally with our Hamiltonian formulation of constrained dynamics and our time-stepping algorithm. However, the norm constraint on the Euler parameters is associated with a lack of strict convexity of the Lagrangian with respect to the generalized velocities, and is accordingly of a distinctly different nature from the imposed constraints of inextensibility and unshearability. As a consequence, our treatment combines elements of the impetus-striction formulation, to handle the material constraints, with Dirac's theory of constraints (cf., e.g., [2], [10]) to handle the Euler parameter norm condition.

We begin, in Section 2, with an introduction to the impetus-striction method within a finite-dimensional context. In Section 3 we summarize a generic model for the mechanics of large deformations in rods. An impetus-striction formulation of the three-dimensional motion of an inextensible and unshearable rod is described in Section 4, and some associated conservation laws and integrals of motion are detailed in Section 5. Our numerical discretization is developed in Section 6. The discretization is an elementary, low-order one that is intended to demonstrate the efficacy of the impetus-striction formulation as the basis for the stable numerical integration of constrained Lagrangian dynamics, and we make no claim as to its efficiency.

Lastly, some results of our numerical simulations of struts are presented in Section 7.

2. The Impetus-Striction Method

The impetus-striction method is a variant of the method of 'vakonomic' mechanics, as described in [2, p. 33], with modifications designed to facilitate applications to partial differential equations. Vakonomics is a treatment of constrained motion for systems governed by ordinary differential equations associated with a constrained Lagrangian action principle. In vakonomics a system of equations is set up to determine the multipliers associated with the constraints, and subsequently a Legendre transform can be performed to pass to the Hamiltonian form of the dynamics. By contrast, in the impetus-striction approach, we immediately construct the Legendre transform of the constrained Lagrangian to obtain a pre-Hamiltonian that depends upon the multipliers, or strictions, in addition to the phase variables. In so doing, a variational principle for the strictions can be recognized. In particular, if the original Lagrangian is a convex function of the generalized velocities, the constraints are enforced, and the Hamiltonian determined, by minimization of the pre-Hamiltonian over the strictions. We believe that this new approach lays bare the convexity and symmetry properties of the resulting Hamiltonian system. The variational principle for the strictions is particularly convenient in the case of Hamiltonian partial differential equations. In our analysis we restrict attention to the case of holonomic constraints. In the case of nonholonomic constraints it is known that the constrained action principle that is our starting point does not always yield the physically appropriate time dynamics.

The essential features of the impetus-striction method are as follows. For simplicity, we consider a system of ordinary differential equations and a single constraint. The extension to partial differential equations and multiple constraints required in Section 4 is not difficult. For a more detailed development, see [8].

Let $L(q, \dot{q}) \in R$ be a Lagrangian in which $q \in R^n$ represents the configuration, $\dot{q} \in R^n$ represents the generalized velocities, and L is strictly convex in $\dot{q}$. We seek to determine corresponding motions $q(\cdot)$ subject to a holonomic constraint of the form

$$f(q(t)) = 0, \qquad \forall t \geq 0. \tag{2.1}$$

The first step leading to an unconstrained Hamiltonian formulation of the dynamics is to replace the constraint (2.1) with its time derivative

$$f_q(q(t)) \cdot \dot{q}(t) = 0, \qquad \forall t \geq 0. \tag{2.2}$$

If the initial data $q(0)$ satisfy the constraint (2.1), the conditions (2.1) and (2.2) are equivalent. Now define the functional

$$\mathcal{L}(q, \dot{q}, \Lambda) \equiv L(q, \dot{q}) + \Lambda \, f_q(q) \cdot \dot{q}. \tag{2.3}$$

The Lagrange multiplier Λ in (2.3) associated with the time-differentiated constraint is termed the *striction* [6], [8]. In order to obtain a Hamiltonian system associated with the Lagrangian (2.3), we define the conjugate variable

$$\gamma \equiv \mathcal{L}_{\dot{q}} = L_{\dot{q}}(q, \dot{q}) + f_q(q) \, \Lambda, \tag{2.4}$$

and the classic (conjugate) momentum

$$p \equiv L_{\dot{q}}(q, \dot{q}), \tag{2.5}$$

so γ and p are related by

$$p = \gamma - f_q(q) \, \Lambda. \tag{2.6}$$

(To simplify notation, we have identified the derivatives $L_{\dot{q}}$ and f_q with the corresponding column vectors.) The conjugate variable γ is thus similar to, but distinct from, the momentum p. To emphasize this distinction, we refer to γ as the *impetus*.

Given strict convexity of L with respect to $\dot{q}$, equation (2.5) can be solved for $\dot{q}$:

$$\dot{q} = v(q, p) = v(q, \gamma - f_q \Lambda). \tag{2.7}$$

Because $\dot{q}$ depends upon the unknown striction Λ in (2.7), a well-defined evolutionary system cannot yet be defined. Instead, we regard the Legendre transform

$$L^*(q, p) \equiv p \cdot v(q, p) - L(q, v(q, p)), \tag{2.8}$$

as defining a *pre-Hamiltonian*

$$M(q, \gamma, \Lambda) \equiv L^*(q, \gamma - f_q \Lambda). \tag{2.9}$$

The crucial step in our formulation of the unconstrained Hamiltonian system is to observe that the expression (2.6) for the momentum p in terms of q and γ implies

$$M_\Lambda = -f_q(q) \cdot v(q, p). \tag{2.10}$$

Provided that (2.7) holds, the constraint (2.2) is equivalent to requiring that $M_\Lambda = 0$. Therefore, we define the striction to be that Λ which *minimizes* the pre-Hamiltonian M for specified values of q and γ. To see that M has a unique minimizer, recall that the Legendre transform L^* is strictly convex in p when L is strictly convex in $\dot{q}$. Thus M is strictly convex in Λ for fixed q and γ, so a unique minimizer $\Lambda = \tilde{\Lambda}(q, \gamma)$ exists. (If the Lagrangian is not a convex function of the velocities, a stationarity condition for the strictions is obtained.)

We now define a Hamiltonian on $R^n \times R^n$ by

$$H(q, \gamma) \equiv \min_\Lambda M(q, \gamma, \Lambda) = M(q, \gamma, \tilde{\Lambda}(q, \gamma)). \tag{2.11}$$

The associated canonical Hamiltonian system yields the evolution equations:

$$\dot{q} = H_\gamma(q, \gamma) = v(q, \tilde{p}(q, \gamma)), \tag{2.12}$$
$$\dot{\gamma} = -H_q(q, \gamma) = L_q(q, \gamma) + \tilde{\Lambda}(q, \gamma) f_{qq}(q) v(q, \tilde{p}(q, \gamma)), \tag{2.13}$$

where $\tilde{p}(q, \gamma) \equiv \gamma - f_q(q) \tilde{\Lambda}(q, \gamma)$. It is perhaps not immediately apparent, but equations (2.12) and (2.13) yield the correct dynamics. Moreover, $f(q)$ is an integral of motion. Therefore, if the initial data q_0 satisfy $f(q_0) = 0$, constraint (2.1) holds for all time.

Given a motion $q(t)$, the momentum p is completely determined by equation (2.5). By contrast, equation (2.6) implies only that the linear combination $\gamma - f_q \Lambda$ is determined by the motion, and individually neither the impetus γ nor the striction Λ is completely specified. However, once an initial value is prescribed for either $\Lambda(0)$ or $\gamma(0)$, the impetus $\gamma(t)$ is determined by solving the differential equation (2.13). We refer to this indeterminacy in the initial data as a *gauge freedom*. The important point is that the gauge freedom has no effect on the evolution $q(t)$ of the configuration. Indeed, if (q, γ) is a solution of the Hamiltonian system, with striction $\tilde{\Lambda}(q, \gamma)$, then for any constant Λ^*, $(q, \gamma + f_q \Lambda^*)$ is also a solution, with striction $\tilde{\Lambda}(q, \gamma) + \Lambda^*$. In fact it is precisely this gauge freedom that generates, via Noether's Theorem, the integral $f(q)$.

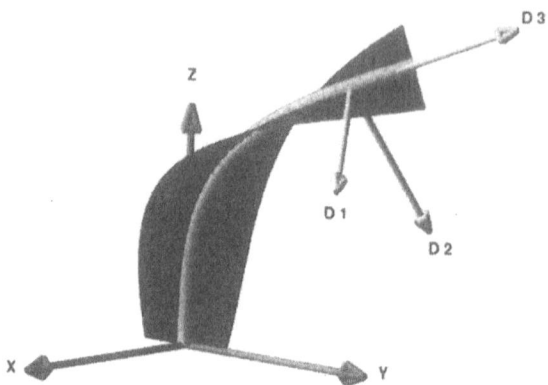

Fig. 3.1. A representative ribbon describing the configuration of a rod. The centerline of the ribbon represents the centerline of the rod and is parameterized by its arc-length s. The attached ribbon illustrates the twisting of the rod. The orthogonal triad of directors d_1, d_2 and d_3 can be thought of as indicating the orientation of a cross-section of the rod, with d_3 tangent to the centerline, and d_1 determining the ribbon.

3. Classical Rod Dynamics

This section provides a brief summary of the rod model that will be employed in this article. For further details, the reader is referred to [1]. We first review the kinematics and balance laws of the director theory of rods, and specify the constitutive law to be considered. We then describe the spatial kinematics of an elastica using Euler parameters.

In the director model for rod kinematics, the dependent variables are a vector function $r(s, t) \in R^3$ and a right-handed orthonormal frame of directors $d_k(s, t) \in R^3, k = 1, 2, 3$. The independent variables are two scalars, namely, undeformed arc-length s that ranges from 0 to length ℓ, and time t. At each time t, the curve $r(\cdot, t)$ can be interpreted as the configuration of a material line in a long, slender elastic body. The triad $\{d_k\}$ can be interpreted as providing information concerning the orientation of the material cross-section of the rod. (See Figure 3.1.)

The kinematics of the rod are encapsulated in the relations

$$d_k' = u \times d_k, \tag{3.1}$$

$$\dot{d}_k = \omega \times d_k, \tag{3.2}$$

$$r' = d_3. \tag{3.3}$$

Here d_k' denotes the partial derivative with respect to arc-length s, while $\dot{d}_k$ denotes the partial derivative with respect to t, etc. Equations (3.1) and (3.2) follow from the orthonormality of the directors $\{d_k\}$ as functions of s and t. The components $u_k \equiv u \cdot d_k$

of the (Darboux) vector u are the strains associated with bending and twist, and the components $\omega_k \equiv \omega \cdot d_k$ of the vector $\omega(s, t)$ can be regarded as the body components of the angular velocity of the material cross-section at s when viewed as a rigid body. We use $\underline{u}$ to denote the set of components (u_1, u_2, u_3), and similarly let $\underline{\omega}$ denote $(\omega_1, \omega_2, \omega_3)$. Equation (3.3) expresses the condition that the rod be inextensible and unshearable. For brevity equation (3.3) is referred to henceforth as the inextensibility condition.

The densities of linear and angular momentum are

$$p(s, t) \equiv \rho(s)\dot{r}(s, t), \tag{3.4}$$

$$\pi(s, t) \equiv \sum_{i,j=1}^{3} I_{ij}(s)\omega_j(s, t)d_i(s, t). \tag{3.5}$$

Here $\rho(s)$ is the mass per unit arc-length associated with the material cross-section at the point s, which is determined entirely by the reference configuration. The quantities $I_{ij}(s)$ are the components of the positive-definite inertia tensor of the material cross-section at s expressed with respect to the triad $\{d_k\}$. It will be assumed that these inertia coefficients are independent of the strains u_k and v_k, and that d_1 and d_2 are the principal axes of cross-section, so that

$$I_{ij} = 0, \quad i \neq j, \qquad I_{33} = I_{11} + I_{22}. \tag{3.6}$$

For simplicity we will denote I_{jj} by I_j.

In a rod model the stresses acting across each material cross-section are reduced to a net force $n(s, t)$ and moment $m(s, t)$ (of the material on the side s^+ acting on the material on the side s^- at the point $r(s, t)$). The balance of linear and angular momentum yields the equations

$$\dot{p} = n' \tag{3.7}$$

and

$$\dot{\pi} = m' + r' \times n. \tag{3.8}$$

It is stipulated that the constraint (3.3) holds for all possible forces and moments. The force is determined by this requirement, and a constitutive equation is specified only for the moments. We assume that the unstressed rod is straight with no twist, and that the components $m_j \equiv m \cdot d_j$ are given by the linear constitutive relations

$$m_i = K_i u_i. \tag{3.9}$$

More general reference configurations and hyperelastic constitutive relations could also be considered in a straightforward manner. The kinetic energy of the rod is given by

$$T(r, \underline{\omega}) = \int_0^\ell \frac{1}{2}\left\{\rho|r_t|^2 + \sum_{j=1}^{3} I_j\omega_j^2\right\} ds, \tag{3.10}$$

and the potential energy is

$$V(\underline{u}) = \int_0^\ell \frac{1}{2}K_j u_j^2 \, ds. \tag{3.11}$$

We must specify the set of boundary conditions which the rod is to satisfy. In this article, we consider a strut, that is, a rod with one end held fixed and the other end free. Specifically, the end $s = 0$ is fixed at the origin with the directors aligned with a reference frame $\{e_k\}$, so that

$$r(0, t) = 0 \tag{3.12}$$

and

$$d_k(0, t) = e_k. \tag{3.13}$$

The end $s = \ell$ experiences no loads, so that

$$n(\ell, t) = 0, \qquad m(\ell, t) = 0. \tag{3.14}$$

To fully describe the three-dimensional motion of a rod, it is convenient to parametrize the directors $\{d_k\}$, which is equivalent to selecting a representation for the group of proper orthogonal transformations $SO(3)$. Euler parameters provide a four-dimensional, global, two-to-one parametrization of $SO(3)$ which involves only ratios of quadratic functions. A set of Euler parameters, also called a unit quaternion, is a quadruple of real numbers $q = (q_1, q_2, q_3, q_4)^T$ that satisfies the identity

$$q \cdot q = q_1{}^2 + q_2{}^2 + q_3{}^2 + q_4^2 = 1. \tag{3.15}$$

By definition, the Euler parameters associated with a rotation through an angle Φ about an axis determined by a unit vector $k = k_j e_j$ are given by $q_j = k_j \sin(\Phi/2)$ for $j = 1, 2, 3$, and $q_4 = \cos(\Phi/2)$. The set of directors $\{d_1, d_2, d_3\}$ can be expressed in terms of the Euler parameters as

$$d_1(q) = \begin{pmatrix} q_1^2 - q_2^2 - q_3^2 + q_4^2 \\ 2(q_1 q_2 + q_3 q_4) \\ 2(q_1 q_3 - q_2 q_4) \end{pmatrix}, \tag{3.16}$$

$$d_2(q) = \begin{pmatrix} 2(q_1 q_2 - q_3 q_4) \\ -q_1^2 + q_2^2 - q_3^2 + q_4^2 \\ 2(q_1 q_4 + q_2 q_3) \end{pmatrix}, \tag{3.17}$$

$$d_3(q) = \begin{pmatrix} 2(q_1 q_3 + q_2 q_4) \\ 2(q_2 q_3 - q_1 q_4) \\ -q_1^2 - q_2^2 + q_3^2 + q_4^2 \end{pmatrix}. \tag{3.18}$$

In particular the set of Euler parameters

$$q(0, t) = (0, 0, 0, 1)^T \tag{3.19}$$

satisfies the boundary conditions (3.13).

The strains u and the angular velocities ω are given in terms of the Euler parameters by

$$u_j = \frac{2q_s \cdot B_j q}{|q|^2}, \qquad j = 1, 2, 3, \tag{3.20}$$

and

$$\omega_j = \frac{2q_t \cdot B_j q}{|q|^2}, \qquad j = 1, 2, 3, \tag{3.21}$$

where $\{B_1, B_2, B_3\}$ is a set of three 4-by-4 skew matrices, satisfying the relations

$$B_1 q = \begin{pmatrix} q_4, & q_3, & -q_2, & -q_1 \end{pmatrix}^T, \tag{3.22}$$

$$B_2 q = \begin{pmatrix} -q_3, & q_4, & q_1, & -q_2 \end{pmatrix}^T, \tag{3.23}$$

$$B_3 q = \begin{pmatrix} q_2, & -q_1, & q_4, & -q_3 \end{pmatrix}^T. \tag{3.24}$$

The skewness of the matrices B_j together with the norm constraint (3.15) imply that the four vectors $\{q, B_1 q, B_2 q, B_3 q\}$ form an orthonormal basis for R^4. Geometrically, the vector $B_j q$ acts as the infinitesimal generator of rotation about the d_j axis.

4. Hamiltonian Formulation

We now obtain a Hamiltonian formulation describing the three-dimensional motion of an inextensible and unshearable rod, represented in terms of Euler parameters. For the partial differential equations describing rod dynamics, we must deal with both the Euler parameter constraint and the more complicated inextensibility and unshearability constraints. It is because of these constraints that we adopt the impetus-striction formulation.

The time derivative of the inextensibility constraint (3.3) is

$$(r_s - d_3)_t = r_{st} - D_3 (q)^T q_t = 0, \tag{4.1}$$

where, for $k = 1, 2, 3$, the 3-by-4 matrices $D_k (q)$ are

$$D_k (q) \equiv \left(\frac{\partial d_k}{\partial q} \right)^T = \frac{2}{|q|^2} \sum_{i,j=1}^{3} \epsilon_{ijk} B_j q \otimes d_i + \frac{2}{|q|^2} q \otimes d_k, \tag{4.2}$$

with $\otimes$ denoting the outer product. (A direct differentiation of equations (3.16)–(3.18) yields a simpler expression for the matrices $D_k (q)$, but equation (4.2) has proved to be a more useful form for analysis.) The time derivative of equation (3.15) is

$$q \cdot q_t = 0. \tag{4.3}$$

Consequently, the Lagrangian we consider is

$$\mathcal{L} = T(r, \underline{\omega}) - V(\underline{u}) + \int_0^\ell \left\{ \Lambda \cdot (r_{st} - D_3 (q)^T q_t) + vq \cdot q_t \right\} ds,$$

$$= \int_0^\ell \left\{ \frac{1}{2}\rho|r_t|^2 + \frac{1}{2}\sum_{j=1}^{3} I_j \omega_j^2 - \frac{1}{2}\sum_{j=1}^{3} K_j u_j^2 + \Lambda \cdot r_{st} - q_t \cdot D_3 (q) \Lambda + vq \cdot q_t \right\} ds, \tag{4.4}$$

where the kinetic energy T is given by (3.10), the potential energy V is given by (3.11), the angular velocities ω_j are given by (3.21), the strains u_j are given by (3.20), and Λ

and v are the Lagrange multipliers associated respectively with the constraints (4.1) and (4.3). It can be shown that the striction Λ appearing in (4.4) is the time anti-derivative of the force n, that is

$$\Lambda_t = n \tag{4.5}$$

(cf. [6]). The impetuses associated with translation and rotation are defined by

$$\chi \equiv \frac{\partial \mathcal{L}}{\partial r_t} = \rho r_t - \Lambda_s, \tag{4.6}$$

$$\gamma \equiv \frac{\partial \mathcal{L}}{\partial q_t} = \sum_{j=1}^{3} I_j \omega_j \frac{2 B_j q}{|q|^2} - D_3 (q) \Lambda + v q. \tag{4.7}$$

Combining equations (3.4) and (4.6), we can rewrite the linear momentum density as

$$p = \rho r_t = \chi + \Lambda_s. \tag{4.8}$$

One can use the orthonormality of the basis $\{q, B_j q\}$, together with the constraint (4.3), to solve equation (4.7) for the multiplier v and the components $\pi_j = \pi \cdot d_j$ of angular momentum density to obtain

$$\pi_j = I_j \omega_j = \tfrac{1}{2} \gamma \cdot B_j q + \sum_{i=1}^{3} \epsilon_{ij3} d_i \cdot \Lambda = \tfrac{1}{2} \gamma \cdot B_j q + d_j \cdot (d_3 \times \Lambda) \tag{4.9}$$

and

$$v = [\gamma + D_3 (q) \Lambda] \cdot \frac{q}{|q|^2} = \frac{\gamma \cdot q}{|q|^2} + 2 d_3 \cdot \Lambda, \tag{4.10}$$

where the second equality in (4.10) follows from the expression (4.2) for $D_3 (q)$. We can then employ the orthonormality of the basis $\{q, B_j q\}$ to invert equation (3.21) for the angular velocities and obtain the expression for q_t, namely,

$$q_t = \sum_{j=1}^{3} \tfrac{1}{2} \omega_j B_j q, \tag{4.11}$$

where ω_j now is to be interpreted as defined in terms of the Hamiltonian variables by (4.9).

It is necessary to invoke the constraint (4.3) in order to solve (4.7) for q_t, for otherwise the component $q \cdot q_t$ is indeterminate. This is a special case of a general theory developed by Dirac [10], which constructs Hamiltonian formulations of Lagrangian systems that are not strictly convex in the velocities. The Dirac theory, and particularly associated gauge freedoms, remains an active area of research (cf. [5] and references therein). Notice that in contrast to the striction Λ, the multiplier v does not appear in the expressions (4.8) and (4.9) for p and π_j. As a consequence, v does not arise in the Hamiltonian dynamics.

The next step in our Hamiltonian formulation is to express the total energy, which is the sum of the kinetic and potential energies, in terms of the Hamiltonian state variable

$z = (r, q, \chi, \gamma)$. Equations (4.8) and (4.9) allow us to rewrite the energy in the form of the pre-Hamiltonian

$$M(z, \Lambda) = \int_0^\ell \frac{1}{2} \left\{ \rho^{-1}|p|^2 + \sum_{j=1}^3 I_j^{-1}\pi_j^2 + \sum_{j=1}^3 K_j u_j^2 \right\} ds, \qquad (4.12)$$

where the linear momentum density p is given by (4.8), the components of angular momentum density π_j are given by (4.9), and the components of strain u_j are given by (3.20). The strictions are determined by the requirement that $\Lambda = \tilde{\Lambda}(z)$ minimize M for given phase $z = (r, q, \chi, \gamma)$, so the strictions satisfy $\delta M/\delta\Lambda = \mathbf{0}$, or

$$\mathbf{0} = -\left[\rho^{-1}(\chi + \Lambda_s) \right]_s - I_1^{-1}\left(\tfrac{1}{2}\gamma \cdot B_1 q - d_2 \cdot \Lambda \right) d_2$$
$$+ I_2^{-1}\left(\tfrac{1}{2}\gamma \cdot B_2 q + d_1 \cdot \Lambda \right) d_1. \qquad (4.13)$$

The boundary conditions connected with the set of elliptic equations (4.13) are the natural boundary condition

$$(\chi + \Lambda_s)(0, t) = \mathbf{0}, \qquad (4.14)$$

corresponding to the fact that the end $s = 0$ is fixed, and

$$\Lambda(\ell, t) = \mathbf{0}, \qquad (4.15)$$

corresponding to the fact that the end $s = \ell$ is free. Using equations (4.8) and (4.9), one can show that the equation (4.13) is simply the enforcement of the inextensibility constraint. The striction $\tilde{\Lambda}(z)$ is uniquely determined by equations (4.13)–(4.15). The proof that the striction $\tilde{\Lambda}$ exists and is unique is given in [8] for planar motions (of an infinite rod) and in [6] for three-dimensional motion.

Henceforth we shall use p, π_j and ω_j respectively to denote the linear momentum density, the components of angular momentum density and the components of angular velocity, each evaluated at $\Lambda = \tilde{\Lambda}(z)$. The resulting Hamiltonian is

$$H(z) = M(z, \tilde{\Lambda}(z)) = \int_0^\ell \frac{1}{2} \left\{ \rho^{-1}|p|^2 + \sum_{j=1}^3 I_j^{-1}\pi_j^2 + \sum_{j=1}^3 K_j u_j^2 \right\} ds, \qquad (4.16)$$

with the associated Hamiltonian system of partial differential equations

$$r_t = \frac{\delta H}{\delta \chi} = \rho^{-1} p, \qquad (4.17)$$

$$q_t = \frac{\delta H}{\delta \gamma} = \sum_{j=1}^3 \tfrac{1}{2}\omega_j B_j q, \qquad (4.18)$$

$$\chi_t = -\frac{\delta H}{\delta r} = \mathbf{0}, \qquad (4.19)$$

$$\gamma_t = -\frac{\delta H}{\delta q} = \sum_{j=1}^3 \omega_j \left(\tfrac{1}{2}B_j \gamma - \sum_{i=1}^3 \epsilon_{ij3} D_i \tilde{\Lambda} \right)$$
$$+ \sum_{j=1}^3 \left[K_j u_j \frac{2B_j q}{|q|^2} \right]_s + \sum_{j=1}^3 K_j u_j \frac{2B_j q_s}{|q|^2} + \sum_{j=1}^3 K_j u_j^2 \frac{2q}{|q|^2}. \qquad (4.20)$$

The dynamics (4.17)–(4.20) are subject to the boundary conditions (3.12) and (3.19). To determine the initial conditions, one must fix the gauge by specifying $\tilde{\mathbf{\Lambda}}(s, 0)$ and $v(s)$. It is convenient to set

$$\tilde{\mathbf{\Lambda}}(s, 0) = \mathbf{0}, \tag{4.21}$$

$$v(s) = \mathbf{0}, \tag{4.22}$$

which satisfy the boundary condition (4.15). Equations (4.21) and (4.22) together with definitions (4.6) and (4.7) yield the initial conditions for the impetuses,

$$\chi(s, 0) = \rho(s)\mathbf{r}_t(s, 0), \tag{4.23}$$

$$\gamma(s, 0) = \sum_{j=1}^{3} I_j(s)\omega_j(s, 0)\frac{2\mathbf{B}_j\mathbf{q}(s, 0)}{|\mathbf{q}(s, 0)|^2}. \tag{4.24}$$

5. Conservation Laws and Integrals of Motion

A number of conservation laws and integrals of motion arise in our model of the rod dynamics. The Lagrangian forms of these conservation laws are discussed in detail in [18]. Whether these conservation laws imply associated integrals of motion depends upon the boundary conditions under consideration. For example, the Hamiltonian system is time-independent, and the Hamiltonian is an integral of motion. On the other hand, the balance law (3.7) for linear momentum is a conservation law, but for the boundary conditions (3.12)–(3.14) appropriate for a strut, the total linear momentum

$$\mathbf{P}(t) \equiv \int_0^\ell \mathbf{p}(s, t)\, ds \tag{5.1}$$

is not an integral of motion.

The Hamiltonian system under consideration also possesses several pointwise invariants. By the impetus-striction construction, the quantities $|\mathbf{q}(s, t)|^2$ and $\mathbf{r}_s(s, t) - \mathbf{d}_3(s, t)$ appearing in the constraints are integrals of motion for each s. Moreover, the variable r is ignorable (i.e., the Hamiltonian does not depend upon r), so its conjugate variable, namely the impetus χ, is also a pointwise invariant.

Although the total linear momentum is not conserved for the strut, the pointwise invariance of the impetus immediately implies that the total linear impetus

$$X(t) \equiv \int_0^\ell \chi(s, t)\, ds \tag{5.2}$$

is also conserved. The expressions (4.8) and (5.1) for the linear momentum density and total linear momentum together with the boundary condition (4.15) can be used to rewrite the total linear impetus as

$$X(t) = \mathbf{P}(t) + \tilde{\mathbf{\Lambda}}(0, t). \tag{5.3}$$

Equation (5.3) illustrates the physical meaning of the strictions within the context of rods: The change in $\tilde{\mathbf{\Lambda}}(0, \cdot)$ over the time interval $[0, t]$ is the negative of the change in the linear momentum of the rod over the same time interval. The negative sign appearing in this interpretation is merely a consequence of the sign convention for Λ in our definition of the Lagrangian (4.4). More generally, because $\chi(s, t)$ is constant for each s, it follows that for any a and b such that $0 \leq a < b \leq \ell$, the quantity

$$\int_a^b p(s, t) \, ds - \tilde{\Lambda}(b, t) + \tilde{\Lambda}(a, t) \tag{5.4}$$

is conserved, so the change in $\tilde{\Lambda}(b, \cdot) - \tilde{\Lambda}(a, \cdot)$ over the time interval $[0, t]$ is the change in the linear momentum of the segment $[a, b]$ over the same time interval. This interpretation of the striction is equivalent to the statement (4.5) that the striction Λ is the time anti-derivative of the force n in the context of rods. One could exploit the gauge freedom in the choice of initial data to specify $\chi(s, 0) = 0$ for all s, and so guarantee that the quantity $\tilde{\Lambda}(b, t) - \tilde{\Lambda}(a, t)$ actually equals the linear momentum of the segment $[a, b]$, for all a, b and for all times t. It is reasonable to regard (5.3) as an integral of motion because it is expressed in terms of quantities that are determined solely by the location in phase space.

6. Numerical Approximation of the Hamiltonian Dynamics

We now shift our focus to the numerical integration of our model equations. Our purpose here is to demonstrate the feasibility of the impetus-striction method as the basis for a stable numerical integration scheme for constrained systems, and in the first instance no efforts were made to optimize the speed of the code. To numerically approximate the partial differential equations arising in our model of rod dynamics we employ the method of lines and first discretize with respect to arc-length, thereby obtaining a system of ordinary differential equations. This first stage is referred to as the semidiscretization. The second stage is then to numerically integrate the system of ordinary differential equations in time.

At both stages, our choice of discretization has been guided by two goals. First, we retain the Hamiltonian structure of the equations. Second, we preserve, as much as is possible, the conservation laws and invariants of the original system of partial differential equations. In particular, we enforce discrete versions of inextensibility condition (3.3) and norm condition (3.15) on the Euler parameters. To achieve these goals, it is simplest to discretize the constrained Lagrangian with respect to arc-length s, and then to apply the impetus-striction method to obtain a Hamiltonian system of ordinary differential equations that can also be recognized as a discretization of the Hamiltonian partial differential equations. In Section 6.1 we describe a finite-difference scheme which is second order in space, and in Section 6.2 we motivate the choice of the midpoint rule as our time-stepping scheme.

6.1. Spatial Discretization

Our first step is to discretize the rod with respect to arc-length over a uniform grid $\{s^n\}$, where n runs from 0 to N, to obtain N segments of length $h \equiv \ell/N$. The segments are numbered by half-integers, running from $1/2$ to $(N - 1/2)$, with segment $(n + 1/2)$ lying between the nodes n and $n + 1$. We find it convenient to associate a position r^n with each node n, but to associate a quaternion $q^{n+1/2}$ with each segment $(n + 1/2)$. In this way, a $O(h^2)$ discrete form of the inextensibility condition (3.3) can be written as

$$\frac{r^{n+1} - r^n}{h} = d_3^{n+\frac{1}{2}},$$

(6.1.1)

where

$$d_j^{n+\frac{1}{2}} \equiv d_j(q^{n+\frac{1}{2}}),$$

(6.1.2)

and the norm condition on the quaternions is

$$q^{n+\frac{1}{2}} \cdot q^{n+\frac{1}{2}} = 1.$$

(6.1.3)

Each segment has an associated striction and multiplier arising from the inextensibility and norm conditions, denoted $\Lambda^{n+1/2}$ and $\nu^{n+1/2}$ respectively. The discrete analogs of the boundary conditions (3.12) and (3.19) are

$$r^0(t) = 0,$$

(6.1.4)

and

$$\frac{1}{2}\left(q^{\frac{1}{2}}(t) + q^{-\frac{1}{2}}(t)\right) = (0, 0, 0, 1)^T.$$

(6.1.5)

Here (6.1.5) is a second-order approximation to (3.19) involving the additional variable $q^{-1/2}(t)$ which also appears in the discretization of the strain u_j^n.

By using the trapezoid rule for variables defined at nodes, and the midpoint rule for variables associated with segments, the continuous Lagrangian (4.4) can be approximated to order $O(h^2)$ by the discretized Lagrangian

$$
\begin{aligned}
\mathcal{L}^N(q, r, \Lambda, \nu) = &\sum_{n=1}^{N-1} \frac{1}{2}\rho^n |\dot{r}^n|^2 + \frac{\rho^N}{4}|\dot{r}^N|^2 + \sum_{n=0}^{N-1}\sum_{j=1}^{3} \frac{1}{2}I_j^{n+\frac{1}{2}}\left(\omega_j^{n+\frac{1}{2}}\right)^2 \\
&- \sum_{j=1}^{3} \frac{K_j^0}{4}\left(u_j^0\right)^2 - \sum_{n=1}^{N-1}\sum_{j=1}^{3} \frac{1}{2}K_j^n\left(u_j^n\right)^2 \\
&+ \sum_{n=0}^{N-1}\Lambda^{n+\frac{1}{2}} \cdot \left[\frac{\dot{r}^{n+1} - \dot{r}^n}{h} - \left(D_3^{n+\frac{1}{2}}\right)^T \dot{q}^{n+\frac{1}{2}}\right] \\
&+ \sum_{n=0}^{N-1}\nu^{n+\frac{1}{2}}q^{n+\frac{1}{2}} \cdot \dot{q}^{n+\frac{1}{2}},
\end{aligned}
$$

(6.1.6)

where the boundary conditions (6.1.4) and $u_j^N = 0$ have been used,

$$\omega_j^{n+\frac{1}{2}} \equiv \frac{2\dot{q}^{n+\frac{1}{2}} \cdot B_j q^{n+\frac{1}{2}}}{|q^{n+\frac{1}{2}}|^2}, \qquad n = 0, \ldots, (N-1),$$

(6.1.7)

and

$$u_j^n \equiv \frac{2}{|q^{n+\frac{1}{2}}||q^{n-\frac{1}{2}}|} \left(\frac{q^{n+\frac{1}{2}} - q^{n-\frac{1}{2}}}{h} \right) \cdot B_j \left(\frac{q^{n+\frac{1}{2}} + q^{n-\frac{1}{2}}}{2} \right)$$

$$= \frac{2 q^{n+\frac{1}{2}} \cdot B_j q^{n-\frac{1}{2}}}{h |q^{n+\frac{1}{2}}||q^{n-\frac{1}{2}}|}, \qquad n = 0, \dots, (N-1), \qquad (6.1.8)$$

with $u_j^N = 0$, and $q^{-1/2}$ defined through boundary condition (6.1.5). The expression (6.1.7) is a natural semidiscrete form of equation (3.21). The expression (6.1.8) is an approximation to the strain (3.20) that is closely related to the quaternion which measures the rotation between the frame $\{d^{n-1/2}\}$ and the frame $\{d^{n+1/2}\}$. However, this choice of the discrete approximation to the strain u_j is by no means unique.

The pre-Hamiltonian can then be written as

$$M^N(q, \chi, \gamma, \Lambda) = \sum_{n=1}^{N-1} \frac{|p^n|^2}{2\rho^n} + \frac{|p^N|^2}{\rho^N} + \sum_{n=0}^{N-1} \sum_{j=1}^{3} \frac{1}{2} \left(I_j^{n+\frac{1}{2}} \right)^{-1} \left(\pi_j^{n+\frac{1}{2}} \right)^2$$

$$+ \sum_{j=1}^{3} \frac{K_j^0}{4} \left(u_j^0 \right)^2 + \sum_{n=0}^{N-1} \sum_{j=1}^{3} \frac{1}{2} K_j^n \left(u_j^n \right)^2. \qquad (6.1.9)$$

Here the linear momentum density associated with each node is

$$p^n = \chi^n + \frac{\Lambda^{n+\frac{1}{2}} - \Lambda^{n-\frac{1}{2}}}{h}, \qquad n = 1, \dots, (N-1), \qquad (6.1.10)$$

with $p^0 = 0$ reflecting boundary condition (6.1.4), $p^N = \chi^N + \frac{\Lambda^{n+\frac{1}{2}} - \Lambda^{n-\frac{1}{2}}}{2h}$, and the components of angular momentum density associated with segment $n + \frac{1}{2}$ (for $n = 0, \dots, (N-1)$) are

$$\pi_j^{n+\frac{1}{2}} = I_j^{n+\frac{1}{2}} \omega_j^{n+\frac{1}{2}} = \frac{1}{2} \gamma^{n+\frac{1}{2}} \cdot B_j q^{n+\frac{1}{2}} + d_j^{n+\frac{1}{2}} \cdot \left(d_3^{n+\frac{1}{2}} \times \Lambda^{n+\frac{1}{2}} \right). \qquad (6.1.11)$$

(Cf. equations (4.8) and (4.9).) The strictions $\tilde{\Lambda}^{n+1/2}$ for $n = 0, \dots, (N-1)$ are chosen such that the semidiscrete inextensibility condition (6.1.1) is enforced, while $\Lambda^{N+1/2}$ is defined through the discretization

$$\frac{1}{2} \left(\Lambda^{N+\frac{1}{2}} + \Lambda^{N-\frac{1}{2}} \right) = 0 \qquad (6.1.12)$$

of boundary condition (4.15). This condition yields a system of linear equations, involving a banded, positive definite symmetric matrix, which must be solved whenever the Hamiltonian is evaluated. To specify the gauge, we employ the discrete analogs of equations (4.21) and (4.22).

Inserting the strictions $\tilde{\Lambda}^{n+1/2}$ into the pre-Hamiltonian M (6.1.9), we obtain the Hamiltonian

$$H^N(z) = M^N(z, \tilde{\Lambda}(z)), \qquad (6.1.13)$$

with associated dynamics

$$\dot{r}^{n+1} = \frac{\partial H^N}{\partial \chi^{n+1}}, \tag{6.1.14}$$

$$\dot{q}^{n+\frac{1}{2}} = \frac{\partial H^N}{\partial \gamma^{n+\frac{1}{2}}}, \tag{6.1.15}$$

$$\dot{\chi}^{n+1} = -\frac{\partial H^N}{\partial r^{n+1}} = \mathbf{0}, \tag{6.1.16}$$

$$\dot{\gamma}^{n+\frac{1}{2}} = -\frac{\partial H^N}{\partial q^{n+\frac{1}{2}}}, \tag{6.1.17}$$

for $n = 0, \ldots, (N-1)$. The semidiscrete Hamiltonian system of ordinary differential equations (6.1.14)–(6.1.17) inherits versions of all but one of the conservation laws of the associated partial differential equations (4.17)–(4.20). The exception is the conservation law for linear impulse, which corresponds to the symmetry of translation in arc-length, and which is therefore lost in the spatial discretization (cf. [18], [9]). The Hamiltonian (6.1.13) is time-independent and is therefore conserved. The semidiscrete forms of the inextensibility constraint (6.1.1) and the norm condition on quaternions (6.1.3) are conserved as a consequence of the impetus-striction formulation. The semidiscrete version of the total linear momentum, analogous to (5.2), is given by

$$P^N(t) \equiv h \sum_{n=1}^{N} p^n(t), \tag{6.1.18}$$

where p^n is given by (6.1.10). Although the total linear momentum is not conserved, the total linear impetus

$$X^N(t) \equiv h \sum_{n=1}^{N} \chi^n(t) = P^N(t) + \tilde{\Lambda}^{\frac{1}{2}}(t), \tag{6.1.19}$$

analogous to (5.3), is an integral of motion.

6.2. Time Discretization

Having discretized the Hamiltonian system with respect to arc-length, we must also se-lect a method to approximate the time evolution associated with the system of ordinary differential equations (6.1.14)–(6.1.17). Our primary criteria in selecting a time inte-gration scheme were to preserve both the Hamiltonian structure of the system and as many of the invariants of motions as possible, but most importantly the inextensibility condition (6.1.1) and the norm condition (6.1.3).

Various integration schemes have been developed that preserve the symplectic struc-ture of a Hamiltonian system. In addition, it is known [20] that any Gaussian Runge-Kutta scheme is symplectic (Theorem 8.1, p. 99) and preserves quadratic invariants (Theorem 10.1, p. 136). The norm condition (6.1.3) is clearly quadratic, and the inextensibil-ity condition (6.1.1) is also quadratic because the directors (3.16)–(3.18) are quadratic functions of the quaternion. The total linear impetus X^N (6.1.19) is actually linear in the state variables. Thus, among the invariants of motion only the Hamiltonian is not quadratic.

In the first instance we chose to implement the simplest Gaussian Runge-Kutta method—namely, the midpoint method, which is a second-order integration scheme. For an ordinary differential equation of the form $\dot{z} = f(z)$, the midpoint method is

$$z^{k+1} = z^k + \tau f \left(\frac{z^{k+1} + z^k}{2} \right), \tag{6.2.1}$$

where τ is the time step, and z^k is the state at time $k\tau$. The midpoint method (6.2.1) is implicit and must be solved iteratively. As a result, the quadratic invariants are in practice preserved only with the accuracy to which the implicit equation (6.2.1) is solved. For the purposes of the simulations presented here, it was sufficient to solve equation (6.2.1) by implementing a rather simple functional iteration scheme.

7. Results of Numerical Experiments

In this section we describe the results of some numerical simulations which were conducted using the algorithms outlined above. For the computations we chose the values of the material coefficients to be $\rho = 1$, $I_1 = 1 = I_2$, $I_3 = I_1 + I_2$, and $K_1 = 16 = K_2 = K_3$, and considered a rod of length 2π. These values of the coefficients were chosen purely for demonstration purposes and have no associated units. Were one to choose values for these coefficients in MKS units corresponding to a thin metallic wire, the moments of inertia I_j would be rather small compared to both ρ and the bending stiffness K_j. One can think of the simulations presented here as corresponding to a set of nondimensionalized and scaled units of a realistic rod, in which the actual real time of the simulation is relatively short. In each case the rod spatial discretization was in 64 segments, so the dynamical system comprises 1024 ordinary differential equations. The time step used was $\tau = 5 \cdot 10^{-3}$, and the duration of the simulation is 600, so the total number of time steps is 120,000. For the initial data taken in Example 1 detailed below, the simulation includes more than 30 large amplitude oscillations. The iterative approximation to the solution of the midpoint method (6.2.1) was considered to have converged when the distance between successive approximations was less than $\epsilon = 10^{-11}$.

Our interest is in the calculation of the dynamics of large deformations of a rod. In Example 1, we consider a naturally straight, elastic rod and initially deform it to an untwisted circle, with zero initial velocities. In Figure 7.1 the first node, which is fixed, lies at the center of the figure on the left side of the initial circle, with the first segment pointing upward. While the initial configuration is closed, the final and initial segments of the rod are not connected, and Figure 7.1 depicts the ensuing motion when the rod is released.

The configurations are shown as snapshots at equal time intervals, during which the rod moves in a generally counterclockwise direction. Figure 7.2 is a continuation of the evolution of the rod, with the first snapshot being the final state shown in Figure 7.1. During the second sequence the tip of the rod moves in a generally clockwise direction.

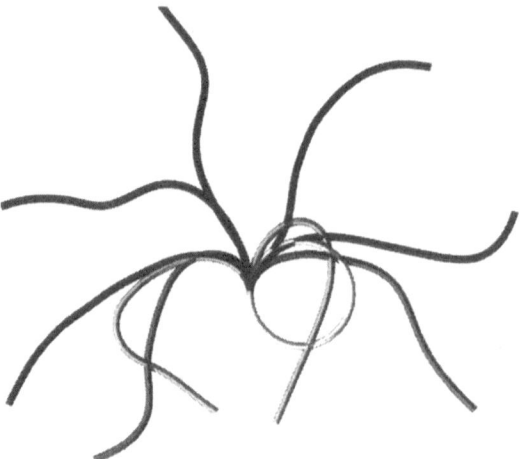

Fig. 7.1. Snapshots of the motion of the rod in Example 1. Although the rod is initially circular in shape, the initial and final segments are not connected. The shading is used to code the sequence with the rod being white initially, becoming darker as the tip moves counterclockwise toward the top of the figure, and then becoming lighter again, until it is white in the last configuration which appears at the bottom left of the figure.

Fig. 7.2. The motion of the rod in Example 1, continued. The same shading scheme is used as in Figure 7.1, with the overall motion being clockwise.

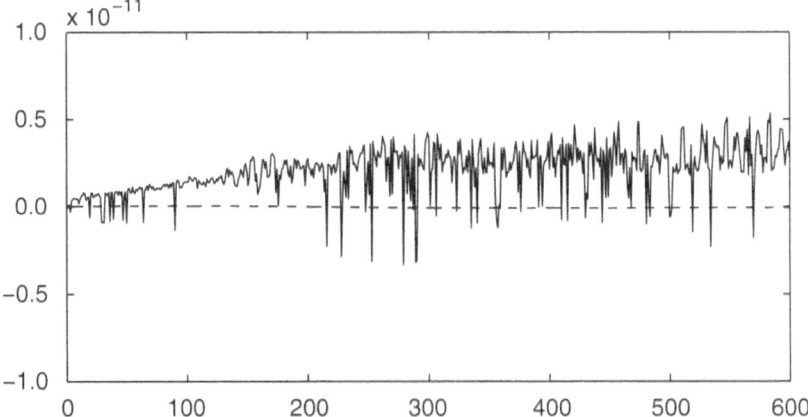

Fig. 7.3. Errors in the constraints for the planar motion in Example 1. The accumulated error in the inextensibility condition (solid curve) is less than 0.6ϵ, where $\epsilon = 10^{-11}$ is the tolerance used in the solution of the implicit midpoint method at each time step. The accumulated error in the quaternion norm (broken curve) is less than 0.01ϵ.

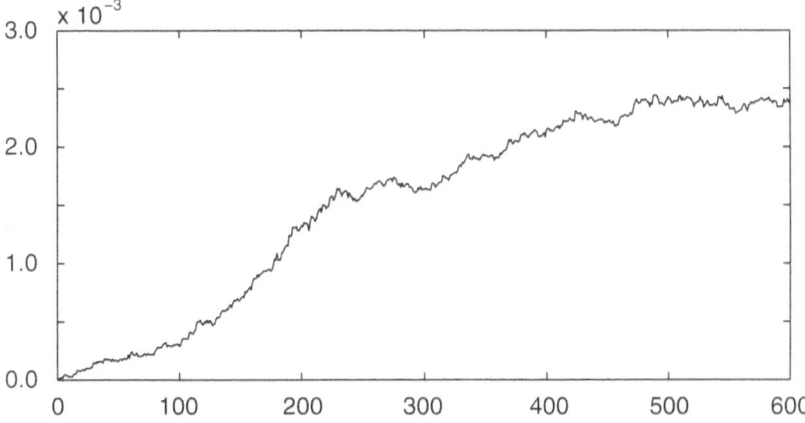

Fig. 7.4. Relative error in the Hamiltonian for Example 1. The error at each integration step is proportional to τ^2, where $\tau = 5 \cdot 10^{-3}$ is the time step.

Throughout the simulation the rod remains planar (as it should), although we actually compute a fully three-dimensional motion. As could have been expected, the motion is generally oscillatory, winding and unwinding from side to side, but the configuration does not come particularly close to resuming its initial circular shape. The elapsed time in Figures 7.1 and 7.2 together is 18 time units, so the entire simulation of 600 time units is the equivalent of approximately 33 oscillations.

Figure 7.3 shows the errors in the constraints for Example 1 over the full length of the simulation, i.e., a total number of $120,000$ time steps. The broken curve represents the maximum error in $|q^{n+1/2}|^2$ among all of the nodes, while the solid curve represents the maximum error in inextensibility among all of the segments. Because these constraints are quadratic functions of the state, the errors in the constraints are governed by the

Fig. 7.5. A sequence of snapshots depicting the three-dimensional motion arising in Example 2. Here the centerline is initially colored white, and becomes darker as the evolution proceeds. The ribbon attached to the centerline of the rod indicates the twisting of the directors.

tolerance used in the solution of the implicit midpoint method at each time step, namely $\epsilon = 10^{-11}$. The accumulated error in the inextensibility condition is less than 0.6ϵ, while the accumulated error in the quaternion norm is less than 0.01ϵ.

Figure 7.4 shows the relative error in the Hamiltonian for Example 1. The Hamiltonian H^N is not a quadratic function of the state. Therefore the error in H^N is proportional to $\tau^2 = 2.5 \cdot 10^{-5}$, the order of accuracy for the midpoint method. Note that the error in the constraints, although increasing, remains several orders of magnitude smaller than the error in the time-stepping scheme. We do not plot the errors in the total linear impetus (6.1.19) because the dynamics (6.1.16) imply that the linear impetus χ^n is conserved *pointwise*. It follows that the total linear impetus X^N is *exactly* (i.e. to machine precision) constant throughout the time interval.

In Example 2, we consider the same initial circular configuration, still with no initial velocities, but now with an additional uniform imposed twist corresponding to one complete rotation at the end. While the initial shape is planar, an unloading of the twist leads to highly nonplanar dynamics. The ensuing three-dimensional motions are much harder to illustrate effectively in snapshot figures, and movies can be found at the WWW site http://lcrmwww.epfl.ch.

Figure 7.5 shows the early evolution of the rod from its initial planar configuration. In this case there is significant coupling between the bending and twisting motions. Nevertheless, Figures 7.6 and 7.7 show that once again the norm condition, the inextensibility condition and the Hamiltonian are all preserved accurately by the time dynamics. As depicted in Figure 7.6 the maximum error in $|q^{n+1/2}|^2$ (broken curve) is less than 0.3ϵ and

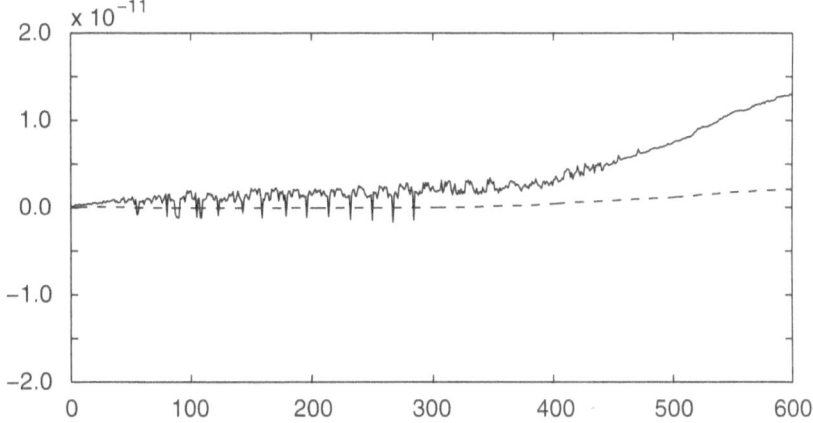

Fig. 7.6. Errors in the constraints for the three-dimensional motion in Example 2. The accumulated error in the inextensibility condition (solid curve) is less than 1.3ϵ, where the tolerance ϵ is the same as in Example 1. The accumulated error in the quaternion norm (broken curve) is less than 0.3ϵ.

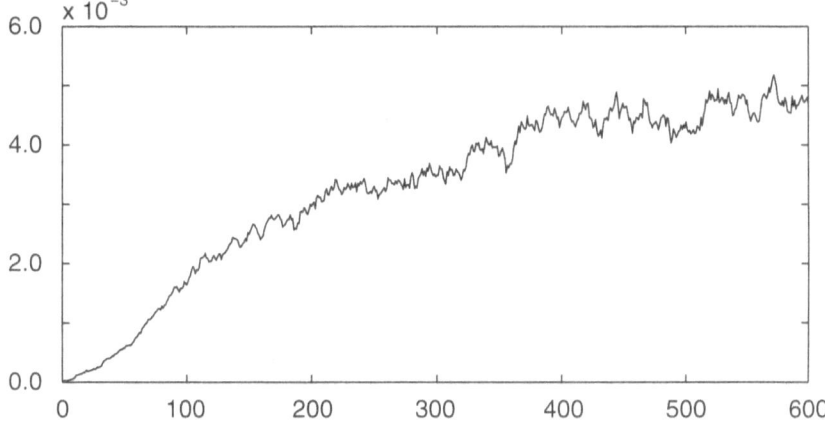

Fig. 7.7. Relative error in the Hamiltonian for Example 2 is again proportional to τ^2 at each integration step, where the time step τ is the same as in Example 1.

the maximum error in inextensibility (solid curve) is less than 1.3ϵ after $120,000$ time steps, both significantly less than the discretization error associated with the solution. Figure 7.7 shows the relative error in the Hamiltonian for Example 2.

Acknowledgments

It is a pleasure for the authors to thank Randy Paffenroth for assistance in the development of the graphics, and Dr. Joseph Sedlak of Computer Sciences Corporation, Professor R. B. Kellogg, Dr. Oscar Gonzalez, Dr. Rob Manning and Dr. Jian-Ming Xu for valuable discussions.

References

[1] S. S. Antman. *Nonlinear Problems of Elasticity* (Springer-Verlag, New York, 1994).

[2] V. I., Arnold, V. V. Kozlov & A. I. Neishtadt. Mathematical aspects of classical and celestial mechanics, in *Dynamical Systems III, Encyclopædia of the Mathematical Sciences Volume 3*, ed. V. I. Arnold (Springer-Verlag, New York, 1988).

[3] O. Bottema & B. Roth. *Theoretical Mechanics* (Dover, 1979).

[4] K. E. Brenan, S. L. Campbell & L. R. Petzold. *Numerical Solution of Initial-Value Problems in Differential-Algebraic Equations* (SIAM, Philadelphia, 1995).

[5] J. M. Charap, ed. *Geometry of Constrained Dynamical Systems* (Publications of the Newton Institute, Cambridge Press, 1995).

[6] D. J. Dichmann. Hamiltonian Dynamics of an Elastica and Stability of Solitary Waves. Ph.D. thesis, University of Maryland (1994).

[7] D. J. Dichmann., Y. W. Li & J. H. Maddocks. Hamiltonian formulations and symmetries in rod mechanics, in *Mathematical Approaches to Biomolecular Structure and Dynamics*, J. P. Mesirov, K. Schulten & D. W. Sumners, eds., IMA Volumes in Mathematics and Its Applications, **82** (Springer-Verlag, New York, 1996), 71–113.

[8] D. J. Dichmann, J. H. Maddocks & R. L. Pego. Hamiltonian dynamics of an elastica and the stability of solitary waves. *Arch. Rat. Mech. Anal.* To appear.

[9] D. J. Dichmann, J. H. Maddocks & J. M. Xu. Three-dimensional Hamiltonian dynamics of an elastica. In preparation.

[10] P. A. M. Dirac. On generalized Hamiltonian dynamics. *Can. J. Math.* **2** (1950) 129–148.

[11] O. Gonzalez. Mechanical systems subject to holonomic constraints: unconstrained formulations and conservative integration. *Physica D*, submitted

[12] O. Gonzalez & J.-C. Simo. Exact energy and momentum conserving algorithms for general models in nonlinear elasticity. *Comput. Methods Appl. Mech. Eng.*, submitted.

[13] J. Junkins & J. Turner. *Optimal Spacecraft Rotational Maneuvers* (Elsevier, Amsterdam, 1986).

[14] I. Klapper. Biological applications of the dynamics of twisted elastic rods. *J. Comput. Phys.* To appear.

[15] F. Klein & A. Sommerfeld. *Über die Theorie des Kriesels* (Johnson Reprint, 1965).

[16] B. Leimkuhler & S. Reich. Symplectic integration of constrained Hamiltonian systems. *Math. Comp.* **63** (1994) 589–605.

[17] B. Leimkuhler & R. Skeel. Symplectic numerical integrators in constrained Hamiltonian systems. *J. Comp. Phys.* **112** (1994) 117–125.

[18] J. H. Maddocks & D. J. Dichmann. Conservation laws in the dynamics of rods. *J. Elasticity* **34** (1994) 83–96.

[19] J. H. Maddocks & R. L. Pego. An unconstrained Hamiltonian formulation for incompressible fluid flow. *Comm. Math. Phys.* **170** (1995) 207–217.

[20] J. M. Sanz-Serna and M. P. Calvo. *Numerical Hamiltonian Problems* (Chapman & Hall, New York, 1994).

[21] J.-C. Simo, N. Tarnow & M. Doblare. Nonlinear dynamics of three-dimensional rods: Exact energy and momentum conserving algorithms. *Int. J. Numer. Methods Eng.* **38** (1995) 1431–1473.

[22] J.-C. Simo & L. Vu-Quoc. A three-dimensional finite-strain rod model. Part II: Computational aspects. *Comput. Methods Appl. Mech. Eng.* **58** (1986) 79–116.

[23] J.-C. Simo & L. Vu-Quoc. On the dynamics in space of rods undergoing large motions: A geometrically exact approach. *Comput. Methods Appl. Mech. Eng.* **66** (1988) 125–161.

[24] M. Tabor & I. Klapper. Dynamics of twist and writhe and the modeling of bacterial fibers, in *Mathematical Approaches to Biomolecular Structure and Dynamics*, J. P. Mesirov, K. Schulten & D. W. Sumners, eds., IMA Volumes in Mathematics and Its Applications, **82** (Springer-Verlag, New York, 1996), 139–160.

A Symplectic Integrator for Riemannian Manifolds

B. Leimkuhler[1] and G.W. Patrick[2]

[1] Department of Mathematics, University of Kansas, Lawrence, KS 66045, USA

[2] Department of Mathematics and Statistics, University of Saskatchewan, Saskatoon, Saskatchewan, S7N 5E6, Canada

Received October 27, 1995; revised manuscript accepted for publication March 13, 1996.
Communicated by Jerrold Marsden and Stephen Wiggins

This paper is dedicated to the memory of Juan-Carlos Simo

Summary. The configuration spaces of mechanical systems usually support Riemannian metrics which have explicitly solvable geodesic flows and parallel transport operators. While not of primary interest, such metrics can be used to generate integration algorithms by using the known parallel transport to evolve points in velocity phase space.

1. Introduction

There exist a number of mechanical systems having configuration and phase spaces that are differentiable manifolds which are not open subsets of Euclidean space. The spherical pendulum, which has configuration space the 2-sphere, and a single rigid body, with configuration space the Lie group $SO(3)$, are two common examples. Integration algorithms, on the other hand, are usually set in the Euclidean context. Use of a covering set of charts is straightforward in principle but can give unsatisfactory results in practice, since (i) it can increase the computational complexity of the algorithm, (ii) although a symplectic method can be used in each chart, between charts the discrete map changes, and this can lead to a systematic energy drift [6], and (iii) the numerical determination of when and how to switch charts can be complicated. If numerical computation must be done with tuples of floating point numbers, how can one proceed without charts?

This question has led to recent research on numerical methods that can be applied directly on various differential geometric objects. Such methods might be called *intrinsic*. Progress cannot be made strictly within the context of a manifold, symplectic or not: one requires the assistance of *some* global structure to construct *any* method. For example, McLachlan and Scovel [11] render configuration spaces as submanifolds of Euclidean space using global embeddings. From that point of view the mechanical system is a constrained system and the main issue is the finding of algorithms that map the constraint set to itself. One finds these ideas also in Barth and Leimkuhler [2] and Reich [14]. Crouch and Grossman [4] posit an independent spanning set of globally defined vector fields on phase space; the flows of these vector fields are used to move through phase space. Intrinsic Runge-Kutta methods, also using global vector fields, are investigated by Munthe-Kaas [12]. Lewis and Simo [7], [8] have constructed conserving algorithms in the context of configuration spaces that are Lie groups, where a multitude of global objects are available.

Riemannian metrics are other global objects worthy of consideration when construct-ing intrinsic methods. Indeed, Riemannian geometry is a subject exactly invented to export common Euclidean notions to the global-manifold context. The main derived objects of a Riemannian metric are its geodesic flow and its parallel transport operators along geodesics. Riemannian geometry may be a promising base for numerical compu-tation since many common diffeotypes of configuration spaces enjoy the presence of a Riemannian metric having explicitly known geodesic flow and parallel transport. Yet, no one seems to have used Riemannian geometry in the context of intrinsic integration algorithms.

In this article we show how such an ambient Riemannian metric can be used to frame the popular "leapfrog" method in the category of Riemannian manifolds, thus creating new intrinsic methods applicable to mechanical systems with configuration spaces as general as homogeneous spaces of semisimple Lie groups. This new method, shown in Figure 1, is implicit, second order, time-reversing, and symplectic, and respects those symmetries of the system which are also isometries of the ambient metric. For arbitrary potential energy, but where the kinetic energy metric of the mechanical system is proportional to the ambient metric, our method is *explicit*, and is equivalent to the splitting method corresponding to the splitting of the Hamiltonian into kinetic and potential energy.

2. Context and Notation

We will consider mechanical systems with configuration space Q, kinetic energy metric g, and potential V, so that the Lagrangian is

$$L(v_q) = \frac{1}{2} g(v_q, v_q) - V(q), \qquad v_q \in T_q Q.$$

Let a Lie group G act on Q by isometries with respect to g and suppose that V is G-invariant. We will use the following standard notations [1]:

$$g^\flat \colon TQ \to T^*Q \quad \langle g^\flat(v_q), w_q \rangle = g(q)(v_q, w_q),$$
$$g^\sharp \colon T^*Q \to TQ \quad g^\sharp = (g^\flat)^{-1},$$
$$\operatorname{grad}_g V = g^\sharp \circ dV.$$

In other words, $g^\flat$ is the Legendre transformation, $g^\sharp$ is its inverse, and $-\operatorname{grad}_g V$ is the force field divided by the "inertia."

Given an initial state $tv_1 \in T_{q_1} Q$, find $w \in T_{q_1} Q$ by iterating

$$w = tv_1 - \frac{t^2}{2} \operatorname{grad}_g V(q_1) + \frac{1}{2} A(q_1, w),$$

where A is defined by

$$g^{\flat}(q_1) A(q_1, w) = \frac{1}{2} \nabla h(q_1) w^2 + h(q_1) \left(w, \nabla\Delta(q_2, q_1) + 2\,\mathrm{Id} \right) - h(q_2) \left(\mathbb{P}_{q_2} w, d\Delta(q_1, q_2) \right),$$

and $q_2 = \exp_{q_1} w$. The convergence of this step is faster for g nearer to $\tilde{g}$, and $A = 0$ for $g = \tilde{g}$, in which case this step is *explicit*. The new configuration is then

$$q_2 = \exp_{q_1} w.$$

The new state is $tv_2 \in T_{q_2} Q$, given by

$$tv_2 = \mathbb{P}_{q_2} w - \frac{t^2}{2} \operatorname{grad}_g V(q_2) + \frac{1}{2} A(q_2, -\mathbb{P}_{q_2} w).$$

If $(Q, \tilde{g})$ is a symmetric space, then

$$g^{\flat}(q_1) A(q_1, w) = \frac{1}{2} \nabla h(q_1) w^2 + h(q_1) \left(w, (2 - f_1/f_2) R_w \right) - h(q_2) \left(\mathbb{P}_{q_2} w \mathbb{P}_{q_2} (1/f_2) R_w \right),$$

where R_w is the curvature operator of $\tilde{g}$, and f_1 and f_2 are the power series

$$f_1(z) = \cos \sqrt{z} = \sum_{k=0}^{\infty} \frac{(-1)^k}{(2k)!} z^k, \qquad f_2(z) = \frac{\sin \sqrt{z}}{\sqrt{z}} = \sum_{k=0}^{\infty} \frac{(-1)^k}{(2k+1)!} z^k.$$

Fig. 1. The algorithm.

Suppose that $f(q_1, q_2)$ is a differentiable function on $Q \times Q$. We use the notations $d_1 f$ and $d_2 f$ for the differential of f in its first and second argument, respectively. Thus, if $q_1(\epsilon)$ is a curve in Q, then

$$\frac{d}{d\epsilon}\Big|_{\epsilon=0} f(q_1(\epsilon), q_2) = d_1 f(q_1(0), q_2) q_1'(0).$$

Suppose that $\tilde{g}$ is another metric on Q, and that the action of G on Q is isometric with respect to $\tilde{g}$. We think of the Riemannian geometry defined by $\tilde{g}$ as being known.

It may be helpful to keep in mind the following example: a particle of mass m moves on the ellipsoid

$$\frac{y_1^2}{a_1^2} + \frac{y_2^2}{a_2^2} + \frac{y_3^2}{a_3^2} = 1$$

in the presence of some potential V, say $V = \mu y_3$, where μ is constant. The diffeotype of the ellipsoid is of course a 2-sphere, and transformation can be made to the 2-sphere by

$$x_1 = \frac{y_1}{a_1}, \qquad x_2 = \frac{y_2}{a_2}, \qquad x_3 = \frac{y_3}{a_3}.$$

Through this transformation, the kinetic energy metric becomes

$$g = m a_1^2 dx_1 \otimes dx_1 + m a_2^2 dx_2 \otimes dx_2 + m a_3^2 dx_3 \otimes dx_3$$

(meaning the restriction of this to the unit 2-sphere) while the potential energy becomes $V = a_3\mu x_3$. If $a_1 = a_2$ then the system admits the S^1 symmetry of rotations about the y_3 axis, and one can take the Lie group G to be S^1. The other metric $\tilde{g}$ is of course

$$\tilde{g} = dx_1 \otimes dx_1 + dx_2 \otimes dx_2 + dx_3 \otimes dx_3.$$

The geometry generated by this metric (spherical geometry) is well-known, and many elements of that geometry have well-known closed formulas. Using the approach developed in this article, the geometry of $\tilde{g}$ can be used to navigate directly on the unit 2-sphere; preservation of constraint $x_1{}^2 + x_2{}^2 + x_3{}^2 = 1$ will not be an issue.

The exponential mapping of $\tilde{g}$ will be denoted by exp, so that the geodesic starting at $q \in Q$ with velocity $v \in T_q Q$ is $\exp_q(tv)$. Parallel translation with respect to $\tilde{g}$ will be denoted by $\mathbb{P}$. Various forms will be used and the distinction will be clear from the context. For example, if $q_1, q_2 \in TQ$, $\mathbb{P}_{q_2,q_1}$ is the parallel translation operator along a previously defined curve joining q_1 to q_2, usually a geodesic. Sometimes q_1 is omitted. If $w \in TQ$, the notation $\mathbb{P}_w$ denotes the parallel translation operator along the geodesic curve $\exp(tw)$, $t \in [0, 1]$. The covariant derivative of a vector field X on Q in the direction $v \in TQ$ is denoted by $\nabla_v X$ and satisfies (or is defined by, depending on how the theory is revealed)

$$\nabla_v X(q) = \frac{d}{d\epsilon}\bigg|_{\epsilon=0} \mathbb{P}_q X(q + \epsilon v),$$

where $q + \epsilon v$ is any curve in ϵ with derivative v at $\epsilon = 0$ and the parallel transport can be along the reverse of that curve. Covariant derivatives of other types of tensors satisfy (or are defined by) analogous formulas.

3. The Coordinate Version, Constant Metric

In this section we derive a special case of our algorithm, in order to introduce it in a universally understood context. On the way, we show that, in this case, our algorithm reduces to the well-known leapfrog algorithm. Specifically, we assume that $Q = \{q^i\}$ is an open subset of Euclidean space. The algorithm will be constructed on the assumption that g is constant. The metric $\tilde{g}$ does not make an appearance here; effectively it is the ordinary Euclidean metric.

3.1. The Generating Function

We aim for a second-order time reversing algorithm, in part so that the Yoshida trick [15] can be applied. For a second-order algorithm from a generating function of type 1, the appropriate generating function is [13]

$$S_t^1(q_1, q_2) \equiv \frac{1}{2t}\left(g_{ij}(q_1)\Delta q^{ij} + \frac{1}{2}g_{ij;k}(q_1)\Delta q^{ijk}\right)$$
$$- t\left(V(q_1) + \frac{1}{2}V_{;i}(q_1)\Delta q^i\right) \tag{1}$$

where

$$\Delta q^{i_1 \cdots i_n} = (q_2^{i_1} - q_1^{i_1})(q_2^{i_2} - q_1^{i_2}) \cdots (q_2^{i_n} - q_1^{i_n}), \tag{2}$$

and $; i$ denotes differentiation by q^i. This generating function is defective since it is not symmetric in q_1 and q_2, and hence does not produce a time-reversing algorithm. To remedy this, one can symmetrize:

$$\frac{1}{2} \left(S_t^1(q_1, q_2) + S_t^1(q_2, q_1) \right) = \frac{1}{4t} A_1 - \frac{t}{2} A_3,$$

where

$$A_1 \equiv \left(g_{ij}(q_1) + g_{ij}(q_2) \right) \Delta q^{ij} + \frac{1}{2} \left(g_{ij;k}(q_1) - g_{ij;k}(q_2) \right) \Delta q^{ijk}$$

and

$$A_3 \equiv \left(V(q_1) + V(q_2) \right) + \frac{1}{2} \left(V_{;i}(q_1) - V_{;i}(q_2) \right) \Delta q^i.$$

By [13], we may retain an order 2 algorithm by discarding terms in A_1 with up to third-order derivatives in q_1 and q_2 zero at $q_1 = q_2$ and terms in A_3 with up to first-order derivatives in q_1 and q_2 zero at $q_1 = q_2$. Thus, the second summands in both A_1 and A_3 may be discarded, and the generating function

$$S_t^2 \equiv \frac{1}{4t} \left(g_{ij}(q_1) + g_{ij}(q_2) \right) \Delta q^{ij} - \frac{t}{2} \left(V(q_1) + V(q_2) \right) \tag{3}$$

defines an order 2 algorithm which is time reversible.

3.2. The Algorithm

The algorithm takes a point (q, p) in canonical phase space and advances by time t to obtain a point (q', p'). The point (q', p') is found by first solving the equation

$$p_k = -\frac{\partial S_t^2}{\partial q_1^k}(q, q') \tag{4}$$

for q' and then calculating

$$p_k' = \frac{\partial S_t^2}{\partial q_2^k}(q, q'). \tag{5}$$

Assuming constant g_{ij}, (4) takes the form

$$-p_k = -\frac{1}{t} g_{ik} \Delta q^i - \frac{t}{2} V_{;k}(q),$$

which has the solution

$$\Delta q^i = t g^{ik} p_k - \frac{t^2}{2} g^{ik} V_{;k}(q),$$

or by setting $v^i = g^{ik} p_k$,

$$q' = q + tv - \frac{t^2}{2} g^{-1} \nabla V(q), \tag{6}$$

where ∇ denotes the common gradient and g is the matrix with entries g_{ij}. Equation (5) becomes

$$\begin{aligned} p'_k &= \frac{1}{t} g_{ik} \Delta q^i - \frac{t}{2} V_{;k}(q') \\ &= p_k - \frac{t}{2} \left(V_{;k}(q) + V_{;k}(q') \right) \end{aligned}$$

or

$$p' = p - \frac{t}{2} \left(\nabla V(q) + \nabla V(q') \right). \tag{7}$$

The algorithm defined by (6) and (7) is the leapfrog algorithm as follows. Starting from the point $(q_0, p_{-\frac{1}{2}})$, the leapfrog algorithm advances to the point $(q_1, p_{\frac{1}{2}})$ by

$$p_{\frac{1}{2}} = p_{-\frac{1}{2}} - t \nabla V(q_0), \tag{8}$$

$$q_1 = q_0 + t g^{-1} p_{\frac{1}{2}}. \tag{9}$$

The next step advances $p_{\frac{1}{2}}$ to $p_{\frac{3}{2}}$ by

$$p_{\frac{3}{2}} = p_{\frac{1}{2}} - t \nabla V(q_1). \tag{10}$$

Set

$$p_0 = \frac{p_{-\frac{1}{2}} + p_{\frac{1}{2}}}{2}, \qquad p_1 = \frac{p_{\frac{1}{2}} + p_{\frac{3}{2}}}{2}. \tag{11}$$

Using (11) to eliminate $p_{-\frac{1}{2}}$ from (8) gives

$$p_{\frac{1}{2}} = p_0 - \frac{t}{2} \nabla V(q_0), \tag{12}$$

and substitution of (12) into (9) gives

$$q_1 = q_0 + t g^{-1} p_0 - \frac{t^2}{2} g^{-1} \nabla V(q_0). \tag{13}$$

Taking the average of (8) and (10) yields

$$p_1 = p_0 - \frac{t}{2} \left(\nabla V(q_0) + \nabla V(q_1) \right). \tag{14}$$

Since (13) and (14) are equivalent to (6) and (7), the algorithm from the generating function (10) is equivalent, for constant g_{ij}, to the leapfrog algorithm.

4. On a Riemannian Manifold

In this section we leave coordinates on Q behind, using instead the Riemannian geometry of the metric $\tilde{g}$. This is accomplished first by replacing coordinate-dependent portions of the generating function (3) by intrinsic (in the $\tilde{g}$ geometry) objects. The algorithm itself is obtained using only intrinsic operations, such as the covariant derivative. The result is an algorithm that advances on Q using parallel translation in the $\tilde{g}$ geometry.

4.1. The Generating Function

The exponential of the standard metric on $\mathbb{R}^n$ is exactly $\exp_q v = q + v$. Given $q_1 \in \mathbb{R}^n$ and $q_2 \in \mathbb{R}^n$, the quantity $\Delta q \equiv q_2 - q_1$ is the tangent vector at q_1 which tells how to get to q_2 using a geodesic. The idea is to replace Δq^i in (3) with $\Delta(q_2, q_1)$, which is defined by the equation

$$\exp_{q_1} \Delta(q_2, q_1) = q_2. \tag{15}$$

We have that $\Delta(q_2, q_1)$ is the same as $\Delta(q_1, q_2, 1)$ of [13], so that $\Delta(q_2, q_1)$ is defined and smooth for q_2 near to q_1. We guess that the function

$$S_t^3 \equiv \frac{1}{4t} \left(g(q_1)\left(\Delta(q_2, q_1), \Delta(q_2, q_1)\right) + g(q_2)\left(\Delta(q_2, q_1), \Delta(q_2, q_1)\right) \right)$$
$$- \frac{t}{2} \left(V(q_1) + V(q_2) \right) \tag{16}$$

generates an order 2 algorithm.

And it does. To check this, take any chart on Q, again with coordinates q^i. Let the Christoffel symbols for the metric $\tilde{g}$ be $\tilde{\Gamma}^i_{jk}$. Then

$$\left[\exp_{q_1} v \right]^i = q_1{}^i + v^i - \frac{1}{2} \tilde{\Gamma}^i_{ab}(q_1) v^a v^b + \cdots = q_2^i$$

gives

$$\left[\Delta(q_2, q_1) \right]^i = \Delta q^i + \frac{1}{2} \tilde{\Gamma}^i_{ab}(q_1) \Delta q^a \Delta q^b + O\left((\Delta q)^3 \right), \tag{17}$$

where $O\left((\Delta q)^3 \right)$ denotes a function that has up to second order derivatives in q_1 and q_2 zero at $q_1 = q_2$. Now substituting (17) into the first part of (16) gives

$$g(q_1)\left(\Delta(q_2, q_1), \Delta(q_2, q_1)\right) + g(q_2)\left(\Delta(q_2, q_1), \Delta(q_2, q_1)\right)$$

$$= g_{ij}(q_1) \left(\Delta q^i + \frac{1}{2} \tilde{\Gamma}^i_{ab}(q_1) \Delta q^a \Delta q^b \right) \left(\Delta q^j + \frac{1}{2} \tilde{\Gamma}^i_{cd}(q_1) \Delta q^c \Delta q^d \right)$$

$$+ g_{ij}(q_2) \left(-\Delta q^i + \frac{1}{2} \tilde{\Gamma}^i_{ab}(q_2) \Delta q^a \Delta q^b \right) \left(-\Delta q^j + \frac{1}{2} \tilde{\Gamma}^i_{cd}(q_2) \Delta q^c \Delta q^d \right)$$

$$+ O\left((\Delta q)^4 \right)$$

$$= \left(g_{ij}(q_1) + g_{ij}(q_2) \right) \Delta q^{ij}$$

$$+ \left(g_{ij}(q_1)\tilde{\Gamma}^i_{ab}(q_1) - g_{ij}(q_2)\tilde{\Gamma}^i_{ab}(q_2) \right) \Delta q^{iab} + O\left((\Delta q)^4 \right)$$

$$= \left(g_{ij}(q_1) + g_{ij}(q_2) \right) \Delta q^{ij} + O\left((\Delta q)^4 \right),$$

as required. Showing that the second part of (16) has the required order of contact with the second part of (3) is even easier.

4.2. The Algorithm

We begin by setting $h = g - \tilde{g}$, so that $g = \tilde{g} + h$. Then [13] writing $\Delta(q_2, q_1)^2$ for a pair of $\Delta(q_2, q_1)$, the generating function

$$S_t^4(q_1, q_2) \equiv \frac{1}{2t}\tilde{g}(q_1)\Delta(q_2, q_1)^2 = \frac{d(q_1, q_2)^2}{2t} = \frac{1}{2t}\tilde{g}(q_2)\Delta(q_1, q_2)^2$$

generates the geodesic flow of $\tilde{g}$, where $d(q_1, q_2)$ is the Riemannian distance between q_1 and q_2. Thus, the generating function (16) can be written as follows:

$$S_t^3(q_1, q_2) = \frac{1}{4t}\left(g(q_1)\Delta(q_2, q_1)^2 + g(q_2)\Delta(q_1, q_2)^2\right) - \frac{t}{2}(V(q_1) + V(q_2))$$

$$= \frac{1}{2t}\tilde{g}(q_1)\Delta(q_2, q_1)^2 - \frac{t}{2}(V(q_1) + V(q_2)) \tag{18}$$

$$+ \frac{1}{4t}\left(h(q_1)\Delta(q_2, q_1)^2 + h(q_2)\Delta(q_1, q_2)^2\right). \tag{19}$$

The algorithm from this generating function takes $\alpha^1 \in T_{q_1}^* Q$ to $\alpha^2 \in T_{q_2}^* Q$, defined by first solving

$$-\alpha^1 = d_1 S_t^3(q_1, q_2) \tag{20}$$

for q_2 and then setting

$$\alpha^2 = d_2 S_t^3(q_1, q_2). \tag{21}$$

In this discussion, we will think of q_2 as $\exp_{q_1} w$, where $w \in T_{q_1} Q$. Since S_t^4 does generate the geodesic flow, we have the identities

$$\tilde{g}^b(q_1)w = -d_1\left(\frac{1}{2t}\tilde{g}(q_1)\Delta(q_2, q_1)^2\right), \tag{22}$$

$$\mathbb{P}_{q_2}\tilde{g}^b(q_1)w = d_2\left(\frac{1}{2t}\tilde{g}(q_1)\Delta(q_2, q_1)^2\right). \tag{23}$$

Using (22), the implicit part (20) of the algorithm becomes

$$-\alpha^1 = -\frac{1}{t}\tilde{g}^b(q_1)w - \frac{t}{2}dV(q_1)$$

$$+ d_1\left(\frac{1}{4t}h(q_1)\Delta(q_2, q_1)^2 + \frac{1}{4t}h(q_2)\Delta(q_1, q_2)^2\right)$$

$$= -\frac{1}{t}\tilde{g}^b(q_1)w - \frac{t}{2}dV(q_1) + \frac{1}{4t}\nabla h(q_1)w^2$$

$$+ \frac{1}{2t}h(q_1)(w, \nabla\Delta(q_2, q_1)) + \frac{1}{2t}h(q_2)(-\mathbb{P}_{q_2}w, d\Delta(q_1, q_2)), \tag{24}$$

where (24) contains various linear operators defined as follows: for $v \in T_{q_1} Q$,

$$\nabla h(q_1) w^2 \cdot v \equiv \nabla_v h(q_1) w^2, \tag{25}$$

$$\nabla \Delta(q_2, q_1) \cdot v \equiv \nabla_v (q_1 \mapsto \Delta(q_2, q_1)), \tag{26}$$

$$d\Delta(q_1, q_2) \cdot v \equiv \left. \frac{d}{d\epsilon} \right|_{\epsilon=0} \Delta(q_1 + \epsilon v, q_2). \tag{27}$$

Here $q_1 + \epsilon v$ is any curve in ϵ having derivative v at $\epsilon = 0$. Equation (24) is calculated in a way that separates, as much as possible, the Riemannian geometry of $\tilde{g}$ from the system-dependent quantity h.

We wish to write (24) as an iterative procedure for calculating w, and that would be possible simply by isolating the term $\tilde{g}^b(q_1)w$. However, the sum of the last two terms of (24) can be expected to fall only linearly with w, and so the iterative procedure would converge just in the case that h is small in comparison to g. This can be seen by considering the approximations

$$d\Delta(q_1, q_2)v \approx v, \qquad \nabla_v \Delta(q_2, q_1) \approx -v, \tag{28}$$

and then the sum of the last two terms of (24) looks like

$$-\frac{1}{t} h(q_1)(w, \cdot).$$

We want an iterative procedure that converges whenever w is small (i.e., whenever q_1 and q_2 are close). So we replace the $\tilde{g}$ in the first term of (24) with $g - h$ to obtain

$$
\begin{aligned}
-\alpha^1 = &-\frac{1}{t} g^b(q_1)w + \frac{1}{t} h(q_1)^b w - \frac{t}{2} dV(q_1) + \frac{1}{4t} \nabla h(q_1) w^2 \\
&+ \frac{1}{2t} h(q_1)(w, \nabla \Delta(q_2, q_1)) + \frac{1}{2t} h(q_2)\left(-\mathbb{P}_{q_2} w, d\Delta(q_1, q_2)\right).
\end{aligned} \tag{29}
$$

If we define the quantity $A(q_1, w)$ by

$$
\begin{aligned}
g^b(q_1) A(q_1, w) = &\frac{1}{2} \nabla h(q_1) w^2 \\
&+ h(q_1)(w, \nabla \Delta(q_2, q_1) + 2\,\mathrm{Id}) \\
&- h(q_2)\left(\mathbb{P}_{q_2} w, d\Delta(q_1, q_2)\right),
\end{aligned} \tag{30}
$$

then (29) becomes

$$-\alpha^1 = -\frac{1}{t} g^b(q_1)w - \frac{t}{2} dV(q_1) + g^b(q_1)\frac{1}{2t} A(q_1, w), \tag{31}$$

and here the approximations (28) suggest that A falls quadratically with w. Isolating w in (31) and making the replacement $\alpha^1 = g^{\sharp}(q_1)v_1$ yields

$$w = tv_1 - \frac{t^2}{2} \mathrm{grad}_g V(q_1) + \frac{1}{2} A(q_1, w), \qquad q_2 = \exp_{q_1} w. \tag{32}$$

For the explicit part of the algorithm that corresponds to (21), we find, using (23) and the fact that interchanging q_1 and q_2 means replacing w with $-\mathbb{P}_{q_2}w$, that

$$
\begin{aligned}
\alpha^2 &= \frac{1}{t}g^{\flat}(q_2)\mathbb{P}_{q_2}w - \frac{t}{2}dV(q_2) \\
&\quad + d_2\left(\frac{1}{4t}h(q_1)\Delta(q_2, q_1)^2 + \frac{1}{4t}h(q_2)\Delta(q_1, q_2)^2\right) \\
&= \frac{1}{t}g^{\flat}(q_2)\mathbb{P}_{q_2}w - \frac{t}{2}dV(q_2) + \frac{1}{2}g^{\flat}(q_2)A(q_2, -\mathbb{P}_{q_2}w).
\end{aligned}
$$

Therefore,

$$
v_2 = \frac{1}{t}\mathbb{P}_{q_2}w - \frac{t}{2}\operatorname{grad}_g V(q_2) + \frac{1}{2t}A(q_2, -\mathbb{P}_{q_2}w). \tag{33}
$$

We note that if g and $\tilde{g}$ are proportional, then $A = 0$, and the algorithm is exactly the iteration of $F_{t/2}G_t F_{t/2}$, where F is the flow of the Hamiltonian consisting just of the potential energy and G is the flow of the Hamiltonian consisting just of the kinetic energy.

4.3. Calculations on a Symmetric Riemannian Manifold

As is apparent, it is not sufficient merely to know the exponential mapping of $\tilde{g}$. Indeed, by (26) and (27), derivatives of the map Δ are required, and Δ, being defined by (15), is in turn the inverse of exp. Thus at least one derivative of exp is required, and the task of calculating $\nabla\Delta$ and $d\Delta$ immediately attracts attention.

In this regard, it is useful to consider *Jacobi fields* and the class of *symmetric Riemannian manifolds*. Let us summarize the relevant facts, for which a general reference is [5]. Let M be a Riemannian manifold and $p \in M$.

1. The curvature operators $R_w: T_pM \to T_pM$, $w \in T_pM$ are defined by

$$
R_w(v) = R(v, w)w, \tag{34}
$$

where R is the Riemann curvature tensor.

2. A Jacobi field over a geodesic $c(t)$ is a vector field $Y(t)$ over $c(t)$ satisfying

$$
\frac{\nabla^2 Y}{dt^2} + R_{\dot{c}(t)}Y(t) = 0.
$$

In appropriate coordinates, this is a nonautonomous, second-order linear ordinary differential equation, and so there is a unique Jacobi field $Y(t)$ for specified $Y(0)$ and $\nabla Y/dt(0)$.

3. Let $p + \epsilon a$ be any curve at p with derivative $a \in T_pM$ and let $b \in T_pM$. From page 113 of [5],[3] the vector field

$$
Y(t) = \left.\frac{d}{d\epsilon}\right|_{\epsilon=0}\exp_{p+\epsilon a}(\mathbb{P}_{p+\epsilon a}wt)
$$

---

[3] In [5] on page 113 the equation for $A(t)$, which is the relevant one for what we require, is in error: A_0 and A_1 must be interchanged.

is the Jacobi field over $\exp_p(wt)$ such that

$$Y(0) = a, \qquad \frac{\nabla Y}{dt}(0) = 0,$$

and the vector field

$$Y(t) = \frac{d}{d\epsilon}\bigg|_{\epsilon=0} \exp_p((w + \epsilon b)t)$$

is the Jacobi field over $\exp_p(wt)$ such that

$$Y(0) = 0, \qquad \frac{\nabla Y}{dt}(0) = b.$$

By definition, M is a symmetric space if it is connected and if for all $p \in M$ there is an isometry σ_p such that $\sigma_p(p) = p$ and $T_p\sigma_p = -\text{Id}$.
Suppose M is a symmetric space, $v, w \in T_pM$. Then

$$\mathbb{P}_{vt} R_w = R_{\mathbb{P}_{vt}w}\mathbb{P}_{vt},$$

that is, R is parallel, and

$$\mathbb{P}_v = T_p\left(\sigma_{\exp_p(v/2)} \circ \sigma_p\right). \qquad (35)$$

Suppose M is a symmetric space, and define the power series

$$f_1(z) = \cos\sqrt{z} = \sum_{k=0}^{\infty} \frac{(-1)^k}{(2k)!} z^k, \qquad f_2(z) = \frac{\sin\sqrt{z}}{\sqrt{z}} = \sum_{k=0}^{\infty} \frac{(-1)^k}{(2k+1)!} z^k.$$

For $a, b \in T_pM$, the Jacobi field $Y(t)$ over $\exp_p(wt)$, such that

$$Y(0) = a, \qquad \frac{\nabla Y}{dt}(0) = b,$$

is

$$Y(t) = \mathbb{P}_{wt}\left(f_1(t^2 R_w)a + f_2(t^2 R_w)b\right).$$

We now calculate $d\Delta$. Let $q_1, q_2 \in Q$, $w = \Delta(q_2, q_1)$ and $\tilde{w} = \mathbb{P}_{q_2}w = -\Delta(q_1, q_2)$. fferentiating the defining equation of Δ, namely

$$\exp_{q_1} \Delta(q_2 + \epsilon v, q_1) = q_2 + \epsilon v, \qquad v \in T_{q_2}Q,$$

c obtain

$$v = \frac{d}{d\epsilon}\bigg|_{\epsilon=0} \exp_{q_1} \Delta(q_2 + \epsilon v, q_1) = \mathbb{P}_{q_2} f_2(R_w)d\Delta(q_2, q_1)v,$$

so that

$$d\Delta(q_2, q_1)v = (f_2(R_w))^{-1} \mathbb{P}_{q_1} v = (1/f_2)(R_w)\mathbb{P}_{q_1} v. \tag{36}$$

Also, for $v \in T_{q_1} Q$,

$$\begin{aligned}
d\Delta(q_1, q_2)v &= (1/f_2)(R_{-\tilde{w}})\mathbb{P}_{q_2} v \\
&= \mathbb{P}_{q_2}\mathbb{P}_{q_1}(1/f_2)(R_{\tilde{w}})\mathbb{P}_{q_2} v \\
&= \mathbb{P}_{q_2}(1/f_2)(R_w)v.
\end{aligned} \tag{37}$$

The calculation of $\nabla\Delta$ is similar, starting with the equation

$$\exp_{q_1+\epsilon v} \Delta(q_2, q_1 + \epsilon v) = q_2, \qquad v \in T_{q_1} Q,$$

and differentiating, we have

$$\begin{aligned}
0 &= \frac{d}{d\epsilon}\bigg|_{\epsilon=0} \exp_{q_1+\epsilon v} \Delta(q_2, q_1 + \epsilon v) \\
&= \frac{d}{d\epsilon}\bigg|_{\epsilon=0} \exp_{q_1+\epsilon v} \left(\mathbb{P}_{q_1+\epsilon v}\mathbb{P}_{q_1} \Delta(q_2, q_1 + \epsilon v)\right) \\
&= \frac{d}{d\epsilon}\bigg|_{\epsilon=0} \exp_{q_1+\epsilon v} \left(\mathbb{P}_{q_1+\epsilon v}\mathbb{P}_{q_1} \Delta(q_2, q_1)\right) + \frac{d}{d\epsilon}\bigg|_{\epsilon=0} \exp_{q_1} \left(\mathbb{P}_{q_1} \Delta(q_2, q_1 + \epsilon v)\right) \\
&= \mathbb{P}_{q_2} \left(f_1(R_w)v + f_2(R_w)\nabla_v\Delta(q_2, q_1)\right).
\end{aligned}$$

Consequently,

$$\nabla_v\Delta(q_2, q_1) = -(f_1/f_2)(R_w)v \tag{38}$$

and for $v \in T_{q_2} Q$,

$$\nabla_v\Delta(q_1, q_2) = -\mathbb{P}_{q_2}(f_1/f_2)(R_w)\mathbb{P}_{q_1} v. \tag{39}$$

Note that since $f_1(z) = 1 + O(z)$ and $f_2(z) = 1 + O(z)$, the approximations (28) are confirmed by (37) and (38). Moreover, since R_w contains w twice, R_w falls quadratically with w. Thus, if the power series for $\nabla\Delta$ and $d\Delta$ are substituted into (30), then the terms that depend on R_w fall cubically with w, suggesting that R_w may not require recalculation on every iteration of (32).

5. Special Cases

5.1. On a 2-Sphere

Throughout this section, we will take

$$Q = \{q \in \mathbb{R}^3 \mid q \cdot q = a^2\}, \qquad \tilde{g}\left((q, v_1), (q, v_2)\right) = \mu^2 v_1 \cdot v_2,$$

where $a > 0$ and $\mu > 0$. We will use the standard identification of $\mathbb{R}^3$ and $so(3)$ by $v \in \mathbb{R}^3 \mapsto v^\wedge \in so(3)$ where

$$v^\wedge = \begin{bmatrix} 0 & -v^3 & v^2 \\ v^3 & 0 & -v^1 \\ -v^2 & v^1 & 0 \end{bmatrix}.$$

Note that $v^\wedge w = v \times w$. We will write the $SO(3)$-exponential mapping of $v \in so(3)$ as $\exp v^\wedge$.

The geodesic through $q \in Q$ in the direction of $v \in T_q Q$ is the great circle along v, so that

$$\exp_q v = \exp\left(\frac{1}{a^2}(q \times v)^\wedge\right) q. \tag{40}$$

By (35) or otherwise,

$$\mathbb{P}_v w = \exp\left(\frac{1}{a^2}(q \times v)^\wedge\right) w. \tag{41}$$

For $\nabla\Delta$ and $d\Delta$, we must calculate $(1/f_2)(R_w)$ and $(f_1/f_2)(R_w)$. From page 101 of [5],

$$R_w v = \frac{1}{a^2}\left(|w|^2 v - (v \cdot w)w\right), \tag{42}$$

and so

$$
\begin{aligned}
R_w^2 v &= \frac{1}{a^2} R_w \left(|w|^2 v - (v \cdot w)w\right) \\
&= \frac{1}{a^4}\left(|w|^2\left(|w|^2 v - (v \cdot w)w\right) - \left(\left(|w|^2 v - (v \cdot w)w\right) \cdot w\right)w\right) \\
&= \frac{|w|^2}{a^4}\left(|w|^2 v - (v \cdot w)w\right) \\
&= \frac{|w|^2}{a^2} R_w v. \tag{43}
\end{aligned}
$$

Now let $f(z)$ be a generic power series in z:

$$f(z) = \sum_{k=0}^\infty c_k z^k.$$

Using (43),

$$
\begin{aligned}
f(R_w) &= c_0 \,\mathrm{Id} + c_1 R_w + c_2 R_w^2 + c_3 R_w^3 + \cdots \\
&= c_0 \,\mathrm{Id} + c_1 R_w + c_2 \frac{|w|^2}{a^2} R_w + c_3 \frac{|w|^4}{a^4} R_w + \cdots \\
&= f(0)\,\mathrm{Id} + \left(f\left(\frac{|w|^2}{a^2}\right) - f(0)\right)\frac{a^2}{|w|^2} R_w.
\end{aligned}
$$

Applying this to $(1/f_2)(z) = \sqrt{z}/\sin\sqrt{z}$ and to $(f_1/f_2)(z) = \sqrt{z}\cos\sqrt{z}/\sin\sqrt{z}$ yields

$$(1/f_2)(R_w) = \mathrm{Id} + g_1\left(\frac{|w|}{a}\right) R_w, \qquad (f_1/f_2)(R_w) = \mathrm{Id} + g_2\left(\frac{|w|}{a}\right) R_w,$$

where

$$g_1(z) = \frac{1}{z^2}\left(\frac{z}{\sin z} - 1\right), \qquad g_2(z) = \frac{1}{z^2}\left(\frac{z\cos z}{\sin z} - 1\right).$$

Care is required to avoid error in the numerical evaluation of these quantities, since $|w|$ is near zero. Recasting them as

$$g_1(z) = \left(\frac{z}{\sin z}\right)\left(\frac{z - \sin z}{z^3}\right)$$

and

$$
\begin{aligned}
g_2(z) &= \frac{1}{z^2 \sin z}(z \cos z - \sin z) \\
&= \frac{1}{z^2 \sin z}(z \cos z - z) + \frac{1}{z^2 \sin z}(z - \sin z) \\
&= \frac{1}{z \sin z}\frac{\cos^2 z - 1}{\cos z + 1} + g_1(z) \\
&= g_1(z) - \frac{\sin z}{z}\frac{1}{1 + \cos z},
\end{aligned}
$$

one sees that it is enough to be able to calculate

$$\frac{\sin z}{z} \quad \text{and} \quad \frac{\sin z - z}{z^3}$$

for small z.

For the calculation of $\nabla h(q)w^2$, we assume that all of g, $\tilde{g}$ and h are the restrictions of tensors (denoted by the same name) on an open subset of $\mathbb{R}^3$ containing Q. Then

$$
\begin{aligned}
\nabla_v h(w, w) &= \frac{d}{d\epsilon}\Big|_{\epsilon=0} h(q + \epsilon v)(\mathbb{P}_{\epsilon v}w, \mathbb{P}_{\epsilon v}w) \\
&= \frac{d}{d\epsilon}\Big|_{\epsilon=0} h(q + \epsilon v)(w, w) + \frac{d}{d\epsilon}\Big|_{\epsilon=0} h(q)(\mathbb{P}_{\epsilon v}w, \mathbb{P}_{\epsilon v}w) \\
&= d\left(h(q)(w, w)\right)v + 2h\left(\Gamma(q, v)w, w\right),
\end{aligned}
$$

where $\Gamma(q, v)$ is

$$
\begin{aligned}
\Gamma(q, v) &= \frac{d}{d\epsilon}\Big|_{\epsilon=0} \exp\left(\frac{\epsilon}{a^2}(q \times v)^\wedge\right)w \\
&= \frac{1}{a^2}(q \times v) \times w \\
&= -\frac{1}{a^2}(w \cdot v)q.
\end{aligned}
$$

Thus,

$$\nabla h(q)w^2 = d\left(h(q)(w, w)\right) - \frac{1}{a^2}h(w, qw^T).$$

Finally, in the calculation of $\text{grad}_g V$ and A, the map $g^\sharp$ is required. So suppose α is a covector in $\mathbb{R}^3$ (i.e., a row vector) representing a covector in Q by $v \mapsto \alpha v$ where $v \in T_q Q \Leftrightarrow v \cdot q = 0$. The vector $z = g^\sharp(\alpha)$ is characterized by $z \cdot q = 0$ and

$g(q)(z, v) = \alpha v$ for all $v \cdot q = 0$. Let G be the matrix of the extension of g at q. Then we seek z such that

$$z^T G v = \alpha v, \qquad z \cdot q = 0.$$

Trying $z^T G = \alpha + \gamma q^T$ for some $\gamma \in \mathbb{R}$ satisfies the first equation. Postmultiplication by G^{-1}, transposing, premultiplying by q^T, and using $z \cdot q = 0$ gives

$$\gamma = -\frac{\alpha G^{-1} q}{q^T G^{-1} q}.$$

Thus,

$$g^\sharp(\alpha) = \alpha - \frac{\alpha G^{-1} q}{q^T G^{-1} q} q.$$

5.2. On a Lie Group or Homogeneous Space

We begin with the specific case where Q is a Lie group upon which $\tilde{g}$ is an bi-invariant metric. Let the Lie algebra of Q be $\mathfrak{q}$. Denote the Lie group exponential mapping of Q by exp; it can be distinguished from the exponential mapping of the bi-invariant metric by context. Identify TQ with $Q \times \mathfrak{q}$ using left translations, so that

$$(q, \tilde{w}) \in Q \times \mathfrak{q} \equiv w = \left.\frac{d}{d\epsilon}\right|_{\epsilon=0} q \exp(\tilde{w}\epsilon) = TL_q \tilde{w},$$

where L is the left multiplication mapping. Then we have the following, from Section (1.7) of [5] or Sections (7.8) and (8.3) of [3].

1. The exponential mapping of the bi-invariant metric is $\exp_q v = q \exp \tilde{v}$.
2. The mappings $\sigma_q x = qx^{-1}q$, $q \in Q$ are isometries making Q into a symmetric space.
3. Parallel translation along the geodesic $\exp_q(wt)$ is the derivative of the map

$$x \mapsto \sigma_{\exp_q(w\epsilon/2)} \sigma_q x$$
$$= \sigma_{q \exp(\tilde{w}\epsilon/2)} q x^{-1} q$$
$$= q \exp(\tilde{w}\epsilon/2) q^{-1} x q^{-1} q \exp(\tilde{w}\epsilon/2)$$
$$= q \exp(\tilde{w}\epsilon/2) q^{-1} x \exp(\tilde{w}\epsilon/2),$$

so that, remembering that the result is to be left translated to the identity, parallel translation becomes

$$\mathbb{P}_{tw} v = TL_{(q \exp(\tilde{w}t))^{-1}} \left.\frac{d}{d\epsilon}\right|_{\epsilon=0} q \exp(\tilde{w}t/2) q^{-1} (q \exp(\tilde{v}\epsilon)) \exp(\tilde{w}t/2)$$
$$= \left.\frac{d}{d\epsilon}\right|_{\epsilon=0} \exp(-\tilde{w}t/2) \exp(\tilde{v}\epsilon) \exp(\tilde{w}t/2)$$
$$= \mathrm{Ad}_{\exp(-\tilde{w}t/2)} \tilde{v}.$$

In particular, $\mathbb{P}_{tw} w = w$, so the parallel translations occuring in the algorithm do not change their arguments.

4. The curvature operator is given by $R_w = -\frac{1}{4} \operatorname{ad}_{\tilde{w}}^2$.
5. Under the identification of TQ with $Q \times \mathfrak{q}$, the tensor h becomes a map from Q to the symmetric bilinear forms on $\mathfrak{q}$. Consequently,

$$
\begin{aligned}
\nabla_v h(q) w^2 &= \left.\frac{d}{d\epsilon}\right|_{\epsilon=0} h\left(q \exp(\tilde{v}\epsilon)\right) \left(\operatorname{Ad}_{\exp(-\tilde{v}\epsilon/2)} \tilde{w}\right)^2 \\
&= \left.\frac{d}{d\epsilon}\right|_{\epsilon=0} \left(h\left(q \exp(\tilde{v}\epsilon)\right) w^2\right) + h(q)(\tilde{w}, \operatorname{ad}_{\tilde{w}} \tilde{v}).
\end{aligned}
\tag{44}
$$

An important special case is obtained by taking $V = 0$ and g to be a left invariant metric; both g and h then determine constant symmetric bilinear forms on $\mathfrak{q}$, which we denote by the same names. One gets, from Equation (30) and the formulas immediately above, and dropping the tilde,

$$
\begin{aligned}
g^\flat A(w) = h\Bigg(w, \frac{1}{2} \operatorname{ad}_w &- \left(2Id - \frac{f_1}{f_2}\left(-\frac{1}{4} \operatorname{ad}_w^2\right)\right. \\
&\left. - \exp\left(\frac{1}{2} \operatorname{ad}_w\right) \frac{1}{f_2}\left(-\frac{1}{4} \operatorname{ad}_w^2\right)\right)\Bigg).
\end{aligned}
\tag{45}
$$

The series expansion for $2 - f_1/f_2 - 1/f_2$ has no constant term and the linear map ad_w is $\tilde{g}$ skew symmetric, so one may replace h in (45) by g. Then

$$
A(w) = g^\sharp g \left(w, 2(1 + Z \cdot Z \coth Z) Z \operatorname{ad}_w\right),
$$

and, begining with the state $tv_1 \in \mathfrak{q}$, the implicit part of the algorithm is the iteration of

$$
w = tv_1 + \frac{1}{2} A(w).
\tag{46}
$$

After the Legendre transform, the dynamics of the system occurs on the cotangent bundle T^*Q, and that dynamics descends, by the quotient map that is left translation to the identity, to the quotient $T^*Q/Q = \mathfrak{q}^*$, where it becomes a *Lie-Poisson* system [9], [10]. On the tangent bundle side, the dynamics descends to $\mathfrak{q}$, again by left translation, where it is equivalent to the Lie-Poisson dynamics of $\mathfrak{q}^*$ by the linear isomorphism $g^\flat$. Since our algorithm respects the symmetry of left translation, it too descends to integrate the dynamics on $\mathfrak{q}$. If one is interested merely in the dynamics on the quotient, then one simply discards the part of the algorithm that updates configurations. An efficiency can be realized by noting that

$$
\frac{1}{2} A(w) + \frac{1}{2} A(-w) = g^\sharp g(w, \operatorname{ad}_w),
$$

and then the explicit part of the algorithm yields the state $tv_2 \in \mathfrak{q}$ given by

$$
\begin{aligned}
tv_2 &= w + \frac{1}{2} A(-w) \\
&= tv_1 + \frac{1}{2} A(w) + \frac{1}{2} A(-w) \\
&= tv_1 + g^\sharp g(w, \operatorname{ad}_w).
\end{aligned}
\tag{47}
$$

Lie-Poisson integrator on $\mathfrak{q}$: Given an initial state $tv_1 \in \mathfrak{q}$, find $w \in \mathfrak{q}$ by iterating

$$w = tv_1 + g^{\sharp}g\left(w, (1 + Z - Z \coth Z)Z = \frac{1}{2} \operatorname{ad}_w\right).$$

The new state is $tv_2 \in \mathfrak{q}$, where

$$tv_2 = tv_1 + g^{\sharp}g(w, \operatorname{ad}_w).$$

Lie-Poisson integrator on $\mathfrak{q}^*$: Given an initial state $t\pi_1 \in \mathfrak{q}^*$, find $\alpha \in \mathfrak{q}^*$ by iterating

$$\alpha = t\pi_1 + \left((1 + Z - Z \coth Z)Z = \frac{1}{2} \operatorname{ad}_w\right)^* \alpha,$$

where $w = g^{\sharp}\alpha$. The new state is $t\pi_2 \in \mathfrak{q}^*$, where

$$t\pi_2 = t\pi_1 + \operatorname{ad}_w^* \alpha.$$

In either case, if the full integration on Q is required, then configurations in Q are updated after each implicit step by left multiplication by $\exp w$.

Fig. 2. The Lie-Poisson algorithm.

Remarkably, the exponential map of the Lie group needs never to be calculated. Another, equivalent algorithm, but directly integrating the Lie-Poisson system on $\mathfrak{q}^*$, is obtained by transforming this algorithm on $\mathfrak{q}$ by the linear map $g^{\flat}$, which amounts to the substitution $\pi = g^{\flat}w$. Both these equivalent *Lie-Poisson integrators* [9] are summarized in Figure 2.

For a different kind of special case, suppose H is a closed subgroup of G and that Q is the homogeneous space of left cosets G/H. Many common diffeotypes are homogeneous spaces: spheres and projective spaces, for example. One gives G/H a Riemannian structure by imposing the condition that the projection $\pi \colon G \to G/H$ be an isometric immersion [5, p. 104], and one can identify TQ with $G \times \mathfrak{h}^{\perp}$, where $\mathfrak{h}$ is the Lie algebra of H, by

$$(g, \tilde{w}) \in G \times \mathfrak{h} \mapsto w = \left.\frac{d}{d\epsilon}\right|_{\epsilon=0} \pi\left(g \exp(\tilde{w}t)\right).$$

The natural implication of this is that we identify Q with G in the sense that $g \in Q \equiv G$ means $\pi(g)$. Then we have the following:

1. The exponential mapping of the metric on G/H is $\exp_q w = q \exp \tilde{w}$.
2. While not necessarily a symmetric space, G/H is symmetric in a local sense [5, p. 157], and this is enough to imply that the Riemannian curvature tensor is parallel, which in turn validates the results of Section 4.3.
3. Parallel translation along the geodesic $\exp_q tw$ is $\mathbb{P}_{tw}v = \operatorname{Ad}_{\exp(-\tilde{w}t/2)} \tilde{v}$. Again, the parallel translations occuring in the algorithm do not change their arguments.
4. The curvature operator may be calculated from page 105 of [5], and is $R_w = -\operatorname{ad}_{\tilde{w}}^2$.
5. With obvious changes in context, $\nabla h w^2$ can be calculated using Equation (44).

Acknowledgments

B. Leimkuhler was supported in part by National Science Foundation Grant No. NSF 9303223. G. W. Patrick was supported in part by NSERC Grant No. OGP0105716.

References

[1] R. Abraham and J. E. Marsden. *Foundations of Mechanics*. Addision-Wesley, Reading, MA, second edition, 1978.

[2] E. Barth and B. Leimkuhler. Symplectic methods for conservative multibody systems. *Fields Inst. Commun.*, 1995. To appear.

[3] W. M. Boothby. *An Introduction to Riemannian Geometry*. Academic Press, New York, 1975.

[4] P. E. Crouch and R. Grossman. Numerical integration of ordinary differential equations on manifolds. *J. Nonlin. Sci.*, 3:1–33, 1993.

[5] W. Klingenberg. *Riemannian Geometry*. Walter de Gruyter, New York, 1982.

[6] B. Leimkuhler and S. Reich. Symplectic integration of constrained Hamiltonian systems. *Math. Comput.*, 63:589–605, 1994.

[7] D. Lewis and J.-C. Simo. Conserving algorithms for the dynamics of Hamiltonian systems of Lie groups. *J. Nonlin. Sci.*, 4:253–299, 1995.

[8] D. Lewis and J.-C. Simo. Conserving algorithms for the n dimensional rigid body. *Fields Inst. Commun.*, 1995. To appear.

[9] J. E. Marsden. *Lectures on Mechanics*, volume 174 of *London Mathematical Society Lecture Note Series*. Cambridge University Press, Cambridge, 1992.

[10] J. E. Marsden and T. S. Ratiu. *Introduction to Mechanics and Symmetry*. Springer-Verlag, New York, 1994.

[11] R. I. McLachlan and C. Scovel. Equivariant constrained symplectic integration. *J. Nonlin. Sci.*, 16:233–256, 1995.

[12] H. Munthe-Kaas. Lie-Butcher theory for Runge-Kutta methods. *BIT*, 35, 1995.

[13] G. W. Patrick. *Two axially symmetric coupled rigid bodies: Relative equilibria, stability, bifurcations, and a momentum preserving symplectic integrator*. Ph.D. thesis, University of California at Berkeley, 1991.

[14] S. Reich. Symplectic integrators for systems of rigid bodies. *Fields Inst. Commun.*, 1995. To appear.

[15] H. Yoshida. Construction of higher order symplectic integrators. *Phys. Lett. A*, 150:262–268, 1990.

Time Integration and Discrete Hamiltonian Systems

O. Gonzalez

Division of Applied Mechanics, Department of Mechanical Engineering, Stanford University, Stanford, CA 94305, USA

Received November 3, 1995; revised manuscript accepted for publication March 4, 1996
Communicated by Stephen Wiggins and Jerrold Marsden

This paper is dedicated to the memory of Juan-Carlos Simo

Summary. This paper develops a formalism for the design of conserving time-integration schemes for Hamiltonian systems with symmetry. The main result is that, through the introduction of a discrete directional derivative, implicit second-order conserving schemes can be constructed for general systems which preserve the Hamiltonian along with a certain class of other first integrals arising from affine symmetries. Discrete Hamiltonian systems are introduced as formal abstractions of conserving schemes and are analyzed within the context of discrete dynamical systems; in particular, various symmetry and stability properties are investigated.

1. Background and Motivation

First integrals or conservation laws for Hamiltonian systems with symmetry are typically lost under numerical integration in time. In some cases, failure to maintain certain conservation laws can lead to physically impossible solutions [3], and in other cases to numerical instability [7], [21]–[24]. For Hamiltonian systems with symmetry it is thus generally desirable that numerical time-integration schemes preserve physically meaningful integrals from the underlying system. These types of integrators are usually referred to as *conserving integrators* and are the subject of this investigation.

This paper develops a formalism for the design of conserving time-integration schemes for Hamiltonian systems with symmetry. The main result is that, through the introduction of a discrete directional derivative, implicit second-order conserving schemes can be constructed for general systems which preserve the Hamiltonian along with quadratic integrals arising from affine symmetries. Discrete Hamiltonian systems are introduced as formal abstractions of conserving schemes and are analyzed within the context of discrete dynamical systems; in particular, various symmetry and stability properties are investigated. It is shown that the proposed class of schemes inherit equi-

libria and relative equilibria from the underlying system along with various notions of
stability.

Only finite-dimensional Hamiltonian systems defined in open sets of Euclidean space
are considered in this paper. However, the framework presented herein easily extends
to infinite-dimensional systems on linear manifolds [6], [8], and can be extended to
canonical systems with holonomic constraints [5]. For other treatments of conserving
schemes, particularly within the context of specific applications, see [2]–[4], [9]–[15],
[17], [19]–[24].

2. Preliminaries

In this section we recall some standard terminology and concepts to be used in the
developments that follow. We refer to Abraham & Marsden [1], Olver [18] or Marsden
& Ratiu [16] for further details not explained here.

2.1. Hamiltonian Differential Equations and First Integrals

Let (P, Ω) denote a symplectic space with P open in m-dimensional Euclidean space
$\mathbb{R}^m$ with points denoted by $z = (z^1, \ldots, z^m)$, and symplectic structure $P \ni z \to \Omega_z \in$
$\mathbb{R}^{m \times m}$, where each Ω_z is viewed as a bilinear form in $T_z P \cong \mathbb{R}^m$. For any $z \in P$ we recall
that Ω_z is skew-symmetric in the sense that $\Omega_z(v, w) = -\Omega_z(w, v)$ for all $v, w \in \mathbb{R}^m$.

To any smooth function $H \colon P \to \mathbb{R}$ we associate a *Hamiltonian vector field* $X_H \colon P \to$
$\mathbb{R}^m$ defined by

$$\Omega_z^\flat(X_H(z)) = DH(z), \tag{2.1}$$

where $DH(z) \in T_z^* P \cong \mathbb{R}^m$ denotes the derivative of H at z. If we denote the com-
ponents of $\Omega_z \in \mathbb{R}^{m \times m}$ by $(\Omega_z)_{ij}$ $(i, j = 1, \ldots, m)$, then $\Omega_z^\flat \colon \mathbb{R}^m \to \mathbb{R}^m$ is defined
in components by $(\Omega_z^\flat(v))_k = (\Omega_z)_{jk} v^j$ where summation on repeated indices is im-
plied. Nondegeneracy conditions on the symplectic structure require that m be even, say
$m = 2n$, and for each $z \in P$ we define $\Omega_z^\sharp \colon \mathbb{R}^m \to \mathbb{R}^m$ to be the inverse of $\Omega_z^\flat$.

Given a Hamiltonian system (P, Ω, H) we will be concerned with the associated
Hamiltonian differential equations

$$\dot{z} = X_H(z) \tag{2.2}$$

where the Hamiltonian vector field X_H is assumed to be smooth. For any $z \in P$ we
note that (2.2) generates a local evolution semigroup $F \colon B \times [0, T] \to P$, where B is a
neighborhood of z and $T > 0$. For any $z_0 \in B$ the curve $\varphi(t) = F(z_0, t) = F_t(z_0)$ is a
solution to (2.2), defined for all $t \in [0, T]$, with initial condition $\varphi(0) = z_0$.

By a (time-independent) *first integral* for the system (P, Ω, H), we mean a smooth
function $f \colon P \to \mathbb{R}$ which is constant along any solution $\psi \colon [0, T] \to P$ of (2.2), i.e.,

$$f(\psi(t)) = f(\psi(0)), \qquad \forall t \in [0, T]. \tag{2.3}$$

Using straightforward arguments it can be shown that f is an integral if and only if the
following *orthogonality condition* is satisfied:

$$Df(z) \cdot X_H(z) = 0, \qquad \forall z \in P. \tag{2.4}$$

Note that the skewness of Ω_z implies that the Hamiltonian $H: P \to \mathbb{R}$ is a first integral for (P, Ω, H).

2.2. Symplectic Actions of Lie Groups and Momentum Maps

Let G be a Lie group with tangent space at the identity denoted by $T_e G$, and let $\Phi: G \times P \to P$ denote a regular symplectic action of G on P. (See, e.g., Olver [18, p. 22] for the definition of a regular action.) Given $\xi \in T_e G$ the *infinitesimal generator* of the G-action corresponding to ξ is a vector field $\xi_P: P \to \mathbb{R}^m$ defined by the relation

$$\xi_P(z) = \frac{d}{ds}\bigg|_{s=0} \Phi(\exp(s\xi), z), \tag{2.5}$$

where $\exp: T_e G \to G$ is the exponential map. For any $z \in P$ we denote by $G \cdot z$ the orbit of z under the action of G, and we denote by $\mathrm{Ad}^*: G \times T_e^* G \to T_e^* G$ the *coadjoint action* of G on $T_e^* G$.

By a *momentum map* for the action of G on P we mean a mapping of the form $J: P \to T_e^* G$ satisfying

$$DJ_\xi(z) = \Omega_z^\flat(\xi_P(z)) \quad \forall \xi \in T_e G, \tag{2.6}$$

where $J_\xi: P \to \mathbb{R}$ is defined by the relation $J_\xi(z) = J(z) \cdot \xi$. We say that J is Ad^*-*equivariant* if

$$J(\Phi(g, z)) = \mathrm{Ad}^*(g^{-1}, J(z)) \tag{2.7}$$

for all $g \in G$ and $z \in P$.

Given $\mu \in T_e^* G$ we denote by $G_\mu \subset G$ the *isotropy group* for μ under the coadjoint action, and we call the quotient space $P_\mu = J^{-1}(\mu)/G_\mu$, induced by the action of G_μ on $J^{-1}(\mu)$, the *reduced phase space* for the momentum value μ. Note that P_μ has the structure of a smooth manifold provided that μ is a regular value for J and G_μ acts regularly on $J^{-1}(\mu)$. In what follows we will assume that the symplectic structure Ω on P induces a well-defined symplectic structure Ω_μ in P_μ, and we will use π_μ to denote the natural projection from $J^{-1}(\mu)$ onto P_μ.

2.3. Symmetry, Conservation Laws and Relative Equilibria

Let (P, Ω) be a symplectic space as described above and let Φ denote the symplectic action of a Lie group G on P. Given a G-*invariant* function $H: P \to \mathbb{R}$, i.e.,

$$H(\Phi(g, z)) = H(z), \qquad \forall g \in G, \ z \in P, \tag{2.8}$$

we call the system (P, Ω, G, H) a Hamiltonian system with *symmetry*. This system has the property that if $\varphi: [0, T] \to P$ is a maximal trajectory for the Hamiltonian vector field X_H, then so is $\Phi_g \circ \varphi$ for any $g \in G$. Here we employ the notation $\Phi_g = \Phi(g, \cdot): P \to P$.

Suppose the action of G possesses a momentum map $J: P \to T_e^* G$. Then J is conserved along trajectories of X_H in the sense that, for any $\xi \in T_e G$, the function $J_\xi = J \cdot \xi: P \to \mathbb{R}$ is an integral for (2.2). To see this result use (2.1) and (2.6) to write

$$DJ_\xi(z) \cdot X_H(z) = -DH(z) \cdot \xi_P(z). \tag{2.9}$$

The result then follows from the G-invariance of H, which implies $DH(z) \cdot \xi_P(z) = 0$ for any $\xi \in T_eG$ and $z \in P$.

For any regular value μ of J we recall that the G-invariance of H implies the existence of a well-defined function H_μ on the reduced phase space P_μ, which we call the *reduced Hamiltonian* associated with H and μ. Thus, given a Hamiltonian system with symmetry as discussed above, and a regular value μ for J, we have a well-defined reduced Hamiltonian system $(P_\mu, \Omega_\mu, H_\mu)$.

Finally, we recall the notion of a relative equilibria for a Hamiltonian system with symmetry. In particular, a point $z_e \in P$ is a *relative equilibrium* if the maximal trajectory of X_H with initial condition z_e, denoted by $\varphi(t)$, satisfies

$$\varphi(t) = \Phi(\exp(t\xi), z_e) \tag{2.10}$$

for some $\xi \in T_eG$. It is well known that, for any regular value μ of J, a point $z_e \in J^{-1}(\mu) \subset P$ is a relative equilibrium if $\pi_\mu(z_e) \in P_\mu$ is a critical point of the reduced Hamiltonian H_μ.

3. Conserving Time Integration

In this section we present a framework for the design and analysis of numerical schemes for (2.2). Our attention will be focused on schemes which inherit underlying integrals. Rather than view an algorithm as a discrete system which approximates a continuous one, we take the point of view that an algorithm defines a discrete system worthy of study in its own right. Hence, we introduce the notion of a discrete Hamiltonian system as a formal abstraction of a conserving scheme.

3.1. A Point of Departure

Given a Hamiltonian system (P, Ω, H) possessing an integral $f: P \to \mathbb{R}$, our goal is to construct a numerical approximation scheme for (2.2) which inherits f as an integral.

As a point of departure, we consider approximating solutions to (2.2) by numerical schemes of the form

$$z_{n+1} - z_n = hX_H(z_n, z_{n+1}), \tag{3.1}$$

where $h > 0$ is a parameter interpreted as the time step and $X_H: P \times P \to \mathbb{R}^m$ is a given smooth map which is viewed as a two-point approximation to the exact vector field X_H, e.g., $X_H(z_n, z_{n+1}) \approx X_H(z_{n+\frac{1}{2}})$ where $z_{n+\frac{1}{2}} = \frac{1}{2}(z_n + z_{n+1})$.

For any $z \in P$ we assume the numerical scheme generates a local evolution semigroup in the sense that there exists a neighborhood B of z, real numbers $h_c, T > 0$, and a mapping $F: B \times [0, h_c] \to P$ such that, for any $z_0 \in B$ and $h \in [0, h_c]$, the sequence (z_n) generated by $F^n(z_0, h) = F_h^n(z_0)$ satisfies (3.1) for all $nh \in [0, T]$. Note that a function $f: P \to \mathbb{R}$ is an integral for (3.1) if for any $z_0 \in P$ we have $f(z_n) = f(z_0)$ for all $nh \in [0, T]$.

The following observations illustrate how (3.1) may be constructed so that it inherits an arbitrary integral from the underlying system. To begin, let f be an integral for (2.2) and assume that, for any x, $y \in P$, there exists a vector $Df(x, y) \in \mathbb{R}^m$ with the property that $Df(x, y) \approx Df(\frac{x+y}{2})$ and

$$Df(x, y) \cdot (y - x) = f(y) - f(x). \tag{3.2}$$

Along any solution sequence of (3.1) we could thus write

$$\begin{aligned} f(z_{n+1}) - f(z_n) &= Df(z_n, z_{n+1}) \cdot (z_{n+1} - z_n) \\ &= hDf(z_n, z_{n+1}) \cdot X_H(z_n, z_{n+1}). \end{aligned} \tag{3.3}$$

Now note that if the approximate vector field X_H satisfied the discrete orthogonality condition

$$Df(x, y) \cdot X_H(x, y) = 0, \qquad \forall x, y \in P, \tag{3.4}$$

then f would be an integral for (3.1).

The preceding arguments suggest that a formalism for constructing conserving schemes can be based on both a discrete derivative operator "D" which allows one to write (3.2) and the discrete orthogonality condition (3.4).

In principle, by projecting $X_H(x, y)$ onto the orthogonal complement of the linear space span$\{Df(x, y)\}$, we could arrange for (3.1) to inherit an arbitrary integral from the underlying system (2.2). For multiple integrals such a projection would likely be inefficient and thus we are interested in simpler ways to satisfy the discrete orthogonality condition. As we will see below, a simplification can be achieved when the integrals of interest are the Hamiltonian and quadratic momentum maps associated with affine symmetries. The preceding ideas are formalized in the next few subsections.

3.2. Definitions

Consider a symplectic space (P, Ω) where the phase space P is an open subset of $\mathbb{R}^m$ and Ω denotes a symplectic structure on P. Motivated by the preceding developments we make the following definition.

Definition 3.1. A *discrete derivative* for a smooth function $f: P \to \mathbb{R}$ is a mapping $Df: P \times P \to \mathbb{R}^m$ with the following properties:

(1) Directionality. $Df(x, y) \cdot v_{xy} = f(y) - f(x)$ for any x, $y \in P$ where $v_{xy} = y - x$.
(2) Consistency. $Df(x, y) = Df(\frac{x+y}{2}) + O(\|y - x\|)$ for all x, $y \in P$ with $\|y - x\|$ sufficiently small. (Here $\| \cdot \|$ denotes the standard Euclidean norm in $\mathbb{R}^m$.)

For any smooth function $H: P \to \mathbb{R}$ we call the system (P, Ω, D, H) a *discrete Hamiltonian system*. We associate with this system a difference equation of the form

$$z_{n+1} - z_n = hX_H(z_n, z_{n+1}), \tag{3.5}$$

where $h \in \mathbb{R}_+$ is a parameter and X_H is a *discrete Hamiltonian vector field* defined by the relation

$$X_H(x, y) = \Omega^\sharp_{(x+y)/2}(DH(x, y)) \tag{3.6}$$

for all x, $y \in P$. Any sequence $(z_n)_{n=0}^N$ in P satisfying (3.5), if it exists, will be called a *trajectory* or *solution sequence* for the discrete system.

We now give some constructive examples of discrete derivatives for functions defined on general inner-product spaces.

3.3. Discrete Derivative: Examples

We begin by considering the general case of functions defined on m-dimensional Euclidean space $\mathbb{R}^m$.

Proposition 3.1. *Let* $f: \mathbb{R}^m \to \mathbb{R}$ *be a smooth function and for any two points* $x, y \in \mathbb{R}^m$ *let* $z = (x + y)/2$ *and* $v = y - x$. *Then a discrete derivative for* f *is defined by the relation*

$$\mathsf{D}f(x, y) = Df(z) + \frac{f(y) - f(x) - Df(z) \cdot v}{\|v\|^2} v, \tag{3.7}$$

where $\| \cdot \|$ *denotes the standard Euclidean norm in* $\mathbb{R}^m$.

Proof. The result follows by direct verification of the directionality and consistency properties.

(1) To verify the directionality condition we apply $\mathsf{D}f(x, y)$ to v and get

$$\begin{aligned}
\mathsf{D}f(x, y) \cdot v &= Df(z) \cdot v + \frac{f(y) - f(x) - Df(z) \cdot v}{\|v\|^2} v \cdot v \\
&= f(y) - f(x).
\end{aligned} \tag{3.8}$$

(2) To verify the consistency condition we examine what happens to (3.7) as v approaches zero. As a first step, given $v = y - x$, we use Taylor's Theorem to write

$$\begin{aligned}
f(y) &= f(z) + \tfrac{1}{2}Df(z) \cdot v + \tfrac{1}{4}D^2 f(z) \cdot (v, v) + \tfrac{1}{8}D^3 f(z) \cdot (v, v, v) \\
&\quad + \tfrac{1}{16}D^4 f(z) \cdot (v, v, v, v) + O(\|v\|^5),
\end{aligned} \tag{3.9}$$

$$\begin{aligned}
f(x) &= f(z) - \tfrac{1}{2}Df(z) \cdot v + \tfrac{1}{4}D^2 f(z) \cdot (v, v) - \tfrac{1}{8}D^3 f(z) \cdot (v, v, v) \\
&\quad + \tfrac{1}{16}D^4 f(z) \cdot (v, v, v, v) + O(\|v\|^5),
\end{aligned} \tag{3.10}$$

which implies

$$f(y) - f(x) - Df(z) \cdot v = \tfrac{1}{4}D^3 f(z) \cdot (v, v, v) + O(\|v\|^5). \tag{3.11}$$

Let $v = y - x = \alpha w$ where $\alpha > 0$ and $w \in \mathbb{R}^m$ is a unit vector. Then the last expression can be written as

$$f(y) - f(x) - Df(z) \cdot v = \tfrac{1}{4}\alpha^3 D^3 f(z) \cdot (w, w, w) + O(\alpha^5). \tag{3.12}$$

Using the above result in (3.7) gives the relation

$$\mathsf{D}f(x, y) = Df(z) + \left(\tfrac{1}{4}\alpha^2 D^3 f(z) \cdot (w, w, w) + O(\alpha^4)\right) w, \tag{3.13}$$

which shows that $\mathsf{D}f(x, y)$ is well defined as $\alpha = \|y - x\| \to 0$. In particular, the expression for $\mathsf{D}f(x, y)$ given in (3.7) satisfies the consistency requirement.   □

Here we note that, for any $x, y \in \mathbb{R}^m$, the construction above yields a discrete derivative which, in the classical sense, is a second-order approximation to the exact derivative at the midpoint $z = \frac{1}{2}(x + y)$. For reference, we now list some (second-order) discrete derivatives for more general situations:

(1) General case. Let $(U, \langle \cdot, \cdot \rangle_U)$ be an inner-product space. Then, for any smooth function $f \colon U \to \mathbb{R}$, a second-order discrete derivative is given by

$$\mathsf{D}f(x, y) = Df(z) + \frac{f(y) - f(x) - \langle Df(z), v_{xy} \rangle_U}{\langle v_{xy}, v_{xy} \rangle_U} v_{xy}, \tag{3.14}$$

where $v_{xy} = y - x$.

(2) Partitioned case. Let $(U, \langle \cdot, \cdot \rangle_U)$ be an inner-product space where $U = U_1 \times \cdots \times U_k$ for some $k \geq 1$, and suppose each U_i $(i = 1, \ldots, k)$ is endowed with an inner-product $\langle \cdot, \cdot \rangle_{U_i}$. Here we would like a discrete derivative which respects the product structure of U. To this end, for any smooth function $f \colon U \to \mathbb{R}$ a second-order discrete derivative is defined by the relation

$$\tilde{\mathsf{D}}f(x, y) \cdot u = \sum_{i=1}^{k} \tfrac{1}{2} \left(\mathsf{D}f_{xy}^i(x_i, y_i) + \mathsf{D}f_{yx}^i(x_i, y_i) \right) \cdot u_i \tag{3.15}$$

for all $u = (u_1, \ldots, u_k) \in U$, where $x = (x_1, \ldots, x_k) \in U$, $y = (y_1, \ldots, y_k) \in U$, and $f_{xy}^i, f_{yx}^i \colon U_i \to \mathbb{R}$ are defined by the relations

$$f_{xy}^i(w) = f(x_1, x_2, \ldots, x_{i-1}, w, y_{i+1}, \ldots, y_k), \tag{3.16}$$

$$f_{yx}^i(w) = f(y_1, y_2, \ldots, y_{i-1}, w, x_{i+1}, \ldots, x_k). \tag{3.17}$$

3.4. The Algorithmic Viewpoint

The interpretation of the above developments within an algorithmic framework should be clear; in particular, we may view the Hamiltonian difference equation (3.5), together with (3.14) or (3.15), as defining an algorithm for the approximation of (2.2). Moreover, the approximation is formally second-order since $X_H(z_n, z_{n+1})$ is a second-order approximation to $X_H(z_{n+\frac{1}{2}})$, where $z_{n+\frac{1}{2}} = \frac{1}{2}(z_n + z_{n+1})$.

To develop the theory for discrete Hamiltonian systems we assume that the algorithm defined by (3.5) generates an evolution semigroup so that, for any $z_0 \in P$, n sufficiently small, we may speak of unique solution sequences $(z_n)_{n=0}^N$. With this in mind, we may then view a discrete trajectory as being generated by a mapping F_h, defined at least locally, such that $z_n = \mathsf{F}_h^n(z_0)$. In particular, F_h^n has the semigroup properties $\mathsf{F}_h^{n+m} = \mathsf{F}_h^n \circ \mathsf{F}_h^m$ and $\mathsf{F}_h^0 = id$. Also, we note that for all fixed n the mapping F_h^n is continuous in h in the sense that $z_n = \mathsf{F}_h^n(z_0)$ and any $z_i = \mathsf{F}_h^i(z_0)$ for $i = 0, \ldots, n - 1$ can be forced to remain in a neighborhood of z_0 for h sufficiently small.

3.5. Discrete Brackets and First Integrals

We next introduce the concept of a discrete bracket which we will use to define integrals for discrete Hamiltonian systems.

Let (P, Ω, D) be a symplectic space with a discrete derivative and, for any smooth function $H: P \to \mathbb{R}$, let X_H denote the associated discrete Hamiltonian vector field. For any $z_0 \in P$ let $(z_n)_{n=0}^N$ be the trajectory generated by X_H for some $h > 0$. We say that a smooth function $f: P \to \mathbb{R}$ is an *integral* for the discrete system $(P, \Omega, \mathsf{D}, H)$ if it is constant along trajectories. That is, f is an integral for X_H if, for any trajectory $(z_n)_{n=0}^N$, we have $f(z_n) = f(z_0)$ for all $n = 0, \ldots, N$.

The condition that f be an integral for X_H may be expressed locally by the condition $\{f, H\} = 0$ where the *discrete bracket* $\{f, H\}: P \times P \to \mathbb{R}$ is defined as

$$\{f, H\}(x, y) = \mathsf{D}f(x, y) \cdot \mathsf{X}_H(x, y) = -\{H, f\}(x, y). \tag{3.18}$$

This is the essence of the following proposition.

Proposition 3.2. *A smooth function $f: P \to \mathbb{R}$ is an integral for a discrete Hamiltonian system $(P, \Omega, \mathsf{D}, H)$ if the discrete bracket of f and H vanishes, i.e.,*

$$\{f, H\}(x, y) = 0, \qquad \forall x, y \in P. \tag{3.19}$$

Proof. For any $z_0 \in P$ let $(z_n)_{n=0}^N$ denote the trajectory generated by X_H for some $h > 0$. By the definitions of the discrete bracket, discrete Hamiltonian vector field and discrete derivative we have

$$
\begin{aligned}
\{f, H\}(z_n, z_{n+1}) &= \mathsf{D}f(z_n, z_{n+1}) \cdot \mathsf{X}_H(z_n, z_{n+1}) \\
&= \mathsf{D}f(z_n, z_{n+1}) \cdot (z_{n+1} - z_n)/h \\
&= (f(z_{n+1}) - f(z_n))/h. \tag{3.20}
\end{aligned}
$$

The result follows. $\qquad\square$

Proposition 3.3 follows from the skew-symmetry property of the discrete bracket.

Proposition 3.3. *The Hamiltonian $H: P \to \mathbb{R}$ is an integral for the discrete Hamiltonian system $(P, \Omega, \mathsf{D}, H)$.*

Remark 3.1. The discrete brackets defined above are motivated by the discrete orthogonality condition (3.4). As defined, these brackets do not satisfy the Jacobi identity and hence are not Poisson brackets. The difficulty lies in the fact that the discrete brackets are defined for functions on P, while the discrete bracket of two functions is a function on $P \times P$. $\qquad\square$

3.6. Symmetry and Conservation Laws

In this section we define a discrete derivative for G-invariant functions and use it to introduce the concept of a discrete Hamiltonian system with symmetry. In what follows we let P be an open set in m-dimensional Euclidean space $\mathbb{R}^m$ and we denote by Φ the symplectic action of a group G on P.

Definition 3.2. A *G-equivariant discrete derivative* for a smooth *G*-invariant function $f: P \to \mathbb{R}$ is a mapping $\mathsf{D}^G f: P \times P \to \mathbb{R}^m$ satisfying the requirements for a discrete derivative together with the following properties:

(1) Equivariance. $\mathsf{D}^G f(\Phi_g(x), \Phi_g(y)) = \left[D\Phi_g(\frac{x+y}{2}) \right]^{-\mathsf{T}} \cdot \mathsf{D}^G f(x, y)$ for all $g \in G$
    and $x, y \in P$. (For any $z \in P$ note that $D\Phi_g(z) \in \mathbb{R}^{m \times m}$.)
(2) Orthogonality Condition. $\mathsf{D}^G f(x, y) \cdot \xi_P(\frac{x+y}{2}) = 0$ for all $\xi \in T_e G$ and $x, y \in P$.

For any smooth *G*-invariant function H we call the system $(P, \Omega, G, \mathsf{D}^G, H)$ a discrete Hamiltonian system with *symmetry*. As before, we associate with this system a difference equation of the form

$$z_{n+1} - z_n = h X_H(z_n, z_{n+1}), \tag{3.21}$$

where $h \in \mathbb{R}_+$ is a parameter and X_H is a discrete Hamiltonian vector field defined by the relation

$$X_H(x, y) = \Omega^{\sharp}_{(x+y)/2}(\mathsf{D}^G H(x, y)) \tag{3.22}$$

for all $x, y \in P$.

Remark 3.2. The equivariance and orthogonality conditions stated above are motivated by properties of the derivatives of *G*-invariant functions.   □

Before giving some constructive examples of *G*-equivariant discrete derivatives, we first summarize some properties of discrete Hamiltonian systems with symmetry.

Proposition 3.4. *Let $(P, \Omega, G, \mathsf{D}^G, H)$ be a discrete Hamiltonian system with symmetry and let Φ denote an affine symplectic action of G on P. Then solution sequences satisfying (3.21) are invariant under G. That is, if $(z_n)_{n=0}^N$ is a solution sequence, then so is $(\Phi_g(z_n))_{n=0}^N$ for any $g \in G$.*

Proof. For arbitrary $z_0 \in P$ let $(z_n)_{n=0}^N$ be the trajectory for X_H defined by (3.21) for some $h > 0$. For any $g \in G$ consider the transformed sequence $(\Phi_g(z_n))_{n=0}^N$. Since by assumption Φ_g is affine, we may write

$$\Phi_g(z_{n+1}) - \Phi_g(z_n) = D\Phi_g(z_{n+\frac{1}{2}}) \cdot (z_{n+1} - z_n)$$
$$= h D\Phi_g(z_{n+\frac{1}{2}}) \cdot X_H(z_n, z_{n+1}). \tag{3.23}$$

The above statement implies that the transformed sequence $(\Phi_g(z_n))_{n=0}^N$ is a trajectory of H if and only if the discrete vector field X_H satisfies the equivariance relation

$$X_H(\Phi_g(z_n), \Phi_g(z_{n+1})) = D\Phi_g(z_{n+\frac{1}{2}}) \cdot X_H(z_n, z_{n+1}). \tag{3.24}$$

The result follows from the fact that (3.24) is equivalent to the equivariance condition on $\mathsf{D}^G H$.   □

Recall that, under certain circumstances, the action of a group G on a phase space P possesses a momentum map $J\colon P \to T_e^*G$. Furthermore, if a momentum map exists, it is conserved by the system (P, Ω, G, H) in the sense that the function $J_\xi = J \cdot \xi$ is an integral for any $\xi \in T_eG$. We now state a similar result for the discrete case.

Proposition 3.5. *Let* $(P, \Omega, G, \mathsf{D}^G, H)$ *be a discrete Hamiltonian system with symmetry and denote by* Φ *a symplectic action of G on P. Suppose this action possesses a momentum map* $J\colon P \to T_e^*G$. *If J is at most quadratic in* $z \in P$, *then J is conserved by the discrete system in the sense that the function* $J_\xi = J \cdot \xi$ *is an integral for any* $\xi \in T_eG$.

Proof. To begin, note that if the map $J\colon P \to T_e^*G$ is at most quadratic, then for any $x, y \in P$ we have

$$J_\xi(y) - J_\xi(x) = DJ_\xi\left(\tfrac{x+y}{2}\right) \cdot (y - x). \tag{3.25}$$

Now let $(z_n)_{n=0}^N$ be any trajectory generated by the discrete system (3.21). For any $\xi \in T_eG$ we use (3.25), (3.21) and (2.6) to write

$$
\begin{aligned}
J_\xi(z_{n+1}) - J_\xi(z_n) &= DJ_\xi(z_{n+\frac{1}{2}}) \cdot (z_{n+1} - z_n) \\
&= h DJ_\xi(z_{n+\frac{1}{2}}) \cdot \mathsf{X}_H(z_n, z_{n+1}) \\
&= h\Omega^\flat_{z_{n+\frac{1}{2}}}(\xi_P(z_{n+\frac{1}{2}})) \cdot \mathsf{X}_H(z_n, z_{n+1}) \\
&= -h\Omega^\flat_{z_{n+\frac{1}{2}}}(\mathsf{X}_H(z_n, z_{n+1})) \cdot \xi_P(z_{n+\frac{1}{2}}) \\
&= -h\mathsf{D}^G H(z_n, z_{n+1}) \cdot \xi_P(z_{n+\frac{1}{2}}),
\end{aligned}
\tag{3.26}
$$

which vanishes in view of the orthogonality condition on $\mathsf{D}^G H$. $\square$

We next give some constructive examples of G-equivariant discrete derivatives.

3.7. Discrete Derivative: G-Equivariant Case

Let (P, Ω, G) be a phase space with symmetry where P is an open set in m-dimensional Euclidean space $\mathbb{R}^m$, and denote by Φ a regular affine symplectic action of G on P. Assume the action of G has orbits of dimension s, so that the quotient or orbit space P/G can be identified locally with $\mathbb{R}^{m-s}$. In particular, let $\pi_i\colon P \to \mathbb{R}\ (i = 1, \ldots, m-s)$ be invariants of G (assumed to be globally defined, for simplicity) so that $P/G \cong \pi(P) \subset \mathbb{R}^{m-s}$ where $\pi\colon P \to \mathbb{R}^{m-s}$ is defined by $\pi = (\pi_1, \ldots, \pi_{m-s})$. With this setup a G-equivariant discrete derivative is contained in the following proposition.

Proposition 3.6. *Let* $f\colon P \to \mathbb{R}$ *be a smooth G-invariant function and denote by* $\tilde{f}\colon \pi(P) \subset \mathbb{R}^{m-s} \to \mathbb{R}$ *the associated reduced function, defined by the expression* $\tilde{f}(\pi(z)) = f(z)$ *for all $z \in P$. Consider any two points $x, y \in P$ and let $z = (x + y)/2$*

and $v = y - x$. If the invariants $\pi_i \colon P \to \mathbb{R}$ are at most quadratic, then a G-equivariant discrete derivative for f is defined by the relation

$$\mathsf{D}^G f(x, y) = \mathsf{D}\tilde{f}(\pi(x), \pi(y)) \circ D\pi(z)$$
$$= [D\pi(z)]^{\mathsf{T}} \cdot \mathsf{D}\tilde{f}(\pi(x), \pi(y)), \qquad (3.27)$$

where on the right-hand side D represents a discrete derivative for functions on $\mathbb{R}^{m-s}$. (For any $z \in P$ note that $D\pi(z) \in \mathbb{R}^{(m-s) \times m}$.)

Proof. The result follows by direct verification of the defining conditions.

(1) To verify the directionality condition we apply $\mathsf{D}^G f(x, y)$ to v and obtain

$$\mathsf{D}^G f(x, y) \cdot v = \mathsf{D}\tilde{f}(\pi(x), \pi(y)) \cdot (D\pi(z) \cdot v). \qquad (3.28)$$

Since by assumption π is at most quadratic we have that $D\pi(z) \cdot v = \pi(y) - \pi(x)$. Hence

$$\mathsf{D}^G f(x, y) \cdot v = \mathsf{D}\tilde{f}(\pi(x), \pi(y)) \cdot (\pi(y) - \pi(x))$$
$$= \tilde{f}(\pi(y)) - \tilde{f}(\pi(x))$$
$$= f(y) - f(x). \qquad (3.29)$$

(2) Consistency follows from the consistency of the discrete derivative for functions defined on $\mathbb{R}^{m-s}$.

(3) To verify the equivariance condition we note that, since π is invariant, i.e., $\pi(\Phi_g(z)) = \pi(z)$ for all $z \in P$ and $g \in G$, we have

$$D\pi(\Phi_g(z)) = D\pi(z) \circ [D\Phi_g(z)]^{-1}. \qquad (3.30)$$

Since $\Phi_g \colon P \to P$ is affine we have $\frac{1}{2}(\Phi_g(x) + \Phi_g(y)) = \Phi_g(z)$, and thus

$$\mathsf{D}^G f(\Phi_g(x), \Phi_g(y)) = \mathsf{D}\tilde{f}(\pi(\Phi_g(x)), \pi(\Phi_g(y))) \circ D\pi(\Phi_g(z))$$
$$= \mathsf{D}\tilde{f}(\pi(x), \pi(y)) \circ D\pi(\Phi_g(z))$$
$$= \mathsf{D}\tilde{f}(\pi(x), \pi(y)) \circ D\pi(z) \circ [D\Phi_g(z)]^{-1}$$
$$= [D\Phi_g(z)]^{-\mathsf{T}} \cdot \left(\mathsf{D}\tilde{f}(\pi(x), \pi(y)) \circ D\pi(z)\right)$$
$$= [D\Phi_g(z)]^{-\mathsf{T}} \cdot \mathsf{D}^G f(x, y). \qquad (3.31)$$

(4) To verify the orthogonality condition we again exploit the invariance of the mapping $\pi \colon P \to \mathbb{R}^{m-s}$. In particular, we have $D\pi(z) \cdot \xi_P(z) = 0$ for all $\xi \in T_e G$. So

$$\mathsf{D}^G f(x, y) \cdot \xi_P(z) = \mathsf{D}\tilde{f}(\pi(x), \pi(y)) \cdot (D\pi(z) \cdot \xi_P(z)) = 0, \qquad (3.32)$$

for all $\xi \in T_e G$. $\qquad \square$

We next give an example to clarify the above ideas.

Example 3.1. Let P be an open set in $\mathbb{R}^3 \times \mathbb{R}^3$ of the form

$$P = \{(q, p) \in \mathbb{R}^3 \times \mathbb{R}^3 \mid q \times p \neq 0\}, \tag{3.33}$$

and let Ω denote the canonical symplectic structure on P. Let $H: P \to \mathbb{R}$ be a smooth function of the form

$$H(q, p) = V(q) + K(p), \tag{3.34}$$

where $V(q) = \hat{V}(\|q\|)$ for some function $\hat{V}: \mathbb{R}_+ \to \mathbb{R}$ and $K(p) = \hat{K}(\|p\|) = \|p\|^2/2m$ for some $m > 0$.

Clearly, the above Hamiltonian system (P, Ω, H) has symmetry under the regular affine action of $G = SO(3)$ on P defined as $\Phi(\Lambda, (q, p)) = (\Lambda q, \Lambda p)$, i.e., the Hamiltonian is invariant under this action. Moreover, this action is symplectic with momentum map $J: P \to T_e^*G \cong \mathbb{R}^3$ given by $J(q, p) = q \times p$, which is called the angular momentum for the system.

To construct an associated discrete system with symmetry we need to construct a G-equivariant discrete derivative for G-invariant functions on P. To do this, we need to find a set of independent invariants of G which are at most quadratic. In particular, since P is of dimension $k = 6$ and the action of G has orbits of dimension $s = 3$, we need to find $k - s = 3$ independent invariants of G. By inspection, we have that

$$\left.\begin{array}{l} \pi_1(q, p) = \|q\|^2 = q \cdot q \\ \pi_2(q, p) = q \cdot p \\ \pi_3(q, p) = \|p\|^2 = p \cdot p \end{array}\right\} \tag{3.35}$$

are a set of independent invariants which are quadratic. Hence we have $P/G \cong \pi(P) \subset \mathbb{R}^3$ where

$$\pi(P) = \{(x_1, x_2, x_3) \in \mathbb{R}^3 \mid x_1 > 0, \ x_3 > 0, \ |x_2| < x_1 x_3\}, \tag{3.36}$$

and the associated reduced function $\tilde{H}: \pi(P) \subset \mathbb{R}^3 \to \mathbb{R}$ for H is

$$\begin{aligned} \tilde{H}(\pi_1, \pi_2, \pi_3) &= \hat{V}(\sqrt{\pi_1}) + \hat{K}(\sqrt{\pi_3}) \\ &= \tilde{V}(\pi_1) + \tilde{K}(\pi_3), \end{aligned} \tag{3.37}$$

where $\tilde{V}(\pi_1) = \hat{V}(\sqrt{\pi_1})$ and $\tilde{K}(\pi_3) = \hat{K}(\sqrt{\pi_3}) = \pi_3/2m$.

Now, for any $x, y \in P$ let $z = (x + y)/2$. Then, using a partitioned discrete derivative for $\tilde{H}$, a G-equivariant discrete derivative for H is

$$\mathrm{D}^G H(x, y) = \mathrm{D}\tilde{V}(\pi_1(x), \pi_1(y)) \circ \mathrm{D}\pi_1(z) + \mathrm{D}\tilde{K}(\pi_3(x), \pi_3(y)) \circ \mathrm{D}\pi_3(z). \tag{3.38}$$

Since $\tilde{V}: \mathbb{R}_+ \to \mathbb{R}$ we have

$$\begin{aligned} \mathrm{D}\tilde{V}(\tau, t) &= \tilde{V}'(\tfrac{\tau+t}{2}) + \frac{\tilde{V}(t) - \tilde{V}(\tau) - \tilde{V}'(\tfrac{\tau+t}{2})(t - \tau)}{|t - \tau|^2}(t - \tau) \\ &= \frac{\tilde{V}(t) - \tilde{V}(\tau)}{t - \tau}. \end{aligned} \tag{3.39}$$

Similarly,

$$D\tilde{K}(\tau, t) = \frac{\tilde{K}(t) - \tilde{K}(\tau)}{t - \tau} = \frac{1}{2m}. \tag{3.40}$$

So

$$D^G H(x, y) = \frac{\tilde{V}(\pi_1(y)) - \tilde{V}(\pi_1(x))}{\pi_1(y) - \pi_1(x)} D\pi_1(z) + \frac{1}{2m} D\pi_3(z). \tag{3.41}$$

If we let $x = (q_n, p_n)$ and $y = (q_{n+1}, p_{n+1})$ then $z = (q_{n+\frac{1}{2}}, p_{n+\frac{1}{2}})$, and we get

$D^G H((q_n, p_n), (q_{n+1}, p_{n+1}))$

$$= \frac{\tilde{V}(\|q_{n+1}\|^2) - \tilde{V}(\|q_n\|^2)}{\|q_{n+1}\|^2 - \|q_n\|^2} (2q_{n+\frac{1}{2}}, 0) + \frac{1}{2m}(0, 2p_{n+\frac{1}{2}})$$

$$= \left(\frac{\hat{V}(\|q_{n+1}\|) - \hat{V}(\|q_n\|)}{\|q_{n+1}\| - \|q_n\|} \frac{q_{n+\frac{1}{2}}}{\frac{1}{2}(\|q_{n+1}\| + \|q_n\|)}, m^{-1}p_{n+\frac{1}{2}} \right). \tag{3.42}$$

With the canonical symplectic structure, we obtain the difference equations for our discrete system with symmetry as

$$\left. \begin{array}{l} q_{n+1} - q_n = hm^{-1}p_{n+\frac{1}{2}} \\[2mm] p_{n+1} - p_n = -h \frac{\hat{V}(\|q_{n+1}\|) - \hat{V}(\|q_n\|)}{\|q_{n+1}\| - \|q_n\|} \frac{q_{n+\frac{1}{2}}}{\frac{1}{2}(\|q_{n+1}\| + \|q_n\|)} \end{array} \right\}, \tag{3.43}$$

where $h > 0$ is a parameter.   □

Remarks 3.3.

(1) Within an algorithmic framework the above system is a second-order, implicit, one-step approximation to the underlying Hamiltonian differential equation which preserves the Hamiltonian and the angular momentum. This scheme is studied in detail in [7]. For an n-body generalization of the above scheme, together with a numerical assessment of performance, see [24].

(2) Generally speaking, the idea of replacing the derivative of a potential with a finite-difference quotient in order to achieve energy and momentum conservation goes back to the work of Greenspan [9] and LaBudde & Greenspan [12]–[14].   □

3.8. Reduced Trajectories

Given a discrete system with symmetry possessing a momentum map J, we can introduce the notion of reduced trajectories as is done for the underlying system. The existence of these reduced trajectories will be crucial when we consider questions of stability in later sections.

Let (P, Ω, G, D^G, H) be a discrete Hamiltonian system with symmetry and let Φ denote a regular affine symplectic action of G on P. Assume this action possesses an Ad*-equivariant momentum map $J: P \to T_e^*G$ which is an integral for the discrete system, and let $\mu \in T_e^*G$ be a regular value for J so that the preimage $J^{-1}(\mu)$ is a smooth manifold in P. Since J is an integral for the system, any trajectory which starts in $J^{-1}(\mu)$ remains there. Hence, given any $z_0 \in J^{-1}(\mu)$, there is a well-defined trajectory $(z_n)_{n=0}^N$ in $J^{-1}(\mu)$, which implies the existence of a well-defined discrete system on $J^{-1}(\mu)$.

As before, let G_μ denote the isotropy subgroup of μ under the coadjoint action of G on T_e^*G, i.e., $G_\mu = \{g \in G \mid \mathrm{Ad}_{g^{-1}}^*(\mu) = \mu\}$. Then $J^{-1}(\mu)$ is invariant under the action of G_μ. Since G_μ is a subgroup of G, we have, by Proposition 3.4, that any trajectory maps to a trajectory under the action of G_μ. In particular, the action of G_μ maps trajectories in $J^{-1}(\mu)$ to trajectories in $J^{-1}(\mu)$. Hence, the restriction of the discrete system to $J^{-1}(\mu)$ is a well-defined system with symmetry.

Since by assumption the action of G_μ is regular, there is a well-defined reduced phase space $P_\mu = J^{-1}(\mu)/G_\mu$ and a natural projection $\pi_\mu : J^{-1}(\mu) \to P_\mu$. If $(z_n)_{n=0}^N$ is a trajectory for the original system lying in $J^{-1}(\mu)$, then $(\pi_\mu(z_n))_{n=0}^N$ is a well-defined trajectory in the reduced space. In particular, trajectories in $J^{-1}(\mu)$ and P_μ differ by some sequence of transformations under the action. More importantly, since the reduced Hamiltonian $H_\mu : P_\mu \to \mathbb{R}$ depends only on the original Hamiltonian H and the momentum value μ, it follows that H_μ is an integral for the reduced trajectory, i.e., $H_\mu(\pi_\mu(z_n)) = H_\mu(\pi_\mu(z_0))$ for all $n = 0, \ldots, N$.

3.9. Fixed Points

Consider a discrete Hamiltonian system $(P, \Omega, \mathsf{D}, H)$ where P is open in m-dimensional Euclidean space $\mathbb{R}^m$. An *equilibrium point* or *equilibria* of the system is a point $z_0 \in P$ for which the constant sequence $(z_0)_{n=0}^\infty$ satisfies the associated Hamiltonian difference equation (3.5). In terms of the discrete vector field, it follows by induction that z_0 is an equilibrium point if and only if $\mathsf{X}_H(z_0, z_0) = 0$.

We can characterize equilibria of general discrete Hamiltonian systems with the following proposition.

Proposition 3.7. *Let* $(P, \Omega, \mathsf{D}, H)$ *be a discrete Hamiltonian system. Then a point* $z_0 \in P$ *is an equilibrium point if and only if* z_0 *is a critical point of the Hamiltonian* H, *i.e.,* $\mathsf{D}_{z_0} H = 0$.

Proof. A point z_0 is an equilibria if and only if $\mathsf{X}_H(z_0, z_0) = 0$. Since $\mathsf{X}_H(z_0, z_0) = \Omega_{z_0}^\sharp(\mathsf{D}_{(z_0, z_0)} H)$, and Ω_{z_0} is nondegenerate, z_0 is an equilibria if and only if $\mathsf{D}_{(z_0, z_0)} H = 0$. The result follows from the fact that $\mathsf{D}_{(z_0, z_0)} H = \mathsf{D}_{z_0} H$. $\qquad\square$

Comparing the discrete system with the underlying system (P, Ω, H) we see that both possess equilibrium points which are critical points of the Hamiltonian function H. In particular, the trajectory $(z_0)_{n=0}^\infty$ is a discrete analog of the equilibrium solution $\varphi(t) = z_0$ for all $t \in \mathbb{R}$ of the underlying system.

We next consider discrete Hamiltonian systems with symmetry and determine whether they inherit discrete analogs of relative equilibria.

3.10. Relative Equilibria

Let $(P, \Omega, G, \mathbf{D}^G, H)$ be a discrete Hamiltonian system with symmetry and denote by Φ a regular affine symplectic action of G on P. Suppose the action has a momentum map $J \colon P \to T_e^* G$ which is conserved along trajectories of this system. We say a point $z_e \in P$ is a *relative equilibria* of the discrete system with symmetry if, for given $h > 0$, the local trajectory $(z_n)_{n=0}^N$ through z_e is of the form

$$z_n = \Phi(g^n, z_e), \tag{3.44}$$

for some $g \in G$ where g^n denotes n products of g. Note that this definition is just a discrete analog of the definition for the underlying system, and is motivated from that definition by considering $z(t_n)$ where $t_n = hn$. Regarding relative equilibria for discrete systems with symmetry, we have the following proposition.

Proposition 3.8. *Let $(P, \Omega, G, \mathbf{D}^G, H)$ be a discrete Hamiltonian system with symmetry and denote by Φ a regular affine symplectic action of G on P. Assume this action possesses an Ad^*-equivariant momentum map $J \colon P \to T_e^* G$ which is an integral for the discrete system. Then, for any regular momentum value μ, a point $z_e \in J^{-1}(\mu) \subset P$ is a relative equilibria if and only if, for given $h > 0$, the local trajectory at z_e projects to a constant trajectory (i.e., fixed point) in the reduced space P_μ.*

Proof. Consider a point $z_e \in J^{-1}(\mu)$ and recall that there exists a neighborhood B of z_e in P, real numbers $h_c, T > 0$, and an evolution semigroup $\mathsf{F} \colon B \times [0, h_c] \to P$ such that, for any $0 < h < h_c$, the local trajectory at z_e is given by $z_n = \mathsf{F}_h^n(z_e)$ for $n = 0, \ldots, N$ where $N \geq 1$ is such that $Nh \leq T$. Furthermore, we have $z_n \in J^{-1}(\mu)$ for all $n = 0, \ldots, N$.

Now, if z_e is a relative equilibria, then $z_n = \Phi(a^n, z_e)$ for all $n = 0, \ldots, N$ for some $a \in G_\mu$. Hence, for each n it follows that $z_n \in G_\mu \cdot z_e$, i.e., z_n is in the orbit of z_e under the action of G_μ. By definition of the projection $\pi_\mu \colon J^{-1}(\mu) \to P_\mu$, we then have $\pi_\mu(z_n) = \pi_\mu(z_e)$ for all $n = 0, \ldots, N$, and the reduced sequence $(\pi_\mu(z_n))_{n=0}^N$ in P_μ is a constant sequence.

Conversely, assume the local trajectory $(z_n)_{n=0}^N$ in $J^{-1}(\mu)$ projects to a constant sequence $(\pi_\mu(z_n))_{n=0}^N$ in P_μ, i.e., $\pi_\mu(z_n) = \pi_\mu(z_e)$ for all $n = 0, \ldots, N$. By definition of π_μ, we must have $z_n \in G_\mu \cdot z_e$ for each n. That is, there exists a sequence $(g_n)_{n=0}^N$ in G_μ such that $z_n = \Phi(g_n, z_e)$. Since $z_n = \mathsf{F}_h^n(z_e)$ for $n = 0, \ldots, N$ and the mappings F_h^n have the semigroup properties $\mathsf{F}_h^0 = id$ and $\mathsf{F}_h^{n+m} = \mathsf{F}_h^n \circ \mathsf{F}_h^m$, we can use properties of the action Φ to deduce that the sequence $(g_n)_{n=0}^N$ must have the properties $g_0 = e$ and $g_{n+m} = g_n g_m$. Let $g_1 = a \in G_\mu$, and for induction assume $g_n = a^n$. Then, since $g_{n+1} = g_n g_1$, it follows that $g_{n+1} = a^{n+1}$. Hence, the sequence $(g_n)_{n=0}^N$ is defined by $g_n = a^n$ for some $a \in G_\mu$. It then follows that z_e is a relative equilibria. $\square$

As with equilibrium points, we would like to be able to characterize relative equilibria of discrete systems in terms of properties of the underlying system. To this end we have the following proposition.

Proposition 3.9. *Let (P, Ω, G, H) be a Hamiltonian system with symmetry with an Ad*-equivariant momentum map J. Assume this system possesses a relative equilibria z_e with a regular momentum value μ. If $\pi_\mu(z_e) \in P_\mu$ is a nondegenerate minima or maxima of the reduced Hamiltonian H_μ, and the parameter $h > 0$ is sufficiently small, then z_e is a relative equilibria for an associated discrete system with symmetry $(P, \Omega, G, \mathsf{D}^G, H)$ provided that the action of G on P is affine and J is an integral for this system.*

Proof. The result follows from Proposition 3.8, together with the observations that there is a well-defined discrete system in P_μ and H_μ is an integral for this system. □

The following proposition follows from the definitions of relative equilibria for both the underlying system and an associated discrete system.

Proposition 3.10. *If z_e is a relative equilibria for both the underlying system and an associated discrete system, then there is a sequence $(g_n)_{n=0}^N$ in G_μ such that the sampled trajectory $\varphi(hn)$ of the underlying system through z_e, and the local trajectory $(z_n)_{n=0}^N$ of the discrete system through z_e, differ by group transformations of the form*

$$\varphi(hn) = \Phi(g_n, z_n), \tag{3.45}$$

for all $n = 0, \ldots, N$.

3.11. Notions of Stability

In analogy with the underlying Hamiltonian system we now introduce the notions of general dynamic stability and stability of equilibria and relative equilibria of an associated discrete system.

3.11.1. General Dynamic Stability. Consider a discrete Hamiltonian system $(P, \Omega, \mathsf{D}, H)$ with P open in $\mathbb{R}^m$ and consider the associated Hamiltonian difference equation (3.5). If the system has symmetry under the affine action of a group G, we suppose this action possesses a momentum map $J: P \to T_e^* G$ which is conserved along trajectories.

For any $z_0 \in P$ let (z_n) denote the maximal trajectory through z_0 for given $h > 0$. We say that the trajectory (z_n) is *dynamically stable* if it is defined for all $n \geq 0$ and if there is a constant $K > 0$, depending on h and z_0, such that $\|z_n\| \leq K$ for all $n \geq 0$. More generally, we say that the system is dynamically stable on a subset B of P if, for each $z_0 \in B$, there is a real number $h > 0$ such that the maximal trajectory through z_0 is dynamically stable.

An elementary criterion for dynamical stability is contained in the following proposition whose proof is straightforward.

Proposition 3.11. *Without loss of generality consider a discrete Hamiltonian system with symmetry $(P, \Omega, G, \mathsf{D}^G, H)$ possessing a momentum map J. Given $z_0 \in P$ such that $H(z_0) = c$ and $J(z_0) = \mu$, the trajectory through z_0 is dynamically stable if the subset $H^{-1}(c) \cap J^{-1}(\mu) \subset P$ is bounded, and the parameter $h > 0$ is sufficiently*

small. In particular, the system is dynamically stable on any bounded subset of the form
$H^{-1}(c) \cap J^{-1}(\mu)$.

3.11.2. Stability of Equilibria and Relative Equilibria.

Our second notion of stability is that of stability of equilibria and relative equilibria, which is concerned with the behavior of solutions with nearby initial conditions. For concreteness consider a discrete Hamiltonian system $(P, \Omega, \mathsf{D}, H)$ where P is open in m-dimensional Euclidean space $\mathbb{R}^m$.

Suppose we are given an equilibrium point $z_0 \in P$. We say that z_0 is *stable* in the sense of Lyapunov if, for any neighborhood U of z_0, there is a neighborhood V of z_0 and a real number $h_{max} > 0$ such that, for any $y \in V$ and $0 < h < h_{max}$, the solution sequence (y_n) at y is defined and satisfies $y_n \in U$ for all $n \geq 0$. Roughly speaking, z_0 is stable if all solution sequences beginning in a neighborhood of z_0 remain in a neighborhood of z_0.

An elementary criterion for the Lyapunov stability of an equilibrium point of a discrete Hamiltonian system is contained in the following proposition whose proof is analogous to the time-continuous version [1].

Proposition 3.12. *Let z_0 be an equilibrium point of a discrete Hamiltonian system $(P, \Omega, \mathsf{D}, H)$. If the bilinear form $D^2 H(z_0)$ is positive- or negative-definite, i.e., $D^2 H(z_0) \cdot (v, v) > 0$ or $D^2 H(z_0) \cdot (v, v) < 0$, respectively, for all nonzero $v \in T_{z_0} P \cong \mathbb{R}^m$, then z_0 is stable in the sense of Lyapunov.*

The conditions of the above proposition are sufficient to guarantee Lyapunov stability of an equilibrium point for a general discrete Hamiltonian system. However, the proposition cannot be applied as is to systems with *symmetry*. The underlying reason is that the conditions of the above proposition imply the equilibrium point is isolated, which in general is not true for systems with symmetry. In particular, since equilibrium points z_0 correspond to critical points of the Hamiltonian, any point in the orbit $G \cdot z_0$ is also a critical point. In this case, the most we can hope for is stability of the set $G \cdot z_0$, i.e., stability up to the group action.

Similar difficulties are encountered when studying the stability of relative equilibria; in particular, for a relative equilibria with momentum value μ the most we can hope for is stability of the set $G_\mu \cdot z_e$. Since equilibrium points are special cases of relative equilibria—in particular, they are relative equilibria with $g = e$ (the group identity) in expression (3.44)—we can discuss their stability together as follows.

Let z_e be a relative equilibrium point with momentum value $J(z_e) = \mu$ and recall, from Proposition 3.8, that the trajectory in P through z_e projects to a fixed point in the reduced phase space P_μ. Also, for any trajectory in $J^{-1}(\mu) \subset P$, recall that the reduced Hamiltonian $H_\mu \colon P_\mu \to \mathbb{R}$ is an integral for the reduced trajectory. With this in mind, we can establish a criterion for the relative stability of a discrete relative equilibrium point z_e. In particular, we will say that z_e is *relatively stable* if the fixed point $\pi_\mu(z_e)$ in P_μ is stable in the sense of Lyapunov. As for the underlying time-continuous system [1], we have the following criterion for relative stability.

Proposition 3.13. *Let z_e be a relative equilibrium point with momentum μ of a discrete Hamiltonian system with symmetry $(P, \Omega, G, \mathsf{D}^G, H)$, and denote by π_μ the canonical projection from $J^{-1}(\mu)$ onto P_μ. If the bilinear form $D^2 H_\mu(\pi_\mu(z_e))$ is positive- or negative-definite, i.e. $D^2 H_\mu(\pi_\mu(z_e)) \cdot (v, v) > 0$ or $D^2 H_\mu(\pi_\mu(z_e)) \cdot (v, v) < 0$, respectively, for all nonzero $v \in T_{\pi_\mu(z_e)} P_\mu$, then z_e is relatively stable.*

4. Concluding Remarks

Using the notion of a discrete Hamiltonian system, this paper has developed a framework for the design and analysis of conserving time-integration schemes for Hamiltonian systems with symmetry. Given a Hamiltonian system defined on an open set of Euclidean space, we have shown that a Hamiltonian-conserving scheme can always be constructed. Furthermore, if the system has symmetry under a group of affine transformations, we have shown how a conserving scheme that inherits this symmetry may be constructed. Regarding qualitative properties, it was shown that conserving schemes which fit within the proposed framework inherit invariant sets in phase space such as equilibria and relative equilibria, along with their stability properties.

The results summarized above were obtained for the case in which the underlying phase space was open in some Euclidean space and equipped with a symplectic structure. However, from the point of view of design, it is easy to see that the framework presented herein extends immediately to Euclidean phase spaces with more general Poisson structures. Moreover, the framework extends to infinite-dimensional systems. In this case, one introduces the idea of discrete functional derivatives, analogous to the discrete derivatives introduced in this paper, and then constructs a system of difference equations using the Poisson structure of the underlying problem (see [6] and [8] for details). For extensions of the ideas presented herein to constrained systems, see [5].

With regards to accuracy, we note that conserving schemes constructed using the discrete derivatives presented in this paper are formally second-order. However, one can employ time substepping procedures such as that proposed in [25] to increase the accuracy of a given conserving scheme.

Acknowledgment

The author gratefully acknowledges the support of the National Science Foundation as a graduate fellow.

References

[1] R. Abraham & J. E. Marsden (1978) *Foundations of Mechanics*, Second Edition, Addison-Wesley, Reading, MA.

[2] M. Austin, P. S. Krishnaprasad & L. S. Wang (1993) Almost Poisson integration of rigid body systems, *J. Computational Physics*, **107**, 105–117.

[3] M. A. Crisfield & J. Shi (1994) A co-rotational element/time-integration strategy for nonlinear dynamics, *International Journal for Numerical Methods in Engineering*, **37**, 1897–1913.

[4] M. Delfour, M. Fortin & G. Payre (1981) Finite difference solutions of a nonlinear Schrödinger equation, *J. Computational Physics*, **44**, 277–288.

[5] O. Gonzalez (1996a) Mechanical systems subject to holonomic constraints: Differential-algebraic formulations and conservative integration, *Physica D*, to appear.

[6] O. Gonzalez (1996b) "Design and analysis of conserving integrators for nonlinear Hamiltonian systems with symmetry," Ph.D. Dissertation, Department of Mechanical Engineering, Division of Applied Mechanics, Stanford University.

[7] O. Gonzalez & J.-C. Simo (1996a) On the stability of symplectic and energy-momentum algorithms for nonlinear Hamiltonian systems with symmetry, *Computer Methods in Applied Mechanics and Engineering*, to appear.

[8] O. Gonzalez & J.-C. Simo (1996b) Exact energy-momentum conserving algorithms for general models in nonlinear elasticity, submitted.

[9] D. Greenspan (1973) *Discrete Models*, Addison-Wesley, Reading, MA.

[10] T. Itoh & K. Abe (1988) Hamiltonian-conserving discrete canonical equations based on variational difference quotients, *J. of Computational Physics*, **77**, 85–102.

[11] T. Itoh & K. Abe (1989) Discrete Lagrange's equations and canonical equations based on the principle of least action, *Applied Mathematics and Computation*, **29**, 161–183.

[12] R. A. LaBudde & D. Greenspan (1974) Discrete mechanics—A general treatment, *J. of Computational Physics*, **15**, 134–167.

[13] R. A. LaBudde & D. Greenspan (1976a) Energy and momentum conserving methods of arbitrary order for the numerical integration of equations of motion. Part I, *Numerisch Mathematik*, **25**, 323–346.

[14] R. A. LaBudde & D. Greenspan (1976b) Energy and momentum conserving methods of arbitrary order for the numerical integration of equations of motion. Part II, *Numerisch Mathematik*, **26**, 1–16.

[15] D. Lewis & J.-C. Simo (1994) Conserving algorithms for the dynamics of Hamiltonian systems on Lie groups, *Journal of Nonlinear Science*, **4**, 253–300.

[16] J. E. Marsden & T. S. Ratiu (1994) *Introduction to Mechanics and Symmetry*, Springer-Verlag, New York.

[17] J. Moser & A. P. Veselov (1991) Discrete versions of some classical integrable systems and factorization of matrix polynomials, *Communications in Mathematical Physics*, **139**, 217–243.

[18] P. J. Olver (1993) *Applications of Lie Groups to Differential Equations*, Second Edition, Springer-Verlag, New York.

[19] J. M. Sanz-Serna (1984) Methods for the numerical solution of the nonlinear Schrödinger equation, *Mathematics of Computation*, **43**, 21–27.

[20] J.-C. Simo & K. K. Wong (1991) Unconditionally stable algorithms for rigid body dynamics that exactly preserve energy and angular momentum, *International J. Numerical Methods in Engineering*, **31**, 19–52.

[21] J.-C. Simo & N. Tarnow (1992) The discrete energy-momentum method. Conserving algorithms for nonlinear elastodynamics, *ZAMP*, **43**, 757–793.

[22] J.-C. Simo & N. Tarnow (1994) A new energy-momentum method for the dynamics of nonlinear shells, *International J. Numerical Methods in Engineering*, **37**, 2525–2550.

[23] J.-C. Simo, N. Tarnow & M. Doblaré (1994) Exact energy-momentum algorithms for the dynamics of nonlinear rods, *International J. Numerical Methods in Engineering*, **38**, 1431–1473.

[24] J.-C. Simo & O. Gonzalez (1993) Assessment of energy-momentum and symplectic schemes for stiff dynamical systems, *American Society of Mechanical Engineers*, ASME Winter Annual Meeting, New Orleans, LA.

[25] N. Tarnow & J.-C. Simo (1994) How to render second order accurate time-stepping algorithms fourth order accurate while retaining the stability and conservation properties, *Computer Methods in Applied Mechanics and Engineering*, **115**, 233–252.

Problems and Progress in Microswimming

J. Koiller,[1] K. Ehlers,[2] and R. Montgomery[2]

[1] Laboratório Nacional de Computacão Científica, R. Lauro Muller 455, Rio de Janeiro, 22290-160, RJ, Brazil
[2] Mathematics Department, University of California at Santa Cruz, Applied Sciences Building, Santa Cruz, CA 95064, USA

Received January 5, 1996; revised manuscript accepted for publication June 1, 1996
Communicated by Jerrold Marsden and Stephen Wiggins

*Dedicated to the memory of Juan-Carlos Simo, a pioneer in the use of geometry
to produce better analytical and numerical methods in mechanics*

Summary. Stokesian swimming is a geometric exercise, a collective game. In Part I, we review Shapere and Wilczek's gauge-theoretical approach for a single organism. We estimate the speeds of organisms moving by propagating small amplitude waves, and we make a conjecture regarding a new inequality for the Stokes' curvature. In Part II, we extend the gauge theory to collective motions. We advocate the influx of nonlinear control theory and subriemannian geometry. Computationally, parallel algorithms are natural, each microorganism representing a separate processor. In the final section, open questions motivated by biology are presented.

Key words. Stokes' flows, geometric phases, nonholonomic control

MSC classification codes (1991): 76D07, 76Z10, 93B29, 93C10, 51P05, 53C05

PACS classification codes (1990): 47.15.Gf, 87.45.-k, 87.10.+e, 02.40.+m, 03.40.Gc

Part I: Bachelor's Life at Low Reynolds Number

... Microwildernesses exist in a handful of soil or aqueous silt collected almost anywhere in the world. They are at least close to a pristine state and still unvisited. Bacteria, protists, nematodes, mites, and other minute creatures swarm around us, an animate matrix that binds Earths surface. They are objects of potentially endless study and admiration, if we are willing to sweep our vision down from the world lined by the horizon to include the world an arm's length away. A lifetime can be spent in a Magellanic voyage around the trunk of a single tree. If I could do it all over again, and relive my vision in the twenty-first century, I would be a microbial ecologist. Ten billion bacteria live in a gram of ordinary soil, a mere pinch held between thumb and forefinger. They represent thousands of species, almost none of which are known to science.

—E. O. Wilson, from the end of his book, *The Naturalist*

1. Introduction

The motions of living beings inspired Leonardo da Vinci's engineering endeavors. Giovanni Borelli's *De motum animalium* (1680) was an early interdisciplinary effort. Two centuries later, biologists, importing ideas from Industrial Revolution scientists [C. Bernard (*Les phénomenes de la vie*, 1878) and L. Frederick (who coined the term "regulatory agencies" in 1885)], showed that control theory applies as well to living machines as for steam engines. More recently, N. Wiener pointed out many areas of contact between applied mathematics and biology. He introduced the word "cybernetics" in 1961.

The earliest life forms appeared 3600 million years ago. The first invertebrates appeared 1600 million years ago. Genus homo appeared about 1 million years ago and civilization arose about .01 million years ago. Bacteria and protozoa, among the first life forms, have had plenty of time to optimize their performance.[1] Paradoxically, very few roboticists or control theorists have looked into microorganism motion and survival strategies. (An exception is R. S. Fearing [33].) We suggest a control-theoretic approach to microorganism motility. These ancient organisms may have used "modern" ideas from nonholonomic motion planning and parallel computing. But a disclaimer is in order. Biofluiddynamics has a long tradition and we are novices, so we ask for indulgence.

We became interested in the problems of microswimmers through the work of A. Shapere and F. Wilczek [75]. In 1989 two of us heard these two talk on their "gauge-theory" explanation for microorganism motion. We felt that their approach paved the way for other ideas from geometric mechanics [1]. We were working on geometric phases, constrained variational problems, and nonholonomic systems. In the Workshop on Geometric Mechanics, Rio de Janeiro, 1993, one of us (JK) outlined a "vakonomic" [3] mechanics approach for microorganism motion. (Juan Simo raised his eyebrows in his ironic but very supportive way.) In the winter quarter of 1993–1994, we organized a seminar on microorganism motion at Santa Cruz. With the help of S. Wiggins and

---

[1] John Cohen, Interdisciplinary talkfest prompts flurry of questions, *Science* **270** (24 Nov 1995), page 1294:

"Bring together 100 of the best and brightest young (under age 45) scientists for 3 days to listen to talks from the cutting edge of a variety of disciplines, and what do you get? An endless barrage of questions, as participants try to slake their intellectual thirst in fields far removed from their own.... In a discussion at the end of the talks, the audience began peppering the engineers with questions and ideas about applying the microlessons of biological systems. Why not model a miniature motor after a flagella, exquisite micromotors that propel cells around the body? How about linking silicon devices to cells?"

R. Murray we went down to Caltech, where we met Professors T. Y. Wu and C. Brennen (who not only did seminal work in the seventies [84], [19], [26], but have encouraged and influenced a great deal of other people's research). K. Ehlers, who was doing his thesis with R. Montgomery, found Professor Berg's beautiful book on bacterial motion. Together with the Brazilian biophysicists studying magnetotactic bacteria (Darci Esquivel, Marcos Farina, Henrique Lins de Barros), we had the privilege of having Prof. Berg visiting Rio in May 1995. Ehlers and Montgomery came down to participate. The list of problems in the last section came from these discussions.

The Stokes' approximation—neglecting the inertial terms in the Navier-Stokes equations—is valid for microorganisms (see, e.g., [81]).[2] Their lives are dominated by two properties of their aqueous environment. These are viscous drag and Brownian motion. Due to drag, inertia plays essentially no role for the microswimmer and this is why the Stokes' equation is used instead of the Navier-Stokes.[3] To illustrate this, H. Berg calculated in his book [4] that an *E. coli* coasts a distance small compared to the diameter of a hydrogen atom when it "throws in the clutch" and stops actively swimming. However, *E. coli* can never actually stop moving due to Brownian motion of the surrounding liquid. It is knocked around on the average of one body-width per second. *E. coli* cannot chase nutrients but relies on diffusion to take the nutrients to it. See the subsection on feeding in the problem list at the end. These two basic properties and various of their consequences are beautifully discussed in Purcell's article "Life at Low Reynold's Number" [71] which we urge all of our readers to read.

It is fashionable nowadays to mention possible applications:[4]

(i) "Peristaltic" (= low Reynolds) pumps. Two of us (RM and KE) met in July 1994 with engineers at the actuators and sensors division of EECS in Berkeley to investigate the possibilities of building a peristaltic pump based on Kurt's ideas. These engineers were in the process of building something very close. Instead of both walls of the pipe vibrating, they had only one wall. There was some disagreement about the underlying physics. We felt that the bulk of the transport might be accounted for by Stokes' equation. They used "acoustic streaming" which mathematically amounts to including part of the nonlinear inertial term of Navier-Stokes as a perturbation. In any case, their tiny pump can be built on a computer chip and could eventually be used for cooling or actuating.

(ii) Oil recovery. Suppose one can construct small, mobile robots, which move about in underground porous media and are able to "sniff" hydrocarbon molecules and follow their gradients. Why not imitate *E. coli*'s use of chemotaxis ("sniffing" [4]) to locate oil patches? In the same vein, an exploration of Mars by a swarm of mini-robots is being planned [34].

(iii) Health. Unfortunately, there are not many studies relating motility to infectivity. But recent work may start to reverse this situation: strategies for vaccine development

---

[2] This fact seems to have been first noticed by Wilhelm Ludwig [57], whose work in the thirties remained forgotten for a long time. (G. Taylor, in his seminal 1951 work [79] did not refer to it.)

[3] It is worth comparing with high Reynolds swimming and flying; see [24], [54], [55].

[4] It is useful to know how to swim at low Reynolds numbers. Prof. H. Berg told us that he has a *Boston Globe* clipping saying that several unfortunate people were drowned when a large tank of molasses broke apart in 1912.

are being attempted that rely on changing the genetic plans for the flagellum. Pathogenic
E. coli, a major cause of infant mortality in third world countries, is the microorganism
of choice.[5]

This paper is organized as follows. In Section 2 we review the gauge theory for one
microswimmer; swimming ellipses and other examples from K. Ehlers' thesis are outlined. A result on motions induced by small amplitude waves is presented in Section 3:
we add a bit of flesh to a general theorem in microswimming first brought to our attention
by A. Shapere's thesis [76]. This is the "tangent plane approximation" for estimating
the net swimming velocity. In Section 4 we outline our main new idea for collective
swimming. It has been observed that there can be an advantage with cooperative microswimmers' behavior. In principle, this can be analyzed by extending Shapere and
Wilczek's gauge-theoretic framework and borrowing ideas from nonholonomic control
theory. In Section 5 we work out some "toy" examples to illustrate this. In Section 6 we
give an annotated list of mathematical problems in microswimming.[6] Most of them are
real biological problems that came out of discussions with H. Berg (misunderstandings
are our responsibility!).

2. The Geometric Description of a Single Swimming Microorganism

> No more pleasant sight has ever yet come before my eye than these many thousands
> of living creatures, seen all alive in a little drop of water, moving among one another,
> each creature having its own proper motion.
>
> —Anton van Leeuwenhoek, 1676

A gauge theory for the motion of an isolated, free-swimming microorganism $\mathcal{R}$ was
proposed by Shapere and Wilczek in 1989 [75]. They considered the "envelope" model
for ciliates, but with appropriate rewording, the same ideas hold for flagellar propulsion
and more realistic models of ciliary motion [16], [17], [18], [84]. In this section we
summarize the main features. For details, see also K. Ehlers' thesis [28].

One distinguishes from the outset abstract (or intrinsic) shapes s and instantaneous
intrinsic shape changes $\dot{s}$, from located shapes $q\colon S^2 \to \mathcal{B} \subset \mathbb{R}^3$, $\mathcal{B} = \partial\mathcal{R}$, and
corresponding boundary velocities $\vec{v}$. The set of located shapes forms our configuration
space $\mathcal{Q}$, and the set of abstract shapes, denoted $\mathcal{S}$, is the quotient of $\mathcal{Q}$ by the group
G of rigid motions. Motion of the body in the fluid is an *indirect* result of the intrinsic
shape changes. A more familiar example is a cat dropped upside down, which is able
to reorient itself doing clever shape maneuvers (the physical constraint of zero angular
momentum is satisfied throughout [62]).

A two-step procedure is commonly used by fluid mechanists. Fix a particular configuration of an isolated organism, and first consider a *trial* boundary condition $\vec{V}$ along
the located boundary of the body, corresponding to the intrinsic infinitesimal shape
deformation.

---

[5] For instance, scientists have been able to use bacteria to introduce genes into animal cells. See [77].

[6] We did some leisure reading too; some interesting sources are [27], [42], [23], [74], [67].

There is an unavoidable ambiguity: $\vec{V}$ plus an infinitesimal rigid motion X (consisting of infinitesimal translation and rotation) is as good as $\vec{V}$ in the sense that intrinsic shape deformation is the same. The correct $\vec{V}$ is picked out of the collection of $\vec{V} = \vec{V}_{\text{trial}} + X$ by adding the physical restriction that $\vec{V}$ induces no net force or torque on the fluid.

2.1. Definitions and Notations

The *propulsion operator* of a *single* microorganism is the linear map $\mathcal{L}$ that associates to a trial vectorfield $\vec{V}$ along the boundary the corresponding total force and torque $(\vec{F}, \vec{T})$ acting on the organism.[7] Restricted to infinitesimal rigid motions, it is called the *resistance* operator $\mathcal{G}$ [40], which is represented by a 6×6 symmetric, positive definite matrix. Properties of $\mathcal{G}$ are summarized in Section 5.

In order to compute $\mathcal{L}$ one needs the first-order germ of the solution of Stokes' equations with boundary condition $\vec{V}$. If the whole solution $\hat{\vec{V}}$ is known, then $\mathcal{L}$ can also be computed from the asymptotics of $\hat{\vec{V}}$ at infinity (see [75], [28]).

The operator

$$A = \mathcal{G}^{-1}\mathcal{L} \tag{1}$$

is called the *Stokes' connection 1-form.* Its range is the Lie-algebra $s E(3)$. For a given intrinsic shape deformation $\dot{s}_s$ and located boundary q, there is a unique vectorfield $\vec{v}$ along the located boundary satisfying $A \cdot \vec{v} = 0$ and consistent with $\dot{s}$. It gives the correct infinitesimal motion of the body in the fluid corresponding to the intrinsic shape deformation $\dot{s}$. The passage from the trial solution to the correct solution may be written (with some abuse of notation) as

$$\vec{v} = (I - A)\vec{V}. \tag{2}$$

2.2. Fluid Dynamics Background

The fluid dynamics of swimming microorganisms are dominated by shear stresses. Inertia plays no role—any motion of the microorganism is communicated to the entire fluid without delay. The swimmer is not subjected to a net force or torque.

The equations of motion, known as Stokes' equations, are

$$\partial_j \sigma_{ij} = 0,$$
$$\partial_i v_i = 0, \tag{3}$$

where

$$\sigma_{ij} = -p\delta_{ij} + \mu \left(\frac{\partial v_i}{\partial x_j} + \frac{\partial v_j}{\partial x_i} \right). \tag{4}$$

Here v_i is the fluid velocity and p the pressure. The boundary conditions are no slip:

$$\vec{v} = \vec{V} \text{ on the boundary,}$$

---

[7] This name was proposed by Purcell.

where $\vec{V} = \vec{V}(x)$ is a given vectorfield on the boundary representing the organism (e.g., the "trial" deformation of the previous section), together with

$$\vec{V} = 0 \text{ at infinity.}$$

The stress tensor, σ_{ij}, represents the ith component of the force per unit area exerted across a plane with normal in the j direction. The total force exerted on the fluid is then given by the formula $F_i = -\int \sigma_{ij} \hat{n}_j \, dA$, where the integral can be taken over any surface enclosing the organism (since the divergence of σ is zero). In short,

$$\vec{F} = -\int \sigma \cdot \hat{n} \, dA. \tag{5}$$

Analogously, the total torque (with respect to an arbitrary origin 0) is given by

$$\vec{T} = -\int \vec{r}_0 \times (\sigma \cdot \hat{n}) \, dA. \tag{6}$$

If we put these together, we get the propulsion operator

$$\mathcal{L}(V) = (\vec{F}, \vec{T}).$$

The resistance operator $\mathcal{G}$ is simply the propulsion operator restricted to infinitesimal isometries. Such an isometry is represented by a vectorfield $\vec{x} \rightarrow \vec{U} + \vec{\omega} \times \vec{x}$ where $\vec{U}$ represents the constant velocity and $\vec{\omega}$ represents the constant angular velocity. Let $V_{\vec{U}, \vec{\omega}}$ be the restriction of this vectorfield to the boundary of the organism. Then

$$\mathcal{G}(\vec{U}, \vec{\omega}) = \mathcal{L}(V_{\vec{U}, \vec{\omega}}).$$

Remark. For those familiar with the mechanical connection associated to the N-body problem, note the similarity. $\mathcal{L}$ plays the role of the total angular momentum and $\mathcal{G}$ plays the role of the locked inertia tensor.

2.3. Geometric Description of a Swimming Stroke

Consider a moving cell membrane. We model it by a one-parameter family $q(t)$ of embeddings of the sphere S^2 into $\mathbb{R}^3$. The set of all embeddings of the two-sphere into three-space forms an infinite-dimensional manifold denoted by $\mathcal{Q}$. It is the underlying configuration space for swimming, and $q(t)$ is a curve in it. The derivative $\dot{q}$ of $q(t)$ at $t = 0$ is then a tangent vector to $\mathcal{Q}$ at the particular located shape $q(0)$. Write $\mathcal{B}$ for the image of the sphere under $q(0)$, that is, for the physical cell membrane at time 0. Then $\vec{V} = \dot{q}$ is a map $\mathcal{B} \rightarrow \mathbb{R}^3$ which represents the infinitesimal deformations of the particles of the cell membrane. The set of all tangent vectors $\vec{V}$ will be denoted by $T\mathcal{Q}$ and forms the tangent bundle to our space of located shapes. They are the possible boundary values for Stokes' equations. (See immediately following equation (4) above.)

If the organism is not allowed to move freely while attempting some infinitesimal deformation $\vec{V}$—for example, if we pin it to the microscope slide by some means—then of necessity it will be subjected to a net force and torque (provided by the pins). These forces and torques are linear functionals of the boundary data $\vec{V}$. Now if we remove the pins and allow the organism to move freely in response to the same infinitesimal deformation $\vec{V}$, then it will rotate and translate infinitesimally by some amount $\vec{X}$. The physics that determine this response $\vec{X}$ are that the organism exerts no net force and torque on the fluid. In this manner we obtain an linear assignment $\vec{V} \mapsto \vec{X} = \vec{A}_s(\vec{V})$ from $T_q \mathcal{Q}$ to $\mathbb{R}^6$, the Lie algebra of infinitesimal Euclidean motions. This map A is called the *Stokes' connection* and is summarized by equation (1).

Now let $\mathcal{S}$ denote the collection of all possible intrinsic shapes which the organism can assume. It is the quotient of $\mathcal{Q}$ by the group G of rigid motions. Its tangent bundle $T\mathcal{S}$ is then the collection of all shapes, together with the infinitesimal deformations. A tangent vector can be thought of as a vectorfield $\vec{V} \colon \mathcal{B} \to \mathbb{R}^3$ *modulo* infinitesimal motions $\vec{X}$. A *swimming stroke* is represented by a closed loop in $\mathcal{S}$:

$$s \colon [0, T] \to \mathcal{S}, \tag{7}$$

with $s(0) = s(T)$. When the organism performs this stroke in the fluid, it will rotate and translate so as to never violate the no net force and torque law. Although the initial and final shapes are the same, with any luck, the initial and final positions are not: the organism has swum! The initial and final *located shape* will differ by some Euclidean motion $g(T) \in G$. *Our first goal is to understand $g(T)$ as a function of the swimming stroke.*

In order to find it, choose some arbitrary "lift" or realization of the unlocated swimming stroke $s(t)$ by some curve $q(t)$ in $\mathcal{Q}$. This curve is not the physically accurate motion; rather, the correct located shape of the organism can be described by $g(t) \cdot q(t)$. The group elements $g(t) \in G$ provide the necessary corrections to our rather arbitrary "gauge" $q(t)$. With a bit of work (see Shapere and Wilczek [75]), one can show that g satisfies the first-order ordinary linear time-dependent differential equation:

$$\frac{dg}{dt} = g(t) \cdot A_{q(t)}(\dot{q}(t)). \tag{8}$$

(Caveat: a minus sign may appear depending on if you define the force and torque exerted by the fluid on the body or vice versa.)

Because the group of Euclidean motions is not Abelian, the solution is not as straightforward as one would hope. Physicists call the solution a "path-ordered exponential":

$$g(T) = \bar{P} \exp \int_{0 < t < T} A(t)\,dt$$

$$= I + \int_{0 < t_1 < T} A(t_1)\,dt_1 + \int_{0 < t_1 < t_2 < T} A(t_1)A(t_2)\,dt_1\,dt_2 + \cdots. \tag{9}$$

For large deformations, $g(T)$ must, in general, be found numerically.[8] The following recent references may be useful for those interested in pursuing numerical work (we apologize for omissions): [83], [35], [31], [69], [72], [73].

---

[8] The authors have little computational background. Juan Simo's work combined deep abstract thinking together with tremendous numerical ability. We think Juan would enjoy this problem.

Let us summarize what we have said so far, introducing the corresponding geometric language as we go. A swimming stroke is a closed loop in the space S of abstract shapes. The resulting distance (and reorientation) suffered as a result this stroke is the *holonomy* of this loop with respect to the Stokes' connection. The physical motion $q(t)$ representing where the cell membrane is in space at any instant t of time is the horizontal lift of the loop in S to a (nonclosed) curve in Q with respect to the Stokes' connection. Either process of lifting or finding the holonomy involves solving the linear differential equation (8), which is also known as the equation of parallel transport.

2.4. The Curvature Approximation Formula

For small swimming motions g can be estimated (to second order in the amplitude of the deformation) using the curvature of the connection form A.

Let $\{v_n\}$ be a basis for the vectorfields on the surface of a given located shape q. Define $\mathcal{F}_{mn}$ to be the infinitesimal Euclidean motion given by the coupling of the modes v_m and v_n (e.g., Fourier modes). The $\mathcal{F}_{mn}$ are nothing more than the components of the curvature two-form $\mathcal{F}$ of the connection form A, evaluated at the shape q contracted with the vectors v_m and v_n. A formula for $\mathcal{F}_{mn}$ is

$$\mathcal{F}_{mn} = A([v_n^h, v_m^h]), \tag{10}$$

where

$$[v_n^h, v_m^h] = (v_n^h \cdot \nabla)\hat{v}_{m|\text{shape}}^h - (v_m^h \cdot \nabla)\hat{v}_{n|\text{shape}}^h \tag{11}$$

is the Lie bracket. The hat indicates the Stokes' extension of the boundary condition to the fluid; the fluid response in a neighborhood of the boundary is necessary in order to compute the derivatives. The superscript h denotes "horizontal projection"—which in practice means subtracting $A_q(v)$ from the input boundary conditions so that their Stokes' extensions lead to no net force or torque on the fluid.

Once we have the components of the curvature calculated at a particular shape we can approximate the connection form in a neighborhood of that shape. Let a_m be the coordinates associated to the v_m. A boundary condition can then be written $v = \sum a_m v_m$, and the a_m are to be thought of as the amplitudes. Then,

$$\mathcal{F}_{|s} = \sum_{m<n} \mathcal{F}_{mn} da_m \wedge da_n$$

$$= \sum_{m<n} d(a_m \mathcal{F}_{mn} da_n). \tag{12}$$

Hence $A \cong \sum a_m \mathcal{F}_{mn} da_n + \text{exact}$. So if a swimming motion is gauge-parameterized by

$$s(t) = q + \sum a_n(t)v_n, \tag{13}$$

where q is a given located shape, then substituting the approximation for A into formula (9), we obtain an approximation for the net motion associated to the periodic swimming stroke:

$$\bar{P}\exp\int_0^1 A(t)\,dt = I + \int \sum_{m<n} \mathcal{F}_{mn} a_m \dot{a}_n\,dt + O(|a|^3). \tag{14}$$

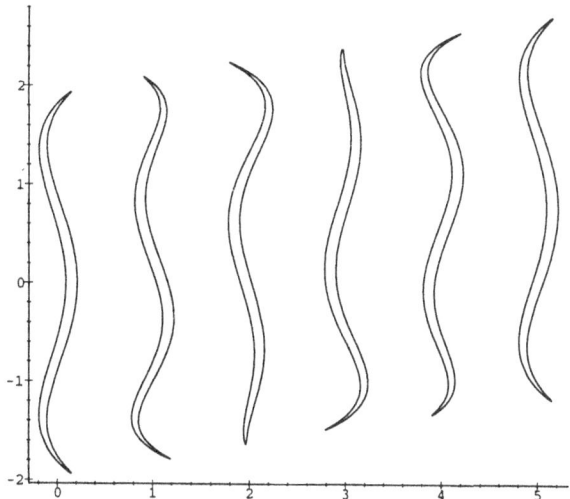

Fig. 1. Tubatrix model at times $t = 0, 0.2, 0.4, 0.6, 0.8$, and 1.0.

So, to describe arbitrary small amplitude swimming motions of a base shape s, we must compute the curvature at a given shape. The resulting matrix encodes the essential dynamics of the problem. To compute the curvature we need the following ingredients:

1. a fixed basis for the vectorfields defined on the exterior of the base shape,
2. solutions to Stokes' equations with boundary conditions specified on the base shape,
3. an expression for the Lie bracket in the appropriate coordinates,
4. an efficient way to compute the net force and torque associated to an arbitrary vectorfield specified on the base shape (the operator $\mathcal{L}$),
5. the vectorfields defined on the exterior of the base shape corresponding to rigid rotations and rigid translations.

2.5. Examples

For two-dimensional problems, complex variable techniques for the biharmonic equation [66], [30] are available. Following the steps above makes the calculations quite algorithmical. One of us (KE) computed swimming strokes for deforming ellipses [28]. In his thesis the curvature for an ellipsoid of revolution and for the axial modes on a cylinder were also investigated. The cylinder problem can be done analytically. Components of the ellipsoid curvature are calculated using Maple©. The curvatures for a number of other examples are described in the next section.

A swimming pattern for a model *Tubatrix aceti* is depicted in Figure 1. To represent this long, thin organism that swims by passing undulatory waves down its body like an eel, one takes a large eccentricity ellipse and deforms it with purely imaginary coefficients. The predicted translation per stroke is 18% of its length. A real *Tubatrix* is slightly slower; it has been observed to translate approximately 16.5% of its length per stroke.

The interested reader may see animation of various swimming models using the software Maple©. Contact the authors by e-mail for the code (kehlers@bert.ucsc.edu, rmont@cats.ucsc.edu, or jair@server02.lncc.br).

3. Traveling Waves on Surfaces and Curvature Asymptotics

Imagine a series of waves of very small amplitude a and very short wavelength λ traveling continuously along a sphere or other surface with wave speed c. With what speed V will the surface translate through the fluid? In this section we will derive the estimate:

$$V \sim \frac{1}{2} c \left(\frac{a}{\lambda} \right)^2 \tag{15}$$

by using the curvature approximation formula. Then we will relate it to the "tangent plane approximation" of Shapere [76].

This estimate for V can be guessed at on dimensional grounds, once we realize that the curvature approximation formula implies that V must be proportional to a^2. It is hidden in formulas of Blake and others for special types of waves on special types of surfaces. (See Brennen [18] for plane waves on planes and Blake for Legendre waves on spheres [16]. These and other examples are worked below.) But we have not found this simple form written down explicitly anywhere. We learned of it from Berg, who in turn learned of it through a recent conversation with Blake. It is a starting point in trying to understand whether it is possible for the mystery swimmer (see section 6.2 below) to swim using this imagined strategy.

The careful reader will have a number of serious questions regarding this whole discussion. What is a "traveling wave on a surface," especially some general asymmetric surface? What is its wave length? Is it a local or global object? Et cetera. We will only touch these questions below. But first we proceed as follows. Represent the surface wave by the vector function $\vec{w}(x, t)$ where x is a material point on the surface boundary and $x + \vec{w}(x, t)$ represents that point at time t. Take some L_2 orthonormal basis $\{\vec{u}_m\}$ for the space of vector functions $\mathcal{T}$ along the surface and expand w in terms of them:

$$\vec{w}(x, t) = \Sigma a_m(t) \vec{u}_m(x),$$

with the condition that

$$\int_0^T a_m(t) \, dt = 0.$$

Also expand the Stokes' curvature in this basis:

$$\mathcal{F} = \Sigma_{m<n} \mathcal{F}_{mn} da_m \wedge da_n.$$

Remember that $\vec{w}$ and hence the a_m are meant to be very small. Then the curvature approximation formula states that to one period T of the wave motion the cell membrane has translated by approximately $\int_0^T \Sigma_{m<n} \mathcal{F}_{mn} a_m(t) \frac{d}{dt} a_n(t) \, dt$. (Hereafter, we only con-

sider the translational part of the curvature.) Hence the mean velocity $\vec{V}$ of its motion is

$$\vec{V} = \frac{1}{T} \int_0^T \Sigma_{m<n} \mathcal{F}_{mn} a_m(t) \frac{d}{dt} a_n(t) \, dt.$$

Imagine a wave where only two modes are coupled:

$$\vec{w}(x,t) = a\{\cos(\omega t)\vec{u}_m(x) + \sin(\omega t)\vec{u}_n(x)\}.$$

We calculate $\Sigma_{m<n} \mathcal{F}_{mn} a_m \frac{d}{dt} a_n = a^2 \omega \mathcal{F}_{mn} \cos^2(\omega t)$. Since the average of $\cos^2$ is $1/2$, we arrive at the velocity formula:

$$\vec{V} = \frac{1}{2} a^2 \omega \mathcal{F}_{mn}.$$

On the other hand, the wave speed c of a true plane wave is related to its wavelength and frequency by $c = \omega \lambda$. Then our original estimate for V becomes $V \sim \frac{1}{2} a^2 \omega \frac{1}{\lambda}$. Comparing formulae, we obtain

$$\mathcal{F}_{mn} \sim \frac{1}{\lambda} \tag{16}$$

relating the curvature components to the effective wavelength λ of the waves in our basis. But which λ do we use, the one for m or the one for n? And again, what do we even mean by the wave number for waves on a surface?

Do we have a guess for the leading term? If we scale the whole surface by an amount ℓ, $\mathcal{F}_{mn}$ scales by ℓ^{-1}. This suggests that the leading term is related to the diameter of the body. (The answer does depend on dimension; for plane swimming $\mathcal{F}_{mn}$ is invariant under rescaling.)

This is a good time to go to the examples.

3.1. The Swimming Slab

(See [24], especially equation (3.14), p. 26, [76, ch. 4], [18], and [28, final section].)

Plane waves of a fixed wavelength $2\pi/k$ travel along an infinite sheet. The sheet is represented by the xy plane and the half-space $z \le 0$ represents the body of the "microorganism." The waves travel in the x direction and we will suppose that they have no components in the y direction. Then an appropriate basis is

$$u_m^+ = \cos(mkx)e_1,$$

$$u_m^- = \sin(mkx)e_1,$$

$$v_m^+ = \cos(mkx)e_3,$$

$$v_m^- = \sin(mkx)e_3.$$

The curvature entries involving different modes m and n with $m \neq n$ are all zero. So the curvature comes in 4×4 blocks indexed by the mode m. Remember the curvature is a vector-valued matrix. In this case its components are all proportional to e_1, the direction of wave propagation, so we will drop the vector index and think of the curvature as scalar-valued. Relative to the basis above this matrix is mk times a matrix whose entries consist solely of 1s, 0s, and -1s.

Take a traveling wave:

$$\bar{w}(x, t) = a \cos(kx - \omega t)e_3$$

representing a vertical deformation. Expanding the cosine we find that $w = a\{\cos(\omega t)v_1^+ (x) + \sin(\omega t)v_1^-(x)\}$, precisely the above form. Using the above formula for the velocity we find that it is

$$V = \frac{1}{2}a^2\omega k.$$

This is a special case of formula (3.14) of Childress's book [24, p. 26].

3.2. Circles and Cylinders

(See [15], [75], and [28].)

The surface is represented by the interior of a circle or cylinder. In the case of the cylinder we only allow deformations that are independent of the axial (z) coordinate and that have no component in this direction. Then the cylinder problem reduces to that of the circle. Identify vectors in the plane with complex numbers. Then an appropriate basis for the deformations is

$$u_m(\theta) = e^{im\theta}$$

and

$$v_m(\theta) = i e^{im\theta}.$$

The only nonzero curvature components are of the form $\mathcal{F}(u_m, v_n)$ with $m - n = \pm 1$, and they are $\pm i2m$. Again $\mathcal{F}_{mn} \sim \frac{1}{\lambda} \sim m$.

3.3. The Sphere

(See [16], [75], and [28].)

The surface is the unit sphere. The appropriate basis for surface deformations is the vector spherical harmonics, written $Y_{J,l,m}$. They form an L_2-orthonormal basis of eigenfunctions for the vector-valued Laplacian. The index J represents the total angular momentum (as in quantum mechanics) and is asymptotic to the square root of the eigenvalue of the spherical Laplacian. The only modes that couple in the curvature tensor are those with the same J and for which one of the indices m and l differ by 1. The corresponding curvature components are asymptotic to a constant times J.

The simplest case is that of purely radial deformations which only depend on the latitudinal angle θ. These can be described by Legendre polynomials P_n. When normalized this yields the basis

$$\left\{ \sqrt{\frac{4\pi}{2n + 1}} P_n(\cos(\theta))e_r \right\}.$$

The only nonzero components are $\mathcal{F}_{n,n+1}$ and these are asymptotic to $\frac{-1}{4\pi}n$ (cf. pp. 60–61 of Shapere's thesis).

3.4. The Ellipse

One of us (Ehlers) found the curvature at an ellipse using complex variable techniques. Under the conformal mapping

$$w(\xi) = R\left(\xi + \frac{\epsilon}{\xi}\right)$$

the unit circle gets mapped to an ellipse with major and minor axes having lengths $2R(1 + \epsilon)$ and $2R(1 - \epsilon)$. We use the basis

$$u_n(w) = e^{in\xi}, \qquad w = w(\xi), \qquad |\xi| = 1,$$

$$v_n(w) = i e^{in\xi}, \qquad w = w(\xi), \qquad |\xi| = 1,$$

for deformations of the ellipse. (Warning: we have shifted our index by 1 compared to [28].) It is convenient to combine the u and v in expressing the curvature using the fact that $v_n = i u_n$. For α, β complex numbers write

$$\mathcal{F}(\alpha u_m, \beta u_n) = \alpha\beta\mathcal{F}_{mn} + \bar\alpha\beta\mathcal{F}_{\bar m n} + \alpha\bar\beta\mathcal{F}_{m\bar n} + \bar\alpha\bar\beta\mathcal{F}_{\bar m\bar n},$$

thus defining the complex curvature components $\mathcal{F}_{mn}, \mathcal{F}_{\bar m n}$, etc. (Of course they are easily related to the usual $\mathcal{F}(u_m, u_n), \mathcal{F}(u_m, v_n)$, etc.) The result is that

$$\mathcal{F}_{m\bar n} = m\epsilon^l, \qquad m - n = 2l + 1 > 0, \qquad m, n > 0.$$

The components with $m - n$ even and the $\mathcal{F}_{mn}, \mathcal{F}_{mn}$ are zero. $\mathcal{F}_{\bar m n}$ can be obtained from $\mathcal{F}_{m\bar n}$ by skew-symmetry. Note that again

$$\mathcal{F}_{mn} \sim \min\{m, n\}.$$

3.5. Discussion and Problems

In all these cases the nonzero entries of the curvature go as

$$|\mathcal{F}_{mn}| \sim |m|$$

for $|m|, |n| \gg 1$, provided the basis $\{\bar u_m\}$ is appropriately normalized and the absolute value $|m|$ is appropriately interpreted (for the sphere $|(J, l, m)| = |J|$). In the symmetric examples of the plane, circle, and sphere, the curvature matrix breaks up into small blocks, no more than 4×4. For the ellipses infinitely many modes enter, but still $|\mathcal{F}_{mn}| \le \min\{|m|, |n|\}$. This suggest that the wavelength λ corresponds to the mode number m. In the symmetric examples above the basis we used was a normalized eigenbasis for the Laplacian on that surface. More precisely, each component of the vectorfield along the surface is an eigenfunction of the Laplacian. The mode numbers were proportional, for large m, to the square root of the corresponding eigenvalue $E(m)$ for the Laplace operator.

3.6. Spectral Methods

In the examples above the eigenbasis for the surface deformations that we used were taken from the Laplacian operator. A more natural operator is the stress tensor σ contracted with the normal vector $\vec{n}$ to the surface boundary. Fix a located boundary shape $\mathcal{B}$, which we think of as a surface in space representing the cell membrane. (See Section 2.3.) Recall (Section 2.3) that the tangent space at $\mathcal{B}$ to our configuration space $\mathcal{Q}$ of all located shapes is the linear space $T_\mathcal{B}\mathcal{Q}$ consisting of all vector-valued maps $\mathcal{B} \to \mathbb{R}^3$. The stress tensor as defined by equation (4) and the Stokes' extension $\vec{v}$ of the deformation $\vec{V}$ defines a linear operator $\mathcal{P}\colon T \to T$ by mapping $\vec{V}$ to

$$\mathcal{P}(\vec{V}) =: \sigma_{|\mathcal{B}} \cdot \vec{n}.$$

Here $\vec{n}$ is the unit normal vector to the surface $\mathcal{B}$. Physically, $\mathcal{P}(\vec{V})(x)$ is the force per unit area that the fluid exerts on the boundary at x, given that the boundary deforms according to $\vec{V}$.

The bilinear form

$$(W, V) = \int_\mathcal{B} \vec{W} \cdot \mathcal{P}\vec{V} \, d^2s$$

defines a Riemannian metric on our configuration space $\mathcal{Q}$. This remarkable fact will be discussed and proved in a forthcoming paper. The horizontal space, or physically allowable (zero force and torque) deformations, and the "vertical space" (span of the infinitesimal rigid motions $\vec{X}$) are orthogonal complements to each other relative to this metric. It follows that the Stokes' connection is a "mechanical" or "Riemannian submersion" induced connection, just like the connection for the falling cat.

A fundamental fact is that the operator $\mathcal{P}$ is symmetric and positive definite. (This result was published by H. A. Lorentz in 1906). See [40, Section 3.5] for a proof in a slightly more general situation. Positive definiteness is seen physically by observing that

$$E(\vec{V}) = \int \vec{V} \cdot \mathcal{P}(\vec{V})$$

represents the mechanical power expenditure due to the deformation $\vec{V}$. The propulsion operator $\mathcal{L}$ (see Section 2.2), which is the basic ingredient of the Stokes' connection, can be written in the form $\mathcal{L}(\vec{V}) = (\vec{F}(\vec{V}), \vec{T}(\vec{V}))$ where

$$\vec{F} = -\int \mathcal{P}(\vec{V}) \, dS \tag{17}$$

represents the total force and

$$\vec{T} = -\int \vec{r}_0 \times \mathcal{P}(\vec{V}) \, dS \tag{18}$$

the total torque. See Section 2.2.

Some thought suggests that a "Poincaré inequality" of the form

$$|\mathcal{F}(\vec{u}, \vec{v})| \leq \|\vec{u}\| \|\mathcal{P}(\vec{v})\| + \|\vec{v}\| \|\mathcal{P}(\vec{u})\|$$

ought to be true. Here the norms of deformation vectorfields, $\vec{u}$, $\vec{v}$, are their L_2 norms defined by integration over the surface. Let $\{\vec{u}_m\}$ be an L_2 *orthonormal basis* for the space of vector functions on the surface, consisting of the eigenvectors of $\mathcal{P}$:

$$\mathcal{P}(\vec{u}_m) = \lambda_m \vec{u}_m,$$

with corresponding real positive eigenvalues λ_m. Such a "Poincaré" inequality would yield the curvature bounds

$$\mathcal{F}_{mn} \leq \lambda_m + \lambda_n$$

for the components of the curvature $\mathcal{F}$ relative to this basis.

Remark. This hypothetical curvature bound and the calculations of the above examples suggest that (at least for those examples) there will be a simple relationship between $\mathcal{P}$ and the "godfather of all operators," the Laplacian.

3.6.1. A Spectral Algorithm. We now present a theoretical algorithm for solving the Stokesian swimming based on the spectral analysis of $\mathcal{P}$. The crux of the Stokesian swimming problem is to find an effective means of computing (17) and (18) for every trial boundary condition on every located boundary shape $\mathcal{B}$. The spectral decomposition of $\mathcal{P}$ may provide such a means, and it is the heart of this algorithm. As far as we know, this proposed algorithm is new.

Step 1. Find $\{\vec{u}_m\}$, an L_2 *orthonormal basis* for the space of vector functions on the surface, consisting of the eigenvectors of $\mathcal{P}$. Also find the corresponding real positive eigenvalues λ_m.

Numerically, this step can be implemented via singularity collocation methods [39], [83]. Peskin's method [69] is another interesting possibiliy, since it yields $\vec{v}_m$, given $\mathcal{P}$.

Step 2. Define the "linear" and "angular momentum" vectors of a given trial boundary vectorfield $\vec{V}$

$$\vec{L}(\vec{V}) = \int \vec{V} \, dS, \tag{19}$$

$$\vec{M}(\vec{V}) = \int \vec{r}_0 \times \vec{V} \, dS, \tag{20}$$

so that

$$\vec{F}_m = \vec{F}(\vec{u}_m) = -\lambda_m \vec{L}\vec{u}_m,$$

$$\vec{T}_m = \vec{T}(\vec{u}_m) = -\lambda_m \vec{M}\vec{u}_m,$$

are the associated forces and torques for our eigenbasis. See Section 2.2.

This illustrates that in some sense the eigenvectorfields are "principal directions" for Stokesian deformations.

Step 3. Expand the given trial vectorfield $\vec{V}$ in terms of the eigenbasis:

$$\vec{V}(x) = \Sigma a_m \vec{u}_m(x). \tag{21}$$

Then, by linearity,

$$\vec{F}(\vec{V}) = \Sigma a_m \vec{F}_m, \qquad \vec{T}(\vec{V}) = \Sigma a_m \vec{T}_m. \tag{22}$$

Step 4. Compute the *resistance matrix* $\mathcal{G}$ of the shape $\mathcal{B}$. See Section 2.2. (Recall that this transformation is obtained by applying $\mathcal{L}$ to infinitesimal translations and rotations restricted to the boundary, so it can be computed, for example, by applying step 3 to these translation and rotation vectorfields as trials.)

Step 5. Invert $\mathcal{G}$ and then either apply it directly to $(\vec{F}, \vec{T})$ of step 3, to obtain $X = A(V)$, or apply it to the $(\vec{F}_m, \vec{T}_m)$ of step 2 to obtain the Lie-algebra elements

$$X_m = \mathcal{G}^{-1}(\vec{F}_m, \ \vec{T}_m) = -\lambda_m \mathcal{G}^{-1}(\vec{L}_m, \ \vec{M}_m) \tag{23}$$

and hence the complete connection one-form at our given shape: $A = \Sigma X_m d\vec{u}_m$. We have the linear relation

$$X = \Sigma a_m X_m.$$

The Stokes' flow for X is the "counterflow" associated to the trial $\vec{V}$ which gives the actual flow: $\vec{v} = \vec{V} - \hat{X}$. *This ends the algorithm.*

Remarks. The mechanical power expenditure of the trial vector (21) is

$$E = \Sigma \lambda_m a_m^2. \tag{24}$$

The power expenditure of the corrected vectorfield is

$$\begin{aligned} E &= \langle \mathcal{P}(\vec{v}), \vec{v} \rangle \\ &= \langle \mathcal{P}(\vec{V}), \vec{V} \rangle - \vec{F} \cdot \vec{U} + \vec{T} \cdot \vec{\omega}, \end{aligned}$$

where $\vec{v} = \vec{V} - X$ is horizontal, $X = (\vec{U}, \vec{\omega})$ is the rigid motion producing the counter-flow, and $\mathcal{L}(\vec{V}) = (\vec{F}, \vec{T})$. Equation (23) seems to indicate that higher modes are more effective for swimming. However, (24) shows they consume more power.

The algorithm may be appended to give a possible alternative to calculating the curvature.

In a future publication we plan to present examples where the spectral decomposition of $\mathcal{P}$ can be obtained analytically. For the sphere of radius a we have found its spectrum to be given by (here $n = -2, -3, \dots$; χ_n, φ_n, p_n are solid spheric harmonics of order n)

Family 1: $\lambda_n^{(1)} = \mu(n - 1)/a$, $v_n = \nabla \times (\chi_n \vec{r})$;
Family 2: $\lambda_n^{(2)} = 2\mu(n - 1)/a$, $v_n^{(2)} = \nabla \varphi_n$;
Family 3: $\lambda_n^{(3)} = \frac{2\mu}{a}(2n + 4 + 3/n)$.

The eigenvector is

$$v_n^{(3)} = \frac{r^2 - a^2}{2\mu(2n + 1)} \nabla p_n + \frac{n r^{2n+3}}{(n + 1)(2n + 1)(2n + 3)} \nabla \left(\frac{p_n}{r^{2n+1}} \right).$$

3.7. The Tangent Plane Approximation: An Open Problem

This is taken from Shapere's thesis. Again, imagine a very small-amplitude, short-wavelength deformation of the surface. Follow the following "algorithm":

- Step 1. Near each point P replace the surface by its tangent plane at P and the wave by the corresponding periodic deformation of the plane.
- Step 2. Calculate the flow field and resultant velocity at infinity, $u(P)$ due to this planar deformation.
- Step 3. Replace the original boundary conditions at the surface with the new boundary condition $u(P)$. Solve Stokes' equations. Find the resulting flow at infinity, thus obtaining the net translation by finding the resulting force.

The rationale in this construction is that the wave is so small that infinity for the plane-wave approximation (steps 1 and 2) is right near the actual original surface. The first step is the problematic one. What do we mean by "the corresponding periodic deformation" of the tangent plane? This only appears to make sense asymptotically. Consider a family

$$\vec{w}(x, t; a, \lambda) = a\vec{W}(x, t; \lambda)$$

with a being the amplitude and λ some measure of wavelength as above. Express $\vec{W}$ relative to Gaussian normal coordinates x centered at an arbitrary point x_0 of the surface. Rescale the coordinate x to focus attention on the point of contact. Thus consider $\vec{W}(x_0 + \lambda x, t; \lambda)$. Now hope (or assume) that in the limit the corresponding function of x converges on compact sets to a plane wave or at least to a periodic wave on the tangent plane. The phase, frequency, and amplitude of this wave will vary as x_0 varies.

Perhaps this can be made more rigorous or palatable by multiplying $\vec{W}$ or its rescaled version by cut-off functions and then analyzing the Fourier transforms as $\lambda \to 0$. Some help here might be buried somewhere in the book *Geometric Asymptotics* [38] or in the larger pseudodifferential calculus literature.

Part II: Social Life at Low Reynolds Number

On the basis of organization we may regard the [protozoan] cell as a huge computer on a molecular scale, with a very broad and complicated program.

—T. L. Jahn and E. C. Bovee

At some unrecorded point in time, perhaps very early in the history of protozoa, certain of the one-celled animals began to form societies of cells—communities in which different cells had different functions—and from these societies the metazoa, the many-celled animals, are believed to have evolved.

—Helena Curtis

Protozoa are 'simple' only in that they are generally tiny and have few observable body structures. Physiologically and behaviorally, they are not simple, carrying out all functions carried out by multicellular forms within a restricted framework. They have mechanoreceptors, chemoreceptors, and photoreceptors ... At the time of conjugation in ciliates, or fusion of gametes in flagellates, two individuals come together, and this might suggest some form of mutual or individual attraction.... For certain peritrichous ciliates, the microconjugant, which is free-swimming, can identify the macroconjugant, which is sessile, by chemicals given off by the latter. When a microconjugant passes within one millimeter of a macroconjugant (a considerable distance in its size scale), it stops and undertakes a search until it finds the other.

—Hubert and Mable Frings

4. Nonholonomic Path Planning Inside Stokes' Flows

It may be a good idea to begin with an example.

4.1. Can Scallops Swim at Zero Reynolds Number?

A *scallop* is any Stokesian free swimmer whose shape can be parameterized by a line segment (one-dimensional). The "scallop theorem"[9] states that scallops are non-swimmers. If the shape is changed and returned to the original by the reverse path (reparameterizations not only allowed but encouraged) the initial and end states coincide. A formal statement can be found in Childress [24].

A basic fact stressed *ad nauseum* is that the shape space of a successful Stokesian swimmer must be at least *two*-dimensional. A swimming sequence of a two-hinged organism is illustrated in [71]. We now know[10] why it swims from left to right (a challenge in Purcell's paper).

It turns out that, after all, scallops may swim *if* there are two or more of them, and *if they cooperate*. This is one of the main points of this second half of our paper. The reader is challenged to provide a choreography. Scallops a (Alice) and b (Bob) are ready to perform their *pas-de-deux* in Figure 2.

4.2. General Framework

Let N organisms $\mathcal{R}_j$ with deformable boundaries $\mathcal{B}_j$ swim at zero Reynolds number.[11] The *grand configuration space* $\mathcal{Q} = \mathcal{Q}_1 \times \cdots \times \mathcal{Q}_N$ consists of N-tuples of maps $q_j: S^2 \to \mathcal{B}_j$ from the sphere to the Euclidean space, describing the located shape of each body. More precisely, the map gives the location of every material particle on the boundary. We mark a special point on every organism (a "nose").

4.2.1. A Fiber Bundle Festival. The Euclidean group $G = SE(3)$ acts on the right on the q_j by composition and hence defines a diagonal action on $\mathcal{Q}$. The quotient space $\mathcal{S}$ is

---

[9] The name is due to Purcell [71].

[10] We thank Rubens Sampaio Filho.

[11] Microorganisms may divide or conjugate, so strictly speaking N may vary in time.

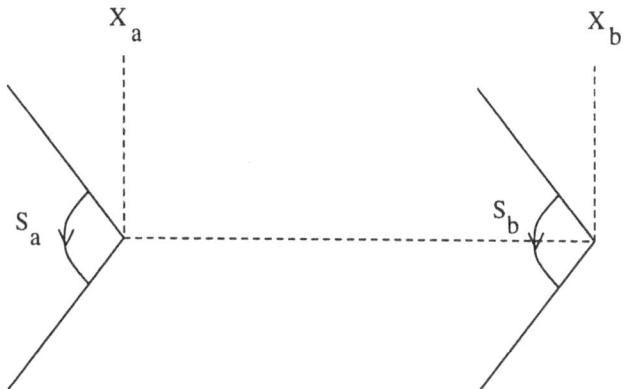

Fig. 2. Scallops Alice and Bob swimming happily.

called the *overall shape space*. We take the diagonal action because the relative positions and orientations of the bodies matter (otherwise there would be no collective swimming effects). Let $S_j = Q_j/SE(3)$ be the individual shape spaces. Notice that S is a principal bundle over $S_1 \times \cdots \times S_N$ with fiber $SE(3)^{N-1}$.

(i) Suppose that there is a collective "intelligence" that simultaneously controls all intrinsic shape changes, $\dot{s} = (\dot{s}_1, \ldots, \dot{s}_N)$. Since we marked a nose on each body, we have a fiber bundle $Q \to S_1 \times \cdots \times S_N$. It is a principal bundle whose fiber is the direct product $SE(3) \times \cdots \times SE(3)$ (N-copies). Each $SE(3)$ acts on its own organism alone. We will see that *collective swimming is not described by a principal bundle connection; rather, it is given by an Ehresman connection*. The distribution has codimension $6N$ as expected, but it is not $(SE(3))^N$ equivariant. It is equivariant only under the diagonal $SE(3)$ action.

(ii) Another point of view is more "individualistic." Each body j controls *only* its own boundary shape deformation $\dot{s}_j$. Ignoring the presence of other bodies, we define N principal bundles $\pi_j : Q_j \to S_j$ with connection forms as in Part 1. The total space Q is the product of the Q_j.

From these individual connections we can reconstruct the global Ehresman connection, using the reflection method [40] to account for the mutual effect of the organisms. We next describe this connection from both points of view, (i) and (ii), and show that they are equal.

4.2.2. The Connection Form: Viewpoint (i).

Associated to a set of trial boundary conditions, there is a unique Stokes' flow external to the union of the domains. The *grand propulsion operator* $\mathcal{L}_{\text{grand}}$ has now a range of N forces and N torques, whose calculation again requires only the first-order jet of the Stokes' flow computed along the boundaries. Restricted to N (generally distinct) infinitesimal rigid motions as boundary conditions, one gets a $6N \times 6N$ grand-resistance matrix $\mathcal{G}_{\text{grand}}$ [40]. The connection form is $A_{\text{grand}} = \mathcal{G}_{\text{grand}}^{-1} \mathcal{L}_{\text{grand}}$. The corrected boundary velocities are given by

$$(\vec{u}_1|_{B_1}, \ldots, \vec{u}_N|_{B_N}) = (I - A_{\text{grand}})(\vec{U}_1|_{B_1}, \ldots, \vec{U}_N|_{B_N}).$$

where the $\vec{U}_j$ are arbitrary trial vectorfields on the boundaries consistent with the intrinsic shape changes.

4.2.3. The Connection Form: Viewpoint (ii).

We now present the connection form, viewpoint (ii). Although it will be equal to (i), its ingredients are useful for comparing the collective swimming against the individual efforts. With several organisms interacting, there are two additional complications: first, the counterflows $\hat{X}_j$ induce different translations and rotations on every body. Second, the "intrinsic" shape deformation performed by each isolated organism must be modified in order to yield a desired translation (due to the influence of flows generated by the others). To address these two issues, we follow the biological rationale that each organism is a separate processor, independently making its "decentralized decisions" (but taking into consideration the "social" environment).

We call $\vec{v}_j|_{B_j}$ a *virtual* located shape deformation (of the jth organism). Solving Stokes' equations for the whole space external to $\mathcal{R}_j$ (ignoring the other organisms, that is filling them in with fluid) produces a fluid flow that we also denote by $\vec{v}_j$. In the geometric language, this defines a connection on the jth bundle $\mathcal{Q}_j \rightarrow \mathcal{S}_j$. We repeat: the idea behind this construction is the following: *if* all the other bodies $k \neq j$ could correctly "guess" the vectorfield $\vec{v}_j$ and choose to change shape accordingly, then the jth organism would not be able to tell (hydrodynamically) that the others were present and could proceed as if they were not.

For microorganisms close enough to interact hydrodynamically, virtual shape deformations are "adjusted" to result in prescribed actual deformations $\vec{u}_k|_{B_k}$. In fact, besides its own self-generated virtual motion, the kth organism is under the influence of ("feels") the flows generated by all others. Hence its *actual* deformation is given by

$$\vec{u}_k|_{B_k} = \vec{v}_k|_{B_k} + \sum_{j \neq k} v_j|_{B_k}. \tag{25}$$

It is important to verify that the off-diagonal terms (i.e., the other organism influences) $j \neq k$ produce no net force and torque on the kth organism. This follows directly from the fact that the stress tensor for $\vec{v}_k$ is divergence-free, on the exterior to $\mathcal{R}_k$, the region of organism k.

Uniqueness theorems guarantee that (i) and (ii) produce the same solution. In practice, the extra work needed to compute the "grand" operators is equivalent (or worse) to the functional inversion involved in the iterative process.

We anticipate at this point a biological question:

Problem. Protozoa have a sense of "touch," i.e., have mechanical pressure transducers. Can they "compute" the virtual deformations (the modified shape deformations) whose implementation will result in true deformations?

4.3. Discussion

Suppose organism 1 wants to translate (and rotate) in the direction X_1 ($X_1 \in se(3)$). Let v_1 be a shape deformation that would yield X_1, *given that no other organisms are present*. Repeat this process for organisms 2 through n, arriving at shape deformations $v_1, \ldots, v_n$.

Due to interactions between the flows the organisms will *not* move in direction X_i. The fields u_i of equation (25) represent the two conditions. They result instead in motions Y_i. (As the distance between organisms tends to infinity, the Y_i tend to the X_i.) On the other hand, v_i is what organism i actually controls. So we are left with the following inversion problem: given that the Stokes' flow u, with $u_{|B} = u_i$, leads to desired motions Y_i, find individual deformations v_i which lead, through equation (25), to the flow u. This is a problem of functional inversion. In control theoretical jargon, the inverse operator is called a *feedback control transformation*. In low Reynolds hydrodynamics literature, *reflection* methods [40] implement the functional inversion by an iterative process. In computer science terminology, each organism acts as a separate processor, working in parallel; interaction is implemented via the off-diagonal terms, or an equivalent domain-decomposition technique.

It remains to be seen whether these metaphors can be biologically meaningful. We summarize them in the words of Andrew Klinger:

> Not only mathematical terms but also problem solving methods are inevitably modeled on Nature. Cooperation may not require direct communication. Domain decomposition could indeed be (our description of) the process, with no need for direct links. Assuming that the time scale for the individual organism 'awareness' to be small compared to the time scale of the actual motion, the collectivity may quite efficiently achieve 'virtual communication' by an iterative process of solving a local problem and then comparing the immediate results with the goals.

4.4. Collective Path Planning in Stokes' Flows

"The study of nonholonomic motion planning is in its infancy." (S. Sastry, R. Murray)

Nature has a 4 billion year expertise on nonholonomic control theory.

—the authors

Nonholonomic motion planning is a recent area of nonlinear control theory stimulated by robotics [53]. Some applications are airport baggage-trailer parking, satellite attitude control, multifingered manipulators, and ultrasonic motors. Recall the key notions. Let y denote n-dimensional states and r m-dimensional control variables ($m = 2$ typically). The states obey an ODE of the form

$$\dot{y} = \sum_{j=1}^{m} r_j Y_j(y). \tag{26}$$

Suppose that for every y, the Lie algebra spanned by the vectorfields Y_j, under Lie bracketing, spans the tangent space at y. By Chow's theorem [25] the system is controllable, in the sense that any state can be reached. One needs more or fewer "zig-zags" to steer toward the goal, depending upon the Lie algebra structure. Recently, several theoretical results have been obtained to quantify this notion and to obtain normal forms useful for engineering practice [53]. Control problems on a principal G-bundle were studied by Montgomery [63], [64].

Our viewpoint is that equation (25) may also be interpreted as a nonlinear control system of type (26), with the individual shape deformations $r_j = \dot{s}_j$ as controls. Thus Nature has a 3.6 (plus) billion year expertise on nonholonomic control theory. The control vectorfields are lifted by the linear connection operator to induced vectorfields along the boundaries of all the bodies, implementing the state changes.

Problem. Do some microorganisms actually perform collective path planning strategies to achieve a desired location? Can optimality criteria be defined? Do they play differential games (predator vs. prey)?

We have here a more general form of control problem on principal bundles. To finish this section, it is appropriate to distill an abstract nonsense summary. The following is a setting for "collective control problems" (generalizing [64]).

Collective Connections and Variational Problems. Let $(G_i \hookrightarrow Q_i \to S_i, A_i)$, $i = 1, \ldots, N$, principal bundles with connection forms A_i. For simplicity take all G_i equal to a same group G. Let $Q \doteq Q_1 \times \cdots \times Q_N$ and π_j the projection over Q_j. Denote by $\text{hor}_i(\dot{s}_i^{\text{virt}})$ the horizontal lift operator of the ith connection. We assume that hor_i can be extended to $\text{Hor}_i(\dot{s}_i^{\text{virt}}) \in T_q Q$, equivariantly under the diagonal action of G, in such a way that $A_j(\pi_{*j}\text{Hor}_i(\bullet)) = 0$. Call $\dot{s}_i^{\text{virt}}$ a *virtual* rate of shape change. Dynamics in the state space Q is governed by the *joint* lift of $\dot{s}_i^{\text{virt}}$, i.e.,

$$\dot{q} = \sum_{i=1}^{N} \text{Hor}_i(\dot{s}_i^{\text{virt}}). \tag{27}$$

The constraints $A_i(\pi_{*i}\dot{q}) = 0$, $i = 1, \ldots, N$, are satisfied, but there is a subtle point here: $\pi_i|_*(\dot{q})$ is in general different from $\dot{s}_i^{\text{virt}}$. So we cannot event start lifting a virtual curve $s^{\text{virt}}(t)$, because the actual shapes would change differently! We circumvent this difficulty by banishing virtual curves and allowing only the symbol $\dot{s}^{\text{virt}}$—regarded as "quasi-controls." A *control protocol* is a curve $s(t) = (s_1(t), \ldots, s_N(t))$. The N equations $\dot{s}_i =: \pi_i|_*(\dot{q})$ bring into stage the *observability* issue. We must assume that the linear mapping $(\dot{s}_1^{\text{virt}}, \ldots, \dot{s}_N^{\text{virt}}) \to (\dot{s}_1, \ldots, \dot{s}_N)$ is invertible. The inverse is called the "feedback transformation," relating the true shape rates to "quasi-controls." Control curve $s(t)$ is lifted to $q(t)$ via (27).

Now, suppose Q has a G-invariant Riemannian metric g. Defining length of curves $(s_1(t), \ldots, s_N(t))$ as $\int \sqrt{g(\dot{q}, \dot{q})}\, dt$ produces a sub-Riemannian metric on Q. It is still sub-Riemannian in Q/G (with the diagonal action).

5. Toy Examples

The reasonable microbiologist adapts himself to the microbial world. The unreasonable one persists in trying to adapt the microbial world to himself. Therefore, all progress depends on the unreasonable microbe.

—D. Mirelman

5.1. Can Scallops Swim at Zero Reynolds Number?

We now sketch some calculations for our problem (Section 4.1). Suppose that scallop a (Alice) is trying to swim along the x-axis. Let s_a be a parameter describing Alice's shape, x_a her position. Alice's velocity is of the form $v_a = \kappa_a(s_a)\dot{s}_a$. The far field generated by Alice is of a dipole, whose associated stokeslets have opposite strength $4\pi\mu\varphi_a(s_a)\dot{s}_a$. Functions κ_a and φ_a encode Alice's hydrodynamic characteristics.

Suppose that another scallop b (Bob) is around. The joint states of Alice and Bob are represented by $y = (s_a, s_b, x_a, x_b)$, and the control system is given by

$$\dot{y} = \dot{s}_a Y_a(y) + \dot{s}_b Y_b(y) \tag{28}$$

with

$$Y_a = \left(1, 0, \kappa_a(s_a), \frac{\varphi_a(s_a)}{(x_b - x_a)^2}\right), \tag{29}$$

$$Y_b = \left(0, 1, \frac{\varphi_b(s_b)}{(x_b - x_a)^2}, \kappa_b(s_b)\right). \tag{30}$$

Exercise. Compute the Lie brackets of Y_a and Y_b and check that the system is controllable.

We observe that a variational problem, leading to a sub-Riemannian geometry is defined by the mechanical power expenditure. This is of the form

$$g = g_a(s_a, \ell)\dot{s}_a^2 + g_b(s_b, \ell)\dot{s}_b^2, \qquad \ell = |x_b - x_a|.$$

5.2. Virtual vs. True Velocities on Stokeslets Fields

Consider two stokeslets "a" and "b", point particles subject to external forces $\vec{F}_a, \vec{F}_b$ and having virtual velocities $\vec{V}_a, \vec{V}_b$ respectively. This means that $\vec{F}_a = -\mu K_a \vec{V}_a$ would relate force and velocity at a, if b were not present. Likewise, $\vec{F}_b = -\mu K_b \vec{V}_b$.

Take the origin at a and the positive y-axis along the line-segment $\overline{ab}$ and let ℓ be the distance between the two stokeslets. Denote by $\vec{U}_a = (U_{ax}, U_{ay})$, $\vec{U}_b = (U_{bx}, U_{by})$ the true velocities. From the analytical expression of a stokeslet field one constructs the *mobility matrix*

$$\mathbf{A} = \begin{pmatrix} 1 & 0 & \frac{K_b}{8\pi\ell} & 0 \\ 0 & 1 & 0 & \frac{K_b}{4\pi\ell} \\ \frac{K_a}{8\pi\ell} & 0 & 1 & 0 \\ 0 & \frac{K_a}{4\pi\ell} & 0 & 1 \end{pmatrix} \tag{31}$$

relating the forces to the true velocities, that is,

$$(U_{ax}, U_{ay}, U_{bx}, U_{by})^\dagger = \mathbf{A}\left(\frac{F_{ax}}{-\mu K_a}, \frac{F_{ay}}{-\mu K_a}, \frac{F_{bx}}{-\mu K_b}, \frac{F_{by}}{-\mu K_b}\right)^\dagger.$$

Inverting the mobility matrix one gets the *resistance matrix*, which gives $(\vec{F}_a, \vec{F}_b)$ in terms of the true velocities. Thus

$$\frac{(F_{ax}, F_{ay})}{\mu K_a} = \left(\frac{-U_{ax} + \frac{K_b}{8\pi\ell}U_{bx}}{1 - \frac{K_a K_b}{(8\pi\ell)^2}}, \quad \frac{-U_{ay} + \frac{K_b}{4\pi\ell}U_{by}}{1 - \frac{K_a K_b}{(4\pi\ell)^2}} \right) \tag{32}$$

(just exchange indices for the force in b). This is precisely equation (6.2-15) in [40], obtained via the reflection method (a tricky calculation, where one sums geometric series).

Exercise. The reader is invited to extend (31) to a more general form (31)' where the line connecting the two stokeslets makes an angle α with the x-axis.

5.2.1. The Hydrodynamical Binary. Suppose that "a" and "b" can interact, but there are no external forces. The condition $\vec{F}_{\text{total}} = \vec{F}_a + \vec{F}_b = 0$ yields, after a simple algebra,

$$U_{ax} = -p(\ell)U_{bx}, \qquad U_{ay} = -q(\ell)U_{by} \tag{33}$$

which

$$p = \frac{A + \frac{B}{\ell}}{1 + \frac{B}{\ell}}, \qquad q = \frac{A + \frac{2B}{\ell}}{1 + \frac{2B}{\ell}},$$

and where the parameters are given by $A = K_a/K_b$, $B = K_b/8\pi$.

In general, the segment $\overline{ab}$ makes an angle α with the x-axis. By equivariance, we may apply (33) if we rotate the velocity vectors by $\exp i(\frac{\pi}{2} - \alpha)$. We get a *distribution of planes* in configuration space $\mathbb{R}^2 \times \mathbb{R}^2 = \{(x_a, y_a, x_b, y_b)\}$ given by

$$\begin{aligned} (\sin\alpha U_{ax} - \cos\alpha U_{ay}) &= -p(\ell) \ (\sin\alpha U_{bx} - \cos\alpha U_{by}), \\ (\cos\alpha U_{ax} + \sin\alpha U_{ay}) &= -q(\ell) \ (\cos\alpha U_{bx} + \sin\alpha U_{by}), \end{aligned} \tag{34}$$

where $\ell\cos\alpha = x_b - x_a$, $\ell\sin\alpha = y_b - y_a$, $\ell^2 = (x_b - x_a)^2 + (y_b - y_a)^2$.

Exercise. If $K_a \neq K_b$, the distribution is nonintegrable and of full rank. The whole configuration space is accessible.

We remark that if $K_a = K_b$ then $A = 1$ and the functions F, G become constant, $p = q \equiv 1$. This implies that the distribution, in this case defined by $\mathbf{U}_a = -\mathbf{U}_b$, is obviously integrable. The midpoint between "a" and "b" remains fixed.

5.3. The Hydrodynamical Connection for N Rigid Bodies

5.3.1. The Setting. Consider N rigid bodies moving in a viscous fluid. We assume that they can interact, say, by hydrodynamically negligible "rods," but there are no external forces. The *configuration space* is $Q = SE(3)^N$ (minus collisions), acted diagonally upon by the group of rigid motions $SE(3)$, yielding a principal bundle whose base is the *shape space* $S = Q/SE(3) = SE(3)^{N-1}$.

The principal bundle for the motion of N isotropic *point* particles is just $(\mathbb{R}^3)^N$ acted upon by the group of rigid motions of the space. For $N \geq 3$ this action is generally free; however this is not so for $N = 2$, because rotations along the axis determined by the two particles do not act. We can make the action free by assuming that the particles move on a plane. The configuration space is now four-dimensional, and the group, three-dimensional. The quotient space is parameterized by the distances $\ell > 0$ between the two particles.[12]

5.3.2. Basic Facts. An isolated rigid body, moving in a Stokes' flow, is characterized by 21 parameters (twelve for two symmetric 3×3 matrices K, Ω and nine for a general matrix C). More generally, the *collective* motion of N rigid bodies is characterized by a $6N \times 6N$ symmetric matrix [40, chapter 8]

$$\mathcal{G} = \begin{pmatrix} \mathbf{K} & \mathbf{C}^\dagger \\ \mathbf{C} & \mathbf{\Omega} \end{pmatrix} \tag{35}$$

called the *grand resistance matrix*. $\mathbf{K}$ is formed by N^2 3×3 blocks K_{ij} (likewise $\mathbf{C}$, $\mathbf{\Omega}$). Therefore

$$K_{ij}^\dagger = K_{ji}, \qquad \Omega_{ij}^\dagger = \Omega_{ji}.$$

$\mathcal{G}$ depends upon (i) the shapes and sizes of the particles, and (ii) the distances and orientations with respect to each other. $\mathcal{G}$ relates linearly the *forces and torques acting on each and every one particle* of the system with their *velocities and spins* with respect to the fluid at rest at infinity:

$$(\vec{F}_1, \ldots, \vec{F}_N, \vec{T}_1, \ldots, \vec{T}_N)^\dagger = \mathcal{G}(\vec{U}_1, \ldots, \vec{U}_N, \vec{\omega}_1, \ldots, \vec{\omega}_N)^\dagger. \tag{36}$$

If G is known, we can obtain the total force $\vec{F}$ and torque $\vec{T}$:

$$\vec{F} = \sum_i \vec{F}_i = \sum_j \left(\sum_i K_{ij} \right) \vec{U}_j + \sum_j \left(\sum_i C_{ij} \right) \vec{\omega}_j, \tag{37}$$

$$\vec{T} = \sum_i \vec{T}_i = \sum_j \left(\sum_i C_{ij}^\dagger \right) \vec{U}_j + \sum_j \left(\sum_i \Omega_{ij} \right) \vec{\omega}_j. \tag{38}$$

5.3.3. An Isoholonomic Problem. The grand-resistance matrix has been mainly used for sedimentation studies, where the applied force is gravity. Here we suggest another application:

Robotics model. The bodies are articulated by "invisible rods" (i.e., hydrodynamically negligible) which can control the shape of the configuration.

---

[12] If you have your feet trapped on tar, follow this advice: make $N = 3$ or 4 by putting your hands in too. You should be able to escape!

Since actual motions inside the fluid satisfy the conditions of zero total force and total torque on the system, these conditions define a connection in the bundle. We call it the *hydrodynamical connection of a system of rigid bodies.*

The power, or instantaneous *mechanical energy dissipation rate* E (i.e., the sum of the instantaneous rates at which the stresses acting over the surfaces of the bodies are doing work upon it) is given by

$$
\begin{aligned}
E &= -\sum_i \int_{S_i} d\vec{n}_i \cdot \sigma \cdot \mathbf{v}_i \\
&= -\sum_i (\vec{U}_i, \vec{\omega}_i) \cdot (\vec{F}_i, \vec{T}_i)^\dagger \\
&= \mu(\mathbf{U}, \omega)\mathcal{G}(\mathbf{U}, \omega)^\dagger.
\end{aligned}
\tag{39}
$$

We get an *isoholonomic variational problem* [63] on configuration space Q with Lagrangian $L = E$ given by (39) subject to the constraints $\vec{F} = 0$ and $\vec{T} = 0$.

5.3.4. Propositions. The following assertions must be true (a verification is in order):

(i) The local holonomy algebra of the hydrodynamical connection is all of $se(3)$.

Digression. Guichardet explicitly identified what some call the "natural mechanical connection" associated to N mass points moving in empty space [37]. The configuration space is $Q = (\mathbb{R}^3)^N$. $G = SE(3)$ acts on Q. Here the physical constraints are zero linear and zero angular momentum. The first leads to the conservation of center of mass, which means that *for the Guichardet's connection no holonomy takes place in the translation subgroup.* The usual reduced Guichardet connection is a connection on the $SO(3)$ principal bundle whose total space is the subspace V of $(\mathbb{R}^3)^N$ of configurations whose center of mass is at the origin. (Remove collinear configurations to obtain a true principal bundle.) Guichardet's connection is defined on the reduced configuration space and its local holonomy algebra is all of $so(3)$. This fact is well known to cats, athletes, and astronauts, which are able to perform interesting rotational maneuvers. Note: For molecular chemistry and robotics it should be important to extend Guichardet's connection to the case where instead of point particles one considers spinning rigid bodies, in which case the configuration space becomes $Q = SE(3)^N$.

Returning to the viscous fluid, if the bodies are so far apart that their interaction can be neglected, then G becomes sparse, corresponding (after a trivial rearrangement) to N 6×6 blocks of individual resistance matrices. More precisely, this six-diagonal form corresponds to writing sequentially the forces and torques of each particle on one side of the equation and the velocities and spins on the other. In other words, we assert that:

(ii) The hydrodynamical connection is asymptotically flat.

(iii) As a consequence, in the isoholonomic variational problem, in order for a system of bodies to translate, it may be an advantage to first cluster together to interact.

5.3.5. Specific Examples. Happel-Brenner's treatise [40] is a treasure chest for those interested in concrete examples. For instance:

(i) *Motions of N point particles on a line.* Smoluchowski's formulas in [40, section 8.3] are a starting point to compute the hydrodynamical connection.

(ii) *Two concentric spheres.* The Stokes' flow is given in [40, section 7.8]. A configuration of the system is $(R_1, R_2) \in SO(3) \times SO(3)$, a moving frame for each sphere. Suppose one can control the *relative* attitude $S = R_1 R_2^{-1}$. Given $S^{-1}\dot{S}$, what are the instantaneous angular velocities of the spheres, viewed externally?

(iii) *Towing with stokeslets.* Interpret the forces $\vec{F}_a$, $\vec{F}_b$ in equation (31)′ as the controls, and positions (x_a, y_a, x_b, y_b) as states. For N point particles the controls are forces $\vec{F}_j$, $j = 1, \ldots, N$, some of which can be inactive. Think of marble balls inside a syrup tank, some of which have iron cores and may be controlled by magnetic fields. Is the system of marbles controllable? Can just one iron ball control the whole system?

Remark. Because of the zero force and torque condition on free-swimming organisms, pure stokeslets have been considered "outlaws." Nevertheless we believe we have found a problem for example (iii).[13] See [59] ("Insects set sail": cover of *Nature*, September 28, 1995). In the next section we advertise other real biological problems—we believe the mathematical treatment can be very interesting.

6. A List of Problems

> *E. coli* is an optimist.
>
> —H. Berg

Some of these problems were collected during the winter 1993–1994 when we had a seminar in Santa Cruz, and some during the discussions we had in May 1995, during Howard Berg's visit to Rio. The first two problems are the ones of most interest to Berg and are probably of the greatest biological import.

6.1. Feeding at Low Reynolds Number

In trying to understand the mechanisms and strategies behind the myriad of feeding methods, one expects to uncover interesting interplays between hydrodynamics and diffusion. Here are some of the basic facts. When a low Reynolds number swimmer approaches another object it pushes it away. Objects tend to slide around each other without touching. Activities such as eating (or sex) become very tricky [27], [67].

Diffusion, on the other hand, brings small objects such as nutrients in contact with a microswimmer's surface according to the laws and time scales of Brownian motion. So, in this realm bacteria and protozoa live different realities. Microorganisms on the scale of *E. coli* and below are completely dependent on diffusion [4]. According to Berg, Purcell estimated in a "back-of-the-envelope" calculation that 1 μm objects would have

---

[13] Mathematicians often have a solution looking for a problem.

to expend enough energy to boil water in order to get much use out of using stirring to feed. However when the organism gets even 10 μm bigger, stirring becomes a viable strategy (the essence of the calculation is that stirring efficiency is proportional to R^6).

Berg showed us a video, called "Death at Low Reynolds Number" (music from "Jaws") made for comic relief at conferences, in which rotifers and other animals hunt and eat *E. coli*. The predatory rotifers have an amazingly intricate structure of cilia which brings the *E. coli* into their gut. There is another predator which curls up into a small ball, and then unfurls itself like a frog tongue to suck up *E. coli*.

Berg also has amazing observations [6], [7] on the information processing and feeding strategies of *E. coli*. Their motions consist of a series of "runs"—nearly straight lines and "tumbles"—random changes of direction. Due to Brownian motion, *E. coli* are turned by ninety degrees about every 10 seconds. Any run that lasts longer than this is no longer close to straight. Because of this it does not make sense for *E. coli* to store information about its environment for any longer than 10 seconds. By subjecting *E. coli* to delta function-like pulses of nutrient concentrations and then observing their response, Berg was able to determine that they store information about chemical gradients for about 4 seconds. They integrate this information and it affects their swimming strategy. Essentially, if a chemical concentration is increasing in time, either due to their motion, or because of the biologist's pipette, then the runs become longer. Larger microorganisms may take advantage of a spatial process as well. Slime molds are able to steer toward nutrients. *C. elegans*, which are organisms composed of about 1000 cells and 300 neurons, follow nutritive gradients along spiral paths.

- *Swimming enhanced diffusion.* Can an animal's swimming stroke effect the rate of diffusive uptake of nutrients by an appreciable amount? Purcell's calculation suggests not. Berg and Purcell estimated the effect of swimming by using a diffusion equation with drift term provided by Stokes' flow past a sphere [4]. However, the flow field produced by a microorganism, seen from far away, is that of a Stokes' dipole, not that of flow past a sphere, which is a stokeslet (monopole). Nowadays, accurate description of the near-field of a bacterial motion is possible [72].

What is the effect of using this more accurate drift term? Are certain swimming strategies much better than others in terms of diffusive uptake? If so, has evolution taken advantage of them?

- *Filter feeding.* Small filter feeders clearly use Stokes' or near-Stokesian flow in order to amplify the diffusive uptake of nutrients. There are many strategies for doing this. A classic example of a filter feeder is the tunicate. They spend their life attached to the sea floor pumping sea water through their gut using a layer of densely packed cilia. Their hearts are reversible peristaltic pumps. See [28] for an analysis of these organisms from our point of view. One problem is to find a reasonable measure of efficiency for this strategy and to formulate the corresponding optimal design problems. The solutions can then be compared to what is found in nature.
- *Collar flagellates.* A mathematical theory was developed by Higdon [41].
- *"Suckers".* The ability of flagellar motion to produce pressure gradients for suction is exemplified in *Giardia lamblia*, a most popular intestinal parasite. Giardia uses its sucking disk for attachment into the mucosa [45].

- *Calanoid copepods.* These are planktonic crustaceans very important for the ecological food chain in oceans and lakes. The physiology of copepod feeding is quite intricate. High-speed movies have shown that copepods' appendages form a "dream team" of low Reynolds number basketball players—they are able to generate flows to direct particles of food into their gut [48]. It may be interesting to solve the idealized problem of controlling a particle using two rigid appendages. The Reynolds number range for the appendages of copepods, 10^{-2} to 1, allows for different flow regimes, depending on the particular task at hand [49].
- *Rotifers.* Rotifers look like small vacuum cleaners as they draw *E. coli* into their gut. They generate currents around their oral cavity using a layer of densely packed cilia. It is possible that an envelope model can be used to describe the flows generated by the cilia surrounding the oral cavity of these organisms.

6.2. Mystery Swimmers

Interesting descriptions of microorganism locomotion can be found in [46], [70].[14] A mystery swimmer is a microorganism for which we have no explanation of how they swim. Find a reasonable explanation that is experimentally testable. Here are the two examples we know of.

- *The Synechococcus.* We learned of this swimmer from Berg. This is a bacterium living in sea water, with an important ecological role (carbon cycle). It is about 1 μm in diameter, and swims fast, at 20 μm per sec. It does not change shape, at least in a manner visible to the eye. It has no internal flagella.[15] Motion by electrical means, and by "jets" have both been ruled out. So far, Berg was not able to glue tiny polystyrene balls on its surface. On a related organism, this was done, and one observes a strip of these balls streaming one way on or close to the organism's surface, and right next to them another strip of the balls streaming the opposite direction, as if they were on "conveyor belts." It seems unlikely that surface motion such as this could account for the swimming speeds of synechococcus as this would require it to transport its entire surface area down its length about 20 times a second. One mechanism which has not yet been ruled out is very small high-frequency oscillations of its surface.[16] See the discussion in Section 3.

Remark. Synechococcus is related to certain "gliding organisms" [51]. These organisms latch onto the water surface and move along at a good clip (2–3 μm/sec); when they are separated from the surface they are unable to swim, they only drift, being bounced around erratically by Brownian motion. Perhaps some gliding motions are done via

---

[14] Reynolds numbers seem to be ignored, as late as 1972, in [46].

[15] *Spirochetes* were once mystery swimmers, until it was discovered that they have internal flagella. As they move inside the body, the external sheath rotates in opposite direction. In consequence, there is a translational velocity as well. It should be interesting to work out this calculation in detail. Here's an example where the tail wags the dog!

[16] While this paper was in the editorial process, KE, HB, and RM realized a very simple explanation that may solve the synechococcus mystery [29].

classical nonholonomic mechanics.[17] Motion of bacteria on solid surfaces or in modified fluids (e.g., suspensions of methyl cellulose and/or ficoll to obtain a "non-Newtonian" environment) requires them knowing other physical principles.

- *"Rio's magnetotactic aggregate"* [32]. A multicellular magnetotatic organism lives in the brackish water near the bottom of Rodrigues de Freitas Lagoon in Rio de Janeiro (a few blocks from Ipanema beach). This organism consists of about 20 cells; the overall size is about 4 μm. It swims at 30–100 μm/sec, in quite a distinct manner. No cilia or flagellar bundles are observed. This aggregate could be a bacterial colony, but behaves as an individual organism. Our conjecture is that its motion may be a biological realization of "holonomy drive." The idealized model is a system of points connected by invisible rods, as discussed in the toy examples.

6.3. Studies on Magnetotactic Bacteria

General information on magnetotaxis can be found in [52]. Nogueira and Lins de Barros worked out an *ab initio* model for motion of bacteria (magnetotactic or not) [68]. Numerical solutions of the ODEs reproduce some finer details, which normally have been averaged out. For instance, the helical pattern around a "guiding-center" type averaged trajectory. We believe that their approach can benefit from some mathematical abstraction. In particular, can it be Hamiltonianized? For two or more bacteria, include interaction terms.

6.4. Cooperative Behavior

- Flagella of adjacent spermatozoa or flagellar bundles of adjacent bacteria tend to beat in unison, and they will go visibly faster.[18] Is this a phase-locking phenomenon? Use Stokes' equations to explain the increase in speed.
- Coordination of bacterial flagella in a bundle seems to be mediated hydrodynamically, but the mechanism is not known. In ciliated protozoa, synchronization producing metachronal waves is most likely mechanically induced. Explain the mechanism. Again, could it be a phase-locking phenomenon?[19]
- *Spirila* is an elongated bacteria (about 50 μm) which swims back and forth. The flagellar bundles at the ends alternate clockwise (CW) and counterclockwise (CCW) rotation. The CCW bundle is in the direction of the body; the bundle undergoing CW motion is tilted. How is the switch coordinated?
- Bacterial flagella are rotated by independent motors [43]. What is the synchronization process by which a majority "vote" for CCW rotation?

---

[17] Imagine placing a small colony of these gliders on a wetted sphere, or pseudosphere. In this manner we could study low Reynolds number life in non-Euclidean geometries.

[18] "Taylor calculated the forces acting to synchronize parallel undulating sheets and found these forces to be large for sheet spacings smaller than the wavelength of the undulations" [5].

[19] Cilia axonemes can be removed from the cells (using triton) and glued on glass. Exposed to ATP, metachronal waves are observed!

- Some ciliated microorganisms have an isolated flagellum (Luis Henrique Leal, high-speed movies). The cilia perform an helicoidal wave, and the beating cycle of the flagellum is sharply in 1 : 2 resonance. Why?

6.5. Patterns and Collective Phenomena

- *Traveling bands of bacteria.* First seen just over hundred years ago by Engelmann and Pfeffer, such experiments can be done in science high-school classes. Traveling bands of magnetotactic bacteria, with different speeds, reminiscent to "solitons" have been observed in dense populations [78]. Organisms in a band travel side by side. A plausible hydrodynamical argument was given by D. C. Guell *et al.* [36] and we believe it can be elaborated in more detail.
- *Spatio temporal patterns, self-organization.* These have striking complexity, offering new opportunities for bifurcation theorists and experts on reaction-diffusion PDEs. Some references are [82], [21], [22]. We just mention some observations in [21]: for *E. coli* grown in a semi-solid agar culture, on a 7.5-cm diameter dish, one observed first a sunflower-like array of spots. At smaller radii the spots tend to appear in circles, at larger radii on intersecting clockwise and counterclockwise spirals; other patterns are radial arrays of spots, spots and stripes, spots with radial tails arrayed in chevrons.

6.6. Energy Expenditure

- *The bacterial motor.* Berg has devoted much of his scientific life to the study of the structure and dynamics of the flagellar rotary motors of *E. coli* which he discovered in the early 1970s. We are not able to pose questions on this topic at this point, but we mention some references on this intriguing subject: [13], [8], [9], [10], [11], [12], [47], [60], [56], [61], [58].
- Microorganisms swim for fun and profit, having no energy consumption problems as far as locomotion is concerned.[20] Their motor is powered directly by proton flux [13]. Efficiency is almost 100% at low speeds (in laboratory preparations), but drops to 1% at cruising speed. If energy consumption and efficiency issues arise, perhaps a *Finsler metric* should be chosen as the functional. If several organisms are involved one expects to obtain a variational problem of isoholonomic type involving the many-body connection and the grand resistance matrix. Is this biologically important? Can the problem be derived from first principles?

6.7. Stochastic Processes in Bacterial Motion

H. Berg developed the tracking microscope in the early 1970s, an instrument capable of following the motion of a single bacterium [14]. He found out that *E. coli* (and most other motile bacteria) move in a series of "runs" (lasting on average 1 sec) and "tumbles" (of about 0.1 sec). Runs are longer when the motion is in a positive gradient of a chemoattractant, but do not get shorter when in a negative direction (acuna-matata)!

---

[20] Purcell says it is like driving a Datsun in Saudi Arabia.

- Consider a swarm of *E. coli*, sufficiently dense to interact hydrodynamically. It is plausible that the collective motion in the positive gradient will be enhanced, while the motions in other directions will be averaged out by the hydrodynamic interaction. How dense must the population be in order that this effect could be observed?
- Simultaneous optimization of two or more functions have been studied in Mathematical Economics. When there are two chemoattractants, can "Pareto-Smale sets" be observed?
- When particles are diffusing by Brownian motion, one gets the usual diffusion (heat) equation. Is it helpful to consider the low Reynolds hydrodynamical interaction? Inside a Stokes' flow, does it make sense to consider a subelliptic "grand-Laplacian"? (Its symbol would belong to a codimension-6 distribution in phase space defined by zero total force and torque.)

We would like to end with another quote from E. O. Wilson's *The Naturalist* (p. 123):

I have evolved a rule that has proved useful for myself and might be for others not born with championship potential: for every level of mathematical ability there exists a field of science poorly enough developed to support original theory.

. . .

The advice I give to students in science is to move laterally and up and down and peer all around. If you have the will there is a discipline in which you can succeed. Look for the ones still thinly populated, . . . be a hunter and explorer, not a problem solver . . .

Acknowledgments

We thank the participants of our seminar in Santa Cruz, especially students Andrew Klinger, Junko Hoshi, and Cesar Castilho; Professors T. Y. Wu and C. Brennen for their encouragement; and Howard Berg, for his beautiful talks and the many conversations in Rio. J. K. would like to thank Prof. Manuel de Leon for his hospitality at CSIC (Madrid, 1992), when some ideas started to take form. We also thank Darci Esquivel, Marcos Farina, and Henrique de Barros for providing an enthusiastic biophysical environment during Prof. Berg's visit. J. K. would like to acknowledge support provided by a Guggenheim Foundation fellowship, January–March 1994, to visit the Mathematics Department, University of California at Santa Cruz.

References

[1] Abraham, R. and Marsden, J. E.: *Foundations of Mechanics*, Addison–Wesley, Reading, MA (1978).
[2] Ambrose, W. and Singer, I. M.: A theorem on holonomy, Trans. AMS **75**, 428–453 (1953).
[3] Arnold, V., Kozlov, V. V., and Neishtadt, A. I.: *Dynamical Systems III* , Encyclopaedia of Mathematical Sciences, Springer–Verlag, New York (1988).
[4] Berg, H.C.: *Random Walks in Biology*, expanded edition, Princeton University Press, Princeton, NJ (1993).

[5] Berg, H. C. and Anderson, R. A.: Bacteria swim by rotating their flagellar filaments, Nature **245**, 380–384 (1973).

[6] Berg, H. C. and Purcell, E. M.: Physics of chemoreception. Biophys. J. **20**, 193–219 (1977).

[7] Berg, H. C.: Physics of bacterial chemotaxis, pp. 19–30 in *Sensory Perception and Transduction in Aneural Organisms*, eds. Colombetti, G., Lenci, F., and Song, P. S., Plenum, New York (1985).

[8] Berg, H.C.: Studies of motile bacteria. In *Physics News*, ed. by Schewe, P. F., American Institute of Physics, New York (1991).

[9] Berg, H. C.: Dynamic properties of bacterial flagellar motors. Nature **249**, 77–79 (1974).

[10] Berg, H. C. and Khan, S.: A model for the flagellar rotary motor. In *Mobility and Recognition in Cell Biology*, pp. 485–497, eds. Sund, H. and Veeger, C., deGruyter, Berlin (1983).

[11] Berg, H. C., Manson, M. D., and Conley, M. P.: Dynamics and energetics of flagellar rotation in bacteria, *Symp. Soc. Exp. Biol.* **35**, 1–31 (1982).

[12] Berg, H. C.: Dynamics and energetics of the bacterial rotary motor. In *Protein Dynamics and Energy Transduction*, pp. 312–344, ed. Ishiwata, S.–I., Taniguchi Foundation (1980).

[13] Berg, H.C.: Torque generation by the flagellar rotary motor, *Biophys. J.* **68**, 163s–167s (1995).

[14] Berg, H. C.: How to track bacteria, *Rev. Sci. Instrum.* **42**, 868–871 (1971).

[15] Blake, J. R.: Self propulsion due to oscillations on the surface of a cylinder at low Reynolds number, *Bull. Austral. Math. Soc.* **3**, 255–264 (1971).

[16] Blake, J. R.: A spherical envelope approach to ciliary propulsion, *J. Fluid Mech.* **46**, 199–208 (1971).

[17] Blake, J. R.: A model for the micro–structure in ciliated organisms, *J. Fluid Mech.* **55**, 1–23 (1972).

[18] Brennen, C.: An oscillating boundary layer theory for ciliary propulsion, *J. Fluid Mech.* **65**, 799–824 (1974).

[19] Brennen, C. and Winnet, H.: Fluid mechanics of propulsion by cilia and flagella, *Ann. Rev. Fluid Mech.* **9**, 339–398 (1977).

[20] Brockett, R. W. and Dai, L: Non-holonomic kinematics and the role of elliptic functions in constructive controllability, in *Nonholonomic Motion Planning*, eds. Li, Z., Canny, J. F., Kluwer, Dordrecht (1993).

[21] Budrene, E. O. and Berg, H. C.: Complex patterns formed by motile cells of Escherichia coli, *Nature* **349**, 630–633 (1991).

[22] Budrene, E. O. and Berg, H. C.: Dynamics of formation of symmetrical patterns of chemotactic bacteria, *Nature* **376**, 49–53 (6 July 1995).

[23] Calleja, G. B.: *Microbial Aggregation*, CRC Press, Boca Raton, FL (1984).

[24] Childress, S.: *Mechanics of Swimming and Flying*, Cambridge University Press (1981).

[25] Chow, W. L.: Uber systeme van linearen partiellen differentialgleichungen ersten ordnung, *Math. Ann.* **117**, 98–105 (1939).

[26] Chwang, A. T. and Wu, T. Y.: A note on the helical movement of micro-organisms, *Proc. Roy. Soc. Lond. B.* **178**, 327–346 (1971).

[27] Curtis, H.: *The Marvelous Animals: An Introduction to the Protozoa*, The Natural History Press, New York (1968).

[28] Ehlers, K. M.: *The Geometry of Swimming and Pumping at Low Reynolds Number*, Ph.D. Thesis, University of California, Santa Cruz (1995).

[29] Ehlers, K. M, Berg, H. C., and Montgomery, R.: Do synechococcus swim using traveling surface waves?, Proc. Natl. Acad. Sci. USA **93**, 8340–8344 (1996).

[30] England, H.: *Complex Variable Methods in Elasticity*, Wiley-Interscience, New York (1971).

[31] Fauci, L. J.: Computational modeling of the swimming of biflagellated algal cells, in *Contemp. Math*, **141**, 91–102, ed. A. Y. Cheer, C. P. van Dam (1993).

[32] Farina, M., Esquivel, D. M. S., and Lins de Barros, H. G. P.: Magnetic iron–sulphur crystals from a magnetotactic microorganism, *Nature* **343**, 6255, 256–258 (1990).

[33] Fearing, R. S.: Control of a micro–organism as a prototype micro–robot, *Second Int. Symp. on Micromachines and Human Sciences, Nagoya, Japan, 1991*.

[34] Flam, F.: Swarms of mini-robots set to take on Mars terrain, *Science* **257**, 1621, (18 Sept. 1992).

[35] Greenbaum, A., Greengard, L., and Mayo, A.: On the numerical solution of the biharmonic equation on the plane, *Physica D* **60**(1–4), 216–225 (1992).

[36] Guell, D. C., Brenner, H., Frankel, R. B., and Hartman, H.: Hydrodynamic forces and band formation in swimming magnetotactic bacteria, *J. Theor. Biol.* **135**, 525–542 (1988).

[37] Guichardet, A.: On rotation and vibration motions of molecules, *Ann. Inst. H. Poincaré, Phys. Theor.* **40**(3), 329–342 (1984).

[38] Guillemin, V. and Sternberg, S.: *Geometric Asymptotics*, American Mathematical Society, Providence, RI (1977).

[39] Hasimoto, H. and Sano, H.: Stokeslets and eddies in creeping flow, *Ann. Rev. Fluid Mech.* **12**, 335–364 (1980).

[40] Happel, J. and Brenner, H.: *Low Reynolds Number Hydrodynamics*, Kluwer Academic, Dordrecht (1991).

[41] Higdon, J. J. L.: The generation of feeding currents by flagellar motions, *J. Fluid Mech.* **94**(2), 305–330 (1979).

[42] Hirsch, P.: Microcolony formation and consortia, in *Microbial Adhesion and Aggregation*, ed. K. C. Marshall, pp. 373–393, Springer-Verlag, New York (1984).

[43] Ishihara, A., Segall, J. E., Block, S. M., and Berg, H. C.: Coordination of flagella on filamentous cells of Escherichia coli, *J. Bacteriol.* **155**, 228–237 (1983).

[44] Jahn, T. L., and Bovee, E. C.: Motile behavior of Protozoa, in *Research in Protozoology*, ed. Tze–Tuan Chen, vol. 1, Pergamon, New York (1967).

[45] Jones, R. D., Lemanski, C., and Jones, T. J.: Theory of attachment in Giardia, *Biophys. J.* **44**, 185–190 (1983).

[46] Jahn, T. L. and Votta, J. J.: Locomotion of protozoa, *Ann. Rev. Fluid Mech.* **4**, 93–116 (1972).

[47] Khan, S., Meister, M., and Berg, H. C.: Constraints on flagellar rotation, *J. Mol. Biol.* **184**, 645–656 (1985).

[48] Koehl, M. A. R.: Feeding at low Reynolds number by Copepods, *Lect. Math. Life Sciences* **14**, 89–117 (1981).

[49] Koehl, M. A. R.: Hairy little legs: feeding, smelling and swimming at low Reynolds numbers, in *Contemp. Math.* **141**, 33–64, Fluid dynamics in biology, eds. Cheer, A. Y., van Dam, C. P. (1993).

[50] Keller, S. T. and Wu, T. Y.: A porous prolate–spheroidal model for ciliated micro–organisms, *J. Fluid Mech.* **80**(2), 259–278 (1977).

[51] Lapidus, R. and Berg, H. C.: Gliding motility of Cytophaga sp.Strain U67, *J. Bacteriology* **151**(1), 384–398 (1982).

[52] Lins de Barros, H. G. P., Esquivel, D. M. S., and Farina, M.: Magnetotaxis, *Sci. Progress Oxford* **74**, 347–359 (1990).

[53] Li, Z. and Canny, J. F. (eds.): *Nonholonomic Motion Planning*, Kluwer, Dordrecht (1993).

[54] Lighthill, J.: Biofluiddynamics: A survey, *Contemp. Math.* **141**, 1–23 (1993), eds. Cheer, A. Y., van Dam, C. P.

[55] Lighthill, L.: *Mathematical Biofluidmechanics*, SIAM (1975).

[56] Lowe, G., Meister, M., and Berg, H. C.: Rapid rotation of flagellar bundles in swimming bacteria, *Nature* **325**, 637–640 (1987).

[57] Ludwig, W.: Zur theorie der flimmerbewegung (dynamik, nutzeffekt, energiebilanz), *Z. vergl. Physiol.* **13**, 397–504 (1930).

[58] Manson, M. D., Tedesco, P. M. and Berg, H. C.: Energetics of flagellar rotation in bacteria, *J. Mol. Biol.* **138**, 541–561 (1980).

[59] Marden, J. H. and Kramer, M. G.: Locomotor performance of insects with rudimentary wings, *Nature* **377**, 332–334 (28 Sept. 1995).

[60] Meister, M., Lowe, G., and Berg, H.C.: The proton flux through the bacterial flagellar motor, *Cell* **49**, 643–650 (1987).

[61] Meister, M. and Berg, H. C.: The stall torque of the bacterial flagellar motor, *Biophys. J.* **52**, 413–419 (1987).

[62] Montgomery, R.: Gauge theory of the falling cat, *Fields Institute Communications* **1**, 193–218 (1993).

[63] Montgomery, R.: Isoholonomic problems and some applications, *Commun. Math. Phys.* **128**, 565–592 (1990).

[64] Montgomery, R.: Nonholonomic control and gauge theory, in *Nonholonomic Motion Planning*, eds. Li, Z., Canny, J. F., Kluwer, Dordrecht (1993).

[65] Murray, R. M. and Sastry, S. S.: Steering nonholonomic control systems using sinusoids, in *Nonholonomic Motion Planning*, ed. Li, Z., Canny, J. F., Kluwer, Dordrecht (1993).

[66] Muskhelishvili, N. I.: *Some Basic Problems of the Mathematical Theory of Elasticity*, P. Noordhoff, Groningen, Holland (1953).

[67] Nisbet, B.: *Nutrition and Feeding Strategies in Protozoa*, Croom Helm, London and Camberra (1984).

[68] Nogueira, F. S. and Lins de Barros, H. G. P.: Study on the motion of magnetotactic bacteria, *Eur. Biophys. J.* **24**, 13–21 (1995).

[69] Peskin, C. S., McQueen, D. M.: Computational biofluid dynamics, in *Contemp. Math.* **141**, 161–186, ed. Cheer, A. Y., van Dam, C. P. (1993).

[70] Pedley, T. J. and Kessler, J. O.: Hydrodynamic phenomena in suspensions of swimming microorganisms, *Ann. Rev. Fluid Mech.* **24**, 313–358 (1992).

[71] Purcell, E.: Life at low Reynolds number, *Am. J. Phys.* **45**, 3–11 (1977).

[72] Ramia, M., Tullock, D. L., and Phan–Thien, N.: The role of hydrodynamic interaction in the locomotion of microorganisms, *Biophys. J.* **65**, 755–778 (1993).

[73] Ramia, M. and Swan, M. A.: The swimming of unipolar cells of *spirillum volutans*: Theory and observations, *J. Exp. Biol.* **187**, 75–100 (1994).

[74] Saier, M. H. and Jacobson, G. R.: *The Molecular Basis of Sex and Differentiation*, Springer-Verlag, New York (1984).

[75] Shapere, A. and Wilczek, F.: Geometry of self–propulsion at low Reynolds number, *J. Fluid Mech.* **198**, 557–585 (1989).

[76] Shapere, A.: *Gauge Theory of Deformable Bodies: A Theory of Something*, Ph.D. thesis, Princeton University Physics Dept., Princeton, NJ (1989).

[77] Sizemore, D. R., Branstrom, A., and Sadoff, J. C.: Attenuatd *Shigella* as a DNA delivery vehicle for DNA-mediated immunization, *Science* **270**, 299–302 (13 October 1995).

[78] Spormann, A. M.: Unusual swimming behavior of a magnetotactic bacterium, *FEMS Microb. Ecol.* **45**, 37–45 (1987).

[79] Taylor, G. I.: Analysis of the swimming of microscopic organisms, *Proc. Roy. Soc. Lond. A* **209**, 447–461 (1951).

[80] Taylor, G. I.: The action of waving cylindrical tails in propelling microscopic organisms, *Proc. Roy. Soc. Lond. A* **211**, 225–239 (1952).

[81] Yates, G.: How microorganisms move through water, *Am. Sci.* **74**, 358–365 (1986).

[82] Woodward, D. E., Tyson, R., Myerscough, M. R., Murray, J. D., Budrene, E. O., and Berg, H. C.: Spatio-temporal patterns generated by *Salmonella typhimurium*, *Biophys. J.* **68** 2181–2189 (1995).

[83] Weinbaum, S. and Ganatos, P.: Numerical multipole and boundary integral equation techniques in Stokes flow, *Ann. Rev. Fluid Mech.* **22**, 275–316, 1990.

[84] Wu, T. Y., Brennen, C., and Brokaw, C. (eds.): *Swimming and Flying in Nature*, vol. 1, Plenum, New York (1975).

Symmetry Methods in Collisionless Many-Body Problems

I. Stewart

Mathematics Institute, University of Warwick, Coventry CV4 7AL, UK

Received January 29, 1996; revised manuscript accepted for publication June 6, 1996
Communicated by Jerrold Marsden and Stephen Wiggins

This paper is dedicated to the memory of Juan-Carlos Simo

Summary. We formulate an appropriate symmetry context for studying periodic solutions to equal-mass many-body problems in the plane and 3-space. In a technically tractable but unphysical case (attractive force a smooth function of squared distance, bodies permitted to coincide) we apply the equivariant Moser-Weinstein Theorem of Montaldi *et al.* to prove the existence of various symmetry classes of solutions. In so doing we expoit the direct product structure of the symmetry group and use recent results of Dionne *et al.* on 'C-axial' isotropy subgroups. Along the way we obtain a classification of C-axial subgroups of the symmetric group. The paper concludes with a speculative analysis of a three-dimensional solution to the $2n$-body problem found by Davies *et al.* and some suggestions for further work.

1. Introduction

This paper is a preliminary report on ongoing work to formulate an appropriate symmetry context for equal-mass multi-body dynamics and use it to investigate a wide class of periodic solutions. We derive the consequences of such a formulation for the simplest cases and speculate on possible future developments. In particular we identify several symmetric configurations whose perturbations may be expected to give rise to interesting dynamics, and we briefly discuss how such perturbations might be analysed. We hope to develop these ideas further in future work.

The dynamical behaviour of systems of point particles interacting through central forces has been of interest since the time of Newton, and has given rise to a considerable body of research. In the hands of Poincaré it gave rise to the qualitative theory of ODEs and modern dynamical systems theory, a recent triumph being the proof by Xia [26] that Arnold diffusion—and hence in particular chaos—occurs in the restricted 3-body problem.

Despite these complexities, many results are known for special kinds of motion of an N-body system, in particular *central configurations*; see for example Meyer [17]. It is fairly common to consider the symmetric case of N equal masses, in which case many of the known solutions also possess a degree of symmetry. The aim of this paper is to set up a symmetry-based context for equal-mass N-body problems, both in the plane $\mathbf{R}^2$ and in 3-space $\mathbf{R}^3$, and to illustrate how some simple general principles for equivariant dynamics might be used to organize the known results and—eventually—derive new ones. Because of technical limitations in current techniques, we will make two rather unorthodox assumptions:

Assumption 1. The potential corresponding to each body is a smooth (C^∞) function of (squared) distance.

Assumption 2. Bodies that collide are free to pass through each other, and several distinct bodies may occupy the same spatial position simultaneously.

Assumption 1 rules out, for example, inverse-square law attraction, but it has the advantage of not creating singularities at collisions. Assumption 2 may appear unphysical, but it is central to the approach adopted here, because we will obtain 'interesting' solutions by perturbing the trivial solution in which all N bodies are coincident and stationary. Many of these perturbations do not involve collisions, and thus have physical significance. Those that do involve collisions can be further perturbed in order to remove them.

No attempt is made to "regularize" collisions: they are simply accepted. A similar assumption is common in the dynamics of low-density distributions of celestial bodies such as globular clusters and galaxies; see Binney and Tremaine [3, p. 190]. Here the problem is often formulated using the collisionless Boltzmann equation; note, however, that "collisionless" there carries the further technical implication that each body moves under the influence of the mean field potential of the others. Our model could—stretching a point—be interpreted as describing a system of N nominally identical globular clusters or galaxies, assuming (again unphysically) that these retain their general form if they pass close to or through each other. The Hamiltonian for such clusters can, in a reasonable model, be assumed to obey the smoothness condition (1), even though individual stars exert inverse-square law gravitational attraction, in the same way that Poisson's equation rather than Laplace's applies within a solid body. However, this suggestion is not intended very seriously.

Assumptions 1 and 2 give this paper a somewhat speculative feel: it is an attempt to explore the implications of equivariant dynamics in a technically tractable case. Many of the solutions whose existence will be proved can be obtained by other methods, even when Assumption 1 is relaxed to allow the inverse-square law. In fact existence is often obvious for the simpler solutions, although it can involve heavy computations for the more complicated ones. However, the resulting framework fits together rather elegantly, and it suggests many directions for future work—including relaxing the current sweeping assumptions.

2. Periodic Orbits for Symmetric Hamiltonians

The basic "local bifurcation" existence theorem for periodic orbits in Hamiltonian dynamics is the Liapunov Centre Theorem [16]. Suppose that H is a Hamiltonian on a $2n$-dimensional symplectic manifold P and that $p \in P$ is an equilibrium, so that $DH|_p = 0$. Since we shall work locally near p there will be no loss of generality in taking P to be $\mathbf{R}^{2n} \equiv \mathbf{C}^n$ with the standard symplectic structure, and $p = 0$, but for the moment we stay in the more general setting. Assume that p is a nondegenerate minimum of H, that is, $DH|_p = 0$ and $D^2 H|_p$ is positive definite. Let L be the linearization of the Hamiltonian vector field at p and let the eigenvalues of L be the purely imaginary pairs $\{\pm\lambda_1, \ldots, \pm\lambda_n\}$. Liapunov [16] proved that if some λ_i is *non-resonant* then there exists a smooth two-dimensional submanifold of P, passing through p and intersecting every energy level near p in a periodic orbit, with period approaching $2\pi/|\lambda_i|$ for orbits near p. For a proof see Abraham and Marsden [1]. By "non-resonant" we mean that λ_j is not an integer multiple of λ_i for $j \neq i$.

Various authors showed that the hypothesis of non-resonance could be relaxed in certain special cases. In a celebrated paper Weinstein [25] proved that even when there is resonance, there must exist at least n families of periodic solutions on each energy level near p. The proof was simplified by Moser [21], and the result has come to be known as the Moser-Weinstein Theorem.

Montaldi *et al.* [19] obtained an equivariant version of the Moser-Weinstein Theorem, and this result will be the basis of this paper. Suppose that a compact Lie group Γ acts symplectically on P, let $p \in P$ be a fixed point for Γ, and suppose that the Hamiltonian H is Γ-invariant. This symmetry may force some of the λ_i to be equal, creating unavoidable resonances.

Let $u(t)$ be a periodic orbit of the flow of H having period T. Let $\mathbf{S}^1$ be the circle group, identified with $\mathbf{R}/2\pi\mathbf{Z}$, and define an action of $\Gamma \times \mathbf{S}^1$ on the *loop space* $C^k(T)$ of k-times differentiable T-periodic functions $u = u(t)$ by

$$(\gamma, \theta).u(t) = \gamma u(t + T\theta/2\pi).$$

Define the *symmetry group* of $u \in C^k(T)$ to be

$$\Sigma_u = \{(\gamma, \theta) \in \Gamma \times \mathbf{S}^1 \mid \gamma.u(t + T\theta/2\pi) = u(t)\}.$$

Recall that for any action of a group G on a set Y the *isotropy subgroup* G_y of a point $y \in Y$ is

$$G_y = \{\sigma \in G \mid \sigma.y = y\},$$

and that for any subgroup Σ of G the *fixed-point set* $\mathrm{Fix}(\Sigma)$ is defined to be

$$\mathrm{Fix}(\Sigma) = \{y \in Y \mid \sigma.y = y \ \forall \sigma \in \Sigma\}.$$

When P is a vector space over $\mathbf{R}$ and G acts linearly, $\mathrm{Fix}(\Sigma)$ is a linear subspace. If P is a symplectic vector space over $\mathbf{R}$ and G acts linearly and symplectically, then $\mathrm{Fix}(\Sigma)$ is a symplectic linear subspace.

Let X be the vector field of H, let L be the linearization of X at p, and assume that the degree of smoothness for loop space is $k = \infty$. Define the *linearized flow* to be the flow generated by the ODE

$$\dot{x} + Lx = 0$$

on the tangent space $V = T_p P$ to P at p. Let λ be a nonzero purely imaginary eigenvalue of L and define the *resonance space* $V_\lambda \subset V$ to be the (real part of the) sum of the generalized eigenspaces of L for eigenvalues $k\lambda$, where $k \in \mathbf{Z}$. Assume the following conditions on H:

1. $D^2 H_p$ is a nondegenerate quadratic form.
2. $D^2 H_p|_{V_\lambda}$ is positive definite.

Condition (1) is equivalent to L being nonsingular, and (2) implies that $L|_{V_\lambda}$ is semisimple (diagonalizable over $\mathbf{C}$).

Clearly L is Γ-equivariant, so V_λ is invariant under the action of Γ. It is also invariant under the linearized flow. Because $L|_{V_\lambda}$ is semisimple, the orbits of the linearized flow are all periodic with period $2\pi/|\lambda|$ and hence define an action of $\mathbf{S}^1$ on V_λ. Explicitly,

$$\theta \cdot v = \exp\left(\frac{\theta}{|\lambda|}L\right)v.$$

This action commutes with the action of Γ, so together they define a $\Gamma \times \mathbf{S}^1$-action on V_λ.

We may now state the Equivariant Weinstein-Moser Theorem:

Theorem 1. *Suppose that the Hamiltonian H satisfies (1) and (2). Then for every isotropy subgroup Σ of the $\Gamma \times \mathbf{S}^1$-action on V_λ, and for all sufficiently small ϵ, there exist at least $\frac{1}{2}\dim \mathrm{Fix}(\Sigma)$ periodic orbits of X with periods near $2\pi/|\lambda|$ and symmetry group containing Σ, on the energy surface $H(x) = H(p) + \epsilon^2$.*

For further details see Montaldi *et al.* [19]. A rather different approach to an equivariant Liapunov Centre Theorem can be found in Dellnitz *et al.* [9].

In practice—though it is more a rule of thumb than a provable theorem—the "primary" isotropy subgroups Σ are those for which $\dim \mathrm{Fix}(\Sigma)$ is small. Because of the symplectic structure, $\dim \mathrm{Fix}(\Sigma)$ is always even. The most important isotropy subgroups, and the most tractable, of all are those for which $\dim \mathrm{Fix}(\Sigma)$ attains its minimum value, namely 2. Following Dionne *et al.* [7, 8] we define a subgroup $\Sigma \subset \Gamma \times \mathbf{S}^1$ to be **C**-*axial* if it is an isotropy subgroup having a two-dimensional fixed-point subspace. This implies in particular that it is a maximal isotropy subgroup (though the converse is not true). Moreover, the equivariant Weinstein-Moser Theorem implies that under the usual hypotheses if Σ is **C**-axial then there exists at least one family of periodic solutions with isotropy group *equal* to Σ. This follows since $\frac{1}{2}\dim \mathrm{Fix}(\Sigma) = 1$ and Σ is a maximal isotropy subgroup, so that "containing Σ" implies "equal to Σ."

The group theory involved in the $\Gamma \times \mathbf{S}^1$-action is identical to that introduced by Sattinger [23] and developed by Golubitsky and Stewart [10] to study equivariant Hopf bifurcation. This is not coincidence, but a consequence of the loop space technique employed. The Hopf bifurcation case has been studied more widely: see Golubitsky *et al.* [12] for a survey of results up to 1989. We will make use of this relationship to import results from equivariant Hopf bifurcation. Usually we restate them in Hamiltonian language.

3. Symmetry of the Planar Problem

We first establish the appropriate symmetry context for the N-body problem in the plane. Consider a system of N equal point particles in $\mathbf{R}^2$. Assume that each pair of particles experiences an attractive force given by a smooth function $F(d^2)$ where d is the distance between them. Let $q^1, \ldots, q^N$ be the position coordinates of the particles, and let $p^1, \ldots, p^N \in \mathbf{R}^2$ be their momentum coordinates. Then the motion is governed by a smooth Hamiltonian

$$H: \mathbf{R}^{2N} \times \mathbf{R}^{2N} \to \mathbf{R}$$

with $2N$ degrees of freedom. Because the force F is attractive there is clearly an equilibrium state with all q^j and p^j zero. We will require this state to be nondegenerate, in a slightly stronger sense than Section 2; that is, there is a purely imaginary eigenvalue λ with $V_\lambda = V = \mathbf{R}^{2N} \times \mathbf{R}^{2N}$, and $D^2 H_0$ is positive definite.

An important special case of the theory developed in Montaldi *et al.* [19] is when the group Γ acts absolutely irreducibly on configuration space and dually on momentum space. That is, the only linear maps that commute with Γ are scalar multiples of the identity. We will see below that this case applies to the N-body system. Results of Montaldi *et al.* [19] imply that by a symplectic change of coordinates we can identify $T_p P$ with $V \oplus V \equiv V \otimes_\mathbf{R} \mathbf{C}$ in such a way that the action of $(\gamma\theta) \in \Gamma \times \mathbf{S}^1$ is given by

$$(\gamma, \theta).(v \otimes z) = \gamma v \otimes e^{i\theta} z$$

for $v \in V$, $z \in \mathbf{C}$. A more concrete way to say this is to complexify V to $V \oplus V$, having the same basis but with complex coefficients — so that in a sense it should be thought of as $V \oplus iV$. Then Γ acts on $V \oplus V$ using the same matrices as for its action on V, and $\theta \in \mathbf{S}^1$ acts as scalar multiplication by $e^{i\theta}$. The crucial point is that the action of $\Gamma \times \mathbf{S}^1$ on $T_p P$ is known up to symplectic isomorphism once that of Γ on configuration space is known.

Let us apply this remark to the N-body model. There are two obvious types of symmetry on configuration space:

- *Label symmetries.* The Hamiltonian H is invariant under permutations of the N particles, hence under permutations of the q^j.
- *Euclidean symmetries.* The Hamiltonian H is invariant if the same element α of the two-dimensional euclidean group is applied simultaneously to all q^j.

There is a reduction of the problem that simplifies the analysis. Without loss of generality we may choose a coordinate system whose origin is at the centre of mass. That is, we may restrict the dynamics to the subspace

$$q^1 + \cdots + q^N = 0,$$
$$p^1 + \cdots + p^N = 0.$$

Then the euclidean symmetries are restricted to elements of $\mathbf{O}(2)$, the orthogonal group in the plane, since the centre of mass remains fixed.

Let $\mathbf{S}_N$ denote the symmetric group, comprising all permutations of $\{1, \ldots, N\}$. Because label and euclidean symmetries commute, the Hamiltonian is invariant under an action of $\Gamma = \mathbf{O}(2) \times \mathbf{S}_N$. Explicitly, this action is as follows:

$$(\gamma, \sigma)(q^1, \ldots, q^N) = (\gamma q^{\sigma^{-1}(1)}, \ldots, \gamma q^{\sigma^{-1}(N)}).$$

Here we assume that $q^1 + \cdots + q^N = 0$, so there is redundancy in the coordinate system. More abstractly we can define this action as follows. Introduce a space $\mathbf{R}^N$ on which $\mathbf{S}_N$ acts by permuting a basis, and identify $\mathbf{R}^{2N} = \{(q^1, \ldots, q^N)\}$ with $\mathbf{R}^2 \otimes_{\mathbf{R}} \mathbf{R}^N$ so that $(0, \ldots, 0, q^j, 0, \ldots, 0)$ is identified with $q^j \otimes e_j$, where $e_j = (0, \ldots, 0, 1, 0, \ldots, 0)$ with the 1 in the jth position. Then $\mathbf{O}(2) \times \mathbf{S}_N$ acts on $\mathbf{R}^2 \otimes_{\mathbf{R}} \mathbf{R}^N$ by

$$(\gamma, \sigma)(v \otimes w) = \gamma v \otimes \sigma w.$$

Define the subspace

$$\mathbf{R}^{N-1} = \left\{ \sum \mu_j e_j \in \mathbf{R}^N \mid \sum \mu_j = 0 \right\}.$$

Then confining attention to the subspace $\{\sum q^j = 0\}$ is equivalent to restricting the action of Γ to the subspace $V = \mathbf{R}^2 \otimes_{\mathbf{R}} \mathbf{R}^{N-1}$.

Lemma 2. *The group* $\Gamma = \mathbf{O}(2) \times \mathbf{S}_N$ *acts absolutely irreducibly on* $V = \mathbf{R}^2 \otimes_{\mathbf{R}} \mathbf{R}^{N-1}$.

Proof. We know that $\mathbf{O}(2)$ acts absolutely irreducibly on $\mathbf{R}^2$ and $\mathbf{S}_N$ acts absolutely irreducibly on $\mathbf{R}^{N-1}$. By Adams [2, Theorem 3.6.5, p. 71], Γ acts irreducibly on V, and by Adams [2, p. 29] that action is absolutely irreducible.   □

By the remarks above, the action of Γ on $\mathbf{R}^{2N-2} \times \mathbf{R}^{2N-2}$ can be identified with that on $V \oplus V \equiv V \otimes_{\mathbf{R}} \mathbf{C}$, in such a way that $\exp(\frac{1}{\omega}\theta L) \in \mathbf{S}^1$ acts as scalar multiplication by $e^{i\theta}$. But

$$V \otimes \mathbf{C} \cong (\mathbf{R}^2 \otimes_{\mathbf{R}} \mathbf{R}^{N-1}) \otimes_{\mathbf{R}} \mathbf{C} \cong \mathbf{C}^2 \otimes_{\mathbf{C}} \mathbf{C}^{N-1}.$$

Thus we have the following

Proposition 3. *The action of* $\mathbf{O}(2) \times \mathbf{S}_n \times \mathbf{S}^1$ *on* $\mathbf{R}^{2N-2} \times \mathbf{R}^{2N-2}$ *can be identified with the action on* $\mathbf{C}^2 \otimes_{\mathbf{C}} \mathbf{C}^{N-1}$, *where*

(a) $\gamma \in \mathbf{O}(2)$ *acts by complexifying its action on* $\mathbf{R}^2$.
(b) $\sigma \in \mathbf{S}_N$ *acts by permuting a basis of* $\mathbf{C}^N \supset \mathbf{C}^{N-1}$ *where*

$$\mathbf{C}^{N-1} = \left\{ z \in \mathbf{C}^N \mid \sum z_j = 0 \right\}.$$

(c) $\theta \in \mathbf{S}^1$ *acts as scalar multiplication by* $e^{i\theta}$.

Note that with this complex structure the action is unitary.

4. Twisted Products

The main group-theoretic results that we shall employ are taken from Dionne *et al.* [7, 8] and were originally developed to study arrays of coupled identical cells that possess both global and internal symmetries. Here we concentrate on the "direct product" case (see Dionne *et al.* [8]) for which the symmetry group of the system decomposes as the direct product $\mathcal{L} \times \mathcal{G}$ of the internal group $\mathcal{L}$ and the global group $\mathcal{G}$. Although motivated by Hopf bifurcation, the main algebraic results are entirely abstract and relate the **C**-axial subgroups of $\mathcal{L} \times \mathcal{G}$ to those of $\mathcal{L}$ or $\mathcal{G}$ separately.

In order to study periodic solutions of symmetric Hamiltonian systems near equilibrium, we first identify the generic (symplectically) irreducible representations of the symmetry group on (the tangent space to) phase space. Let Γ be a compact Lie group, acting linearly and symplectically on a finite-dimensional symplectic vector space V, and let $H: V \rightarrow \mathbf{R}$ be a Γ-invariant C^∞ Hamiltonian. Let λ be a purely imaginary eigenvalue, and define the resonance space $Y = V_\lambda$. Let $\Gamma^* = \Gamma \times \mathbf{S}^1$ act on Y as in Section 2.

By an argument similar to that of Golubitsky *et al.* [12, XVI, Proposition 1.4], but applied to invariant functions rather than equivariant mappings, it follows that the resonance space Y is generically Γ-*simple*, that is, either

$Y \cong W \oplus W$ where W is absolutely irreducible under Γ, or
Y is non-absolutely irreducible under Γ.

We can use the $\mathbf{S}^1$-action to give Y the structure of a complex vector space. To do this, let $z \in \mathbf{C}$, where $z = re^{i\theta}$, $r \geq 0$, and $\theta \in \mathbf{S}^1$. Define

$$zy = \theta \cdot (ry) = r(\theta \cdot y). \tag{1}$$

Because the $\mathbf{S}^1$-action is fixed-point free, it follows that if $z, y \neq 0$ then $zy \neq 0$. The remaining properties of a complex vector space are easily verified. We call this complex vector space $Y_c \cong W \otimes \mathbf{C}$.

The following statements are equivalent:

(a) Y is Γ-simple as a real representation.
(b) $W \otimes \mathbf{C}$ is a real irreducible representation of $\Gamma \times \mathbf{S}^1$ of complex type.
(c) Y_c is a complex irreducible representation of Γ.

For proofs see Golubitsky *et al.* [12, XVI, Proposition 3.5].

We can now redefine **C**-axial subgroups using the complex structure on Y_c. A subgroup Σ is **C**-*axial* if it is an isotropy subgroup with a complex one-dimensional fixed-point subspace. Note that **C**-axial subgroups are maximal isotropy subgroups.

It is shown in Golubitsky *et al.* [12] that, in the case of Γ-simple center subspaces, isotropy subgroups of Γ^* always have the form of a *twisted subgroup*, that is, a subgroup $\Sigma = A^\phi \subset \Gamma \times \mathbf{S}^1$ where $A \subset \Gamma$ is the projection of Σ into Γ, the map $\phi: A \rightarrow \mathbf{S}^1$ is a homomorphism, and

$$A^\phi = \{(a, \phi(a)): a \in A\}.$$

In a twisted subgroup there are no elements of the form $(1, \theta)$ where $\theta \neq 0$ and this point follows from the generic assumption that the resonance space is Γ-simple.

We now specialize to the case of interest in this paper, groups of the form $\Gamma = \mathcal{L} \times \mathcal{G}$ acting on $X = U \otimes V$ as before. Generically the action of $\mathcal{L} \times \mathcal{G}$ on the imaginary eigenspace Y_c is a complex irreducible representation of $\mathcal{L} \times \mathcal{G}$, and we henceforth assume this. A crucial simplification occurs in this case: as a complex representation of $\Gamma = \mathcal{L} \times \mathcal{G}$,

$$Y_c \cong U' \otimes_c V', \tag{2}$$

where U' is a **C**-irreducible representation of $\mathcal{L}$ and V' is a **C**-irreducible representation of $\mathcal{G}$.

Recall that $\mathcal{L}^* = \mathcal{L} \times \mathbf{S}^1$ and similarly for $\mathcal{G}^*$. There is a simple way to combine twisted subgroups $A^\phi \subset \mathcal{L}^*$ and $B^\psi \subset \mathcal{G}^*$ to get the *twisted product* subgroup $A^\phi \dot{\times} B^\psi \subset \mathcal{L} \times \mathcal{G} \times \mathbf{S}^1$ as follows. Twist the direct product $A \times B \subset \mathcal{L} \times \mathcal{G}$ using the homomorphism $\phi + \psi$ defined by $(\phi + \psi)(a, b) = \phi(a) + \psi(b)$. Equivalently,

$$A^\phi \dot{\times} B^\psi = \Theta(A^\phi \times B^\psi), \tag{3}$$

where

$$\Theta: \Omega \to \mathcal{L} \times \mathcal{G} \times \mathbf{S}^1$$

$$\Theta(\ell, \phi, g, \psi) = (\ell, g, \phi + \psi).$$

Lemma 4.

$$\dim_\mathbf{C} Fix_{U \otimes_c V}(A^\phi \dot{\times} B^\psi) = \dim_\mathbf{C} Fix_U(A^\phi) \cdot \dim_\mathbf{C} Fix_V(B^\psi).$$

This gives a simple way to obtain **C**-axial subgroups of Ω from **C**-axial subgroups of $\mathcal{L}^*$ and $\mathcal{G}^*$:

Proposition 5. *Suppose that $A^\phi \subset \mathcal{L}^*$ and $B^\psi \subset \mathcal{G}^*$ are **C**-axial. Then $P = A^\phi \dot{\times} B^\psi \subset \mathcal{L} \times \mathcal{G} \times \mathbf{S}^1$ is **C**-axial.*

The above proposition does not yield all **C**-axial subgroups. In principle (and often in practice) they can all be found using the following representation-theoretic criterion for **C**-axiality. In order to state it, let

$$\Omega = \mathcal{L}^* \times \mathcal{G}^*$$

and let

$$\Pi_{\mathcal{L}^*}: \Omega \to \mathcal{L}^* \quad \text{and} \quad \Pi_{\mathcal{G}^*}: \Omega \to \mathcal{G}^*$$

be the canonical projections. Given a subgroup $Q \subset \Omega$, we define representations of Q on U and V as follows:

$$\rho_U = \eta_U \circ \Pi_{\mathcal{L}^*} \quad \text{and} \quad \rho_V = \eta_V \circ \Pi_{\mathcal{G}^*},$$

where η_U and η_V are the given representations of $\mathcal{L}^*$ and $\mathcal{G}^*$ on U and V, respectively.

Proposition 6. *Let $Q \subset \Omega$ be a maximal isotropy subgroup. The Q has a (complex) one-dimensional fixed-point subspace if and only if there is precisely one irreducible representation of the action ρ_U of Q on U that is isomorphic to precisely one irreducible representation of the action ρ_V of Q on V.*

5. C-Axials for the Symmetric Group

In order to apply the results of Section 4 we classify the C-axial subgroups of the symmetric group S_n. We observe in passing that this classification has direct applications to Hopf bifurcation in a system of ODEs with S_n symmetry, as well as to periodic solutions of S_n-invariant Hamiltonian systems. The appropriate action of $S_n \times S^1$ is the "natural" action on

$$\mathbf{C}^{n,0} = \{(z_1, \ldots, z_n) \in \mathbf{C}^n \mid z_1 + \cdots + z_n = 0\},$$

in which $\sigma \in S_n$ acts by permutations of coordinates and $\theta \in S^1$ acts as multiplication by $e^{i\theta}$:

$$(\sigma, \theta).z = e^{i\theta}(z_{\sigma(1)}, \ldots, z_{\sigma(n)}). \tag{4}$$

The permutation representation of S_n on $\mathbf{R}^{n,0}$ is absolutely irreducible, and the representation (4) is isomorphic to $\mathbf{R}^{n,0} \otimes \mathbf{C}$ and is S_n-simple. For calculations it is often convenient to extend the action (4) to an action on $\mathbf{C}^n$ and to impose the condition $z_1 + \cdots + z_n = 0$ at a later stage.

We now classify the C-axial subgroups of S_n acting on $\mathbf{C}^{n,0}$ as in (4). Such subgroups are of the form H^θ where $H \subset S_n$, and $\theta : H \to S^1$ is a group homomorphism. Moreover, they are maximal with respect to fixing a complex line $\mathbf{C}z = \{\mu z \mid \mu \in \mathbf{C}\}$, where $z \neq 0$.

We classify possible H^θ by asking under what conditions on z the isotropy subgroup Σ_z in $S_n \times S^1$ fixes *only* $\mathbf{C}z$. Call such a z an *axis*.

Our aim is to prove:

Theorem 7. *Suppose that $n \geq 2$. Then the axes of $S_n \times S^1$ acting on $\mathbf{C}^{n,0}$ have orbit representatives as follows:*

Type I: *Let $n = qk + p$ where $2 \leq k \leq n, q \geq 1, p \geq 0$. Let $\zeta = e^{2\pi i/k}$, and set*

$$z = (\underbrace{1, \ldots, 1}_{q}; \underbrace{\zeta, \ldots, \zeta}_{q}; \underbrace{\zeta^2, \ldots, \zeta^2}_{q}; \cdots; \underbrace{\zeta^{k-1}, \ldots, \zeta^{k-1}}_{q}; \underbrace{0, \ldots, 0}_{p}).$$

Type II: *Let $n = q + p, 1 \leq q < \frac{n}{2}$, and set*

$$z = (\underbrace{1, \ldots, 1}_{q}; \underbrace{a, \ldots, a}_{p})$$

where $a = -q/p$.

Proof. Suppose that z is an axis and $\Sigma_z = H^\theta$. If $(\sigma, \theta(\sigma)) \in H^\theta$ then by (4)

$$(\sigma, \theta(\sigma)) \cdot (z_1, \ldots, z_n) = e^{i\theta(\sigma)}(z_{\sigma(1)}, \ldots, z_{\sigma(n)}). \tag{5}$$

Suppose $\mathrm{im}(\theta) = \mathbf{Z}_k = \langle \zeta \rangle$, where $\zeta = e^{2\pi i/k}$. Then every element $(\sigma, \theta(\sigma)) \in H^\theta$ permutes the coordinates of z and multiplies the result by ζ^m for some m that may depend on σ. By renumbering coordinates if necessary (a conjugacy in $\mathbf{S}_n$) we can break z up into blocks

$$z = (z_1, \ldots, z_{b_1}; z_{b_1+1}, \ldots, z_{b_2}; \cdots; z_{b_{p-1}+1}, \ldots, z_{b_p})$$

such that for each j all the complex numbers

$$z_{b_j+1}, \ldots, z_{b_{j+1}}$$

lie in the same $\mathbf{Z}_k$-orbit (for the standard action of $\mathbf{Z}_k$ on $\mathbf{C}$). Define

$$B_j = \{b_j + 1, \ldots, b_{j+1}\} \subset \{1, \ldots, n\}.$$

Since $z \in \mathrm{Fix}(H^\theta)$, equation (1) implies that for all $\sigma \in H$ we have $\sigma : B_j \to B_j$ for $j = 1, \ldots, p$. Moreover, the $\mathbf{Z}_k$-action also preserves the block structure because the blocks are defined via $\mathbf{Z}_k$-orbits. Thus we may write $\mathbf{C}^n$ as a direct sum

$$\mathbf{C}^n = \bigoplus_{j=1}^{p} \mathbf{C}_j \tag{6}$$

where $\mathbf{C}_j \cong \mathbf{C}^{b_j - b_{j-1}}$ corresponds to the entries of z whose subscripts are in the block B_j (and we take $b_0 = 0$). We know that H^θ leaves each $\mathbf{C}_j$ invariant. It follows that if more than one component in (6) contains nonzero elements whose coordinates sum to zero, then H^θ is not $\mathbf{C}$-axial on $\mathbf{C}^{n,0}$. For if there are two nonzero blocks (without loss of generality B_1 and B_2) whose elements sum to zero, then H^θ fixes

$$(\alpha(z_1, \ldots, z_{b_1}); \beta(z_{b_1+1}, \ldots, z_{b_2}); \cdots)i \in \mathbf{C}^{n,0}$$

for all $\alpha, \beta \in \mathbf{C}$; but this contradicts H^θ being $\mathbf{C}$-axial.

Suppose first that $k \geq 2$. Consider a block B_j and consider the H^θ-action on the corresponding $\mathbf{C}_j$. Let z^j be the component of z in this block. By re-ordering the subscripts and normalizing the first component to 1 we can decompose z according to powers of ζ, so that

$$z = (1, \ldots, 1; \zeta, \ldots, \zeta; \zeta^2, \ldots, \zeta^2; \cdots; \zeta^{k-1}, \ldots, \zeta^{k-1})$$

where there are $q_j \geq 0$ successive occurrences of ζ^j.

Let $m = b_j - b_{j-1}$ be the dimension of the block $\mathbf{C}_j$. Decompose B_j into sub-blocks C_ℓ according to which power of ζ occurs. Let $K = \ker(\theta) \lhd H$, so that $K = H \cap \mathbf{S}_n$. Then K acts by pure permutations. Choose $\rho \in H$ such that $\theta(\rho) = \zeta$. Then

$$H = \langle K, \rho \rangle = K \langle \rho \rangle$$

and $\rho^k \in K$.

Now (5) implies that the ρ-action cycles the k blocks C_ℓ so that $C_1 \to C_2 \to \cdots \to C_k \to C_1$ under ρ. We deduce from (5) applied to $(\rho, \theta(\rho))$ that

$$q_1 = q_2 = \cdots = q_k = q$$

for some fixed q.

It now follows that the sum of the coordinates of z^j on block $\mathbf{C}_j$ is $q(1 + \zeta + \cdots + \zeta^{k-1}) = 0$. Therefore there can be at most one nonzero block $\mathbf{C}_j$.

We assume for the moment that there is no zero block, so $n = qk$. From (5) we have

$$K \supset \mathbf{S}_q^1 \times \cdots \times \mathbf{S}_q^k$$

where $\mathbf{S}_q^j$ is the full permutation group on block C_j. Moreover, we may replace ρ by any k-cycle of the form

$$\rho = (a_1 \cdots a_k) \qquad a_\ell \in C_\ell$$

by maximality of H^θ; and now K, ρ between them generate *all* permutations that cycle the blocks C_j. Those that fix each block belong to K, and the remainder are generated by K and ρ. Moreover, ρ is twisted into $\mathbf{S}$ by ζ, and this defines the map θ on the whole of H. (Any permutation that sends block C_1 to C_{s+1} is twisted by ζ^s.)

It is easy to check that the only elements fixed by such a group are the complex scalar multiples of z, so z is an axis and H^θ is its isotropy subgroup.

Next we suppose that $k = 1$ and there is no zero block. Suppose that z decomposes into s blocks

$$(z_1, \ldots, z_1; z_2, \ldots, z_2; \cdots; z_s, \ldots, z_s), \tag{7}$$

where the z_j are distinct, each z_j occurs a_j times, and

$$a_1 z_1 + \cdots + a_j z_j = 0.$$

We claim that maximality of H^θ implies that only two blocks occur in (7), so that $s = 2$. The reason is that the isotropy subgroup of (7) is $\mathbf{S}_{a_1} \times \cdots \times \mathbf{S}_{a_s}$. However, if $s > 2$ this is contained in $\mathbf{S}_{a_1} \times \mathbf{S}_{n-a_1}$ which fixes $(z_1, \ldots, z_1; z_2, \ldots, z_2)$ with a_1 occurrences of z_1 and $n - a_1$ of z_2. The sum of the entries is zero provided we take $z_2 = \frac{a_1}{a_1 - n} z_2$. But this contradicts maximality of H^θ.

Finally we consider how the presence of a zero block modifies the above conclusions. If there is a zero block, then decompose $\{1, \ldots, n\}$ into $\{1, \ldots, q\} \cup \{q + 1, \ldots, n\}$, where $1 \leq q < n$. Let

$$z = (z_1, \ldots, z_q; 0, \ldots, 0)$$

with $z_j \neq 0$ for $j = 1, \ldots, q$. Then the first block leads to a twisted subgroup H^θ of $\mathbf{S}_q \times \mathbf{S}^1$ as just described, while the second is fixed by $\mathbf{S}_p$ where $p = n - q$. If $p \neq 0$ and $k = 1$ then this construction leads to a non-maximal group because the zero block can be merged with a nonzero block in the same way that we proved only two blocks occur when $k = 1$ and there is no zero block. $\qquad \square$

Next we consider the corresponding isotropy subgroups. For type I we have

$$\Sigma_z = \mathbf{S}_q \widetilde{\wr} \mathbf{Z}_k \times \mathbf{S}_p \overset{\text{def}}{=} \Sigma_{q,p}^1.$$

Here $\wr$ denotes the wreath product (see Hall [13, p. 81]) and the tilde indicates that $\mathbf{Z}_k$ is twisted into $\mathbf{S}^1$. In more explicit terms, let

$$K = \ker(\theta) = \mathbf{S}_q^1 \times \cdots \times \mathbf{S}_q^k \times \mathbf{S}_p,$$

where $\mathbf{S}_q^j$ is the symmetric group on $B_j = \{(j-1)q+1, \ldots, jq\}$ and $\mathbf{S}_p$ is the symmetric group on $B_0 = \{kq+1, \ldots, n\}$. Define α to be the k-cycle

$$\alpha = (1, q+1, 2q+1, \ldots, (k-1)q+1).$$

Then $\Sigma_z = \langle K, (\alpha, \zeta) \rangle$. The interpretation for a many-body problem is that the solution comprises k clumps of q coincident bodies whose oscillations are identical except for phase shifts by $\frac{mT}{k}$ where $m = 0, \ldots, k-1$ and T is the period, together with one clump of p coincident bodies.

In type II the isotropy subgroup is

$$\Sigma_z = \mathbf{S}_q \times \mathbf{S}_p \overset{\text{def}}{=} \Sigma_q^{\text{II}}$$

where the respective factors are the symmetric groups on $\{1, \ldots, q\}$ and $\{q+1, \ldots, n\}$. In a many-body interpretation this correponds to two clumps consisting of q and p coincident bodies respectively.

Note that the statement of Theorem 7 excludes the case $p = q$, which occurs only when n is even. The reason is that then

$$z = (\underbrace{1, \ldots, 1}_{q}; \underbrace{-1, \ldots, -1}_{q})$$

which is of type I with $k = 2, q = n/2, p = 0$. So this case is already included (and there is an extra symmetry interchanging the two blocks, with a π twist, which is why we do not allow this case in type II).

It remains only to check that no groups defined in the above manner are contained in conjugates of others, but this is routine.

6. C-Axials for the Planar Problem

In this section we apply the above theory to the N-body problem in the plane, seeking periodic solutions arising by symmetry-breaking from the (extremely!) trivial equilibrium in which all particles are coincident and stationary.

First we recall the C-axial subgroups of $\mathbf{O}(2)$ in its standard action on $\mathbf{R}^2$. From the Hopf bifurcation results of Golubitsky and Stewart [10] and Golubitsky et al. [12] there are two C-axial subgroups A_1 and A_2, where

$$\begin{aligned}
A_1 &= \widetilde{\mathbf{SO}(2)} \overset{\text{def}}{=} \{(\theta, \theta) \in \mathbf{O}(2) \times \mathbf{S}\}, \\
A_2 &= \mathbf{D}_1^{(\kappa,0)} \times \mathbf{Z}_2^{(\pi,\pi)}.
\end{aligned}$$

The C-axial subgroups of $\mathbf{S}_N$ have been classified in Section 5. Using Proposition 5 we can combine the results for the two groups to obtain C-axial subgroups of $\mathbf{S}_N \times \mathbf{O}(2)$, and by Theorem 1 these are the isotropy subgroups of periodic solutions near the trivial equilibrium of the N-body system (under the usual technical hypotheses). In order to illustrate the method we first tackle a few small values of N.

6.1. N = 3

We begin with the case $N = 3$. Because $\mathbf{S}_3 \cong \mathbf{D}_3$, this case reduces to a system with $\mathbf{D}_3 \times \mathbf{O}(2)$ symmetry, which is worked out in detail as an example in Dionne *et al.* [8].

Recall that the **C**-axial subgroups of $\mathbf{O}(2)$ are A_1, A_2 given by (8). By Theorem 7 $\mathbf{S}_3$ has precisely three **C**-axial subgroups, namely

$$\Sigma^I_{1,0} \quad \Sigma^I_{1,1} \quad \Sigma^{II}_1.$$

Therefore by Theorem 5 $\mathbf{S}_3 \times \mathbf{O}(2)$ has (at least) the following six **C**-axial subgroups:

$$\Sigma^I_{1,0} \times A_1 \quad \Sigma^I_{1,1} \times A_1 \quad \Sigma^{II}_1 \times A_1,$$
$$\Sigma^I_{1,0} \times A_2 \quad \Sigma^I_{1,1} \times A_2 \quad \Sigma^{II}_1 \times A_2.$$

The complete analysis of Dionne *et al.* [8] shows that there is precisely one additional **C**-axial subgroup, namely

$$\widetilde{\mathbf{D}}_3 = \langle (\rho, R_{2\pi/3}), (\sigma, \kappa) \rangle,$$

where $\rho = (123)$ is a 3-cycle, $\sigma = (12)$ is a transposition, $R_{2\pi/3}$ is rotation through an angle $2\pi/3$, and κ is a flip.

Figure 1 illustrates the solution types corresponding to these seven **C**-axial subgroups.

Observe that solutions of the form $\Sigma \times A_1$ are "rotating waves," whereas those of the form $\Sigma \times A_2$ are restricted to a line in configuration space. The isotropy group $\widetilde{\mathbf{D}}_3$ corresponds to a "collapsing triangle." (Note that in this paper all such figures are schematic, serving only to indicate the symmetries of the solutions. Other structure in the figure is not necessarily indicative of general features. However, orbits that appear circular *are* circular, because the rotating wave nature of $\mathbf{O}(2)$ solutions with isotropy subgroup A_1 forces circularity.)

6.2. N = 4, 5

When $N \geq 4$ the literature does not contain a complete classification of the **C**-axial subgroups of $\mathbf{S}_N \times \mathbf{O}(2)$. We make a start on this question by listing those **C**-axial subgroups whose existence follows easily from the general theory described above.

To provide specific examples we here consider the cases $N = 4, 5$. We make some general remarks in the next subsection.

When $N = 4$ the **C**-axial subgroups of $\mathbf{S}_4$ are listed in Table 1.

As for $N = 3$, these five groups combine via the twisted product with the two **C**-axial subgroups for $\mathbf{O}(2)$ to create ten **C**-axial subgroups of $\mathbf{S}_4 \times \mathbf{O}(2)$.

In addition, a simple calculation using Proposition 6 shows that there is also a **C**-axial subgroup of $\mathbf{S}_4 \times \mathbf{O}(2)$ of the form

$$\widetilde{\mathbf{D}}_4 = \langle (\rho, R_{\pi/2}), (\sigma, \kappa) \rangle,$$

where now $\rho = (1234)$. This is a "collapsing square." The eleven corresponding solutions of the 4-body problem are shown in Figure 2.

When $N = 5$ we find seven **C**-axial subgroups of $\mathbf{S}_5$, listed in Table 2.

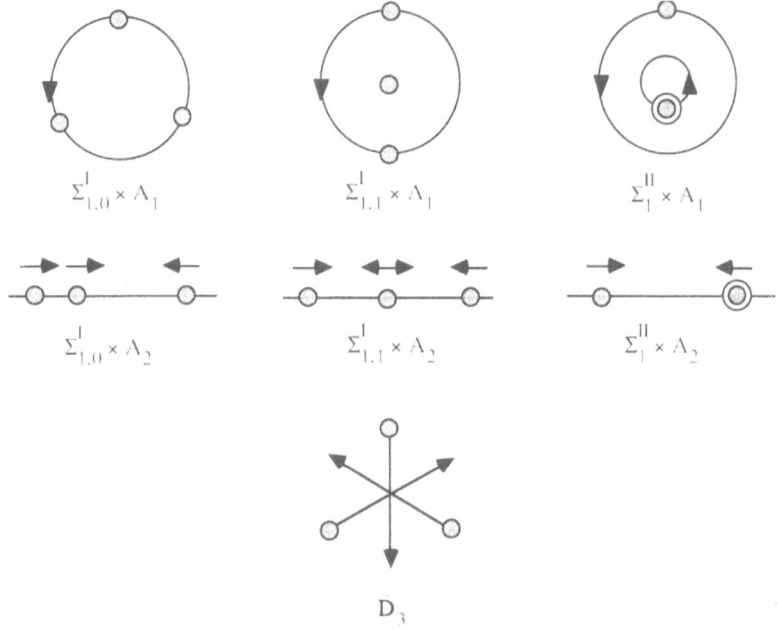

Fig. 1. The seven C-axial solutions for three bodies (schematic).

Table 1.

axis	q	p	k	Σ
$(1, -1, 0, 0)$	1	2	2	$\Sigma^I_{1.2}$
$(1, 1, -1, -1)$	2	0	2	$\Sigma^I_{2.0}$
$(1, \omega, \omega^2, 0)$	1	1	3	$\Sigma^I_{1.1}$
$(1, i, -1, -i)$	1	0	4	$\Sigma^I_{1.0}$
$(3, -1, -1, -1)$	1	3		Σ^{II}_1

Here $\xi = e^{2\pi i/5}$ is a primitive fifth root of unity. These seven groups combine via the twisted product with the two C-axial subgroups for $\mathbf{O}(2)$ to create 14 C-axial subgroups of $\mathbf{S}_5 \times \mathbf{O}(2)$. In addition, Prosopition 6 shows that there is also a C-axial subgroup of $\mathbf{S}_5 \times \mathbf{O}(2)$ of the form

$$\widetilde{\mathbf{D}}_5 = \langle (\rho, R_{2\pi/5}), (\sigma, \kappa) \rangle,$$

where now $\rho = (12345)$. This is a "collapsing pentagon." The fifteen corresponding solutions of the 5-body problem are analogous to those in Figures 2 and 3.

6.3. General N

The same arguments for general N yield $[\frac{N}{2}] + [\frac{N}{3}] + \cdots + [\frac{N}{N}]$ C-axial subgroups of type I and $[\frac{N-1}{2}]$ of type II. Via the twisted-product construction, each yields two C-axial

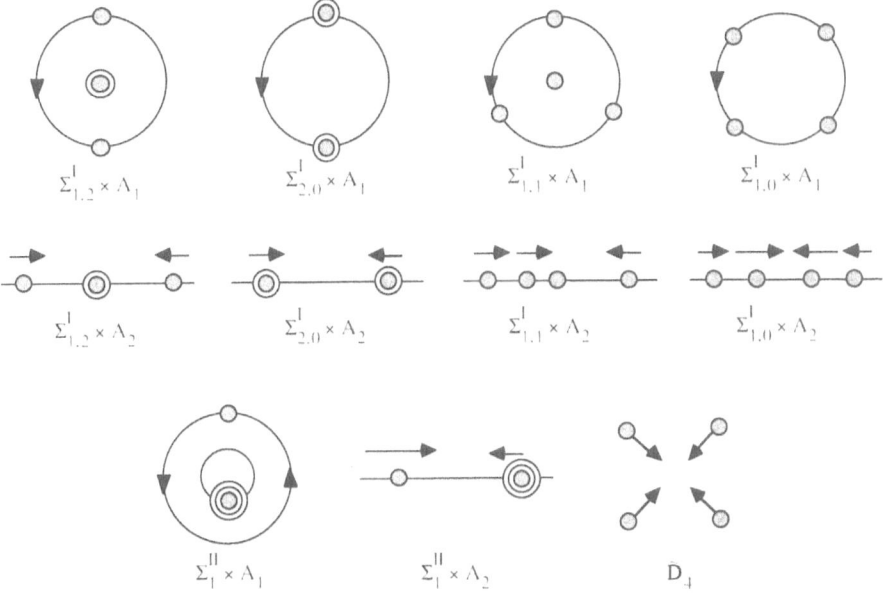

Fig. 2. Eleven **C**-axial solutions for four bodies (schematic).

Table 2.

axis	q	p	k	Σ
$(1, -1, 0, 0, 0)$	1	3	2	$\Sigma^I_{1,3}$
$(1, 1, -1, -1, 0)$	2	1	2	$\Sigma^I_{2,1}$
$(1, \omega, \omega^2, 0, 0)$	1	2	3	$\Sigma^I_{1,2}$
$(1, i, -1, -i, 0)$	1	1	4	$\Sigma^I_{1,1}$
$(1, \xi, \xi^2, \xi^3, \xi^4)$	1	0	5	$\Sigma^I_{1,0}$
$(4, -1, -1, -1, -1)$	1	4		Σ^{II}_1
$(3, 3, -2, -2, -2)$	2	3		Σ^{II}_2

subgroups of $\mathbf{S}_N \times \mathbf{O}(2)$. There is also a "collapsing N-gon" with isotropy subgroup $\widetilde{\mathbf{D}}_N$ defined in the obvious way. Thus we have found

$$2\left(\left[\frac{N}{2}\right] + \left[\frac{N}{3}\right] + \cdots + \left[\frac{N}{N}\right] + \left[\frac{N-1}{2}\right]\right) + 1$$

C-axial subgroups of $\mathbf{S}_N \times \mathbf{O}(2)$.

There are collinear solutions corresponding to the one-dimensional problem, solutions in which a number of coincident bunches formed from equal numbers of point masses form a rotating regular polygon (often with another bunch of coincident point masses stationary at its centre), together with the collapsing polygon. The solutions will become more interesting when we consider possible perturbations: see Section 8.

Table 3.

Σ	$\chi[V]$	$\chi[\mathbf{g}/\mathbf{g}_z \otimes_R \mathbf{C}]$	$\chi[\mathcal{R}_\Sigma]$
$\Sigma^I_{1,0} \times A_1$	$\chi_{00} + \chi_{01} + \chi_{-20} + \chi_{-21}$	χ_{00}	$\chi_{01} + \chi_{-20} + \chi_{-21}$
$\Sigma^I_{1,1} \times A_1$	$\chi_{00} + \chi_{01} + \chi_{-20} + \chi_{-21}$	χ_{00}	$\chi_{01} + \chi_{-20} + \chi_{-21}$
$\Sigma^{II}_1 \times A_1$	$\chi_{00} + \chi_{01} + \chi_{-20} + \chi_{-21}$	χ_{00}	$\chi_{01} + \chi_{-20} + \chi_{-21}$
$\Sigma^I_{1,0} \times A_2$	$\chi_{000} + \chi_{100} + \chi_{010} + \chi_{110}$	χ_{000}	$\chi_{100} + \chi_{010} + \chi_{110}$
$\Sigma^I_{1,1} \times A_2$	$\chi_{000} + \chi_{100} + \chi_{010} + \chi_{110}$	χ_{000}	$\chi_{100} + \chi_{010} + \chi_{110}$
$\Sigma^{II}_1 \times A_2$	$\chi_{000} + \chi_{100} + \chi_{010} + \chi_{110}$	χ_{000}	$\chi_{100} + \chi_{010} + \chi_{110}$
$\tilde{\mathbf{D}}_3$	$\chi_0 + \chi_1 + \chi_2$	$\chi_0 + \chi_1$	χ_2

6.4. Linearized Stability

Montaldi *et al.* [19, Theorem 5.1, p. 264] provides a formalism for computing the linearized stability of C-axial, solutions with isotropy subgroup Σ in terms of the *residual space*

$$\mathcal{R}_\Sigma = (K \cap E)/K \cap \mathbf{g} \cdot u(0)$$

where $u(t)$ is the associated periodic solution, $\mathbf{g}$ is the Lie algebra of the symmetry group Γ of the Hamiltonian H, the space E is (a suitable deformation of) the resonance space V_λ, equal to $\mathbf{R}^{2N-2} \times \mathbf{R}^{2N-2}$ in this case, and

$$K = \ker \mathbf{D}\Phi_{u(0)}$$

for the momentum map $\Phi: P \to \mathbf{g}^*$. In particular they identify a class of solutions, said to be *cyclospectral*, for which the action of Σ on $\mathcal{R}$ forces linearized stability (ellipticity).

They also show in Proposition 6.8, p. 275, that it is possible to compute the character of this representation using the formula

$$\chi[\mathcal{R}_\Sigma] = \chi[V_\lambda] - \chi[\mathbf{g}_\mu/\mathbf{g}_z \otimes_R \mathbf{C}] - \chi[\mathbf{g}/\mathbf{g}_\mu]. \tag{8}$$

Here $\chi[U]$ denotes the character of the representation of Σ on the space U, and $z = u(0)$, $\mu = \Phi(z)$.

When G^0 is abelian (as here because $G = \mathbf{O}(2) \times \mathbf{S}_N$) then $\mathbf{g}_\mu = \mathbf{g}$, and when Σ is finite then $\mathbf{g}_z = 0$. Moreover, here V_λ is the whole space. So (8) reduces to

$$\chi[\mathcal{R}_\Sigma] = \chi[\mathbf{R}^{2N-2} \times \mathbf{R}^{2N-2}] - \chi[\mathbf{g}/\mathbf{g}_z \otimes_R \mathbf{C}]. \tag{9}$$

When $N = 3$ we have used this formula to compute the character corresponding to $\mathcal{R}_\Sigma$. The result is shown in Table 3. Here we use the notation $\alpha = (123) \in \mathbf{S}_3$ and $\beta = (12) \in \mathbf{S}_3$. Our notation for characters is as follows. If an element $(\psi, \alpha) \in \mathbf{SO}(2) \times \mathbf{S}_3$ acts as multiplication by $e^{p\mathbf{i}\psi} e^{2q\pi \mathbf{i}/3}$ then the corresponding character is χ_{pq}. If the element $\kappa \in \mathbf{O}(2)$ acts as multiplication by $e^{p\mathbf{i}\pi}$, the element $\pi \in \mathbf{SO}(2)$ acts as multiplication by $e^{q\mathbf{i}\pi}$, and $\alpha \in \mathbf{S}^1_3$ acts as multiplication by $e^{r\mathbf{i}\pi}$, then the corresponding character is χ_{pqr}.

Applying Montaldi *et al.* [19, Theorem 2.3, p. 250] we find that the only cyclospectral case is $\Sigma_{1,0}^I \times A_1$, the rotating equilateral triangle. (Note that for this case χ_{01} is of complex type, whereas for $\Sigma_{1,1}^I \times A_1$ and $\Sigma_1^{II} \times A_1$ it is of real type.) Other solutions *may* be stable or not, depending on normal form coefficients.

7. The Three-Dimensional Problem

Turning to the three-dimensional version of the problem, we find by similar reasoning that the appropriate symmetry group is $O(3) \times S_N \times S^1$ acting on $R^{3N-3} \times R^{3N-3}$, which can be identified with $C^3 \otimes_c C^{N-1}$ where

(a) $\Gamma \in O(3)$ acts by complexifying its action on R^3.
(b) $\sigma \in S_N$ acts by permuting a basis of $C^N \supset C^{N-1}$ where

$$C^{N-1} = \left\{ z \in C^N \mid \sum z_j = 0 \right\}.$$

(c) $\theta \in S^1$ acts as scalar multiplication by $e^{i\theta}$.

Again we can use Proposition 5 to find C-axial subgroups by combining C-axial subgroups for $O(3)$ and S_N. By Golubitsky *et al.* [12, p. 341] the C-axial subgroups for $O(3)$ are

$$\widetilde{SO}(2) \times O(1)$$

and

$$Z_2^c \times O(2),$$

where

$$\widetilde{SO}(2) = \{(\theta, \theta) \mid \theta \in SO(2)\} \subset O(3) \times S^1,$$
$$Z_2^c = \langle -\text{id} \rangle \subset O(3).$$

These are the natural extensions to $O(3)$ of the corresponding subgroups of $O(2)$, and we therefore obtain just the $O(2)$ (planar) solutions again, but now embedded in R^3.

However, for suitable N we can also obtain solutions that are genuinely three-dimensional. Let T, O, I be the tetrahedral, octahedral, and icosahedral subgroups of $O(3)$ (as in Golubitsky *et al.* [12, p. 104] but adding in the extra generator $-\text{id}$). Then, for example, when $N = 4$ we expect to find a "collapsing tetrahedron" with isotropy subgroup

$$\widetilde{T} = \{(\gamma, \alpha(\gamma))\},$$

where γ runs through T and $\alpha \colon S_4 \to O(3)$ is an isomorphism onto T. By Proposition 6 this subgroup is C-axial. Similar remarks apply to $\widetilde{O}$ when $N = 6, 8$ (collapsing cube and collapsing octahedron) and $\widetilde{I}$ when $N = 12, 20$ (collapsing dodecahedron and collapsing icosahedron).

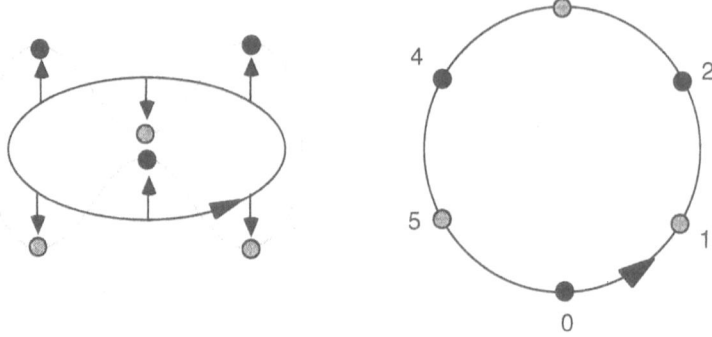

Fig. 3. Type of three-dimensional motion found by Davies *et al.*, illustrated for six bodies (schematic). Left: front view. Right: top view.

8. Perturbations and Speculations

None of the solutions obtained from the **C**-axial subgroups that we have found are especially surprising—indeed those that are confined to a line are arguably more surprising than the planar ones. However, we would not expect **C**-axial subgroups to produce particularly exotic solutions, since they are in particular maximal isotropy subgroups and the solutions therefore possess a large amount of symmetry. The above analysis should be seen as a preliminary step that provides a strong grip on a much more interesting class of solutions, which can be obtained by perturbation from suitable **C**-axial solutions. In this final section we sketch out a few ideas in this direction. Some could be pinned down with little extra work, others seem to need rather more, but either way no details will be given here.

The original inspiration for this paper was an attempt to understand, in symmetry terms, the work of Davies *et al.* [6], who found a genuinely three-dimensional solution to the equal mass $2n$-body problem (with inverse-square law attraction). We call this the *DTW-solution*. Meyer and Schmidt [18] have reformulated the derivation of the DTW-solution by exploiting its symmetry. The next few paragraphs explain how the analogue of this solution, for the kind of system considered here, fits into the framework set out in this paper.

The DTW-solution is illustrated in Figure 3 in the case $2n = 6$. Take coordinates x, y in the "equatorial plane" and z at right angles to this. The six bodies divide into two sets of three, and each set forms an equilateral triangle. When projected into the xy-plane the six bodies lie at the vertex of a regular hexagon, and we number the bodies 0–5 anticlockwise round this hexagon. At any given instant t, bodies 0, 2, 4 have the same z-coordinate $z(t)$, while bodies 1, 3, 5 have z-coordinate $-z(t)$. The motion is a combination of a periodic oscillation in the z-direction (in which bodies 1, 3, 5 are half a period out of phase with bodies 0, 2, 4) with a steady rotation in the xy-plane. In addition, however, the size of the hexagon in the xy-plane expands and contracts periodically.

For a periodic solution the periodic mode in the z-direction (which we call the *z-mode*) must be in resonance with the rotational *xy-mode*. The most natural case is when the z-mode is in $3 : 1$ resonance with the xy-mode, so that after one complete period of the

z-mode body 0 has rotated to the original position of body 2. In this case all six bodies follow the same trajectory in 3-space, separated at intervals of one-sixth of a period. Different resonances give rise to more intricate "braided" motion. In a coupled oscillator interpretation this solution is rather like a rotating version of the "tripod gait" found in models of insect locomotion; see Collins and Stewart [5] and Kroon and Stewart [15].

A similar solution occurs for $2n$ bodies, and now the most natural case is the $n : 1$ resonance where again all $2n$ bodies follow the same trajectory in 3-space. As hinted by the description above, we can think of such motions as "mode interactions" between two solutions of the following type:

(a) Two sets, each comprising n coincident bodies, oscillating half a period out of phase with each other along the z-axis.
(b) A rotating rigid $2n$-gon in the xy-plane.

One way to find such solutions is to look in the fixed-point space of the appropriate symmetry group, which we would expect to be whichever isotropy subgoup is maximal with respect to being contained in (suitable conjugates of) the isotropy subgroups of the two separate modes. Here the simplest way to do this is to work out the isotropy subgroup Σ_{DTW} of the DTW-solution and compute its fixed-point space V_{DTW}. It turns out that V_{DTW} has dimension 2 over $\mathbf{C}$ and that Σ_{DTW} is indeed contained in the isotropy subgroups of both component modes. In order to demonstrate the existence of such a solution it is therefore sufficient to perform a Liapunov-Schmidt reduction on to V_{DTW} in the case where the two primary modes are resonant. (Alternatively the system can be put in Birkhoff normal form and restricted to this fixed-point space, but the usual technical problems with truncation must be dealt with.)

The element $-\mathrm{id} \in \mathbf{O}(3)$ obviously normalizes Σ_{DTW} but does not lie in that group, so there is a $\mathbf{Z}_2$ normalizer symmetry induced on V_{DTW}. Define a *Liapunov mode* to be a periodic oscillation that can be obtained by applying the Liapunov centre theorem when the resonance is slightly detuned. Then, because the complex dimension of V_{DTW} is 2, the dynamics on V_{DTW}, in the abstract, is (the Liapunov-Schmidt reduction of) the generic resonance between two Liapunov modes with this $\mathbf{Z}_2$ normalizer symmetry (see Roberts [22] or Montaldi *et al.* [20, §6]). This symmetric resonance behaves in a very similar manner to the standard asymmetric resonance between two Liapunov modes; see for example Dionne *et al.* [9] and the wealth of references contained therein. These sources prove that for open sets of coefficients in the normal form there exists a resonant periodic motion that is different from the two Liapunov modes; that is, the resonant mode interaction can lead to a periodic mixed mode. This is clearly the DTW-solution. The different kinds of phase-locking found by Davies *et al.* [6] depend on the choice of resonance in this procedure.

Another way to obtain interesting solutions using mode interactions is to start with a solution in which bodies are grouped in coincident clumps, and break up the clumps by perturbing them according to a second mode. Again, provided the motions are resonant, we expect to obtain a reduction onto a low-dimensional fixed-point space (with luck two-dimensional over $\mathbf{C}$) on which we find a (generic) asymmetric or $\mathbf{Z}_2$-symmetric resonance. Figure 4 illustrates one such solution that may exist for ten bodies: each clump of two coincident bodies is perturbed into a "binary pair."

A third possibility is to perturb collapsing solutions. If these are elliptic then by the

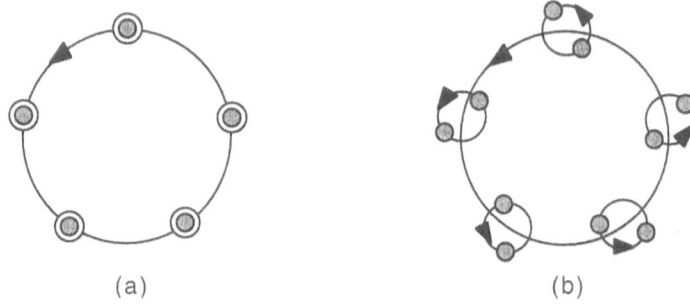

Fig. 4. Left: a rotating pentagon formed by clumped pairs. Right: perturbing it to yield a collision-free solution. (Schematic.)

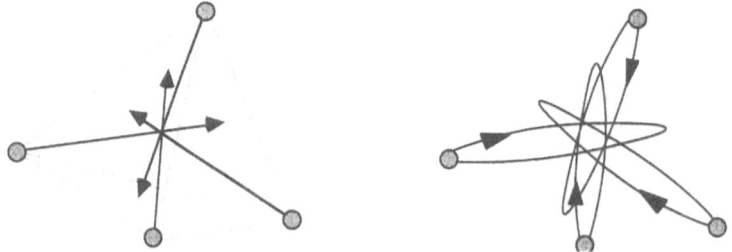

Fig. 5. Left: a collapsing tetrahedron. Right: perturbing it to yield a collision-free solution. (Schematic.)

KAM theorem they may support nearby tori, and on these tori we may find resonant periodic motions. Figure 5 indicates the general idea for a collapsing tetrahedron in $\mathbf{R}^3$ with four bodies. Variations include starting with a collapsing set of clumps and perturbing either the clumps (to break them up), the collapsing solutions, or both. Interesting quasiperiodic solutions might also be located using such procedures.

Incomplete and speculative though these suggestions are, they indicate that a systematic symmetry-based study of the equal-mass N-body problem would be of definite interest. Even more interesting—but probably technically harder—would be to apply the methods of this paper to singular potentials, especially Newtonian ones. For this, radically new methods seem to be required.

References

[1] R. Abraham & J. E. Marsden. *Foundations of Mechanics*. Benjamin/Cummings, Reading, MA, 1985.

[2] J. F. Adams, *Lectures on Lie Groups*. Benjamin/Cummings, New York, 1969.

[3] J. Binney & S. Tremaine. *Galactic Dynamics*. Princeton Unuversity Press, Princeton, NJ, 1987.

[4] T. Bröcker & T. tom Dieck. *Representations of Compact Lie Groups*. Springer-Verlag, New York, 1985.

[5] J. J. Collins & I. Stewart. Hexapodal gaits and coupled nonlinear oscillator models, *Biol. Cybernet.* **68** (1993) 287–298.

[6] I. Davies, A. Truman, & D. Williams. Classical periodic solutions of the equal-mass $2n$-body problem, $2n$-ion problem, and the n-electron atom problem, *Phys. Lett.* A **99** (1983) 15–18.

[7] B. Dionne, M. Golubitsky, & I. Stewart. Coupled cells with internal symmetry. Part I: wreath products, *Nonlinearity*, **9** (1996) 559–574.

[8] B. Dionne, M. Golubitsky, & I. Stewart. Coupled cells with internal symmetry. Part 2: direct products, *Nonlinearity*, **9** (1996) 575–599.

[9] M. Golubitsky, J. E. Marsden, I. Stewart & M. Dellnitz. The constrained Liapunov-Schmidt procedure and periodic orbits, *Fields Inst. Commun.* **4** (1995) 81–127.

[10] M. Golubitsky & I. Stewart. Hopf bifurcation in the presence of symmetry, *Arch. Ratl. Mech. Anal.* **87** (1985) 107–165.

[11] M. Golubitsky, I. Stewart, & B. Dionne. Coupled cells: wreath products and direct products, in *Dynamics, Bifurcation, and Symmetry*, ed. P. Chossat. Proceedings, Cargèse 1993, NATO ASI Series C **437**, Kluwer, Dordrecht, 1994, 127–138.

[12] M. Golubitsky, I. Stewart, & D. G. Schaeffer. *Singularities and Groups in Bifurcation Theory*, Vol. 2. Springer-Verlag, New York, 1988.

[13] M. Hall. *The Theory of Groups*. Macmillan, New York, 1959.

[14] A. A. Kirillov. *Elements of the Theory of Representations*. Springer-Verlag, Berlin, 1976.

[15] M. Kroon & I. N. Stewart. Detecting the symmetry of attractors for six oscillators coupled in a ring, *Int. J. Bifurcations Chaos* **5** (1995) 209–229.

[16] A. M. Liapunov. The general problems of the stability of motion, Doctoral Dissertation, University of Kharkhov 1892, published by Kharkhov Math. Soc. English transl. (transl. and ed. A. T. Fuller), Taylor and Francis, London, 1992.

[17] K. R. Meyer. Periodic solutions of the N-body problem, *J. Diff. Eq.* **39** (1981) 2–38.

[18] K. R. Meyer & D. S. Schmidt. Librations of central configurations and braided Saturn rings, *Celest. Mech. Dyn. Astron.* **55** (1993) 289–303.

[19] J. A. Montaldi, R. M. Roberts, & I. Stewart. Periodic solutions near equilibria of symmetric Hamiltonian systems, *Phil. Trans. R. Soc. Lond.* A**325** (1988) 237–293.

[20] J. A. Montaldi, R. M. Roberts, & I. Stewart. Existence of nonlinear modes of symmetric Hamiltonian systems, *Nonlinearity* **3** (1990) 695–730.

[21] J. Moser. Periodic orbits near equilibrium and a theorem by Alan Weinstein, *Commun. Pure Appl. Math.* **29** (1976) 727–747.

[22] R. M. Roberts. Nonlinear normal modes of the spring pendulum, in *Papers Presented to Christopher Zeeman*, unpublished duplicated notes, Math. Inst. U. Warwick, June 1988, 207–216.

[23] D. H. Sattinger. Branching in the presence of symmetry, *CBMS-NSF Conference Notes* **40**, SIAM, Philadelphia, 1983, pp. 1-73.

[24] V. S. Varadarajan. *Lie Groups, Lie Algebras, and Their Representations*, Graduate Texts in Math. **102**. Springer-Verlag, New York, 1984.

[25] A. Weinstein. Normal modes for nonlinear Hamiltonian systems, *Invent. Math.* **20** (1973) 47–57.

[26] Z. Xia. Arnold diffusion and oscillatory solutions in the planar three-body problem, *J. Diff. Eq.* **110** (1994) 289–321.

Mathematical Analysis of Sideband Instabilities with Application to Rayleigh-Bénard Convection

A. Mielke

Institut für Angewandte Mathematik, Universität Hannover, W3000 Hannover 1, Germany

Received September 25, 1995; revised manuscript accepted for publication September 25, 1996
Communicated by Jerrold Marsden and Stephen Wiggins

This paper is dedicated to the memory of Juan-Carlos Simo

Summary. We introduce a new method for the analysis of sideband instabilities which are important for periodic patterns appearing in systems close to the instability threshold. The method relies on a two-fold application of the Liapunov–Schmidt reduction procedure, a first application to the nonlinear bifurcation problem and a second application to the linear spectral problem. We obtain rigorous results on the spectrum of the associated linearization in spaces allowing for general sideband perturbations by treating the sideband vector and the spectral parameter as small bifurcation parameters.

We apply the theory to the small roll solutions in the Rayleigh–Bénard convection and derive domains in Rayleigh, Prandtl, and wave number space where the rolls are unstable. We recover the Eckhaus, zigzag, and skew-varicose instabilities obtained earlier by formal methods.

Key words. Sideband instability, convection roll, Navier Stokes equation, Eckhaus zigzag varicose instability

MSC numbers. Primary 76D05, 35B35

1. Introduction

The development of modern bifurcation theory led to the understanding of many nonlinear phenomena in the natural sciences, notably problems involving the spontaneous formation of patterns from a homogeneous state. While the bifurcation of these new patterns is often very well understood mathematically, there is still a need for tools in order to study the associated stability problems. Here we are concerned with problems which are posed on physically extended domains which generate patterns of much smaller wave length. Classical bifurcation and stability theory, e.g., the Liapunov–Schmidt reduction or the center manifold technique, can easily treat the stability of such patterns with respect to the perturbations which have the same type of periodicity as the basic pattern. However, many problems involve instabilities with respect to more general perturbations, such as quasiperiodic or localized ones.

It was Wiktor Eckhaus [Eck65] who introduced a first systematic approach to such sideband instabilities for problems which have one unbounded direction such as the Couette–Taylor problem between infinitely long cylinders. From this starting point a vast body of research emanated in order to understand more complicated situations in hydrodynamic stability theory, e.g., involving mean flow effects or two unbounded spatial directions (cf. [Bu71], [SdP78], [BuB84], [Be94], and for a recent review see [NPL93], [DP94]). Yet, all of these theories lack a sound mathematical basis, since they are built on formal methods involving the method of amplitude equations or the three (or higher) mode interaction theory.

The first rigorous approaches to the Eckhaus instability were given in [MT84], [CE90], where the one-dimensional Swift–Hohenberg problem was considered. A more general mathematical approach to sideband instabilities was developed in [BM95], [BM96], which can detect instabilities of small bifurcating periodic solutions for problems on infinitely long cylindrical domains. The method relies on the theory of spatial center manifolds in which the elliptic nonlinear problem as well as the eigenvalue problem are reduced to ordinary differential equations. In [BM95] a first rigorous proof of the Benjamin–Feir instability of gravity surface waves on a fluid layer of finite depth could be derived.

Here we introduce a new method, which was already outlined in [Mi95], where the sideband instabilities of the roll solutions in the two-dimensional Swift–Hohenberg equation were studied. This method will be generalized here and applied in detail to the sideband instabilities of convection rolls in the Rayleigh–Bénard problem with stress-free boundary conditions. In contrast to [BM95] we base our theory on the Liapunov–Schmidt reduction (LSR) rather than the spatial center manifold reduction. The key observation is that the spectral problem obtained from linearization at the given steady state can also be reduced by the LSR to a finite dimensional eigen-value problem which then contains all the needed information on possible unstable sidebands.

To be more specific let us consider a system of PDEs for a vector-valued function $u(t, x, \tilde{x})$ on the domain $\mathbb{R}^d \times \Sigma$ where $x \in \mathbb{R}^d$, $d \geq 1$, denotes the unbounded variable whereas $\tilde{x} \in \Sigma \subset \mathbb{R}^m$, $m \geq 0$, denotes the bounded cross-sectional variables, namely

$$M \partial_t u = \mathcal{N}(\mu, \partial_x, u) \overset{\text{def}}{=} A_\mu(\partial_x)u + N(\mu, u). \tag{1.1}$$

Here $A_\mu(\partial_x)$ is an elliptic operator on $\mathbb{R}^d \times \Sigma$ where only the derivatives with respect to the unbounded variables x are indicated for later convenience. For the present purposes it suffices to assume that M is a constant bounded linear operator, but generalizations are possible. Moreover, $N(\mu, u) = \mathcal{O}(|u|^2)$, μ is a vector of parameters with $|\mu|$ small, and $u \equiv 0$ is the trivial homogeneous solution.

The bifurcation of steady, spatially periodic patterns can be treated by considering the time-independent problem on a function space Z with given periodicity in $x \in \mathbb{R}^d$. The basic assumption is that the bifurcation problem $\mathcal{N}(\mu, \partial_x, u) = 0$ can be reduced by the LSR, i.e., $A_0(\partial_x): Z \to Y$ has a finite-dimensional kernel Z_0 and co-range Y_0, such that the projections $Q\colon Z \to Z_0$ and $P\colon Y \to Y_0$ are bounded. Let $u = u_0 + u_1$ with $u_0 = Qu$ and consider

$$P\mathcal{N}(\mu, \partial_x, u_0 + u_1) = 0,$$

$$(I - P)\mathcal{N}(\mu, \partial_x, u_0 + u_1) = 0. \tag{1.2}$$

The basic idea is to solve the second equation in (1.2) by the implicit function theorem for $u_1 = \mathcal{U}(\mu, u_0)$ in order to obtain the finite-dimensional bifurcation problem

$$0 = n(\mu, u_0) \overset{\text{def}}{=} P\mathcal{N}(\mu, \partial_x, u_0 + \mathcal{U}(\mu, u_0)). \tag{1.3}$$

Using this algebraic problem in Z_0 the existence of nontrivial patterns can be studied. We refer to [CH88] and [GS85], Ch. VII, for the general theory of the LSR.

Assume now that we have found a family $(\mu, u_\mu) = (\mu, u_{\mu 0} + \mathcal{U}(\mu, u_{\mu 0}))$ of steady states of (1.1). To study the stability of u_μ we consider the linearization of the full problem (1.1) and assume that the perturbation is of the form $v(t, x) = e^{\lambda t} w(x)$. We obtain

$$M(\lambda w) = A_\mu(\partial_x) w + DN(\mu, u_\mu) w. \tag{1.4}$$

This resolvent problem again has to be solved in some function space which should include general enough perturbations, e.g., $w \in L^\infty(\mathbb{R}^d, L^2(\Sigma))$. It is obvious that u_μ is (linearly) unstable if there exists (λ, w) with $\operatorname{Re}\lambda > 0$ and a nontrivial bounded w.

Following the formal approaches to sideband instability (see, e.g., [Bu71], [BuB84], [Be94]), it is natural to look for perturbations in the form $v(t, x) = e^{\lambda t + i\sigma \cdot x} W(x)$ with $\sigma \in \mathbb{R}^d$ and $W \in Z$. Here σ is called the sideband vector and it will be small in our analysis. Hence, in general v is quasiperiodic in $x \in \mathbb{R}^d$ and need not lie in Z. The problem for W reads

$$0 = \mathcal{B}(\mu, \sigma, \lambda) W \overset{\text{def}}{=} \left[D_u \mathcal{N}(\mu, i\sigma + \partial_x, u_\mu) - \lambda M \right] W, \tag{1.5}$$

where $\mathcal{B}(\mu, \sigma, \lambda)$ is a bounded linear operator from Z into Y. Note the appearance of $i\sigma + \partial_x$, which also explains why we display the differential with respect to the unbounded variable. Since $\mathcal{B}(0, 0, 0) = A_0(\partial_x)$, it is obvious that nontrivial solutions of (1.5) can be obtained via LSR in the form $W = W_0 + W_1$ with $W_1 = \mathcal{W}(\mu, \sigma, \lambda) W_0$; and (1.5) is equivalent to the *reduced spectral problem*

$$0 = b(\mu, \sigma, \lambda) W_0 \overset{\text{def}}{=} P\mathcal{B}(\mu, \sigma, \lambda) [W_0 + \mathcal{W}(\mu, \sigma, \lambda) W_0], \tag{1.6}$$

for small (μ, σ, λ) and the same P as in (1.2). Instability of the steady periodic pattern (μ, u_μ) of (1.1) is shown as soon as we find (σ, λ) such that $\det b(\mu, \sigma, \lambda) = 0$ and $\operatorname{Re}\lambda > 0$. This reduction of the spectral problem (1.4) leads to the *Principle of Reduced Instability* (cf. [BM95], [BM96], and [Mi95]) since it is clear that any unstable solution (i.e., $\operatorname{Re}\lambda > 0$) of the reduced spectral problem (1.6) leads to instability of u_μ in the full problem (1.1). However, the opposite conclusion need not be valid, since our reduction requires λ and σ to be small, such that instabilities with large λ or σ are not detected.

Thus, the question of sideband instability has been reduced to a sequence of two related LSRs: (i) the nonlinear bifurcation problem and (ii) the linear spectral problem. The functional analytic set up in both reductions is the same (i.e., Z_0, Z_1, Y_0, Y_1 do not change); in fact in Theorem 2.2 we show

$$b(\mu, 0, 0) = D_{u_0} n(\mu, u_{\mu 0}) \qquad \text{and} \qquad b(0, \sigma, \lambda) = P(A_0(i\sigma + \partial_x) - \lambda M)Q.$$

There are several papers on the so-called *Principle of Reduced Stability* which states that the stability of a steady state $u_\mu \in Z$ of the problem $\partial_t u = \mathcal{N}(\mu, u)$ can be determined from the stability of the steady state $u_{\mu 0} \in Z_0$ of the problem $\partial_t u_0 = n(\mu, u_0)$, cf. [GS85], Ch. I, Thm. 4.1. However, in general the desired result does not hold (see [KL83]) and a very subtle theory was developed to derive sufficient conditions for the validity of the principle [Va84], [Re89], [Da91]. Our principle of reduced instability works the other way round. We construct a slightly more complicated reduced stability problem (involving λ nonlinearly) which contains the exact stability information.

The main part of this paper is the application of the abstract theory to the Bénard problem for a fluid layer between two horizontal plates at distance π, that is, for the flow in the domain $\mathcal{Q}_\infty = \mathbb{R}^2 \times (0, \pi)$. Let (u, θ, p) be nondimensionalized velocity, temperature, and pressure, $\text{Ra} = \mathcal{R}^2$ the Rayleigh, and $\mathcal{P}$ the Prandtl number; then the problem reads

$$\left. \begin{aligned} \partial_t u + (u \cdot \nabla)u + \nabla p - \Delta u - \mathcal{R}\theta e_3 &= 0 \\ \mathcal{P}(\partial_t \theta + (u \cdot \nabla)\theta) - \Delta\theta - \mathcal{R}u_3 &= 0 \\ -\operatorname{div} u &= 0 \end{aligned} \right\} \quad \text{in } \mathcal{Q}_\infty,$$

$$\partial_{x_3} u_1 = \partial_{x_3} u_2 = u_3 = \theta = 0 \quad \text{on } \partial\mathcal{Q}_\infty.$$

The bifurcation parameter is $\mathcal{R} - \mathcal{R}_0$ with the critical Rayleigh number $\text{Ra}_{\text{crit}} = \mathcal{R}_0^2 = 27/4$. The critical wave number is $k_0 = 1/\sqrt{2}$. We have assumed stress-free boundary conditions which enables us to perform all calculations in the reduction process analytically. Of course the problem with no-slip boundary conditions (i.e., $u = 0$ on $\mathcal{Q}_\infty$) is also tractable with the method developed here; however, it is not known how to give the relevant eigenfunctions explicitly. Before going into the analysis we give a definition of instability for the Bénard problem, which is somewhat tricky since the evolution problem cannot be posed in $L^\infty(\mathcal{Q}_\infty)$, as we shall see in Section 3.

The existence of steady Bénard convection rolls is classical and we give, for completeness, an independent proof in which important information for the stability question is prepared. For all $\mathcal{R} > \widetilde{\mathcal{R}}(k, 0) = (1+k^2)^{3/2}/k$ (with $|k-k_0|$ and $\mathcal{R} - \widetilde{\mathcal{R}}(k, 0)$ sufficiently small) there exists a solution which is two-dimensional ($u_2 \equiv 0$, $\partial_{x_2}(u, \theta, p) \equiv 0$) and has period $2\pi/k$ in the x_1-direction:

$$z_{\rho,k}(x_1, x_3) = \begin{pmatrix} u_1 \\ u_2 \\ u_3 \\ \theta \\ p \end{pmatrix} = \rho \begin{pmatrix} \sqrt{2}\cos(kx_1)\cos x_3 \\ 0 \\ \sin(kx_1)\sin x_3 \\ \sqrt{3}\sin(kx_1)\sin x_3 \\ -3\sin(kx_1)\cos x_3 \end{pmatrix} + \mathcal{O}(\rho^2),$$

where $k \approx k_0 = 1/\sqrt{2}$ and ρ is defined such that $\mathcal{R} = (1+k^2)^{3/2}/k + \sqrt{3}\mathcal{P}^2\rho^2/16 + \mathcal{O}(\rho^4)$.

Rescale the periodicity variable $y = kx_1$ with $y \in S^1 = \mathbb{R}/2\pi\mathbb{Z}$. Our construction in Section 4 is built on the basic spaces

$$Y = L^2(\Omega) \times H^1(\Omega)$$

and

$$Z = \{ (u, \theta, p) \in \left[H^2(\Omega)\right]^4 \times H^1(\Omega): \partial_{x_3}u_1 = \partial_{x_3}u_2 = u_3 = \theta = 0 \text{ on } \partial\Omega \},$$

where $\Omega = S^1 \times (0, \pi)$ is the bounded periodicity domain.

The linear operator $A_0: Z \to Y$ of the steady problem at $(\mathcal{R}, k) = (\mathcal{R}_0, k_0)$ is symmetric with respect to the $L^2(\Omega)^5$ scalar product and has a five-dimensional kernel which is spanned by $\xi_1, \ldots, \xi_5$ given by

$$\begin{pmatrix} 1 \\ 0 \\ 0 \\ 0 \\ 0 \end{pmatrix}, \begin{pmatrix} 0 \\ 1 \\ 0 \\ 0 \\ 0 \end{pmatrix}, \begin{pmatrix} 0 \\ 0 \\ 0 \\ 0 \\ 1 \end{pmatrix}, \begin{pmatrix} \sqrt{2}\cos y \cos x_3 \\ 0 \\ \sin y \sin x_3 \\ \sqrt{3}\sin y \sin x_3 \\ -3\sin y \cos x_3 \end{pmatrix}, \begin{pmatrix} \sqrt{2}\sin y \cos x_3 \\ 0 \\ -\cos y \sin x_3 \\ -\sqrt{3}\cos y \sin x_3 \\ 3\cos y \cos x_3 \end{pmatrix}.$$

Note that ξ_1 and ξ_2 describe mean flow effects in x_1 and x_2 direction while ξ_3 corresponds to the trivial addition of a constant to the pressure.

Letting $z_0 = \sum \eta_j\xi_j \in Z_0$, the LSR of the steady problem leads to

$$\tilde{n}(\mathcal{R}, k, \eta) = \begin{pmatrix} g_1(\mathcal{R}, k, \eta_1^2, r)\eta_1 \\ 0 \\ 0 \\ g_4(\mathcal{R}, k, \eta_1^2, r)\eta_4 - g_5(\mathcal{R}, k, \eta_1^2, r)\eta_1\eta_5 \\ g_5(\mathcal{R}, k, \eta_1^2, r)\eta_1\eta_4 + g_4(\mathcal{R}, k, \eta_1^2, r)\eta_5 \end{pmatrix} = 0,$$

where $r = \eta_4^2 + \eta_5^2$. The special structure of $\tilde{n}$ follows from the symmetry properties of the Bénard problem given in Section 3. Since $g_5(\mathcal{R}_0, k_0, 0, 0) \neq 0$, the expansion

$$g_4(\mathcal{R}, k, \eta_1^2, r) = -\frac{\sqrt{3}}{2}(\mathcal{R} - \mathcal{R}_0) + 3(k - k_0)^2 - \frac{7\mathcal{P}^2+34\mathcal{P}-23}{200}\eta_1^2 + \frac{3\mathcal{P}^2}{32}r + \cdots$$

yields immediately the unique solution family $\eta_{\rho,\tau} = (0, \alpha, \beta, \rho\cos\tau, \rho\sin\tau)^T$, $\alpha, \beta \in \mathbb{R}$, $\tau \in [0, 2\pi)$, with $\rho = \sqrt{r}$ and $\mathcal{R} = \tilde{\mathcal{R}}(k, \rho^2)$.

In Section 5 we analyze the stability of the roll solution $z_{\rho,k}$ associated with $\eta = (0, 0, 0, \rho, 0)^T$. We investigate perturbations in the form

$$\hat{z}(t, x_1, x_2, x_3) = e^{\lambda t + i(\sigma_1 x_1 + \sigma_2 x_2)} W(kx_1, x_3) \quad \text{with } W \in Z,$$

which leads to a $(\rho, k, \sigma, \lambda)$-depending linear problem which is treated again by the LSR. Using $W_0 = \sum \omega_j\xi_j \in Z_0$ we find the reduced problem $\tilde{b}(\rho, k, \sigma, \lambda)\omega = 0$ where the matrix $\tilde{b} \in \mathbb{C}^{5\times 5}$ has the form

$$\tilde{b} = \begin{pmatrix} \gamma_{11} & -\sigma_1\sigma_2\gamma_{12} & i\sigma_1 & i\sigma_1\rho\gamma_{14} & \rho\gamma_{15} \\ -\sigma_1\sigma_2\gamma_{21} & \gamma_{22} & i\sigma_2 & i\sigma_2\rho\gamma_{24} & -\sigma_1\sigma_2\rho\gamma_{25} \\ -i\sigma_1 & -i\sigma_2 & 0 & 0 & 0 \\ i\sigma_1\rho\gamma_{41} & i\sigma_2\rho\gamma_{42} & 0 & \gamma_{44} & i\sigma_1\gamma_{45} \\ \rho\gamma_{51} & -\sigma_1\sigma_2\rho\gamma_{52} & 0 & i\sigma_1\gamma_{54} & \gamma_{55} \end{pmatrix},$$

where all γ_{jl} are analytical functions of $(\rho^2, k, \sigma_1^2, \sigma_2^2, \lambda)$. This special form of $\tilde{b}$ exploits the reflection symmetries of the Bénard rolls. Using the structure of the third column and the third row, the problem $\tilde{b}\omega = 0$ can be reduced to $b^*\omega^* = 0$ for an equivalent 3×3 matrix b^* which then corresponds to the matrices which were derived in [BuB84] and [Be94] by purely formal methods. Expanding the eigenvalue equation

$$0 = \Lambda(\rho, k, \sigma, \lambda) \stackrel{\text{def}}{=} \det b^*(\rho, k, \sigma, \lambda),$$

we obtain instability of $z_{\rho,k}$ whenever there exists (σ, λ) such that $\Lambda(\rho, k, \sigma, \lambda) = 0$ and $\operatorname{Re}\lambda > 0$. To the lowest order this equation is

$$0 = |\sigma|^2 \left(\frac{9}{16}(\mathcal{P}+1)^2\lambda^3 + p_2(\rho^2, \sigma_1^2, \sigma_2^2)\lambda^2 + p_1(\rho^2, \sigma_1^2, \sigma_2^2)\lambda + p_0(\rho^2, k, \sigma_1^2, \sigma_2^2) \right),$$
(1.7)

which can be analyzed by Hurwitz's stability criterium [Hu895]:

$$p_j \geq 0 \quad \text{for } j = 0, 1, 2 \quad \text{and} \quad p_1 p_2 - \frac{9}{16}(\mathcal{P}+1)^2|\sigma|^2 p_0 \geq 0. \quad (1.8)$$

These inequalities are necessary and sufficient for all three roots to satisfy $\operatorname{Re}\lambda \leq 0$, and thus any violation gives rise to an instability.

From this expansion the classical instabilities will be derived; the Eckhaus instability with $\sigma = (\sigma_1, 0)$, the zigzag instability with $\sigma = (0, \sigma_2)$, the monotonic skew-varicose (MSV) instability, where $\lambda > 0$ and $\sigma_1\sigma_2 \neq 0$, and the oscillatory skew-varicose (OSV) instability where $|\operatorname{Im}\lambda| \operatorname{Re}\lambda \neq 0$ and $\sigma_1\sigma_2 \neq 0$. The main conclusion is that small convection rolls existing for the parameter values $(\mathcal{R}, k) \approx (\mathcal{R}_0, k_0)$ are unstable if $\mathcal{P} < \mathcal{P}_{\text{MSV}} \approx 0.78197$. For larger $\mathcal{P}$ the rolls are unstable with respect to the MSV instability if

$$\mathcal{R} > \max\{\tilde{\mathcal{R}}(k, 0), \mathcal{R}_{\text{MSV}}(k)\} \quad \text{where } \mathcal{R}_{\text{MSV}}(k) = \mathcal{R}_0 - \frac{6\sqrt{6}}{7}(k - k_0) + \mathcal{O}(|k - k_0|).$$

Moreover, small rolls are unstable with respect to the OSV instability for $k < k_0$ and $\tilde{\mathcal{R}}(k, 0) \leq \mathcal{R} \leq \mathcal{R}_{\text{OSV}}(k)$ where

$$\mathcal{R}_{\text{OSV}}(k) = \mathcal{R}_0 - \frac{6\sqrt{6}\,(\mathcal{P}+1)^2}{3\mathcal{P}^2\sqrt{(\mathcal{P}+1)(\mathcal{P}+5)} + (\mathcal{P}+3)(3\mathcal{P}^2+2\mathcal{P}+2)}(k - k_0) + \mathcal{O}(|k - k_0|).$$

A more exact statement is given in Theorem 5.8.

The instability of all rolls in the case $\mathcal{P} < \mathcal{P}_{\text{MSV}}$ was already reported in [ZS83], but this was neither recovered by the analysis in [BuB84], [Be94] nor by the numerical approach in [BoB85] where only the bounds $\mathcal{R}_{\text{MSV}}$ and $\mathcal{R}_{\text{OSV}}$ were given. This discrepancy led to a considerable discussion in the physics literature ([Be94], [DP94]) usually concluding that the results in [ZS83] are incorrect. Our work reconciles the two analyses: In both cases the Hurwitz criterion (1.8) was only evaluated in certain limits leading to two different sufficient conditions for instability, and the union of these two conditions provides the best answer. For more details see our discussion in Section 6.

We believe that our method leads to calculations which are completely equivalent to most of the formal approaches. However, our sound mathematical basis allows us to validate that certain effects seen in lower order terms are really stable with respect to higher order corrections. As an example we mention the role of Galilean invariance which appears here as a crucial mechanism (see Lemma 5.3). Moreover, the systematic approach based on the LSR enables us to do many calculations automatically either by numerical codes or by computer algebra.

Finally, we warn the reader that, up to now, only sufficient conditions for instability are derived and that proofs of (linear) stability are still missing. In particular, we do not study instabilities which appear with respect to large vectors σ, like the cross-roll instability, cf. [NPL93], [DP94].

Recently, in [KvW96] a first stability result was obtained for the case of the two-dimensional Rayleigh–Bénard problem. Since no dependence on x_2 is present (i.e., $\sigma_2 = 0$), the eigenvalue equation $\Lambda(\rho, k, (\sigma_1, 0), \lambda) = 0$ simplifies greatly, and the only remaining instability is the Eckhaus instability. Between the Eckhaus bounds the stability of the roll solutions is then proved. The stability of rolls in the Swift–Hohenberg problem (see (2.9)) with two unbounded directions (i.e., $\sigma \in \mathbb{R}^2$) is established in [Ku96], [Mi96].

A general theory for the linearized stability of spatially periodic steady states in the Navier-Stokes equations was recently developed in [Sc95a], [Sc95b] and [ScW98].

2. General Theory

We consider systems of partial differential equations which are posed over unbounded domains of the form $\mathcal{Q}_\infty = \mathbb{R}^d \times \Sigma$ with variables $(x, \tilde{x}) \in \mathbb{R}^d \times \Sigma$. We assume for simplicity the form

$$M\partial_t u = A_\mu(\partial_x)u + N(\mu, \partial_x, u) \quad \text{in } \mathcal{Q}_\infty, \tag{2.1}$$

where $A_\mu(\partial_x)$ is a linear differential operator in the $(x, \tilde{x})$ variables and suitable boundary conditions on $\partial \mathcal{Q}_\infty = \mathbb{R}^d \times \partial \Sigma$ are specified. The vector $\mu \in \mathbb{R}^p$ collectively denotes all the parameters of the problem. Note that the problem is translational invariant (no x-dependence) while dependence on the cross-sectional variable $\tilde{x}$ and appearance of derivatives $\partial_{\tilde{x}}$ is allowed but not explicitly displayed. For notational simplicity we have assumed that M is a constant operator; however, it is easy to see that the whole theory goes through for more general $M = M(\mu, \partial_x, u)$.

A periodic pattern $\tilde{u}$ is characterized by a lattice group $\mathcal{L} \subset \mathbb{R}^d$ such that $\tilde{u}(x+\ell, \cdot) = \tilde{u}(x, \cdot)$ for all $\ell \in \mathcal{L}$ and an additional symmetry group $\Gamma \subset O(d)$ which corresponds to rotations and reflections of the pattern. For simplicity of this presentation, we neglect the latter symmetry in this general part and refer to [GSS88] (especially Case Study 4) for a discussion. However, in the application to the Bénard problem given below, we heavily use the underlying symmetries.

The dual lattice group $\mathcal{L}^* \subset \mathbb{R}^d$ is given by

$$\mathcal{L}^* = \{ h \in \mathbb{R}^d : h \cdot \ell \in 2\pi\mathbb{Z} \quad \text{for all} \quad \ell \in \mathcal{L} \},$$

and the periodicity domain $\Omega = \mathcal{T} \times \Sigma \subset \mathbb{R}^d \times \Sigma$ is $\mathcal{Q}_\infty / \mathcal{L} = \mathcal{T} \times \Sigma$, where $\mathcal{T} = \mathbb{R}^d / \mathcal{L}$ is a torus of dimension less than or equal to d. Here we assume that $\mathcal{L}$ contains at least d linearly independent vectors. If the connected component of $\mathcal{L}$ which contains 0 is

$\tilde{d}$-dimensional, then $\mathcal{T}$ is a $(d - \tilde{d})$-dimensional torus. Under this condition on $\mathcal{L}$, the dual lattice $\mathcal{L}^*$ is discrete, and a pattern $\tilde{u}$ with lattice $\mathcal{L}$ can be written as a generalized Fourier series $\tilde{u}(x, \tilde{x}) = \sum_{h \in \mathcal{L}^*} e^{ih \cdot x} b_h(\tilde{x})$ where

$$b_h(\tilde{x}) = \frac{1}{|\mathcal{T}|} \int_{\mathcal{T}} e^{-ih \cdot x} \tilde{u}(x, \tilde{x}) \, dx. \tag{2.2}$$

Varying the parameter μ the lattice of a pattern might slightly change under a smooth family of affine transformation. To fix the lattice group we may rescale $x = T(\mu)y + b(\mu)$ such that the y-variable lives in the constant domain $\mathcal{T}$ associated with the constant group $\mathcal{L}$. Hence, without loss of generality we assume constant $\mathcal{L}$ from now on.

We define now function spaces Z and Y over $\mathcal{Q}_\infty$ which contain $\mathcal{L}$-periodic functions, or, equivalently we consider Z and Y as function spaces over the periodicity domain $\Omega = \mathcal{T} \times \Sigma$. Moreover, the space Z contains the boundary conditions on $\partial \Omega = \mathcal{T} \times \partial \Sigma$ associated with (2.1). Our main assumption is that $A_0(\partial_x)\colon Z \to Y$ is a Fredholm operator of index 0. That means that $A_0(\partial_x)$ is a bounded linear operator from Z to Y, such that $\text{kernel}[A_0(\partial_x)] = \{u \in Z\colon A_0(\partial_x)u = 0\}$ is finite dimensional and that $\text{range}[A_0(\partial_x)] = \{A_0(\partial_x)u\colon u \in Z\} \subset Y$ is a closed subspace of Y with finite codimension which is equal to $\dim(\text{kernel}[A_0(\partial_x)])$. Moreover we assume $M\colon Z \to Y$ and $N(\cdot, \partial_x, \cdot) \in C^m(\mathbb{R}^p \times Z, Y)$ for some $m \geq 1$.

In typical applications the spectral problem

$$\lambda M w = A_0(\partial_x)w, \qquad w \in Z,$$

will have a discrete spectrum (since the domain $\Omega = \mathcal{T} \times \Sigma$ has a compact closure), but this is not necessary for the present method. Of course we are interested in instabilities of bifurcating steady solutions; hence we assume that $\lambda = 0$ is an isolated eigenvalue, i.e., $\dim(\text{kernel}[A_0(\partial_x)]) > 0$. Moreover, the trivial solution should not be linearly unstable; otherwise all bifurcating solutions would inherit this instability. Thus, our method is useful only if $\lambda M - A_0(\partial_x)\colon Z \to Y$ is invertible for all λ with $\text{Re}\,\lambda > 0$. Eigenvalues on the imaginary axis are allowed, but instabilities arising from there will not be captured by our method.

Since $A_0(\partial_x)\colon Z \to Y$ has Fredholm index 0 there exist bounded projections $Q\colon Z \to Z$ and $P\colon Y \to Y$ such that $Z_0 = QZ$ and $Y_0 = PY$ have the same finite dimension. Let $Z_1 = (I - Q)Z$ and $Y_1 = (I - P)Y$ and decompose the steady problem into

$$P[A_\mu(\partial_x)(u_0 + u_1) + N(\mu, \partial_x, u_0 + u_1)] = 0,$$

$$(I - P)[A_\mu(\partial_x)(u_0 + u_1) + N(\mu, \partial_x, u_0 + u_1)] = 0, \tag{2.3}$$

where $u_j \in Z_j$. By construction $A_0\colon Z_1 \to Y_1$ has a bounded inverse which allows us to solve the second equation in (2.3) locally by the implicit function theorem $u_1 = \mathcal{U}(\mu, u_0)$. This solution can be inserted into the first equation to derive the *reduced problem*

$$0 = n(\mu, u_0) \stackrel{\text{def}}{=} P\left[A_\mu(\partial_x)(u_0 + \mathcal{U}(\mu, u_0)) + N(\mu, \partial_x, u_0 + \mathcal{U}(\mu, u_0))\right], \tag{2.4}$$

which is also called the *bifurcation equation*. This reduction of the problem on Z to the finite-dimensional problem on Z_0 is called the *Liapunov–Schmidt reduction* (LSR) which has been a well-established tool in bifurcation theory for more than thirty years, cf. [CH88], [GS85].

We now want to proceed further and investigate the stability of any of the bifurcated periodic patterns u_μ. The linearization around u_μ reads

$$M \partial_t v = A_\mu(\partial_x) v + D_u N(\mu, \partial_x, u_\mu)[v]. \tag{2.5}$$

We call u_μ *linearly unstable* if there exists an initial condition v_0 which is bounded over Q_∞ such that the solution $v = v(t, x)$ of (2.5) with $v(0, x) = v_0(x)$ exists and is not bounded uniformly for $(t, x) \in [0, \infty) \times \mathbb{R}^d$.

A first step is to consider the stability of u_μ with respect to perturbations in Z, i.e., with the same periodicity class. This could be done by using center manifold techniques or by the *principle of reduced stability* which was developed for the case $M = I$. The latter theory is intimately related to the LSR; see [GS85], Ch. I, Thm. 4.1 as well as [KL83], [Va84]. However, as discussed in [Re89], [Da91], the conditions for general situations are very complicated. Thus, we prefer our method which will include this theory as a special case.

As our problem is posed on all of $Q_\infty = \mathbb{R}^d \times \Sigma$ we are allowed to consider perturbations of the form $v(t, x) = e^{i\sigma \cdot x} w(t, x)$ for any sideband vector $\sigma \in \mathbb{R}^d$ and $w(t, \cdot) \in Z$. Since the problem has no time dependence we can also look for special perturbations growing exponentially in time: $v(t, x) = e^{\lambda t + i\sigma \cdot x} W(x)$ with $W \in Z$. Inserting this Ansatz into (2.5) yields a solution whenever W satisfies the spectral problem

$$0 = \mathcal{B}(\mu, \sigma, \lambda) W \overset{\text{def}}{=} \left[A_\mu(i\sigma + \partial_x) + D_u N(\mu, i\sigma + \partial_x, u_\mu) - \lambda M \right] W. \tag{2.6}$$

Note that $A_\mu(i\sigma + \partial_x) W = e^{-i\sigma \cdot x} \left(A_\mu(\partial_x) [e^{i\sigma \cdot x} W] \right)$ and similarly for the term $D_u N$. (Here we see the reason for displaying ∂_x explicitly.)

The key idea of our method is to find nontrivial solutions of this linear problem by the same method as we used for the nonlinear problem, namely the LSR. Since $\mathcal{B}(0, 0, 0) = A_0(\partial_x)$ we can use the same projections P and Q as for (2.3) and let $W = W_0 + W_1$ with $W_0 = QW$.

$$P\mathcal{B}(\mu, \sigma, \lambda)[W_0 + W_1] = 0,$$

$$(I - P)\mathcal{B}(\mu, \sigma, \lambda)[W_0 + W_1] = 0. \tag{2.7}$$

Again the second equation can be solved locally for small (μ, σ, λ) as $W_1 = \mathcal{W}(\mu, \sigma, \lambda) W_0$; and we obtain the reduced spectral problem

$$0 = b(\mu, \sigma, \lambda) W_0 \overset{\text{def}}{=} P\mathcal{B}(\mu, \sigma, \lambda)[W_0 + \mathcal{W}(\mu, \sigma, \lambda) W_0]. \tag{2.8}$$

Thus nontrivial solutions W_0 can be found if and only if $\det b(\mu, \sigma, \lambda) = 0$. We summarize the result as follows.

Theorem 2.1 (Principle of Reduced Instability). *Let (μ, u_μ) be a small steady solution of (2.1) obtained by LSR as indicated above. Let $b(\mu, \sigma, \lambda): Z_0 \to Y_0$ in (2.8) describe the reduced spectral problem around (μ, u_μ). Then, u_μ is spectrally unstable if there exists (σ, λ) with $\operatorname{Re} \lambda > 0$ such that $\det b(\mu, \sigma, \lambda) = 0$.*

The result is now obvious, since the assumption of the theorem implies the existence of a nontrivial solution W_0 for (2.8) which immediately leads to the perturbation

$$v(t, x) = e^{\lambda t + i\sigma \cdot x}(W_0(x) + (\mathcal{W}(\mu, \sigma, \lambda)W_0)(x)),$$

which is bounded over $Q_\infty = \mathbb{R}^d \times \Sigma$ and grows exponentially in time.

The name "sideband instability" comes about as follows. The basic pattern u_μ consists of modes which have wave vectors $h \in \mathcal{L}^*$, while the perturbation $v(t, \cdot)$ has modes with wave vectors $h + \sigma$. Of course, the name sideband is only justified if $|\sigma|$ is small compared to the distances within $\mathcal{L}^*$. When treating $\sigma \in \mathbb{R}^d$ as a real parameter the set $S(\mu) \subset \mathbb{R}^d$ of instability points σ for u_μ is an open set

$$S(\mu) = \{\sigma \in \mathbb{R}^d: \text{ there exists } \lambda \in \mathbb{C} \text{ with } \operatorname{Re}\lambda > 0 \text{ and } \det b(\mu, \sigma, \lambda) = 0\}.$$

If $\sigma = 0$ is contained in $S(\mu)$ then we already have instability with respect to periodic perturbations in Z. More important is the case $0 \notin S(\mu)$, which has two subcases: (i) if $0 \in \overline{S(\mu)}$ (closure of $S(\mu)$ in $\mathbb{R}^d$), we have an *attached* sideband of unstable modes, and (ii) if $0 \notin \overline{S(\mu)}$, we have a *detached* sideband instability.

Example. For a simple application of this theory we refer to [Mi95]. There the Swift-Hohenberg equation

$$\partial_t u = -(1 + \Delta)^2 u + \varepsilon^2 u - u^3, \qquad x \in \mathbb{R}^2, \tag{2.9}$$

is considered which has x_2-independent roll solutions $u_{\varepsilon,k}(x_1) = \rho\cos(kx_1) + \mathcal{O}(\rho^2)$ where $\rho = \sqrt{\frac{8}{3}(\varepsilon^2 - (1 - k^2)^2)}$. The space $Z_0 = Y_0 = \operatorname{span}\{\cos(kx_1), \sin(kx_1)\}$ is two-dimensional and the reduced spectral problem leads to the 2×2-matrix

$$\widetilde{b}(\varepsilon, k, \sigma_1, \sigma_2, \lambda) = \begin{pmatrix} v(k, \sigma) + c(\varepsilon, k) - \lambda & i\delta(k, \sigma) \\ -i\delta(k, \sigma) & v(k, \sigma) - \lambda \end{pmatrix}$$

$$+ \begin{pmatrix} \mathcal{O}(\rho^4(|\sigma|^2 + |\lambda|)) & \mathcal{O}(\rho^4(|\sigma| + |\lambda|)) \\ \mathcal{O}(\rho^4(|\sigma| + |\lambda|)) & \mathcal{O}(\rho^4(|\sigma|^2 + |\lambda|)) \end{pmatrix},$$

with coefficients $c(\varepsilon, k) = -\frac{3}{4}\rho^2 + \mathcal{O}(\rho^4)$, $v(k, \sigma) = -4\sigma_1^2 + 2(1 - k^2)|\sigma|^2 - |\sigma|^4$, and $\delta(k, \sigma) = 4\sigma_1 k(1 - k^2 - |\sigma|^2)$. As a result one finds two small eigenvalues having the expansion

$$\lambda_1 = c(\varepsilon, k) + \mathcal{O}(|\sigma|^2),$$

$$\lambda_2 = \frac{8}{c(\varepsilon, k)}(3(1 - k^2)^2 - \varepsilon^2 + \mathcal{O}(\rho^2))\sigma_1^2 + \frac{2}{c(\varepsilon, k)}(1 - k^2 + \mathcal{O}(\rho^2))\sigma_2^2 + \mathcal{O}(|\sigma|^4).$$

The second eigenvalue is positive for sufficiently small $\sigma \neq 0$ if one of the coefficients of σ_j^2 is positive. Hence, we find the *Eckhaus instability* (with $\sigma = (\sigma_1, 0)$) for $|1 - k^2| > \varepsilon/\sqrt{3} + \mathcal{O}(\varepsilon^2)$ and the *zigzag instability* (with $\sigma = (0, \sigma_2)$) for $k < 1 + \mathcal{O}(\varepsilon^2)$. Moreover, in [Mi95] the sets $S(\varepsilon, k)$ of all unstable sideband vectors are characterized.

A similar instability theory was developed in [BM95], [BM96]; however there the spatial center manifold reduction (SCMR) was used instead of the LSR. Although the use of SCMR and the use of LSR are very similar, there are certain restrictions and advantages for each method. The LSR applies only to periodic patterns but is technically simpler and available for problems with several unbounded directions. The SCMR is restricted to one unbounded spatial variable but allows for more general solutions; additionally it gives more information on the spatial dynamics.

In both cases it is important to note that only sufficient conditions for instability are derived. Even if we show that for all small (μ, σ) the function $\det b(\mu, \sigma, \lambda)$ has no small solutions λ with $\operatorname{Re} \lambda > 0$ we cannot conclude stability. There might be unstable modes with either σ or λ large which are not captured by the local reduction methods. Only if more global information is at hand, which allows us to conclude that these local modes are the most unstable ones, then a stability result can be derived; see [Ku96], [KvW96], [Mi96].

One of the major advantages of the unified treatment of the nonlinear bifurcation problem and the linear stability problem is that all the functional analytic framework and much of the computational effort have to be done only once. As in the above example it is often sufficient to treat the linear stability problem in only two steps. First we can calculate the LSR of the linear problem only (linearization at $u \equiv 0$), which has a simple structure due to the translational invariance of the problem. Second we can find the lowest order influence of the nonlinearity (i.e., $D_u N(\mu, \partial_x, u_\mu)$) by the following theorem which was already used in [Mi95], Thm. 2.1 (see also [BM95] for the analogue in the case of the SCMR).

Theorem 2.2. *Let $u_1 = \mathcal{U}(\mu, u_0)$ be obtained from the LSR of (2.3) such that $n(\mu, u_0) = 0$ (i.e. (2.4)) is the reduced problem. Then the reduction $W_1 = \mathcal{W}(\mu, \sigma, \lambda) W_0$ of the spectral problem (2.6) at $(\mu, u_\mu) = (\mu, u_{\mu 0} + \mathcal{U}(\mu, u_{\mu 0}))$ satisfies*

$$\mathcal{W}(\mu, 0, 0) = D_{u_0} \mathcal{U}(\mu, u_{\mu 0}) \quad in \ \mathcal{L}(Z_0, Y_1),$$

and

$$b(\mu, 0, 0) = D_{u_0} n(\mu, u_{\mu 0}) \quad in \ \mathcal{L}(Z_0, Y_0).$$

Proof. We have $n(\mu, u_0) = P\mathcal{N}(\mu, u_0 + \mathcal{U}(\mu, u_0))$ and $(I - P)\mathcal{N}(\mu, u_0 + \mathcal{U}(\mu, u_0)) = 0$. Taking the u_0-derivative at $(\mu, u_{\mu 0})$ in both equations gives

$$D_{u_0} n(\mu, u_{\mu 0})[W_0] = P D_u \mathcal{N}(\mu, u_\mu)\big(W_0 + D_{u_0} \mathcal{U}(\mu, u_{\mu 0})[W_0]\big),$$

$$(I - P) D_u \mathcal{N}(\mu, u_\mu)\big(W_0 + D_{u_0} \mathcal{U}(\mu, u_{\mu 0})[W_0]\big) = 0.$$

However, by the uniqueness of the reduction functions $\mathcal{U}$ and $\mathcal{W}$ we find $\mathcal{W}(\mu, 0, 0) = D_{u_0} \mathcal{U}(\mu, u_{\mu 0})$, and thus $b(\mu, 0, 0) = D_{u_0} n(\mu, u_{\mu 0})$ follows. $\square$

Recalling that we started out with a translationally invariant problem and noting that u_μ really depends nontrivially on $x \in \mathbb{R}^d$, we see that some of the derivatives $\partial_{x_j} u_\mu$ are nontrivial and satisfy $D_u \mathcal{N}(\mu, u_\mu) W = 0$. These vectors as well as associated Jordan blocks lead to nontrivial null vectors and Jordan blocks in the reduced problem. This will be important in the Bénard problem where the Galilean invariance enforces a Jordan block of length 2.

Theorem 2.3. *(a) If $D_u \mathcal{N}(\mu, u_\mu)\tilde{w} = 0$, then $b(\mu, 0, 0)Q\tilde{w} = 0$.*
    (b) If, in addition, $M\tilde{w} = D_u \mathcal{N}(\mu, u_\mu)\tilde{v}$, then $b(\mu, 0, 0)Q\tilde{v} + \frac{\partial}{\partial\lambda}b(\mu, 0, 0)Q\tilde{w} = 0$.

Proof. (a) We obviously have $\tilde{w}_1 = (I - Q)\tilde{w} = \mathcal{W}(\mu, 0, 0)\tilde{w}_0$ where $\tilde{w}_0 = Q\tilde{w}$. Hence, the result is clear.

(b) The relation of $\tilde{v}$ and $\tilde{w}$ can also be written as $\mathcal{B}(\mu, 0, 0)\tilde{v} + \frac{\partial}{\partial\lambda}\mathcal{B}(\mu, 0, 0)\tilde{w} = 0$, which is analogous to the result for the reduced problem. For all λ we have

$$(D_u \mathcal{N}(\mu, u_\mu) - \lambda M)(\lambda\tilde{v} + \tilde{w}) + \lambda^2 M\tilde{v} = 0.$$

Now choose any bounded linear operator K with $K\tilde{v} = 0$ and $K\tilde{w} = M\tilde{v}$ and consider the LSR for the linear problem

$$(D_u \mathcal{N}(\mu, u_\mu) - \lambda M + \delta K)W = 0,$$

with the same splitting $W = W_0 + W_1$ with $W_0 = QW$ as above. For small $(\lambda, \delta) \in \mathbb{C}^2$ we obtain a reduced problem $m(\mu, \lambda, \delta)W_0 = 0$. By Theorem 2.2, we know that $m(\mu, 0, 0)$ as well as $b(\mu, 0, 0)$ coincide with $D_{u_0}n(\mu, u_{\mu 0})$.

By construction $(D_u \mathcal{N} - \lambda M + \lambda^2 K)(\lambda\tilde{v} + \tilde{w}) \equiv 0$, and hence after reduction we have $m(\mu, \lambda, \lambda^2)Q(\lambda\tilde{v} + \tilde{w}) \equiv 0$. Obviously the relation $b(\mu, 0, \lambda) = m(\mu, \lambda, \lambda^2) + \mathcal{O}(|\lambda|^2)$ holds. Hence, $b(\mu, 0, 0)Q\tilde{w} = 0$ and $b(\mu, 0, 0)Q\tilde{v} + \partial_\lambda b(\mu, 0, 0)Q\tilde{w} = 0$, which is the desired result. $\square$

We mention another important consequence of the translational invariance of the problem. The linear problem

$$(A_0(i\sigma + \partial_x) - \lambda M)w = 0$$

has x-independent coefficients. Define Z^Σ and Y^Σ to be the Banach spaces of those functions on Σ which can be obtained as Fourier coefficients in (2.2). Then, $\mathcal{B}(0, \sigma, \lambda) = A_0(i\sigma + \partial_x) - \lambda M$ maps the function $u = e^{ih\cdot x}\phi$ with $\phi \in Z^\Sigma$ onto $e^{ih\cdot x}\psi$, where

$$\psi = [A_0(i\sigma + ih) - \lambda M]\phi \in Y^\Sigma.$$

Thus, we are left with a differential operator on the domain Σ only, which degenerates to an algebraic problem if Σ is zero-dimensional as in the example of the Swift–Hohenberg equation (2.9). Now the kernel Z_0 of $\mathcal{B}(0, 0, 0)$ has the representation

$$Z_0 = \left\{ \sum_{j=0}^{p} e^{ih^{(j)}\cdot x} \left(\sum_{l=1}^{m_j} a_{jl}\phi_{jl} \right) : a_{jl} \in \mathbb{C} \text{ with } a_{p-j,l} = \overline{a_{jl}} \right\},$$

where $h^{(j)} = -h^{(p-j)} \in \mathcal{L}^*$, $\phi_{jl} = \overline{\phi_{p-j,l}} \in Z^\Sigma$, and $\dim Z_0 = m_0 + \cdots + m_p$.

This shows that the reduced problem $b(0, \sigma, \lambda)$ decouples into $(p+1)$ blocks of lengths $m_0, m_1, \ldots, m_p$, respectively. Write the projections $P\colon Y \to Y_0$ and $Q\colon Z \to Z_0$ in the form

$$Pv = \sum_{j=0}^{p} \sum_{l=1}^{m_j} \left\langle \int_T e^{-ih^{(j)} \cdot y} v(y)\, dy, \widetilde{\psi}_{jl} \right\rangle_\Sigma e^{ih^{(j)} \cdot x}\, \psi_{jl},$$

$$Qw = \sum_{j=0}^{p} \sum_{l=1}^{m_j} \left\langle \int_T e^{-ih^{(j)} \cdot y} w(y)\, dy, \widetilde{\phi}_{jl} \right\rangle_\Sigma e^{ih^{(j)} \cdot x}\, \phi_{jl},$$

where $\langle \cdot, \cdot \rangle_\Sigma$ is the L^2-scalar product on Σ. The block of $b(0, \sigma, \lambda)$ which is associated with $h^{(j)}$ is constructed as follows. For $l = 1, \ldots, m_j$, solve the following linear problem for $(W_{jl}, \alpha_{jl1}, \ldots, \alpha_{jlm_j}) \in Z^\Sigma \times \mathbb{C}^{m_j}$:

$$(A_0(i\sigma + ih^{(j)}) - \lambda M)[\phi_{jl} + W_{jl}] = \sum_{q=1}^{m_j} \alpha_{jlq} \psi_{jq},$$

$$\langle W_{jl}, \widetilde{\psi}_{jq} \rangle = 0 \text{ for } q = 1, \ldots, m_j.$$

For sufficiently small (σ, λ) there is a unique solution, since the bordered operator

$$\begin{pmatrix} A_0(ih^{(j)}) & \psi_{j1} & \cdots & \psi_{jm_j} \\ \langle \cdot, \widetilde{\psi}_{j1} \rangle & 0 & \cdots & 0 \\ \vdots & \vdots & & \vdots \\ \langle \cdot, \widetilde{\psi}_{jm_j} \rangle & 0 & \cdots & 0 \end{pmatrix} \quad \text{from } Z^\Sigma \times \mathbb{C}^{m_j} \text{ into } Y^\Sigma \times \mathbb{C}^{m_j}$$

has a bounded inverse due to the construction according to Fredholm's alternative. With the solution $W_{jl}(\sigma, \lambda)$ and $\alpha_{jlq}(\sigma, \lambda)$ the reduced linear problem satisfies the relations

$$\mathcal{W}(0, \sigma, \lambda) W_0 = e^{ih^{(j)} \cdot x} \sum_{l=1}^{m_j} a_l W_{jl}(\sigma, \lambda),$$

$$b(0, \sigma, \lambda) W_0 = e^{ih^{(j)} \cdot x} \sum_{q=1}^{m_j} \left[\sum_{l=1}^{m_j} a_l \alpha_{jlq}(\sigma, \lambda) \right] \psi_{jq},$$

for $W_0 = e^{ih^{(j)} \cdot x}(\sum_{l=1}^{m_j} a_l \phi_{jl})$. Hence, the functions $\alpha_{jlq}(\sigma, \lambda)$ appear as matrix entries when $b(0, \sigma, \lambda)$ is represented in the given bases of Z_0 and Y_0.

It may be also convenient to use a real basis for Z_0 and Y_0. Consider the case, to be used in the Bénard problem below, that $h^{(j)} \neq 0$ and $m_j = 1$. The associated complex basis elements are $\zeta_j = e^{ih^{(j)} \cdot x} \phi_{j1}$ and $\zeta_{p-j} = \overline{\zeta}_j$ in Z_0 and $\kappa_j = e^{ih^{(j)} \cdot x} \psi_{j1}$ and $\kappa_{p-j} = \overline{\kappa}_j$ in Y_0, such that $b(0, \sigma, \lambda)\zeta_j = \alpha_{j,1,1}(\sigma, \lambda)\kappa_j$ and $b(0, \sigma, \lambda)\zeta_{p-j} = \alpha_{p-j,1,1}(\sigma, \lambda)\kappa_{p-j}$.

Lemma 2.4. *For any $(c_1, c_2) \in \mathbb{R}^2$ we have*

$$b(0, \sigma, \lambda) \left(c_1 \operatorname{Re} \zeta_j + c_2 \operatorname{Im} \zeta_j \right) = d_1 \operatorname{Re} \kappa_j + d_2 \operatorname{Im} \kappa_j$$

with

$$\begin{pmatrix} d_1 \\ d_2 \end{pmatrix} = \begin{pmatrix} v(\sigma, \lambda) & -i\delta(\sigma, \lambda) \\ i\delta(\sigma, \lambda) & v(\sigma, \lambda) \end{pmatrix} \begin{pmatrix} c_1 \\ c_2 \end{pmatrix},$$

where $2v(\sigma, \lambda) = \alpha_{j11}(\sigma, \lambda) + \overline{\alpha_{j11}(-\sigma, \overline{\lambda})}$ *and* $2\delta(\sigma, \lambda) = \alpha_{j11}(\sigma, \lambda) - \overline{\alpha_{j11}(-\sigma, \overline{\lambda})}$.

Proof. As we started with a real problem we have $\overline{\mathcal{B}(0, \sigma, \lambda)} = \mathcal{B}(0, -\sigma, \overline{\lambda})$ which leads by the uniqueness of the LSR to $\overline{\mathcal{W}(0, \sigma, \lambda)} = \mathcal{W}(0, -\sigma, \overline{\lambda})$ and hence to $\overline{b(0, \sigma, \lambda)} = b(0, -\sigma, \overline{\lambda})$. In particular, this implies $\alpha_{p-j,11}(-\sigma, \overline{\lambda}) = \overline{\alpha_{j11}(\sigma, \lambda)}$ which leads to the desired result. $\qquad\qquad\qquad\qquad\qquad\qquad\qquad\qquad\qquad\qquad\qquad\qquad\qquad\qquad$ □

The above considerations show that the reduced spectral problem $b(\mu, \sigma, \lambda) W_0 = 0$ can be calculated to the lowest order with little additional effort compared to the effort for obtaining the reduced bifurcation problem. Assume that $b(0, \sigma, \mu)$ and $n(\mu, u_0)$ are known (to a sufficiently high order), then $b(0, 0, 0) = 0$ and Theorem 2.2 implies

$$b(\mu, \sigma, \lambda) = b(0, \sigma, \lambda) + D_{u_0}n(\mu, u_{\mu 0}) + \mathcal{O}(|\mu|^{\varepsilon_1}(|\sigma|^{\varepsilon_2} + |\lambda|^{\varepsilon_3})), \qquad (2.10)$$

with $\varepsilon_1 = \varepsilon_2 = \varepsilon_3 = 1$. Often additional properties of the problem (e.g., symmetries) provide even larger ε_j. The order of approximation in (2.10) turns out to be sufficient in the Swift–Hohenberg problem, yet for the Bénard problem treated below we have to calculate a few more coefficients in order to study the relevant instabilities.

3. The Rayleigh–Bénard Problem

The Rayleigh–Bénard convection is described by the Navier–Stokes equations for the divergence-free velocity field u of the fluid and the temperature distribution θ, which generates a buoyancy force

$$\left.\begin{array}{rcl} \partial_t u + (u \cdot \nabla)u + \nabla p - \mathcal{R}\theta e_3 - \Delta u &=& 0, \\ \mathcal{P}(\partial_t \theta + u \cdot \nabla\theta) - \mathcal{R}u_3 - \Delta\theta &=& 0, \\ \nabla \cdot u &=& 0, \end{array}\right\} \text{ for } x \in \mathbb{R}^2 \times (0, \pi); \qquad (3.1)$$

with stress-free boundary conditions

$$\partial_3 u_1 = \partial_3 u_2 = u_3 = \theta = 0 \quad \text{for } x_3 = 0 \quad \text{or} \quad x_3 = \pi.$$

The parameter $\mathcal{R}$ is related to the Rayleigh number via $\text{Ra} = \mathcal{R}^2$ and used as a bifurcation parameter. The Prandtl number $\mathcal{P} > 0$ is treated as fixed but arbitrary and therefore normally the dependence on $\mathcal{P}$ will be omitted. For additional physical background we refer to [BuB84], [Be94] and the references therein. We have used the special scaling of [SvW92] which makes the linearized operator symmetric in the standard L^2 scalar product. We will use the abbreviation $z = (u, \theta, p)^T \in \mathbb{R}^5$.

The system has a rich symmetry structure that plays an important role in the subsequent analysis. The Euclidean group of $\mathbb{R}^2$ is generated by translations by $\tau = (\tau_1, \tau_2) \in \mathbb{R}^2$

and rotations through the angle $\alpha \in S^1 = \mathbb{R}/_{2\pi \mathbb{Z}}$

$$
\begin{pmatrix} u \\ \theta \\ p \end{pmatrix}(x) \mapsto \begin{pmatrix} R(-\alpha)u \\ \theta \\ p \end{pmatrix}(R(\alpha)x + (\tau_1, \tau_2, 0)),
$$

where

$$
R(\alpha) = \begin{pmatrix} \cos\alpha & -\sin\alpha & 0 \\ \sin\alpha & \cos\alpha & 0 \\ 0 & 0 & 1 \end{pmatrix}.
$$

We have reflection symmetries $K^{(j)}$, $j = 1, 2$, corresponding to reflections of the x_j-coordinate

$$
K^{(1)}: z(x) \mapsto (-u_1, u_2, u_3, \theta, p)(-x_1, x_2, x_3),
$$

$$
K^{(2)}: z(x) \mapsto (u_1, -u_2, u_3, \theta, p)(x_1, -x_2, x_3).
$$

There is a further reflection symmetry about the midplane which arises from the Boussinesq approximation:

$$
K^{(3)}: z(x) \mapsto (u_1, u_2, -u_3, -\theta, p)(x_1, x_2, \pi - x_3).
$$

See also [GSS88], Case Study 4, for a discussion of these symmetries. However, there is an additional (generalized) symmetry, which comes from the Galilean invariance associated with a moving frame with speed $v = (v_1, v_2, 0)^T$. This transformation is given explicitly below in (3.2), and it will play an important role in the discussion of the sideband instabilities (see Lemma 5.3).

When trying to set up an instability theory for the Navier–Stokes problem in unbounded domains one has to be aware of the problems at infinity when no decay assumptions on the flow are imposed. Since it is our aim to allow for periodic solutions (which are in $W^{k,\infty}(\mathbb{R}^2 \times (0, \pi))$) we have to impose some conditions at infinity in order to control the mass flux there. If $z = (u, \theta, p)$ solves the Bénard problem and $\tilde{x} = (\tilde{x}_1(t), \tilde{x}_2(t), 0)$ and $q = q(t) \in \mathbb{R}$ are any smooth functions, then

$$
\begin{aligned}
\tilde{u}(t, x) &= u(t, x - \tilde{x}(t)) + \dot{\tilde{x}}(t), \\
\tilde{\theta}(t, x) &= \theta(t, x - \tilde{x}(t)), \\
\tilde{p}(t, x) &= p(t, x - \tilde{x}(t)) + q(t) - \ddot{\tilde{x}}(t) \cdot x,
\end{aligned} \tag{3.2}
$$

solves the Bénard problem as well. For $\tilde{x}(t) = (v_1 t, v_2 t, 0)$ this is exactly the Galilean transformation into a moving frame with constant speed v.

A rigorous theory can be developed when periodicity in (x_1, x_2) on some periodicity domain is assumed. For simplicity we restrict our analysis to the case of a large rectangle $T_\ell = (0, \ell_1) \times (0, \ell_2)$. The periodicity of velocity and temperature is reasonable; however there is a choice for the pressure, namely to impose (i) periodicity on p, or (ii) periodicity in ∇p only. According to the above nonuniqueness, case (ii) needs an extra condition on the mean value of the first two components of u. Since there is no horizontal forcing

in the problem, it is easy to see that the mean values of the horizontal components of u
satisfy

$$\frac{d}{dt} \int_{T_\ell \times (0,\pi)} \begin{pmatrix} u_1 \\ u_2 \end{pmatrix} dx + \int_{T_\ell \times (0,\pi)} \begin{pmatrix} \partial_{x_1} p \\ \partial_{x_2} p \end{pmatrix} dx = 0,$$

in case (i) and (ii). Hence, while the mean flow is constant in case (i), the pressure
gradient has zero mean in case (ii). From this we conclude that case (ii) is a special case
of (i), namely that with mean flow zero.

Thus, we can use the following spaces. Let $\mathcal{Q}_\ell = T_\ell \times (0, \pi)$ and

$$\mathcal{H}_\ell = \{ (u, \theta) \in H^1(\mathcal{Q}_\ell)^4 \colon \operatorname{div} u \equiv 0, u_3 = 0 \text{ on } T_\ell \times \{0, \pi\} \}.$$

The problem (3.1) can be written as an evolution problem

$$\frac{d}{dt} \begin{pmatrix} u \\ \theta \end{pmatrix} = L_{\ell, \mathcal{R}} \begin{pmatrix} u \\ \theta \end{pmatrix} + F(\mathcal{R}, u, \theta) \tag{3.3}$$

in $\mathcal{H}_\ell$ after applying the suitable Stokes projection. In [SvW92] it is shown that this
initial value problem is locally well posed for any ℓ.

Now assume we have found a periodic steady state $\widetilde{U} = (\widetilde{u}, \widetilde{\theta}) \in \mathcal{H}_\ell$. We may
investigate its spectral properties in the space $\mathcal{H}_\ell$ by classical methods from functional
analysis. Here we want to take into account more general perturbations, but keeping the
restriction to periodicity. We solve this apparent dilemma by considering the solution $\widetilde{z}$
as a solution on all larger domains which are compatible with the period of the pattern
$\widetilde{z}$. Thus, our instability definition reads as follows.

Definition 3.1. *Assume that $\widetilde{U}$ is a steady solution of the Bénard problem satisfying*
$\widetilde{U}(x + (n_1 \ell_1, n_2 \ell_2, 0)) = \widetilde{U}(x)$ *for all $x \in \mathcal{Q}_\infty = \mathbb{R}^2 \times (0, \pi)$ and all $n_1, n_2 \in \mathbb{Z}$, where*
$\ell_1, \ell_2 > 0$ *are fixed.*

*Then, the solution $\widetilde{U}$ is called (linearly) **unstable** on $\mathcal{Q}_\infty$ if there exists any $n_1, n_2 > 0$*
such that $\widetilde{U}$ is a (linearly) unstable equilibrium point of (3.3) when considered in the
function space $\mathcal{H}_{(n_1\ell_1, n_2\ell)}$.

In this definition the case in which (n_1, n_2) is very large is of particular interest. A
sideband instability corresponds to the case where (neutral) stability holds for small
(n_1, n_2) but instability appears for large n. As developed above we will study the instability problem by introducing the sideband vector $\sigma = (\sigma_1, \sigma_2)$ as a small continuous
parameter. However, according to the definition we have to consider perturbations in
$\mathcal{H}_{(n_1\ell_1, n_2\ell_2)}$, namely in the form $v(t, x) = e^{\lambda t + i\sigma \cdot x} W(x)$ with $W \in \mathcal{H}_\ell$. Thus, σ should
satisfy $e^{i\sigma \cdot (n_1\ell_1, n_2\ell_2)} = 1$, or, which is equivalent, lie in the discrete set

$$\Sigma_{(n_1\ell_1, n_2\ell_2)} = \{ (\tfrac{2\pi m_1}{n_1\ell_1}, \tfrac{2\pi m_2}{n_2\ell_2}) \in \mathbb{R}^2 \colon m_1, m_2 \in \mathbb{Z} \}.$$

Hence, if our analysis with continuous $\sigma \in \mathbb{R}^2$ provides any open set $S(\mathcal{R}, \ell) \subset \mathbb{R}^2$
in which nontrivial solutions with $\operatorname{Re} \lambda > 0$ exists, then by making n sufficiently large
we can find points of $\Sigma_{(n_1\ell_1, n_2\ell_2)}$ lying inside of $S(\mathcal{R}, \ell)$. Thus instability in the sense of
the above definition can be concluded. Using this philosophy, we are able to establish
rigorous instability results for the Bénard rolls on suitably large periodicity domains.

4. The Linear Problem and Bifurcation of Steady Rolls

We look for the solutions of the linear problem (linearized at the trivial flow $u = 0$). We use a Fourier Ansatz in all three spatial directions, where the x_3-direction has boundary conditions which make the Fourier spectrum discrete:

$$z = e^{i\sigma \cdot x + \lambda t} \varphi_n(\alpha)(x_3) \quad \text{with } \varphi_n(\alpha)(x_3) = \begin{pmatrix} \alpha_1 \cos(nx_3) \\ \alpha_2 \cos(nx_3) \\ \alpha_3 \sin(nx_3) \\ \alpha_4 \sin(nx_3) \\ \alpha_5 \cos(nx_3) \end{pmatrix}, \tag{4.1}$$

$\sigma = (\sigma_1, \sigma_2, 0)^T$, $\alpha \in \mathbb{C}^5$, and $\alpha_3 = \alpha_4 = 0$ for $n = 0$. There are nontrivial solutions of the linear problem if α satisfies $F_n(\mathcal{R}, \sigma_1, \sigma_2, \lambda)\alpha = 0$, where $F_n(\mathcal{R}, \sigma_1, \sigma_2, \lambda) =$

$$\begin{pmatrix} \lambda + n^2 + |\sigma|^2 & 0 & 0 & 0 & i\sigma_1 \\ 0 & \lambda + n^2 + |\sigma|^2 & 0 & 0 & i\sigma_2 \\ 0 & 0 & \lambda + n^2 + |\sigma|^2 & -\mathcal{R} & -n \\ 0 & 0 & -\mathcal{R} & \mathcal{P}\lambda + n^2 + |\sigma|^2 & 0 \\ -i\sigma_1 & -i\sigma_2 & -n & 0 & 0 \end{pmatrix}, \tag{4.2}$$

and $n = 1, 2, \ldots$. Calculating the determinant we find

$$\det F_n = (\lambda + n^2 + |\sigma|^2)\Big[\mathcal{R}^2|\sigma|^2 - (\lambda + n^2 + |\sigma|^2)(\mathcal{P}\lambda + n^2 + |\sigma|^2)(n^2 + |\sigma|^2)\Big],$$

and, from $\det F_n = 0$, the eigenvalues are

$$\lambda_1 = -n^2 - |\sigma|^2, \qquad \lambda_{2,3} = -\frac{1}{2\mathcal{P}}(\mathcal{P}+1)(n^2 + |\sigma|^2)\Big(1 \pm \sqrt{1 - 4\mathcal{P}\delta/(\mathcal{P}+1)^2}\Big),$$

where $\delta = 1 - \mathcal{R}^2|\sigma|^2/(n^2 + |\sigma|^2)^3$. In particular, for small δ, we obtain

$$\lambda_3 = -\frac{n^2 + |\sigma|^2}{\mathcal{P}+1}\delta + \mathcal{O}(\delta^2) \quad \text{for } \delta \to 0.$$

Thus, for $n = 0$ and $\sigma = 0$ we have a three-dimensional neutral direction (recall $\alpha_3 = \alpha_4 = 0$) which corresponds to translations in the (x_1, x_2)-directions and to adding a constant to the pressure. Eigenvalues with a positive real part exist only when δ becomes negative, which is the case for $\mathcal{R}^2 > 27n^2/4$. Hence unstable directions occur first for $n = 1$ and are associated with $\mathcal{R} > \mathcal{R}_0 = 3\sqrt{3}/2$. The relevant wave vector $\sigma = (\sigma_1, \sigma_2, 0)^T$ then satisfies $|\sigma| \approx 1/\sqrt{2}$. Our further study will be restricted to Rayleigh numbers with $\mathcal{R} \in (\mathcal{R}_0, \mathcal{R}_0 + \varepsilon)$; then the only unstable modes can appear for $n = 1$, but $n = 0$ also contains neutral modes.

Our main interest lies in roll patterns which are defined by periodicity in one unbounded direction while they are independent of the other direction. Without loss of generality (using the rotational invariance of the problem) we may assume that the rolls are independent of x_2 and periodic in x_1, namely, $\partial_2 z = 0$ and $z(x_1 + 2\pi/k, x_3) = z(x_1, x_3)$, where k is the wave number of the periodic pattern.

Thus, we introduce the scaled variable $y = kx_1 \in S^1 = \mathbb{R}/_{2\pi\mathbb{Z}}$ and the basic domain $\Omega = S^1 \times (0, \pi)$. Moreover, we define the spaces

$$Y = \left(L^2(\Omega)\right)^4 \times H^1(\Omega),$$

$$Z = \left\{(u, \theta, p) \in (H^2(\Omega))^4 \times H^1(\Omega): \partial_3 u_1 = \partial_3 u_2 = u_3 = \theta = 0 \text{ on } \partial\Omega \right\},$$

$$\langle z, \tilde{z} \rangle = \frac{1}{2\pi^2} \int_\Omega z(y, x_3) \cdot \tilde{z}(y, x_3) \, dy \, dx_3. \tag{4.3}$$

Note that the spaces $H^j(\Omega)$ automatically include the assumption of periodicity in y. For all integrals over Ω we will drop the integration variables $dy \, dx_3$ whenever there is no danger of confusion. We further use the scaled versions of the gradient and Laplace operator on Ω,

$$\nabla_k = (k\partial_y, 0, \partial_3)^T, \qquad \Delta_k = k^2\partial_y^2 + \partial_3^2.$$

The steady roll patterns satisfy the nonlinear problem

$$\mathcal{N}(\mathcal{R}, k, z) \stackrel{\text{def}}{=} \begin{pmatrix} -\Delta_k u - \mathcal{R}\theta e_3 + \nabla_k p + (u \cdot \nabla_k)u \\ -\Delta_k \theta - \mathcal{R}u_3 + \mathcal{P}u \cdot \nabla_k\theta \\ -\nabla_k \cdot u \end{pmatrix} = 0. \tag{4.4}$$

The map $\mathcal{N}$ is analytic from Z into Y, since the nonlinear terms are quadratic with both factors in $H^1(\Omega) \subset L^4(\Omega)$ by Sobolev's embedding theorem (recall that Ω is two-dimensional).

Note that the second component of $\mathcal{N} = 0$ reads $-\Delta_k u_2 + u_1 k\partial_y u_2 + u_3\partial_3 u_2 = 0$, which is a linear problem having the trivial solutions $u_2 = a_2 = $const. Multiplying this equation by u_2 and integrating by parts shows that these are the only solutions:

$$\int_\Omega |\nabla_k u_2|^2 = -\int_\Omega u_2 \Delta_k u_2 = -\int_\Omega u_2(u \cdot \nabla_k)u_2 = -\int_\Omega (u \cdot \nabla_k)\left[\frac{1}{2}u_2^2\right]$$

$$= \int_\Omega (\nabla_k \cdot u)\frac{1}{2}u_2^2 = 0,$$

where all boundary terms cancel due to $z \in Z$. Moreover, note that u_2 does not enter in any of the other four equations. Thus, the u_2 component decouples; nevertheless we keep it in the following calculations since this decoupling will not be valid when the instability is investigated.

We are interested in the bifurcations of nontrivial patterns from the trivial solution $z = 0$. The linear theory tells us that $z = 0$ remains stable as long as the Rayleigh number stays less than 27/4, or in our notation $\mathcal{R} \leq \mathcal{R}_0 = 3\sqrt{3}/2$. When $\mathcal{R}$ passes through $\mathcal{R}_0$, the waves with wave number k close to $k_0 = 1/\sqrt{2}$ become unstable. Hence, the basic linear operator we have to study is

$$A_0 \stackrel{\text{def}}{=} D_z\mathcal{N}(\mathcal{R}_0, k_0, 0): \begin{pmatrix} u \\ \theta \\ p \end{pmatrix} \mapsto \begin{pmatrix} -\Delta_{k_0} u - \mathcal{R}_0\theta e_3 + \nabla_{k_0} p \\ -\Delta_{k_0}\theta - \mathcal{R}_0 u_3 \\ -\nabla_{k_0} \cdot u \end{pmatrix}, \tag{4.5}$$

and it is easy to see that A_0 is symmetric with respect to the scalar product given above, i.e., $\langle A_0 z, \tilde{z} \rangle = \langle z, A_0\tilde{z} \rangle$ for all $z, \tilde{z} \in Z$.

This operator can be analyzed by expanding the elements in Z and Y into Fourier series as follows:

$$z(y, x_3) = \sum_{n \in \mathbb{N}_0} \sum_{m \in \mathbb{Z}} e^{imy} \varphi_n(\alpha^{(n,m)})(x_3),$$

where φ_n was defined in (4.1) and $\alpha^{(n,-m)} = \overline{\alpha}^{(n,m)} \in \mathbb{C}^5$.

The five-dimensional spaces $Z_{(n,m)} = \{e^{imy}\varphi_n(\alpha) \in Z : \alpha \in \mathbb{C}^5\}$ are left invariant by the application of A_0 since

$$A_0 \sum_{n,m} e^{imy} \varphi_n(\alpha^{(n,m)}) = \sum_{n,m} e^{imy} \varphi_n(F^0(n, m)\alpha^{(n,m)}), \tag{4.6}$$

where $F^0(n, m) = F_n(\mathcal{R}_0, k_0 m, 0, 0) \in \mathbb{C}^{5 \times 5}$ as defined in (4.2). Obviously the kernel of A_0 is found as a direct sum from the kernels of the $F^0(n, m)$.

Lemma 4.1. *The operator $A_0 : Z \subset Y \to Y$ is symmetric in Y with respect to the scalar product $\langle \cdot, \cdot \rangle$. Moreover, $A_0 : Z \to Y$ is a Fredholm operator of index zero with a five-dimensional kernel:* $\mathrm{span}\{\xi_1, \ldots, \xi_5\}$ *where*

$$\xi_1 = \begin{pmatrix} 1 \\ 0 \\ 0 \\ 0 \\ 0 \end{pmatrix}, \; \xi_2 = \begin{pmatrix} 0 \\ 1 \\ 0 \\ 0 \\ 0 \end{pmatrix}, \; \xi_3 = \begin{pmatrix} 0 \\ 0 \\ 0 \\ 0 \\ 1 \end{pmatrix}, \quad \begin{array}{l} \xi_4 = \mathrm{Re}\,\zeta, \\ \xi_5 = \mathrm{Im}\,\zeta, \end{array} \quad \text{with } \zeta = e^{iy} \begin{pmatrix} \sqrt{2}\cos x_3 \\ 0 \\ -i\sin x_3 \\ -i\sqrt{3}\sin x_3 \\ 3i\cos x_3 \end{pmatrix}.$$

Note that $\xi_4 = K^{(2)}\xi_4 = -K^{(1)}K^{(3)}\xi_4$ and $\xi_5(y) = \xi_4(y - \pi/2)$.

Proof. The result on the kernel follows easily by studying the kernels of each $F^0(n, m)$ separately and then using (4.6).

The symmetry of A_0 follows by an explicit calculation. The Fredholm property can be derived from classical elliptic regularity theory (see, e.g., [RR92] Ch. 8) or by writing $A_0 z = g \in Y$ in its Fourier representation $F^0(n, m)\alpha^{(n,m)} = \gamma^{(n,m)}$ and estimating $\alpha^{(n,m)}$ directly. □

The question of whether nontrivial roll patterns exist is now investigated by applying the LSR to the nonlinear equation $\mathcal{N}(\mathcal{R}, k, z) = 0$. Therefore, we rewrite the mapping $\mathcal{N}$ as $\mathcal{N}(\mathcal{R}, k, z) = A_0 z + N(\mathcal{R}, k, z)$ with $N(\mathcal{R}, k, z) = \mathcal{O}(\varepsilon\|z\|_Z + \|z\|_Z^2)$ where $\varepsilon = |\mathcal{R} - \mathcal{R}_0| + |k - k_0|$. Moreover, we let $Z_0 = Y_0 = \mathrm{kernel}[A_0]$ and define the orthogonal projection P from Y onto Y_0 which also projects Z onto Z_0. The kernels of P are denoted by Y_1 and Z_1, respectively. Thus, we decompose $z = z_0 + z_1$ with $z_j \in Z_j$ and, since $P A_0 z = 0$ and $(I - P)A_0(z_0 + z_1) = A_0 z_1$, the nonlinear problem (4.4) can be written as

$$P N(\mathcal{R}, k, z_0 + z_1) = 0,$$

$$A_0 z_1 + (I - P)N(\mathcal{R}, k, z_0 + z_1) = 0. \tag{4.7}$$

The second equation in (4.7) has a locally unique solution $z_1 = \mathcal{Z}(\mathcal{R}, k, z_0) \in Z_1$ which depends smoothly (in fact, analytically) on $(\mathcal{R}, k, z_0)$ and is such that

$$\mathcal{Z}(\mathcal{R}, k, z_0) = \mathcal{O}(\varepsilon\|z_0\| + \|z_0\|^2).$$

Substituting this relation into the first equation of (4.7) we obtain the bifurcation equation

$$0 = n(\mathcal{R}, k, z_0) \overset{\text{def}}{=} PN(\mathcal{R}, k, z_0 + \mathcal{Z}(\mathcal{R}, k, z_0)), \tag{4.8}$$

which is equivalent to

$$0 = \tilde{n}(\mathcal{R}, k, \eta) \overset{\text{def}}{=} \left(\left\langle \xi_j, n\left(\mathcal{R}, k, \sum_{l=1}^{5} \eta_l \xi_l \right) \right\rangle \right)_{j=1,\dots,5} \in \mathbb{R}^5. \tag{4.9}$$

The function $\tilde{n}$ is the coordinate representation of the function n, which has values in Y_0, with respect to the basis $\{\xi_1, \dots, \xi_5\}$. Hence, finding nontrivial roll patterns is equivalent to finding $\eta \in \mathbb{R}^5$ with $\tilde{n}(\mathcal{R}, k, \eta) = 0$.

Before giving the calculations of the relevant lowest order terms in an expansion of $\tilde{n}$, we discuss the general structure of $\tilde{n}$ which is enforced by the symmetries of the problem. We have invariance of the problem under adding multiples of ξ_2 and ξ_3 to z. Hence, $\mathcal{Z}$ and thus $\tilde{n}$ do not depend on (η_2, η_3). Moreover, we have equivariance of the problem under translation in y-direction

$$T_\tau: z(\cdot) \mapsto z(\cdot - \tau),$$

and under the reflections $K^{(j)}$, $j = 1, 2, 3$. These symmetries induce symmetries on the bifurcation equation relative to the following action on $\mathbb{R}^5$:

$$
\begin{aligned}
\widehat{K}^{(1)}&: \eta \mapsto \operatorname{diag}(-1, 1, 1, -1, 1)\eta, \\
\widehat{K}^{(2)}&: \eta \mapsto \operatorname{diag}(1, -1, 1, 1, 1)\eta, \\
\widehat{K}^{(3)}&: \eta \mapsto \operatorname{diag}(1, 1, 1, -1, -1)\eta,
\end{aligned}
\qquad
\widehat{T}_\tau: \eta \mapsto
\begin{pmatrix}
\eta_1 \\
\eta_2 \\
\eta_3 \\
\cos(\tau)\eta_4 - \sin(\tau)\eta_5 \\
\sin(\tau)\eta_4 + \cos(\tau)\eta_5
\end{pmatrix}. \tag{4.10}
$$

Notice that $\widehat{K}^{(3)} = \widehat{T}_\pi$.

Lemma 4.2. *The function $\tilde{n}$ in (4.9) takes the form*

$$
\tilde{n}(\mathcal{R}, k, \eta) =
\begin{pmatrix}
g_1(\mathcal{R}, k, \eta_1^2, r)\eta_1 \\
0 \\
0 \\
g_4(\mathcal{R}, k, \eta_1^2, r)\eta_4 - g_5(\mathcal{R}, k, \eta_1^2, r)\eta_1\eta_5 \\
g_5(\mathcal{R}, k, \eta_1^2, r)\eta_1\eta_4 + g_4(\mathcal{R}, k, \eta_1^2, r)\eta_5
\end{pmatrix},
\qquad \text{where } r = \eta_4^2 + \eta_5^2.
$$

Proof. As mentioned above, $\tilde{n}$ is independent of η_2 and η_3. Moreover, the equivariance with respect to $\widehat{K}^{(2)}$ implies that $\tilde{n}_2$ is odd in η_2. This is only possible if $\tilde{n}_2 \equiv 0$. The third component is zero since $\tilde{n}_3 = \langle \xi_3, \mathcal{N}(\mathcal{R}, k, z) \rangle = -\int_\Omega \nabla_k \cdot u = 0$ due to the boundary conditions for elements in Z.

The equivariance with respect to $\widehat{T}_\tau$ implies

$$
\begin{pmatrix} \widetilde{n}_1 \\ \widetilde{n}_4 \\ \widetilde{n}_5 \end{pmatrix} = \begin{pmatrix} \widetilde{g}_1 \\ 0 \\ 0 \end{pmatrix} + \widetilde{g}_4 \begin{pmatrix} 0 \\ \eta_4 \\ \eta_5 \end{pmatrix} + \widetilde{g}_5 \begin{pmatrix} 0 \\ -\eta_5 \\ \eta_4 \end{pmatrix},
$$

where each $\widetilde{g}_j$ depends on $(\mathcal{R}, k, \eta_1, r)$ (see Proposition VIII,2.3 in [GS85]). Using $\widehat{K}^{(1)}$ we find that $\widetilde{g}_1$ and $\widetilde{g}_5$ must be odd in η_1 whereas $\widetilde{g}_4$ is even.   □

A nice consequence of the special form of $\widetilde{n}$ is that all the functions g_j can be calculated by setting $\eta_2 = \eta_3 = \eta_5 = 0$. In particular, we find

$$
\mathcal{Z}(\mathcal{R}, k, \eta_1 \xi_1 + \eta_4 \xi_4) = \eta_1 \eta_4 Z_{14} + \eta_4^2 Z_{44} + \eta_4 (k - k_0) Z_{4k} + \eta_4 (\mathcal{R} - \mathcal{R}_0) Z_{4\mathcal{R}} + \cdots.
$$

For explicit expressions for Z_{14}, Z_{44}, Z_{4k}, and $Z_{4\mathcal{R}}$ see Appendix A. For the further analysis it will only be important to calculate a finite number of coefficients, namely c_j for $j = 1, \ldots, 5$ in

$$
g_1(\mathcal{R}_0, k_0, 0, r) \doteq c_1 r + \mathcal{O}(r^2),
$$

$$
g_4(\mathcal{R}, k, 0, r) = c_2(\mathcal{R} - \mathcal{R}_0) + c_3(k - k_0)^2 + c_4 r + \cdots,
$$

$$
g_5(\mathcal{R}_0, k_0, 0, 0) = c_5. \tag{4.11}
$$

The calculation of these coefficients is given in Appendix A, and the result is

$$
c_1 = -3(\mathcal{P} + 1)/20, \qquad c_2 = -\sqrt{3}/2, \qquad c_3 = 3,
$$

$$
c_4 = 3\mathcal{P}^2/32, \qquad c_5 = -3\sqrt{2}(\mathcal{P} + 1)/8.
$$

We conclude from $c_5 \neq 0$ that small solutions can only exist for $\eta_1 = 0$, which means that the mean flow in x_1-direction is zero. In this case, the solutions are given by $\eta = (0, \eta_2, \eta_3, \rho \cos(\tau), \rho \sin(\tau))^T$ where $\rho = \sqrt{r}$ solves the scalar equation $g_4(\mathcal{R}, k, 0, \rho^2) = 0$. As $\mathcal{R} = (1 + k^2)^{3/2}/k$ is the line of neutral stability, we know that $g_4((1 + k^2)^{3/2}/k, k, 0, 0) = 0$. Using $c_4 > 0$ we conclude that nontrivial solutions exist exactly for those $(\mathcal{R}, k) \approx (\mathcal{R}_0, k_0)$ with $g_4(\mathcal{R}, k, 0, 0) < 0$. Thus, we have proved the following (well-known) result (cf. [GSS88], Case Study 4 and the references therein).

Theorem 4.3. *For all $(\mathcal{R}, k)$ close to $(\mathcal{R}_0, k_0)$ and satisfying $\mathcal{R} > (1 + k^2)^{3/2}/k$ there exists a roll solution which is unique up to a shift in x_1-direction. Let $\mathcal{R} = \widetilde{\mathcal{R}}(k, r)$ be the unique solution of $g_4(\mathcal{R}, k, 0, r) = 0$ and $\rho = \sqrt{r}$, then*

$$
z_{\rho, k} = \rho \xi_4 + \mathcal{Z}(\widetilde{\mathcal{R}}(k, \rho^2), k, \rho \xi_4) \in Z
$$

is the unique roll solution of (4.4) with u_1 an even function in y and $u_1 > 0$ at $(y, x_3) = (0, 0)$.

We will continue to use (ρ, k) as parameters for our problem and express $\mathcal{R}$ as

$$
\mathcal{R} = \widetilde{\mathcal{R}}(k, \rho^2) = \widetilde{\mathcal{R}}(k, 0) - \frac{c_4}{c_2}\rho^2 + \mathcal{O}(|k - k_0|\rho^2 + \rho^4),
$$

$$
= \mathcal{R}_0 + 2\sqrt{3}(k - k_0)^2 + \frac{\sqrt{3}\mathcal{P}^2}{16}\rho^2 + \mathcal{O}(|k - k_0|^3 + |k - k_0|\rho^2 + \rho^4).
$$

As a basis of our following stability analysis we choose the roll solution $z_{\rho,k} \in Z$ of Theorem 4.3. Note that $z_{\rho,k}$ is associated with the solution $\eta = \eta_\rho = (0, 0, 0, \rho, 0)^T$ of (4.9), or with the solution $z_0 = \rho\xi_4$ of the reduced problem (4.8), respectively. Moreover, we have the expansion

$$z_{\rho,k} = \rho\xi_4 + \rho^2 Z_{44} + \rho^3 \left(Z_{444} - \frac{c_4}{c_2} Z_{4\mathcal{R}} \right)$$

$$+ \rho(k - k_0)Z_{4k} + \mathcal{O}(\rho(\rho^3 + |k - k_0|^2)). \tag{4.12}$$

The additional term $Z_{4\mathcal{R}}$ is present since $\mathcal{R} = \widetilde{\mathcal{R}}$ is considered as a function of ρ.

5. Sideband Instabilities

We assume that $\widehat{z}$ is a roll solution as constructed above, i.e., $\widehat{z} = z_{\rho,k}$. We want to study the instability of this solution as a steady state of the full Rayleigh–Bénard problem (3.1) in the sense of Definition 3.1. The linearization is obtained by letting $z = \widehat{z} + \varepsilon w$, with $w = (v, \vartheta, q)$, and considering only terms which are linear in ε:

$$\left.\begin{array}{rcl} \partial_t v + (\widehat{u} \cdot \nabla)v + (v \cdot \nabla)\widehat{u} + \nabla q - \widetilde{\mathcal{R}}\vartheta e_3 - \Delta v & = & 0, \\ \mathcal{P}(\partial_t \vartheta + \widehat{u} \cdot \nabla\vartheta + v \cdot \nabla\widehat{\theta}) - \widetilde{\mathcal{R}}v_3 - \Delta\vartheta & = & 0, \\ \nabla \cdot v & = & 0, \end{array}\right\} \text{ for } x \in \mathbb{R}^2 \times (0, \pi);$$

$$\partial_3 v_1 = \partial_3 v_2 = v_3 = \vartheta = 0 \text{ for } x_3 = 0 \text{ or } x_3 = \pi.$$
$$\tag{5.1}$$

As discussed above we study perturbations of the form

$$w(t, x) = e^{\lambda t + i(\sigma_1 x_1 + \sigma_2 x_2)} W(y, x_3), \text{ where } y = kx_1 \text{ and } W \in Z.$$

Inserting this Ansatz into (5.1) we obtain a linear problem for $W \in Z$ which depends on the additional parameters λ and σ. In fact, we have complexified the linear problem such that $W \in Z^{\mathbb{C}} = Z \oplus iZ$; however, for notational convenience we omit the superscript $\mathbb{C}$ in the following.

The problem for $W = (v, \vartheta, q) \in Z$ reads as follows:

$$\left.\begin{array}{rcl} \lambda v + (\widehat{u} \cdot \nabla_{k,\sigma})v + (v \cdot \nabla_k)\widehat{u} + \nabla_{k,\sigma} q - \widetilde{\mathcal{R}}\vartheta e_3 - \Delta_{k,\sigma} v & = & 0, \\ \mathcal{P}(\lambda\vartheta + \widehat{u} \cdot \nabla_{k,\sigma}\vartheta + v \cdot \nabla_k\widehat{\theta}) - \widetilde{\mathcal{R}}v_3 - \Delta_{k,\sigma}\vartheta & = & 0, \\ -\nabla_{k,\sigma} \cdot v & = & 0, \end{array}\right\} \text{ in } \mathbb{R}^2 \times (0, \pi),$$
$$\tag{5.2}$$

with boundary conditions

$$\partial_3 v_1 = \partial_3 v_2 = v_3 = \vartheta = 0 \text{ for } x_3 = 0 \text{ or } x_3 = \pi.$$

We continue to use the abbreviations

$$\nabla_{k,\sigma} = \begin{pmatrix} i\sigma_1 + k\partial_y \\ i\sigma_2 \\ \partial_3 \end{pmatrix}, \qquad \Delta_{k,\sigma} = k^2\partial_y^2 + \partial_3^2 + 2i\sigma_1 k\partial_y - |\sigma|^2 = \Delta_k + 2i\sigma_1 k\partial_y - |\sigma|^2.$$

It is important to note that the gradient operator $\nabla_{k,\sigma}$ is applied to the linear perturbation W while the steady solution $\widehat{z}$ is acted upon by $\nabla_k = \nabla_{k,0}$.

We can write this system as a general linear problem

$$\mathcal{B}(\rho, k, \sigma, \lambda)W = 0, \qquad (5.3)$$

where $\mathcal{B}(\cdots)$ is a bounded operator from Z into Y depending smoothly on the variables $(\rho, k, \sigma, \lambda)$. It can be written as

$$\mathcal{B}(\rho, k, \sigma, \lambda) = D_z \mathcal{N}(\widetilde{\mathcal{R}}(k, \rho^2), k, \nabla_{k,\sigma}, z_{\rho,k}) - \lambda M,$$

where $M = \mathrm{diag}(-1, -1, -1, -\mathcal{P}, 0)$. Thus, we have $\mathcal{B}(\rho, k, 0, 0) = D_z \mathcal{N}(\rho, k, z_{\rho,k})$ and so $\mathcal{B}(0, k_0, 0, 0) = A_0$.

The existence of nontrivial solutions W for (5.3) can again be studied with the help of the LSR. Again we use the splitting $W = W_0 + W_1$ with $W_j \in Z_j$ and (5.3) is thus equivalent to

$$P\mathcal{B}(\rho, k, \sigma, \lambda)(W_0 + W_1) = 0,$$

$$(I - P)\mathcal{B}(\rho, k, \sigma, \lambda)(W_0 + W_1) = 0. \qquad (5.4)$$

The second equation is solvable in the form $W_1 = \mathcal{W}(\rho, k, \sigma, \lambda)W_0$ by the implicit function theorem for $(\rho, k, \sigma, \lambda)$ sufficiently close to $(0, k_0, 0, 0)$.

Using the coordinates $\omega = (\omega_1, \ldots, \omega_5)^T \in \mathbb{C}^5$ in Z_0 via $W_0 = \sum_{j=1}^5 \omega_j \xi_j$ we obtain $W_1 = \sum_{j=1}^5 \omega_j \mathcal{W}_j(\rho, k, \sigma, \lambda)$. Introducing this into the first equation of (5.4) the reduced linear problem takes the form

$$\widetilde{b}(\rho, k, \sigma, \lambda)\omega = 0, \quad \text{where } \widetilde{b}(\cdots) \in \mathbb{C}^{5 \times 5}, \qquad (5.5)$$

and $\widetilde{b}(\cdots)_{jl} = \langle \mathcal{B}(\cdots)[\xi_l + \mathcal{W}_l(\cdots)], \xi_j \rangle$.

Thus, the problem of linearized instability is reduced to the study of the matrix $\widetilde{b}$. For a given (ρ, k) we can now prove spectral instability of the roll pattern $z_{\rho,k}$ by finding a solution $(\sigma, \lambda, \omega) \in \mathbb{R}^2 \times \mathbb{C} \times \mathbb{C}^5$ with $\mathrm{Re}\,\lambda > 0$ and $\omega \notin \mathbb{C}e_3$. (This last condition on ω is necessary to exclude solutions which only grow exponentially in the mean value of the pressure.)

As in the case of the nonlinear bifurcation problem, we can give more structure to the matrix $\widetilde{b}$ by using the special form and the symmetries of the problem. This will minimize the amount of calculations for the relevant coefficients in the lowest order expansion.

Lemma 5.1. *The matrix $\widetilde{b}$ takes the form*

$$\widetilde{b} = \begin{pmatrix} \gamma_{11} & -\sigma_1\sigma_2\gamma_{12} & i\sigma_1 & i\sigma_1\rho\gamma_{14} & \rho\gamma_{15} \\ -\sigma_1\sigma_2\gamma_{21} & \gamma_{22} & i\sigma_2 & i\sigma_2\rho\gamma_{24} & -\sigma_1\sigma_2\rho\gamma_{25} \\ -i\sigma_1 & -i\sigma_2 & 0 & 0 & 0 \\ i\sigma_1\rho\gamma_{41} & i\sigma_2\rho\gamma_{42} & 0 & \gamma_{44} & i\sigma_1\gamma_{45} \\ \rho\gamma_{51} & -\sigma_1\sigma_2\rho\gamma_{52} & 0 & i\sigma_1\gamma_{54} & \gamma_{55} \end{pmatrix}, \qquad (5.6)$$

where all γ_{jl} are analytic functions of $(\rho^2, k, \sigma_1^2, \sigma_2^2, \lambda)$.

Proof. First we note that the third row and the third column of $\widetilde{b}$ can be given exactly. The reason for this is two-fold. On the one hand the third component of $\widetilde{b}\,\omega \in \mathbb{C}^5$ reads

$$\langle \xi_3, \mathcal{B}(\cdots)W \rangle = -\int_\Omega \nabla_{k,\sigma} v = -\int_\Omega (i\sigma \cdot v + k\partial_y v_1 + \partial_3 v_3) = -i\sigma \cdot (\omega_1, \omega_2)^T,$$

since the integrals over the derivatives cancel due to the boundary conditions. On the other hand, $\omega_3 = \frac{1}{2\pi^2}\int_\Omega q$ enters $\mathcal{B}(\cdots)W$ only in the first and second equation, namely as $(i\sigma_1\omega, i\sigma_2\omega, 0, 0, 0)^T$. Hence, $(I - P)\mathcal{B}(\cdots)(W_0 + W_1)$ is independent of ω_3 and so is $W_1 = \mathcal{W}(\cdots)W_0$. This proves the form of $\widetilde{b}_{jl}$ when either j or l equals 3.

Next we establish the symmetry properties of $\widetilde{b}$. Since imaginary coefficients only appear through $i\sigma$ in the full problem $\mathcal{B}W = 0$, we find that

$$\overline{\widetilde{b}(\rho, k, \sigma, \lambda)} = \widetilde{b}(\rho, k, -\sigma, \bar{\lambda}).$$

The symmetry $K^{(2)}$: $x_2 \mapsto -x_2$ leads to the equivariance

$$\widehat{K}^{(2)}\widetilde{b}(\rho, k, \sigma_1, -\sigma_2, \lambda)\widehat{K}^{(2)} = \widetilde{b}(\rho, k, \sigma, \lambda),$$

where the matrices $\widehat{K}^{(j)} \in \mathbb{R}^{5\times5}$ are defined in (4.10). The basic roll pattern $z_{\rho,k}$ is still invariant under the transformation $K^{(1)}K^{(3)}$: $(y, x_3) \mapsto (-y, \pi - x_3)$ which amounts to

$$\widehat{K}^*\widetilde{b}(\rho, k, -\sigma_1, \sigma_2, \lambda)\widehat{K}^* = \widetilde{b}(\rho, k, \sigma, \lambda) \text{ with } \widehat{K}^* = \widehat{K}^{(1)}\widehat{K}^{(3)} = \operatorname{diag}(-1, 1, 1, 1, -1).$$

In addition, we may change the sign of ρ, since $\mathcal{R} = \widetilde{\mathcal{R}}(k, \rho^2)$ and $z_{-\rho,k}$ is obtained from $z_{\rho,k}$ by the reflection $K^{(3)}$: $x_3 \mapsto \pi - x_3$. Hence,

$$\widehat{K}^{(3)}\widetilde{b}(-\rho, k, \sigma, \lambda)\widehat{K}^{(3)} = \widetilde{b}(\rho, k, \sigma, \lambda).$$

These four symmetries imply the existence of the functions $\gamma_{jl}(\rho^2, k, \sigma_1^2, \sigma_2^2, \lambda)$. $\square$

The structure of the third row and column of $\widetilde{b}$ allows us to reduce the problem to a three-dimensional one for the variables $(\omega_0, \omega_4, \omega_5) \in \mathbb{C}^3$, where $\omega_0 = \sigma_2\omega_1 - \sigma_1\omega_2$. In fact, this reduction is analogous to the classical reductions of Navier–Stokes problems by enforcing the incompressibility $\nabla \cdot u = 0$ and projecting out the pressure. Moreover, the 3×3-matrix b^* defined below corresponds to the matrices in the previous studies [BuB84], equations (2.9) and (2.11), and [Be94], equation (3.11).

Lemma 5.2. *Let $\widetilde{b}$ be given as in (5.6) and define $b^* =$*

$$\begin{pmatrix} \sigma_2^2\gamma_{11}+\sigma_1^2\gamma_{22}+\sigma_1^2\sigma_2^2(\gamma_{12}+\gamma_{21}) & \sigma_1\sigma_2\rho(\gamma_{14}-\gamma_{24}) & -i\sigma_2\rho(\gamma_{15}+\sigma_1^2\gamma_{25}) \\ \sigma_1\sigma_2\rho(\gamma_{42} - \gamma_{41}) & \gamma_{44} & i\sigma_1\gamma_{45} \\ i\sigma_2\rho(\gamma_{51} + \sigma_1^2\gamma_{52}) & i\sigma_1\gamma_{54} & \gamma_{55} \end{pmatrix}.$$

$$(5.7)$$

Then we have the relation $\det \widetilde{b}(\rho, k, \sigma, \lambda) = -\det b^*(\rho, k, \sigma, \lambda)$.

Proof. Obviously, $\det \widetilde{b} = \det b^* = 0$ for $\sigma = 0$ due to (5.6) and the definition of b^*. For $\sigma \neq 0$ define

$$T = \begin{pmatrix} 0 & i\sigma_1 & i\sigma_2 & 0 & 0 \\ 0 & i\sigma_2 & -i\sigma_1 & 0 & 0 \\ 1 & 0 & 0 & 0 & 0 \\ 0 & 0 & 0 & 1 & 0 \\ 0 & 0 & 0 & 0 & 1 \end{pmatrix},$$

which has $\det T = |\sigma|^2$. We find that

$$C = \overline{T}^T \widetilde{b} T = \begin{pmatrix} 0 & |\sigma|^2 & 0 & 0 & 0 \\ |\sigma|^2 & * & * & * & * \\ 0 & * & & & \\ 0 & * & & b^* & \\ 0 & * & & & \end{pmatrix},$$

which has $\det C = -|\sigma|^4 \det b^*$. This proves the result for $\sigma \neq 0$. $\qquad\square$

We apply our abstract result from Theorem 2.2 to control the terms obtained from the nonlinearity in the case $(\sigma, \lambda) = 0$:

$$\widetilde{b}(\rho, k, 0, 0) = D_\eta \widetilde{n}(\widetilde{\mathcal{R}}(k, \rho^2), k, \eta_\rho) = \begin{pmatrix} g_1 & 0 & 0 & 0 & 0 \\ 0 & 0 & 0 & 0 & 0 \\ 0 & 0 & 0 & 0 & 0 \\ 0 & 0 & 0 & 2\rho^2 \partial g_4/\partial r & 0 \\ \rho g_5 & 0 & 0 & 0 & 0 \end{pmatrix}. \qquad (5.8)$$

Here we have used $\eta_\rho = (0, 0, 0, \rho, 0)^T$ and $g_4(\widetilde{\mathcal{R}}, k, \eta_\rho) = 0$ by definition of ρ. In this formula all g_j are evaluated at $(\widetilde{\mathcal{R}}(k, \rho^2), k, \eta_\rho)$. This yields additional information on the following functions γ_{jl} (recall that $r = \rho^2$):

$$\gamma_{15}(r, k, 0, 0, 0) = \gamma_{22}(r, k, 0, 0, 0) = \gamma_{55}(r, k, 0, 0, 0) = 0,$$

$$\gamma_{11}(r, k, 0, 0, 0) = g_1 = c_1 r + \mathcal{O}(r^2), \quad \gamma_{44}(r, k, 0, 0, 0) = 2r\frac{\partial g_4}{\partial r} = 2c_4 r + \mathcal{O}(r^2),$$

$$\gamma_{51}(r, k, 0, 0, 0) = g_5 = c_5 + \mathcal{O}(r).$$

Moreover, employing Lemma 2.4 we can construct the part arising from the linearization at the trivial flow $\rho = 0$. In particular, the terms from the mean flows decouple from those arising from $e^{\pm iy}$. Thus, we obtain the block structure

$$\widetilde{b}(0, k, \sigma, \lambda) = \begin{pmatrix} \widetilde{b}^{(123)}(k, \sigma, \lambda) & 0 \\ 0 & \widetilde{b}^{(45)}(k, \sigma, \lambda) \end{pmatrix},$$

where the individual blocks have the form

$$\widetilde{b}^{(123)}(k, \sigma, \lambda) = \begin{pmatrix} \lambda + |\sigma|^2 & 0 & i\sigma_1 \\ 0 & \lambda + |\sigma|^2 & i\sigma_2 \\ -i\sigma_1 & -i\sigma_2 & 0 \end{pmatrix}, \qquad (5.9)$$

and

$$\tilde{b}^{(45)}(k, \sigma, \lambda) = \begin{pmatrix} \nu(k, \sigma_1^2, \sigma_2^2, \lambda) & -i\sigma_1\tilde{\delta}(k, \sigma_1^2, \sigma_2^2, \lambda) \\ i\sigma_1\tilde{\delta}(k, \sigma_1^2, \sigma_2^2, \lambda) & \nu(k, \sigma_1^2, \sigma_2^2, \lambda) \end{pmatrix}. \qquad (5.10)$$

The following expansion for $\tilde{b}^{(45)}$ is derived in Appendix B:

$$\nu(k, \sigma_1^2, \sigma_2^2, \lambda) = \tfrac{3}{4}(\mathcal{P}+1)\lambda + 3\sigma_1^2 + 3\sqrt{2}(k - k_0)\sigma_2^2 + \tfrac{3}{2}\sigma_2^4 + \tfrac{7\mathcal{P}^2+34\mathcal{P}-23}{100}\lambda^2 + \cdots,$$

$$\tilde{\delta}(k, \sigma_1^2, \sigma_2^2, \lambda) = \tfrac{\sqrt{2}}{15}(45\sqrt{2}(k - k_0) + 31\sigma_1^2 + 45\sigma_2^2 + 9(\mathcal{P}+1)\lambda + \cdots).$$

We can draw a few helpful conclusions about some of the γ_{jl} at $\rho = 0$:

$$\gamma_{jl}(0, k, \sigma_1^2, \sigma_2^2, \lambda) = 0 \text{ except for } (j, l) \in \{(1, 1), (2, 2), (4, 4), (4, 5), (5, 4), (5, 5)\}.$$

The two pieces of information can be put together by the general relation

$$\gamma_{jl}(\rho^2, k, \sigma_1^2, \sigma_2^2, \lambda) = \gamma_{jl}(0, k, \sigma_1^2, \sigma_2^2, \lambda) + \gamma_{jl}(\rho^2, k, 0, 0, 0) + \mathcal{O}(\rho^2(|\sigma|^2 + |\lambda|)),$$

which follows from the general fact that $f(a, b) = f(a, 0) + f(0, b) + \mathcal{O}(|ab|)$ for a C^2 function satisfying $f(0, 0) = 0$, cf. also (2.10).

According to the Principle of Reduced Instability (Theorem 2.1) it remains to analyze the solution set of the eigenvalue equation

$$0 = \Lambda(\rho, k, \sigma, \lambda) \overset{\text{def}}{=} \det b^*(\rho, k, \sigma, \lambda),$$

where $(\rho, k, \sigma, \lambda) \approx (0, k_0, 0, 0)$. If for a given (ρ, k) we can find a solution (σ, λ) of $\Lambda(\rho, k, \sigma, \lambda) = 0$ with $\text{Re }\lambda > 0$, then, in view of the discussion in Section 3, instability of the roll solution $z_{\rho,k}$ in the sense of Definition 3.1 is proved.

A complete discussion of $\Lambda = 0$ turns out to be difficult and will not be given here. We treat the cases $\sigma = (\sigma_1, 0)$ and $\sigma = (0, \sigma_2)$ exactly, whereas for $\sigma_1\sigma_2 \neq 0$ only certain scaling limits are considered.

5.1. The Case $\sigma = 0$

Before going into detail we derive a useful result for the eigenvalue problem in the case of $\sigma = 0$. Hence, stability is only calculated with respect to perturbations with the same period. In that case we have to treat the matrix $\tilde{b} \in \mathbb{R}^{5\times5}$ directly without using the reduction associated with Lemma 5.2:

$$\tilde{b} = \begin{pmatrix} \gamma_{11} & 0 & 0 & 0 & \rho\gamma_{15} \\ 0 & \gamma_{22} & 0 & 0 & 0 \\ 0 & 0 & 0 & 0 & 0 \\ 0 & 0 & 0 & \gamma_{44} & 0 \\ \rho\gamma_{51} & 0 & 0 & 0 & \gamma_{55} \end{pmatrix},$$

where

$$\gamma_{11} = \lambda + g_1(\tilde{\mathcal{R}}, k, 0, r) + \mathcal{O}(|\lambda|r), \qquad \gamma_{22} = \lambda(1 + \cdots),$$
$$\gamma_{44} = \tfrac{3}{4}(\mathcal{P}+1)\lambda + 2c_4\rho^2 + \cdots, \qquad \gamma_{51} = c_5 + \cdots,$$
$$\gamma_{15} = c_{15}\lambda + \cdots, \qquad \gamma_{55} = \tfrac{3}{4}(\mathcal{P}+1)\lambda + \cdots.$$

Eliminating the third column and row which is due to the pressure indeterminacy, we obtain a four-dimensional problem, where the diagonal entries γ_{22} and γ_{44} decouple. We immediately obtain the eigenvalue $\lambda_2 = 0$ from $\gamma_{22} = 0$ which is due to the mean flow in the x_2-direction. From $\gamma_{44} = 0$, we obtain the eigenvalue λ_4 which is strictly negative.

It remains to investigate the relation $0 = \gamma_{11}\gamma_{55} - \rho^2\gamma_{15}\gamma_{51}$. Of course one eigenvalue has to be $\lambda = 0$, since we have the translational mode in the x_1-direction. In fact, by Galilean invariance the zero $\lambda = 0$ has to be double. This is the content of the following lemma which is based on Theorem 2.3.

Lemma 5.3. (a) For $(\sigma, \lambda) = (0, 0)$ we have $\widetilde{b}(\rho, k, 0, 0)e_5 = 0$, or equivalently

$$\gamma_{15}(\rho^2, k, 0, 0) = \gamma_{55}(\rho^2, k, 0, 0) = 0 \quad \text{for all small } (\rho, k - k_0).$$

(b) There is a smooth function $\alpha: \mathbb{R}^2 \to \mathbb{R}$ with $\alpha(\rho, k) = \rho(1 + \mathcal{O}(\rho^2 + (k-k_0)^2))$ such that the relation $\widetilde{b}(\rho, k, 0, 0)e_1 + k\alpha(\rho, k)\partial_\lambda\widetilde{b}(\rho, k, 0, 0)e_5 = 0$ holds, or equivalently

$$\gamma_{11} + k\alpha(\rho, k)\rho\partial_\lambda\gamma_{15} = 0, \qquad \gamma_{51} + k\alpha(\rho, k)\partial_\lambda\gamma_{55} = 0,$$

for all small $(\rho, k - k_0)$ and $\sigma = \lambda = 0$.

Proof. We apply parts (a) and (b) of Theorem 2.3 using the fact that $\widetilde{w} = -\partial_y z_{\rho,k}$ satisfies $\mathcal{B}(\rho, k, 0, 0)\widetilde{w} = D_z\mathcal{N}(\widetilde{\mathcal{R}}, k, z_{\rho,k})\widetilde{w} = 0$ and $Q\widetilde{w} = \alpha(\rho, k,)\xi_5$. The last result is easy to see since $\widetilde{w} = \partial_y z_{\rho,k}$ has an average of zero, i.e., $\langle \xi_j, \widetilde{w} \rangle = 0$ for $j = 1, 2, 3$. The case $j = 4$ follows from symmetry considerations, namely $K^{(1)}\xi_4 = \xi_4$ and $K^{(1)}\widetilde{w} = -\widetilde{w}$. Moreover, we have $\widetilde{w} = \rho\xi_5 + \mathcal{O}(\rho^2)$ which proves part (a).

To establish (b), we use the Galilean invariance (3.2). Consider the time-dependent solution $z(t, \cdot) = \tau\xi_1 + T_{k\tau t}z_{\rho,k}$ (recall that $T_{k\tau}z = z(\cdot - k\tau)$) of Bénard's problem $M\partial_t z = \mathcal{N}(\widetilde{\mathcal{R}}, k, z)$, which is just the same roll as before but considered in a frame moving with constant velocity τ in x_1-direction. As a consequence we find

$$\mathcal{N}(\tau\xi_1 + T_{k\tau t}z_{\rho,k}) = M\partial_t\big(\tau\xi_1 + T_{k\tau t}z_{\rho,k}\big) = k\tau M T_{k\tau t}\widetilde{w},$$

for all $\tau \in \mathbb{R}$. Evaluating the derivative with respect to τ at $\tau = 0$ we arrive at

$$D_z\mathcal{N}(\widetilde{\mathcal{R}}, k, z_{\rho,k})[\xi_1 + kt\widetilde{w}] - kM\widetilde{w} = 0.$$

However, this is just the assumption of part (b) in Theorem 2.3, and the result follows with $\widetilde{v} = \xi_1 = Q\xi_1$.   □

5.2. Eckhaus Instability: $\sigma = (\sigma_1, 0)$

The first case we study is the two-dimensional problem (that is, with no x_2-dependence) which will lead to the Eckhaus instability. We let $\sigma_2 = 0$ and obtain

$$b^*(\rho, k, \sigma_1, 0, \lambda) = \begin{pmatrix} \sigma_1^2\gamma_{22} & 0 & 0 \\ 0 & \gamma_{44} & i\sigma_1\gamma_{45} \\ 0 & i\sigma_1\gamma_{54} & \gamma_{55} \end{pmatrix}.$$

The decoupling yields a first eigenvalue $\lambda_1 = -\sigma_1^2 + \cdots$ which is always stable.

The 2×2 block admits the expansion

$$\begin{pmatrix} \frac{3}{4}(\mathcal{P}+1)\lambda + 3\sigma_1^2 + 2c_4\rho^2 + \cdots & -i\sigma_1(6(k-k_0)+\cdots) \\ i\sigma_1(6(k-k_0)+\cdots) & \frac{3}{4}(\mathcal{P}+1)\lambda + 3\sigma_1^2 + \cdots \end{pmatrix}.$$

We immediately find that the second eigenvalue is always less than $-\frac{8}{3}c_4\rho^2/(\mathcal{P}+1)$ while the third eigenvalue is positive if and only if $\widehat{E}(\rho, k, \sigma_1) = \left[\gamma_{44}\gamma_{55} - \sigma_1^2\gamma_{45}\gamma_{54}\right]_{\lambda=0}$ is negative. Since $\widehat{E}$ has the form $\widehat{E}(\rho, k, \sigma_1) = \sigma_1^2 E(\rho^2, k, \sigma_1^2)$ with

$$E(r, k, s) = 3(3s + 2c_4r) - 36(k - k_0)^2 + \mathcal{O}(|k - k_0|(|k - k_0|^2 + r + s) + r^2 + s^2),$$

negative values of $\widehat{E}$ (for $s = \sigma_1^2 > 0$) can occur if and only if $\rho^2 \le r_E(k)$. The function r_E is obtained from solving $E(r, k, 0) = 0$ for $r = r_E(k)$ and has the expansion $r_E(k) = 6(k - k_0)^2/c_4 + \mathcal{O}(|k - k_0|^3)$.

Transforming back into the $(\mathcal{R}, k)$ coordinates, we let $\mathcal{R}_{\text{Eckhaus}}(k) \overset{\text{def}}{=} \widetilde{\mathcal{R}}(k, r_E(k))$ and find existence of rolls which are *Eckhaus unstable* for

$$\widetilde{\mathcal{R}}(k, 0) < \mathcal{R} < \mathcal{R}_{\text{Eckhaus}}(k) = \mathcal{R}_0 + 3 \cdot 2\sqrt{3}(k - k_0)^2 + \mathcal{O}(|k - k_0|^3),$$

$$= \mathcal{R}_0 + 3(\widetilde{\mathcal{R}}(k, 0) - \mathcal{R}_0) + \mathcal{O}(|k - k_0|^3).$$

We summarize the result as follows.

Theorem 5.4 (Eckhaus instability). *The roll solution $z_{\rho,k}$ is unstable with respect to two-dimensional sideband perturbations (that is, $v(x) = e^{i\sigma_1 x_1} W(kx_1, x_3)$ with $W \in Z$) if $\widetilde{\mathcal{R}}(k, \rho^2) < \mathcal{R}_{\text{Eckhaus}}(k)$.*

In [KvW96] the two-dimensional problem (no dependence on x_2) is analyzed in more detail. There stability of $z_{\rho,k}$ in the sense of Definition 3.1 is established for the case $\widetilde{\mathcal{R}}(k, \rho^2) \ge \mathcal{R}_{\text{Eckhaus}}(k)$.

5.3. Zigzag Instabilities: $\sigma = (0, \sigma_2)$

The second case relates to three-dimensional perturbations with $\sigma_1 = 0$. Again, the matrix b^* decouples into

$$b^* = \begin{pmatrix} \sigma_2^2\gamma_{11} & 0 & -i\sigma_2\rho\gamma_{15} \\ 0 & \gamma_{44} & 0 \\ i\sigma_2\rho\gamma_{51} & 0 & \gamma_{55} \end{pmatrix}.$$

From $\gamma_{44} = 0$ we obtain the second eigenvalue

$$\lambda_2(\rho^2, k, \sigma_2^2) = -\frac{4}{3(\mathcal{P}+1)}\left(2c_4\rho^2 + 3\sqrt{2}(k - k_0)\sigma_2^2 + \frac{3}{2}\sigma_2^4 + \cdots\right).$$

For $k > k_0$ this eigenvalue is always negative, while for $k < k_0$ the maximum eigenvalue is achieved for finite $\sigma_2^2 = S_1(\rho^2, k)$ with $S_1(\rho^2, k) = -\sqrt{2}(k - k_0) + \cdots$. This maximum is positive if and only if

$$\rho^2 < r_Z^{(1)}(k) = \frac{3}{2c_4}(k - k_0)^2 + \mathcal{O}(|k - k_0|^3) \qquad \text{and} \qquad k < k_0.$$

Here $r_Z^{(1)}$ is defined by solving $\lambda_2(r, k, S_1(r, k)) = 0$ for r.

This sideband instability is detached: For very small $|\sigma_2|$ the modes are always stable as well as for large values. Moreover, this instability only occurs where the Eckhaus instability is already present, since it appears in the regime

$$\widetilde{\mathcal{R}}(k, 0) < \mathcal{R} < \mathcal{R}_{zigzag}^{(1)}(k) = \mathcal{R}_0 + \frac{3}{2}(\widetilde{\mathcal{R}}(k, 0) - \mathcal{R}_0) + \mathcal{O}(|k - k_0|^3) \qquad \text{and} \qquad k < k_0,$$

where $\mathcal{R}_{zigzag}^{(1)}(k) \overset{\text{def}}{=} \widetilde{\mathcal{R}}(k, r_Z^{(1)}(k)) = \mathcal{R}_0 + 3\sqrt{3}(k - k_0)^2 + \mathcal{O}(|k - k_0|^3).$

The remaining 2×2 block can now be studied separately. Since we consider only $\sigma_2 \neq 0$ we may divide out σ_2 and are left with

$$\begin{pmatrix} \gamma_{11} & \rho\gamma_{15} \\ \rho\gamma_{51} & \gamma_{55} \end{pmatrix}$$
$$= \begin{pmatrix} \lambda + \sigma_2^2 + c_1\rho^2 + \cdots & \rho(-\lambda\sqrt{2}\,c_1 + d_2\sigma_2^2 + \cdots) \\ \rho(c_5 + \cdots) & \frac{3(\mathcal{P}+1)}{4}\lambda + 3\sqrt{2}(k - k_0)\sigma_2^2 + \frac{3}{2}\sigma_2^4 + \cdots \end{pmatrix},$$

where $c_1 = -\frac{3}{20}(\mathcal{P} + 1)$, $c_5 = -\frac{3}{4\sqrt{2}}(\mathcal{P} + 1)$, and $d_2 = 1/\sqrt{2}$ (cf. Appendix C). This leads to the eigenvalue equation

$$\frac{3(\mathcal{P}+1)}{4}\lambda^2 + \left(\frac{3(\mathcal{P}+1)}{4}\sigma_2^2 + 3\sqrt{2}(k - k_0)\sigma_2^2 + \frac{3}{2}\sigma_2^4\right)\lambda$$
$$+ (\sigma_2^2 + c_1\rho^2)(3\sqrt{2}(k - k_0)\sigma_2^2 + \frac{3}{2}\sigma_2^4) - \rho^2 c_5 d_2\sigma_2^2 + \cdots = 0.$$

Since $d_2 > 0$, for small σ_2 the eigenvalues are not real with $\text{Im}\,\lambda = \mathcal{O}(\rho\sigma_2)$ and $\text{Re}\,\lambda \le -\frac{1}{2}\sigma_2^2 + \cdots$. However, for larger σ_2 (viz., $\sigma_2^2 > 16\sqrt{2}\rho^2/(3(\mathcal{P} + 1))$) the eigenvalues become real, and a positive eigenvalue appears if the coefficient of λ^0 is negative. This coefficient is $\widetilde{F}(\rho, k, \sigma_2) = [\gamma_{11}\gamma_{55} - \rho^2\gamma_{15}\gamma_{51}]_{\lambda=0}$, and it takes the form $\widetilde{F}(\rho, k, \sigma_2) = \sigma_2^2 F(\rho^2, k, \sigma_2^2)$. The minimum of F is attained for $\sigma_2^2 = S_2(\rho^2, k) = -\sqrt{2}(k - k_0) - \frac{c_1}{2}\rho^2 + \cdots$, and testing for negativity leads to the instability region $\rho^2 < r_Z^{(2)}(k) = \frac{16}{\mathcal{P}+1}(k - k_0)^2 + \mathcal{O}(|k - k_0|^3)$. Thus, we have zigzag unstable rolls in the region

$$\widetilde{\mathcal{R}}(k, 0) < \mathcal{R} < \mathcal{R}_{zigzag}^{(2)}(k) = \mathcal{R}_0 + \left(1 + \frac{\mathcal{P}^2}{4(\mathcal{P}+1)}\right)2\sqrt{3}(k - k_0)^2 + \mathcal{O}(|k - k_0|^3) \text{ and } k < k_0.$$

This is again a detached sideband instability.

We summarize the results in the case $\sigma = (0, \sigma_2)$ as follows.

Theorem 5.5 (Zigzag instability). *The roll solution $z_{\rho,k}$ is unstable with respect to zigzag perturbations (that is, $v(t, x) = e^{i\sigma_2 x_2} W(kx_1, x_3)$ with $W \in Z$) if $\widetilde{\mathcal{R}}(k, \rho^2) < \max\{\mathcal{R}_{zigzag}^{(1)}(k), \mathcal{R}_{zigzag}^{(2)}(k)\}$.*

5.4. Skew-Varicose Instabilities: $\sigma_1 \sigma_2 \neq 0$

Finally we investigate instabilities which arise due to fully three-dimensional perturbations, that is, $\sigma_1 \neq 0 \neq \sigma_2$. The discussion of $\Lambda(\rho, k, \sigma, \lambda) = 0$ is now very complicated and not completely understood. To derive necessary conditions for instability the following procedure is used. We introduce a scaling

$$\rho = \varepsilon^{\nu_1} \widetilde{\rho}, \qquad k - k_0 = \varepsilon^{\nu_2} \widetilde{\kappa},$$

$$\sigma_1 = \varepsilon^{\nu_3} \widetilde{\sigma}_1, \qquad \sigma_2 = \varepsilon^{\nu_4} \widetilde{\sigma}_2, \qquad \lambda = \varepsilon^{\nu_5} \widetilde{\lambda}, \quad \text{with } \nu_j \in \mathbb{N},$$

where ε is a small positive parameter and $\widetilde{\rho}, \widetilde{\kappa}, \widetilde{\sigma}$, and $\widetilde{\lambda}$ are considered to be of order one compared to ε. Inserting this scaling into Λ gives

$$\Lambda(\varepsilon^{\nu_1} \widetilde{\rho}, k_0 + \varepsilon^{\nu_2} \widetilde{\kappa}, \varepsilon^{\nu_3} \widetilde{\sigma}_1, \varepsilon^{\nu_4} \widetilde{\sigma}_2, \varepsilon^{\nu_5} \widetilde{\lambda}) = \varepsilon^{\nu^*} \Lambda^*(\widetilde{\rho}, \widetilde{\kappa}, \widetilde{\sigma}, \widetilde{\lambda}) + \mathcal{O}(\varepsilon^{\nu_6}),$$

where ν^* is chosen such that Λ^* is nontrivial and $\nu^* < \nu_6$. Note that Λ^* is a polynomial.

Assume now that $\Lambda^*(\cdots) = 0$ has a solution $(\widehat{\rho}, \widehat{\kappa}, \widehat{\sigma}, \widehat{\lambda})$ such that $\widehat{\sigma} \neq 0$, $\operatorname{Re} \widehat{\lambda} > 0$, and $\widehat{\lambda}$ is a simple root of $\Lambda^*(\widehat{\rho}, \widehat{\kappa}, \widehat{\sigma}, \cdot)$. Then by standard perturbation arguments there exists an $\varepsilon_0 > 0$ such that for all $\varepsilon \in (0, \varepsilon_0)$ the equation $\Lambda(\varepsilon^{\nu_1} \widehat{\rho}, k_0 + \varepsilon^{\nu_2} \widehat{\kappa}, \varepsilon^{\nu_3} \widehat{\sigma}_1, \varepsilon^{\nu_4} \widehat{\sigma}_2, \lambda)$ $= 0$ has a simple root $\lambda = \lambda(\varepsilon)$ which has the expansion $\lambda(\varepsilon) = \varepsilon^{\nu_5}(\widehat{\lambda} + \mathcal{O}(\varepsilon^{\nu_6 - \nu^*}))$. From $\operatorname{Re} \widehat{\lambda} > 0$ we thus conclude sideband instability of the rolls $z_{\rho, k}$ along the curve $(\rho, k) = (\varepsilon^{\nu_1} \widehat{\rho}, k_0 + \varepsilon^{\nu_2} \widehat{\kappa})$, $\varepsilon \in (0, \varepsilon_0)$.

It is important to realize that for different scaling vectors $\nu = (\nu_1, \ldots, \nu_5)$ one may obtain different instability criteria. The results on the Eckhaus and zigzag instabilities are obtained with $\nu = (2, 2, 2, 1, 4)$, which is also the scaling used in [ZS83].

Here we follow the approach in [BuB84], [Be94] with $\nu = (1, 2, 1, 1, 2)$, or explicitly

$$\rho = \varepsilon \widetilde{\rho}, \qquad k - k_0 = \varepsilon^2 \widetilde{\kappa}, \qquad \sigma_1 = \varepsilon \widetilde{\sigma}_1, \qquad \sigma_2 = \varepsilon \widetilde{\sigma}_2, \qquad \lambda = \varepsilon^2 \widetilde{\lambda}. \quad (5.11)$$

Using the information on the functions $\gamma_{jk}(\rho, k, \sigma_1^2, \sigma_2^2, \lambda)$ from above we obtain the following expansion for the reduced stability matrix $b^* =$

$$\begin{pmatrix} \varepsilon^4 b_{11}^0 + \mathcal{O}(\varepsilon^6) & \varepsilon^3 \sigma_1 \sigma_2 \rho (d_0 + \mathcal{O}(\varepsilon^2)) & -i\varepsilon^4 \sigma_2 \rho \left(b_{13}^0 + \mathcal{O}(\varepsilon^2)\right) \\ \varepsilon^3 \sigma_1 \sigma_2 \rho (d_3 + \mathcal{O}(\varepsilon^2)) & \varepsilon^2 b_{22}^0 + \mathcal{O}(\varepsilon^4) & -i\varepsilon^3 \sigma_1 (b_{23}^0 + \mathcal{O}(\varepsilon^2)) \\ i\varepsilon^2 \sigma_2 \rho \left[\frac{-3(\mathcal{P}+1)}{4\sqrt{2}} + \mathcal{O}(\varepsilon^2)\right] & i\sigma_1 \mathcal{O}(\varepsilon^3) & \varepsilon^2 b_{33}^0 + \mathcal{O}(\varepsilon^4) \end{pmatrix},$$

$$(5.12)$$

where

$$b_{11}^0 = |\widetilde{\sigma}|^2 (\widetilde{\lambda} + |\widetilde{\sigma}|^2) + c_1 \widetilde{\rho}^2 \widetilde{\sigma}_2^2,$$

$$b_{13}^0 = -c_1 \sqrt{2} \widetilde{\lambda} + d_1 \widetilde{\sigma}_1^2 + d_2 \widetilde{\sigma}_2^2,$$

$$b_{22}^0 = \frac{3(\mathcal{P}+1)}{4} \widetilde{\lambda} + 3\widetilde{\sigma}_1^2 + 2c_4 \widetilde{\rho}^2,$$

$$b_{23}^0 = \frac{\sqrt{2}}{15}(31\widetilde{\sigma}_1^2 + 45\widetilde{\sigma}_2^2 + 9(\mathcal{P}+1)\lambda + 45\sqrt{2}\widetilde{\kappa}) + d_4 \widetilde{\rho}^2,$$

$$b_{33}^0 = \frac{3(\mathcal{P}+1)}{4} \widetilde{\lambda} + 3\widetilde{\sigma}_1^2 = b_{22}^0 - 2c_4 \widetilde{\rho}^2.$$

There are four new coefficients d_j which are needed for the subsequent stability analysis. They are calculated in Appendix C:

$$d_0 = 1, \qquad d_1 = \frac{\sqrt{2}}{6}, \qquad d_2 = 1/\sqrt{2}, \qquad d_3 = -\frac{3\mathcal{P}+8}{10}, \qquad d_4 = \frac{11\sqrt{2}\,\mathcal{P}^2}{80}.$$

The scaling of Λ gives $v^* = 8$, $v_6 = 10$, and

$$\Lambda^*(\widetilde{\rho}, \widetilde{\kappa}, \widetilde{\sigma}, \widetilde{\lambda}) = b_{33}^0[b_{11}^0 b_{22}^0 - \widetilde{\sigma}_1^2 \widetilde{\sigma}_2^2 \widetilde{\rho}^2 d_0 d_3] + \frac{3(\mathcal{P}+1)}{4\sqrt{2}}\widetilde{\sigma}_2^2 \widetilde{\rho}^2 [b_{22}^0 b_{13}^0 - \widetilde{\sigma}_1^2 d_0 b_{23}^0].$$

The components of the sideband vector $\widetilde{\sigma}$ and the amplitude $\widetilde{\rho}$ appear only as squares; hence we let $r = \widetilde{\rho}^2$, $\widetilde{\sigma}_1^2 = s(1-\delta)$, and $\widetilde{\sigma}_2^2 = s\delta$ where $|\widetilde{\sigma}| = \sqrt{s}$ and $\delta \in [0,1]$. After dividing $\Lambda^* = 0$ by s we obtain the following equation:

$$0 = p(\widetilde{\lambda}) \overset{\text{def}}{=} p_3\widetilde{\lambda}^3 + p_2(r, \delta, s)\widetilde{\lambda}^2 + p_1(r, \delta, s)\widetilde{\lambda} + p_0(r, \kappa, \delta, s),$$

where $\kappa = 48\sqrt{2}\widetilde{\kappa}/(r\mathcal{P}^2)$. The coefficients p_j have the explicit representation

$$p_3 = \frac{9}{16}(\mathcal{P}+1)^2,$$

$$p_2 = \frac{9(\mathcal{P}+1)}{64}\left[4(\mathcal{P}+9-8\delta)s + \mathcal{P}^2 r\right],$$

$$p_1 = \frac{9}{2}(1-\delta)(\mathcal{P}+3-2\delta)s^2$$
$$+ \frac{1}{64}\left[24(\mathcal{P}+1)(2\mathcal{P}+1)\delta^2 - 2\mathcal{P}(33\mathcal{P}^2+18\mathcal{P}+2)\delta + 9\mathcal{P}^2(\mathcal{P}+5)\right]rs,$$

$$p_0 = 9(1-\delta)^2 s^3 + \frac{1}{16}(1-\delta)\left[8(\mathcal{P}-2)\delta^2 - 2(13\mathcal{P}+1)\delta + 9\mathcal{P}^2\right]rs^2$$
$$+ \frac{3\mathcal{P}^2(\mathcal{P}+1)}{128}\left[\delta(10\delta-7) - 2\kappa(\delta-\delta^2)\right]r^2 s.$$

The Routh-Hurwitz criterion (see [Ro877] and [Hu895]) states that $p(\widetilde{\lambda}) = 0$ has all roots in the left complex half-plane including the imaginary axis if and only if the coefficients satisfy the inequalities

$$p_j \geq 0 \quad \text{for } j = 0, \ldots, 3 \qquad \text{and} \qquad p_1 p_2 - p_0 p_3 \geq 0.$$

It is easy to see that p_3 and p_2 are nonnegative for all $s, r \geq 0$ and $\delta \in [0,1]$.

The discussion of the remaining conditions is more difficult. In the case the coefficient p_0 changes sign a simple real eigenvalue crosses the imaginary axis which is called *monotonic skew-varicose* (MSV) instability. If p_0 remains positive but p_1 or $p_1 p_2 - p_0 p_3$ become negative, a pair of complex conjugate eigenvalues crosses the imaginary axis which is called *oscillatory skew-varicose* (OSV) instability. In the two lemmata below we give regions where we can guarantee that these instabilities occur. A schematic view of the regions in the $(\mathcal{P}, \kappa)$-plane is given in Figure 5.1.

Lemma 5.6 (Monotonic skew-varicose instability). *Let*

$$\mathcal{P}_{\text{MSV}} = (3 - \sqrt{8})(1 + \sqrt{7 + \sqrt{32}}) \approx 0.78197,$$

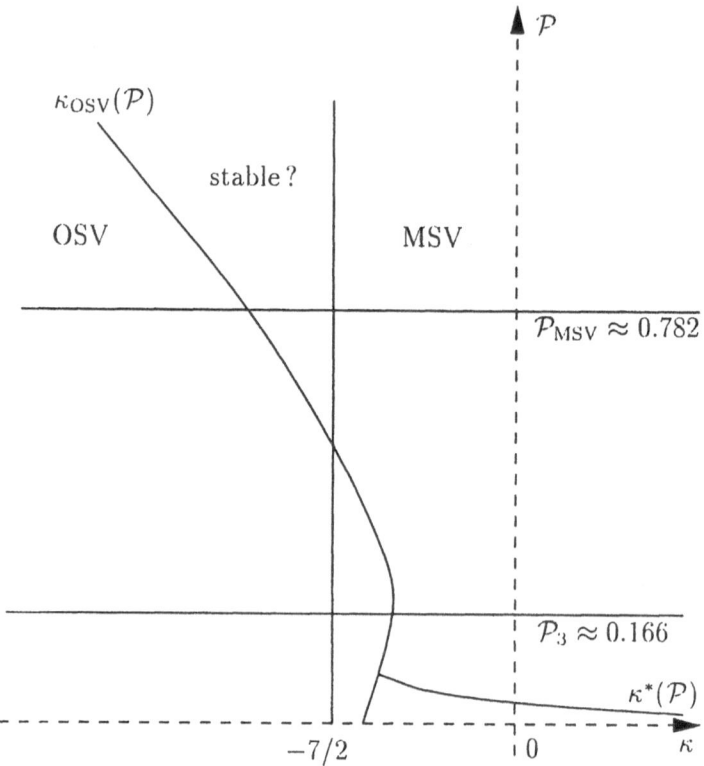

Fig. 5.1. The skew varicose stability regions as functions of the Prandtl number $\mathcal{P}$, where $\kappa = 48\sqrt{2}(k - k_0)/(\mathcal{P}^2 \rho^2)$.

then $p_0(r, \kappa, s, \delta)$ attains negative values for some $(s, \delta) \in (0, \infty) \times [0, 1]$ if and only if $(\mathcal{P}, \kappa)$ satisfies

$$\mathcal{P} \in (0, \mathcal{P}_{MSV}) \quad or \quad \kappa > -7/2.$$

Proof. Note that p_0 has the form $s(a_2 s^2 + a_1 rs + a_0 r^2)$ which remains positive as long as $a_2 \geq 0$, $a_0 \geq 0$, and $a_1 + 2\sqrt{a_0 a_2} \geq 0$. From the definitions of a_2 and a_0 it is easy to see that a_2 is always nonnegative and that a_0 is nonnegative for all $\delta \in [0, 1]$ and $\mathcal{P} > 0$ if and only if $\kappa \leq -7/2$.

The condition $a_1 + 2\sqrt{a_0 a_2} \geq 0$ is analyzed only for $\kappa \leq -7/2$. We rewrite the problem in the form

$$\alpha(P, \delta) \stackrel{\text{def}}{=} \frac{8(\mathcal{P}-2)\delta^2 - (26\mathcal{P}+2)\delta + 9\mathcal{P}^2}{6\mathcal{P}\sqrt{6(\mathcal{P}+1)}}$$

$$\geq \beta(\kappa, \delta) \stackrel{\text{def}}{=} -\sqrt{\delta\big[(10+2\kappa)\delta - (7+2\kappa)\big]}.$$

Here β is independent of $\mathcal{P}$ and α is a quadratic polynomial in δ which is independent of κ. For $\mathcal{P} \leq 2$ the function α is concave in δ while β is convex. Hence the condition is satisfied if and only if $\alpha(P, 0) \geq \beta(\kappa, 0) = 0$ and $\alpha(P, 1) \geq \beta(\kappa, 1) = -\sqrt{3}$. The inequality at $\delta = 0$ is always valid while the condition at $\delta = 1$ holds exactly for $P \geq P_{MSV}$.

The case $P > 2$ is easily estimated, since α is now convex and lies above its tangent $T(P, \delta) = \alpha(P, 1) + \partial_\delta \alpha(P, 1)(\delta - 1)$ in the point $\delta = 1$. Moreover, $\beta(\kappa, \delta) \leq \beta(-7/2, \delta) = -\sqrt{3}\delta$, and an easy check yields $T(P, \delta) \geq -\sqrt{3}\delta$ for $P > 2$ and $\delta \in [0, 1]$. $\qquad\square$

From the proof of this lemma we see that the instability for $\kappa > -7/2$ occurs for $\delta \in (0, \delta_0)$ and $s \in (0, s_0 r)$, which is an attached sideband instability with $|\sigma_2| \ll |\sigma_1| = \mathcal{O}(\rho)$ (like the Eckhaus instability). The case $\kappa < -7/2$ and $P < P_{MSV}$ leads to negative values for $p_0(r, \kappa, \delta, s)$ when $\delta \in (\delta_1, 1)$ and $s \in (s_1 r, s_2 r)$ with $0 < s_1 < s_2$ fixed. Thus, this is a detached sideband instability with $|\sigma_1| \ll |\sigma_2|$ and $|\sigma_1| \in (\sqrt{s_1}\rho, \sqrt{s_2}\rho)$ (like the zigzag instabilities).

Lemma 5.7 (Oscillatory skew-varicose instability). *(a) There is a value $P_3 \approx 0.1654$ such that $p_1(r, s, \delta)$ attains negative values for some $(s, \delta) \in (0, \infty) \times [0, 1]$ if and only if $P \in [0, P_3)$.*

(b) There is a value $P_{OSV} \approx 0.050712$ and a function $\kappa^* = \kappa^*(P)$ such that the function $p_1 p_2 - p_0 p_3$ attains negative values for some $(s, \delta) \in (0, \infty) \times [0, 1]$ if and only if (P, κ) satisfies $\kappa < \kappa^*(P)$. Letting

$$\kappa_{OSV}(P) = -\frac{1}{2(P+1)^2}\left((P+3)(3P^2 + 2P + 2) + 3P^2\sqrt{(P+1)(P+5)}\right),$$

we have $\kappa^*(P) = \kappa_{OSV}(P)$ for $P \geq P_{OSV}$. In the region $0 < P < P_{OSV}$ we have $\kappa^*(P) > \kappa_{OSV}(P)$ and $\kappa^*(P) = \gamma/P^2 + \mathcal{O}(1/P)$ for $P \to 0$ with $\gamma \approx 0.0006665$.

Proof. (a) For $p_1 = a_2 s^2 + a_1 rs \geq 0$ we need $a_2, a_1 \geq 0$. Obviously, $a_2 \geq 0$ and a_1 is a quadratic polynomial in δ which attains its minimum value $\tilde{a}(P)$ for some $\delta \in (0, 1)$. From $\tilde{a}(P) = (48P^5 + 191P^4 + 252P^3 + 62P^2 - 12P - 1)/(8(P + 1)(2P + 1))$ we find the desired result.

(b) As in the previous lemma we may write $p_1 p_2 - p_0 p_3 = \frac{4096}{27} s(b_2 s^2 + b_1 rs + b_0 r^2)$ which is nonnegative as long as b_0, b_2, and $b_1 + 2\sqrt{b_0 b_2}$ are nonnegative. We have

$$b_2 = 384(1 - \delta)(P + 5 - 4\delta)^2,$$

$$b_1 = \frac{4(P + 1)}{3}\left[-32(P + 1)(11P + 8)\delta^3 + 16(3P^3 + 65P^2 + 55P + 20)\delta^2 \right.$$
$$\left. - 2(87P^3 + 605P^2 + 109P + 23)\delta + 9P^2(P + 5)(P + 13)\right],$$

$$b_0 = P^2(P + 1)\left[4(P + 1)^2(\delta - \delta^2)\kappa - 4(P + 1)(P + 3)\delta^2 \right.$$
$$\left. - 4(2P^2 - 4P - 3)\delta + 3P^2(P + 5)\right],$$

which immediately gives $b_2 \geq 0$. Moreover, we see that $b_0(P, \kappa, \delta)$ increases monotonically with κ and satisfies

$$b_0(P, \kappa, 0) = 3P^4(P + 1)(P + 5), \qquad b_0(P, \kappa, 1) = 3P^4(P + 1)^2.$$

For sufficiently large κ the minimum of b_0 over $\delta \in [0, 1]$ will be positive and there is a unique $\kappa = \kappa_{OSV}(\mathcal{P})$ such that the minimum is negative if and only if $\kappa < \kappa_{OSV}(\mathcal{P})$. The value is characterized by a double zero of the quadratic polynomial $b_0(\mathcal{P}, \kappa, \cdot)$. Using the values of b_0 at $\delta \in \{0, 1\}$, this double zero has to occur at $\delta = \sqrt{\mathcal{P} + 5}/(\sqrt{\mathcal{P} + 1} + \sqrt{\mathcal{P} + 5})$. Equating $b_0(\mathcal{P}, \kappa, \widetilde{\delta})$ to zero, we find exactly the given expression $\kappa = \kappa_{OSV}(\mathcal{P})$.

It remains to consider the third condition $b_3 = b_1 + 2\sqrt{b_0 b_2} \geq 0$. As a first estimate, consider b_1 divided by $\mathcal{P} + 1$ as a polynomial in $\mathcal{P}$ and minimize each coefficient over $\delta \in [0, 1]$. This gives $b_1(\mathcal{P}, \delta) \geq \frac{4}{3}(\mathcal{P} + 1)(9\mathcal{P}^4 + 36\mathcal{P}^3 + 63\mathcal{P}^2 - 15\mathcal{P} - 2)$ which implies that b_1, and hence b_3, is positive for all $(\mathcal{P}, \delta) \in [0.3, \infty) \times [0, 1]$. However, at $\mathcal{P} = 0$ we obtain $b_3(0, \kappa, \delta) = -8\delta(23 - 128\delta + 160\delta^2)/3$, which attains negative values for some δ and every κ. Hence, κ^* has to grow to infinity for $\mathcal{P} \to 0$. We substitute $\kappa = \mu/\mathcal{P}^2$ into b_3 and obtain

$$\widetilde{b}_3(\mathcal{P}, \delta, \mu) = -8\delta(23 - 128\delta + 160\delta^2)/3 + 32\sqrt{6\mu\delta}\,(1 - \delta)(5 - 4\delta) + \mathcal{O}(\mathcal{P})\ \text{for}\ \mathcal{P} \to 0.$$

Thus, $\widetilde{b}_3$ has a positive minimum if and only if $\mu \geq \widetilde{\mu}(\mathcal{P}) = \gamma + \mathcal{O}(\mathcal{P})$. The constant γ can be found numerically by looking for a double zero.

The point where the curves $\kappa_{OSV}(\mathcal{P})$ and $\kappa^*(\mathcal{P})$ separate can again be found numerically by searching for a double zero in δ of the function $b_3(\mathcal{P}, \kappa_{OSV}(\mathcal{P}), \delta)$ which leads to the value $\mathcal{P}_{OSV}$.   □

We note that $\kappa_{OSV}(\mathcal{P}) \leq -7/2$ if and only if $\mathcal{P} \geq \mathcal{P}_2 \approx 0.54398$.

We may translate these stability boundaries back into the $(\mathcal{R}, k)$-plane by using $\mathcal{R} = \mathcal{R}_0 + 2\sqrt{3}(k - k_0)^2 + \frac{\sqrt{3}\mathcal{P}^2}{16}\rho^2 + \cdots$ and (5.11), viz., $\rho = \varepsilon\sqrt{r}$ and $k = k_0 + \varepsilon^2 \frac{\mathcal{P}^2}{48\sqrt{2}}\kappa r$. Without loss of generality we may assume $r = 1$ which means that $\varepsilon = \rho$. Thus, we arrived at the main instability result.

Theorem 5.8 (Skew-varicose instability). *Assume that*

$$(\mathcal{P}, \kappa) \in \Gamma \overset{\text{def}}{=} \{\, (\mathcal{P}, \kappa) \in (0, \infty) \times \mathbb{R}\colon \mathcal{P} < \mathcal{P}_{MSV}, \ \text{or}\ \kappa > -7/2, \ \text{or}\ \kappa < \kappa_{OSV}(\mathcal{P}) \,\}.$$

Then, there exists a number $\rho_0 = \rho_0(\mathcal{P}, \kappa) > 0$ *such that for all* $\rho \in (0, \rho_0)$ *the roll solution* $z_{\rho, k(\rho)}$, *where* $k(\rho) = k_0 + \frac{\mathcal{P}^2}{48\sqrt{2}}\kappa\rho^2$, *is unstable in the sense of Definition 3.1. (Here,* $\rho_0(\mathcal{P}, \kappa)$ *can be chosen such that it remains bounded away from zero on any compact subsets of* Γ.)

The lines $(\rho, k(\rho))$ given in the theorem correspond to curves in the $(\mathcal{R}, k)$-plane which have the form $k = k_0 + \kappa(\mathcal{R} - \mathcal{R}_0)/(3\sqrt{6}) + \mathcal{O}(|\mathcal{R} - \mathcal{R}_0|^2)$. Hence, for a given $\mathcal{P} > \mathcal{P}_{MSV}$ we may vary κ suitably to obtain a family of curves which fill a region

$$A_{SV}(\mathcal{P}) =$$

$$\left\{ (\widetilde{\mathcal{R}}(k, \rho^2), k)\colon \kappa = \tfrac{48\sqrt{2}(k - k_0)}{\mathcal{P}^2\rho^2} \in \mathbb{R} \setminus \left[\kappa_{OSV}(\mathcal{P}), -\tfrac{7}{2}\right] \text{ and } \rho \in (0, \rho_0(\mathcal{P}, \kappa)) \right\},$$

which is the set of parameters $(\mathcal{R}, k)$ for which we have proved instability. It consists of two drop-shaped regions, one for the MSV and one for the OSV instability. The drops have their tip in the point $(\mathcal{R}_0, k_0)$ with tangents $\mathcal{R} = \mathcal{R}_0$ and $\mathcal{R} = \mathcal{R}_0 + \frac{3\sqrt{6}}{\kappa}(k - k_0)$, where $\kappa = -7/2$ and $\kappa = \kappa_{\text{OSV}}(\mathcal{P})$ for MSV and OSV, respectively. See also Figure 6.1.

This result is somewhat weaker than in the case of the Eckhaus and the zigzag instabilities. There we obtained exact curves $\mathcal{R} = \mathcal{R}_{\text{Eckhaus}}(k)$ and $\mathcal{R} = \mathcal{R}_{\text{zigzag}}^{(j)}(k)$ which bound the region of unstable rolls. Here the boundary of the instability region $A_{\text{SV}}(\mathcal{P})$ has no algebraic characterization; it depends on $\rho_0(\mathcal{P}, \kappa)$ which can only be estimated.

6. Discussion

We have developed a mathematically rigorous and computationally efficient method for the study of sideband instabilities of small periodic patterns bifurcating from a homogeneous state. The theory can detect all instabilities which arise through perturbations having small sideband vectors; however, other instabilities need not be seen. The instability theory is based on a Liapunov-Schmidt reduction (LSR) for the linear spectral problem which yields a complete analysis of the unfolding of the critical eigenvalues as functions of the sideband vector and other bifurcation parameters.

We believe that most of the formal methods for the study of sideband instabilities using multiple scalings or amplitude equations are in fact formally equivalent to our method, i.e., the method presented here leads to computations which are completely analogous to those arising in the other formal theories. The major difference here is our mathematically rigorous background which leads to several principal advantages even without discussing the necessity of proofs:

(1) For more involved problems it will be almost impossible to get a secure result without having a clean algorithm. Our method provides a simple standardized approach for the nonlinear bifurcation problem as well as for the sideband instability theory. The LSR can be implemented as numerical code or as a computer algebra tool. Using these tools the analysis of sideband instabilities can be made very efficient.

(2) Moreover, the knowledge that expansions obtained formally with a few terms are really an approximation of a nice smooth function allows us to apply perturbations arguments which show that effects seen in the lower order terms really persist for the full problem.

(3) Applying our method we are forced to set up a proper functional analytic framework. As a reward we obtain rigorous statements on the spectral properties of the linearization around the basic periodic pattern.

The usefulness of the method is established by the application to the three-dimensional Bénard problem. We were led to a LSR with a five-dimensional bifurcation equation, where the mean flows, the mean pressure, and the two convective modes are taken into account. We note that the whole theory can be restricted to the two-dimensional case, then leading to a four-dimensional reduced stability problem (just delete the second row and column in $\tilde{b}$ and set $\sigma_2 = 0$). There the only instability is the so-called Eckhaus instability which we have studied in Section 5.2. A more complete treatment is given in [KvW96] where stability is also shown.

Further instabilities arise through the presence of three-dimensional perturbations of the form $v(t, x) = e^{\lambda t + i(\sigma_1 x_1 + \sigma_2 x_2)} W(kx_1, x_3)$ where $\sigma_2 \neq 0$. We have two types of zigzag

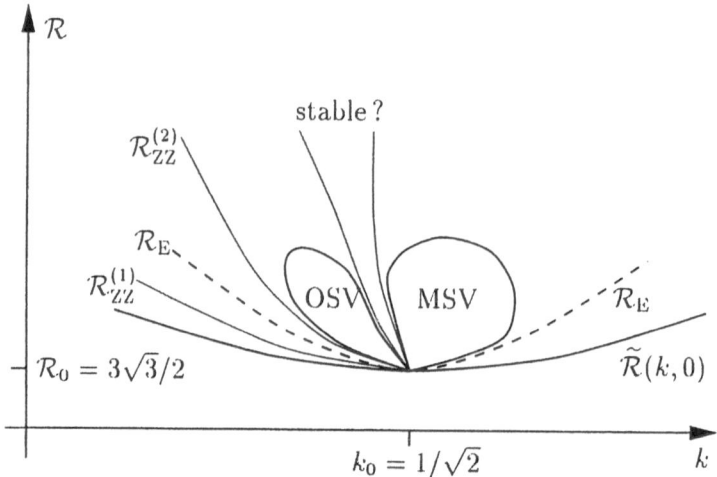

Fig. 6.1. Instabilities for the Bénard convection rolls $\mathcal{P} > 17$. Eckhaus instability $\mathcal{R}_E$, zigzag instability $\mathcal{R}_{ZZ}^{(j)}$, skew-varicose instabilities: monotonic $\mathcal{R}_{MSV}$ and oscillatory $\mathcal{R}_{OSV}$.

instabilities ($\sigma_1 = 0$), the monotonic and the oscillatory skew-varicose instability. For large Prandtl numbers ($\mathcal{P} \geq 17$), our instability regions are depicted in Figure 6.1.

We note that all bifurcating rolls are unstable for $\mathcal{P} < \mathcal{P}_{MSV} \approx 0.782$ while for larger values for $\mathcal{P}$ there is a sector in the $(\mathcal{R}, k)$-plane where the rolls might be stable. We do not claim stability since we only derived sufficient conditions for *instability*.

Another aspect of our analysis is that the Galilean invariance obtains a central role; see Lemma 5.3. It guarantees that all roll solutions have, for $\sigma = 0$, a three-fold eigenvalue zero. The understanding of this degeneracy is necessary for the proper unfolding of the critical eigenvalues as functions of σ. The complexity of the instability problem for the Bénard roll is mainly due to this fact.

Our results compare nicely with the previous work on the formal level as was derived in [ZS83] and [BuB84], [Be94]. In fact, our work reconciles the discrepancies between the first and the two latter works. In both cases, only sufficient conditions for instability were derived, and our analysis implies that the union of both instability criteria is correct.

The instability for $\mathcal{P} < \mathcal{P}_{MSV}$ was not detected in [BuB84] and [Be94] since there the analysis of p_0 was restricted to the case where $|\sigma| \ll \rho$. Thus, only attached sideband instabilities were detected whereas the instability for $\mathcal{P} < \mathcal{P}_{MSV}$ is due to a detached sideband instability with $|\sigma_1| \ll |\sigma_2|$ and $|\sigma_1| \in (\sqrt{s_0}\rho, \sqrt{s_1}\rho)$.

In [ZS83] the analysis is based on the scaling

$$\rho = \varepsilon^2 \widetilde{\rho}, \qquad k - k_0 = \varepsilon^2 \widetilde{\kappa}, \qquad \sigma_1 = \varepsilon^2 \widetilde{\sigma}_1, \qquad \sigma_2 = \varepsilon \widetilde{\sigma}_2, \qquad \lambda = \varepsilon^4 \widetilde{\lambda}, \qquad (6.1)$$

which is the classical scaling for the Newell-Whitehead-(Segel) equation (cf. [ZS83], [NPL93], [Mi96]), which in turn was developed to study slow modulations of roll patterns. The resulting scaled eigenvalue problem $\Lambda_{ZS}^*(\widetilde{\rho}, \widetilde{\kappa}, \widetilde{\sigma}, \widetilde{\lambda}) = 0$ is then a quadratic polynomial in $\widetilde{\lambda}$, where unstable eigenvalues are always real. Thus, no oscillatory skew-varicose instability was detected. The reason is that the eigenvalues λ associated with the oscillatory instability have an imaginary part of order $\mathcal{O}(\rho|\sigma_2|)$ (cf. equations (3.8) and (3.9) in [BuB84]) which is incompatible with the scaling (6.1).

Appendix A. The Nonlinear Problem

In order to present the calculations of the coefficients $c_1, \ldots, c_5$ in (4.11) most efficiently, we just give the result which is obtained from applying the LSR inductively on the terms which are needed. We define the order parameter $\varepsilon = \sqrt{\mathcal{R} - \mathcal{R}_0} + |k - k_0| + |\eta|$ and note that it is sufficient to calculate $\mathcal{Z}(\mathcal{R}, k, \sum \eta_j \xi_j)$ up to terms of order $\mathcal{O}(\varepsilon^m)$ in order to find all coefficients of the bifurcation equation $\tilde{n}(\mathcal{R}, k, \eta) = 0$ up to terms of order $\mathcal{O}(\varepsilon^{m+1})$. This is a general fact which holds since the term $A_0 \mathcal{Z}$ does not enter into $\tilde{n}$ since $P A_0 \mathcal{Z} = 0$.

Define the vectors

$$
Z_{14} = \begin{pmatrix} \frac{38-2P}{75} \sin y \cos x_3 \\ 0 \\ \frac{7+47P}{75\sqrt{2}} \cos y \sin x_3 \\ \frac{(17+7P)\sqrt{3}}{75\sqrt{2}} \cos y \sin x_3 \\ \frac{24P-6}{75\sqrt{2}} \cos y \cos x_3 \end{pmatrix}, \quad
Z_{44} = \begin{pmatrix} 0 \\ 0 \\ 0 \\ \frac{-P\sqrt{3}}{8} \sin(2x_3) \\ \frac{8+9P}{32} \cos(2x_3) - \frac{1}{2} \cos(2y) \end{pmatrix},
$$

$$
Z_{4k} = \begin{pmatrix} \frac{2}{15} \cos y \cos x_3 \\ 0 \\ \frac{16\sqrt{2}}{15} \sin y \sin x_3 \\ \frac{2\sqrt{6}}{5} \sin y \sin x_3 \\ \frac{4\sqrt{2}}{5} \sin y \cos x_3 \end{pmatrix}, \quad
Z_{4\mathcal{R}} = \begin{pmatrix} \frac{2\sqrt{6}}{75} \cos y \cos x_3 \\ 0 \\ \frac{-28\sqrt{3}}{75} \sin y \sin x_3 \\ \frac{-18}{25} \sin y \sin x_3 \\ \frac{-26\sqrt{3}}{75} \sin y \cos x_3 \end{pmatrix}, \quad Z_{11} = 0;
$$

then all Z_{jl} lie in $Z_1 = (I - Q)Z$. (For the calculation of the coefficients c_j the vector $Z_{4\mathcal{R}}$ is not needed but we use it in Appendix C.) By the symmetry arguments in Section 4 it is enough to consider the case $\eta = (\eta_1, 0, 0, \eta_4, 0)$, and we let

$$
z^* = \eta_1 \xi_1 + \eta_4 \xi_4 + \mathcal{Z}^*
$$
$$
\text{with } \mathcal{Z}^* = \eta_1^2 Z_{11} + \eta_1 \eta_4 Z_{14} + \eta_4^2 Z_{44} + \eta_4 (k - k_0) Z_{4k} \in Z_1
$$
$$
= (I - Q)Z,
$$

and define $\mathcal{N}^* = \mathcal{N}(\mathcal{R}, k, z^*)$. We obtain

$$
g_1 \eta_1 = \langle \mathcal{N}^*, \xi_1 \rangle + \mathcal{O}(\varepsilon^4) = \frac{-3(P+1)}{20} \eta_1 \eta_4^2 + \mathcal{O}(\varepsilon^4),
$$

$$
g_4 \eta_4 = \langle \mathcal{N}^*, \xi_4 \rangle + \mathcal{O}(\varepsilon^4)
$$
$$
= \left(-\frac{\sqrt{3}}{2}(\mathcal{R} - \mathcal{R}_0) + 3(k - k_0)^2 - \frac{7P^2 + 34P - 23}{200} \eta_1^2 + \frac{3P^2}{32} \eta_4^2 \right) \eta_4 + \mathcal{O}(\varepsilon^4),
$$

$$
g_5 \eta_1 \eta_4 = \langle \mathcal{N}^*, \xi_5 \rangle = -\left(\frac{3\sqrt{2}(P+1)}{8} + \frac{27(P+1)}{20}(k - k_0) \right) \eta_1 \eta_4 + \mathcal{O}(\varepsilon^4), \tag{A.1}
$$

where all g_j are evaluated at $(\mathcal{R}, k, \eta_1^2, \eta_4^2)$.

To check that the given approximation Z^* of $Z(\mathcal{R}, k, z_0)$ is in fact sufficient we have to calculate the error $M = N^* - \langle N^*, \xi_1 \rangle \xi_1 - \frac{4}{15}(\langle N^*, \xi_4 \rangle \xi_4 + \langle N^*, \xi_5 \rangle \xi_5)$. By explicit calculation we find that $M = \mathcal{O}(\varepsilon^3)$ which is enough to guarantee $Z(\mathcal{R}, k, \eta_1 \xi_1 + \eta_4 \xi_4) = Z^* + \mathcal{O}(\varepsilon^3)$.

Using the equivariance $Z(\mathcal{R}, k, T_\tau z_0) = T_\tau Z(\mathcal{R}, k, z_0)$ the expansion with $\eta_5 \neq 0$ is

$$Z\left(\mathcal{R}, k, \sum_{l=1}^{5} \eta_l \xi_l\right) = \eta_1\left(\eta_4 Z_{14} + \eta_5 Z_{15}\right) + \eta_4^2 Z_{44} + \eta_4 \eta_5 Z_{45} + \eta_5^2 Z_{55}$$

$$+ (k - k_0)\left(\eta_4 Z_{4k} + \eta_5 Z_{5k}\right) + \mathcal{O}(\varepsilon^3),$$

where $Z_{15} = T_{\pi/2} Z_{14}$, $Z_{55} = T_{3\pi/2} Z_{44}$, $Z_{5k} = T_{\pi/2} Z_{4k}$, and

$$Z_{45} = T_{5\pi/4} Z_{44} - T_{7\pi/4} Z_{44} = (0, 0, 0, 0, -\sin(2y))^T.$$

Appendix B. The Linear Problem

Here we establish the form (5.9) and (5.10) of $\widetilde{b}(0, k, \sigma, \lambda)$. To this end we use the abstract Lemma 2.4. First consider the functions in the span of ξ_1, ξ_2, and ξ_3. Since $\mathcal{B}(0, k, \sigma, \lambda)$ leaves this three-dimensional subspace invariant we find that the LSR is just the restriction to this subspace. Thus, the form of $\widetilde{b}^{(123)}$ is clear.

For the reduction in the case of ξ_4 and ξ_5 we can use the fact that $\mathcal{B}(0, k, \sigma, \lambda)$ leaves the subspace $Z_{(1,1)} = \{e^{iy}\varphi_1(\beta) \in Z : \beta \in \mathbb{C}^5\}$ invariant, where

$$\mathcal{B}(0, k, \sigma, \lambda)e^{iy}\varphi_1(\beta) = e^{iy}\varphi_1(F_1(\widetilde{\mathcal{R}}(k, 0), \sigma, \lambda)\beta),$$

and φ_1 and F_1 are defined in (4.1) and (4.2), respectively.

We first calculate the LSR of the matrix F_1 by solving $F_1(\widetilde{\mathcal{R}}(k, 0), \sigma, \lambda)[\chi_4 + V] = \alpha \chi_4$ under the side condition $\chi_4^T V = 0$, where $\chi_4 = (\sqrt{2}, 0, -i, -i\sqrt{3}, 3i)^T$ is chosen such that $\zeta = \xi_4 + i\xi_5 = \varphi_1(\chi_4)$. From this linear problem for $(\alpha, V) \in \mathbb{C}^6$ we obtain

$$\alpha = \frac{P+1}{5}\lambda + \frac{4}{5}\sigma_1^2 + \frac{4\sqrt{2}}{5}(k - k_0)\sigma_2^2 + \frac{4}{15}\sigma_1^2\sigma_2^2 + \frac{2}{5}\sigma_2^4 + \frac{16(P+1)}{75}\sigma_2^2\lambda$$

$$+ \frac{4\sqrt{2}}{45}\left(5\sqrt{2}(k - k_0) + \frac{31}{9}\sigma_1^2 + 5\sigma_2^2 + (P+1)\lambda\right)\sigma_1 + \mathcal{O}(\widetilde{\varepsilon}^4),$$

and

$$V = \begin{pmatrix} \frac{2}{15}\sigma_1 + \frac{\sqrt{2}}{225}\left(15\sqrt{2}(k-k_0) - 224\sigma_1^2 - 180\sigma_2^2 + (6P-114)\lambda\right) \\ 2\sigma_2 - \frac{28\sqrt{2}}{15}\sigma_1\sigma_2 \\ \frac{-16i\sqrt{2}}{15}\sigma_1 - \frac{i}{225}\left(240\sqrt{2}(k-k_0) + 346\sigma_1^2 + 270\sigma_2^2 + (141P+21)\lambda\right) \\ \frac{-2i\sqrt{6}}{5}\sigma_1 - \frac{i\sqrt{3}}{225}\left(90\sqrt{2}(k-k_0) + 196\sigma_1^2 + 120\sigma_2^2 + (21P+51)\lambda\right) \\ \frac{-4i\sqrt{2}}{5}\sigma_1 - \frac{2i}{25}\left(10\sqrt{2}(k-k_0) + 9\sigma_1^2 + 5\sigma_2^2 + (4P-1)\lambda\right) \end{pmatrix} + \mathcal{O}(\widetilde{\varepsilon}^{9/4}),$$

where $\widetilde{\varepsilon}^2 = |k - k_0| + \sigma_1^2 + |\sigma_2|^{3/2} + |\lambda|$. The apparently strange scaling defined through $\widetilde{\varepsilon}$ stems from the fact that we have to use a special scaling in Section 5.3 where the zigzag instability is considered. There the term $\frac{2}{5}\sigma_2^4$ occurring in the function α is needed.

According to Lemma 2.4 we find the functions v and $\widetilde{\delta}$ in (5.10) to be given by

$$v(k, \sigma, \lambda) = \frac{15}{8}[\alpha(k, \sigma, \lambda) + \overline{\alpha(k, -\sigma, \overline{\lambda})}],$$

$$\widetilde{\delta}(k, \sigma, \lambda) = \frac{15}{8\sigma_1}[\alpha(k, \sigma, \lambda) - \overline{\alpha(k, -\sigma, \overline{\lambda})}]. \tag{B.1}$$

The additional factor 15/4 appears because $\langle \xi_4, \xi_4 \rangle = \langle \xi_5, \xi_5 \rangle = 15/4$. Moreover, the function $\mathcal{W}_j(0, k, \sigma, \lambda)$ satisfying $(I - P)\mathcal{B}(0, k, \sigma, \lambda)[\xi_j + \mathcal{W}_j] = 0$ can be calculated explicitly:

$$\mathcal{W}_j(0, k, \sigma, \lambda) = 0 \quad \text{for } j = 1, 2, 3,$$

$$\mathcal{W}_4(0, k, \sigma, \lambda) = \frac{1}{2}\left[e^{iy}\varphi(V(k, \sigma, \lambda)) + e^{-iy}\varphi(\overline{V(k, -\sigma, \overline{\lambda})})\right],$$

$$\mathcal{W}_5(0, k, \sigma, \lambda) = \frac{-i}{2}\left[e^{iy}\varphi(V(k, \sigma, \lambda)) - e^{-iy}\varphi(\overline{V(k, -\sigma, \overline{\lambda})})\right]. \tag{B.2}$$

Appendix C. The Stability Problem

It remains to calculate the coefficients $d_0, \ldots, d_4$ which are important in studying the zigzag and the skew-varicose instability. To find the expansion of b^* as given in (5.12) we recall the definition in (5.7) and then we find

$$d_0 = \gamma_{14}(0, k_0, 0, 0, 0) - \gamma_{24}(0, k_0, 0, 0, 0),$$

$$d_1 = \partial_{(\sigma_1^2)}\gamma_{15}(0, k_0, 0, 0, 0) + \gamma_{25}(0, k_0, 0, 0, 0),$$

$$d_2 = \partial_{(\sigma_2^2)}\gamma_{15}(0, k_0, 0, 0, 0),$$

$$d_3 = \gamma_{42}(0, k_0, 0, 0, 0) - \gamma_{41}(0, k_0, 0, 0, 0),$$

$$d_4 = -\partial_{(\rho^2)}\gamma_{45}(0, k_0, 0, 0, 0).$$

As in Section 5.4 we let $\varepsilon = \rho + |\sigma| + \sqrt{|\lambda| + |k - k_0|}$ and find that we have to calculate the columns of $\widetilde{b}$ up to different orders in ε: namely, up to ε^2 for the columns 1, 2, and 4, and up to ε^3 for column 5. Before we treat these cases separately we recall that Theorem 2.2 allows us to reduce the computational effort for the solution of $(I - P)\mathcal{B}(\rho, k, \sigma, \lambda)[\xi_j + \mathcal{W}_j] = 0$ considerably. In fact, we will employ the results from Appendix A and B in the calculations below to find the relevant terms in the expansions of $\mathcal{W}_j$ for either $(\sigma, \lambda) = (0, 0)$ or $\rho = 0$, respectively. Recall also the expansion (4.12):

$$z_{\rho,k} = \rho\xi_4 + \rho^2 Z_{44} + \rho^3\left(Z_{444} - \frac{c_4}{c_2}Z_{4R}\right) + \rho(k - k_0)Z_{4k} + \mathcal{O}(\varepsilon^4).$$

To find $\widetilde{b}(\rho, k, \sigma, \lambda)e_j$ up to order $\mathcal{O}(\varepsilon^m)$, we calculate $\mathcal{W}_j$ up to order $\mathcal{O}(\varepsilon^{m-1})$. Moreover, we have to be aware that $z_{\rho,k}$ enters the problem linearly from the derivative of the quadratic nonlinearity and it is multiplied with the order 1 term ξ_j. Hence, we use an expansion of $z_{\rho,k}$ up to order $\mathcal{O}(\varepsilon^m)$.

Columns 1 to 4 of $\widetilde{b}$

We only need to calculate the terms up to and including order ε^2. Thus, it is sufficient to expand only the term of order ε in $\mathcal{W}_j$, $j = 1, \ldots, 4$, namely

$$\sum_{j=1}^{4} \omega_j \mathcal{W}_j = \omega_1\left(\rho Z_{14} + \mathcal{O}(\varepsilon^2)\right) + \omega_2 \mathcal{O}(\varepsilon^2) + \omega_3 \mathcal{O}(\varepsilon^2)$$

$$+ \omega_4\left(2\rho Z_{44} + i\sigma_1 W_{41}^0 + i\sigma_2 W_{42}^0 + \mathcal{O}(\varepsilon^2)\right),$$

where W_{4j}^0 is the corresponding lowest order expansion coefficent of $\mathcal{W}_4(0, k, \sigma, \lambda)$ from Appendix B:

$$W_{41}^0 = \frac{\sqrt{2}}{15}\begin{pmatrix} \sqrt{2}\sin y \cos x_3 \\ 0 \\ -16\cos y \sin x_3 \\ -6\sqrt{3}\cos y \sin x_3 \\ -12\cos y \cos x_3 \end{pmatrix}, \qquad W_{42}^0 = \begin{pmatrix} 0 \\ 2\sin y \cos x_3 \\ 0 \\ 0 \\ 0 \end{pmatrix}.$$

For the roll solution we use the approximation $z_{\rho,k} = \rho\xi_4 + \rho^2 Z_{44} + \mathcal{O}(\varepsilon^3)$. Inserting all this into $\widetilde{b}_{jl} = \langle \mathcal{B}(\rho, k, \sigma, \lambda)[\xi_l + \mathcal{W}_l], \xi_j\rangle$, we find

$$\widetilde{b} = \begin{pmatrix} \lambda + |\sigma|^2 - \frac{3(\mathcal{P}+1)}{20}\rho^2 & 0 & i\sigma_1 & i\sigma_1\rho & * \\ 0 & \lambda + |\sigma|^2 & i\sigma_2 & 0 & * \\ -i\sigma_1 & -i\sigma_2 & 0 & 0 & 0 \\ \frac{3\mathcal{P}+8}{10}i\sigma_1\rho & 0 & 0 & \frac{3(\mathcal{P}+1)}{4}\lambda + 3\sigma_1^2 + \frac{3\mathcal{P}^2}{16}\rho^2 & * \\ -\frac{3(\mathcal{P}+1)}{4\sqrt{2}}\rho & 0 & 0 & 0 & * \end{pmatrix} + \mathcal{O}(\varepsilon^3),$$

which gives $d_0 = 1$ and $d_3 = -(3\mathcal{P} + 8)/10$. Moreover, an explicit calculation yields

$$\mathcal{B}(\cdots)[\xi_l + \mathcal{W}(\cdots)_l] - \sum_{j=1}^{5}\frac{\widetilde{b}_{jl}(\cdots)}{\langle\xi_j,\xi_j\rangle}\xi_j = \mathcal{O}(\varepsilon^2), \quad \text{for } l = 1, \ldots, 4,$$

as desired.

Column 5 of $\widetilde{b}$

We now need a higher order approximation. To avoid the calculation of Z_{444} we replace (4.12) with

$$z_{\rho,k} = \rho\xi_4 + \rho^2 Z_{44} + \rho(k - k_0)Z_{4k} + \mathcal{O}(\rho^3 + \varepsilon^4).$$

For W_5 we have

$$W_5 = \rho Z_{45} + i\sigma_1 W_{51}^0 + i\sigma_2 W_{52}^0 + \rho^2\left(\tfrac{2c_4}{\sqrt{3}} Z_{5\mathcal{R}} + Z_{445}\right) + i\rho\left(\sigma_1 W_{51}^1 + \sigma_2 W_{52}^1\right)$$
$$+ \lambda W_{5\lambda}^0 + (k - k_0) W_{5k}^0 - \sigma_1^2 W_{511}^0 - \sigma_1\sigma_2 W_{512}^0 - \sigma_2^2 W_{522}^0 + \mathcal{O}(\varepsilon^3).$$

Here the superscript l in W^l stands for the associated power of ρ, hence all coefficients W^0 can be easily obtained from V and (B.2). Further we have $W_{52}^1 = 0$, $Z_{5\mathcal{R}} = T_{\pi/2} Z_{4\mathcal{R}}$,

$$Z_{445} = \frac{\mathcal{P}^2}{600}\begin{pmatrix} 2\sqrt{2}\sin y\cos x_3 \\ 0 \\ -47\cos y\sin x_3 \\ -7\sqrt{3}\cos y\sin x_3 \\ -24\cos y\cos x_3 \end{pmatrix} + \frac{\mathcal{P}^2}{27328}\begin{pmatrix} 27\sqrt{2}\sin y\cos(3x_3) \\ 0 \\ -9\cos y\sin(3x_3) \\ -361\sqrt{3}\cos y\sin(3x_3) \\ 513\cos y\cos(3x_3) \end{pmatrix},$$

and

$$W_{51}^1 = \frac{1}{240}\begin{pmatrix} 0 \\ 0 \\ 0 \\ 44\mathcal{P}\sqrt{6}\sin(2x_3) \\ 16\sqrt{2}\cos(2y) - (99\mathcal{P}+128)\sqrt{2}\cos(2x_3) \end{pmatrix}.$$

Hence, $\tilde{b}_{j5} = \langle\mathcal{B}(\cdots)[\xi_5 + W_5], \xi_j\rangle$ leads to

$$\begin{pmatrix} \sqrt{2}\left(\tfrac{2}{3}\sigma_1^2 + \tfrac{1}{2}\sigma_2^2 + \tfrac{3}{20}(\mathcal{P}+1)\lambda\right)\rho \\ \tfrac{\sqrt{2}}{2}\rho\sigma_1\sigma_2 \\ 0 \\ \tfrac{-i\sqrt{2}}{240}\left(33\mathcal{P}^2\rho^2 + 720\sqrt{2}(k-k_0) + 496\sigma_1^2 + 720\sigma_2^2 + 144(\mathcal{P}+1)\lambda\right)\sigma_1 \\ \tfrac{3(\mathcal{P}+1)}{4}\lambda + 3\sigma_1^2 \end{pmatrix} + \mathcal{O}(\rho^3 + \varepsilon^4).$$

From our general considerations in Section 5 (cf. (5.8)), we know that the last column of $\tilde{b}$ cannot contain a term ρ^3. Thus, we have established the desired expansion and found the coefficients $d_1 = \sqrt{2}/6$, $d_2 = 1/\sqrt{2}$, and $d_4 = 11\sqrt{2}\mathcal{P}^2/80$. It may be verified that

$$\mathcal{B}(\cdots)[\xi_5 + W(\cdots)_5] - \sum_{j=1}^{5}\frac{\tilde{b}_{j5}(\cdots)}{\langle\xi_j,\xi_j\rangle}\xi_j = \mathcal{O}(\varepsilon^3),$$

when the given approximations for $z_{\rho,k}$ and W_5 are inserted.

Acknowledgments

The author is grateful to F. H. Busse for stimulating discussions and for helping to find a computational error in an earlier version of this paper.

References

[Be94]   A.J. Bernoff. Finite amplitude convection between stress-free boundaries; Ginzburg-Landau equations and modulation theory. *Eur. J. Appl. Math.* **5** (1994) 267–282.

[BM95]   T.J. Bridges, A. Mielke. A proof of the Benjamin-Feir instability. *Arch. Rat. Mech. Anal.* **133** (1995) 145–198.

[BM96]   T.J. Bridges, A. Mielke. Instability of spatially-periodic states for a family of semilinear PDE's on an infinite strip. *Math. Nachrichten* **179** (1996) 5–25.

[BoB85] E.W. Bolton, F.H. Busse. Stabilities of convection rolls in a layer with stress-free boundaries. *J. Fluid Mech.* **150** (1984) 487–498.

[Bu71]   F.H. Busse. Stability regions of cellular fluid flow. In *Instability of Continuous Systems*, H. Leipholz (ed.), Proc. of the IUTAM Symposium in Bad Herrenalb 1969, Berlin: Springer-Verlag, 1971, 41–47.

[BuB84] F.H. Busse, E.W. Bolton. Instabilities of convection rolls with stress-free boundaries near threshold. *J. Fluid Mech.* **146** (1984) 115–125.

[CE90]   P. Collet, J.-P. Eckmann. *Instabilities and Fronts in Extended Systems*. Princeton, NJ: Princeton University Press, 1990.

[CH88]   S.-N. Chow, J.K. Hale. *Methods of Bifurcation Theory.* New York: Springer-Verlag, 1985.

[Da91]   K. Damerow. A sufficient condition for the validity of the principle of reduced stability. *Math. Nachrichten* **154** (1991) 243–252.

[DP94]   W. Decker, W. Pesch. Order parameter and amplitude equations for the Rayleigh-Bénard convection. *J. Phys. II France* **4** (1994) 419–438.

[Eck65] W. Eckhaus. *Studies in Non-Linear Stability Theory.* Berlin: Springer-Verlag. Springer Tracts in Nat. Phil. Vol. **6**, 1965.

[GS85]   M. Golubitsky, D.G. Schaeffer. *Singularities and Groups in Bifurcation Theory Vol. I.* New York: Springer-Verlag, 1985.

[GSS88] M. Golubitsky, I. Stewart, D.G. Schaeffer. *Singularities and Groups in Bifurcation Theory Vol. II.* New York: Springer-Verlag, 1988.

[Hu895] A. Hurwitz. Über die Bedingungen unter welchen eine Gleichung nur Wurzeln mit negativen reellen Teilen besitzt. *Math. Ann.* **46** (1895) 273–284.

[KL83]   H. Kielhöfer, R. Lauterbach. On the principle of reduced stability. *J. Functional Anal.* **53** (1983) 99–111.

[Ku96]   M. Kuwamura. The stability of roll solutions of the 2-D Swift-Hohenberg equation and the phase diffusion equation. *SIAM J. Math. Anal.* **27** (1996) 1311–1335.

[KvW96] Y. Kagei, W. von Wahl. Stability of convection roll solutions of Boussinesq equations in two space dimensions. *Int. J. Non-Linear Mechanics* (1996). To appear.

[Mi95]   A. Mielke. A new approach to sideband instabilities using the principle of reduced instability. In *Nonlinear Dynamics and Pattern Formation in the Natural Environment.* A. Doelman, A. van Harten (eds.), Pitman Research Notes in Mathematics Series **335.** Longman, 1995. 206–222.

[Mi96]   A. Mielke. Instability and stability of rolls in the Swift-Hohenberg equation. Preprint Universität Hannover, 1996. Submitted to *Comm. Math. Physics.*

[MT84]   B. Malomed, M.I. Tribel'skii. Bifurcations in distributed kinetic systems with aperiodic instability. *Physica D* **14** (1984) 67–87.

[NPL93] A.C. Newell, T. Passot, J. Lega. Order parameter equations for patterns. *Annu. Rev. Fluid Mech.* **25** (1993) 399–453.

[Re89]   L. Recke. On linear stability of bifurcating equilibria. *Math. Nachrichten* **140** (1989) 59–68.

[RR92]   M. Renardy, R.C. Rogers. *An Introduction to Partial Differential Equations.* New York: Springer-Verlag, 1992.

[SdP78] J.T. Stuart, R.C. di Prima. The Eckhaus and the Benjamin–Feir resonance. *Proc. Royal Soc. London,* **A 362** (1978) 27–41.

[SvW92] B.J. Schmitt, W. von Wahl. Decomposition of solenoidal fields into poloidal, toroidal and the mean flow. Applications to the Boussinesq equation. In *The Navier-*

Stokes Equation II—Theory and Numerical Methods. Proceedings Oberwolfach 1991. J.G. Heywood, K. Masuda, R. Rautmann, S.A. Solonnnikov (eds.). Lecture Notes in Math. **1530**, Berlin: Springer-Verlag, 1992, 291–305.

[Va84] A. Vanderbauwhede. Stability of bifurcating equilibria and the principle of reduced stability. In *Bifurcation Theory and Applications*, Proceedings of the CIME meeting in Montecatini 1983, L. Salvadori (ed.). Lecture Notes in Math. **1057**. Berlin: Springer-Verlag, 1984, 209–223.

[ZS83] A. Zippelius, E.D. Siggia. Stability of finite-amplitude convection. *Phys. Fluids* **26** (1983) 2905–2915.

[Ro877] E. J. Routh. *A Treatise on the Stability of a Given State of Motion (Adams Price Essay)*. Macmillan, London, 1877.

[Sc95a] B. Scarpellini. $\mathcal{L}^2$-perturbations of periodic equilibria of Navier-Stokes. *Zeits. Anal. Andwendungen* **14** (1995) 779–828.

[Sc95b] B. Scarpellini. The principle of linearized stability for the space periodic equilibria of Navier-Stokes on an infinite plate. *Analysis* **15** (1995) 359–394.

[ScW98] B. Scarpellini, W. von Wahl. Stability properties of the Boussinesq equations. *Zeits. Angew. Math. Physik* **49** (1998) 294–321.

KAM Theory Near Multiplicity One Resonant Surfaces in Perturbations of A-Priori Stable Hamiltonian Systems

M. Rudnev[1] and S. Wiggins[2]

[1] Department of Applied Mathematics 217-50, California Institute of Technology, Pasadena, CA 91125, USA
[2] Departments of Applied Mechanics and Control and Dynamical Systems 116-81, California Institute of Technology, Pasadena, CA 91125, USA

Received January 15, 1996; manuscript accepted for publication July 8, 1996
Communicated by Jerrold Marsden

This paper is dedicated to the memory of Juan-Carlos Simo

Summary. We consider a near-integrable Hamiltonian system in the action-angle variables with analytic Hamiltonian. For a given resonant surface of multiplicity one we show that near a Cantor set of points on this surface, whose remaining frequencies enjoy the usual diophantine condition, the Hamiltonian may be written in a simple normal form which, under certain assumptions, may be related to the class which, following Chierchia and Gallavotti [1994], we call *a-priori unstable*. For the a-priori unstable Hamiltonian we prove a KAM-type result for the survival of whiskered tori under the perturbation as an infinitely differentiable family, in the sense of Whitney, which can then be applied to the above normal form in the neighborhood of the resonant surface.

Key words. KAM theory, multiplicity-one resonance surface, whiskered tori, normal form, integrability on Cantor sets

MSC numbers. 58F05, 58F07, 58F27, 58F30, 58F36, 70H05, 70K30

1. Introduction

This paper is concerned with certain aspects of multiplicity-one resonant dynamics for a generic n-degree of freedom a-priori stable system, i.e., a near-integrable nondegenerate Hamiltonian system expressed in the action-angle variables whose Hamiltonian is real-analytic in some open domain. The number of degrees of freedom n is greater than 2. We consider a Cantor set of points on a codimension-one resonant plane $\omega_1 = 0$ in the frequency space, whose $n - 1$ remaining frequencies satisfy a classical diophantine condition. This Cantor set remains fixed throughout our analysis. The main theorem we prove establishes local integrability near this set, if several assumptions are satisfied. We show that one may take a comparably large neighborhood of this Cantor set, namely a union of open n-balls centered at every point of this set, whose size will be defined by the norm of the perturbation, the parameters of the problem, and the number of degrees of freedom, where a simple normal form may be derived. The latter under certain assumptions may be regarded as a perturbed a-priori unstable system, namely a system whose integrable part possesses a separatrix. The a-priori stable systems allowing such a construction we call *hyperbolic in the first order*.

For this reason at the beginning we devote a separate section to the a-priori unstable systems. We present a KAM-type proof to establish the existence of a differentiable, in the sense of Whitney, family of whiskered tori which survive a perturbation. Generally we follow the well established iterative scheme due to Arnold. Our approach relies heavily on Pöschel [1982], [1993]. Further we apply this result to the neighborhood of the resonant surface for the initial a-priori stable system. There we argue that the one-parameter families of $n - 1$-dimensional invariant tori, corresponding to the multiplicity-one resonances for the unperturbed problem, break up under perturbation and yield hyperbolic $n \doteq 1$-tori with their stable and unstable manifolds (whiskers). In this way we build up a global structure in the neighborhood of the entire resonant surface as a differentiable family of whiskered tori, thus relating different whiskered tori to each other, which is further expected to enable us to advance in the search for the heteroclinic intersection of their whiskers and to establish locally a possibility of drift along the resonant surface, where the density of the whiskered tori is sufficiently high.

Related results can be found in Graff [1974], where a novel *nonsingular* extension of the KAM theory to systems with hyperbolicity was set up and much more, and later in de la Llave and Wayne [1990], Treschev [1991], and Chierchia and Gallavotti [1994], where the terms "a-priori stable" and "a-priori unstable" were coined.

However, our work differs from the aforementioned references both in its spirit and technique. We consider it the first building block in our research project concerning the so-called Arnold's diffusion problem, or rather *Arnold's mechanism*, suggested in Arnold [1964], the latter being a much less general and simpler problem than the former, as pointed out by Lochak [1995], though far from being completely understood. The main purpose of this piece is to suggest a useful Normal form, globally valid in the neighborhood of a multiplicity-one resonant surface that describes simultaneously all the weakly hyperbolic tori with frequencies satisfying some diophantine condition, and straightens out their local stable and unstable manifolds, the kind of Normal form that can further be used as the starting point for analysis of the homoclinic splitting. In this connection all the quoted references except for Chierchia and Gallavotti [1994] deliver only local results that explicitly pertain to only one single torus, and are consequently insufficient for being immediately used in search of the heteroclinic orbits; the last one (namely the part of it which proves the analogous result for the a-priori unstable systems) is close in its philosophy to what we suggest, but we believe that the technique we use, in particular Lie transforms and what we call the Absolute Norm (see Appendix A), a modification of the weighted norm introduced by Pöschel [1993], makes it less unwieldy and essentially easier to conceive and to use. Withal, we are a bit more concerned with the sharpness of estimates, performed in the course of the proof, which becomes important for further considering the splitting of separatrices, since the latter problem is exclusively sensitive with respect to different parameters and their relation to each other; generally we believe that most of the dependencies we suggest are in a way optimal.

We want to emphasize that, although in the section devoted to the a-priori unstable systems we pay tribute to the tradition founded by Poincaré in his "Méthodes nouvelles..." by introducing an "independent" perturbation parameter μ apart from ε, throughout the whole presentation we are dealing with the *singular* problem (see also Treschev [1991]), which turns out to be very similar to a nonsingular one as far as the construction of the Normal form is concerned with all the modifications included into the smallness condition for the perturbation parameter.

In concluding this section we also want to point out that our approach differs from the one used by Graff [1974] and Treschev [1991] in about the same sense that the ways to prove the standard KAM theorem, suggested by Kolmogorov [1954] and Arnold [1963], are different. The Whitney-smoothness of the family of the perturbed whiskered tori along the resonant surface that we establish is the essential fact for further study of the heteroclinic intersections between their whiskers.

2. Main Results

2.1. Set-up and Notation

We study a Hamiltonian system whose Hamiltonian has the form

$$H(I, \varphi, \varepsilon) = H^0(I) + \varepsilon H^1(I, \varphi, \varepsilon), \tag{1}$$

in the action-angle variables $(I, \varphi) \in D \times T^n$, where D is an open subset of R^n, and T^n is an n-torus, obtained from R^n by identifying points whose coordinates differ by integer multiples of 2π, and is analytic in some complex neighborhood of the above domain, and smooth enough in the small real non-negative perturbation parameter ε. We will assume that H^1 is uniformly bounded for all small values of ε and that its Taylor series in ε has a nontrivial zero order term. H^1 is assumed to be 2π-periodic in each of the angle variables φ_i, $i = 1, 2, \ldots, n$, where $n > 2$.

The standard KAM theorem argues that for the Hamiltonian (1) in the action-angle variables, which for $\varepsilon = 0$ describes simple dynamics on tori with constant frequencies $\omega(I) = H^0_I$, the introduction of the perturbation does not affect most of these tori (having asymptotically full measure as $\varepsilon \to 0$), causing only their slight deformation, unless their frequencies are near resonances, which are described by the condition that $k \cdot \omega = 0$ for some nontrivial integer vector $k \in Z^n$.

Here we consider a multiplicity-one resonance, i.e., there is only one independent integer vector satisfying $k \cdot \omega = 0$. Geometrically, this relation defines a hyperplane in the frequency space which we refer to as a resonant plane. By a suitable linear symplectic transformation of the coordinate system this resonant plane can be transformed to $\omega_1 = 0$. We will assume this has been done. Hence, a hyperplane

$$P \equiv \{\omega \colon \omega_1 = 0\} \tag{2}$$

in the *frequency space* Ω we call the *resonant plane*. Its image in the action space

$$\Pi \equiv \{I: \omega_1(I) = 0\} \tag{3}$$

we call the *resonant surface*. Further we'll need several nondegeneracy assumptions on the unperturbed *frequency map* Ψ_0: $\omega(I) = H_I^0$. Qualitatively, we assume that it is *locally invertible* near the whole resonant surface (otherwise we would have to restrict ourselves to only a part of it). This will allow us to consider along with the action-angle variables (I, φ) the frequency-angle variables (ω, φ), though noncanonical. The frequency-angle variable formulation used by Pöschel [1982] allows certain conceptual simplifications that prove convenient.

Henceforth we will extend the action of the frequency map Ψ and its inverse Ψ^{-1} over all the other variables and parameters upon which the Hamiltonian may depend apart from I as the identity transformation:

$$\Psi(I, \cdot) = (\omega(I), \cdot), \qquad \Psi^{-1}(\omega, \cdot) = (I, \cdot).$$

For functions of the frequency-angle variables we will use calligraphic letters. In particular, for the Hamiltonian (1), expressed in these variables, we write

$$\mathcal{H}(\omega, \varphi, \varepsilon) = H \circ \Psi^{-1}. \tag{4}$$

This will be helpful, since in the frequency space the resonant condition holds on a hyperplane; besides, the construction we make is quite simple in this representation.

We shall sometimes admit a slight abuse of rigor, simply saying "the frequency variables" or "the action variables" without referring to the other (angle or hyperbolic) variables that the Hamiltonian in consideration may depend on, basically meaning transiting back and forth under the action of the just generalized transformations Ψ and Ψ^{-1}.

On a resonant plane P we identify a Cantor set Ω_γ of "good" diophantine frequencies, and let I_γ^0 be its image through the inverse of the frequency map. Quantitatively,

$$\Omega_\gamma \equiv \{\omega: \omega_1 = 0, |\tilde{\omega} \cdot k| \geq \gamma |k|^{-\tau}, \forall k \in Z^{n-1} \setminus \{0\}\}, \tag{5}$$

where $\tilde{\omega} \equiv (\omega_2, \ldots, \omega_n)$ and $\gamma > 0$, $|k| = \sum_{i=1}^{n-1} |k_i|$; $\tau \geq n - 2$ is kept fixed. For a small γ the relative measure of Ω_γ on the resonant plane is big if $\tau > n - 2$. Further we will often write ω, instead of $\tilde{\omega}$, just keeping in mind that for the given a-priori stable system we always stay close to the plane P.

Remark. In fact, we shall consider the intersection of D and a neighborhood of Π which is the image in the action space of some full-dimensional neighborhood of Ω_γ through the inverse of the frequency map, or, equivalently, a subset $\tilde{\Omega}_\gamma \subseteq \Omega_\gamma$, such that the image of some of its full-dimensional neighborhood lies in D. To avoid the unnecessary notational complications we will always assume that D itself and its complex extension will contain, respectively, all the real sets in the action space and their complex extensions that we will be dealing with, including Π.

At this point we will introduce a necessary piece of notation. Here we assume that apart from the action-angle variables (I, φ) the Hamiltonian also depends on two extra variables (p, q), to which we refer as the *hyperbolic variables*, given in an open real ball $\bar{B}_\kappa^2 \equiv \{(p, q) \in R^2 : |p|^2 + |q|^2 < \kappa^2\}$ with radius $\kappa > 0$ near the origin, namely,

$$H = H(I, \varphi, p, q, \varepsilon). \tag{6}$$

We will need this in the following discussion about the a-priori unstable systems. If there is no dependence on (p, q), it must just be ignored in the notation for domains and norms by skipping the indices corresponding to these variables.

Further, $r, s, \kappa, \rho, \sigma, \eta$ will stand for some positive parameters.

Henceforward, the notation $\bar{B}_\kappa^l$ (B_κ^l) will stand for a real (complex) ball in R^l (C^l) of radius κ. For instance,

$$B_\kappa^2 \equiv \{(p, q) \in C^2 : |p|^2 + |q|^2 < \kappa^2\}.$$

Also if S is an arbitrary set in R^l, which stands for either the action or the frequency space, then for an arbitrary dimension l we write the complex extension of the domain of definition of (6) as

$$S + (r, s, \kappa) \equiv S + (r) \times W_s T^l \times B_\kappa^2 \subseteq C^l \times C^l \times C^2,$$

where

$$S + (r) \equiv \{z \in C^l : |z - S| < r\}$$

in the *sup*-norm, and

$$W_s T^l \equiv \{\varphi \in C^l : |\Im \varphi| \leq s\}$$

in the *sup*-norm. Equivalently, sometimes we write $S + (r, s, \kappa)$ as $S_{r,s,\kappa}$. Also we reserve the notation $S + (r, \kappa)$, or $S_{r,\kappa}$, for the direct product

$$S + (r, \kappa) \equiv S + (r) \times B_\kappa^2 \subseteq C^l \times C^2.$$

For a complex set $S \subseteq C^{l_0}$ which, like $S_{r,s,\kappa}$, is the direct product of three complex bounded sets $S_1 \subseteq C^{l_1}$, $S_2 \subseteq C^{l_2}$, $S_3 \subseteq C^{l_3}$, where $l_1 + l_2 + l_3 = l_0$ we will write $S - (\rho, \sigma, \eta)$ when we shrink S, cutting off the border layers of thickness ρ, σ, η from the corresponding sets S_i, $i = 1, 2, 3$ of S and taking the direct product.

Now we will define the Absolute norm we will be using. For more information the reader is referred to Appendix A which contains a summary of the properties of this norm and its relation to a more traditional *sup*-norm. Given the function $u(\cdot, p, q)$, analytic in $p, q \in B_\kappa^2$, we define as $\overline{u(\cdot, p, q)}$ the absolute sum of its Taylor series in p, q coordinates near $(p, q) = (0, 0)$. Namely, if

$$u(\cdot, p, q) = \sum_{i,j=0}^{\infty} u(\cdot)_{ij} p^i q^j,$$

then

$$\overline{u(\cdot, p, q)} \equiv \sum_{i,j=0}^{\infty} |u(\cdot)_{ij} p^i q^j|.$$

Then, given an analytic in $S_{r,s,\kappa}$ function $u(x, p, q, \varphi)$, where $x \in S + (r)$ stands for either the action or the frequency variables, with a Fourier series

$$u(\cdot, \varphi) = \sum_{k \in Z^{n-1}} u^k(\cdot) e^{ik \cdot \varphi},$$

we define its absolute norm as

$$|u|_{S_{r,s,\kappa}} \equiv \sup_{(x,p,q) \in S_{r,\kappa}} \sum_{k \in Z^{n-1}} \overline{u^k(x, p, q)} e^{|k|s}.$$

Also, for norms of functions given on the complex extension S of the fixed domain C in the action or the frequency space $S = S_{r,s,\kappa}$, we will often write $| \cdot |_S$, or $| \cdot |_{r,s,\kappa}$ instead of $| \cdot |_{S,r,s,\kappa}$. For an arbitrary vector-function v the corresponding norm will be $|v|_{r,s,\kappa} \equiv \max_i |v_i|_{r,s,\kappa}$; for an arbitrary matrix A the corresponding norm $|A|_{r,s,\kappa} \equiv \max_{i,j} |A_{ij}|_{r,s,\kappa}$. Often in the notation for domains and norms we will omit the indices (in which we won't be interested at the moment), where it does not lead to confusion.

To finish with the introduction, below we give several more definitions and pieces of notation we will be using. For an analytic function $u(\cdot, p, q)$ we will write $\mathcal{D}u$ to denote the part of its Taylor series in (p, q), which depends only on the product pq, namely

$$\mathcal{D}u(\cdot, p, q) \equiv \sum_{i=0}^{\infty} u(\cdot)_{ii} (pq)^i.$$

We will use the $\mathcal{P}$ symbol for a norm $|I, \varphi, p, q|_{\mathcal{P}} \equiv \max(|I|, |\varphi|, |p|, |q|)$ and for the induced operator norm.

If $f = \sum_{k \in Z^n} f_k e^{ik \cdot \varphi}$ denotes the Fourier expansion of a function, then we define $T_K f \equiv \sum_{k \in Z_K^n} f_k e^{ik \cdot \varphi}$, where $Z_K^n \equiv \{k \in Z^n \mid |k| \leq K\}$. We refer to $T_K f$ as the *ultraviolet cutoff* of f.

We assume that for some open domain $D \subseteq R^n$ in the action space the Hamiltonian (1) is real analytic in $D + (r_0, s_0)$ for some positive parameters r_0, s_0. Besides, suppose, for all small positive values of ε for $|H^1|_{r_0, s_0} \leq 1$.

2.2. Main Theorem and General Strategy

Let's consider the unperturbed resonant surface Π and the image I_γ^0 of the Cantor set Ω_γ on it through the inverse of the frequency map Ψ^{-1}. For the unperturbed problem each point of this set corresponds to an n-dimensional multiplicity-one resonant torus, which is foliated into a one-parameter family of tori of dimension $n - 1$. Under certain assumptions, as we'll show, for this family the perturbation brings in hyperbolicity in the first order in ε. We will call perturbed systems of this kind *hyperbolic in the first order*. More technically speaking, by applying once a canonical near-identity transformation $\mathcal{R}_0$ which will be provided by our Main Lemma, we derive a simple normal form for the Hamiltonian (1), valid in some full-dimensional open neighborhood of the set I_γ^0, whose truncation up to the first-order terms in ε for $\varepsilon \neq 0$ is integrable, and may be considered as an a-priori unstable system. We will show that (I_1, φ_1) may be considered as the "hyperbolic variables." We will redefine them as (p, q) and reserve the notation (I, ω, φ) for the rest of the actions, frequencies, and angles.

A well-known result (Moser [1956], Chierchia and Gallavotti [1994]) says that locally in the neighborhood of a hyperbolic equilibrium point the integrable Hamiltonian of the a-priori unstable system by applying some canonical transformation $\mathcal{R}$ may be written in some new canonical coordinates (I', φ', p', q') such that $I = I'$, $J = p'q'$ simply as a function of (I, J). So we can apply this result to the truncation of the above normal form and, if we proceed thinking of it as the "unperturbed" Hamiltonian, and the remainder as a perturbation, the question basically boils down to the KAM theory for the a-priori unstable systems.

The part of the discussion devoted to the a-priori unstable systems will be presented in a self-contained fashion, in particular the notation will be entirely independent. The basic strategy of the iterative KAM-type proof for the a-priori unstable systems will be as follows. At every step, numbered by $i = 0, 1, \ldots$, the Hamiltonian H_i will be the sum of two components: the "unperturbed Hamiltonian" $h(I, J)_i$ plus a perturbation $f(I, \varphi, p, q)_i$ which we target to make smaller and smaller. The way we position the index i is also to point out that at every step we are actually dealing with different coordinate systems. The "frequency map" on each step will be defined as Ψ_i: $(I, \cdot) \rightarrow (\omega = \partial_I h_i, \cdot)$; as mentioned in Section 2.1 we consider its action upon all the other variables and parameters as an identity. Then we can write the Hamiltonian in the frequency variables simply as $\mathcal{H}(\omega, \varphi, p, q)_i = H_i \circ \Psi_i^{-1}$. When we apply to H_i the Main Lemma, which will be our technical tool to seek the canonical changes of variables, thus reducing the size of the perturbation, the unperturbed Hamiltonian picks up a part $\mathcal{D}(T_K \bar{f})_i$ of the Fourier series for the perturbation, truncated with some ultraviolet cutoff parameter K, growing bigger from step to step, averaged in the "fast" angles, which depends only on the actions I and the product pq. So, the new unperturbed Hamiltonian becomes $h_{i+1} = h_i + \mathcal{D}(T_K \bar{f})_i$, and it defines a new frequency map Ψ_{i+1}, which is close to Ψ_i.

To accomplish this we use Lie transforms and the canonical variables (I, φ, p, q). In the Main Lemma that we use on each step we construct a Hamiltonian flow $X^t_{\phi_i}$ with Hamiltonian ϕ_i and define the transformation we are looking for in the *canonical* variables as $\Xi_i = X^1_{\phi_i}$. Then in the frequency variables the corresponding transformation will be $\Phi_i = \Psi_i \Xi_i \Psi_{i+1}^{-1}$. It is in terms of this transformation that we formulate the result of the Main Lemma and work out all the estimates for the Lie transform Hamiltonian, using the standard Cauchy inequalities. We do it to make sure that the projection of the range of Φ_i onto the frequency space lies within a union of the open balls $\Omega_\gamma + (r_i)$ with radius r_i around Ω_γ, the latter being fixed once and for all; then the projection of its domain must be the union of balls of smaller size r_{i+1}.

Thus, we construct a sequence of open domains $\{\Omega_\gamma + (r_i)\}$, where the sequence $\{r_i\}$ rapidly goes to zero; then the limit for the above sequence of the domains is a Cantor set Ω_γ. The transformation $\Phi = \Phi_0 \circ \Phi_1 \circ \cdots$ then acts from $\overline{\Omega_\gamma + (0, s_*, \kappa/2)}$ into $\Omega_\gamma + (r, s, \kappa)$ for some initial values (r, s, κ) of the analyticity parameters, and $s_* < s$; it is C^∞ in the frequency variables in the sense of Whitney and analytic in (φ, p, q). Finally we define the limit frequency map as Ψ_∞: $(I, \cdot) \rightarrow (\lim_{i \to \infty} \partial_I h_i, \cdot)$ and then the transformation $\Xi \equiv \Psi_0^{-1} \Phi \Psi_\infty$ will be canonical on a closed set $\Psi_\infty^{-1} \Omega_\gamma$. The latter set is certainly not as palpable as Ω_γ.

Finally taking the composition $\Xi_* = \mathcal{R}_0 \circ \mathcal{R} \circ \Xi$ we arrive at our main result which is formulated in our Main Theorem.

At this point we want to quote the Whitney Extension Theorem which allows us to speak about the differentiable family of the surviving tori. Using the fact that Ω_γ is closed, Whitney's theorem enables one to extend the above transformation Φ to the open domain in the frequency space containing the Cantor set Ω_γ in such a way that being restricted to Ω_γ, this extension $\mathcal{E}\Phi$ will boil down to the transformation Φ itself.

Theorem 2.1 (The Extension Theorem). *For any closed set $F \in R^n$ there exists a linear extension operator $\mathcal{E} \colon C^\infty(F) \to C^\infty(R^n)$, for $u \in C^\infty(F)$ $u \to U \equiv \mathcal{E}u$, such that for all the derivatives of u in the sense of Whitney $D^k U|_F = u^{(k)}$, $k = 0, 1, \ldots$ and $|U|_{R^n} \le c|u|_F$ in the C^∞-norm.*

Remark. The extension $\mathcal{E}\,\Xi$ will be canonical *only* by being restricted to a Cantor set $\Psi_\infty^{-1}\Omega_\gamma$ by construction.

For definitions of Whitney differentiation the reader is referred to Pöschel [1982].

During our treatment, following Pöschel [1982], we will fix the value of the parameter γ in the diophantine condition (5) equal to 1, thus dealing with the Cantor set Ω_1 rather than with Ω_γ. The result we prove then can be easily reformulated for any $\gamma < 1$, because Ω_1 may be obtained from Ω_γ by "blowing up" the frequencies by a factor γ^{-1} and multiplying the Hamiltonian by γ^{-2}.

Applying the KAM theory for the a-priori unstable systems to the neighborhood of the multiplicity-one resonant surface of the a-priori stable system, which is hyperbolic in the first order, we arrive at our main result, which we formulate now; the rest of this paper is devoted to its proof. The formulation we give here is at one point qualitative, as far as the exact definition of hyperbolicity in the first order is concerned. Expressing everything rigorously will require the quantitative formulation of the assumptions we've mentioned, which we do in Section 4.

Theorem 2.2 (Main Theorem). *For the nondegenerate Hamiltonian system (1) hyperbolic in the first order there exists ε_0 small enough depending only on the analyticity and nondegeneracy parameters of the problem and the parameter $\tau \ge 1$, proportional to γ^2, such that if*

$$|H^1|_{r_0, s_0} \le 1$$

for all $0 < \varepsilon \le \varepsilon_0$, there exists a C^∞-differentiable family in the Whitney sense of invariant whiskered tori $\mathcal{T}_\varepsilon$, whose projection on the action space is ε-close to the Cantor set $I_\gamma^0 = \Psi_0^{-1}\Omega_\gamma$.

Moreover, there exists a transformation Ξ_*, such that the projection of its range on the action space lies in the small neighborhood of the resonant surface Π, and whose action on ε is trivial:

$$\Xi_* \colon (p, I', q, \varphi', \varepsilon) \to (I, \varphi, \varepsilon),$$

where (I', φ') are the new $n - 1$ action-angles and (p, q) are the hyperbolic variables.

The transformation Ξ_* casts the Hamiltonian (1) into the form

$$H \circ \Xi_* = H'(I', pq, \sqrt{\varepsilon}),$$ (7)

which is as smooth in $\sqrt{\varepsilon}$ as (1) is in ε for $0 \le \varepsilon \le \varepsilon_0$.

The new Hamiltonian (7) defines the new frequency map Ψ_*: $(I', \cdot) \to (\omega' = \partial_{I'} H'(I', J, \sqrt{\varepsilon}), \cdot)$, such that its inverse Ψ_*^{-1}: $(\omega', \cdot) \to (I'(\omega', J, \sqrt{\varepsilon}), \cdot)$ is real analytic in J for $J = pq$, smooth in $\sqrt{\varepsilon}$ and C^∞ in the $n-1$ frequencies ω'.

The maps Ξ_* and $\Phi_* = \Psi_0 \circ \Xi_* \circ \Psi_*^{-1}$ are real analytic in the variables $\varphi' \in W_{s_*} T^{n-1}$, $(p, q) \in B_{\kappa_*}^2$ for some small positive parameters $s_* < s_0$ and $\kappa_* = O(\sqrt{\varepsilon})$. Besides, the transformation Φ_* is C^∞ in the $n-1$ frequency variables ω' on the whole resonant plane P; also, for real J the transformation Ξ_* will be C^∞ in the $n-1$ action variables I'.

Moreover, the restriction of Ξ_* in the action space to the Cantor set $\Psi_*^{-1}\Omega_\gamma$ is canonical with respect to the standard symplectic structure in the new variables (I', φ', p, q),

$$\omega^2 = dI' \wedge \varphi' + dp \wedge dq.$$

In the new variables the equations of motion for $I' \in I'_\gamma \equiv \Psi_*^{-1}\Omega_\gamma$ are given by

$$
\begin{aligned}
I' &= I'_0, \\
\varphi' &= \varphi'_0 + \omega't, \quad \omega' \in \Omega_\gamma, \\
p &= p_0 e^{-\lambda't}, \\
q &= q_0 e^{\lambda't},
\end{aligned}
$$ (8)

where $\lambda' = \partial_J H'(I', J, \sqrt{\varepsilon}) = O(\sqrt{\varepsilon})$ is a real analytic function of J, smooth in $\sqrt{\varepsilon}$, such that its real part at $J = 0$ is strictly positive for $\varepsilon > 0$.

The following corollary is the restatement of the part of this theorem describing the transformation Ξ_*, so its proof is immediate.

Corollary 1. *The tori T_ε are a part of invariant n-dimensional surfaces, which can be analytically parameterized by $(p, q) \in B_{\kappa_*}^2$ and $\varphi' \in W_{s_*} T^{n-1}$ as*

$$
\begin{aligned}
I_i &= I'_i(pq, \omega', \sqrt{\varepsilon}) + \varepsilon\Theta_1(I', \varphi', p, q, \sqrt{\varepsilon}), \quad i = 2, \ldots, n, \\
\varphi_i &= \varphi'_i + \Theta_2(I', \varphi', p, q, \sqrt{\varepsilon}), \quad i = 2, \ldots, n, \\
I_1 &= \Theta_3(I', \varphi', p, q, \sqrt{\varepsilon}), \\
\varphi_1 &= \Theta_4(I', \varphi', p, q, \sqrt{\varepsilon}),
\end{aligned}
$$ (9)

where $I'(pq, \omega', \sqrt{\varepsilon}) \in \Psi_*^{-1}\Omega_\gamma$, and the smoothness class of functions I', Θ_i, $i = 1, \ldots, 4$ in $\sqrt{\varepsilon}$ is the same as originally for H in (1) in ε.

Remark. In fact, one can say more about the functions $\Theta_2, \Theta_3, \Theta_4$ in the statement of the above Corollary, since as we've mentioned before the transformation Ξ_* is constructed in several steps. The reader can readily retrieve these details, which we choose not to mention in the general formulation, after studying the paper.

3. A-priori Unstable Systems

This section contains the proof of the self-contained result, namely the KAM-type theorem on local integrability of the perturbed a-priori unstable system on a Cantor set Ω_γ in the frequency space. The proof consists of the infinite repetitive use of the Main Lemma, which will be our principal technical tool. This result will subsequently be applied to the neighborhood of the resonant plane P of the a-priori stable system from our earlier discussion.

Nevertheless, the following discussion may be regarded independently. In particular, functions and quantities that stand behind the notation H, n, r, s, ε, etc., are generally different from those of the previous section.

3.1. The Set-up for A-priori Unstable Systems

The focus of this section is a system with Hamiltonian which, apart from the action-angle variables (I, φ), depends on an extra pair of variables (p, q), which we refer to as the *hyperbolic variables*,

$$H(p, I, q, \varphi, \varepsilon, \mu) = h_0(I, \varepsilon) + \tilde{P}(p, I, q, \varepsilon) + \tilde{f}(p, I, q, \varphi, \varepsilon, \mu). \tag{10}$$

This Hamiltonian will be real analytic on some complex domain $G + (r, s, \kappa)$ in the notation of Section 2.1. Here G is some open real domain in the action space: $I \in G \subseteq R^{n-1}$, functions of the angles φ are assumed to be 2π-periodic in these variables; $(p, q) \in B_\kappa^2$. We will consider ε, μ as two independent small parameters such that μ is complex, $|\mu| \ll \varepsilon$, and ε is real and positive. We will assume that the Hamiltonian (10) is analytic in μ for its values in the complex ball $B_{\mu_0}^1$ near the origin with a positive radius μ_0, and smooth in ε. Mostly in this section we will omit the dependence of (10) in ε, which is not important unless the conditions of the lemmata developed below are violated. However, we shall keep in mind that all the nondegeneracy and analyticity parameters of the unperturbed (corresponding to $\mu = 0$) problem are in fact ε-dependent. Speaking about the coordinate transformations, we will imply that their action on the parameters ε, μ is trivial.

We will assume that the Taylor series for $\tilde{f}$ in μ begins with the first-order terms. Also for $\mu = 0$ we assume that for all I, ε $(p, q) = (0, 0)$ is a hyperbolic equilibrium point, with the real part of its Lyapunov exponent λ,

$$\lambda^2(I, \varepsilon) \equiv (\partial_{pq}^2 \tilde{P})^2 - \partial_{pp}^2 \tilde{P} \partial_{qq}^2 \tilde{P} \Big|_{(p,q)=(0,0)},$$

being strictly positive for all I, ε. Without loss of generality we may also assume that for all I, ε $\tilde{P}(0, I, 0, \varepsilon) = 0$. Otherwise we can always achieve this by adjusting h_0. We will often refer to the hyperbolic part of the system (10) given by $\tilde{P}$ as a "pendulum" and the remaining action-angle variables as "rotors." Following Chierchia and Gallavotti [1994], if the Hamiltonian (10) satisfies the basic nondegeneracy assumptions, which we will shortly formulate quantitatively, the Hamiltonian system (10) is called *a-priori unstable*. In this case we can bring such a system into a Normal Form near the equilibrium; see, for example, Moser [1956] or Chierchia and Gallavotti [1994]. In fact, the proof of the following lemma is very much similar to the proof of Theorem 3.1 that follows, although much simpler, since the unperturbed Hamiltonian does not depend on the angles, thus depending on I as a parameter, and the reader can simply synthesize it from the proof of Theorem 3.1. Besides, in the case when P does not contain I, which is usually considered as the application (for instance the so-called Thirring model; see, e.g., Gallavotti [1994] for an alternative consideration), the analysis would not contain the I variable at all.

Lemma 3.1. *For every $I^* \in G$, any $\varepsilon > 0$, and $|\mu|$ small enough, there is a complex ball $B_{r'}^{n-1}$ centered in I^* in the action space with the radius $r'(I^*)$ such that there exists a real analytic canonical transformation $\mathcal{R}$: $(I', \varphi', p', q') \in B_{r'}^{n-1} \times W_{s'} T^{n-1} \times B_{\kappa'}^2 \to (I, \varphi, p, q) \in B_r^{n-1} \times W_s T^{n-1} \times B_\kappa^2$ for some smaller values of the analyticity parameters $r' \leq r$, $s' < s$, $\kappa' < \kappa$, where κ' may be made proportional to the minimum of the real part of the Lyapunov exponent $\lambda(I, \varepsilon)$. $\mathcal{R}$ is smooth in ε, and its action on I, ε is trivial. Moreover, the transformed Hamiltonian becomes*

$$
\begin{aligned}
H \circ \mathcal{R} &= h_0(I', \varepsilon) + P(I', p'q', \varepsilon) + f(p', I', q', \varphi', \varepsilon, \mu) \\
&\equiv h(I', J', \varepsilon) + f(p', I', q', \varphi', \varepsilon, \mu),
\end{aligned}
\tag{11}
$$

where $h = h_0 + P$.

The change of variables brought by $\mathcal{R}$ may be written as follows:

$$
\begin{aligned}
I &= I', \\
\varphi &= \varphi' + \Theta_2^0(I', \varphi', p', q', \varepsilon), \\
p &= \Theta_3^0(I', \varphi', p', q', \varepsilon), \\
q &= \Theta_4^0(I', \varphi', p', q', \varepsilon).
\end{aligned}
\tag{12}
$$

In addition, the Lyapunov exponent remains unchanged under the transformation, and one has $\lambda(I', \varepsilon) = \partial_J' h|_{J'=0}$ for $J' = p'q'$.

Remark. If $\tilde{P}$ does not contain the I-dependence, then obviously the result of this Lemma holds globally for all I and $s' = s$.

For simplicity we will further assume that the transformation $\mathcal{R}$ exists globally for $I \in G$; otherwise the domain G may be taken as small as $B_{r'}^{n-1}$ above. We will further take the Hamiltonian (11) for granted as the initially given Hamiltonian and omit primes in the notation for variables and parameters.

So, in fact we start with the Hamiltonian

$$
\begin{aligned}
H &= h_0(I, \varepsilon) + P(I, pq, \varepsilon) + f(p, I, q, \varphi, \varepsilon, \mu) \\
&\equiv h(I, J, \varepsilon) + f(p, I, q, \varphi, \varepsilon, \mu),
\end{aligned}
\tag{13}
$$

where $J = pq$, analytic in $G_{r,s,\kappa}$ for some positive values of the analyticity parameters r, s, κ.

The unperturbed Hamiltonian equations, given by (13) for $\mu = 0$, imply that for every I, ε there exists an invariant $n - 1$-torus

$$
I = const, \qquad \varphi = \varphi_0 + \omega t, \qquad (p, q) = (0, 0),
$$

where $\omega = \partial_I h(I, 0, \varepsilon)$. This torus is a part of an invariant n-dimensional surface, described by the equations above plus the equations for the stable and unstable manifolds, or whiskers,

$$p = p_0 e^{-\lambda t}, \qquad q = q_0 e^{\lambda t}.$$

In the next section, applying the results of this section to a-priori stable systems, we will be dealing with the situation when $\mu = \varepsilon^p$ for $1 < p < 2$ and will show that the assumptions of the theorem may be satisfied for any $p \geq \frac{3}{2}$ if $\varepsilon > 0$ is small enough.

3.2. Survival of the Quasiperiodic Tori under Perturbations

In Section 2 we defined a Cantor set of "good" frequencies Ω_γ. We will use the same notation for the set with the same diophantine properties in the frequency space of the rotors. The domain where the Hamiltonian is defined and analytic in the action space of the rotors must include the image of the union of open balls with some fixed positive radius, centered at points of Ω_γ. To formulate the result we need several assumptions.

Assumption 1.

• *Uniformly on $G_{r,\kappa}$ for all positive ε and some positive quantity $R(\varepsilon)$ the sup-norms of the Hessian matrix $A = \partial^2_{II} h(I, J, \varepsilon)$ and its inverse are bounded as*

$$|A|^\infty, \qquad |A^{-1}|^\infty \leq R(\varepsilon). \tag{14}$$

• *The real part of the Lyapunov exponent $\lambda \equiv \partial_J h(I, J, \varepsilon)|_{J=0}$ of the pendulum is bounded away from zero uniformly in $G_{r,\kappa}$ for all positive ε by a positive quantity $\Lambda(\varepsilon)$.*

In particular, from the definition of R it's clear that $R \geq 1$, since it's a maximum of a norm of a matrix and its inverse.

Throughout this section we'll mostly be performing our construction in the frequency space of the rotors rather than in their action space. We'll keep in mind though that the frequency-angle coordinates are generally nonsymplectic.

Remark. As in the previous section, we fix the value of the parameter γ in the diophantine condition (5) equal to 1 to ease the proof. Indeed, if we stretch the actions and the hyperbolic variables as $I \to \gamma I, p \to \gamma p, q \to \gamma q$ and multiply the Hamiltonian by γ^{-2}, Ω_γ gets blown up into Ω_1 whereas the Hessian matrix and the Lyapunov exponent remain the same.

By redefining r as $\frac{r}{R}$ we insure that the image of $\Omega_1 + (r, s, \kappa)$ through the inverse of the frequency map lies within the initial domain of analyticity.

We'll denote the Hamiltonian (13), written in the frequency-angle variables of the rotors, as

$$\mathcal{H}(\omega, p, \varphi, q, \varepsilon, \mu) \equiv H \circ \Psi^{-1} = \mathcal{H}^0(\omega, J, \varepsilon) + \mathcal{F}(\omega, p, \varphi, q, \varepsilon, \mu). \tag{15}$$

The Hamiltonian (15) will then be analytic in $\Omega_1 + (r, s, \kappa)$, as well as (13) in the image of this domain through the inverse of the frequency map.

Henceforward, for functions of the frequencies, denoted by calligraphic letters, $|\cdot|_{r,s,\kappa}$ will stand for the Absolute Norm in $\Omega_1 + (r, s, \kappa)$ for some positive values of analyticity parameters r, s, κ.

In addition, we shall assume for convenience that $r, s, \kappa \leq 1$, although the proof shows that this certainly is not a crucial necessity.

Now we will formulate and prove the main theorem of this section. This formulation will be more technical; in particular, it deals with $\gamma = 1$. Further, regarding our Remark, we'll point to the amendments one has to make to generalize the result for the set Ω_γ for an arbitrary $\gamma < 1$ and reformulate the theorem in more general terms.

Theorem 3.1. *Suppose the Hamiltonian (13) written in the action variables or, equivalently, (15), written in the frequency variables, is analytic in $\Omega_1 + (r, s, \kappa)$ and $\mu \in B^1_{\tilde{\mu}_0}$ for some positive $\tilde{\mu}_0$, and the Assumption 1 holds.*
If in (15)

$$|\mathcal{F}|_{r,s,\kappa} \leq |\mu|$$

for all $\mu \in B^1_{\tilde{\mu}_0}$, then given $s_* < s$ one can find some small constant $r_0 = r_0(r, s - s_*, \tau, R, \kappa, \Lambda)$, so that there exists a small positive $\mu_0 = \mu_0(R, \Lambda, r_0, s - s_*, \kappa, \tilde{\mu}_0, \tau, \varepsilon)$, such that for all $\mu: |\mu| \leq \mu_0$ there exists a symplectic near identity map Ξ, casting the Hamiltonian (13) into the form

$$H \circ \Xi = H_+(I, J, \varepsilon, \mu),$$

which is defined on a Cantor set in the action space $\Psi_+^{-1}\Omega_1$, where $\Psi_+^{-1}: (\omega, \cdot) \to (I(\omega, J, \mu), \cdot)$ is the inverse of the new frequency map Ψ_+ given as $\Psi_+: (I, \cdot) \to (\partial_I H_+, \cdot)$.
The map Ξ is real analytic in φ, p, q, μ for $\varphi \in W_{s_*} T^{n-1}$, $p, q \in B^2_{\frac{r}{2}}$, $\mu \in B^1_{\mu_0}$, and for real J, μ it is C^∞ in the sense of Whitney on the set $\Psi_+^{-1}\Omega_1$.
The transformation Ξ induces a transformation $\Phi = \Psi \circ \Xi \circ \Psi_+^{-1}$ in the frequency variables $\Phi: \Omega_1 + (0, s_*, \kappa/2) \to \Omega_1 + (r, s, \kappa)$, real analytic in the variables (φ, p, q) and C^∞—in the sense of Whitney—in the frequency variables, casting the Hamiltonian (15) on its new domain into the form

$$\mathcal{H} \circ \Phi = \mathcal{H}_+(\omega, J, \varepsilon, \mu).$$

In particular, one can take

$$r_0 = \min\left(\frac{1}{2}(4\tau + 20)^{-(\tau+1)}\left(\frac{s - s_*}{3}\right)^{\tau+1}, \frac{1}{2}\Lambda^{\frac{\tau+1}{\tau}}, r\right) \tag{16}$$

and

$$\mu_0 \leq \min\left(\frac{2^{-2\tau-13}r_0^{\frac{2\tau+1}{\tau+1}}(s - s_*)}{27 R^2}, \frac{2^{-2\tau-14}r_0^{\frac{\tau}{\tau+1}}\kappa_0^2}{27}, \frac{2^{-11}r_0^2}{9R^5}, \frac{\Lambda_0\kappa^2}{300}, \tilde{\mu}_0\right). \tag{17}$$

With such a choice of parameters the norm of the Hessian matrix $A_+ = \partial_{II}^2 H_+$ *and the real part of the Lyapunov exponent* $\lambda_+ = \partial_J H_+|_{J=0}$ *for Hamiltonian* H_+ *are bounded as in Assumption 1, but with* $1 + \frac{|\mu|}{\mu_0} R$ *instead of* R *and* $\Lambda(1 - \frac{|\mu|}{2\mu_0})$ *instead of* Λ.

Remark. This theorem as well as Theorem 2.2 and the following Theorem 3.2 can in fact be formulated in terms of existence of the transformations Ψ_+^{-1} and Φ, as it was done in Chierchia and Gallavotti [1994], but we prefer a more traditional formulation, since the whole iterative procedure underlying the proof has the canonical nature.

Proof. The proof is inductive and consists in the repetitive application of the Main Lemma. We construct Φ as a limit of successive transformations $\Phi = \Phi_0 \circ \Phi_1 \circ \Phi_2 \circ \cdots \circ \Phi_i \circ \cdots$.

In order to be consistent we shall write s_0, R_0, Λ_0 for, respectively, s, R, Λ.

Take a sequence

$$\sigma_i = \frac{s_0 - s_*}{3} \cdot 2^{-i}, \qquad i = 1, 2, \ldots, \tag{18}$$

$$s_i = s_0 - 3 \sum_{j=1}^{i} \sigma_j, \qquad i = 1, 2, \ldots, \tag{19}$$

and

$$\eta_i = \frac{\kappa_0}{6} \cdot 2^{-i}, \qquad i = 1, 2, \ldots, \tag{20}$$

and

$$\kappa_i = \kappa_0 - 3 \sum_{j=1}^{i} \eta_j, \qquad i = 1, 2, \ldots, \tag{21}$$

and

$$r_i = K_r \sigma_{i+1}^{\tau+1} = r_0 2^{-i(\tau+1)}, \qquad i = 0, 1, 2, \ldots, \tag{22}$$

with

$$r_0 = 2^{-\tau-1} K_r \left(\frac{s_0 - s_*}{3}\right)^{\tau+1}, \tag{23}$$

$$\rho_{i+1} = \frac{r_i}{6}, \qquad i = 0, 1, 2, \ldots, \tag{24}$$

and

$$\mu_i = \mu_0 2^{-i(2\tau+7)}, \qquad i = 0, 1, 2, \ldots, \tag{25}$$

where K_r, μ_0 are some positive constants, to be determined.

All the above sequences vanish geometrically, their denominators being some powers of 2.

We will formally define four more sequences:

$$K_i = \left(\frac{1}{2r_i}\right)^{\frac{1}{\tau+1}}, \qquad i = 0, 1, 2, \ldots, \tag{26}$$

$$Y_i = \frac{1}{2K_i^\tau} = \frac{1}{2}(2r_i)^{\frac{\tau}{\tau+1}}, \qquad i = 0, 1, 2, \ldots, \tag{27}$$

$$R_i = R_0 \cdot 2^i, \qquad i = 0, 1, 2, \ldots, \tag{28}$$

and

$$\Lambda_i = \Lambda_0 \cdot \left(\frac{3}{4}\right)^{-i}, \qquad i = 0, 1, 2, \ldots. \tag{29}$$

We want to show by induction that, given s_0, $\kappa_0 \leq 1$ for some choice of the initial values of $r_0 < 1$, or equivalently the parameter K_r, and μ_0, the above sequences will satisfy the conditions of the Main Lemma, and we can apply it with parameters r_i, s_i, κ_i, μ_i, R_i, Λ_i, K_i, and ρ_{i+1}, σ_{i+1}, η_{i+1} for $i = 0, 1, \ldots$ to yield a transformation Ψ_i, and show that the size of the remainder will enjoy the estimate

$$|\mathcal{F}|_{r_{i+1}, s_{i+1}, \kappa_{i+1}} \leq \mu_{i+1}.$$

Intuitively the above choice of the last four sequences is rather clear: The sequence (26) will be our choice of the ultraviolet cutoff parameter on each step to obey (41) in the Main Lemma; after each application of the Main Lemma the norm of the frequency map cannot exceed twice its original value and the bound on the real part of the Lyapunov exponent cannot decrease by more than one fourth of itself.

To satisfy for $i = 0$ the claim (42) of the Main Lemma on the Lyapunov exponent we will need

$$\Lambda_0 \geq 2Y_0 = (2r_0)^{\frac{\tau}{\tau+1}},$$

so we shall require

$$r_0 \leq \frac{1}{2}\Lambda_0^{\frac{\tau+1}{\tau}}. \tag{30}$$

With our choice of the sequences (22), (29) it's clear that (30) ensures that the corresponding condition of the Main Lemma will be satisfied for all i.

Checking the condition (43) on the smallness of the perturbation μ_0, for all i with

$$\xi_i^2 = \min\left(\frac{r_i\sigma_{i+1}}{12R_i^2}, \frac{\kappa_i^2}{144}\right), \tag{31}$$

we see that it will be satisfied for all i if satisfied for $i = 0$, i.e.,

$$\mu_0 \leq \frac{Y_0\xi_0^2}{2} \leq \min\left(\frac{r_0^{\frac{2\tau+1}{\tau+1}}(s_0 - s_*)}{144R_0^2}, \frac{r_0^{\frac{\tau}{\tau+1}}\kappa_0^2}{576}\right). \tag{32}$$

Additionally, we need $\mu_0 \leq \frac{\rho_i^2}{8R_0^3}$ to satisfy the condition on the frequency map in the Main Lemma, so regarding (32) we demand

$$\mu_0 \leq \min \left(\frac{r_0^{\frac{2\tau+1}{\tau+1}}(s_0 - s_*)}{144R_0^2}, \frac{r_0^2}{288R_0^5}, \frac{r_0^{\frac{\tau}{\tau+1}}\kappa_0^2}{576} \right). \tag{33}$$

By our choice of denominators for the geometric series (22), (28), (25) the fact that (33) holds for $i = 0$ will ensure that the analogous condition of the Main Lemma will be fulfilled for all i.

So we can apply the Main Lemma at least for the first time for $i = 0$, and we can apply it as many times as we please if we prove that the size of the remainder is consistent with the choice of μ_i in (25). So, suppose we are doing the ith application, then the remainder on the ith step will be given by

$$|\mathcal{F}_+|_{r_i - 3\rho_{i+1}, s_{i+1}, \kappa_{i+1}} \leq \left(1 - \frac{2\mu_i}{\xi_i^2 Y_i}\right)^{-1} \left(\frac{3\mu_i}{\xi_i^2 Y_i} + e^{-K_i \sigma_{i+1}}\right) \mu_i. \tag{34}$$

Evidently, $r_{i+1} \leq r_i - 3\rho_{i+1} = \frac{r_i}{2}$. By the choice of K_i we have $K_i > 1$. Moreover, from (18), (22), (26) we have

$$K_i \sigma_{i+1} = (2K_r)^{-\frac{1}{\tau+1}},$$

and we may choose K_r so that $e^{-K_i \sigma_{i+1}} \leq e^{-2\tau - 10} \leq 2^{-2\tau - 10}$ by letting

$$K_r \leq \frac{1}{2}(2\tau + 10)^{-(\tau+1)}. \tag{35}$$

Suppose the *second* term in the second bracket is larger than the first one. Then (34) will imply:

$$|\mathcal{F}_+|_{r_{i+1}, s_{i+1}, \kappa_{i+1}} \leq 8e^{-K_i \sigma_{i+1}} \mu_i \leq 2^{-2\tau - 7} \mu_i = \mu_{i+1}.$$

Suppose the *first* term dominates in the second bracket. Then using (18), (22), (25), (27), (28), (21), (31) we will have for all $i = 0, 1, 2, \ldots$,

$$\frac{\mu_i}{\xi_i^2 Y_i} \leq \frac{\mu_0}{\xi_0^2 Y_0},$$

and then (34) implies

$$|\mathcal{F}_+|_{r_{i+1}, s_{i+1}, \kappa_{i+1}} \leq \frac{12\mu_i}{\xi_i^2 Y_i}. \tag{36}$$

That's why if we require

$$\min \left(\frac{12 \cdot 288\mu_0}{\kappa_0^2 r_0^{\frac{\tau}{\tau+1}}}, \frac{12 \cdot 36 \cdot 4\mu_0 R_0^2}{r_0^{\frac{2\tau+1}{\tau+1}}(s_0 - s_*)} \right) \leq 2^{-2\tau - 7},$$

where the left-hand side is just an estimate for (34) using (36), this will imply

$$|\mathcal{F}_+|_{r_{i+1}, s_{i+1}, \kappa_{i+1}} \leq \mu_{i+1}.$$

So, to be consistent with our assumptions it's enough to ask

$$\mu_0 \leq \min \left(\frac{2^{-2\tau-13} r_0^{\frac{2\tau+1}{\tau+1}} (s_0 - s_*)}{27 R_0^2}, \frac{2^{-2\tau-14} \kappa_0^2 r_0^{\frac{\tau}{\tau+1}}}{27} \right),$$

which is a stronger constraint compared to (33).

Adding together (30), (35) and recalling that

$$r_0 = K_r \left(\frac{s_0 - s_*}{3} \right)^{\tau+1} \cdot 2^{-\tau-1},$$

we choose

$$r_0 = \min \left(\frac{1}{2} (4\tau + 20)^{-(\tau+1)} \left(\frac{s_0 - s_*}{3} \right)^{\tau+1}, \frac{1}{2} \Lambda_0^{\frac{\tau+1}{\tau}}, r \right)$$

to establish one of the claims of this theorem.

At the end we will go back and compute how much the parameters R_i and Λ_i that one has after applying the Main Lemma i times really differ from their initial values R_0, Λ_0 for $i = 1, 2, \ldots$ (don't mix this up with (28) and (29) which were the formal sequences and virtually delivered the rough upper bounds for R_i and the lower bounds for Λ_i). We can assume inductively $R_i < 2R_0$ for all i. For $i = 1$ this is true, according to the Main Lemma. Then (see in the Proof of Main Lemma below the estimates (50), (51) for the norm of the Hessian matrix and its inverse) we will have the following estimate (with the inductive assumption that this is true for step $0, 1, \ldots, i$):

$$R_{i+1} \leq \frac{R_i}{1 - \frac{4R_i^5 \mu_i}{\rho_{i+1}^2}} \leq R_0 \prod_{i=0}^{\infty} \left(1 + 4 \cdot 2 \cdot 32 \cdot 36 R_0^5 \frac{\mu_i}{r_i^2} \right)$$

$$\leq R_0 \exp \left(\frac{2^{10} \cdot 9 R_0^5 \mu_0}{r_0^2} \sum_{i=0}^{\infty} 32^{-i} \right),$$

where we've used that for $x \geq 0$ $e^x \geq 1 + x$. If we want the exponent to be not bigger than $\log 2 > \frac{16}{31}$ to prove our inductive assumption, then

$$\mu_0 \leq \frac{r_0^2}{2^{11} \cdot 9 R_0^5}, \tag{37}$$

and it is more strict than the second term in brackets in (33).

For the Lyapunov exponent the same inductive procedure can be used to prove that for any $i = 1, 2, \ldots, \Lambda_i \geq \frac{\Lambda_0}{2}$. Indeed, for $i = 1$ this is true by the Main Lemma. Besides, one evidently has

$$\Lambda_{i+1} \geq \Lambda_0 - \sum_{i=0}^{\infty} \frac{\mu_i}{\eta_{i+1}^2} \leq \Lambda_0 - \frac{144 \mu_0}{\kappa_0^2} \sum_{i=0}^{\infty} 32^{-i},$$

so $\Lambda_i \geq \frac{\Lambda_0}{2}$ if

$$\mu_0 \leq \frac{\Lambda_0 \kappa_0^2}{300}. \tag{38}$$

The limit $\Phi = \Phi_0 \circ \Phi_1 \circ \Phi_2 \circ \cdots$ exists on $\Omega_1 + (0, s_*, \kappa_0/2)$ (one can easily check that the above sequence of transformations is fundamental) and is real analytic in (φ, p, q, μ). For $\mu = 0$ it becomes identity. It gives us a C^∞—in the sense of Whitney—family of the surviving tori, corresponding to frequencies in the set Ω_1, since each single transformation Φ_i is analytic, so C^∞ in the frequency variables on its domain. The fact that Φ is near identity simply follows from our choice of the series and the estimates on $\mathcal{P}$-norm of the Main Lemma. Indeed, we just have to calculate the geometric sums $\sum_i \frac{\mu_i}{r_i}, \sum_i \frac{\mu_i}{\sigma_i}, \sum_i \frac{\mu_i}{\kappa_i}$, which by our choice of the sequences all converge not slower than 32^{-i}.

The last remark to make is that Φ has actually been constructed as a sequence of *symplectic* transformations in the action-angle variables: $\Phi_i = \Psi_i \Xi_i \Psi_{i+1}^{-1}$ (see the proof of the Main Lemma). Then there exists a near identity transformation $\Xi = \Psi_0^{-1} \Phi \Psi_\infty$, where $\Psi_\infty = \lim_{i \to \infty} \Psi_i$, analytic in (φ, p, q, μ), which acts on some (generally complex) Cantor set $I_1' = \Psi_\infty^{-1} \Omega_1$, μ-close to the image I_1 in the action variables of Ω_1 through the unperturbed inverse frequency map, $\Psi(I_1) \to \Omega_\gamma$, and depending on the perturbation and J. The map Ξ is C^∞ in I in the sense of Whitney for real J, μ, and the action of this map on I_1' is symplectic. (Since one can pass to a limit in the Poisson bracket operation, then the limit of a sequence of symplectic transformations is symplectic as well.)

After all, we can use the Whitney extension theorem to extend the inverse of the frequency map Ψ_+^{-1} over an open domain in the frequency space, say $\Omega_1 + (r_0)$, since the estimates we have for its norm are formally valid on a Cantor set. Then the right-hand side in (17) should be multiplied by some constant, resulting from the Extension theorem, which will be close to 1 for small r_0 and which we shall omit. This "smoothening" makes differentiation of H_+ by I more palpable in the sense that it all can be expressed in terms of differentiation by ω through the frequency map Ψ_+ and its inverse. $\qquad\square$

Remark. The difference $s - s_*$ may be rather small, say, depending on ε; that becomes important, as pointed out by Jorba and Simo [1994], for rigorous treatment of homoclinic splitting with more than two degrees of freedom; in our proof it explicitly figures in the smallness condition for μ, namely $\mu_0 \sim (s - s_*)^{2\tau+2}$ and does not cause any extra difficulties. In general, we believe that the transformation Φ has a certain type of singularity when $|\Im\varphi|$ approaches s, which depends on the diophantine properties of the frequency vector ω. In particular, when $\omega \in \Omega_\gamma$, this singularity is a pole of a finite order. It is well known that (5) is not the most lax arithmetic condition under which the KAM theorem can be proved, being a particular case of the Bryuno condition (see Bryuno [1989]), or an *approximating function* (see Pöschel [1989] for a readable proof and references therein). Nevertheless, if one pursues optimality in this sense, one encounters the necessity to require the quantity $s - s_*$ to be $O(1)$, which suggests the presence of an essential singularity of Φ when $|\Im\varphi|$ reaches s.

Theorem 3.1 can be reformulated for a fixed $\gamma < 1$. The only changes one has to make is to include γ in the smallness conditions.

Theorem 3.2. *Suppose the Hamiltonian (13), written in the action variables, or, equivalently, (15) in the frequency variables is analytic in $\Omega_\gamma + (r, s, \kappa)$, and $\mu \in B^1_{\tilde\mu_0}$ for some positive $\tilde\mu_0$, and Assumption 1 holds.*

Given $r, s, s_* < s, \kappa, R, \Lambda, \tau, \varepsilon$, one can find some $\mu_0(r, s - s_*, \kappa, R, \Lambda, \tau, \varepsilon) \leq \tilde\mu_0$, proportional to γ^2, such that if

$$|\mathcal{F}|_{r,s,\kappa} \leq \mu \leq \mu_0,$$

then there exists a near identity map Ξ

$$\Xi \colon (I', \varphi', p', q', \varepsilon, \mu) \to (I, \varphi, p, q, \varepsilon, \mu),$$

real analytic in (φ', p', q', μ), acting trivially on ε, μ, that transforms the Hamiltonian function (13) into the form

$$H \circ \Xi = H_+(I', J', \varepsilon, \mu),$$

with $J' = p'q'$, thus defining a new frequency map $\Psi_+ \colon (I', \cdot) \to (\omega' = \partial_{I'} H_+, \cdot)$, such that its inverse $\Psi_+^{-1} \colon (\omega', \cdot) \to (I'(\omega', J', \mu), \cdot)$ is C^∞ in the frequency variables $\omega' \in \Omega$ for some open real domain Ω, such that $\Omega_\gamma \subseteq \Omega \subseteq R^{n-1}$.

The induced transformation $\Phi = \Psi \circ \Xi \circ \Psi_+^{-1}$ acts as:

$$\Phi \colon \Omega + \left(0, s_*, \frac{\kappa}{2}\right) \to \Omega + (r, s, \kappa).$$

In particular,

$$\Omega_\gamma + \left(0, s_*, \frac{\kappa}{2}\right) \to \Omega_\gamma + (r, s, \kappa).$$

The transformation Φ transforms the Hamiltonian function (15) written in the frequency variables into

$$\mathcal{H} \circ \Phi = \mathcal{H}_+(\omega', J', \varepsilon, \mu).$$

It is real analytic in the variables $(\varphi', p', q') \in W_{s_*} T^{n-1} \times B^2_{\frac{\kappa}{2}}$, $\mu \in B^1_{\mu_0}$ and C^∞ in $\omega' \in \Omega$.

The restriction of the transformation Ξ on the set $\Psi_+^{-1}\Omega_\gamma$ is symplectic.

Equivalently, the restriction of the transformation Φ on the set $\Omega_\gamma + (0, s_*, \kappa_*)$ defines a C^∞-family of n-dimensional invariant surfaces, the equations of motion whereupon are simply

$$\omega' = const \in \Omega_\gamma, \qquad \varphi' = \varphi'_0 + \omega' t, \qquad p' = p'_0 e^{-\lambda' t}, \qquad q' = q'_0 e^{\lambda' t}, \qquad (39)$$

where $\lambda' = \partial_{J'} H_+$. These surfaces are C^∞ μ-close to the unperturbed whiskered tori in the coordinates of Lemma 3.1, and may be parameterized in the initial coordinates as

$$
\begin{aligned}
I &= I'(\omega', p'q', \varepsilon, \mu) + \Theta^1_1(I', \varphi', p', q', \varepsilon, \mu), \\
\varphi &= \varphi' + \Theta^0_2(I', \varphi', p', q', \varepsilon) + \Theta^1_2(I', \varphi', p', q', \varepsilon, \mu), \\
p &= \Theta^0_3(I', \varphi', p', q', \varepsilon) + \Theta^1_3(I', \varphi', p', q', \varepsilon, \mu), \\
q &= \Theta^0_4(I', \varphi', p', q', \varepsilon) + \Theta^1_4(I', \varphi', p', q', \varepsilon, \mu),
\end{aligned}
\qquad (40)
$$

where $I'(\omega', J', \varepsilon, \mu) \in \Psi_+^{-1}\Omega_\gamma$, the functions Θ_i^0, $i = 1, \ldots, 4$ are as in (12) of Lemma 3.1, and the Taylor series in μ of the functions Θ_i^1, $i = 1, \ldots, 4$ begin with terms of the first order in μ.

Moreover $|\Theta_i^1 \circ \Psi_+^{-1}|_{\Omega.0.s_*.\kappa_*} \leq 2\mu$, $i = 1, \ldots, 4$. For the new Hamiltonian H_+ the norm of the frequency map Ψ_+ and its inverse is bounded by $(1 + \frac{|\mu|}{\mu_0})R$ in the sup-norm, the real part of the new Lyapunov exponent $\lambda'|_{J'=0}$ is not smaller than $(1 - \frac{|\mu|}{2\mu_0})\Lambda$.

Proof. The way we've included γ follows from the previous remarks. Indeed, for $0 < \gamma < 1$ if we scale $I \to \gamma I$, $p \to \gamma p$, $q \to \gamma q$ and multiply the Hamiltonian by γ^{-2}, then Ω_γ blows up into Ω_1. This modified Hamiltonian to which we apply Theorem 3.1 has the same nondegeneracy parameters as before rescaling; additionally, we will have to divide r and κ by γ, thus increasing them, since $\gamma < 1$. So, the choice of (16) will still remain the same and the right-hand side of (17) should be multiplied by γ^2.

Also, the only actual smallness parameter in the proof of Theorem 3.1 was r_0, or equivalently K_r, and it played the main role in the smallness condition for μ_0. Since we are free to take r_0 as small as we please, then according to Theorem 3.1 for $\mu \leq \mu_0(r_0)$ we can construct a transformation Φ, acting on a Cantor set Ω_γ in the frequency space.

By construction, the transformation Φ in Theorem 3.1 will be defined on the *closed* set $\Omega_\gamma \times \overline{W_{s_*}T^{n-1} \times B_{\frac{s}{2}}^2 \times B_{\mu_0}^1}$. So we can use the Extension theorem and consider a C^∞-extension $\mathcal{E}\Phi$ of Φ over $\Omega + (0, s_*, \frac{\kappa}{2})$ for some open real domain Ω in the frequency space, which contains the Cantor set Ω_γ: $\Omega_\gamma \subset \Omega \subseteq R^{n-1}$ this extension $\mathcal{E}\Phi$ will be analytic in the variables (φ', p', q') and C^∞ in ω'; if restricted to Ω_γ, it will give us the surviving tori.

Moreover, the induced transformation $\mathcal{E}\Xi = \Psi^{-1} \circ \mathcal{E}\Phi \circ \Psi_+$ will be defined on the set $\Phi_+^{-1}\Omega$, and for real J', μ will be C^∞ in the action variables $I' \in \Psi_+^{-1}\Omega$.

The only change in the smallness condition for μ_0 of Theorem (3.1) that the above extension will incur will be multiplying the right-hand side of (17) by some constant, coming from the Extension theorem. $\qquad \square$

3.2.1. The Main Lemma.

This part is purely technical. It contains the formulation and proof of the Main Lemma which is used to establish Theorem 3.2.

Lemma 3.2 (The Main Lemma). *Suppose we are given positive parameters $r, s, \kappa < 1$, and ρ, σ, η such that $3\rho < r$, $3\sigma < s$, $3\eta < \kappa$. Let $\xi^2 \equiv \min(\frac{\rho\sigma}{2R^2}, \eta^2)$, $Y \equiv \frac{1}{2K^\tau}$. Suppose Assumption 1 holds, and also*

$$r \leq \frac{1}{2K^{\tau+1}}, \tag{41}$$

$$\Lambda \geq \frac{1}{K^\tau}. \tag{42}$$

If

$$|f|_{r,s,\kappa} \leq \min\left(\frac{Y\xi^2}{2}, \frac{\rho^2}{8R^5}\right), \tag{43}$$

then there exists a real analytic symplectic transformation Ξ with the range in the domain of the Hamiltonian (13), which casts this Hamiltonian into the form $H \circ \Xi = h + \mathcal{D}\bar{f} + f_+ \equiv h_+ + f_+$.

The new frequency map is defined by Ψ_+: $(I, \cdot) \to (\partial_I h_+, \cdot)$ with the new Hessian matrix $A_+ = \partial^2_{II} h_+$, the new Lyapunov exponent as $\lambda_+ = \partial_J h_+|_{J=0}$, and they both satisfy Assumption 1 with constants $2R, \frac{3}{4}\Lambda$. The transformation $\Phi = \Psi \circ \Xi \circ \Psi_+^{-1}$ is real analytic and acts as follows: Φ: $\Omega_1 + (r - 3\rho, s - 3\sigma, \kappa - 3\eta) \to \Omega_1 + (r, s, \kappa)$. It casts the Hamiltonian (15) into the form $\mathcal{H} \circ \Phi = \mathcal{H}^0 + \mathcal{D}\bar{\mathcal{F}} + \mathcal{F}_+$ and

$$|\mathcal{F}_+|_{r-3\rho, s-3\sigma, \kappa-3\eta} \leq \left(1 - \frac{2|\mathcal{F}|_{r,s,\kappa}}{Y\xi^2}\right)^{-1} \left(\frac{3|\mathcal{F}|_{r,s,\kappa}}{Y\xi^2} + e^{-K\sigma}\right) |\mathcal{F}|_{r,s,\kappa}.$$

Moreover, $|W(\Phi - \mathrm{id})|_{\mathcal{P}} \leq \frac{1}{Y\xi^2}|\mathcal{F}|_{r,s,\kappa}$ uniformly on $\Omega_1 + (r - 3\rho, s - 3\sigma, \kappa - 3\eta)$, where $W = \mathrm{diag}(\rho^{-1}I_{n-1}, \sigma^{-1}I_{n-1}, \eta^{-1}I_2)$, I_l denoting the $l \times l$ identity matrix, $l = n - 1, 2$.

Proof. Let $\mathcal{S}$ be the image of $\Omega_1 + (r, s, \kappa)$ through the inverse Ψ^{-1} of the frequency map. We use the Lie transform method to seek the transformation from some subset $\mathcal{S}'$ into $\mathcal{S}$ which, being interpreted in the frequency variables, has the desired properties. Let X_ϕ^t be the Hamiltonian flow generated by the Hamiltonian ϕ. Ξ denotes the time-1 map obtained from X_ϕ^1. We consider the ultraviolet cutoff of the Fourier expansion of f with parameter K; then $f = T_K f + (f - T_K f)$. We denote $H_0 = h + T_K f$. From Taylor's formula then we have:

$$H_0 \circ \Xi = h + \{h, \phi\} + T_K f + \int_0^1 \{(1 - t)\{h, \phi\} + T_K f, \phi\} \circ X_\phi^t \, dt,$$

where as usual $\{\cdot, \cdot\}$ stands for the Poisson bracket of two smooth functions u, v of (I, φ, p, q),

$$\{u, v\} = (\partial_I u, \partial_\varphi v) - (\partial_I v, \partial_\varphi u) + \partial_p u \partial_q v - \partial_p v \partial_q u,$$

and $(\cdot, \cdot)$ denotes the standard inner product. We define $g \equiv \{h, \phi\} + T_K f$. To calculate the Poisson bracket we use the relations $\omega = \partial_I h$ and $\partial_p h = q \partial_J h = q\lambda$; in the same fashion $\partial_q h = p\lambda$. Then

$$g = -(\omega, \phi_\varphi) - \lambda(q\phi_q - p\phi_p) + T_K f. \tag{44}$$

The transformed Hamiltonian $H \circ \Xi = H_0 \circ \Xi + (f - T_K f) \circ \Xi$ will be given by

$$H \circ \Xi = h + g + f_+, \tag{45}$$

with

$$f_+ = \int_0^1 \{f_t, \phi\} \circ X_\phi^t dt + (f - T_K f) \circ X_\phi^1, \tag{46}$$

and

$$f_t \equiv (1 - t)g + t T_K f.$$

We represent the perturbation as a Taylor series in the (p, q) variables near $(0, 0)$ and a Fourier series in the angle variables,

$$f = \sum_{i,j \geq 0, \; k \in Z^{n-1}} f_{ij}^k p^i q^j e^{\iota(k,\varphi)},$$

where ι stands for the imaginary unit.

A formal solution of (44) which puts the Hamiltonian in the desired form is given by

$$g = \mathcal{D}\bar{f},$$

$$\phi = \sum_{\substack{i,j \\ k \in Z_K^{n-1} \setminus \{0\}}} \frac{f_{ij}^k p^i q^j e^{\iota(k,\varphi)}}{\iota(k,\omega) + \lambda(j - i)} + \sum_{i \neq j} \frac{f_{ij}^0 p^i q^j}{\lambda(j - i)}, \tag{47}$$

where g clearly depends only on I and the product $J = pq$ of the hyperbolic variables. Indeed, ω, λ are functions of I, J only, so they are in the kernel of the Hamilton-Jacobi type operator $p\partial_p - q\partial_q$; that is why (47) formally satisfies (44).

In order to obtain the estimate on the norm of ϕ we need to estimate the small divisors in the sum given in (47). To do this we use the inequality for r given in the statement of the lemma and notice that for the case $i = j$ we have by our choice of r,

$$|(\omega, k)| \geq |(\omega^*, k)| - K|\omega - \omega^*| \geq K^{-\tau} - Kr \geq \frac{1}{2K^\tau},$$

where $\omega^* \in \Omega_1$ is the center of the ball. Otherwise, since ω is real on Ω_1, then $|\Im(\omega, k)| \leq Kr \leq \frac{1}{2K^\tau}$. For the case $i \neq j$ we use the assumption (42) on $\Re\lambda$ given in the statement of the lemma. Denoting $\delta = |f|_D = |\mathcal{F}|_{r,s,\kappa}$, we then easily obtain the following uniform estimates for ϕ and g:

$$|g|_D \leq \delta, \tag{48}$$

$$|\phi|_D \leq \frac{\delta}{Y}, \tag{49}$$

where $Y = \frac{1}{2K^\tau}$, as defined in the statement of the lemma.

Using Lemma 3 of Appendix A, we obtain the following estimates for the derivatives of ϕ:

$$|\phi_I|_{D-(\bar{\rho},0,0)}^\infty \leq \frac{\delta}{Y\bar{\rho}}, \qquad |\phi_{p,q}|_{D-(0,0,\eta)}^\infty \leq \frac{\delta}{Y\eta}, \qquad |\phi_\varphi|_{D-(0,\sigma,0)}^\infty \leq |\phi|_{D-(0,\sigma,0)}^1 \leq \frac{\delta}{Y\sigma},$$

where σ, η are as in the formulation of the lemma, and $\bar{\rho}$ is a parameter for denoting shrinking in the domain of the action variables, which we will define shortly. Here and further on we use the simple fact that the Absolute Norm of a function is never smaller than its infinity-norm; see Appendix A, Lemma 1.

Furthermore, if $\Xi: S' \to S$, then in the frequency variables we can define an induced transformation $\Phi = \Psi \circ \Xi \circ \Psi_+^{-1}$, where Ψ_+ is the new frequency map defined as $\Psi_+ \equiv \partial_I h_+$ by the new unperturbed Hamiltonian $h_+ = h + g$. Using Cauchy inequalities we get the following bounds on its Hessian matrix $A_+ = \partial_{II}^2 h_+$:

$$|A_+ - A|^\infty \leq \frac{\delta}{\bar{\rho}^2}. \tag{50}$$

Using the equality

$$A^{-1} - A_+^{-1} = A_+^{-1}(A_+ - A)A^{-1},$$

from which by rearranging the terms and transiting to an arbitrary norm we derive

$$|A_+^{-1}|(1 - |A_+ - A|)|A^{-1}| \leq |A^{-1}|;$$

in particular, in the infinity-norm,

$$|A_+^{-1}| \leq \frac{|A^{-1}|}{1 - |A_+ - A||A^{-1}|} \leq \frac{R}{1 - \frac{\delta}{\bar{\rho}^2}R}, \tag{51}$$

provided that the denominator there is positive, which we will ensure by our choice of $\bar{\rho}$. Since one always has $R \geq 1$, fulfillment of (51) will guarantee (50); so to justify our claim for the norm of the new frequency map it's enough to ask

$$\delta \leq \frac{\bar{\rho}^2}{2R}.$$

Thus, if we define $\bar{\rho} = \frac{\rho}{2R^2}$, we can always be sure that the induced loss of analyticity in the frequency space is not greater than ρ, and the above inequality will transfigure into

$$\delta \leq \frac{\rho^2}{8R^5},$$

which is part of (43).

In addition, for the new Lyapunov exponent $\lambda_+ = \partial_J h_+|_{J=0}$ regarding (42) and (43) we will have

$$|\Re(\lambda - \lambda_+)| \leq \frac{\delta}{\eta^2} \leq \frac{\Lambda}{4},$$

which easily follows from (42) and (43). This proves the claim on the new Lyapunov exponent.

Now we want to restrict the domain of the Lie transform Hamiltonian ϕ in (49) to make sure that Φ maps $\Omega_1 + (r - 3\rho/2, s - 3\sigma/2, \kappa - 3\eta/2)$ into $\Omega_1 + (r - \rho, s - \sigma, \kappa - \eta)$, where the estimate (49) is valid. So we require

$$\frac{2R^2\delta}{Y\rho} \leq \sigma/2, \qquad \frac{\delta}{Y\eta} \leq \eta/2, \qquad \frac{2R^2\delta}{Y\sigma} \leq \rho/2.$$

Then if we define $\xi^2 \equiv \min(\frac{\rho\sigma}{2R^2}, \eta^2)$, these three requirements boil down to

$$\delta \leq \frac{\xi^2 Y}{2}. \tag{52}$$

To put it in other terms, if $W = diag(\rho^{-1}I_{n-1}, \sigma^{-1}I_{n-1}, \eta^{-1}I_2)$, I_l denoting the $l \times l$ identity matrix, $l = n - 1, 2$, then $|W(\Phi - id)|_{\mathcal{P}} \leq \frac{1}{Y\xi^2}|f|_{r.s.\kappa}$ uniformly on $\Omega_1 + (r - 3\rho/2, s - 3\sigma/2, \kappa - 3\eta/2)$. This proves the claims for Φ.

It remains to estimate f_+. Evidently for $0 \leq t \leq 1$,

$$|f_t|_{r,s,\kappa} \leq \delta.$$

Using this estimate along with (49), Lemma 4 of Appendix A gives us

$$|\{f_t, \phi\}|_{r-\bar{\rho},s-\sigma,\kappa-\eta} \leq \frac{3}{\xi^2 Y} \delta^2.$$

Lemma 5 of Appendix A gives us a general estimate for the time-t map X_ϕ^t for $0 \leq t \leq 1$,

$$|u \circ X_\phi^t|_{r-2\rho,s-2\sigma,\kappa-2\eta} \leq \left(1 - \frac{2}{\xi^2}|\phi|_{r,s,\kappa}\right)^{-1} |u|_{r,s,\kappa}.$$

We also have the estimate $|f - T_K f|_{r,s-\sigma,\kappa} \leq e^{-K\sigma}|f|_{r,s,\kappa}$. Assembling all these pieces together we obtain the estimate

$$|f_+|_{r-3\rho,s-3\sigma,\kappa-3\eta} \leq \left(1 - \frac{2|f|_{r,s,\kappa}}{\xi^2 Y}\right)^{-1} \left(\frac{3|f|_{r,s,\kappa}}{\xi^2 Y} + e^{-K\sigma}\right) |f|_{r,s,\kappa},$$

which will be the same for $\mathcal{F}_+ = f_+ \circ \Psi_+^{-1}$, as stated in the lemma.   □

Remark. This lemma and its estimates may be used in an obvious way when there is no hyperbolicity, simply by ignoring all that involves the dependence of the Hamiltonian (13) on (p, q). Moreover, the proof can be carried out if the elliptic part of the system has more degrees of freedom than $n - 1$. Indeed, suppose the Hamiltonian (13) depends on m more action-angle pairs, and Ω_1 is then a point set on the $n - 1$-dimensional hyperplane in the $n - 1 + m$-dimensional frequency space, defined, for example, by m linearly independent resonant relations. Then g, being the average only in $n - 1$ angles whose corresponding frequencies are known to be nonresonant at points of Ω_1, will also depend on these additional action-angle variables. The only difference is that in this case $\Omega_1 + (r)$ would mean a full-dimensional neighborhood of a lower dimensional Cantor set Ω_1.

4. Multiplicity-One Resonant Dynamics for the A-priori Stable Systems

In this section we will apply the theory of the a-priori unstable systems to the system (1). To introduce hyperbolicity we bring our system into a simple normal form near the resonant surface, and define what we call hyperbolicity in the first order. Then we apply the KAM theory of the a-priori unstable systems, developed in the previous section, to this normal form and give the proof of Theorem 2.2.

4.1. A Simple Normal Form

Here we present a simple normal form valid near the resonant surface, which can be constructed under rather general assumptions. This is the transitional step to introduce the hyperbolic variables, thus reducing the problem to an a-priori unstable system. The normal form is obtained by a single application of the Main Lemma of the previous section in a union of open balls of full dimension in the frequency space near Ω_1. At this point we come up with the first set of assumptions necessary to go on with our discussion.

Assumption 2 (Nondegeneracy).

- Suppose for any $I \in D_{r_0}$ the infinity-norm of the Hessian matrix $Q = \partial_{II}^2 H^0$ and its inverse are uniformly bounded by some positive constant R_0,

$$|Q|^\infty, |Q^{-1}|^\infty \le R_0. \tag{53}$$

- Suppose the resonant surface Π can be represented as a graph over variables $I^* \equiv (I_2, \ldots, I_n)$, namely

$$\Pi: I_1 = F(I^*),$$

where $F(I^*)$ is analytic in D_{r_0}.

For convenience we shall assume that $r_0, s_0 < 1$.

Consider a complex set $\Omega_\gamma + (r, s)$, where $r \le \frac{r_0}{R_0}, s \le s_0$. Then the Hamiltonian (1), considered as a function of the frequency-angle variables,

$$\mathcal{H} = H \circ \Psi^{-1} = \mathcal{H}^0(\omega) + \varepsilon \mathcal{H}^1(\omega, \varphi, \varepsilon), \tag{54}$$

is analytic in this set. Also suppose that uniformly $\varepsilon |\mathcal{H}^1|_{r,s} \le \varepsilon \ll 1$.

As we mentioned before, we start by "blowing up" the frequency space; namely, we rescale $I \to \gamma I$ and multiply the Hamiltonian by γ^{-2}. Then Ω_γ becomes Ω_1, and instead of r we will have $r\gamma^{-1}$.

Let's suppose that the perturbation H^1 in (1) is such that its zero-order coefficient in its Taylor series expansion in ε is not identically zero. In the above domain we can apply the version of the Main Lemma proved above for some ultraviolet cutoff parameter K. Following the Remark after its proof, the averaging in the Main Lemma must be made only throughout the *fast* angles $\varphi_2, \ldots, \varphi_n$, and the part of the perturbation that is added to H^0 will be $\varepsilon T_K P_1 H^1(I, \varphi, \varepsilon)$, where P_1 stands for the projection of the Fourier series onto the first angle variable. Our purpose is to choose the values of the parameters in the Main Lemma such that the remainder will be of the order $O(\varepsilon^p)$ for $1 < p < 2$. Recall the notation and constraints of the Main Lemma. We can choose, e.g., $\sigma = s/6$, $K = -\frac{6 \log \varepsilon}{s}$. To make sure that $K > 1$ it's enough to ask $\varepsilon < e^{-1}$. Besides, we require that ε is small enough to make sure that $r \ge \frac{1}{2} K^{-\tau-1}$ to obey (41), e.g.,

$$\varepsilon \le e^{-\frac{R_0}{12r_0}},$$

since we've agreed that $r, s \le 1$. Then we can redefine r as

$$r = \frac{1}{2} \left(\frac{s}{6 \log \varepsilon^{-1}} \right)^{\tau+1}. \tag{55}$$

Hence, with $\xi^2 = \frac{\sigma\rho}{2R_0^2}$, $Y = \frac{1}{2K^\tau}$ the remainder $\mathcal{H}_+^1$ will be bounded as

$$|\mathcal{H}_+^1|_{r/2, s/2} \le \left(1 - \frac{2\varepsilon}{Y\xi^2}\right)^{-1} \left(\frac{3\varepsilon}{Y\xi^2} + e^{-K\sigma}\right)\varepsilon. \tag{56}$$

If we express r, Y in terms of K, then to achieve our goal we must obey both the inequalities in (43), for which it will be enough to ask $144 \cdot 8s^{-1} R_0^5 K^{2\tau+2}\varepsilon^{2-p} \le 1$, which can certainly be achieved by taking ε small enough; e.g., for any δ such that $0 < \delta < 2 - p$, we can choose

$$\varepsilon \le \left(\left(\frac{s}{6}\right)^{2\tau+2} \frac{1}{1152 R_0^5} \frac{1}{\max_{0<x<1} x^{2-p-\delta} \log^{2\tau+2} x^{-1}}\right)^{\frac{1}{\delta}}. \tag{57}$$

With the above choice of ε the norm of the remainder is guaranteed to be smaller than ε^p. Needless to say, neither here, nor elsewhere, do we pursue the sharpest threshold smallness estimates for ε. So, in the new variables the Hamiltonian will be analytic in a smaller domain (namely in $\Omega_y + (r/2, s/2)$), and according to the Main Lemma in the action-angles it will look as follows:

$$H = H^0(I) + \varepsilon T_K P_1 H^1(I, \varphi_1, \varepsilon) + O(\varepsilon^p), \tag{58}$$

where P_1 stands for the projection operator on the first angle variable (which is equivalent to averaging in $\varphi_2, \ldots, \varphi_n$). This Hamiltonian, truncated at the first order in ε, may be written as

$$\tilde{h}(I, \varphi, \varepsilon) = H^0(I) + \varepsilon \tilde{H}^1(I, \varphi_1), \tag{59}$$

where $\tilde{H}^1 = T_K P_1 H^1(I, \varphi, 0)$. Here we've attempted to make the dependence of the new "unperturbed" Hamiltonian in ε the simplest possible.

Henceforth we'll assign $p = 3/2$; then if originally $H^1(I, \varphi, \varepsilon)$ is analytic in ε, the transformed Hamiltonian (58) will be analytic in $\sqrt{\varepsilon}$.

At this point we'll change our notation. We will use symbols (p, q) instead of (I_1, φ_1), and (I, ω, φ) will stand for the rest of the actions, frequencies, and angles in an obvious way. Then we may rewrite (59) as

$$\tilde{h}(I, p, q, \varepsilon) = H^0(I, p) + \varepsilon \tilde{H}^1(I, p, q). \tag{60}$$

With this notational change (59) describes the integrable system whose equations of motion are

$$\begin{aligned}
\dot{I} &= 0, \\
\dot{\varphi} &= \omega, \quad \omega \in \Omega_1 + (r/2), \\
\dot{p} &= -\varepsilon \tilde{H}_q^1(I, p, q), \\
\dot{q} &= H_p^0(I, p) + \varepsilon \tilde{H}_p^1(I, p, q).
\end{aligned} \tag{61}$$

In addition, regarding (60) as the new "unperturbed Hamiltonian" we get the new frequency map $\omega = \tilde{h}_I$ which acts as an identity upon the rest of the variables, and the *sup*-norm of its Hessian matrix and its inverse will by construction be bounded by a constant $R = 2R_0$. We will call the perturbed Hamiltonian (1) *hyperbolic in the first order* if Equations (61) describe an a-priori unstable system. The following assumption formalizes this notion.

Assumption 3 (Hyperbolicity in the first order).

- *For $\varepsilon \neq 0$ suppose, for any I from the analyticity domain of the Hamiltonian (60), that the two last equations in (61) possess a hyperbolic equilibrium point (p^*, q^*), $p^* = p^*(I)$, $q^* = q^*(I)$, and the real part of the Lyapunov exponent $\lambda(I, \varepsilon)$,*

$$\lambda^2(I, \varepsilon) = (\partial^2_{pq}\tilde{h})^2 - \partial^2_{pp}\tilde{h}\partial^2_{qq}\tilde{h}\Big|_{p=p^*, q=q^*},$$

at this point is bounded away from zero for all I and some positive constant λ_0 as $\lambda(I, \varepsilon) \leq \sqrt{\varepsilon}\lambda_0$.

In fact, this assumption requires a certain type of nondegeneracy from the perturbation H^1, namely that it shall remove the degeneracy on the resonant surface Π in the system of equations (61) for $\varepsilon = 0$. Also it becomes clear from the view of $\tilde{h}$ that we can't introduce hyperbolicity in the first order for complex ε in a ring near the origin, because of our claim on the real part of λ. Nevertheless one can see from (61) that this Assumption is naturally satisfied in plenty of examples. In particular the genericity of this Assumption becomes clear if one does the more traditional local analysis near a fixed point on a resonant surface in the action space, whose remaining frequencies are diophantine (see e.g. Treschev [1991]), which can actually be done with (61) by fixing some I^* and Taylor expanding inside its small neighborhood.

In fact, from equations (61) we can easily derive that

$$p^*(I) = F(I) + O(\varepsilon),$$

where $F(I)$ is the same as in Assumption 2, and

$$q^*(I) = G(I) + O(\varepsilon),$$

where $G(I)$ is defined implicitly by

$$\partial_q \tilde{H}^1(I, F(I), G(I)) = 0.$$

Also, the square of the Lyapunov exponent of the truncated "unperturbed" Hamiltonian (60) equals

$$\lambda^2(I, \varepsilon) = \varepsilon^2 \partial^2_{pq}\tilde{H}^1(I, p^*, q^*) -$$

$$\varepsilon\partial^2_{qq}\tilde{H}^1(I, p^*, q^*) \cdot \partial^2_{pp}\left(H^0(I, p^*) + \varepsilon\tilde{H}^1(I, p^*, q^*)\right) = O(\varepsilon),$$

if in (1) the term $H^1(I, \varphi, \varepsilon)$ contains φ_1-dependence in its lower Fourier modes.

If we attribute all we've done with the original Hamiltonian (1) to the action of some "normalizing transformation" $\mathcal{R}_0$ which will certainly be symplectic and well-defined, then finally we arrive at Hamiltonian

$$H \circ \mathcal{R}_0 = H^0(I, p) + \varepsilon \tilde{H}^1(I, p, q) + \varepsilon^{\frac{3}{2}} B^{-1} \tilde{f}(I, \varphi, p, q, \sqrt{\varepsilon}), \qquad (62)$$

which satisfies Assumption 2 with $R = 2R_0$ and Assumption 3, and thus represents an a-priori unstable system with $R = 2R_0$, $\Lambda = \sqrt{\varepsilon}\lambda_0$. From the way we've constructed the above normal form, $|\tilde{f}| \leq 1$ in the Absolute Norm on the domain $\mathcal{R}_0$.

Then without loss of generality (see Chierchia and Gallavotti [1994]) the equilibrium may be chosen as $(p^*, q^*) = (0, 0)$. Indeed, it may be brought into this form by a canonical transformation, defined by the generating function

$$(q - q^*(I'))p' + p^*(I')\sin(q - q^*(I')) + \varphi \cdot I'$$
$$= (q - G(I'))p' + F(I')\sin(q - G(I')) + \varphi \cdot I' + O(\varepsilon), \qquad (63)$$

where the $O(\varepsilon)$ term does not contain φ-dependence, and so the transformation $\mathcal{R}_1$ defined by (63) does not affect the actions.

Also by adjusting the (p, q)-independent part of it, we may rewrite (60) as

$$\tilde{h}(I, p, q, \varepsilon) = h_0(I, \varepsilon) + \tilde{P}(I, p, q, \varepsilon), \qquad (64)$$

with the additional assumption that $\tilde{P}(I, 0, 0, \varepsilon) = 0$.

Finally, as for our estimate for the size of the remainder, we may introduce a new small parameter $\mu = B^{-1}\varepsilon^{\frac{3}{2}}$, so that (62) may be rewritten as

$$H(I, p, q, \sqrt{\varepsilon}) = h_0(I, \varepsilon) + \tilde{P}(I, p, q, \varepsilon) + \mu \tilde{f}(I, \varphi, p, q, \sqrt{\varepsilon}), \qquad (65)$$

which describes an a-priori unstable system and will be analytic for $I \in D^* + (\frac{r}{2})$, $\varphi \in W_{\frac{s_0}{2}} T^{n-1}$, $p \in B_{\frac{1}{2}}$, $q \in B_{\frac{s_0}{2}}$ in the notation of Section 2.1, where r is given by (55) with s_0 instead of s.

To finish this section, we'll redefine $\mathcal{R}_0$ as $\mathcal{R}_1 \circ \mathcal{R}_0$ and call it a normal form transformation. By construction, its impact on the action variables is near-identity.

4.2. Proof of the Main Theorem

At this point we have all the necessary building blocks to prove Theorem 2.2. To begin with, after applying the "simple normal form" transformation $\mathcal{R}_0$, Assumption 3 has quantified the notion of hyperbolicity in the first order, and the entire previous subsection has prepared the ground for using the KAM theory of the a-priori unstable systems.

Our first step will be to apply Lemma 3.1, which establishes the transformation $\mathcal{R}$ that casts the "unperturbed" Hamiltonian (64) into the normal form

$$\tilde{h} \circ \mathcal{R} = h(I, pq, \sqrt{\varepsilon}) = h_0(I, \sqrt{\varepsilon}) + P(I, pq, \sqrt{\varepsilon}), \qquad (66)$$

and the full Hamiltonian (65) into

$$H(I, p, q, \sqrt{\varepsilon}) = h(I, pq, \sqrt{\varepsilon}) + \mu f(I, \varphi, p, q, \sqrt{\varepsilon}). \qquad (67)$$

Evidently, if ε is small enough then by construction of the above normal form and by Lemma 3.1 this Hamiltonian will be analytic in some domain $G_{r',s',\kappa}$ for some $0 < r' < r$, $0 < s' < \frac{s_0}{2}$, where s' does not depend on ε, and in particular we can take $\kappa = \Lambda = \lambda_0 \sqrt{\varepsilon}$.

Besides, for any $B > 1$ we can find ε small enough as in (57) such that

$$|f|_{r,s,\kappa} \leq B^{-1}\varepsilon^{\frac{3}{2}} = \mu.$$

We only have to check whether all the conditions of Theorem 3.1 will be satisfied with the choice of some large-enough value of a constant B. If we assign, for instance, $s_* = s'/2$ and assume $\tau \geq 2$ in the statement of Theorem 3.1, then the condition (17) boils down to choosing

$$B > B_1(r_0, s_0, R_0, \lambda_0, \tau),$$

where B_1 does not depend on ε, since in (17) (see also (16)) we will have at most $\mu_0 = B_1^{-1}\varepsilon^{\frac{3}{2}}$. That's why we always can find B large enough to exceed B_1; then (57) will give us the smallness condition on ε.

So, applying Theorem 3.1 and its generalization Theorem 3.2, we in fact arrive at the statement of the Main Theorem and the Corollary following it. Indeed, Theorems 3.1, 3.2 give us a symplectic transformation Ξ that establishes integrability of the corresponding a-priori unstable system on a Cantor set. Thus we can define $\Xi_* = \mathcal{R}_0 \circ \mathcal{R} \circ \Xi$ which transforms the Hamiltonian (1) in the way it is stated in the Main Theorem. Since Ξ differs from identity by $O(\varepsilon^{\frac{3}{2}})$ and $\mathcal{R}$ does not affect the action variables, then Ξ_* itself is near identity projected on the actions $I_2, \ldots, I_n$.

The spurious element is that we've had to require $\tau \geq 2$ to be able to apply Theorem 3.1 (see (16), (17)) to the normal form of Section 4.1. The reason for it was our choice of $p = \frac{3}{2}$ in Section 4.1. By choosing $\frac{1}{2} < p < 2$ we can extend our result for $\tau > 1$. To make it valid for all $\tau \geq 1$, which may be of interest, we should have applied the Main Lemma in Section 4.1 not once but twice, thus bringing the remainder down in size to $O(\varepsilon^2)$, which would be enough to apply Theorem 3.1. But this would not change almost anything as far as Assumption 3 or the system of equations (61) are concerned, since the added terms would be of higher order in ε. With this final remark we conclude the proof of the Main Theorem. $\qquad\square$

Appendix A. The Absolute Norm

We call the norm we use here the *absolute norm*. We introduce more notation for an arbitrary function $u(x, \varphi, p, q)$, analytic in the domain $C_{r,s,\kappa}$ for some open set $C \subseteq R^l$. As before $x \in C$ stands for either the action or the frequency variables, which are related through the globally invertible frequency map.

For $(x, p, q) \in C_{r,\kappa}$ we set

$$|u|_{x,p,q,s} \equiv \sum_{k \in Z^{n-1}} \overline{u_k(x, p, q)}e^{|k|s};$$

in the notation of Section 2.1,

$$|u|_{x,\kappa,s} \equiv \sup_{B_\kappa^2} |u|_{x,p,q,s}.$$

Lemma 1 (Lemma A1). *For the sup-norm $|\cdot|^\infty_{r,s,\kappa}$ on $C_{r,s,\kappa}$ for all $\sigma > 0$, $\eta > 0$, one has*

$$|u|^\infty_{r,s,\kappa} \leq |u|_{r,s,\kappa} \leq \frac{(\kappa + \eta)^2}{\eta^2} \coth^{n-1} \sigma |u|^\infty_{r,s+2\sigma,\kappa+\eta}.$$

Proof. The first inequality is obvious. To prove the second one let's consider the absolute sum of the Taylor series of a function $u(\cdot, p, q)$ analytic in p, q in $B^2_{\kappa+\eta}$,

$$u(\cdot, p, q) = \sum_{p,q=0}^{\infty} u_{ij}(\cdot) p^i q^j.$$

Clearly for a monomial $u_{ij} p^i q^j$ the coefficient equals

$$u_{ij} = \frac{1}{i!j!} \frac{\partial^{i+j}}{\partial p^i \partial q^j} u(0,0) \leq \frac{1}{(\kappa + \eta)^{i+j}} |u|^\infty_{r,s,\kappa+\eta};$$

here we skip the $(\cdot)$-dependence of u, and so

$$\overline{u(p,q)} \leq |u|^\infty_{\kappa+\eta} \left(1 + \frac{p}{\kappa + \eta} + \frac{p^2}{(\kappa + \eta)^2} + \cdots\right) \cdot \left(1 + \frac{q}{\kappa + \eta} + \frac{q^2}{(\kappa + \eta)^2} + \cdots\right)$$

$$\leq \frac{(\kappa + \eta)^2}{\eta^2} |u|^\infty_{\kappa+\eta}.$$

Then using the familiar estimate for the Fourier coefficients $|u_k(x, p, q)| \leq e^{-|k|s} |u|^\infty_{r,s,\kappa}$ of the analytic functions and the identity $\sum_{k \in \mathbb{Z}^{n-1}} e^{-2|k|\sigma} = \coth^{n-1} \sigma$, we come to the claimed result. $\qquad\square$

Lemma 2. *For an arbitrary $x \in C_r$ $|uv|_{x,s,\kappa} \leq |u|_{x,s,\kappa} |v|_{x,s,\kappa}$ for arbitrary functions u, v, analytic in $C_{r,s,\kappa}$.*

Proof. For any $(x, p, q) \in C_{r,s,\kappa}$ we have

$$|uv|_{x,p,q,s} = \sum_k \overline{(uv)_k(x, p, q)} e^{|k|s} \leq \sum_k \sum_l \overline{u_{k-l}(x, p, q)} e^{|k-l|s} \overline{v_l(x, p, q)} e^{|l|s}$$

$$= |u|_{x,p,q,s} |v|_{x,p,q,s}. \qquad\square$$

Lemma 3. *For $0 < \rho < r$, $0 < \sigma < s$, $0 < \eta < \kappa$, and $(x, p, q) \in C_{r-\rho,\kappa-\eta}$,*

$$|u_\varphi|^1_{x,p,q,s-\sigma} \equiv \sum_{i=1}^{n-1} |u_{\varphi_i}|_{x,p,q,s-\sigma} \leq \frac{1}{e\sigma} |u|_{r,s,\kappa}, \quad \max_{1 \leq i \leq n-1} |u_{x_i}|_{x,p,q,s} \leq \frac{1}{\rho} |u|_{r,s,\kappa},$$

$$\max(|u_p|_{x,p,q,s}, |u_q|_{x,p,q,s}) \leq \frac{1}{\eta} |u|_{r,s,\kappa}.$$

Proof. For the φ-derivatives we have

$$
\sum_{i=1}^{n-1} |u_{\phi_i}|_{x,p,q,s-\sigma} = \sum_{i=1}^{n-1} \sum_k |k_i| \overline{|u_k(x, p, q)|} e^{|k|(s-\sigma)}
$$

$$
= \sum_k |k| \overline{|u_k(x, p, q)|} e^{|k|(s-\sigma)}
$$

$$
\leq \sup_{t \geq 0} t e^{-t\sigma} \cdot |u|_{x,p,q,s} = \frac{1}{e\sigma} |u|_{x,p,q,s}.
$$

The inequalities for the x, p, q-derivatives immediately follow from the Cauchy formula. Here is how it follows for the x-derivatives; for the p, q-derivatives it's just the same argument.

If (x, p, q) is a point in $C_{r-\rho,\kappa-\eta}$, then

$$
u_{x_i} = \frac{1}{2\pi i} \int_\gamma \frac{u(x + \zeta, \cdot)}{\zeta} d\zeta,
$$

where γ is a circle with radius ρ in the x-space. Then it follows that

$$
|u_{x_i}|_{x,s,\kappa} \leq \frac{1}{2\pi} \int_\gamma \frac{|u|_{x+\zeta,s,\kappa-\eta}}{|\zeta|} |d\zeta| \leq \frac{1}{\rho} |u|_{r,s,\kappa}
$$

uniformly for all $x \in C_{r-\rho}$ and all i. $\qquad\square$

Lemma 4. *For* $0 < r - \rho < r_0$, $0 < s - \sigma < s_0$, $0 < \kappa - \eta < \kappa_0$ *for* u *analytic in* C_{r_0,s_0,κ_0} *and* v *analytic in* $C_{r,s,\kappa}$:

$$
|\{u, v\}|_{r-\rho,s-\sigma,\kappa-\eta}
$$

$$
\leq \left(\frac{1}{e} \left(\frac{1}{(r_0 - r + \rho)\sigma} + \frac{1}{(s_0 - s + \sigma)\rho} \right) + \frac{2}{(\kappa_0 - \kappa + \eta)\eta} \right) |u|_{r_0,s_0,\kappa_0} |v|_{r,s,\kappa}.
$$

Proof. The proof directly follows from the previous lemma. We just write for any $(x, p, q) \in C_{r-\rho,\kappa-\eta}$,

$$
|\{u, v\}|_{r-\rho,s-\sigma,\kappa-\eta} \leq \sum_{1 \leq i \leq n-1} |u_{x_i}|_{x,p,q,s-\sigma} |v_{\varphi_i}|_{x,p,q,s-\sigma}
$$

$$
+ \sum_{1 \leq i \leq n-1} |v_{x_i}|_{x,p,q,s-\sigma} |u_{\varphi_i}|_{x,p,q,s-\sigma}
$$

$$
+ \frac{2}{(\kappa_0 - \kappa + \eta)\eta} |u|_{x,p,q,s} |v|_{x,p,q,s}
$$

$$\leq \sum_{i=1}^{n-1} |u_{\varphi_i}|_{x,p,q,s-\sigma} \max_{1\leq i\leq n-1} |v_{x_i}|_{x,p,q,s}$$

$$+ \sum_{i=1}^{n-1} |v_{\varphi_i}|_{x,p,q,s-\sigma} \max_{1\leq i\leq n-1} |u_{x_i}|_{x,p,q,s}$$

$$+ \frac{2}{(\kappa_0 - \kappa + \eta)\eta} |u|_{x,p,q,s} |v|_{x,p,q,s}$$

$$\leq \left(\frac{1}{e} \left(\frac{1}{(r_0 - r + \rho)\sigma} + \frac{1}{(s_0 - s + \sigma)\rho} \right) + \frac{2}{(\kappa_0 - \kappa + \eta)\eta} \right)$$

$$\times |u|_{r_0,s_0,\kappa_0} |v|_{r,s,\kappa}.$$

Here we've used the fact that u is analytic in a larger domain.   □

Lemma 5. *If $|\phi|_{r_0,s_0,\kappa_0} < \xi^2/2$, where ξ comes from the Main Lemma, then*

$$|u \circ X_\phi^1|_{r-2\rho,s-2\sigma,\kappa-2\eta} \leq \left(1 - \frac{2}{\xi^2} |\phi|_{r_0,s_0,\kappa_0} \right)^{-1} |u|_{r,s,\kappa},$$

for $0 < \rho < r \leq r_0 - \rho, 0 < \sigma < s \leq s_0 - \sigma, 0 < \eta < \kappa \leq \kappa_0 - \eta$.

Proof. Consider the Lie series expansion,

$$u \circ X_\phi^1 = \sum_{h\geq 0} \frac{1}{h!} ad_\phi^h u,$$

where

$$ad_\phi^0 u = u, \quad ad_\phi^h u = \{ad_\phi^{h-1}u, \phi\}, \quad h \geq 1.$$

For $h \geq 1$ let's introduce $\tilde{\rho} = 2\rho/h, \tilde{\sigma} = 2\sigma/h, \tilde{\eta} = 2\eta/h, \tilde{\xi}^2 = 4\xi^2/h^2$. Let $|\cdot|_i \equiv |\cdot|_{r-i\tilde{\rho},s-i\tilde{\sigma},\kappa-i\tilde{\eta}}$ for $1 \leq i \leq h$. Then we'll have

$$|ad_\phi^i u|_i \leq \left(\frac{1}{e\tilde{\rho}(s_0 - s + i\tilde{\sigma})} + \frac{1}{e\tilde{\sigma}(r_0 - r + i\tilde{\rho})} + \frac{2}{\tilde{\eta}(\kappa_0 - \kappa + i\tilde{\eta})} \right)$$

$$\times |\phi|_{r_0,s_0,\kappa_0} |ad_\phi^{i-1} u|_{i-1}$$

$$\leq \frac{3}{\tilde{\xi}^2} \frac{1}{h+i} |\phi|_{r_0,s_0,\kappa_0} |ad_\phi^{i-1} u|_{i-1}.$$

Thus,

$$|ad_\phi^h u|_{r-2\rho,s-2\sigma,\kappa-2\eta} \leq \left(\frac{3}{\tilde{\xi}^2} \right)^h \frac{h!}{(2h)!} |\phi|_{r_0,s_0,\kappa_0}^h |u|_{r,s,\kappa}.$$

Using a version of Stirling's formula, $(2h)! \leq (2h)^{2h} e^{-2h}$, we arrive at

$$\left(\frac{3}{\tilde{\xi}^2} \right)^h \frac{1}{(2h)!} \leq \left(\frac{3h^2}{4\xi^2} \right)^h \frac{e^{2h}}{(2h)^{2h}} \leq \left(\frac{3e^2}{16\xi^2} \right)^h \leq \left(\frac{2}{\xi^2} \right)^h.$$

So, we end up with

$$|u \circ X_\phi^1|_{r-2\rho,s-2\sigma,\kappa-2\eta} \leq \sum_{h \geq 0} \frac{1}{h!} |ad_\phi^h u|_{r-2\rho,s-2\sigma,\kappa-2\eta}$$

$$\leq \sum_{h \geq 0} \left(\frac{2}{\xi^2} |\phi|_{r_0,s_0,\kappa_0}\right)^h |u|_{r,s,\kappa}$$

$$= \left(1 - \frac{2}{\xi^2} |\phi|_{r_0,s_0,\kappa_0}\right)^{-1} |u|_{r,s,\kappa}. \qquad \square$$

Acknowledgments

S. Wiggins would like to acknowledge research support by the National Science Foundation, DMS-9403691.

References

[1] Arnold, V.I. [1963] Proof of A. N. Kolmogorov's theorem on the preservation of quasiperiodic motions under small perturbations of the Hamiltonian, *Russ. Math. Surv.*, **18**(5), 9–36.

[2] Arnold, V. I. [1964] Instability of dynamical systems with several degrees of freedom, *Sov. Math. Dokl.*, **5** (1964), 581–585.

[3] Bryuno, A. D. [1989] *Local Methods in Nonlinear Differential Equations.* Springer-Verlag: New York, Heidelberg, Berlin.

[4] Chierchia, L., and Gallavotti, G. [1994] Drift and diffusion in phase space. *Ann. Inst. H. Poincaré Phys. Th.*, **60**(91), 1–144.

[5] Gallavotti, G. [1994] Twistless KAM tori, quasi flat homoclinic intersections, and other cancellations in the perturbation series of certain completely integrable Hamiltonian systems. A review. *Rev. Math. Phys.*, **6**, 343–411.

[6] Graff, S. [1974] On the construction of hyperbolic invariant tori for Hamiltonian systems, *J. Differential Eqns.*, **15** (1974), 1–69.

[7] Jorba, A., and Simo, C. [1994] On Quasiperiodic Perturbations of Elliptic Equilibrium Points [1994]. #94-310 Deposited in the archive mp_arc. To get a TeX version of it, send an empty e-mail message to mp_arc@math.utexas.edu.

[8] de la Llave, R., and Wayne, C.E. [1990] Whiskered and low dimensional tori in nearly integrable Hamiltonian systems, University of Texas, Austin, preprint.

[9] Kolmogorov, A. N. [1954] On conservation of conditionally periodic motions under small perturbations of the Hamiltonian. *Dokl. Akad. Nauk USSR*, **98**(4), 527–530.

[10] Lochak, P. [1995] Arnold diffusion: A compendium of remarks and questions. Report.

[11] Moser, J. [1956] The analytic invariants of an area-preserving mapping near a hyperbolic fixed point. *Commun. Pure Appl. Math.* **9**, 673–692.

[12] Pöschel, J. [1982] Integrability of Hamiltonian systems on Cantor sets. *Commun. Pure Appl. Math.*, **35**, 653–695.

[13] Pöschel, J. [1989] On elliptic lower dimensional tori in Hamiltonian systems. *Math. Z.*, **213**, 559–608.

[14] Pöschel, J. [1993] Nekhoroshev estimates for quasi-convex Hamiltonian systems. *Math. Z.*, **202**, 187–216.

[15] Treschev, D.V. [1991] The mechanism of destruction of resonant tori of Hamiltonian systems, *Math. USSR Sb.*, **68**, 181.

Constrained Euler Buckling

G. Domokos[1], P. Holmes[2], and B. Royce[3]*

[1] Department of Strength of Materials, Technical University of Budapest, H-1521 Budapest, Hungary
[2] Program in Applied and Computational Mathematics and Department of Mechanical and Aerospace Engineering, Princeton University, Princeton, NJ 08544, USA
[3] Department of Mechanical and Aerospace Engineering, and Princeton Materials Institute, Princeton University, Princeton, NJ 08544, USA

Received June 27, 1996; revised manuscript accepted for publication September 11, 1996
Communicated by Jerrold Marsden and Stephen Wiggins

Dedicated to the memory of Juan-Carlos Simo

Summary. We consider elastic buckling of an inextensible beam confined to the plane and subject to fixed end displacements, in the presence of rigid, frictionless side-walls which constrain overall lateral displacements. We formulate the geometrically nonlinear (Euler) problem, derive some analytical results for special cases, and develop a numerical shooting scheme for solution. We compare these theoretical and numerical results with experiments on slender steel beams. In contrast to the simple behavior of the unconstrained problem, we find a rich bifurcation structure, with multiple branches and concomitant hysteresis in the overall load-displacement curves.

Key words. Elastic buckling, bifurcation, nonlinear boundary value problem

MSC numbers. 34B15, 58F14, 73K05

1. Introduction

Buckling is important in applications as diverse as automotive crash protection and fiber preparation of nonwoven fabrics. The work described in this paper was motivated by the latter, specifically, a desire to understand plastic buckling of polypropylene fibers in a "stuffer box" process. In this manufacturing environment, many approximately parallel fibers forming a "tow" emerge from rapidly rotating rollers into a confined half-space in

---

* **Note:** Since this article was published, two further papers on constrained buckling have appeared [HSD], [DHRS].

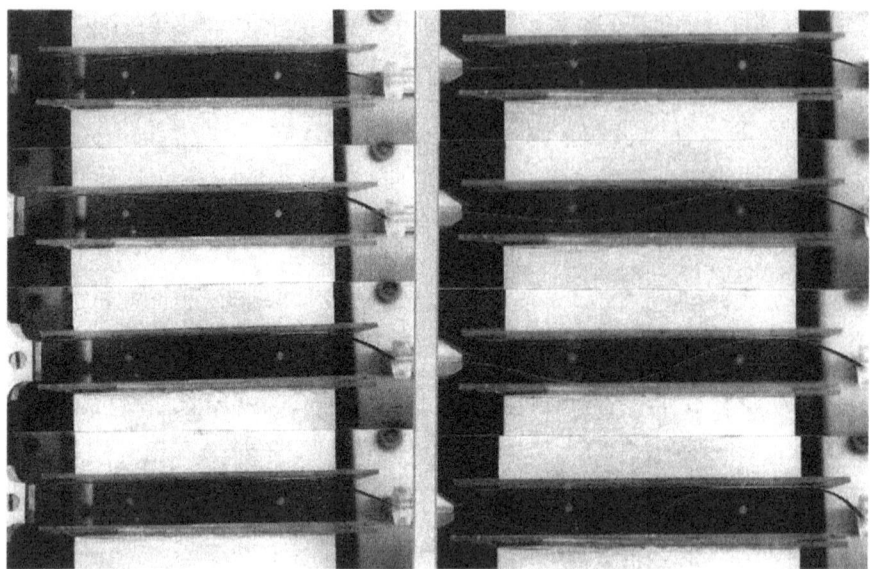

Fig. 1. Some typical buckled shapes. Left column: pinned-pinned beam; right column: clamped-pinned beam. See Section 4 for description of asymmetrical shapes.

which they suffer large velocity changes as a result of collisions with previously slowed and crimped material. At the rollers, the tow approximates a close-packed cylinder configuration, and during crimping the fibers are able to expand into a less close-packed form. The final configuration of the crimped tow suggests that cooperative deformation of many fibers occurs during this process. Our desire to better understand the basic mechanics has led us to consider a simpler, but still interesting problem more amenable to theoretical, numerical, and experimental study. In this paper, we address the planar *elastic* buckling of an inextensible lamella (beam) confined between rigid, parallel, frictionless walls. The specimen shape confines buckling to the plane normal to the lamella's major axis. Geometries more characteristic of the manufacturing process, with wedge-shaped and circular constraint walls, will be treated in future work. Here our main goal is to understand the basic mechanics of the process, including buckling mode selection and overall end load-displacement relations.

In a typical experiment of the type to be described in Section 4, the end positions of a slender steel lamella are constrained and their separation is gradually decreased, leading to buckling and contact with one or both side-walls. Point contact can become line contact as the end load increases; secondary buckling of the flattened portion and jumping between different modes may then ensue. Figure 1 shows two sequences of photographs taken during experimental runs, illustrating multiple contacts, mode jumping, and flattening of the beam against the constraint walls: all phenomena to be investigated below.

The model developed in Section 2 is based on the classical one of Euler [Eul44]. It is fully nonlinear geometrically [Ant95] but employs a linear constitutive relation, adequate for the slender lamellae and modest forces and moments used in our experimental work.

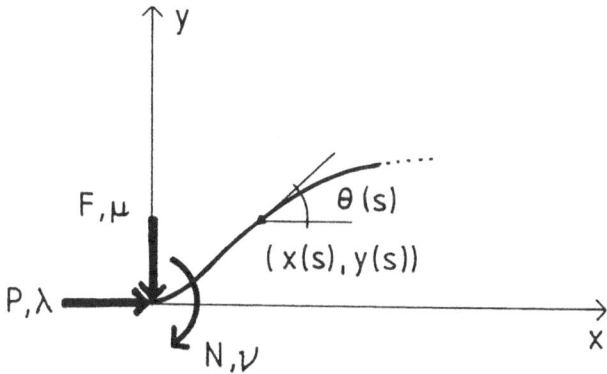

Fig. 2. A free body diagram for the buckled rod.

The unconstrained problem was originally posed and solved essentially completely by Euler, who, as Antman notes [Ant95], thus has a prior claim to Poincaré for invention of the qualitative theory of differential equations; he discovered many properties of elliptic functions before Jacobi.

In Section 3 we describe a numerical scheme developed for solution of rather general constrained problems, and give some examples to illustrate the analysis of Section 2. Then in Section 4 we discuss experimental results and compare them with the predictions of the model. Section 5 contains brief conclusions and identifies problems for future research.

The literature on nonlinear buckling and postbuckling is vast, but there is relatively little work on contact problems. A linear study of the pinned-pinned rod treated below appears in a "problems" book of Feodosyev [Feo77]; in Section 2.3 we show that our results reduce to his in the limit of small displacements. Keller and his colleagues have studied self-contact problems for rings [KFR72], the elastica [KF73], and more recently, thin sheets descending from a spool onto a rough surface [MK95]. The elastica work, in which the authors consider self-contact of the planar rod with clamped-clamped ends in its first mode, and with pinned-pinned ends in the third mode, is closest to that described below. In [KF73] a similarity solution for the shape of the rod between the contact points is found, load versus displacement bifurcation diagrams are computed numerically, and asymptotic expressions are developed for the contact point locations in terms of load.

2. Formulation of the Model

We recall the classical planar Euler buckling problem (cf. [Ant95], [Lov27]). Consider a deformed arc of an initially straight, uniform rod of flexural rigidity EI and length L subject to axial and lateral loads P, F, and moment N at the end $S = 0$, where S measures arclength. Taking moments at any point $S \geq 0$ and referring to the Cartesian coordinate system (x, y) and force sign convention of Figure 2, we have

$$EI \frac{d}{dS}\theta(S) + Py(S) + Fx(S) - N = 0, \tag{1}$$

where $\theta(S)$ denotes the slope at the point S and clockwise moments are positive. The position $(x(S), y(S))$ at S is obtained from the integral relations,

$$x(S) = \int_0^S \cos\theta(\sigma)\,d\sigma, \quad \text{and} \quad y(S) = \int_0^S \sin\theta(\sigma)\,d\sigma. \tag{2}$$

Differentiating (1), using (2), and defining nondimensional loads and moment via $\lambda = L^2 P/EI$, $\mu = L^2 F/EI$, and $\nu = LN/EI$, we obtain

$$\theta'' + \lambda\sin\theta + \mu\cos\theta = 0, \tag{3}$$

where $\theta'' = \frac{d^2\theta}{ds^2}$ and $s = S/L$ is the nondimensional arclength. This ordinary differential equation (ODE) can be posed as either an initial (IVP) or a boundary value problem (BVP). In the IVP, (3) is augmented by

$$\theta(0) = \theta_0, \quad \theta'(0) = \nu, \tag{4}$$

and in view of standard ODE theory, it has a unique smooth solution $\theta(s) = \theta(s; \theta_0, \nu, \lambda, \mu)$, with $s \in [0, 1]$ corresponding to the physically relevant part of the solution. As noted by Kirchhoff (cf. [Lov27]), this is precisely analogous to the dynamical problem of the planar pendulum, with a shifted coordinate θ when $\mu \neq 0$. We appeal to this in developing a numerical scheme to construct solutions of the constrained problem in Section 3.

Equation (3) has the first integral

$$\frac{\theta'^2}{2} - \lambda\cos\theta + \mu\sin\theta = \text{const}, \tag{5}$$

on level sets of which its solutions necessarily lie. This permits the direct construction of phase portraits, but before giving examples, we introduce a sequence of BVPs from which we shall assemble the governing equations and solutions of the constrained problem. As noted in Section 1, we restrict ourselves to the case of symmetric, frictionless walls at $y = \pm h$, parallel to the line joining the endpoints.

There are five "elementary" BVPs from which full solutions will be built as sequences of solutions applying between end and/or contact points. The relevant boundary conditions (BCs) to supplement equation (3) are listed below. Three cases involve contact and admit special limits in which internal moments at contact points vanish.

As we have noted, the IVP solutions can be uniquely specified by the four scalars θ_0, θ_0', λ, μ. In the case of a BVP we need the same number of scalar conditions, although they may not specify a solution uniquely. We list three of these conditions for each case between arbitrary endpoints $s = s_a$ and $s = s_b$:

(a) Pinned-pinned without contact:

$$\left.\begin{array}{rl} 1. & \theta'(s_a) = 0, \\ 2. & \theta'(s_b) = 0, \\ 3. \quad y(s_b) - y(s_a) = 0, \\ \text{or} & \mu = 0. \end{array}\right\} \tag{6}$$

(b) Clamped-pinned without contact:

$$
\left.\begin{array}{ll}
1. & \theta(s_a) = 0, \\
2. & \theta'(s_b) = 0, \\
3. & y(s_b) - y(s_a) = 0.
\end{array}\right\} \tag{7}
$$

(c) Pinned-contact:

$$
\left.\begin{array}{ll}
1. & \theta'(s_a) = 0, \\
2. & \theta(s_b) = 0, \\
3. & y(s_b) - y(s_a) = h.
\end{array}\right\} \tag{8}
$$

(d) Clamped-contact:

$$
\left.\begin{array}{ll}
1. & \theta(s_a) = 0, \\
2. & \theta(s_b) = 0, \\
3. & y(s_b) - y(s_a) = h.
\end{array}\right\} \tag{9}
$$

(e) Contact-contact, same wall:

$$
\left.\begin{array}{ll}
1. & \theta(s_a) = 0, \\
2. & \theta(s_b) = 0, \\
3. & \mu = 0.
\end{array}\right\} \tag{10}
$$

The fourth condition is identical in all cases and can be defined as

$$
x(1) = d \qquad \text{or} \qquad \lambda = \lambda_0, \tag{11}
$$

corresponding to the displacement (hard) boundary problem and the traction (soft or dead-load) problem, respectively.

We show typical element shapes and corresponding phase portraits in Figure 3, those for the degenerate zero internal moment cases being shown at far right. Note that the lateral load, μ, can take either sign in some cases, and the sense specified above is arbitrary. Correspondingly, for $\mu \neq 0$ the vertical axis of symmetry in the phase plane is shifted left or right from $\theta = 0$. A sixth case, contact-contact between opposite walls, is essentially identical to case (d), and (d), via symmetry about the inflection point, is in turn similar to case (c). In the explicit solutions developed below, solutions involving integer multiples of quarter circuits in the phase plane ((a) and (e)) are expressed in terms of complete elliptic integrals; the general cases ((b)–(d)) require incomplete elliptic integrals.

At a contact point the axial load and moment acting on an element must balance those on the adjacent element, but the lateral force μ may suffer a jump due to the net constraint force. This leads to a jump in shear force θ'', so that the resulting solutions $\theta(s)$ are only once-differentiable, the moment (curvature) $\theta'(s)$ being continuous.

In the physical experiments we are primarily interested in the displacement problem, in which the BVP for $s \in [0, 1]$ is controlled by the overall constraint

$$
x(1) = \int_0^1 \cos \theta(s) \, ds = d \in (-1, 1], \tag{12}
$$

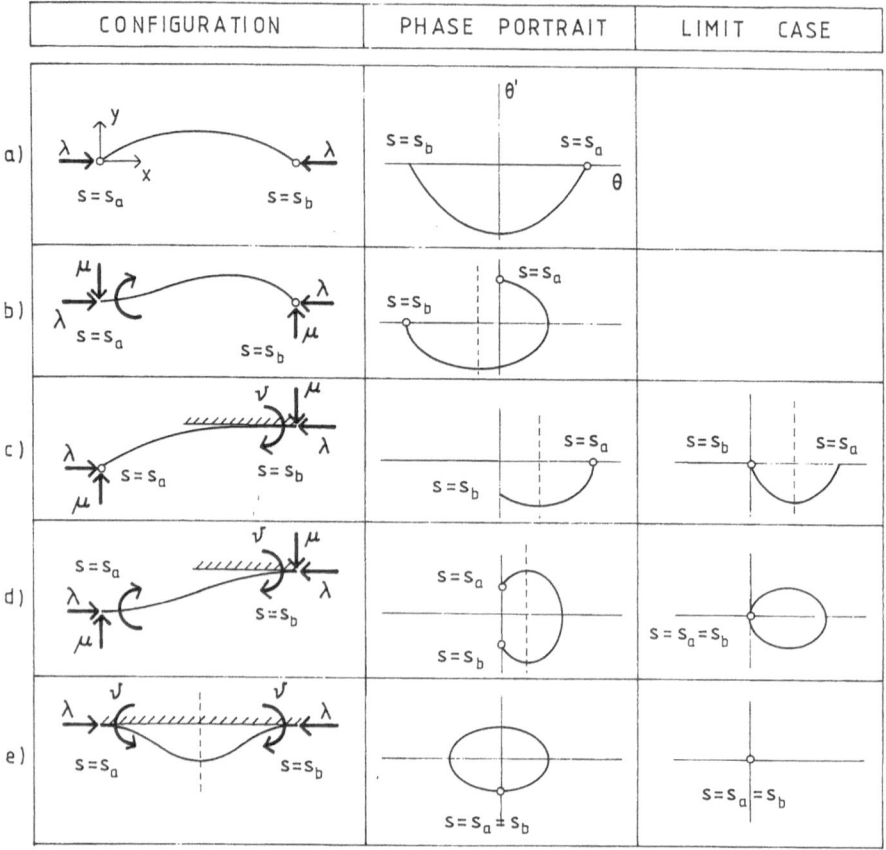

Fig. 3. Elemental BVPs and phase portaits for the constrained rod: (a) pinned-pinned (no contact); (b) clamped-pinned (no contact); (c) pinned-contact; (d) clamped-contact; (e) contact-contact (same wall). Limiting contact cases (c)–(e) shown in right column.

and the loads λ, μ are regarded as unknown, to be implicitly determined as part of the solution. However, it will be convenient to consider also the traction problem in which λ, μ are given and the displacements solved for directly. A third alternative is provided by path continuation, which our numerical approach adopts, as described in Section 3. This approach, in contrast with the previous two, does not admit jumps or snaps.

2.1. The Pinned-Pinned Case

Following the experimental protocol, we imagine monotonically decreasing the distance $d = x(1)$ between the endpoints, with fixed symmetric side-walls at a distance $\pm h$ from the centerline. Initially there is no contact, lateral forces and moments are zero, and the first half-wave is described by the elementary BVP (3)–(6), with $s_a = 0$; $s_b = 1/n$. Adopting $\mu = 0$, the lateral constraint $y(1) - y(0) = 0$ is satisfied automatically. Imposing reflectional symmetry about the midpoint of each half-wave implies that solutions for all mode numbers n can be found at once. (For completeness, we mention that the alternative third boundary condition, $y(1) - y(0) = 0$, yields $\mu = 0$ except for the degenerate case at $x(1) = x(0)$, excluded from this paper; see, e.g., [Mad84] and [Dom94].) This classical problem has been solved many times since Euler's work, but in the appendix we recall some explicit details, since they do not seem readily available. We summarise key conclusions below.

There is a unique solution $\theta(s)$, λ to the displacement problem specified by (6) and (12) with $s_a = 0$, $s_b = 1$, for each $d \in (-1, 1)$ and mode n (denoting the number of half-waves). The exact solution is given in terms of elliptic integrals, and a simple approximation relating the end load and displacement may be found via their power series:

$$\lambda = n^2 \pi^2 \left[1 + \frac{(1-d)}{2} + \frac{9(1-d)^2}{32} + \mathcal{O}(|1-d|^3) \right]. \tag{13}$$

The exact analysis results in two critical loads and end displacements, the first being the classical Euler buckling load,

$$\lambda_n^b = n^2 \pi^2, \qquad d_n^b = 1, \tag{14}$$

and the second the condition at which point contact with a side-wall first occurs,

$$\lambda_n^p = 4[n K(k)]^2, \qquad d_n^p = \left[\frac{2E(k)}{K(k)} - 1 \right], \qquad \text{where } K(k) = \frac{k}{nh}. \tag{15}$$

Here $K(k)$ and $E(k)$ are the complete elliptic integrals of the first and second kinds respectively, and $k \in [0, 1]$ is the elliptic modulus. See [AS65], but note that the modulus defined there is $m = k^2$. In (15) $k = \sin(\theta(0)/2)$ and $\theta(0)$ is the slope at the endpoint. When $\lambda = \lambda_n^p$ we denote the corresponding value $\theta(0) = \theta_n^p$: This is used in the input data in Section 3.1.

We note that h can also be regarded as a control parameter, for example with $d < 1$ fixed. Equations (15) can then be used to determine critical values $h_n^p(d)$; in this case the elliptic moduli are found from the second equality and substituted into the third. In our experiments the value of h was always fixed. However, the critical values $h_n^p(d)$, interpreted as a function of d, have a global maximum, and this maximal value (which, for the sake of simplicity will be denoted by h_n^p) is of great interest both from the experimental and numerical point of view: If $h > h_n^p$ then contact is impossible for *any* d, while for $h < h_n^p$ (15) has two solutions k_n^p corresponding respectively to first contact and loss of contact in the nth mode with decreasing d (the latter may occur in the everted state $d < 0$).

The value h_n^p is found by simultaneous solution of the last equality of (15) and its derivative with respect to k,

$$\frac{d}{dk} K(k) = \frac{1}{k} \left[-K(k) + \frac{E(k)}{1 - k^2} \right] = \frac{1}{nh}, \tag{16}$$

yielding the numerical values

$$k = 0.83745296\ldots \Rightarrow nh_n^p = 0.40314018\ldots. \tag{17}$$

It is interesting to note that this value of k, which characterises the unconstrained pinned-pinned beam at maximum lateral deflection, corresponds to an end angle of $\theta(0) = 113.74°$ and a displacement of $d = 0.19469\ldots$, i.e., neither to $\theta(0) = \pi/2$, nor $d = 0$, which might seem plausible conjectures, but to a state between these two. (Also see [Lov27], Section 263.)

Once contact is made, a lateral force develops at the contact point, which must be balanced by lateral forces at the pins. The resulting elemental problem for the first quarter wave of the n-th mode is now completed by (8), with $s_a = 0$, $s_b = 1/2n$. The explicit solution involves incomplete elliptic integrals and is awkward to work with, except in a second limiting case when the internal moment at the contact point drops to zero. This occurs when the resultant of the axial and lateral forces passes through pin and contact point, leading to a third critical load and displacement given by,

$$\lambda_n^l = (1 - 2k^2)\,[4nK(k)]^2, \qquad d_n^l = \frac{nh(1 - 2k^2)}{k\sqrt{1 - k^2}}, \tag{18}$$

with corresponding end angle θ_n^l. Here the elliptic moduli satisfy

$$\frac{nh}{k\sqrt{1 - k^2}} = \left[\frac{2E(k)}{K(k)} - 1\right]. \tag{19}$$

As above, (19) can have two, one, or no solutions depending on h, the critical values $h_n^l < h_n^p$ being

$$k = 0.48967120\ldots \;\Rightarrow\; nh_n^l = 0.32107049\ldots, \tag{20}$$

derived by matching (19) and its derivative with respect to k.

If $h < h_n^l$, when d first drops below d_n^l, a (straight) segment of the rod makes contact with the wall (ultimately, for smaller d, line contact is lost again). The equilibrium conditions and the fact that the wall is flat and frictionless implies that the lateral force is now two equal point loads at the ends $s = s_1$ and $s = s_2 = \frac{1}{n} - s_1$ of a contact region throughout which the solution $\theta(s) \equiv 0$. Thus the end moments $\theta'(s_j) = 0$, and the resultant forces in each end element remain colinear with pins and contact points. In this régime the full solution is assembled from solutions to (8) with $s_a = 0$, $s_b = s_1$, followed by (10) with $s_a = s_1$, $s_b = s_2$, and again (8) with $s_a = 1$, $s_b = s_2$. In the phase space $[\theta, \theta']$ the full BVP consists of solutions to (3)–(8) running between $(\theta, \theta') = (2\theta_a, 0)$ and $(\theta, \theta') = (0, 0)$, and $(0, 0)$ and $(-2\theta_a, 0)$, along with flat segments $(\theta, \theta') \equiv (0, 0)$: the trivial solution of (3)–(10).

Here the slope at the end of the rod $\theta(0) = 2\theta_a$; indeed, as long as the internal moments at the ends of the contact line remain zero, we have

$$\tan(\theta(0)/2) = h/\Delta = \mu/\lambda, \tag{21}$$

where $\Delta = \int_0^{s_1} \cos\theta(s)\,ds$ is the axial distance subtended by the end segment. See Figure 3(c) and Equation (65) below. Again, the awkward exact expressions involving elliptic functions, as given in the appendix, can be approximated conveniently using the Taylor series, yielding the force-displacement relation,

$$\lambda = \frac{4\pi^2}{9n^2h^4}(1 - d)^2\left[1 - \frac{(1 - d)^2}{2n^2h^2} + \mathcal{O}(|1 - d|^4)\right]. \tag{22}$$

When each flat segment's length and axial load are such that it can buckle as a clamped-clamped beam in the first mode, as in Figure 3(e), a fourth critical load and displacement emerge, given by

$$\lambda_n^{sb} = \frac{4n^2\pi^2}{(1 - 2ns_1)^2}, \qquad d_n^{sb} = 1 + 2n(\Delta - s_1), \tag{23}$$

where

$$s_1 = \frac{hK(k)}{2k\sqrt{1 - k^2}[2E(k) - K(k)]} \qquad \text{and} \qquad \Delta = \frac{h(1 - 2k^2)}{2k\sqrt{1 - k^2}},$$

and the elliptic moduli are solutions of

$$nh\left[\frac{\pi}{2\sqrt{1 - 2k^2}K(k)} + 1\right] = k\sqrt{1 - k^2}\left[\frac{2E(k)}{K(k)} - 1\right]. \tag{24}$$

Once more, (24) has two, one, or no solutions depending on whether $h < h_n^{sb}$, $= h_n^{sb}$, or $> h_n^{sb}$, where

$$k = 0.43505967\ldots \Rightarrow nh_n^{sb} = 0.14317822\ldots. \tag{25}$$

Following secondary buckling, the end element of each segment of length $\frac{1}{n}$ is governed by (8) and the central one by (10), the solutions of which must be matched via the internal moment $\theta'(s_1)$. A solution of this type persists until the central buckled portion first touches the opposite wall, a condition coinciding with first contact for the $3n$-th mode.

2.2. The Clamped-Pinned Case

This problem does not admit solutions as explicit as those obtained above, but the qualitative phase plane analysis is very similar. Again we start with the unconstrained problem (7), with $s_1 = 0$, $s_2 = 1$. The desired solution in the first mode is sketched in the phase plane of Figure 3(b); the second and higher modes simply make more circuits, but due to displacement of the center of symmetry, cannot be found as directly as in the pinned-pinned case. Nonetheless, as above we may identify several critical loads and displacements, the first being the classical clamped-pinned buckling load,

$$\tan\sqrt{\lambda_n^b} = \sqrt{\lambda_n^b} \Rightarrow \lambda_1^b \approx 2.045\pi^2, \ldots, \qquad d = 1, \tag{26}$$

obtained from the linearized ODE BVP derived from (3) and (7).

The second load λ_n^p, at which contact with a side-wall first occurs, is given by simultaneous solution of the following four equations for the unknown loads $\lambda = \lambda_n^p$ and $\mu = \mu_n^p$, the critical axial displacement $d = d_n^p$, and the parameter k (the elliptic modulus):

$$\sqrt{\bar{\lambda}} = (2n + 1)K(k) - F(\phi_a; k), \tag{27}$$

$$\lambda d = 2\sqrt{\tilde{\lambda}}[(2n+1)E(k) - E(\phi_a; k)] - \tilde{\lambda}, \tag{28}$$

$$\mu d = \sqrt{2(2k^2 - 1)\tilde{\lambda} + 2\lambda}, \tag{29}$$

$$\frac{\tilde{\lambda}^{\frac{3}{2}} h}{4n} = \mu[2E(k) - K(k)], \tag{30}$$

where $\tilde{\lambda} = \sqrt{\lambda^2 + \mu^2}$, $\tan\theta_a = \frac{\mu}{\lambda}$, $\sin\phi_a = \sin(\frac{\theta_u}{2})/k$, and $F(\phi; k)$ and $E(\phi; k)$ are the incomplete elliptic integrals of the first and second kinds, respectively [AS65]. As above, a critical value h_n^p exists, which determines if point contact can occur in specific cases. As for the pinned-pinned case, for $h < h_n^p$ no contact occurs, otherwise two critical loads (or displacements) are obtained, bounding an interval in which contact is maintained. Once contact is made at $s = s_1$ in the first mode, the BVP can be assembled from the elementary BCs (9) with $s_a = 0$, $s_b = s_1$, followed by (8) with $s_a = 1$, $s_b = s_1$.

As before, necessary conditions for onset of a flat segment on the wall, and secondary buckling of that segment, are obtained by requiring that internal moments vanish at the ends of the contact region. To avoid yet more unwieldly formulae, we restrict ourselves to the first buckling mode, in which only a single contact region occurs. Thus we have equivalent pinned-pinned rods of lengths $B_j = \sqrt{\Delta_j^2 + h^2}$, $j = 1, 2$, where $\Delta_1 = \int_0^{s_1} \cos\theta(s)\,ds$ and $\Delta_2 = \int_{s_2}^1 \cos\theta(s)\,ds$ are the axial distances subtended by the end segments, and we can appeal to the pinned-pinned theory. Note that the first element ($j = 1$) is in the *second* buckling mode while the second ($j = 2$) is in the *first* mode and the entire beam lies on one side of the central axis. As shown in the appendix, this leads to critical loads and overall axial displacements specified by simultaneous solution of

$$\lambda_1^l = (1 - 2k_1^2)\left[\frac{4K(k_1)}{s_1}\right]^2 = (1 - 2k_2^2)\left[\frac{2K(k_2)}{1 - s_1}\right]^2, \tag{31}$$

$$d_1^l = \Delta_1 + \Delta_2, \tag{32}$$

$$s_1 = \frac{hK(k_1)}{2k_1\sqrt{1 - k_1^2}\,[2E(k_1) - K(k_1)]}$$

$$= 1 - \frac{hK(k_2)}{2k_2\sqrt{1 - k_2^2}\,[2E(k_2) - K(k_2)]}, \qquad \Delta_j = \frac{h(1 - 2k_j^2)}{2k_j\sqrt{1 - k_j^2}}, \tag{33}$$

and

$$\lambda_1^{sb} = \frac{4\pi^2}{(s_2 - s_1)^2} = (1 - 2k_1^2)\left[\frac{4K(k_1)}{s_1}\right]^2 = (1 - 2k_2^2)\left[\frac{2K(k_2)}{1 - s_2}\right]^2, \tag{34}$$

$$d_1^{sb} = \Delta_1 + \Delta_2 + (s_2 - s_1), \tag{35}$$

$$s_1 = \frac{h K(k_1)}{2k_1\sqrt{1 - k_1^2}\,[2E(k_1) - K(k_1)]}, \tag{36}$$

$$s_2 = 1 - \frac{h K(k_2)}{2k_2\sqrt{1 - k_2^2}\,[2E(k_2) - K(k_2)]}, \qquad \Delta_j = \frac{h(1 - 2k_j^2)}{2k_j\sqrt{1 - k_j^2}}. \tag{37}$$

The second pair of equations in (31) and equations (33) define three conditions which may be solved simultaneously for k_1, k_2, and s_1 (and hence Δ_1, Δ_2), given h. The critical load λ_1^l and displacement $d = d_n^l$ then follow from the first of (31) and (32). Similarly the last two equalities of (34) and equations (36)–(37) may be solved for $k_1, k_2, \Delta_1, \Delta_2$, s_1, and s_2, and the critical load and displacement found afterwards. As in Section 2.1, λ_1^l denotes the load at which the contact region may change from point to line, and λ_1^{sb} that at which secondary buckling can occur on the contact line segment. Recall that we only give results for the first mode $n = 1$ here. As in the pinned-pinned case, maximal critical values h_1^l and h_1^{sb} could be derived to characterize the possible contact behavior.

After developing these criteria for onset of line contact and secondary buckling, we performed the experiments described in Section 4. These revealed the presence of numerous additional (unsymmetrical) branches of solutions, and in the clamped-pinned case (Section 4.2), they showed that, while moments must indeed drop to zero at the contact point as d decreases (provided $h < h_1^l$), the line contact path is not followed; rather the moment changes sign at the clamped end and the beam crosses the center-line, thus developing a second inflection point and eventually a second contact point. See Sections 4 and 5 for further discussions.

The critical loads and displacements identified above and the analysis in the appendix provide checks for the numerically constructed solution branches in Section 3. In turn, the numerical data is compared with experiments in Section 4.

2.3. Linearized Results

Here we briefly show that the critical loads of Section 2.1 for the pinned-pinned problem coincide, in the appropriate limit, with those derived in a linearized analysis by Feodosyev ([Feo77], Problem 121). The limit is that of small lateral displacements: $y \leq h \ll d \approx 1$ and small slopes $\theta \ll 1$, although θ depends more sensitively on the mode number than y. We observe that assuming $h \ll 1$ is equivalent to assuming that h lies below *all* critical values discussed earlier; thus, in the linear theory the number of contact points and segments can increase indefinitely.

Assuming end slopes $\theta_0 \ll 1$, we are concerned with the limit $h \to 0, k = \sin(\frac{\theta_0}{2}) \to 0$, with $\lim_{k \to 0} h/k$ finite, in which case $K(k) \to \frac{\pi}{2}$. Using this, we find that λ_n^p of (15) converges to the buckling load,

$$\lambda_n^p \to \lambda_n^b = \lambda_n^p(\text{linear}) = n^2\pi^2. \tag{38}$$

Applying the same limit in (18), we find

$$\lambda_n^l \to \lambda_n^l(\text{linear}) = 4n^2\pi^2, \tag{39}$$

and in (23)–(24),

$$\lambda_n^{sb} \to \lambda_n^{sb}(\text{linear}) = 16n^2\pi^2, \qquad s_1, \Delta \to \frac{1}{4n}. \tag{40}$$

In all cases linearization restricts $d = 1$, and so gives no meaningful predictions of end displacements.

These results agree with those obtained by direct solution of the standard linearized equation for the lateral displacement y in [Feo77], but an interesting point emerges. Feodosyev assumes that displacements follow the first buckling mode ($n = 1$) until, for $\lambda > \lambda_1^{sb}(\text{linear})$, the central flat region detaches and the rod switches to its third mode. This in turn changes to the ninth mode at $\lambda_3^{sb}(\text{linear})$, etc. Thus Feodosyev assumes that reflectional symmetry about the midpoint is preserved, and only modes $n = 1, 3, 9, \ldots$ are tabulated in his solution. In unloading, from say $\lambda \approx 4 \cdot 3^2\pi^2$, he assumes that the third mode persists to its linear buckling load $3^2\pi^2$, lower than the load $\lambda_1^{sb}(\text{linear}) = 16\pi^2$ at which it appeared, thus predicting hysteresis. Hysteresis is indeed observed in the experiments described below. However, other solutions having different symmetry, including those for $n = 2$, are also possible in this load range and are indeed observed in the experiments of Section 4. One cannot determine which branches are physically relevant without addressing the issue of stability.

2.4. Stability

In this paper we do not consider stability of equilibria in detail, but a few remarks are in order. The total stored energy is given by

$$E[\theta] = \int_0^1 \left[\frac{\theta'(s)^2}{2} + \lambda[\cos\theta(s) - 1] - \mu\sin\theta(s) \right] ds, \tag{41}$$

in which the first term is the strain energy due to curvature and the second two, energies associated with work done by the loading device. We have chosen the "constants" in these terms such that $E[\theta] = 0$ as long as no deflections occur, even if $\lambda \neq 0$. An equilibrium state $\bar{\theta}(s)$ is said to be *stable* if $E[\theta] \geq E[\bar{\theta}]$ for all nearby states θ, in a physically appropriate norm. Note that $\bar{\theta}$ need not be an absolute or global minimizer of E; local minimizers are also stable.

To illustrate a simple case, take the pinned-pinned beam without side-wall constraints ($\mu \equiv 0$) and subject to fixed (dead) end-load λ, suppose that $\bar{\theta}$ is an equilibrium, and consider $E[\bar{\theta} + \eta]$ for a small variation η satisfying the same boundary conditions. Substituting $\theta = \bar{\theta} + \eta$ into (41), expanding in η, and eliminating the $\mathcal{O}(\|\eta\|)$ term via integration by parts and use of the equilibrium equation (3), we arrive at

$$E[\theta] = E[\bar{\theta}] + \int_0^1 \left[\frac{\eta'(s)^2}{2} - \lambda\cos\bar{\theta}(s)\frac{\eta(s)^2}{2} \right] ds + \mathcal{O}(\|\eta\|^3). \tag{42}$$

Taking into account the boundary conditions (6), which imply the Poincaré inequality $\int_0^1 \eta'^2 ds \geq \pi^2 \int_0^1 \eta^2 ds$ for admissible variations, the quadratic form (second variation) is positive-definite if $\max|\lambda\cos\bar{\theta}(s)| < \pi^2$. It follows that the trivial solution up to

$\lambda = \pi^2$ is stable. The first mode equilibria, which bifurcate at $\lambda = \pi^2$, are also stable and remain so until $d = 0$. All higher mode branches are unstable. Analogous observations apply to the clamped-pinned and clamped-clamped beams without side-wall constraints. See Maddocks [Mad84], [Mad87] and Domokos and Gáspár [Dom94]. In fact Maddocks [Mad87] considers a related problem in which the $\theta(1)$ versus λ bifurcation diagram can "double back." He shows that the backward-going portions (λ decreasing as $\theta(1)$ increases) are unstable under dead loading, but stable under displacement (hard) boundary conditions of the type used in the experiments reported below.

In the presence of lateral forces due to side-wall contact or clamped boundaries, a calculation analogous to that above yields

$$E[\theta] = E[\bar\theta] + \int_0^1 \left[\frac{\eta'(s)^2}{2} - \left(\lambda \cos\bar\theta(s) - \mu \sin\bar\theta(s)\right) \frac{\eta(s)^2}{2} \right] ds + \mathcal{O}(\|\eta\|^3), \quad (43)$$

which must be constructed by piecing together the appropriate elementary equilibrium solutions, noting that the sign and value of μ may change as one moves from element to element. For displacement boundary conditions, the loads λ and μ must also be allowed to change when calculating the second variation. Unfortunately, the results of [Mad87] do not apply directly, since as one follows a given branch and contact is gained and/or lost, the governing equations change via jumps in the piecewise-constant lateral force $\mu(s)$. For example, it is well known that the second mode of the unconstrained pinned-pinned beam is unstable, while it is experimentally clear that the same mode is stable for the beam with two point contacts. The corresponding branch in the load-displacement graph is monotonically increasing in the neighborhood of (λ_2^p, d_2^p) where point contact is made, so it is not obvious that stability should change here. In fact, a secondary (asymmetrical) branch bifurcates at this point. See Section 5 (Figure 11) for further discussion.

3. Numerical Method

Elliptic functions are awkward to work with and must typically be evaluated numerically, so rather than using the analytical solutions of Section 2 directly, in practice we solve the BVP by integrating a sequence of IVPs $i = 1, \ldots, N$. We regard the unknown loads λ_i, μ_i, matching points s_i, and the undetermined initial condition $\theta(0) = \theta_0$ (for a pinned end) or $\theta'(0) = \nu$ (for a clamped end) as parameters to be varied in a shooting problem determined by the overall displacement and contact constraints. After outlining the numerical scheme, detailed in [DG95], we proceed as in Section 2, describing its application to the different phases of the loading process.

As already described, the problem can be reduced to a sequence of conventional BVPs on subsegments of the total arclength $[0, 1]$, along with matching conditions. Each can be regarded as a parameter-dependent IVP starting at $s = s_a$ and ending at $s = s_b$, or as an input-output device. The input data are the initial conditions $\theta(s_a)$, $\theta'(s_a)$, parameters λ, μ, and the integration constants $x(s_a)$, $y(s_a)$. The output data are the same six scalars at $s = s_b$. (On each segment, λ and μ remain constant.) Using any convergent integrator for the IVP, the second set of six scalars can be determined with arbitrary precision as functions of the first six. Of course, certain initial conditions are fixed by the boundary conditions at $s = s_a$, so the output depends on fewer parameters in any particular BVP.

We will call the nonconstant initial values *variables* and remark that the location of a BVP's endpoint may also appear as a variable. The variables can be regarded as unknown quantities, needed to integrate the global IVP from $s = 0$ to $s = 1$. If we consider a sequence of IVPs, then the input for the subsequent IVP derives partly from the output of the previous one, and partly from new variables to be added. The matching conditions at intermediate points and the boundary conditions at $s = 1$ can be formulated as *functions* of the *output* scalars, thus, as functions of the variables. Hence, using the numerical integrator, our multipoint BVP can be reduced to a system of nonlinear algebraic equations.

In Sections 3.1 and 3.2 we specify variables and functions, respectively V and F in number, for the pinned-pinned and clamped-pinned elemental problems. As we will see, this procedure, although similar in many cases, requires some insight into the mechanical problem.

Fixing all four boundary conditions for each elementary BVP (3)–(6)—(3)–(10), we obtain isolated solutions. However, we are interested in continuous sequences of equilibria and we therefore omit the fourth condition (11), which is identical for each BVP. Recall that this condition prescribes either the load λ or the end displacement $x(1)$. With *one* condition relaxed, the equations yield a *one*-parameter family of solutions, called the *equilibrium path*. In practice, as control parameter we take neither λ nor $x(1)$ since, in general, the path may have turning points in both parameters.

The choice of algorithm depends largely on the question addressed. We posed the following (modest) question: "Given one equilibrium configuration (solution), provide an arbitrarily long sequence of equilibria, each of which is arbitrarily close to the adjacent ones." This task can be completed by a path-continuation method. In structural mechanics most path-continuation algorithms are based on extrapolation along the equilibrium path. Such algorithms are often referred to as incremental-iterative methods, since after each extrapolation step corrective iteration is necessary [Rik79].

The convergence properties of the iteration are often sensitive and need special attention. Here we applied the iteration-free "simplex" method of [DG95], which, in turn, is based on the so-called PL (piecewise-linear) algorithm (see [AG90]), as well as ideas from mechanics. The space spanned by the variables is subdivided into (small) simplices, the function values are computed at the vertices and the equilibrium path is linearly interpolated inside the simplices. Thus the method uses interpolation instead of extrapolation and provides the (approximate) equilibria via direct recursion. It delivers precisely one subsequent point. Since continuity (C^0) of the equilibrium paths is preserved in the piecewise-linear approximation, when approaching a bifurcation point on the primary path, the algorithm selects one of the (two) "exit" possibilities effectively at random and continues along the selected path. We note that the simplex method is also suitable for a more general task: finding *all* globally admissible equilibria in a given domain [Dom94].

To run the simplex method, for each member in the sequence of elementary BVPs we must specify:

- *V variables*, i.e., those initial conditions and parameters for all elementary BVPs which are not fixed by the boundary conditions at the initial points and not inherited from the previous segment, and in addition, those endpoint locations which are not fixed in advance,

- F *functions*, i.e., the matching conditions at intermediate points and the boundary conditions at $s = 1$, and
- the values of the variables at one equilibrium configuration.

In the following subsections we describe how this information can be obtained for our problem. Since we expect one-parameter families of solutions, the number V of variables has to exceed the number F of functions by one.

One further remark is relevant. When contact is established or lost as one follows a branch, the BVP, and hence the numbers V and F, change. The initial configuration of the new model is inherited from the final configuration of the previous one. We found that precision of 10^{-4} in initial parameters suffices to continue the computation. This is relatively easily maintained for force, moment, and angle; however, the contact point precision is limited by the mesh size of the discretization. Small mesh sizes increase computation time and lead to loss of accuracy via truncated Taylor series approximations of trignometric functions. In our computational environment (an IBM PC with Microsoft QuickBasic v 4.5 and DOS 6.2), mesh size and hence contact point accuracy could not be decreased below 10^{-3}. We therefore had to interpolate carefully in changing models. See [DG95] for general remarks on limitations of the simplex method.

3.1. The Pinned-Pinned Case

For simplicity we discuss only the first mode; higher modes can be treated in a similar manner. Starting with the contact-free configuration, the BCs are (6) with $s_a = 0$, $s_b = 1$. Adopting $\mu = 0$ as the third condition in (6) and fixing the initial point at $x(0) = y(0) = 0$, there remain two variables: $\theta_0 = \theta(0)$ and λ. The single boundary condition to be met at $s = 1$ is (6/2). Thus $V = 2$, $F = 1$, and consequently, $V - F = 1$. The initial configuration is specified by $(\theta_0, \lambda) = (0, 0)$. This case is illustrated in Figure 4(a). While following this equilibrium path the maximal transversal displacement has to be monitored. The symmetry with respect to $s = 0.5$ enforces $y_{max} = y(0.5)$. At $y(0.5) = h$ the constraint is violated and this model becomes inapplicable.

Contact is established at this configuration with $\lambda = \lambda_1^p$. Symmetry reduces the computation to the first segment, described by (8) with $s_a = 0$, $s_b = 0.5$. Since a lateral force now acts, there are now three variables—θ_0, λ, and μ—and two boundary conditions to be met at $s = 0.5$—(8/2) and (8/3). Thus we have $V = 3$, $F = 2$, and consequently, $V - F = 1$. The initial configuration is specified by $(\theta_0, \lambda, \mu) = (\theta_1^p, \lambda_1^p, 0)$, the first two parameters being inherited from the last contact-free configuration. (Recall the definitions of λ_n^p and θ_n^p in and directly following Equation (15).) This situation is shown in Figure 4(b). The point-contact model is valid until $\theta'(0.5)$ changes sign, since this would imply an inflexion point at $s = 0.5$ and thus penetration of the boundary at $y = h$.

If $h < h_1^i$ then in the next phase line contact develops. Since symmetry with respect to $s = 0.5$ is still maintained, it is sufficient to compute the shape through the second contact point. Assuming that the first contact point is at $s = s_1$, the BCs are (8) with $s_a = 0$, $s_b = s_1$, followed by (10) with $s_a = s_1$, $s_b = 1 - s_1$. In (8) we now have four variables: Besides the previous three (θ_0, λ, μ), the location s_1 also appears as an input datum. In turn, we must satisfy three constraints: (8/2), (8/3), and (10/2). Thus we have $V = 4$, $F = 3$, and again $V - F = 1$. The initial configuration is specified

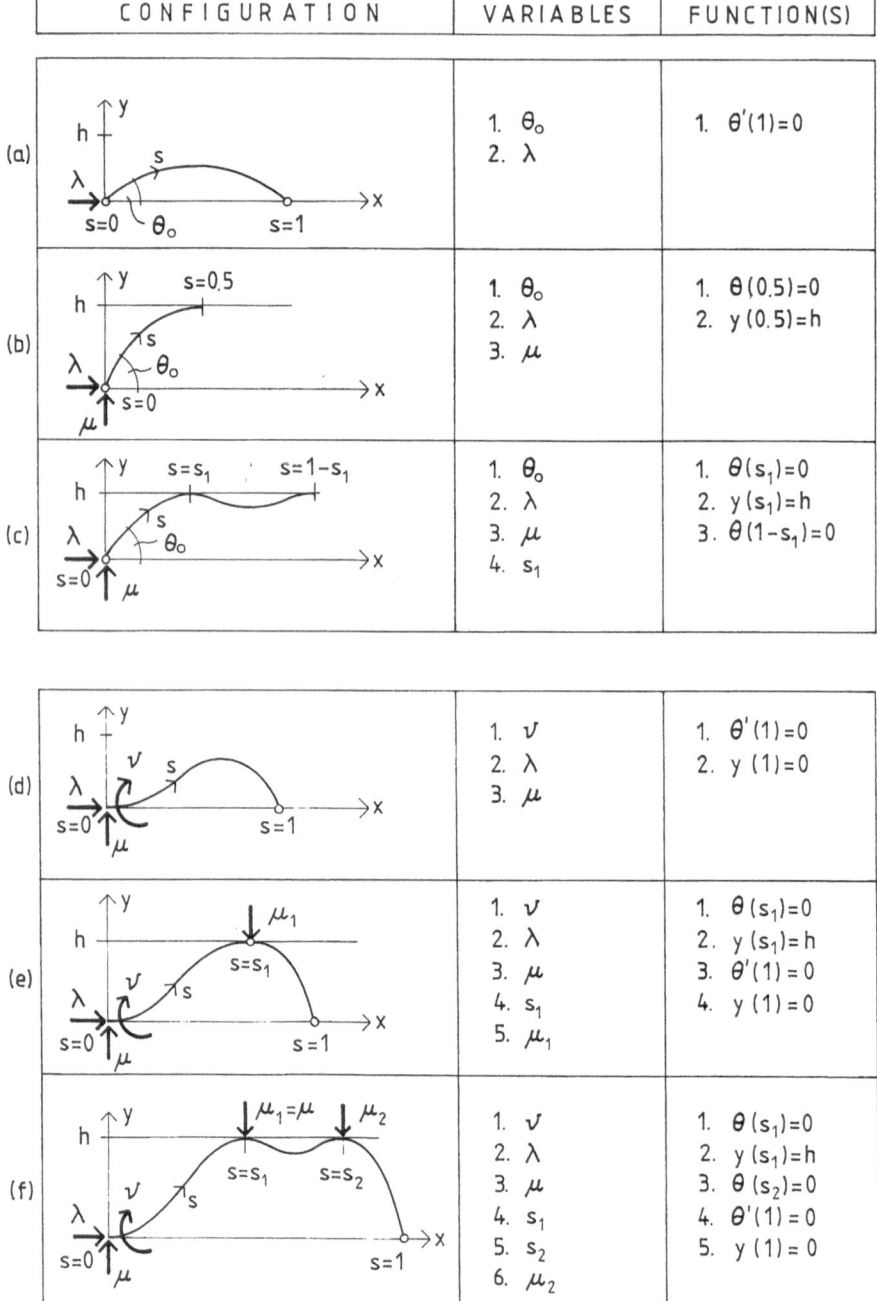

Fig. 4. Variables and functions used in the simplex method: (a) pinned-pinned (no contact); (b) pinned-pinned (point contact); (c) pinned-pinned (line contact); (d) clamped-pinned (no contact); (e) clamped-pinned (point contact); (f) clamped-pinned (line contact).

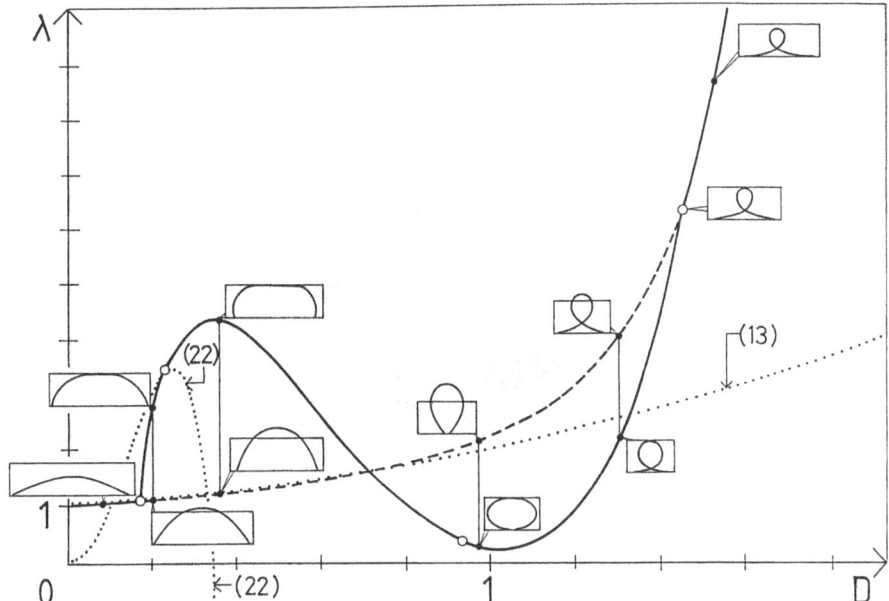

Fig. 5. Load-displacement curve for $h_1^{sb} < h = 0.25 < h_1^l$ in the pinned-pinned case. Constraint-free behavior plotted with dashed line. Onset (loss) of point and line contact indicated by open circles. Dotted curves indicate Taylor series approximations of (13) and (22); also see text.

by $(\theta_0, \lambda, \mu, s_1) = (\theta_1^l, \lambda_1^l, \mu_1^l, 0.5)$; the first four parameters are inherited from the last configuration of the point-contact-phase. Equation (18) also defines and allows direct computation of λ_1^l, θ_1^l.

From the computational viewpoint, secondary buckling of the flat, central portion requires no change in this model. We expect this phenomenon if $h < h_1^{sb}$. As mentioned before, at bifurcation points our algorithm follows one of the nontrivial branches (unless instructed otherwise). This configuration (illustrated in Figure 4(c)) remains valid until new contact is established at $s = 0.5$ with the opposite constraint wall. Then a new model has to be used, again with a different number of variables. This process can be continued; however, as the number of variables grows, the numerical scheme becomes slower and more sensitive.

In sharp contrast to the linearized model (cf. [Feo77], Subsection 2.3), in the nonlinear case the number of contact points (and the number of variables) does not grow indefinitely. Rather, after rising initially, in the limit $d \to -1$ all contact is lost and the model reverts to the classical Euler problem. To illustrate this, in Figure 5 we present a load-displacement curve computed for a case $h_1^{sb} < h < h_1^l$ ($h = 0.25$). For comparison, we also show the analogous curve for a beam without lateral constraint. Physical configurations with equal end-displacements are displayed in the figure. The numerical data agree remarkably well with the critical values from the theory of Section 2.1. Equation (15) yields for λ_1^p the values 10.815 and 63.646 versus the numerical values of 10.814 and 63.229, respectively. From Equation (18) we obtain 34.531 and 3.956 for λ_1^l versus the numerical data 34.508

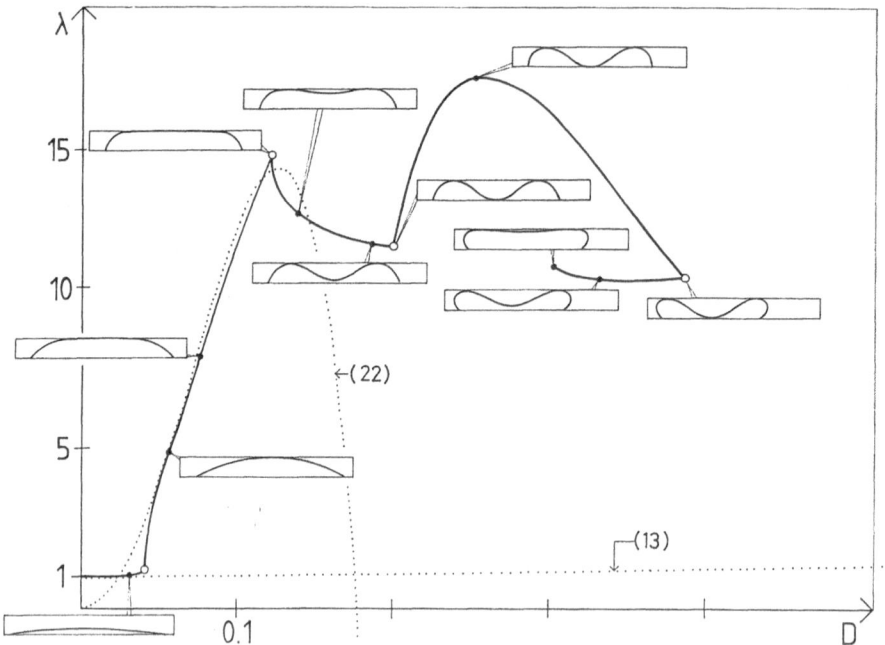

Fig. 6. Load-displacement curve for $h = 0.125 < h_1^{sb}$ and $h = 0$ in the pinned-pinned case (asymmetrical walls). Note horizontal and vertical tangencies. Onset of point contact and secondary buckling indicated by open circles. Dotted curves indicate Taylor series approximations of (13) and (22); also see text.

and 4.001. Agreement of the critical values for d is equally good. The unconstrained and line-contact segments predicted by the Taylor series expansions of (13) and (22) are shown on this and the following figure: They provide acceptable approximations for h smaller than about 0.125 (also see Figure 9).

Note that, here and henceforth, in place of the end displacement d we plot $D = 1 - d$ to agree with conventional load displacement curves of the engineering literature. We also specify all loads in units of $\lambda_1^b = \pi^2$, i.e., nondimensionalized on the pinned-pinned buckling load.

As a second example, in Figure 6 we show a load-displacement diagram and characteristic physical configurations for $h < h_1^{sb}$. In this numerical experiment the constraining walls are set asymmetrically: the upper one at $h = 0.125$, the lower one at $h = 0$. After following the previous scenario, the middle segment buckles and line contact is lost. It is worth noting that neither the load λ nor the end displacement d is monotonic along this equilibrium path, indicating that the sequence of equilibria can be realized experimentally only by switching from displacement to load control. The corresponding critical values are $\lambda_1^p = 10.069$, $\lambda_1^l = 38.499$, and $\lambda_1^{sb} = 146.641$, 110.051, compared with the numerically computed data 10.069, 38.475, and 145.72, 109.51. (We did not follow the branch far enough to observe the second values of λ_1^p and λ_1^l.)

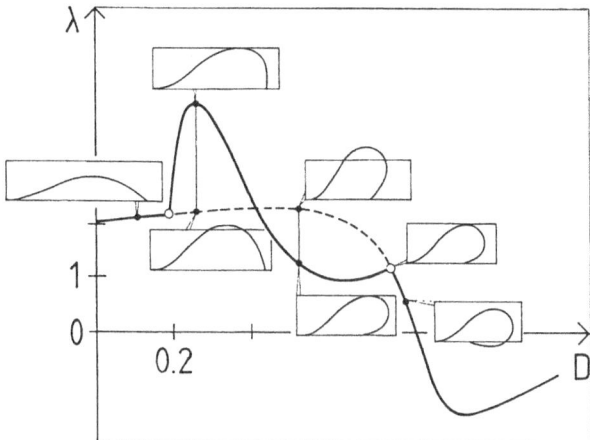

Fig. 7. Load-displacement curve for $h_1^l < h = 0.25 < h_1^p$ in the clamped-pinned case. Constraint-free behavior plotted with dashed line. Onset (loss) of point contact indicated by open circles.

3.2. The Clamped-Pinned Case

As in the previous subsection, we restrict our discussion to the first mode, although computations for higher modes have been done (see Section 4, Figure 10). In the initial, contact-free phase the BCs are (7) with $s_a = 0$, $s_b = 1$. After fixing the initial point at $x(0) = y(0) = 0$ there remain three variables: $v = \theta'(0)$, λ, and μ. The two boundary conditions to be met at $s = 1$ are (7/2) and (7/3); thus we have $V = 3$, $F = 2$, and, as before, $V - F = 1$. The initial configuration is specified by $(v, \lambda, \mu) = (0, 0, 0)$. This case is illustrated in Figure 4(d). This model remains valid until $y_{\max} = h$. Since there is no midpoint reflection symmetry, the location s_1 of the contact point is a priori unknown.

If $h < h_1^p$ contact is established at $\lambda = \lambda_1^p$, the location of the contact point being $s = s_1^p$. In this phase we need two elementary BVPs completed by (9) with $s_a = 0$, $s_b = s_1^p$ followed by a "reversed" (8) with $s_b = s_1^p$, $s_a = 1$; in this case $s_b < s_a$ and we integrate from s_b to s_a. There are five free variables: Besides v, λ, μ, the location s_1, and the transversal contact force μ_1 are unknown when starting the IVP. Four constraints must be fulfilled: (9/2), (9/3), (8/1), and (8/3). The initial configuration is specified by $(v, \lambda, \mu, s_1, \mu_1) = (\theta_1^p, \lambda_1^p, \mu_1^p, s_1^p, 0)$. The first four parameters are inherited from the last configuration of the previous, contact-free phase, λ_1^p can also be determined by the solutions of Equations (27)–(30). This configuration is illustrated in Figure 4(e). As indicated in Section 2.2, the point-contact model remains valid until $\theta'(s_1)$ changes sign. Figure 7 illustrates a load-displacement curve with some characteristic physical shapes. Here $h = 0.25$; thus $h_1^l < h < h_1^p$. For comparison the unconstrained case with identical axial end displacement is also plotted. It is worth mentioning that the contact point moves back and forth as one proceeds along the branch: starting at $s_1^p = 0.593$, increasing to $s_1 = 0.642$, and returning to $s_1 = 0.483$.

If $h < h_1^l$ then the contact point splits into two, $s = s_1$, s_2, and the segment $s \in [s_1, s_2]$ lies flat against the constraint surface. Now we have three elementary BVPs: (3) with

(7) and $s_a = 0$, $s_b = s_1$, followed by (10) and $s_a = s_1$, $s_b = s_2$, and again a "reversed" (8) and $s_a = 1$ and $s_b = s_2$. Since in (10) $\mu = 0$, the contact force μ_1 at s_1 is identical to μ. When integrating the IVP from $s = 0$ the following six variables are unknown: $\nu, \lambda, \mu, s_1, s_2, \mu_2$, the last one denoting the contact force at $s = s_2$. Using these variables five conditions must be met: (9/2), (9/3), (10/2), (8/1), and (8/3). The configuration is shown in Figure 4(f). As in the pinned-pinned case, this path could be followed further, with an increasing number of variables as the number of contact points increases; moreover, as noted in Section 2.2, another branch, with a second critical point, bifurcates from this branch at the point where the moment at the clamped end drops to zero; see Section 4.2 and Figure 10.

4. Experimental Results

Buckling measurements were made using long, slender rectangular stainless steel beams with dimensions of circa $0.05 \times 6.25 \times 100$ mm, shape-constrained to deform in a plane. Buckling was further constrained by rigid walls on either one or both sides placed parallel to and at controlled distances from the beam's center-line. The forces required to maintain a given deformed state were measured using a modified Ohaus CT 200 strain-gauge balance assembly in contact with one end of the beam, contact being with a shallow vertical vee groove approximating a pinned boundary condition. The other end was either supported by a rotary jewelled bearing, or clamped parallel to the deformation axis, depending on the desired boundary conditions.

Both balance assembly and support could be moved with either displacement or displacement rate under control. Balance motion was effected by a Newport motor-micrometer used in either the jog or continuous modes. In the former, step sizes were varied between 2 and 500 microns. Since the strain-gauge balance deflected with increasing load, it was necessary to correct to obtain the true beam end displacement. This was achieved by measuring the force recorded by the balance as it was advanced into a rigid constraint and correcting all displacements for balance deflections at their measured load values. After correction, the loading device approximates hard (displacement) boundary conditions. The experiment was mounted on an optical breadboard table to provide a stable base. The apparatus is shown in Figure 8.

To minimize the effects of friction between beam and side-walls, the latter were made from glass microscope slides treated with a molybdenum disulfide solid lubricant. In jog mode the system was vibrated after each position step using an electric bell motor to minimise hysteresis due to mechanical backlash in the motion stage and friction bewtween beam and constraint. The resulting force/position data pairs were recorded, along with video images and other information regarding the beam's shape.

Typically, the beam is inserted with its long axis parallel to the motion axis and the larger cross-sectional axis vertical, so that buckling occurs in a horizontal plane, thus minimizing gravitational effects. The stationary end is placed in the pivot, or clamp, and the balance end advanced using the motor micrometer until the beam just contacts the vee groove in the thrust pad. In this condition there is a pre-load of about 0.5g. The vertical glass side-walls are placed at equal distances from the center-line and parallel to it. The balance end displacement is then incrementally increased (decreasing the axial

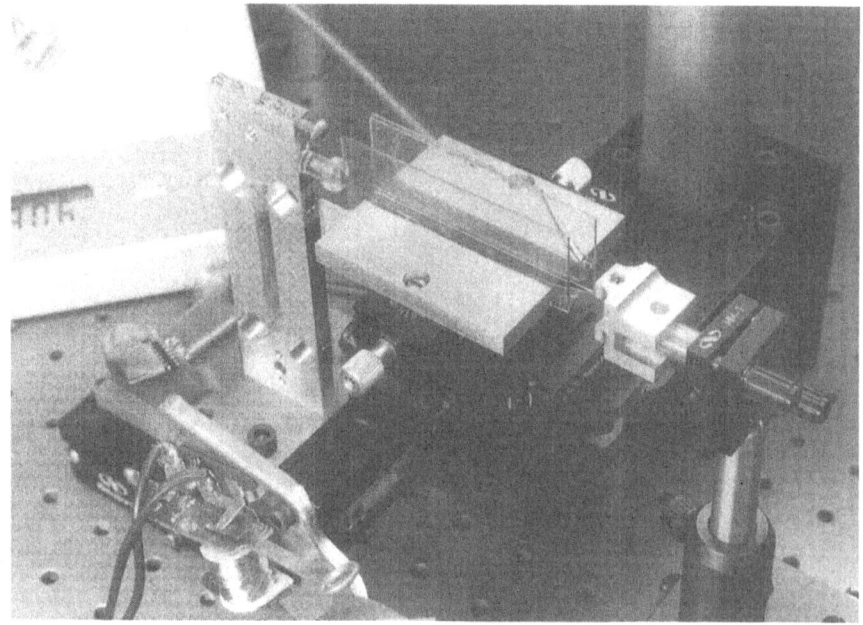

Fig. 8. The experimental apparatus.

length d) in jog mode with data taken following shakedown after each step. Such hard loading allows one to investigate regions in which the corresponding equilibria under soft (traction) loading are unstable.

After maximum displacement (minimum d) is achieved, the process is reversed to detect hysteresis and mode jumping to other equilibrium branches. Throughout, the system remains in the elastic range, but several contact events may occur. Prior to such runs, preliminary surveys of beam response were recorded on videotape under constant strain-rate conditions, and regions for detailed study in the stepping mode were selected from this data.

The flexural rigidity (EI) of the samples is a critical experimental parameter. It was estimated by clamping the beam at one end and measuring the frequency of small amplitude cantilever vibrations as a function of free length. Beam deflection was measured by shining a HeNe laser on the surface and imaging the laser spot on a photo-diode through an interference filter so that it was bisected by a knife edge. The resulting (slowly decaying) periodic signal was recorded on a digital oscilloscope that also determined the frequency. A least squares fit of the data, plotted as frequency squared against the inverse fourth power of free length according to simple linear theory (e.g., [Tho65], Section 8.5) provided the product EI. This method proved more accurate than direct measurement of the cross-sectional dimensions and use of tabulated Young's moduli.

In the following subsections we compare experimental results, in the form of load-displacement curves, with the theoretical and numerical calculations. The two specific cases illustrated in Figures 9 and 10 are typical of several runs.

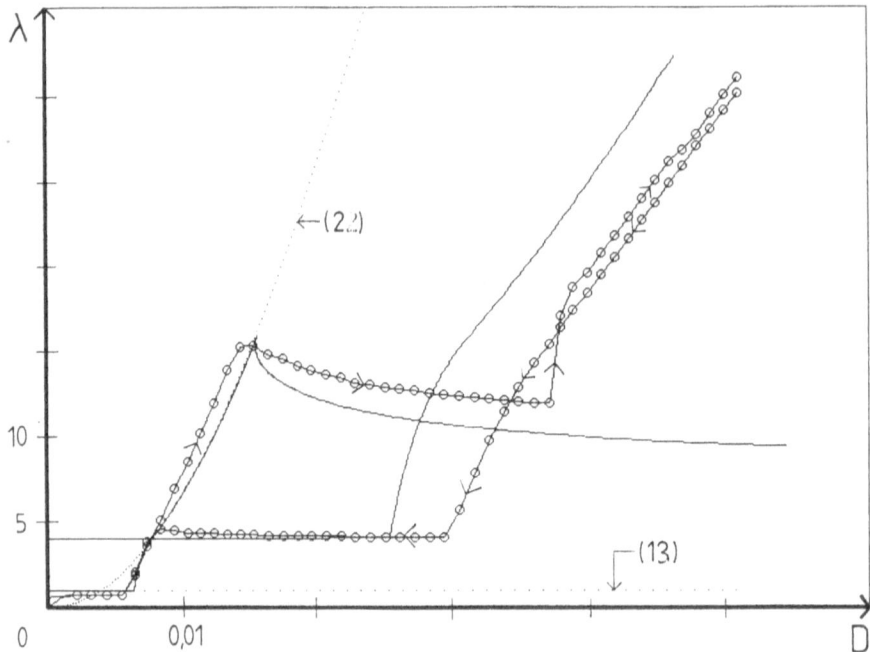

Fig. 9. Load-displacement curves for a pinned-pinned beam: $h = \pm 0.05$; solid curves with circles correspond to experiments. Arrows indicate direction of loading history. Numerical results are shown as solid curves. Taylor expansions of equations (13) and (22) are shown as dotted lines.

4.1. The Pinned-Pinned Case

Figure 9 shows results for a pinned-pinned beam with symmetric side-walls at nondimensionalized distances $h = \pm 0.05$. The load is normalized by the first Euler buckling load $\pi^2 EI/L^2$, with EI estimated as noted above, and the displacement has been corrected for balance flexibility and similarly normalized by the length L. As in the previous figures, we plot the end displacement ($D = 1 - d$). The comparison among analytical, numerical, and experimental critical values is summarized in Table 4.1. The beam has some unavoidable initial curvature (imperfection), with the result that (D) increases continually rather than exhibiting a sudden bifurcation. Nonetheless, it closely follows the theoretical prediction, with the load remaining essentially constant after initial buckling until the beam's center contacts one side-wall. At this point there is an abrupt change in the slope of the load/displacement curve, which then increases monotonically with the beam contacting and subsequently flattening against the wall. (The change from point to line contact is hard to detect visually, but appears to coincide with a slight upturn in slope in some experimental runs.)

This continues until the central section undergoes secondary buckling, after which the load decreases as the amplitude of the center buckle grows. At $D \approx 0.038$, before the central section contacts the opposite side-wall, a jump to the second mode occurs. The two contact portions then flatten and increase in length as D continues to increase until

Table 4.1. Critical values for the pinned-pinned case, $h = 0.05$.

	Analytical	Numerical	Experimental
Point Contact			
λ_1^p	1.003	1.003	0.7836
D_1^p	0.0062	0.00619	0.0053
λ_2^p	4.052	4.050	4.080
D_2^p	0.0251	0.0251	0.0258
Line Contact			
λ_1^l	3.985	3.983	3.620
D_1^l	0.075	0.0755	0.072
λ_2^l	15.75	15.76	14.23
D_2^l	0.0355	0.0309	0.0354
Secondary Buckling			
λ_1^{sb}	15.875	15.837	15.254
D_1^{sb}	0.0155	0.0154	0.0151

the limit of travel of the motor is reached. Reversing displacements, the beam follows the second mode path down past the line/point contact transition until contact is lost at one wall, after which it remains close to the almost horizontal unconstrained second mode branch, with $\lambda \approx 4\pi^2$. Along this branch the experimental shape is a mixture of first and second modes, initially predominantly second, but with the relative contribution of the first mode growing so that the contact point moves toward the center until the path rejoins the first mode branch, which it then follows back to $(0, 0)$.

Comparison with theory is good on the first mode branch, including the portion on which the load drops after secondary buckling occurs. The second branch, including points (d_2^p, λ_2^p) and (d_2^l, λ_2^l), is however significantly displaced to the left of the experimental curve, so that, while the λ values compare well, the d values are in error. We believe that this is primarily due to asymmetry of the specimen. Recalling the sensitive dependence on h revealed in the denominator of Equation (22), it seems reasonable that initial curvature should have a large effect via its influence on the effective value of h; indeed, a series of experiments with identically spaced constraint walls laterally shifted with respect to the beam centerline revealed variations of up to 15% in the position of this portion of the second mode branch.

4.2. The Clamped-Pinned Case

Figure 10 shows typical load-displacement data from a clamped-pinned experiment. Here $h = 0.053 < h_2^l$, and we observe both single and double point contact and line contact configurations. As noted in Section 2.2, after point contact is established in the first mode at $D \approx 0.0075$, the solution follows a smooth branch in which the moment $\theta'(0)$ at the clamped end drops to zero and then changes sign, so that a second critical point, nearer the clamped end, appears at $D \approx 0.011$, in the almost flat portion of the experimental curve. This second critical point first contacts the wall on its side at

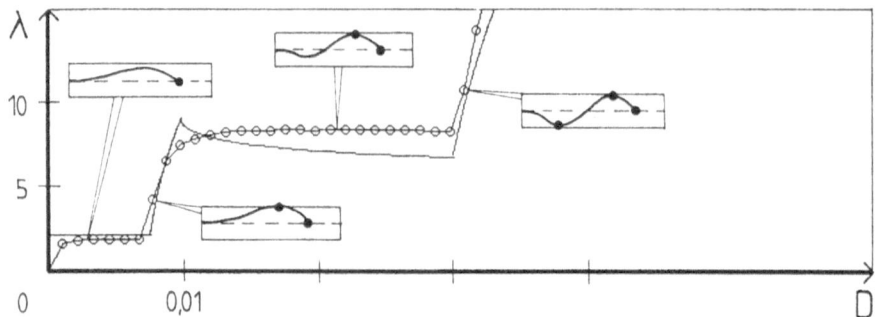

Fig. 10. Load-displacement curves for a clamped-pinned beam: $h = \pm 0.053$. Solid curves with circles correspond to experiments. Numerical results are shown as solid curves. Filled circles on inset diagrams indicate contact points and pin support.

$D \approx 0.030$, after which the beam's shape is essentially that of the second clamped-pinned mode. Thereafter the curve rises steeply, with line contact being established at both contact locations above about $D = 0.035$. Unloading, the beam follows the same path without hysteresis.

Also shown on Figure 10 are numerically computed load-displacement curves. The portions of the curve below $D \approx 0.01$ come from the first mode computation described in Section 3.2, and correspond to the theory of Section 2.2. Plotting of this branch terminates at $D, \lambda \approx 0.011, 8.86$, where the moments at the clamped end and the contact point drop to zero. As described in Section 2.2, the branch continues with a region of line contact, but, taking our cue from the experiment, here we switch to an "asymmetrical" second mode branch with two critical points, which follows a gently descending path to establishment at $D, \lambda \approx 0.029, 6.83$ of the second contact point nearer the clamped end, after which the second mode branch is followed. Key critical values from experiment and numerical computation from the model are summarised in Table 2 (closed form theoretical results are not available for this case).

To understand how the asymmetric first/second mode branch in Figure 10 differs from the "pure" second mode branch, it is necessary to note that the latter contains, at $D, \lambda \approx$

Table 4.2. Critical values for the clamped-pinned case, $h = 0.053$.

	Numerical	Experimental
Point Contact		
λ_1^p	2.051	1.866
D_1^p	0.0073	0.0065
λ_2^p	6.828	8.198
D_2^p	0.0293	0.0294
Line Contact		
λ_1^l	8.86	7.93
D_1^l	0.011	0.0095

6.089, 0.01602, a point where contact first occurs at the critical point nearest the clamped end, whereas along the branch shown contact is always at the critical point nearest the pinned end. The pure second mode branch is intersected at D, $\lambda \approx 6.089, 0.01602$ by a further asymmetric branch containing "unphysical" states with point contact maintained by forces directed *towards* the walls. The complete bifurcation diagram therefore contains at least three double-critical-point branches, only one of which—the one we believe relevant to experiment—is shown in Figure 10.

5. Conclusions

The results reported above demonstrate an acceptable match between experiment and theory. The major "uncontollable" factors in the experiments, omitted from our model, are friction with the side-walls and imperfections (asymmetries) of the specimens. A simple static friction model would give lateral forces, proportional to the normal contact forces μ, which could clearly affect equilibrium states and critical loads and displacements, especially those corresponding to secondary buckling of the flat segment in pinned-pinned experiments. We therefore believe that the primary responsibility for experimental loads generally exceeding theoretical ones lies with friction. We have already commented in Section 4.1 (Figure 9) on the influence of asymmetries in shifting load-displacement curves laterally.

In comparing experiment and theory above, we have included only stable and "physical" configurations, since unstable equilibria cannot be realised experimentally. However, unstable and even unphysical equilibria, such as those requiring *negative* constraint forces (forces directed *towards* the walls) and those involving penetration of walls near points of tangency, are relevant in developing the full global understanding of the interrelations and bifurcations of branches in the load-displacement diagram for any given case. The model predicts *all* equilibria and it is therefore of interest to compute as many as possible and determine their stability properties in order to interpret observations such as mode-jumping and hysteresis.

To this end, Figure 11 shows a large number of secondary branches captured numerically for the pinned-pinned beam. These branches originate from two points labelled 11 and 14: where the primary second and third mode branches make simultaneous multiple contacts with the constraint walls. Segments of some of these branches (1, 2, 3, 4, 10, 11, 12) were shown earlier in Figure 9; another one (13) was observed in another experiment. These branches, corresponding to asymmetric equilibria, are absent in the linearized model discussed in Section 2.3. As $h \to 0$, the number of asymmetric branches grows (they become admissible for higher and higher buckling modes); however, the difference between them decreases. Finally, as we reach the limit corresponding to the linear model, all those belonging to the same mode collapse to a single symmetric solution. Similar observations apply to the clamped-pinned case. It is interesting to ask, which of these asymmetrical branches are stable and hence physically relevant?

The difficulty of stability analysis arises from the fact that, while the problem is continuous, the relevant functions are nonsmooth due to contact forces. Thus the standard tools for stability analysis of continua are useless. Nor can techniques for discrete structures be applied directly either. We envisage a mixed approach to this problem, which we

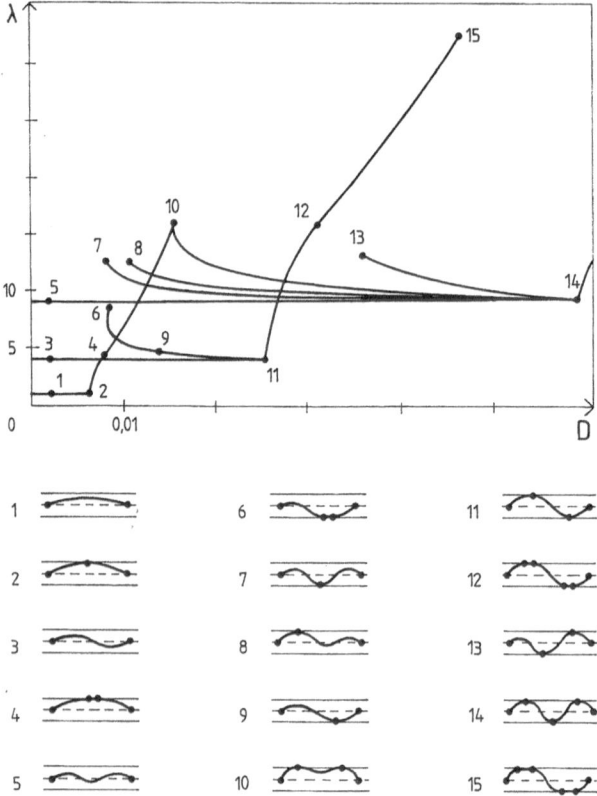

Fig. 11. Numerical load-displacement curves for the pinned-pinned beam: $h = \pm 0.05$; Numbered points on branches correspond to physical shapes indicated.

hope to treat in a forthcoming paper. Here we outline briefly an algorithm which yields some insight even without performing any computations.

As in the model developed in Section 2, we regard the beam as a structure composed of elements, starting and ending at contact points and the endpoints of the beam. At contact points a *node* connects two adjacent elements. A *necessary* condition for the stability of the structure is the stability of each element individually, with nodes fixed. If any element is in an unstable configuration, the structure as a whole is also unstable. The stability of each element can be investigated by standard tools of continuum mechanics, and even without detailed computation, in some cases it can be derived by observation.

If the necessary condition for stability is satisfied, one can seek a sufficient condition for global stability. In this step we regard the beam as an assembly of nodes connected by nonlinear springs, transmitting forces as well as moments, whose stiffnesses are determined via the elemental BVPs. This structure has $2n$ degrees of freedom, n denoting the number of contact points, where each point can *slide* and/or *roll* along the constraining walls. Stability of this discrete structure is determined by the eigenvalues of its $2n \times 2n$

Hessian matrix, which can be found numerically. Positive-definiteness of the Hessian supplies the *sufficient* condition for stability. It is important to note that analysis of the Hessian is meaningful only if the individual segments are stable.

Based on the necessary condition (stability of individual segments), in Figure 11 we can tell that branches 3, 5, and 8 are unstable (for 3 and 5 this is a trivial observation, in the case of 8, a nontrivial one). The stability of the remaining equilibria requires calculation of the Hessian, a substantial task. We hope to address this, and the stability question in general, in a future paper. We also intend to investigate the interrelations of asymmetrical and unphysical branches of equilibria such as those shown in Figure 11 for both pinned-pinned and clamped-pinned cases, and to perturb the symmetric, parallel side-walls considered thus far, in an attempt to derive more complete unfolding and bifurcation diagrams.

In closing, we remark that it is difficult to make a clean distinction between "analytical" and "numerical" techniques in this problem. The theory of Section 2 results in nonlinear algebraic equations containing elliptic integrals. The numerical technique of Section 3 also leads to systems of nonlinear algebraic equations containing special functions. In the latter case the function values are obtained by integrating an IVP for a finite segment; the solutions of the relevant ODEs being, naturally, elliptic functions. Thus, while the two approaches presented here emphasize different aspects, it seems difficult to place them in entirely different categories.

Appendix: Computational Details

The Pinned-Pinned Case

We sketch some details of explicit solutions to the BVPs (3), (6)–(10). We deal first with the pinned-pinned case of (6), for which the first integral (5) reduces to

$$\frac{\theta'^2}{2} - \lambda \cos \theta = -\lambda \cos \theta_0, \tag{44}$$

where $\theta_0 = \theta(0) = -\theta(1)$, by symmetry of reflection about the midpoint $s = \frac{1}{2}$. From (44) derives the quadrature,

$$\int_{\theta_0}^{\theta(s)} \frac{d\theta}{2\sqrt{\lambda[\sin^2(\frac{\theta_0}{2}) - \sin^2(\frac{\theta}{2})]}} = \int_0^s d\sigma, \tag{45}$$

which, via the transformation $\sin(\frac{\theta}{2}) = \sin(\frac{\theta_0}{2}) \sin(\phi)$, may be evaluated as

$$\int_{\phi(s)}^{\frac{\pi}{2}} \frac{d\phi}{\sqrt{1 - k^2 \sin^2 \phi}} = F(\pi/2; k) - F(\phi(s); k) = \sqrt{\lambda} s, \tag{46}$$

where F denotes the incomplete elliptic integral of the first kind and $k = \sin(\frac{\theta_0}{2}) \in [0, 1]$ is the elliptic modulus [AS65]. To satisfy the BVP (3) and (6) we require $\theta'(1) = 0$, implying, via symmetry, that $\theta(s)$ has internal zeros at $s = \frac{1+2j}{2n}$; $j = 0, \ldots, n - 1$

for the n-th buckling mode. When $\theta = 0$ (resp. θ_0), $\phi = 0$ (resp. $\frac{\pi}{2}$), and, from (46), it therefore follows that

$$K(k_n) = \frac{\sqrt{\lambda}}{2n},$$ (47)

where $K(k) = F(\frac{\pi}{2}; k)$ is the complete elliptic integral of the first kind and $\theta_{0n} = 2\sin^{-1}(k_n)$ is the slope at the end of the beam subject to axial load λ, in the n-th mode. From the limit $K(0) = \frac{\pi}{2}$ we derive (14).

Appealing to symmetry and using the relation

$$\frac{d\theta}{ds} = 2\sqrt{\lambda\left[\sin^2\left(\frac{\theta_0}{2}\right) - \sin^2\left(\frac{\theta}{2}\right)\right]}$$

from (44), the displacement constraint (12) may be written as

$$d = \frac{2n}{\sqrt{\lambda}}\int_0^{\theta_0} \frac{\cos\theta \, d\theta}{2\sqrt{\sin^2(\frac{\theta_0}{2}) - \sin^2(\frac{\theta}{2})}}.$$ (48)

After trignometrical manipulation and use of the same transformation as above, this integral evaluates as

$$\int_0^{\frac{\pi}{2}}\left[2\sqrt{1 - k^2\sin^2\phi} - \frac{1}{\sqrt{1 - k^2\sin^2\phi}}\right]d\phi = [2E(k) - K(k)],$$ (49)

where $E(k)$ is the complete elliptic integral of the second kind. From (47), we then obtain the equation

$$d = \left[\frac{2E(k)}{K(k)} - 1\right],$$ (50)

from which $d = d_n$ may be found directly for each $k = k_n$ satisfying equation (47). Alternatively, since $E(k)$ monotonically decreases and $K(k)$ monotonically increases with k, (50) may be solved uniquely for $k = k_d$ in terms of $d \in (-1, 1)$, and an infinite sequence of loads

$$\lambda_n(d) = 4n^2(K(k_d))^2$$ (51)

consistent with this displacement in each given mode, may be found from (47). These provide direct solutions to the displacement BVP. Note that, as $\theta_0 \to 0$ ($k \to 0$), E and K both approach $\frac{\pi}{2}$, giving $d = 1$, as expected for the undeformed rod. As $\theta_0 \to \pi$ ($k \to 1$), $E \to 1$ and $K \to \infty$, so that $d \to -1$, again as expected for the "everted" state (cf. [Lov27]). It is interesting to note that, from (50), the end angle $\theta_0 = 2\sin^{-1}(k_d)$ for a given displacement d is identical in *every* mode.

For relatively small k (θ_0), the truncated power series for the elliptic integrals

$$K(k) = \frac{\pi}{2}\left[1 + \frac{k^2}{4} + \frac{9k^4}{64} + \mathcal{O}(k^6)\right] \quad \text{and} \quad E(k) = \frac{\pi}{2}\left[1 - \frac{k^2}{4} - \frac{3k^4}{64} + \mathcal{O}(k^6)\right]$$
(52)

provide the simple approximation

$$d = 1 - k^2 - \frac{k^4}{8} + \mathcal{O}(k^6)$$

to (50). Using this in (47), we obtain

$$\lambda = n^2 \pi^2 \left[1 + \frac{(1-d)}{2} + \frac{9(1-d)^2}{32} + \mathcal{O}(|1-d|^3) \right]. \tag{53}$$

The condition for first contact in the n-th mode is derived from (3) and (8) with $\mu = 0$, $s_a = 0$, and $s_b = \frac{1}{2n}$. Substituting for $\sin \theta$ from the ODE and integrating gives

$$\left| y \left(\frac{1}{2n} \right) \right| = \left| \int_0^{\frac{1}{2n}} \sin \theta(s) \, ds \right| = \left| \int_0^{\frac{1}{2n}} \frac{\theta''(s)}{\lambda} \, ds \right| = \left| \frac{\theta'(\frac{1}{2n})}{\lambda} \right| = h, \tag{54}$$

or, via (44),

$$\sin \left(\frac{\theta_{0n}}{2} \right) = k_n = \frac{\sqrt{\lambda} h}{2}. \tag{55}$$

Satisfying (55) and (47) simultaneously and using (50) yields (15).

The pinned-contact problem with $\mu \neq 0$ can be reformulated by letting $\psi = \theta - \theta_a$, $\tan \theta_a = \frac{\mu}{\lambda}$, so that (8) becomes

$$\psi'' + \tilde{\lambda} \sin \psi = 0; \qquad \tilde{\lambda} = \sqrt{\lambda^2 + \mu^2},$$
$$\psi'(0) = 0, \qquad \psi(\tfrac{1}{2n}) = -\theta_a, \tag{56}$$

with displacement constraints

$$x(1) = \int_0^{\frac{1}{2n}} \cos(\psi + \theta_a) \, ds = \frac{\lambda}{\tilde{\lambda}} \int_0^{\frac{1}{2n}} \cos \psi \, ds - \frac{\mu}{\tilde{\lambda}} \int_0^{\frac{1}{2n}} \sin \psi \, ds = \frac{d}{2n}, \tag{57}$$

$$y(1) = \int_0^{\frac{1}{2n}} \sin(\psi + \theta_a) \, ds = \frac{\lambda}{\tilde{\lambda}} \int_0^{\frac{1}{2n}} \sin \psi \, ds + \frac{\mu}{\tilde{\lambda}} \int_0^{\frac{1}{2n}} \cos \psi \, ds = h, \tag{58}$$

and we may appeal to the theory for the pinned-pinned case. Following the computations of (45)–(46) and (48)–(49), and recognising that the desired solutions make between one and two quarter turns, we obtain

$$K(k) + F(\phi_a; k) = \frac{\sqrt{\tilde{\lambda}}}{2n} \tag{59}$$

and

$$\int_0^{\frac{1}{2n}} \cos \psi \, ds = \frac{1}{\sqrt{\tilde{\lambda}}} [2E(k) - K(k) + 2E(\phi_a; k) - F(\phi_a; k)], \tag{60}$$

$$\int_0^{\frac{1}{2n}} \sin \psi \, ds = -\frac{\psi'(s)}{\tilde{\lambda}} \Big|_0^{\frac{1}{2n}} = \frac{2k}{\sqrt{\tilde{\lambda}}} \left[\cos \phi \, \Big|_{\frac{\pi}{2}}^{\phi_a} \right] = \frac{2k}{\sqrt{\tilde{\lambda}}} \cos \phi_a$$

$$= \frac{2}{\tilde{\lambda}} \sqrt{\frac{(2k^2 - 1)\tilde{\lambda} + \lambda}{2}}, \tag{61}$$

where $F(\phi_a; k)$ and $E(\phi_a; k)$ are incomplete elliptic integrals of the first and second kinds, $k = \sin(\frac{\psi_0}{2})$, and $\sin \phi_a = \sin(\frac{\theta_a}{2})/\sin(\frac{\psi_0}{2})$. Hence, from (57)–(58) we have

$$\lambda d + 2n\mu h = 4n\sqrt{\tilde{\lambda}}[E(k) + E(\phi_a; k)] - \tilde{\lambda} \text{ and} \tag{62}$$

$$2n\lambda h - \mu d = 4n\sqrt{\frac{(2k^2 - 1)\tilde{\lambda} + \lambda}{2}}. \tag{63}$$

Simultaneous solution of (59) and (62)–(63) yields k, λ, and μ in terms of h, d and n. Note that, when $\mu = 0$, these expressions reduce to (15).

When the resultant of the axial and lateral forces passes through pin and contact point, we have

$$d\mu = 2nh\lambda \quad \text{and} \quad \tan \theta_a = \tan \psi_0 = \frac{2nh}{d}, \tag{64}$$

so that $\phi_a = \frac{\pi}{2}$, $\tilde{\lambda} = \lambda\sqrt{d^2 + 4n^2h^2}/d$, and (59) and (62)–(63) reduce to (18)–(19).

When line contact occurs, we use the geometric condition analogous to (64),

$$\tan \theta_a = \frac{h}{\Delta} = \frac{\mu}{\lambda}, \tag{65}$$

where $\Delta = \int_0^{s_1} \cos \theta(s) \, ds$ is the axial distance subtended by the end segment. As above, this implies that

$$\sin\left(\frac{\theta_a}{2}\right) = k = \sqrt{\frac{B - \Delta}{2B}}; \quad B = \sqrt{\Delta^2 + h^2}. \tag{66}$$

We also have the overall displacement constraint

$$d = 1 + 2n(\Delta - s_1). \tag{67}$$

Treating the end segment as the first buckling mode of a pinned-pinned beam of length s_1, from (50) we have

$$\sqrt{\Delta^2 + h^2} = \left[\frac{2E(k)}{K(k)} - 1\right] s_1, \tag{68}$$

and, via (65),

$$\lambda = \frac{\Delta}{\sqrt{\Delta^2 + h^2}} \left[\frac{2K(k)}{s_1}\right]^2; \quad \mu = \frac{h}{\sqrt{\Delta^2 + h^2}} \left[\frac{2K(k)}{s_1}\right]^2. \tag{69}$$

(Note that the constraint $\int_0^{s_1} \sin \theta(s) \, ds = h$ is satisfied automatically in this formulation.) Given d, h, and n, (66)–(68) may be solved uniquely for s_1 and Δ and the loads λ and μ found from (69). In this way the branch of solutions may be followed (with decreasing s_1) from $s_1 = 1/2n$.

As above, we may use the approximations (52) in (68)–(69) to estimate λ and other quantities such as s_1 in terms of d. After some manipulation, we obtain

$$s_1 \approx \frac{h}{2k(1 - k^2)^{3/2}}, \quad k \approx \frac{1 - d}{3nh},$$

and

$$\lambda = \frac{4\pi^2}{9n^2h^4}(1-d)^2\left[1 - \frac{(1-d)^2}{2n^2h^2} + \mathcal{O}(|1-d|^4)\right].$$ (70)

Secondary buckling of the central straight segment is governed by (9), with critical load

$$\lambda_n^{sb} = \frac{4\pi^2}{(s_2 - s_1)^2}.$$ (71)

Using $s_2 = \frac{1}{n} - s_1$, matching the axial load with (69), and manipulation of (66) and (68) lead to the expressions in (23)–(24).

The Clamped-Pinned Case

We now turn to the clamped-pinned case described by (7) and (12). This problem is reformulated much as the pinned-contact case (8) treated above. Letting $\psi = \theta + \theta_a$, $\tan\theta_a = \frac{\mu}{\lambda}$, (7) becomes

$$\psi'' + \tilde{\lambda}\sin\psi = 0; \qquad \tilde{\lambda} = \sqrt{\lambda^2 + \mu^2},$$
$$\psi(0) = \theta_a, \qquad \psi(1) = 0,$$ (72)

with displacement constraints

$$x(1) = \int_0^1 \cos(\psi - \theta_a)\,ds = \frac{\lambda}{\tilde{\lambda}}\int_0^1 \cos\psi\,ds + \frac{\mu}{\tilde{\lambda}}\int_0^1 \sin\psi\,ds = d,$$ (73)

$$y(1) = \int_0^1 \sin(\psi - \theta_a)\,ds = \frac{\lambda}{\tilde{\lambda}}\int_0^1 \sin\psi\,ds - \frac{\mu}{\tilde{\lambda}}\int_0^1 \cos\psi\,ds = 0.$$ (74)

From Figure 3(c), for buckling in the n-th mode, we are interested in solutions which make between $2n$ and $2n + 1$ quarter-circuits of the origin in the phase plane. Consequently, as above, we deduce from (46) that

$$(2n + 1)K(k) - F(\phi_a; k) = \sqrt{\tilde{\lambda}},$$ (75)

and

$$\int_0^1 \cos\psi\,ds = \frac{1}{\sqrt{\tilde{\lambda}}}\{(2n + 1)[2E(k) - K(k)]$$
$$- [2E(\phi_a; k) - F(\phi_a; k)]\},$$ (76)

$$\int_0^1 \sin\psi\,ds = \frac{2}{\tilde{\lambda}}\sqrt{\frac{(2k^2 - 1)\tilde{\lambda} + \lambda}{2}}.$$ (77)

These expressions in turn lead to:

$$\lambda d = 2\sqrt{\tilde{\lambda}}[(2n+1)E(k) - E(\phi_a; k)] - \tilde{\lambda}, \tag{78}$$

$$\mu d = 2\sqrt{\frac{(2k^2 - 1)\tilde{\lambda} + \lambda}{2}}, \tag{79}$$

where $k = \sin(\frac{\psi_0}{2})$ and $\sin\phi_a = \sin(\frac{\theta_a}{2})/\sin(\frac{\psi_0}{2})$, as before.

The condition for first contact is obtained from the relations above by additionally requiring

$$\int_0^{s_1} \sin(\psi - \theta_a)\, ds = \frac{\lambda}{\tilde{\lambda}}\int_0^{s_1} \sin\psi\, ds - \frac{\mu}{\tilde{\lambda}}\int_0^{s_1} \cos\psi\, ds = h, \tag{80}$$

where $s = s_1$ denotes the first extremum $\theta'(s) = 0$. Referring to the phase plane of Figure 3(b), we compute

$$\int_0^{s_1} \cos\psi\, ds = \frac{2}{\sqrt{\tilde{\lambda}}}\{[2E(k) - K(k)] - [2E(\phi_a; k) - F(\phi_a; k)]\}, \tag{81}$$

$$\int_0^{s_1} \sin\psi\, ds = -\frac{4k}{\sqrt{\tilde{\lambda}}}[\cos\phi \,|_{\phi_a}^{\frac{\pi}{2}}] = \frac{4}{\tilde{\lambda}}\sqrt{\frac{(2k^2 - 1)\tilde{\lambda} + \lambda}{2}}. \tag{82}$$

Substituting from (75) and (78)–(79), (80)–(82) lead to the condition

$$\mu[2E(k) - K(k)] = \frac{\tilde{\lambda}^{\frac{3}{2}} h}{4n}. \tag{83}$$

Equations (27)–(30) are simply (75), (78)–(79), and (83).

When internal moments are zero at the contact point (or throughout the contact region), the resultant of axial and lateral end loads must pass through the boundary points $s = 0$, s_1 and $s = s_2$, 1. We define the geometrical conditions

$$\tan\theta_j = \frac{h}{\Delta_j} = \frac{\mu_j}{\lambda}, \tag{84}$$

where $\Delta_1 = \int_0^{s_1} \cos\theta(s)\, ds$ and $\Delta_2 = \int_{s_2}^1 \cos\theta(s)\, ds$. Assembling the elements of the limiting cases of figures 3(d), (e), and (c) and appealing to the pinned-pinned theory as in (66)–(69), we obtain

$$K(k_1) = \frac{\sqrt{\tilde{\lambda}_1} s_1}{4}; \qquad K(k_2) = \frac{\sqrt{\tilde{\lambda}_2}(1 - s_2)}{2}; \tag{85}$$

$$B_1 = \left[\frac{2E(k_1)}{K(k_1)} - 1\right] s_1; \qquad B_2 = \left[\frac{2E(k_2)}{K(k_2)} - 1\right](1 - s_2), \tag{86}$$

and

$$k_j = \sqrt{\frac{B_j - \Delta_j}{2B_j}}; \qquad B_j = \sqrt{\Delta_j^2 + h^2}; \qquad \tilde{\lambda}_j = \sqrt{\lambda^2 + \mu_j^2}, \tag{87}$$

for $j = 1, 2$, along with the overall axial displacement constraint,

$$d = \Delta_1 + \Delta_2 + (s_2 - s_1). \tag{88}$$

Note that the clamped-contact element is effectively in the second pinned-pinned mode, and the contact-pinned element in the first. From (84) and (88), we have

$$\tilde{\lambda}_j = \lambda \frac{B_j}{\Delta_j}, \tag{89}$$

and using this and the other definitions of (88), we may regard (85)–(86) and (88) as five equations to be solved simultaneously for the unknown s_j, Δ_j and λ, given d and h.

The two critical loads occur when $s_1 = s_2$ and when

$$\lambda = \frac{4\pi^2}{(s_2 - s_1)^2}, \tag{90}$$

leading to the conditions specified in (31)–(36). Here, as in Section 2.2, we give expressions for only the first mode.

Acknowledgments

This work was supported by OTKA grants T015851 and F021307 and MKM grant 815 (GD), AMFK Grant 483/96, DoE Grant DE-FG02-95ER25238 (PH), and an educational grant from Hercules (BR). Tim Healey introduced us to Feodosyev's work. Ralf Wittenberg coerced Maple into correctly solving some of the elliptic function relations in Section 2. John Maddocks provided several useful comments.

References

[AG90] E. L. Allgower and K. Georg. *Numerical Continuation Methods: An Introduction.* Springer-Verlag, Berlin, 1990.

[Ant95] S. S. Antman. *Nonlinear Problems of Elasticity.* Springer-Verlag, New York, 1995.

[AS65] M. Abramowitz and I. A. Stegun. *Handbook of Mathematical Functions.* Dover Publications, New York, 1965.

[DG95] G. Domokos and Zs. Gáspár. A global, direct algorithm for path-following and active static control of elastic bar structures. *Int. J. Struct. Mach.*, 23 (4):549–571, 1995.

[Dom94] G. Domokos. Global description of elastic bars. *Zeitschr. Angew. Math. Mech.*, 74 (4):T289–T291, 1994.

[Eul44] L. Euler. Additamentum I de curvis elasticis, methodus inveniendi lineas curvas maximi minimivi proprietate gaudentes. In *Opera Omnia I, vol. 24*, 231–197. Bousquet, Lausanne, 1744.

[Feo77] V. I. Feodosyev. *Selected Problems and Questions in Strength of Materials.* Mir, Moscow, 1977. Translated from the Russian by M. Konyaeva.

[DHRS] P. Holmes, G. Domokos, J. Schmitt, and I. Szeberényi. Constrained Euler buckling: An interplay of computation and analysis. *Comp. Meth. Appl. Mech. Eng.*, 170 (3–4):175–207, 1999.

[HSD] P. Holmes, J. Schmitt, and G. Domokos. Constrained Euler buckling: Line contact solutions. *Solid Mechanics and its Applications*, vol. 63, pp. 149–158, ed. by F. C. Moon.

Kluwer Academic Publishers, 1999. *Proc. IUTAM Symposium on New Applications of Nonlinear and Chaotic Dynamics in Mechanics*, Cornell University, July 27–August 1, 1997.

[KF73] J. B. Keller and J. E. Flaherty. Contact problems involving a buckled elastica. *SIAM J. Appl. Math.*, 24:215–225, 1973.

[KFR72] J. B. Keller, J. E. Flaherty, and S. I. Rubinow. Post buckling behavior of elastic tubes and rings with opposite sides in contact. *SIAM J. Appl. Math.*, 23:446–455, 1972.

[Lov27] A. E. H. Love. *A Treatise on the Mathematical Theory of Elasticity*. Cambridge University Press, Cambridge, UK, 1927. Reprinted by Dover Publications, New York.

[Mad84] J. H. Maddocks. Stability of nonlinearly elastic rods. *Arch. Rat. Mech. Anal.*, 85:311–354, 1984.

[Mad87] J. H. Maddocks. Stability and folds. *Arch. Rat. Mech. Anal.*, 99:301–328, 1987.

[MK95] L. Mahadevan and J. B. Keller. Periodic folding of thin sheets. *SIAM J. Appl. Math.*, 55:1609–1624, 1995.

[Rik79] E. Riks. An incremental approach to the solution of snapping and buckling problems. *Int. J. Solids Struct.*, 15:529–551, 1979.

[Tho65] W. T. Thomson. *Vibration Theory and Applications*. Prentice Hall, New York, 1965.

Continuity Properties and Global Attractors of Generalized Semiflows and the Navier-Stokes Equations

J.M. Ball
Mathematical Institute, University of Oxford, 24–29 St. Giles, Oxford OX1 3LB, UK

Received 28 October 1996; revised manuscript accepted for publication 16 January 1997
Communicated by Jerrold Marsden

Summary. A class of semiflows having possibly nonunique solutions is defined. The measurability and continuity properties of such generalized semiflows are studied. It is shown that a generalized semiflow has a global attractor if and only if it is pointwise dissipative and asymptotically compact. The structure of the global attractor in the presence of a Lyapunov function, and its connectedness and stability properties are studied. In particular, examples are given in which the global attractor is a single point but is not Lyapunov stable.

The existence of a global attractor for the 3D incompressible Navier-Stokes equations is established under the (unproved) hypothesis that all weak solutions are continuous from $(0, \infty)$ to L^2.

This paper is dedicated to the memory of Juan-Carlos Simo

1. Introduction

Generalized semiflows are an abstraction of autonomous dynamical systems for which there may be more than one solution corresponding to given initial data. The need for a theory of such systems arises for various reasons. First, there may be genuine nonuniqueness of solutions. Second, solutions may not be known to be unique (as, for example, for certain semilinear wave equations with high power nonlinearities, or for the incompressible Navier-Stokes equations in three space dimensions, an example studied in some detail in the paper). Third, there may be free parameters or controls that are not specified and lead to various possible solutions. The paper discusses the measurability and continuity properties of generalized semiflows on a metric space X, and their global attractors.

There are various possible ways of abstracting dynamical systems with non-unique solutions. One method (see Sell [38]) is to recover uniqueness of solutions by working in a space of semitrajectories $\varphi: [0, \infty) \to X$ and defining a corresponding semiflow $T(\cdot)$

by $T(t)\varphi = \varphi^t$, for $t \geq 0$, where $\varphi^t(\tau) := \varphi(t + \tau)$. An interesting example of the use of this method is the recent proof by Sell [39] of the existence of a global attractor for the 3D incompressible Navier-Stokes equations. (For further results see Chepyzhov & Vishik [14].) However, a disadvantage is that the direct connection with the evolution of the system in the 'physical' state space is lost. (In fact, the existence of a global attractor for the Navier-Stokes equations in the original phase space remains an open problem; the existence is proved under a continuity hypothesis on solutions in this paper.) A second method is to consider a set-valued trajectory $t \mapsto T(t)z$ in which $T(t)z$ consists of all possible points reached at time t by solutions with initial data z. This approach has been taken, for example, by Barbashin [10], Budak [13], Bronstein [12], Minkevic [30], Roxin [36], [35], Szego & Treccani [41], Babin & Vishik [3], Babin [2], and Mel'nik [29]. However, the disadvantage of this method is that it is not phrased directly in terms of solutions; in fact in some situations one has to recover solutions by selecting suitable regular paths from the sets $T(t)z$.

The approach taken in this paper is more closely related to the second method than the first, and takes as the primitive objects the solutions themselves; it is an adaptation of that in [8], [9]. A generalized semiflow is defined in Section 2 to be a family of maps $\varphi: [0, \infty) \to X$ satisfying axioms relating to existence, time translation, concatenation, and upper-semicontinuity with respect to initial data. It is shown in Theorem 2.1 that, under a mild technical hypothesis, for generalized semiflows strong measurability of solutions with respect to time implies their continuity on $(0, \infty)$, and hence (Theorem 2.2) that the upper-semicontinuity with respect to initial data is uniform on compact subsets of $(0, \infty)$. These theorems generalize corresponding results for semiflows due to Chernoff & Marsden [16] and the author [6], [7].

In Section 3 global attractors for generalized semiflows are studied. It is shown in Theorem 3.3 that a generalized semiflow has a global attractor if and only if it is point dissipative and asymptotically compact. This result generalizes those for semiflows of Hale [21] and of Ladyzhenskaya [27]. Related results in the context of set-valued semiflows have recently been announced by Mel'nik [29]. In Section 4 the connectedness of the global attractor is proved (Corollary 4.3) provided X is connected and Kneser's property holds; that is, the set $T(t)\{z\}$, consisting of all points $\varphi(t)$ for solutions φ with $\varphi(0) = z$, is connected. In Section 5 the case of an asymptotically compact generalized semiflow with a Lyapunov function is considered. As for semiflows the point dissipativeness can be verified by showing that the set of rest points is bounded (Theorem 5.1).

Section 6 is motivated by results of Sell & You [40] on the Lyapunov stability of attractors. In order to treat the case of a semiflow whose solutions are not necessarily continuous up to $t = 0$, Sell & You were led to change the usual definition of Lyapunov stability. We show by means of two examples, one finite- and the other infinite-dimensional, that without such a change in the definition the global attractor of such a semiflow need not be Lyapunov stable, even if the global attractor consists of a single point. We also give a positive result (Theorem 6.1) giving hypotheses under which the global attractor of a generalized semiflow is in fact stable.

The theory is applied in Section 7 to the case of the 3D incompressible Navier-Stokes equations. The main results are that (Proposition 7.4) weak solutions satisfying an energy inequality form a generalized semiflow, in the usual phase-space H consisting of L^2 vector-fields with zero divergence, if and only if all weak solutions are continuous from $(0, \infty) \rightarrow H$, and that (Theorem 7.6) under this hypothesis there is a global attractor in H. Since weak solutions of the Navier-Stokes equations are not known to be continuous in time, these results may turn out to be vacuous. However, it is notable that we assume neither additional regularity nor the uniqueness of weak solutions.

A further application of the theory to damped semilinear wave equations (see Example 2.3) will appear in [5]. This example in fact motivated this paper. This is a more substantial application of the theory (though to a perhaps less interesting example), since the corresponding generalized semiflows are asymptotically compact but not compact, whereas the generalized semiflow for the Navier-Stokes equations (under the assumed continuity of solutions) is compact. For earlier work on attractors for these equations also not assuming uniqueness of solutions, see Babin & Vishik [3].

2. Generalized Semiflows

Let X be a metric space (not necessarily complete) with metric d. We write $B(a, r)$ for the open ball centre $a \in X$ and radius r. If $C \subset X$ and $b \in X$, we set $\rho(b, C) := \inf_{c \in C} d(b, c)$. If $B \subset X, C \subset X$, we set

$$\text{dist}(B, C) := \sup_{b \in B} \rho(b, C),$$

and define the *Hausdorff distance* $d_H(B, C)$ by

$$d_H(B, C) = \max\{\text{dist}(B, C), \text{dist}(C, B)\}.$$

If $C \subset X$ and $\varepsilon > 0$, we write

$$N_\varepsilon(C) := \{z \in X \colon \rho(z, C) < \varepsilon\}$$

for the open ε-neighbourhood of C.

Definition 2.1. A *generalized semiflow* G on X is a family of maps $\varphi \colon [0, \infty) \rightarrow X$ (called *solutions*) satisfying the hypotheses:

(H1) (*Existence*) For each $z \in X$ there exists at least one $\varphi \in G$ with $\varphi(0) = z$.

(H2) (*Translates of solutions are solutions*) If $\varphi \in G$ and $\tau \geq 0$, then $\varphi^\tau \in G$, where $\varphi^\tau(t) := \varphi(t + \tau), t \in [0, \infty)$.

(H3) (*Concatenation*) If $\varphi, \psi \in G, t \geq 0$, with $\psi(0) = \varphi(t)$, then $\theta \in G$, where

$$\theta(\tau) := \begin{cases} \varphi(\tau) & \text{for } 0 \leq \tau \leq t, \\ \psi(\tau - t) & \text{for } t < \tau. \end{cases}$$

(H4) (*Upper semicontinuity with respect to initial data*) If $\varphi_j \in G$ with $\varphi_j(0) \rightarrow z$, then there exist a subsequence φ_μ of φ_j and $\varphi \in G$ with $\varphi(0) = z$ such that $\varphi_\mu(t) \rightarrow \varphi(t)$ for each $t \geq 0$.

If for each $z \in X$ there is exactly one $\varphi \in G$ with $\varphi(0) = z$, then G is called a *semiflow*. Equivalently, via the correspondence $S(t)z = \varphi(t)$, a semiflow can be defined as a family of continuous maps $S(t): X \to X, t \geq 0$, satisfying the semigroup properties

(a) $S(0) = $ identity,
(b) $S(s + t) = S(s)S(t)$ for all $s, t \geq 0$.

The simplest examples of generalized semiflows are those generated on $\mathbf{R}^n$ by autonomous ordinary differential equations of the form $\dot{u} = f(u)$ for appropriate continuous $f: \mathbf{R}^n \to \mathbf{R}^n$ (see Sell [37]).

We consider the following additional measurability and continuity assumptions that may be satisfied by G. Recall that a map $f: (0, \infty) \to X$ is *strongly measurable* if there exists a sequence f_j of measurable countably-valued maps converging almost everywhere to f on $(0, \infty)$.

(C0) Each $\varphi \in G$ is strongly measurable from $(0, \infty)$ to X.
(C1) Each $\varphi \in G$ is continuous from $(0, \infty)$ to X.
(C2) If $\varphi_j \in G$ with $\varphi_j(0) \to z$ then there exists a subsequence φ_μ of φ_j and $\varphi \in G$ with $\varphi(0) = z$ such that $\varphi_\mu(t) \to \varphi(t)$ uniformly for t in compact subsets of $(0, \infty)$.
(C3) Each $\varphi \in G$ is continuous from $[0, \infty)$ to X.
(C4) If $\varphi_j \in G$ with $\varphi_j(0) \to z$, then there exist a subsequence φ_μ of φ_j and $\varphi \in G$ with $\varphi(0) = z$ such that $\varphi_\mu(t) \to \varphi(t)$ uniformly for t in compact subsets of $[0, \infty)$.

We illustrate these definitions with some examples:

Example 2.1. Let $X = [-1, 1]$. Define $S(0)\tau = \tau$, and for $t > 0$

$$S(t)\tau = \begin{cases} 1 - e^{-t} & \text{if } -1 \leq \tau \leq 0, \\ 1 - e^{-t}(1 - \tau) & 0 < \tau \leq 1. \end{cases}$$

Then it is easily verified that $S(t)$ is a semiflow satisfying (C1) and (C4) but not (C3).

Example 2.2 (*The one-dimensional heat equation*). For $f \in L^\infty(0, 1)$ define $S(t)f$ to be the unique solution $u(\cdot, t)$ of the problem

$$u_t = u_{xx}, \qquad 0 < x < 1, \qquad t > 0,$$
$$u = 0 \quad \text{at} \quad x = 0, 1,$$
$$u(x, 0) = f(x).$$

Then $S(t)$ is a semiflow on $L^\infty(0, 1)$ satisfying (C1) (since u is smooth for $t > 0$) and (C4) (since $\|S(t)f\|_\infty \leq \|f\|_\infty$), but not (C3) (since otherwise we would have $u(\cdot, t) \to f$ in $L^\infty(0, 1)$ as $t \to 0+$, implying that f is continuous).

Example 2.3 (*A semilinear wave equation*). Let $\Omega \subset \mathbf{R}^n, n \geq 3$, be bounded and open with boundary $\partial\Omega$. Consider the damped semilinear wave equation

$$u_{tt} + \beta u_t - \Delta u + f(u) = 0 \qquad \text{in } \Omega, \tag{2.1}$$

with boundary condition

$$u|_{\partial\Omega} = 0, \tag{2.2}$$

and initial conditions

$$u(x, 0) = u_0(x), \qquad u_t(x, 0) = u_1(x), \tag{2.3}$$

where $\beta > 0$ is a constant and $f \colon \mathbf{R} \to \mathbf{R}$ is continuous and satisfies the growth condition

$$|f(u)| \le c_0(|u|^{\frac{n}{n-2}} + 1),$$

for some constant $c_0 > 0$, and sign condition

$$\liminf_{|u| \to \infty} \frac{f(u)}{u} \ge -\lambda_1,$$

where λ_1 is the first eigenvalue of $-\Delta$ with the boundary condition (2.2). Let $X = H_0^1(\Omega) \times L^2(\Omega)$. Then given $\{u_0, u_1\} \in X$ there exists at least one weak solution $\varphi = \{u, u_t\}$ on $[0, \infty)$ to (2.1)–(2.3), and the set G of all such weak solutions is a generalized semiflow satisfying (C3) and (C4). This is proved in [5]. Note that we do not assume any Lipschitz condition on f, so that there is no reason to suppose that solutions are unique.

Our first result is a simple extension to generalized semiflows of a result of [7]. We say that G *has unique representatives* if whenever $\varphi, \psi \in G$ with $\varphi(t) = \psi(t)$ for a.e. $t > 0$ we have $\varphi(t) = \psi(t)$ for all $t > 0$; it is easily seen that this property holds for semiflows.

Theorem 2.1. *Let G have unique representatives and satisfy* (C0). *Then G satisfies* (C1).

Proof. Let $\varphi \in G$. Following the proof in [7], which uses an argument of Auerbach [1], let $0 < a < a + \delta < \infty$ and let I, J denote the open intervals $(a, a + \delta)$ and $(a + \delta/3, a + 2\delta/3)$, respectively. It suffices to show that φ is continuous in J. Since φ is strongly measurable, by a version of Lusin's theorem (see Oxtoby [31] and [7]), there exists in I a closed set F_j of measure greater than $\delta - 1/j^2$ on which the restriction of φ is continuous. The continuity being uniform, there exists $\eta_j \in (0, \delta/3)$ such that $t, t + h \in F_j$ and $|h| < \eta_j$ imply that $d(\varphi(t + h), \varphi(t)) < 1/j$.

Suppose for contradiction that there exist $t_0 \in J$ and a sequence $h_j \to 0$ with $\varphi(t_0 + h_j) \not\to \varphi(t_0)$. Extracting a subsequence, we may assume that

$$d(\varphi(t_0 + h_j), \varphi(t_0)) > \varepsilon \tag{2.4}$$

for some $\varepsilon > 0$ and all j, and that $|h_j| < \eta_j$ for all j. Let

$$\begin{aligned} E_j &= \{t \in J \colon t, t + h_j \in F_j\} \\ &= F_j \cap (F_j - h_j) \cap J. \end{aligned}$$

Then $\mathrm{meas}(J \backslash F_j) \le 1/j^2$, $\mathrm{meas}(J \backslash (F_j - h_j)) = \mathrm{meas}((J + h_j) \backslash F_j) \le 1/j^2$. Thus $\mathrm{meas}(J \backslash E_j) \le 2/j^2$. Hence $\mathrm{meas}(J \backslash \liminf_{j \to \infty} E_j) = 0$, and so $\varphi(t + h_j) \to \varphi(t)$ for a.e. $t \in J$. In particular there exists $t_1, t_2 \in J$, $t_1 < t_0 < t_2$, with $\varphi(t_i + h_j) \to \varphi(t_i)$ for

$i = 1, 2$. By (H2) $\varphi^{t_1+h_j}$ is a solution, and since $\varphi^{t_1+h_j}(0) \to \varphi(t_1)$, by (H4) there exists a subsequence $\varphi^{t_1+h_\mu}$ and a solution ψ with $\varphi^{t_1+h_\mu}(t) \to \psi(t)$ for all $t \geq 0$. But then $\psi(t) = \varphi(t + t_1)$ for a.e. $t \in (0, a + 2\delta/3 - t_1)$. Now define $\tilde{\psi}$ by

$$\tilde{\psi}(t) = \begin{cases} \varphi(t + t_1) & \text{for } 0 \leq t \leq t_2 - t_1, \\ \psi(t) & \text{for } t > t_2 - t_1. \end{cases}$$

By (H2), (H3) $\tilde{\psi} \in G$, and $\tilde{\psi}(t) = \psi(t)$ for a.e. $t > 0$. Since G has unique representatives it follows that $\tilde{\psi}(t) = \psi(t)$ for all $t > 0$. In particular $\tilde{\psi}(t_0 - t_1) = \psi(t_0 - t_1)$, and hence $\varphi(t_0 + h_\mu) \to \varphi(t_0)$, contradicting (2.4). $\square$

Next, we extend a result for semiflows of Chernoff & Marsden [16] (see also [6]).

Theorem 2.2. *Let G satisfy (C1). Let φ_j, φ be solutions with $\varphi_j(t) \to \varphi(t)$ for all $t > 0$. Then $\varphi_j(t) \to \varphi(t)$ uniformly for t in compact subsets of $(0, \infty)$. In particular G satisfies (C2).*

Proof. Let $0 < a < b < \infty$, and for $\varepsilon > 0, n = 1, 2, \ldots$, set

$$S_{n,\varepsilon} = \{t \in [a, b] \colon j \geq n \text{ implies } d(\varphi_j(t), \varphi(t)) \leq \varepsilon\}.$$

$S_{n,\varepsilon}$ is closed by (C1), and by assumption $\cup_{n=1}^{\infty} S_{n,\varepsilon} = [a, b]$. By the Baire Category theorem, some $S_{r,\varepsilon}$ contains an open interval. Since we may apply this argument to any $[a, b] \subset (0, \infty)$, there exists a dense open subset S_ε of $(0, \infty)$ such that if $t_0 \in S_\varepsilon$ there exist an open neighbourhood $N_\varepsilon(t_0)$ of t_0 and $r_\varepsilon(t_0)$ with $d(\varphi_j(t), \varphi(t)) \leq \varepsilon$ whenever $j \geq r_\varepsilon(t_0), t \in N_\varepsilon(t_0)$.

Let $K = \cap_{i=1}^{\infty} S_{1/i}$. Clearly $\varphi_j(t_j) \to \varphi(t)$ whenever $t_j \to t$ and $t \in K$. Again by the Baire Category theorem, K is dense in $(0, \infty)$.

Now let $t > 0$ be arbitrary and $t_j \to t$, and suppose for contradiction that $\varphi_j(t_j) \not\to \varphi(t)$, and without loss of generality that

$$d(\varphi_j(t_j), \varphi(t)) > \delta \tag{2.5}$$

for all j and some $\delta > 0$. Now let $s \in K, s < t$, and consider the solutions $\psi_j = \varphi_j^{t_j+s-t}$, which are well defined for j large enough. Since $s \in K$, $\psi_j(0) \to \varphi(s)$, and so by (H4) there exist a subsequence ψ_μ and a solution ψ with $\psi_\mu(\tau) \to \psi(\tau)$ for all $\tau \geq 0$. But if $s + \tau \in K$ then $\psi_j(\tau) = \varphi_j(t_j + s - t + \tau) \to \varphi(s + \tau)$. Since K is dense and φ, ψ continuous on $(0, \infty)$, it follows that $\psi(\tau) = \varphi(s + \tau)$ for all $\tau \geq 0$. Hence $\varphi_\mu(t_\mu) = \psi_\mu(t - s) \to \psi(t - s) = \varphi(t)$, contradicting (2.5). $\square$

If X is locally compact then for semiflows (C3) implies (C4), a result first proved by Dorroh [19]. We next show that the same result holds for generalized semiflows, adapting the simple proof by Chernoff [15] of Dorroh's result.

Theorem 2.3. *Let X be locally compact. Let G satisfy (C3). Then G satisfies (C4).*

Proof. Let $\varphi_j \in G$ with $\varphi_j(0) \to z$. By Theorem 2.2 there exist a subsequence φ_μ of φ_j and $\varphi \in G$ with $\varphi(0) = z$ and $\varphi_\mu(t) \to \varphi(t)$ uniformly for t in compact subsets of $(0, \infty)$. It thus suffices to show that if $t_\mu \to 0$ then $\varphi_\mu(t_\mu) \to z$. Suppose this is not true. Then there exists a further subsequence, which we do not relabel, such that $d(\varphi_\mu(t_\mu), z) \geq \varepsilon$ for some $\varepsilon > 0$. We may assume also that $d(\varphi_\mu(0), z) < \varepsilon$. Hence by (C3) there exists $s_\mu \in [0, t_\mu]$ with $d(\varphi_\mu(s_\mu), z) = \varepsilon$. Since X is locally compact we may assume further that $\varphi_\mu(s_\mu) \to y$, where $d(y, z) = \varepsilon$. Thus by (H3),(H4) there exists $\psi \in G$ such that for a further subsequence $\varphi_\mu(s_\mu + t) \to \psi(t)$ for all $t \geq 0$. But for $t > 0$ we have $\varphi_\mu(s_\mu + t) \to \varphi(t)$, and so $\psi(t) = \varphi(t)$ for all $t > 0$. Letting $t \to 0$ we deduce from (C3) that $y = z$, a contradiction. $\qquad\square$

Chernoff [15] gives an example of a semiflow on a Hilbert space satisfying (C3) but not (C4); we give another more explicit example in Section 6.2.

3. Existence of Global Attractors

We first extend to generalized semiflows various standard definitions for semiflows.

Let G be a generalized semiflow and let $E \subset X$. Define for $t \geq 0$

$$T(t)E = \{\varphi(t) : \varphi \in G \text{ with } \varphi(0) \in E\}, \tag{3.1}$$

so that $T(t): 2^X \to 2^X$, where 2^X is the space of all subsets of X. It follows from (H2), (H3) that $\{T(t)\}_{t\geq 0}$ defines a semigroup on 2^X, i.e., (a), (b) hold for $T(t)$.[1]

Note that (H4) implies that $T(t)\{z\}$ is compact for each $z \in X, t \geq 0$.

The *positive orbit* of $\varphi \in G$ is the set $\gamma^+(\varphi) = \{\varphi(t) : t \geq 0\}$. If $E \subset X$ then the *positive orbit* of E is the set

$$\gamma^+(E) = \bigcup_{t\geq 0} T(t)E$$
$$= \bigcup\{\gamma^+(\varphi) : \varphi \in G \text{ with } \varphi(0) \in E\}.$$

If $\tau \geq 0$ we set

$$\gamma^\tau(E) = \bigcup_{t\geq \tau} T(t)E = \gamma^+(T(\tau)E).$$

The *ω-limit set* of $\varphi \in G$ is the set

$$\omega(\varphi) = \{z \in X : \varphi(t_j) \to z \text{ for some sequence } t_j \to \infty\}.$$

---

[1] This semigroup has various interesting properties; for example, it is monotone with respect to the partial order of set inclusion (i.e., $E \subset F$ implies $T(t)E \subset T(t)F$ for all $t \geq 0$) and its rest points are the invariant sets of G. When restricted, for example, to the space $K(X)$ of compact subsets of X endowed with the Hausdorff metric, it inherits from (H4) the upper semicontinuity property that $K_j \to K$ implies that $\mathrm{dist}(T(t)K_j, T(t)K) \to 0$ for all $t \geq 0$. If G is a semiflow then we have the stronger property that $K_j \to K$ implies $T(t)K_j \to T(t)K$ for all $t \geq 0$, so that $T(\cdot)$ is a semiflow on $K(X)$.

A *complete orbit* is a map $\psi: \mathbf{R} \to X$ such that for any $s \in \mathbf{R}$, $\psi^s \in G$. If ψ is a complete orbit then the α-*limit set* of ψ is the set

$$\alpha(\psi) = \{z \in X: \psi(t_j) \to z \text{ for some sequence } t_j \to -\infty\}.$$

If $E \subset X$ the ω-*limit set* of E is the set

$$\omega(E) = \{z \in X: \text{ there exist } \varphi_j \in G \text{ with } \varphi_j(0) \in E, \ \varphi_j(0) \text{ bounded,}$$
$$\text{and a sequence } t_j \to \infty \text{ with } \varphi_j(t_j) \to z\}.$$

(When E is unbounded this definition differs from the usual one, in which it is not assumed that the $\varphi_j(0)$ are bounded.)

The subset $A \subset X$ *attracts* a set E if $\text{dist}(T(t)E, A) \to 0$ as $t \to \infty$, and is *locally attracting* if A attracts a neighbourhood of A.

We say that A is *positively invariant* if $T(t)A \subset A$ for all $t \geq 0$, that A is *quasi-invariant* if for each $z \in A$ there exists a complete orbit ψ with $\psi(0) = z$ and $\psi(t) \in A$ for all $t \in \mathbf{R}$, and that A is *invariant* if $T(t)A = A$ for all $t \geq 0$. Note that if A is quasi-invariant then $A \subset T(t)A$ for all $t \geq 0$ (this is taken as the definition of quasi-invariance by Barbashin [10]); from this it follows easily that A is invariant if and only if A is positively invariant and quasi-invariant. Note that even for semiflows a set A may be invariant but there may be solutions $\varphi \in G$ with $\varphi(0) \notin A$ and $\varphi(t) \in A$ for some $t > 0$.

The subset A is a *global attractor* if A is compact, invariant, and attracts all bounded sets.

The generalized semiflow G is *eventually bounded* if, given any bounded $B \subset X$, there exists $\tau \geq 0$ with $\gamma^\tau(B)$ bounded.

G is *point dissipative* if there is a bounded set B_0 such that, for any $\varphi \in G$, $\varphi(t) \in B_0$ for all sufficiently large t.

G is *asymptotically compact* if, for any sequence $\varphi_j \in G$ with $\varphi_j(0)$ bounded, and for any sequence $t_j \to \infty$, the sequence $\varphi_j(t_j)$ has a convergent subsequence.

G is *compact* if, for any sequence $\varphi_j \in G$ with $\varphi_j(0)$ bounded, there exists a subsequence φ_μ such that $\varphi_\mu(t)$ is convergent for each $t > 0$.

Proposition 3.1. *Let G be asymptotically compact. Then G is eventually bounded.*

Proof. Let $a \in X$, let $B \subset X$ be bounded, and suppose for contradiction that $\gamma^\tau(B)$ is unbounded for all $\tau \geq 0$. Then there exist $\varphi_j \in G$ with $\varphi_j(0) \in B$ and $t_j \to \infty$ with $d(\varphi_j(t_j), a) \to \infty$. But $\varphi_j(t_j)$ has a convergent subsequence by asymptotic compactness. $\qquad\square$

Proposition 3.2. *Let G be eventually bounded and compact. Then G is asymptotically compact.*

Proof. Let $\varphi_j \in G$ with $\varphi_j(0)$ bounded, and let $t_j \to \infty$. Since G is eventually bounded, $\varphi_j^{t_j-1}(0)$ is bounded. Since G is compact, for some subsequence $\varphi_\mu^{t_\mu-1}(1) = \varphi_\mu(t_\mu)$ is convergent. $\qquad\square$

Theorem 3.3. *A generalized semiflow G has a global attractor if and only if G is point dissipative and asymptotically compact. The global attractor A is unique and given by*

$$A = \bigcup\{\omega(B): \ B \ a \ bounded \ subset \ of \ X\} = \omega(X). \qquad (3.2)$$

Furthermore A is the maximal compact invariant subset of X.

Theorem 3.3 generalizes a corresponding result for semiflows given in Hale [21] and Ladyzhenskaya [27] and having antecedents in the work of Billotti & LaSalle [11] and Hale, LaSalle & Slemrod [26]. Note, however, that we make no assumption that the positive orbits of bounded sets are bounded. A closely related result in the context of a set-valued semiflow $T(t)$ is announced in Mel'nik [29]; the main differences are that in [29] (i) the global attractor A is not asserted to be invariant, but only to satisfy the property $A \subset T(t)A$ for all $t \geq 0$ (this can be traced to the weaker hypothesis made in [29] that $T(t)$ satisfies $T(s + t)\{z\} \subset T(s)T(t)\{z\}$ for all $z \in X$, $s, t \geq 0$, whereas our concatenation hypothesis $(H3)$ implies equality); (ii) the definition of point dissipative is stronger; and (iii) the semiflow is assumed to be eventually bounded. For other results on global attractors and applications see Temam [42] and Babin & Vishik [4].

In order to prove Theorem 3.3 we need to suitably modify the corresponding arguments for semiflows. Our treatment is closest to that of Ladyzhenskaya [27].

Lemma 3.4. *Let G be asymptotically compact.*

(i) Let $B \subset X$ be nonempty and bounded. Then $\omega(B)$ is nonempty, compact, quasi-invariant, and attracts B. If $T(t_0)\omega(B) \subset B$ for some $t_0 \geq 0$, then $\omega(B)$ is invariant.

(ii) If $\varphi \in G$ then $\omega(\varphi)$ is nonempty, compact, quasi-invariant, and $\lim_{t\to\infty} \rho(\varphi(t), \omega(\varphi)) = 0$.

(iii) If ψ is a bounded complete orbit then $\alpha(\psi)$ is nonempty, compact, and quasi-invariant, and $\lim_{t\to-\infty} \rho(\psi(t), \alpha(\psi)) = 0$.

Proof. (i) Let $v \in B$. By (H1) there exists some $\varphi \in G$ with $\varphi(0) = v$. By the asymptotic compactness $\varphi(j)$ has a convergent subsequence, and so $\omega(B)$ is nonempty. Since $\omega(B) \subset \overline{\gamma^\tau(B)}$ for any $\tau \geq 0$, and since G is eventually bounded, $\omega(B)$ is bounded. It is easily seen that $\omega(B)$ is also closed.

Let $z \in \omega(B)$. By definition there exist $\varphi_j \in G$ with $\varphi_j(0) \in B$ and a sequence $t_j \to \infty$ such that $\varphi_j(t_j) \to z$. By (H2), $\varphi_j^{t_j} \in G$. Since $\varphi_j^{t_j}(0) \to z$, by (H4) there exist a subsequence, which we do not relabel, and a solution ψ_0 with $\psi_0(0) = z$, such that $\varphi_j^{t_j}(t) \to \psi_0(t)$ for all $t \geq 0$. Clearly $\psi_0(t) \in \omega(B)$ for all $t \geq 0$. Now consider the sequence $\varphi_j^{t_j-1}$. Since $\varphi_j^{t_j-1}(0) = \varphi_j(t_j - 1)$, by the asymptotic compactness and (H4) we have (after extraction of a further subsequence) that $\varphi_j^{t_j-1}(t) \to \psi_1(t)$ for all $t \geq 0$, where $\psi_1 \in G$. Clearly $\psi_1^1 = \psi_0$. Proceeding inductively, we find for each $r = 1, 2, \ldots$ a solution ψ_r such that $\psi_r^1 = \psi_{r-1}$ and $\psi_r(t) \in \omega(B)$ for all $t \geq 0$. Given $t \in \mathbf{R}$ define $\psi(t)$ to be the common value of $\psi_r(t + r)$ for $r \geq -t$. Then ψ is a complete orbit with $\psi(0) = z$ and $\psi(t) \in \omega(B)$ for all $t \in \mathbf{R}$. Hence $\omega(B)$ is quasi-invariant.

Now suppose $z_k \in \omega(B)$. By the quasi-invariance $z_k = \psi_k(k)$ where $\psi_k \in G$ and

$\psi_k(0) \in \omega(B)$. By the boundedness of $\omega(B)$ and the asymptotic compactness z_k has a convergent subsequence. Hence $\omega(B)$ is compact.

Suppose $\omega(B)$ does not attract B. Then there exist $\varepsilon > 0$, $\varphi_j \in G$ with $\varphi_j(0) \in B$ and a sequence $t_j \to \infty$ with $\varphi_j(t_j) \notin N_\varepsilon(\omega(B))$. But by asymptotic compactness $\varphi_j(t_j)$ has a convergent subsequence, and the limit belongs to $\omega(B)$, a contradiction.

If $T(t_0)\omega(B) \subset B$ for some $t_0 \geq 0$ then by the quasi-invariance of $\omega(B)$ we have $\omega(B) \subset B$. Let $\varphi \in G$ with $\varphi(0) \in \omega(B)$ and let $t \geq 0$. By the quasi-invariance and concatenation we have that, for each $k \geq 0$, $\varphi(t) = \psi_k(k)$ for some $\psi_k \in G$ with $\psi_k(0) \in \omega(B)$. But then $\psi_k(0) \in B$ and so $\varphi(t) \in \omega(B)$. Thus $\omega(B)$ is invariant.

(ii) The proof is similar to (i) but easier.

(iii) If $t_j \to -\infty$ then $\psi(t_j) = \psi^{2t_j}(-t_j)$ and $\psi^{2t_j}(0)$ is bounded, so that by asymptotic compactness $\psi(t_j)$ has a convergent subsequence. The rest of the proof is as for (ii). $\square$

Lemma 3.5. *Let G be pointwise dissipative and asymptotically compact. Then there exists a bounded set B_1 such that given any compact $K \subset X$ there exist $\varepsilon = \varepsilon(K) > 0$, $t_1 = t_1(K) > 0$, such that $T(t)N_\varepsilon(K) \subset B_1$ for all $t \geq t_1$.*

Proof. Let $\delta > 0$. Since by Proposition 3.1 G is eventually bounded, there exists $\tau \geq 0$ such that $B_1 := \gamma^\tau(N_\delta(B_0))$ is bounded. Suppose for contradiction that there exist a compact set K and sequences $\varepsilon_j \to 0$, $t_j \to \infty$, $\varphi_j \in G$ with $\varphi_j(0) \in N_{\varepsilon_j}(K)$ and $\varphi_j(t_j) \notin B_1$. Since $\varphi_j(t_j) = \varphi_j^t(t_j - t)$, it follows that $\varphi_j^t(0) = \varphi_j(t) \notin N_\delta(B_0)$ for $0 \leq t \leq t_j - \tau$. We may also assume that $\varphi_j(0) \to z \in K$. Hence by (H4) there is a subsequence φ_μ and $\varphi \in G$ with $\varphi_\mu \to \varphi$ pointwise, $\varphi(0) = z$, and $\varphi(t) \notin B_0$ for all $t \geq 0$. This contradicts the point dissipativeness of G. $\square$

Proof of Theorem 3.3. Let A be a global attractor for G and let $B_0 = N_\delta(A)$ for some $\delta > 0$. Given $\varphi \in G$ the set consisting of the single point $\{\varphi(0)\}$ is bounded and thus attracted to A. Hence $\varphi(t) \in B_0$ for t sufficiently large, and hence G is point dissipative. If $\varphi_j \in G$ with $\varphi_j(0)$ bounded, the set $\{\varphi_j(0)\}$ is attracted to A, and thus if $t_j \to \infty$ we have $\rho(\varphi_j(t_j), A) \to 0$. Since A is compact, this implies that $\varphi_j(t_j)$ has a convergent subsequence, and thus G is asymptotically compact.

Conversely, let G be point dissipative and asymptotically compact. Let B_1 be as in Lemma 3.5 and let $A = \omega(B_1)$. By Lemma 3.4 A is compact and attracts B_1. We show that A attracts bounded sets. Let B be bounded and let $K = \omega(B)$. By Lemma 3.4 K is compact and attracts B. Let $\varepsilon(K)$, $t_1 = t_1(K)$ be as in Lemma 3.5, and let $0 < \varepsilon < \varepsilon(K)$. Since K attracts B, $T(t_0)B \subset N_\varepsilon(K)$ for some $t_0 > 0$. Hence $T(t_0 + t_1)B = T(t_1)T(t_0)B \subset T(t_1)N_\varepsilon(K) \subset B_1$. Thus $T(t_0 + t_1 + t)B \subset T(t)B_1$ for all $t \geq 0$, and since B_1 is attracted to A so is B. Since by Lemma 3.5 we also have that $T(t_2)\omega(B_1) \subset B_1$ for some $t_2 \geq 0$, it follows from Lemma 3.4 that A is invariant. This proves that A is a global attractor, and that $\omega(B) \subset A$ for any bounded B, so that (3.2) holds.

Suppose A_1 is compact and invariant. Then $\omega(A_1) = A_1$ and so $A_1 \subset A$ by (3.2). Hence A is the maximal compact invariant subset of X. $\square$

4. Connectedness

Proposition 4.1. *Let G be asymptotically compact and satisfy $(C1)$. If $\varphi \in G$, then $\omega(\varphi)$ is connected. If ψ is a complete orbit then $\alpha(\psi)$ is connected.*

Proof. This is standard. By Lemma 3.4 $\omega(\varphi)$ and $\alpha(\psi)$ are compact. If $\omega(\varphi)$, say, were not connected, then $\omega(\varphi) = A_1 \cup A_2$ for nonempty disjoint compact sets A_1, A_2. Let U_1, U_2 be disjoint open sets with $A_1 \subset U_1$, $A_2 \subset U_2$. By $(C1)$ there exists $t_j \to \infty$ with $\varphi(t_j) \notin U_1 \cup U_2$, and by asymptotic compactness this implies that there exists $z \in \omega(\varphi) \backslash (A_1 \cup A_2)$, a contradiction. $\square$

We say that G has *Kneser's property* if $T(t)\{z\}$ is connected for each $z \in X, t \geq 0$. Any semiflow has Kneser's property since $T(t)\{z\}$ is a point.

Theorem 4.2. *Let G satisfy $(C1)$. If G has Kneser's property and if $E \subset X$ is connected then $\omega(E)$ is connected.*

For a related result see Mel'nik [29].

Proof. Suppose E is connected but $\omega(E)$ is not. Then $\omega(E) = A_1 \cup A_2$, for nonempty sets A_1, A_2, where $A_1 \cap \overline{A_2} = A_2 \cap \overline{A_1} = \emptyset$. Since X is a metric space, it is completely normal (see [23], p. 42), so that there exist disjoint open sets U_1, U_2 with $A_1 \subset U_1$, $A_2 \subset U_2$. (If $\omega(E)$ is compact this conclusion is obvious.) For $i = 1, 2$ let $E_i = \{z \in E : \operatorname{dist}(T(t)\{z\}, A_i) \to 0 \text{ as } t \to \infty\}$. We claim that E_1, E_2 are disjoint, nonempty, relatively open subsets of E with $E_1 \cup E_2 = E$. Since E is connected this is a contradiction.

To show that the E_i are disjoint, note that if $z \in E_1 \cap E_2$ then $T(t)\{z\} \subset U_1$ for t large enough and $T(t)\{z\} \subset U_2$ for t large enough, which is impossible since $U_1 \cap U_2 = \emptyset$.

To show that $E_1 \cup E_2 = E$ suppose $z \in E$. Then $\operatorname{dist}(T(t)\{z\}, \omega(E)) \to 0$ as $t \to \infty$, and so there exists $T > 0$ such that $T(t)\{z\} \subset U_1 \cup U_2$ for all $t > T$. By Kneser's property we thus have that for each $t > T$ either $T(t)\{z\} \subset U_1$ or $T(t)\{z\} \subset U_2$. But if $T(r)\{z\} \subset U_1$, $T(s)\{z\} \subset U_2$ for $T < r < s$ then there exists $\varphi \in G$ with $\varphi(0) = z$, $\varphi(r) \in U_1$, $\varphi(s) \in U_2$, and $\varphi(t) \in U_1 \cup U_2$ for $t \in [r, s]$. This is impossible by $(C1)$. Hence $z \in E_1 \cup E_2$.

To show that E_2, say, is nonempty, suppose that $E = E_1$. Let $a \in A_2$, so that there exists $\varphi_j \in G$ with $\varphi_j(0) \in E$, $\varphi_j(0)$ bounded and $t_j \to \infty$ with $\varphi_j(t_j) \to a$. Let $B = \{\varphi_j(0)\}$. Since B is bounded, by Lemma 3.4 B is attracted to $\omega(B) \subset \omega(E)$. Hence there exists $T > 0$ with $\varphi_j(t) \in U_1 \cup U_2$ for all $t > T$ and all j. But since $\varphi_j(0) \in E_1$, $\varphi_j(t) \in U_1$ for all j and all $t > T$. This contradicts $a \in A_2$.

Finally, to show that E_2, say, is relatively open, let $z \in E_2$ and suppose for contradiction that there exist $z_j \to z$ with $z_j \in E_1$ for all j. Thus there exist $\varphi_j \in G$ with $\varphi_j(0) = z_j$, and for each j and for all sufficiently large t we have $\varphi_j(t) \in U_1$. Since $\{\varphi_j(0)\}$ is bounded, as argued above we have that there exists $T > 0$ with $\varphi_j(t) \in U_1 \cup U_2$ for all $t > T$, and thus $\varphi_j(t) \in U_1$ for all j and all $t > T$. But by (H4) we may assume that $\varphi_j(t) \to \varphi(t)$ for all $t > 0$, where $\varphi \in G$ with $\varphi(0) = z$. Since $z \in E_2$, there exists $\tau > T$ with $\varphi(\tau) \in U_2$, and hence $\varphi_j(\tau) \in U_2$ for j sufficiently large, a contradiction. $\square$

Corollary 4.3. *Let X be connected, and let G satisfy (C1) and have Kneser's property. If A is a global attractor, then A is connected.*

Proof. This follows because $A = \omega(X)$.   □

For similar but not identical results for semiflows, see Ladyzhenskaya [27] and Sell & You [40].[2]

5. Lyapunov Functions

A complete orbit $\psi \in G$ is *stationary* if $\psi(t) = z$ for all $t \in \mathbf{R}$ for some $z \in X$. Each such z is called a *rest point*. (Note that in general, if z is a rest point there may also exist nonconstant $\psi \in G$ with $\psi(0) = z$.) We denote the set of rest points of G by $Z(G)$. It follows easily from (H4) that $Z(G)$ is closed.

We say that $V: X \to \mathbf{R}$ is a *Lyapunov function* for G provided

(*i*) V is continuous,
(*ii*) $V(\varphi(t)) \leq V(\varphi(s))$ whenever $\varphi \in G$ and $t \geq s \geq 0$.
(*iii*) if $V(\psi(t)) = $ constant for some complete orbit ψ and all $t \in \mathbf{R}$ then ψ is stationary.

Since a global attractor A is quasi-invariant, given any $a \in A$ there exists a complete orbit ψ with $\psi(0) = a$. In the presence of a Lyapunov function the behaviour of such complete orbits can be characterized.

Theorem 5.1. *Let G be asymptotically compact, let (C1) hold, and suppose there exists a Lyapunov function V for G. Suppose further that Z(G) is bounded. Then G is point dissipative, so that there exists a global attractor A. For each complete orbit ψ lying in A the limit sets $\alpha(\psi), \omega(\psi)$ are connected subsets of Z(G) on which V is constant. If Z(G) is totally disconnected (in particular, if Z(G) is countable) the limits*

$$z_- = \lim_{t \to -\infty} \psi(t), \qquad z_+ = \lim_{t \to \infty} \psi(t)$$

exist and z_-, z_+ are rest points; furthermore, $\varphi(t)$ tends to a rest point as $t \to \infty$ for every $\varphi \in G$.

Proof. Let $\varepsilon > 0$, $B_0 = N_\varepsilon(Z(G))$. If $\varphi \in G$, by properties (i), (ii) of V and Lemma 3.4 we have that $V(z) = \lim_{t \to \infty} V(\varphi(t)) \in \mathbf{R}$ for all $z \in \omega(\varphi)$. Since $\omega(\varphi)$ is quasi-invariant, by property (iii) $\omega(\varphi) \subset Z(G)$ and hence $\varphi(t) \in B_0$ for t sufficiently large. Thus G is point dissipative. If ψ is a complete orbit in A then we have by the above and

---

[2] However, the proof in [27] that the global attractor is connected if X is connected makes use of the incorrect remark that a bounded subset B of a connected metric space X is contained in a bounded connected subset of X; a counterexample is provided by the metric subspace X of $\mathbf{R}^2$ given by the union of the sets $C_j \setminus Q_j$, $j = 1, 2, ...,$ where C_j is the circle centre $(0, j)$ of radius j and Q_j is the square $(0, 1/j)^2$. The intersection B of X with the ball $B(0, 2)$ is not connected, but the only connected set containing B is X itself.

Proposition 4.1 that $\omega(\psi)$ is a connected subset of $Z(G)$, and the corresponding result for $\alpha(\psi)$ holds similarly. The rest of the theorem is then obvious.   □

6. Stability

The subset $A \subset X$ is *Lyapunov stable* if given $\varepsilon > 0$ there exists $\delta > 0$ such that if $E \subset X$ with $\mathrm{dist}(E, A) < \delta$ then $\mathrm{dist}(T(t)E, A) < \varepsilon$ for all $t \geq 0$. It is easily seen that a subset A is Lyapunov stable if and only if given $\varphi_j \in G$ with $\rho(\varphi_j(0), A) \to 0$ and $t_j \geq 0$ we have $\rho(\varphi_j(t_j), A) \to 0$. We say that A is *uniformly asymptotically stable* if A is Lyapunov stable and is locally attracting.

Since a global attractor is compact, invariant, and locally attracting, the following theorem gives in particular conditions under which a global attractor is uniformly asymptotically stable.

Theorem 6.1. *Let G satisfy* (C1) *and* (C4), *and let A be a compact invariant set that is locally attracting. Then A is uniformly asymptotically stable.*

Proof. We must show that A is Lyapunov stable. Suppose it is not. Then there exist $\varphi_j \in G$ and $t_j \geq 0$ such that $\rho(\varphi_j(0), A) \to 0$ but $\rho(\varphi_j(t_j), A) \geq \varepsilon$ for some $\varepsilon > 0$. Since A is compact, we may suppose that $\varphi_j(0) \to z \in A$, and thus by (H4) that $\varphi_j(t) \to \varphi(t)$ for all $t \geq 0$ and for some $\varphi \in G$ with $\varphi(0) = z$. Since A is invariant, $\varphi(t) \in A$ for all $t \geq 0$. We may also suppose that either $t_j \to \infty$ or $t_j \to t \geq 0$.

If $t_j \to \infty$, then, since A is locally attracting we have $\rho(\varphi_j(t_j), A) \to 0$, a contradiction. If $t_j \to t > 0$ then, since G satisfies (C1), by Theorem 2.2 we have $\varphi_j(t_j) \to \varphi(t) \in A$, a contradiction. So it remains to consider the case $t_j \to 0$. By (C4) we have, after the possible extraction of a further subsequence, that $d(\varphi_j(t_j), \varphi(t_j)) \to 0$. Since A is invariant there exists $\theta \in G$ with $\theta(1 + \tau) = \varphi(\tau)$ for all $\tau \geq 0$, and so by (C1) $\varphi(t_j) \to \varphi(0)$. Hence $\varphi_j(t_j) \to \varphi(0) \in A$, a contradiction.   □

Corollary 6.2. *Let X be locally compact, let G satisfy* (C3), *and let A be a compact invariant set that is locally attracting. Then A is uniformly asymptotically stable.*

Proof. This follows immediately from Theorems 2.3 and 6.1.   □

Let G satisfy (C1) and let A be a compact invariant set that is locally attracting. Then the proof of Theorem 6.1 shows that given $\varepsilon > 0$, $\tau > 0$, there exists $\delta > 0$ such that if $E \subset X$ with $\mathrm{dist}(E, A) < \delta$ then $\mathrm{dist}(T(t)E, A) < \varepsilon$ for all $t \geq \tau$. For semiflows, Sell & You [40] take this conclusion as the definition of Lyapunov stability and prove an equivalent result.

We now give two examples of semiflows $\{S(t)\}_{t \geq 0}$ on a metric space X satisfying (C1) for which the global attractor A is a single point but is not Lyapunov stable.

6.1. An Example with $X = \mathbf{R}^2$

In the first example $X = \mathbf{R}^2$ is locally compact. Thus by Corollary 6.2 (C3) cannot be satisfied.

Consider the differential equations in $\mathbf{R}^2$,

$$\dot{r} = -r^{\frac{1}{2}} h(\theta), \tag{6.1a}$$

$$\dot{\theta} = -2r^{-\frac{1}{2}} \sin(\theta/2) h(\theta), \tag{6.1b}$$

where (r, θ) are plane polar coordinates with $0 \le \theta < 2\pi$ and where

$$h(\theta) := \theta^{-2}(2\pi - \theta)^{-2}. \tag{6.2}$$

Writing $x = r \cos\theta$, $y = r \sin\theta$, we see that (6.1) defines a C^∞ vector field in $U = \mathbf{R}^2 \backslash \bar{L}$, where $L = \{(x, 0): x > 0\}$ denotes the positive semi x-axis. Note that the integral curves of (6.1) are given by

$$r = C \tan(\theta/4) \tag{6.3}$$

where $C > 0$ is a constant. These curves do not intersect L, and approach the origin tangent to it from the first quadrant (see Fig. 1a). Since $h(\theta) \ge \pi^{-4}$, we have that

$$r^{\frac{1}{2}}(t) \le r^{\frac{1}{2}}(0) - \frac{1}{2}\pi^{-4}t. \tag{6.4}$$

Hence for any $z \in U$, the solution $(x(t), y(t))$ of (6.1) with initial data $(x(0), y(0)) = z$ reaches the origin in a finite time $t_c = t_c(z) > 0$. We define for $z \in U$

$$R(t)z = \begin{cases} (x(t), y(t)) & \text{if } 0 \le t < t_c, \\ (0, 0) & \text{if } t \ge t_c. \end{cases} \tag{6.5}$$

For $z = (x, 0)$ with $x > 0$ we define

$$R(t)z = \begin{cases} z & \text{if } t = 0, \\ (0, 0) & \text{if } t > 0. \end{cases} \tag{6.6}$$

Thus $R(t): \mathbf{R}^2 \to \mathbf{R}^2$ is defined for all $t \ge 0$, and clearly $R(0) = \text{identity}$, $R(s + t) = R(s)R(t)$ for all $s, t \ge 0$. We show that $R(t)$ is continuous for each $t > 0$. Let $t > 0$ and $z_j \to z$. We must show that $R(t)z_j \to R(t)z$. This is easily proved if $z \in U$ or $z = 0$, using the fact that by (6.4) the origin is stable, so we assume that $z \in L$. By (6.6) we may also assume that $z_j \notin L$ for all j. The result then follows provided we can show that $t_c(z_j) \to 0$, since then $R(t)z_j = R(t)z = 0$ for sufficiently large j.

Let (r_j, θ_j) be the solutions of (6.1) corresponding to the initial data z_j. We first claim that given $\varepsilon > 0$ there exists $J(\varepsilon)$ such that, if $j \ge J(\varepsilon)$ and $\tau \in [0, t_c(z_j))$, then either $r_j(\tau) \le \varepsilon$ or $h(\theta_j(\tau)) \ge \varepsilon^{-1}$. If not there would exist a subsequence j_k and $\tau_k \in [0, t_c(z_{j_k}))$ with $r_{j_k}(\tau_k) > \varepsilon$ and $h(\theta_{j_k}(\tau_k)) < \varepsilon^{-1}$. From (6.3) we have that

$$\frac{r_{j_k}(\tau_k)}{\tan(\theta_{j_k}(\tau_k)/4)} = \frac{r_{j_k}(0)}{\tan(\theta_{j_k}(0)/4)}, \tag{6.7}$$

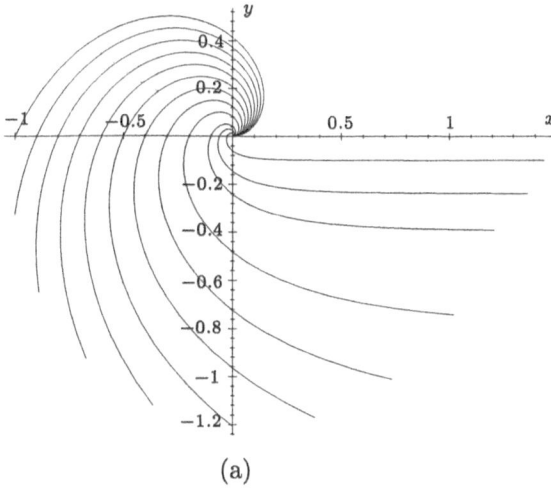

(a)

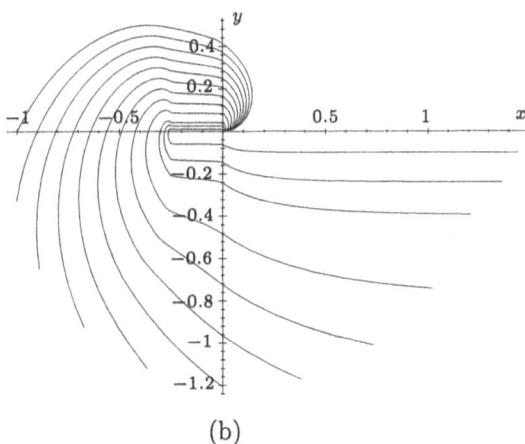

(b)

Fig. 1. (a) Phase portrait for the semiflow $R(t)$. (b) Phase portrait for the semiflow $S(t)$.

and the left-hand side of (6.7) is bounded away from zero and infinity. But this is not true for the right-hand side, since $z \in L$, establishing the claim. Now for $\varepsilon > 0$ let $t^\varepsilon(z_j) = \inf\{s > 0: r_j(s) = \varepsilon\}$. From (6.4) applied to $r(t) = r_j(t + t^\varepsilon(z_j))$ we have that

$$t_c(z_j) - t^\varepsilon(z_j) \le 2\pi^4 \varepsilon^{\frac{1}{2}}. \tag{6.8}$$

But for $j \ge J(\varepsilon)$, $\tau \in [0, t^\varepsilon(z_j))$ we have $h(\theta_j(\tau)) \ge \varepsilon^{-1}$ and thus $\dot{r}_j(\tau) \le -r_j(\tau)^{\frac{1}{2}}\varepsilon^{-1}$. Hence

$$\varepsilon^{\frac{1}{2}} = r_j(t^\varepsilon(z_j))^{\frac{1}{2}} \le r_j(0)^{\frac{1}{2}} - \frac{1}{2}t^\varepsilon(z_j)\varepsilon^{-1}. \tag{6.9}$$

Combining (6.8) and (6.9), we deduce that for $j \geq J(\varepsilon)$,

$$t_c(z_j) \leq 2\pi^4 \varepsilon^{\frac{1}{2}} + 2\varepsilon (r_j(0)^{\frac{1}{2}} - \varepsilon^{\frac{1}{2}}),$$

and letting $\varepsilon \to 0$ we obtain $t_c(z_j) \to 0$ as required.

We have thus shown that $\{R(t)\}_{t\geq 0}$ is a semiflow on $\mathbf{R}^2$. Also, the map $t \mapsto R(t)z$ is clearly continuous on $(0, \infty)$ for any $z \in \mathbf{R}^2$.

We now modify $R(t)$ using a map $P \colon \mathbf{R}^2 \to \mathbf{R}^2$ defined by

$$P(x, y) = (x - f(x, y), y),$$

where $f \in C^1(\mathbf{R}^2 \backslash \{0\})$ satisfies $f(x, y) = 0$ if $(x, y) \notin R := (-1, 0) \times (-1, 1)$, $\lim_{x \to 0-} f(x, 0) = a$, where $0 < a < 1$, and $f_x(x, y) \leq \frac{1}{2}$ for $(x, y) \neq (0, 0)$. Since $\partial_x P(x, y) = (1 - f_x(x, y), 0)$ for $(x, y) \neq (0, 0)$, P is monotone on lines $y = \mathrm{const.}$, and hence P restricted to $\mathbf{R}^2 \backslash \{0\}$ is a diffeomorphism with range $\mathbf{R}^2 \backslash I$, where $I := \{(x, 0) \colon -a \leq x \leq 0\}$. We define $S(t) \colon \mathbf{R}^2 \to \mathbf{R}^2$ by

$$
\begin{aligned}
S(t)z &= P(R(t)P^{-1}z) && \text{if } z \notin I, t \geq 0, \\
S(0)(x, 0) &= (x, 0) && \text{if } (x, 0) \in I, \\
S(t)(x, 0) &= (0, 0) && \text{if } (x, 0) \in I, t > 0.
\end{aligned}
$$

(See Figure 1b, where we have chosen

$$
f(x, y) = \begin{cases}
\frac{1}{4}(x + 1)^2(1 - y^2)^2 & \text{if } -1 < x < -y^2, \\
\frac{1}{4}(x + 1)^2(1 - y^2)^2 \sin^2(\pi x/2y^2) & \text{if } -y^2 \leq x < 0, \\
0 & \text{otherwise,}
\end{cases}
$$

for which $a = \frac{1}{2}$.)

It is easily checked that $S(0) = \mathrm{identity}$, $S(s + t) = S(s)S(t)$ for all $s, t \geq 0$. Each solution $t \mapsto S(t)z$ is continuous on $(0, \infty)$. The only case that is not immediately obvious is when $z \notin I$ and $P^{-1}z \notin L$. But then $R(t)P^{-1}z$ is continuous on $(0, \infty)$ and is zero for $t \geq t_c = t_c(P^{-1}z)$. So we just need to show that $\lim_{t \to t_c-} P(R(t)P^{-1}z) = 0$. But this follows since $R(t)P^{-1}z$ belongs to the first quadrant for $t \in (t_c - \varepsilon, t_c)$ for some $\varepsilon > 0$, and there $P = \mathrm{identity}$.

We now prove that each $S(t)$ is continuous from $\mathbf{R}^2 \to \mathbf{R}^2$. Let $t > 0$, $z_j \to z$. We must show that $S(t)z_j \to S(t)z$. There are two cases.

(i) Suppose $z \in I$, so that $S(t)z = 0$. We may assume that either $z_j \in I$ for all j or $z_j \notin I$ for all j, and in the former case $S(t)z_j = 0$ and we are done. If $z_j \notin I$ for all j then we may assume that $P^{-1}z_j \to w$, and clearly $w = 0$. But then by (6.4) $R(t)P^{-1}z_j = 0$ for j large enough, and hence $S(t)z_j = 0$ for j large enough.

(ii) Suppose $z \notin I$, so that $P^{-1}z \neq 0$ and $P^{-1}z_j \to P^{-1}z$. Then $R(t)P^{-1}z_j \to R(t)P^{-1}z$, so that if $R(t)P^{-1}z \neq 0$ we have $S(t)z_j \to S(t)z$. If $R(t)P^{-1}z = 0$ there are two subcases. If $P^{-1}z \in L$ then we may assume that either $P^{-1}z_j \in L$ for all j or $P^{-1}z_j \notin L$ for all j. In the first situation we have $S(t)z_j = S(t)z = 0$. In the second we already showed that $t_c(P^{-1}z_j) \to 0$. Hence $R(t)P^{-1}z_j = 0$ and $S(t)z_j = 0$ for j large enough. The second subcase is when $P^{-1}z \notin L$. Then

$P^{-1}z_j \notin L$ for j large enough, and since $R(t)P^{-1}z$ approaches 0 from the first quadrant, so $R(t)P^{-1}z_j$ belongs to the closed first quadrant for j large enough. Hence $S(t)z_j = R(t)P^{-1}z_j \to 0 = S(t)z$.

It remains to show that $\{0\}$ is a global attractor for $\{S(t)\}_{t \geq 0}$ that is not Lyapunov stable. To show that $\{0\}$ is a global attractor we just need to prove that it attracts bounded sets. Let $M \geq 2$ and $|z| \leq M$. If $z \notin I$ then, since $P = $ identity outside $B(0,2)$, $|P^{-1}z| \leq M$. Hence, by (6.4) $R(t)P^{-1}z = 0$ for $t \geq 2M^{\frac{1}{2}}\pi^4$, and so $S(t)z = 0$ for such t. If $z \in I$ then $S(t)z = 0$ for $t > 0$. Hence $S(t)B(0, M) = \{0\}$ for $t \geq 2M^{\frac{1}{2}}\pi^4$ and thus $S(\cdot)$ attracts bounded sets.

To see that $\{0\}$ is not Lyapunov stable, let $z_j = (j^{-1}, -j^{-1})$. Then by (6.3) there exists $t_j > 0$ such that $R(t_j)z_j = (-\sqrt{2}/(j \cot 7\pi/16), 0)$. Then $|S(t_j)z_j| = |P(R(t_j)z_j)| \geq a$.

6.2. An Example with X a Hilbert Space

In this example we take $X = H$ to be a real Hilbert space with inner product $(,)$, norm $\| \cdot \|$, and orthonormal basis consisting of the vectors $e_i, \hat{e}_i, i = 1, 2, \ldots$. We construct a semiflow $\{S(t)\}_{t \geq 0}$ on H satisfying (C3) (i.e., $t \mapsto S(t)w$ is continuous from $[0, \infty) \to H$ for each $w \in H$) and such that $\{0\}$ is a global attractor which is not Lyapunov stable. By Corollary 6.2 such an example cannot occur for X locally compact. By Theorem 6.1 the semigroup $S(t)$ cannot satisfy (C4), and so in particular we obtain an explicit example of a semigroup on a Hilbert space satisfying (C3) but not (C4); the example of Chernoff [15] is not explicit and is based on a nontrivial result of infinite-dimensional topology.

Fix β with $0 < \beta < 1$. For each i we let $L_i = \text{span}\{e_i, \hat{e}_i\}$ and

$$K_i := \{u + v \colon u \in L_i, v \in L_i^{\perp}, \|v\|^2 < \beta\|u\|^2\},$$

where $L_i^{\perp}$ denotes the orthogonal complement of L_i. Let $S = \{w \in H \colon \|w\| = 1\}$. We have the following elementary result.

Lemma 6.3.

$$\text{dist}(K_i \cap S, K_j \cap S) = c(\beta)$$

for $i \neq j$, where $c(\beta) := \sqrt{2}(1 - \sqrt{\beta})/\sqrt{1 + \beta} > 0$.

Proof. Let $w_i \in K_i \cap S$, $w_j \in K_j \cap S$. Then we have

$$w_i = u_i + z_j + v,$$
$$w_j = u_j + z_i + \bar{v},$$

with $u_i, z_i \in L_i$, $u_j, z_j \in L_j$, $v, \bar{v} \in (L_i \oplus L_j)^{\perp}$, and

$$\|u_i\|^2 + \|z_j\|^2 + \|v\|^2 = \|u_j\|^2 + \|z_i\|^2 + \|\bar{v}\|^2 = 1, \tag{6.10}$$
$$\|z_j\|^2 + \|v\|^2 < \beta\|u_i\|^2, \qquad \|z_i\|^2 + \|\bar{v}\|^2 < \beta\|u_j\|^2. \tag{6.11}$$

It follows from (6.10) and (6.11) that

$$\|u_i\|^2 > \frac{1}{1+\beta}, \qquad \|u_j\|^2 > \frac{1}{1+\beta},$$

$$\|z_j\|^2 < \frac{\beta}{1+\beta}, \qquad \|z_i\|^2 < \frac{\beta}{1+\beta}.$$

Hence

$$\begin{aligned}
\|w_i - w_j\|^2 &= \|u_i - z_i\|^2 + \|u_j - z_j\|^2 + \|v - \bar{v}\|^2 \\
&\geq (\|u_i\| - \|z_i\|)^2 + (\|u_j\| - \|z_j\|)^2 \\
&> \frac{2(1 - \sqrt{\beta})^2}{1+\beta},
\end{aligned}$$

and so $\text{dist}(K_i \cap S, K_j \cap S) \geq c(\beta)$.

The opposite inequality follows on choosing $v = \bar{v} = 0$, $u_i = (1/\sqrt{1+\beta})e_i$, $u_j = (1/\sqrt{1+\beta})e_j$, $z_i = (\sqrt{\beta}/\sqrt{1+\beta})e_i$, $z_j = (\sqrt{\beta}/\sqrt{1+\beta})e_j$, and noting that $w_i \in K_i \cap S$, $w_j \in K_j \cap S$. $\qquad\square$

Corollary 6.4. *The sets K_i are disjoint.*

Proof. If $w \in K_i \cap K_j$ for $i \neq j$, then $w/\|w\| \in K_i \cap K_j \cap S$. $\qquad\square$

Let $\eta \in C^\infty([0, \infty))$ satisfy $\eta \geq 0$ and

$$\eta(\tau) = \begin{cases} 1 & \text{if } \tau = 0, \\ 0 & \text{if } \tau \geq \frac{1}{3}c(\beta). \end{cases}$$

Define $h: S \to \mathbf{R}$ by

$$h(\theta) = 1 + \sum_{i=1}^{\infty}(i - 1)\eta(\text{dist}(\theta, K_i \cap S)).$$

Given $\theta \in S$, all the terms in the sum vanish in a neighbourhood of θ except perhaps one. Thus h is locally Lipschitz. Clearly $h \geq 1$ and $h(\theta) = i$ for $\theta \in K_i \cap S$.

We construct $S(t)$ through an ordinary differential equation

$$\dot{w} = F(w) \tag{6.12}$$

on $H \setminus \{0\}$, where $F: H \setminus \{0\} \to H$. For $w \notin \bigcup_{i=1}^{\infty} K_i$ we define

$$F(w) = -h\left(\frac{w}{\|w\|}\right)w.$$

To define F on each K_i we first define F on $L_i \setminus \{0\}$. Let $\psi_i \in C_0^\infty(\mathbf{R})$ with $0 \leq \psi_i \leq 1$, $\text{supp } \psi_i \subset (1/i, 3/i)$, $\psi_i(2/i) = 1$, and let $\theta_i \in C_0^\infty(\mathbf{R})$ with $0 \leq \theta_i \leq 1$, $\text{supp } \theta_i \subset (1/i, 4)$, $\theta_i(2/i) = 1$, $\theta_i' > -1$. Define $P_i: L_i \to L_i$ by

$$P_i(xe_i + y\hat{e}_i) = xe_i + (y + \psi_i(x)\theta_i(y))\hat{e}_i.$$

Since

$$\frac{\partial}{\partial y}(y + \psi_i(x)\theta_i(y)) = 1 + \psi_i(x)\theta_i'(y) > 0,$$

it is easily seen that P_i is a diffeomorphism satisfying $P_i = $ identity if $x \notin (1/i, 3/i)$ or $y \notin (1/i, 4)$. Consider the ordinary differential equation

$$\dot{p} = -ip \tag{6.13}$$

on L_i, whose trajectories are given by

$$p(t) = \exp(-it)p_0, \qquad p_0 \in L_i,$$

and the ordinary differential equation

$$\dot{q} = f_i(q)$$

on L_i whose trajectories are given by

$$q(t) = P_i(p(t)). \tag{6.14}$$

Differentiating (6.14) and using (6.13), we see that

$$f_i(q) = -iDP_i(P_i^{-1}(q))P_i^{-1}(q).$$

Note that f_i has the form

$$f_i(xe_i + y\hat{e}_i) = -ixe_i + g_i(x, y)\hat{e}_i,$$

where g_i is smooth with $g_i(x, y) = -iy$ if $x \notin (1/i, 3/i)$ or $y \notin (1/i, 4)$.

Let $w \in K_i$, so that $w = xe_i + y\hat{e}_i + v$ with $(v, e_i) = (v, \hat{e}_i) = 0$ and $\|v\|^2 < \beta(x^2 + y^2)$. Define

$$F(w) = -ixe_i + h_i(w)\hat{e}_i - iv,$$

where

$$h_i(w) := \left(1 - \frac{\|v\|^2}{\beta(x^2 + y^2)}\right) g_i(x, y) - \frac{\|v\|^2}{\beta(x^2 + y^2)} iy.$$

Note that $F(w) = f_i(w)$ if $w \in L_i\backslash\{0\}$, that $F(w) = -iw$ if $x \notin (1/i, 3/i)$ or $y \notin (1/i, 4)$, and that $F(w) + iw \to 0$ as $w \to \partial K_i\backslash\{0\}$.

We have thus defined $F\colon H\backslash\{0\} \to H$, and it is easily seen that F is locally Lipschitz on $H\backslash\{0\}$. If $w_0 \in H\backslash\{0\}$ there thus exists a unique continuous solution $w(t)$ of (6.12) satisfying $w(0) = w_0$, defined and remaining in $H\backslash\{0\}$ on a maximal time interval $[0, t_c)$, where $0 < t_c \leq \infty$. If $w_0 \notin \bigcup_{i=1}^\infty K_i$ then this solution is given by

$$w(t) = \exp\left(-h\left(\frac{w_0}{\|w_0\|}\right)t\right)w_0, \tag{6.15}$$

and so $t_c = \infty$. Also, since $h \geq 1$,

$$\|w(t)\| \leq \exp(-t)\|w_0\|. \tag{6.16}$$

Suppose $w_0 = x_0 e_i + y_0 \hat{e}_i + v \in K_i$. We have that

$$w(t) = x(t)e_i + y(t)\hat{e}_i + \exp(-it)v,$$

where $x(t) = \exp(-it)x_0$ and $\dot{y}(t) = h_i(w(t))$. By the backwards uniqueness of the solution (6.15) $w(t)$ cannot belong to $\partial K_i \setminus \{0\}$ for any t. Also $\dot{w}(t) = -iw(t)$ if $\|w(t)\| \leq \sqrt{2}/i$ or if $\|w(t)\| \geq 4\sqrt{2}$, and so $t_c = \infty$. If $x_0 \leq 1/i$ then $x(t) \leq 1/i$ for all $t \geq 0$ and thus $w(t) = \exp(-it)w_0$. So let $x_0 > 1/i$. Then $x(t_0) = 1/i$ where $t_0 = i^{-1}\ln(ix_0)$, and therefore

$$y(t) = \exp(-i(t-t_0))y(t_0) \qquad \text{for all } t \geq t_0.$$

But $h_i(w(t)) = -iy(t)$ if $|y(t)| \geq 4$, and so

$$|y(t_0)| \leq \max\{|y_0|, 4\}.$$

Hence, if $t \geq i^{-1}\ln(ix_0)$,

$$|y(t)| \leq \exp(-it)ix_0 \max\{|y_0|, 4\},$$

and thus

$$\begin{aligned}
\|w(t)\|^2 &\leq \exp(-2it)(|x_0|^2 + i^2|x_0|^2 \max\{|y_0|^2, 16\} + \|v_0\|^2) \\
&\leq i^2 \exp(-2it)(17|x_0|^2 + |x_0|^2|y_0|^2 + \|v_0\|^2) \\
&\leq i^2 \exp(-2it)(17\|w_0\|^2 + \|w_0\|^4) \\
&\leq \frac{1}{(te)^2}(17\|w_0\|^2 + \|w_0\|^4).
\end{aligned} \qquad (6.17)$$

For $t \geq 0$ we define $S(t)w_0 = w(t)$ if $w_0 \neq 0$ and $S(t)0 = 0$. Then the map $t \mapsto S(t)w_0$ is continuous on $[0, \infty)$ for all $w_0 \in H$, and from standard properties of ordinary differential equations we also have that $S(t)w_{0j} \to S(t)w_0$ whenever $w_{0j} \to w_0 \neq 0$. To show that $S(t)$ is continuous we must therefore show that $w_{0j} \to 0$ implies that $S(t)w_{0j} \to 0$. But if $t > 0$ and $w_0 = xe_i + y\hat{e}_i + v \in H$ with $\|w_0\| \leq te$, we have that

$$\|w(t)\| \leq \exp(-t)\|w_0\|,$$

if $w_0 \notin \bigcup_{i=1}^{\infty} K_i$ (by (6.16)) or if $w_0 \in K_i$ with $x_0 \leq 1/i$, and that (6.17) holds if $w_0 \in K_i$ with $x_0 > 1/i$, since $i^{-1}\ln(ix_0) \leq i^{-1}\ln(ite) \leq t$. Hence $S(t)w_{0j} \to 0$ if $w_{0j} \to 0$.

The same inequalities clearly imply that $\{0\}$ attracts bounded sets, and so it remains to show that the global attractor $A = \{0\}$ is not Lyapunov stable. Let $w_{0i} = 4i^{-1}(e_i + \hat{e}_i)$, $t_i = i^{-1}\ln 2$. Then by (6.14)

$$\begin{aligned}
S(t_i)w_{0i} &= P_i\left(2i^{-1}(e_i + \hat{e}_i)\right) \\
&= 2i^{-1}e_i + (2i^{-1} + 1)\hat{e}_i,
\end{aligned}$$

and thus $\|S(t_i)w_{0i}\| > 1$ for all i.

7. The Incompressible Navier-Stokes Equations

Let $\Omega \subset \mathbf{R}^3$ be a bounded open set with boundary $\partial\Omega$. Let $f \in L^2(\Omega)^3$ and consider the incompressible Navier-Stokes equations

$$u_t + (u \cdot \nabla)u = \nu\Delta u - \nabla p + f, \tag{7.1a}$$
$$\operatorname{div} u = 0, \tag{7.1b}$$

with boundary condition

$$u|_{\partial\Omega} = 0, \tag{7.2}$$

where $\nu > 0$ is a constant. (Similar results to those below can be established for the more realistic case of the nonzero boundary condition $u|_{\partial\Omega} = U$ provided Ω is of class C^2 and that $U = \operatorname{curl} V$ for a sufficiently smooth V, using the well-known device of Hopf [25].) As is customary, we use the function spaces

$$\mathcal{V} = \{u \in C_0^\infty(\Omega)^3;\, \operatorname{div} u = 0\},$$

$$H = \text{closure of } \mathcal{V} \text{ in } L^2(\Omega)^3,$$

$$V = \{u \in H_0^1(\Omega)^3;\, \operatorname{div} u = 0\}.$$

We denote by V' the dual space of V, and by H_w the space H endowed with its weak topology. We denote respectively by $(\cdot, \cdot)$ and $\|\cdot\|$ the inner product and norm in $L^2(\Omega)^3$, and for $u, v \in V$ we write

$$(Du, Dv) = \int_\Omega \sum_{i,j=1}^3 \frac{\partial u_i}{\partial x_j}\frac{\partial v_i}{\partial x_j}dx, \qquad \|Du\| = (Du, Du)^{\frac{1}{2}}.$$

For $u, v, w \in V$, we let

$$b(u, v, w) = \int_\Omega \sum_{i,j=1}^3 u_j v_{i,j} w_i dx.$$

We say that $u: [0, \infty) \to H$ is a *weak solution* of (7.1), (7.2) if $u \in C([0, T]; H_w) \cap L^2(0, T; V)$, $du/dt \in L^1(0, T; V')$ for all $T > 0$; if

$$\left(\frac{du}{dt}, v\right) + \nu(Du, Dv) + b(u, u, v) = (f, v) \qquad \text{for a.e. } t > 0, \tag{7.3}$$

for all $v \in V$, and if u satisfies the energy inequality

$$V(u)(t) \le V(u)(s) \qquad \text{for all } t \ge s, \tag{7.4}$$

for a.e. $s \in (0, \infty)$ and for $s = 0$, where

$$V(u)(t) := \frac{1}{2}\|u(t)\|^2 + \nu\int_0^t \|Du(\tau)\|^2 d\tau - \int_0^t (f, u(\tau))\, d\tau. \tag{7.5}$$

Standard theory [24], [28], [17], [42], [43] shows that given any $u_0 \in H$ there exists at least one weak solution with $u(0) = u_0$, constructed via a Galerkin method. Let G_{NS} denote the set of all weak solutions. G_{NS} satisfies (H1), (H3) but it is not known whether (H4) holds. (H2) would hold if $V(u)(t)$ were nonincreasing, but this does not follow directly from (7.4) which is consistent, for example, with the behaviour $V(u)(t) = 1$ for $t \in [0, 1)$, $V(u)(1) = 0$, $V(u)(t) = a$ for $t \in (1, \infty)$, where $0 < a \leq 1$. This undesirable behaviour cannot be eliminated simply by redefining the weak solution on a set of times of measure zero, since we have already chosen a representative which is continuous from $[0, \infty) \to H_w$.

In Proposition 7.4 below we show that G_{NS} is a generalized semiflow on H if and only if each weak solution u is continuous from $(0, \infty) \to H$. In preparation for this result we give in Proposition 7.3 a consequence of the energy inequality (7.4) that is well known to hold for any weak solution constructed via the Galerkin method; however, our proof does not assume that the weak solution is constructed in this way. We need two lemmas.

Lemma 7.1. *Let $\rho \in L^1_{loc}(0, \infty)$. Then the following conditions are equivalent:*
   (i) ρ has a nonincreasing representative $\bar{\rho}$: $(0, \infty) \to \mathbf{R}$,
   (ii) $\dot{\rho} \leq 0$ in $\mathcal{D}'(0, \infty)$.
If in addition ρ: $[0, \infty) \to \mathbf{R}$ is lower semicontinuous and continuous at zero, then (i) and (ii) are equivalent to
   (iii) $\rho(t) \leq \rho(s)$ for all $t \geq s$, for a.e. $s \in (0, \infty)$ and for $s = 0$.

Proof. The equivalence of (i) and (ii) is standard. For (i) $\Rightarrow$ (ii), one takes $\varphi \in \mathcal{D}(0, \infty)$, $\varphi \geq 0$, and passes to the limit $h \to 0+$ in

$$\int_0^\infty \frac{\rho(t) - \rho(t + h)}{h} \varphi(t) \, dt = \int_0^\infty \rho(t) \frac{\varphi(t) - \varphi(t - h)}{h} \, dt \geq 0,$$

which is valid for $h > 0$ sufficiently small, while (ii) $\Rightarrow$ (i) follows from mollifying ρ.

Suppose now that ρ: $[0, \infty) \to \mathbf{R}$ is lower semicontinuous, continuous at zero, and satisfies (i). Then $\rho(\tau) = \bar{\rho}(\tau)$ for all $\tau \notin N$, where N is a null set. Let $s > 0$, $s \notin N$, $t > s$, and $t_j \to t$ with $t_j \notin N$. Then

$$\rho(t) \leq \liminf_{j \to \infty} \rho(t_j) = \liminf_{j \to \infty} \bar{\rho}(t_j) \leq \bar{\rho}(s) = \rho(s), \tag{7.6}$$

while if $s = 0$ we obtain $\rho(t) \leq \rho(0)$ by passing to the limit $s_k \to 0$ in (7.6) with $s_k \notin N$. Hence (iii) holds.

Conversely, if (iii) holds, then $\bar{\rho}(t) := \sup_{\tau \geq t} \rho(\tau)$ defines a nonincreasing representative of ρ. $\qquad\square$

Lemma 7.2. *Let θ: $[0, \infty) \to \mathbf{R}$ be lower semicontinuous, continuous at zero, $\theta \in L^1(0, T)$ for all $T > 0$, and let θ satisfy, for some constant $c \geq 0$,*

$$\theta(t) + c \int_0^t \theta(\tau) \, d\tau \leq \theta(s) + c \int_0^s \theta(\tau) \, d\tau, \tag{7.7}$$

for all $t \geq s$, for a.e. $s > 0$, and for $s = 0$. Then

$$\theta(t)e^{ct} \leq \theta(s)e^{cs}, \tag{7.8}$$

for all $t \geq s$, for a.e. $s > 0$, and for $s = 0$.

Proof. Let $\rho(t) = \theta(t) + c \int_0^t \theta(\tau) \, d\tau$. Then $\rho \in L^1_{loc}(0, \infty)$, is lower semicontinuous, and continuous at zero. Hence by Lemma 7.1, $\dot{\rho} \leq 0$ in $\mathcal{D}'(0, \infty)$. Hence $\dot{\theta} + c\theta \leq 0$ in $\mathcal{D}'(0, \infty)$ and so $\frac{d}{dt}(\theta e^{ct}) \leq 0$ in $\mathcal{D}'(0, \infty)$. The result then follows from Lemma 7.1 applied to θe^{ct}. $\qquad\square$

Let λ_1 denote the lowest eigenvalue for the Stokes operator on Ω; thus,

$$\|Dv\|^2 \geq \lambda_1 \|v\|^2 \qquad \text{for all } v \in V. \tag{7.9}$$

Proposition 7.3. *Let u be a weak solution. Then*

$$\|u(t)\|^2 - \frac{1}{(\nu\lambda_1)^2}\|f\|^2 \leq e^{-\nu\lambda_1 t}\left(\|u(0)\|^2 - \frac{1}{(\nu\lambda_1)^2}\|f\|^2\right), \tag{7.10}$$

for all $t \geq 0$.

Proof. By (7.4)

$$\frac{1}{2}\|u(t)\|^2 + \nu \int_s^t \|Du(\tau)\|^2 d\tau \leq \frac{1}{2}\|u(s)\|^2 + \int_s^t \|f\| \cdot \|u(\tau)\| \, d\tau \tag{7.11}$$

for all $t \geq s$, for a.e. $s > 0$ and for $s = 0$. Using (7.9) and the inequality

$$\|f\| \cdot \|u(\tau)\| \leq \frac{1}{2}\left(\nu\lambda_1\|u(\tau)\|^2 + \frac{1}{\nu\lambda_1}\|f\|^2\right)$$

it follows that $\theta(t) := \|u(t)\|^2 - \frac{1}{(\nu\lambda_1)^2}\|f\|^2$ satisfies (7.7) with $c = \nu\lambda_1$. Further, since $u: [0, \infty) \to H_w$ is continuous, it follows that θ is lower semicontinuous and, hence also, from (7.7) with $s = 0$, continuous at zero. The result then follows from Lemma 7.2, taking $s = 0$ in (7.8). $\qquad\square$

Proposition 7.4. *The following conditions are equivalent:*
(i) G_{NS} is a generalized semiflow on H.
(ii) Each weak solution u is continuous from $(0, \infty)$ to H.
(iii) Each weak solution u is continuous from $[0, \infty)$ to H.

Proof. (i) $\Rightarrow$ (ii). Suppose G_{NS} is a generalized semiflow on H. Clearly G_{NS} has unique representatives and so by Theorem 2.1 we just need to show that each weak solution u is strongly measurable. But u is weakly continuous, and hence weakly measurable, and so by a well-known result [22, p. 73] u is strongly measurable.

(ii) $\Rightarrow$ (iii). Let u be a weak solution that is continuous from $(0, \infty) \rightarrow H$. Let $t_j \rightarrow$ 0+. Since $u \in C([0, T]; H_w)$ for all $T > 0$ we have that $\|u(0)\| \leq \liminf_{j \rightarrow \infty} \|u(t_j)\|$. But from (7.4) with $s = 0$ we have that $\limsup_{j \rightarrow \infty} \|u(t_j)\|^2 \leq \|u(0)\|^2$, and so $\|u(t_j)\| \rightarrow \|u(0)\|$. Hence $u(t_j) \rightarrow u(0)$ in H strongly, as required.

(iii) $\Rightarrow$ (i). Suppose each weak solution is continuous from $[0, \infty)$ to H. Then $V(u)(t)$ is continuous for $t \geq 0$ and hence $V(u)(t)$ is nonincreasing. In particular, (H2) holds. Also, if we define $\tilde{V}(u)(t) := \frac{1}{2}\|u(s)\|^2 \cdot \int^t (f, u(z))dz$, then $\tilde{V}(u)(t)$ is continuous for $t \geq 0°$ and nonincreasing in t. Let $u^{(j)}$ be a sequence of weak solutions with $u^{(j)}(0) \rightarrow u_0$ in H. By Proposition 7.3 $u^{(j)}$ is bounded in $L^\infty(0, \infty; H)$, and thus from the energy inequality (7.4) $u^{(j)}$ is bounded in $L^2(0, T; V)$ for every $T > 0$. A standard estimate then shows that

$$du^{(j)}/dt \quad \text{is bounded in } L^{4/3}(0, T; V') \tag{7.12}$$

for every $T > 0$, and hence using the usual compactness results that for a diagonal subsequence, which we do not relabel, there exists $u: [0, \infty) \rightarrow H$ with $u \in C([0, T]; H_w) \cap L^2(0, T; V)$ for all $T > 0$ such that

$$u^{(j)}(t) \rightharpoonup u(t) \text{ in } H \text{ for all } t \geq 0, \tag{7.13}$$

$$u^{(j)} \rightharpoonup u \text{ in } L^2(0, T; V) \text{ for all } T > 0, \tag{7.14}$$

$$du^{(j)}/dt \rightharpoonup du/dt \text{ in } L^{4/3}(0, T; V') \text{ for all } T > 0, \tag{7.15}$$

$$u^{(j)} \rightarrow u \text{ strongly in } L^2(0, T; H) \text{ for all } T > 0. \tag{7.16}$$

It follows from (7.16) that extracting a further subsequence we have that

$$u^{(j)}(t) \rightarrow u(t) \text{ in } H, \text{ a.e. } t > 0. \tag{7.17}$$

From (7.13)–(7.17) we deduce that u satisfies (7.3) and that $u(0) = u_0$. Also $\tilde{V}(u^{(j)})(s) \rightarrow \tilde{V}(u)(s)$ for a.e. $s > 0$ and for $s = 0$. If $t \geq s$ then since $V(u^{(j)})(t)$ is nonincreasing, by (7.13) and weak lower semicontinuity, we have that u satisfies the energy inequality (7.4). Hence u is a weak solution. Since, therefore, each $\tilde{V}(u^{(j)})(t)$ and $\tilde{V}(u)(t)$ are nonincreasing and continuous it follows that $\tilde{V}(u^{(j)})(t) \rightarrow \tilde{V}(u)(t)$ for all $t \geq 0$. But this implies that $\|u^{(j)}(t)\| \rightarrow \|u(t)\|$ and so $u^{(j)}(t) \rightarrow u(t)$ in H. Hence (H4) holds and G_{NS} is a generalized semiflow. $\qquad\square$

Corollary 7.5. *If G_{NS} is a generalized semiflow then it satisfies (C4).*

Proof. Let $u^{(j)}$ be as in the proof of Proposition 7.4. By Theorem 2.2 we have to show that if $t_j \rightarrow 0+$ then $u^{(j)}(t_j) \rightarrow u_0$ in H. We first show that $u^{(j)}(t_j) \rightharpoonup u_0$ in H. For $v \in V$ we have that

$$(u^{(j)}(t_j) - u^{(j)}(0), v) = \int_0^{t_j} \left(\frac{du^{(j)}}{dt}, v\right) d\tau. \tag{7.18}$$

Since $du^{(j)}/dt$ is bounded in $L^{4/3}(0, T; V')$, $(du^{(j)}/dt, v)$ is bounded in $L^{4/3}(0, T)$, and hence by Hölder's inequality the right-hand side of (7.18) tends to zero. Hence

$(u^{(j)}(t_j), v) \to (u_0, v)$ for all $v \in V$, and since $u^{(j)}(t_j)$ is bounded in H it follows that $u^{(j)}(t_j) \rightharpoonup u_0$ in H.

To prove strong convergence, note that in fact the convergence of $\tilde{V}(u^{(j)})(t)$ to $\tilde{V}(u)(t)$ is uniform on compact subsets of $[0, \infty)$, and so $\tilde{V}(u^{(j)})(t_j) \to \tilde{V}(u_0)$. From this it follows that $\limsup_{j \to \infty} \|u^{(j)}(t_j)\| \leq \|u_0\|$, and the strong convergence follows. $\qquad \square$

As an application of Theorem 3.3 we prove

Theorem 7.6. *Under the hypothesis that G_{NS} is a generalized semiflow there exists a global attractor for G_{NS}.*

Proof. By Proposition 7.4 G_{NS} is a generalized semiflow if and only if all weak solutions are continuous from $[0, \infty) \to H$. It then follows from Proposition 7.3 that G_{NS} is point dissipative. By Proposition 7.3 we also have that G_{NS} is eventually bounded. Using Proposition 3.2, to show that G_{NS} is asymptotically compact we thus need only show that G_{NS} is compact. But this follows from the argument in Proposition 7.4 (note that we do not need $u^{(j)}(0) \to u_0$ to conclude that $u^{(j)}(t) \to u(t)$ for all $t > 0$). $\qquad \square$

The connectedness of the attractor for G_{NS}, assuming that G_{NS} is a generalized semiflow, depends on whether Kneser's property holds. This does not seem easy to prove.

It follows from Theorem 6.1, Corollary 7.5 that the attractor for G_{NS} (assuming that G_{NS} is a generalized flow) is asymptotically stable. However, this does not use the full strength of Theorem 6.1 since in this case (C3) holds.

It might still be true that there is a global attractor in H if there exist weak solutions that are not continuous from $(0, \infty) \to H$. However, such a global attractor would not exist if there was a complete orbit that was bounded but not continuous, since such an orbit would have to be contained in the global attractor and would not be relatively compact on account of the continuity of weak solutions into H_w. It might be possible to eliminate any discontinuous solutions from $(0, \infty) \to H$ by means of some unknown admissibility criterion; if this could be done in such a way that the resulting family of solutions $\hat{G}_{NS} \subset G_{NS}$ formed a generalized semiflow, then the above methods would guarantee the existence of a global attractor for $\hat{G}_{NS}$ in H.

The existence and properties of a global attractor have been studied by Constantin, Foias & Temam [18] under the a priori assumption that all solutions with initial data in V remain bounded in V for all finite times. The only results on the existence of a global attractor in H which do not make unsubstantiated assumptions on the solutions seem to be those of Raugel & Sell [32], [33], [34] for suitable thin 3D domains.

Foias & Temam [20] have introduced the concept of a *universal attractor* $\tilde{A}$ for the 3D Navier-Stokes equations. In the definition of a weak solution they drop the requirement that the energy equation (7.4) hold with $s = 0$ (as does [39]). With this definition translates of weak solutions are weak solutions. Let us call the corresponding (possibly larger) set of weak solutions $\tilde{G}_{NS}$. For $\tilde{G}_{NS}$ the statements of our theorems need slight modification. Proposition 7.3 holds only for solutions that are continuous at zero, and Proposition 7.4 is replaced by the assertions (i) that $\bar{G}_{NS}$ is a generalized semiflow implies that each weak solution is continuous from $(0, \infty) \to H$, and (ii) that if each weak solution is continuous from $[0, \infty) \to H$ (i.e., (C3) holds) then $\tilde{G}_{NS}$ is a generalized semiflow. Corollary 7.5 is dropped, and Theorem 7.6 is replaced by the statement that a global attractor exists for $\tilde{G}_{NS}$ provided (C3) holds. Foias & Temam then define $\tilde{A}$ as the set of all $u_0 \in H$ through which there passes a complete orbit $u: \mathbf{R} \to H$. If (C3) holds, then it is easily seen that $\tilde{A}$ coincides with the global attractor A in Theorem 7.6. It does not seem to be obvious without making a priori assumptions on solutions that $\tilde{A}$ attracts bounded sets in the sense of weak convergence in H; i.e., that if $u^{(j)}$ are weak solutions with $u^{(j)}(0)$ bounded, and if $t_j \to \infty$, then $u^{(j)}(t_j)$ has a subsequence converging to a point of $\tilde{A}$. This is because the estimate (7.10) is not proved, so that it is not even clear that $u^{(j)}(t_j)$ is bounded. This weak attraction property would hold, however, if we knew that $V(u)(t)$ were nonincreasing for all weak solutions.

Acknowledgements

I am grateful to G.R. Sell and Y. You for providing me with prepublication versions of [40], and to P. Constantin, J.K. Hale, and J. Howie for helpful comments. This version of the paper incorporates a modification to the proofs of Proposition 7.4 and Corollary 7.5 (see Erratum, J. Nonlinear Science 8 (1998) p 233. I am grateful to G. Francfort for querying this point.

References

[1] H. Auerbach. Sur la relation $\lim_{h_n \to \infty} f(x + h_n) = f(x)$. *Fund. Math.*, 11:193–197, 1928.

[2] A.V. Babin. Attractor of the generalized semigroup generated by an elliptic equation in a cylindrical domain. *Izv. Russ. Akad. Nauk.*, 58, 1994. English translation in *Russian Acad. Sci. Izv. Math.* 44:207–223, 1995.

[3] A.V. Babin and M.I. Vishik. Maximal attractors of semigroups corresponding to evolution differential equations. *Math. Sbornik*, 126:397–419, 1985. English translation in *Math. USSR Sbornik*, 54:387–408, 1986.

[4] A.V. Babin and M.I. Vishik. *Attractors of Evolution Equations*. Nauka, Moscow, 1989. English translation, North-Holland, 1992.

[5] J.M. Ball. Global attractors for damped semilinear wave equations. In preparation.

[6] J.M. Ball. Continuity properties of nonlinear semigroups. *J. Funct. Anal.*, 17:91–102, 1974.

[7] J.M. Ball. Measurability and continuity conditions for nonlinear evolutionary processes. *Proc. Amer. Math. Soc.*, 55:353–358, 1976.

[8] J.M. Ball. On the asymptotic behaviour of generalized processes, with applications to nonlinear evolution equations. *J. Differential Eqns.*, 27:224–265, 1978.

[9] J.M. Ball, J. Carr, and O. Penrose. The Becker-Döring cluster equations; Basic properties and asymptotic behaviour of solutions. *Commun. Math. Phys.*, 104:657–692, 1986.

[10] E.A. Barbashin. On the theory of generalized dynamical systems. *Moskov. Gos. Ped. Inst. Učen. Zap.*, 2:110–133, 1948. English translation by U.S. Department of Commerce, Office of Technical Services, Washington D.C. 20235.

[11] J.E. Billotti and J.P. LaSalle. Periodic dissipative processes. *Bull. Amer. Math. Soc.*, 6:1082–1089, 1971.

[12] I.U. Bronstein. On dynamical systems without uniqueness as semigroups of non-singlevalued mappings of a topological space. *Izv. Akad. Nauk. Moldav. SSR*, 1:3–17, 1963. English translation in *Amer. Math. Soc. Transl.*, 97:205–225, 1970.

[13] B.M. Budak. The concept of motion in a generalized dynamical system. *Moskov Gos. Ped. Učen. Zap.*, 5:174–194, 1952. English translation in *Amer. Math. Soc. Transl.*, 97:205–225, 1970.

[14] V.V. Chepyzhov and M.I. Vishik. Trajectory attractors for evolution equations. *Comptes Rendus Acad. Sci. I*, 321:1309–1314, 1995.

[15] P.R. Chernoff. A note on continuity of semigroups of maps. *Proc. Amer. Math. Soc.*, 53:318–320, 1975.

[16] P.R. Chernoff and J.E. Marsden. On continuity and smoothness of group actions. *Bull. Amer. Math. Soc.*, 76:1044–1049, 1970.

[17] P. Constantin and C. Foias. *Navier-Stokes equations*. University of Chicago Press, Chicago, 1989.

[18] P. Constantin, C. Foias, and R. Temam. Attractors representing turbulent flows. *Mem. Amer. Math. Soc.*, 53(314):1–67, 1985.

[19] J.R. Dorroh. Semi-groups of maps in a locally compact space. *Can. J. Math.*, 19:688–696, 1967.

[20] C. Foias and R. Temam. The connection between the Navier-Stokes equations, dynamical systems, and turbulence theory. In M.G. Crandall, P.H. Rabinowitz, and R.E.L. Turner, editors, *Directions in Partial Differential Equations*, pages 55–73. Academic Press, Boston, 1987.

[21] J.K. Hale. *Asymptotic Behavior of Dissipative Systems*, Amer. Math. Soc., Providence, RI, 1988.

[22] E. Hille and R.S. Phillips. *Functional Analysis and Semi-groups*, volume 31 of *Colloq. Publ. Col.*, Amer. Math. Soc., Providence, RI, 1957.

[23] J.G. Hocking and G.S. Young. *Topology*. Addison-Wesley, Reading, MA, 1961.

[24] E. Hopf. Uber die aufangswertaufgabe für die hydrodynamischen grundgliechungen. *Math. Nachr.*, 4:213–231, 1951.

[25] E. Hopf. On nonlinear partial differential equations. In *Lecture Series of the Symposium on Partial Differential Equations, Berkeley, 1955*, pages 1–29. The University of Kansas, 1957.

[26] J.K. Hale, J.P. LaSalle, and M. Slemrod. Theory of a general class of dissipative processes. *J. Math. Anal. Appl.*, 39:177–191, 1972.

[27] O. Ladyzhenskaya. *Attractors for semigroups and evolution equations*. Cambridge University Press, Cambridge, 1991.

[28] J-L. Lions. *Quelques méthodes de résolution des problèmes aux limites nonlinéaires*. Dunod, Paris, 1969.

[29] V.S. Mel'nik. Multivalued semiflows and their attractors. *Dokl. Akad. Nauk.*, 343:302–305, 1995. English translation in *Dokl. Math.*, 52:36–39, 1995.

[30] M.I. Minkevic. A theory of integral funnels for dynamical systems without uniqueness. *Moskov. Gos. Ped. Inst. Učen. Zap.*, 135:134–151, 1948. English translation in *Amer. Math. Soc. Transl.*, 95:11–34, 1970.

[31] J.C. Oxtoby. *Measure and Category*. Springer-Verlag, New York, 1971.

[32] G. Raugel and G.R. Sell. Navier-Stokes equations on thin 3D domains. I. Global attractors and global regularity of solutions. *J. Amer. Math. Soc.*, 6:503–568, 1993.

[33] G. Raugel and G.R. Sell. Navier-Stokes equations on thin 3D domains. II. Global regularity of spatially periodic solutions. In *Nonlinear Partial Differential Equations and Their Applications*, College de France Seminar, Vol. XI, 205–247, Pitman Research Notes in Mathematics, 299, Longman, 1994.

[34] G. Raugel and G.R. Sell. Navier-Stokes equations on thin 3D domains. III. Global and local attractors. In *Turbulence in Fluid Flows: A Dynamical Systems Approach*, IMA Volumes in Mathematics and Its Applications, Vol. 55, 137–163, Springer-Verlag, New York, 1993.

[35] E. Roxin. On generalized dynamical systems defined by contingent equations. *J. Differential Eqns.*, 1:188–205, 1965.

[36] E. Roxin. Stability in general control systems. *J. Differential Eqns.*, 1:115–150, 1965.

[37] G.R. Sell. On the fundamental theory of ordinary differential equations. *J. Differential Eqns.*, 1:370–392, 1965.

[38] G.R. Sell. Differential equations without uniqueness and classical topological dynamics. *J. Differential Eqns.*, 14:42–56, 1973.

[39] G.R. Sell. Global attractors for the three-dimensional Navier-Stokes equations. *J. Dyn. Differential Eqns.*, 8:1–33, 1996.

[40] G.R. Sell and Y. You. *Dynamics of Evolutionary Equations*. Book in preparation, Springer-Verlag, New York.

[41] G.P. Szego and G. Treccani. *Semigruppi di Trasformazioni Multivoche*, volume 101 of *Lecture Notes in Mathematics*. Springer-Verlag, Berlin, 1969.

[42] R. Temam. *Infinite-Dimensional Dynamical Systems in Mechanics and Physics*, volume 68 of *Applied Mathematical Sciences*. Springer-Verlag, New York, 1988.

[43] W. Von Wahl. *The Equations of Navier-Stokes and Abstract Parabolic Equations*. Fried. Vieweg & Sohn, Braunschweig, 1985.

Stacked Lagrange Tops

D. Lewis

Mathematics Board, University of California at Santa Cruz, Santa Cruz, CA 95064, USA
e-mail: lewis@slinky.ucsc.edu

Received July 25, 1996; revised manuscript received July 7, 1997; accepted for publication July 21, 1997
Communicated by Jerrold Marsden

This paper is dedicated to the memory of Juan-Carlos Simo

Summary. We investigate the relative equilibrium and stability conditions for systems of Lagrange tops coupled by universal joints at points on their axes of symmetry and moving under the influence of gravity. We demonstrate the existence of large classes of T^{n+1}-orbitally stable, steady motions of n Lagrange tops for arbitrary n. The symmetry and stability properties of the relative equilibria are closely related to the structure of an $n \times n$ matrix arising in the equilibrium and stability analyses. This structure plays a crucial role in the analysis of the classes of relative equilibria identified here.

1. Introduction

The Lagrange top is one of the best known and most thoroughly analysed mechanical systems. Classic references include Routh [1905] and Klein and Sommerfeld [1910]. Systems of multiple axisymmetric tops connected by universal joints have also been studied, although not as thoroughly as the single Lagrange top. Wittenburg [1977] has studied the dynamics of a wide variety of coupled rigid-body systems and developed computationally efficient formulations of the equations of motion of such systems. Vassileva and Lilov [1987] have derived variational characterizations of the relative equilibrium and formal stability conditions for systems of an arbitrary number of Lagrange tops and identified a simple class of stable relative equilibria existing for arbitrary numbers of tops.[1] Patrick [1989] has analysed in detail the relative equilibria of a pair of coupled Lagrange tops moving in the absence of gravity.

---

[1] I learned of the paper of Vassileva and Lilov [1987] (in Russian) after the present paper was completed. I have kept the derivations of the equilibrium and stability conditions given in the appendix for the sake of clarity and completeness.

The goal of the present work is to develop systematic procedures for the solution of the relative equilibrium equations and for the analysis of the formal and nonlinear stability of the relative equilibria that can be applied efficiently to arbitrarily large numbers of linked Lagrange tops. We do not focus on the derivation of the general relative equilibrium and stability conditions (i.e., the characterization of a relative equilibrium of a Hamiltonian system as a critical point of an appropriate function, and a stable relative equilibrium as a local minimum modulo symmetries of that function), but on the structure of the equalities and inequalities resulting from such a characterization for systems of coupled tops. Whereas the initial formulations of the equilibrium and stability conditions are derived using the techniques of geometric mechanics, the derivation of the final simplified formulations of these conditions relies almost entirely on a few basic results from linear algebra. By exploiting the structure of a family of matrices that appear in both the equilibrium and stability analyses, we obtain simple criteria for stable, steady motion that can be verified easily by hand for relatively small numbers of tops and are well suited for numerical treatment even for very large numbers of tops.

A crucial feature of the steady motions of systems of tops coupled in sequence by universal joints is that the relative equilibrium conditions can be expressed as a pair of homogeneous linear equations and a set of uncoupled nonlinear equations. If the axes of symmetry $s_1, \ldots, s_n$ of the tops have coordinates $s_i = (h_{1i}, h_{2i}, v_i)$, then a configuration with horizontal projections $h_1 = (h_{11}, \ldots, h_{1n})$ and $h_2 = (h_{21}, \ldots, h_{2n})$ determines a precessing relative equilibrium for some choice of spin rates if there exists $\sigma \in \mathbb{R}^n$ such that h_1 and h_2 are both null-vectors of the symmetric matrix $K(\sigma) = \text{diag} [\sigma] - K$. The entries of K are fixed parameters of the system determined by the masses of the tops and the positions of the centers of mass and the joints on the axes of symmetry of the tops. The matrix $K(\sigma)$ has a particularly convenient structure that significantly simplifies several aspects of the analysis. Specifically, it is the sum of a triangular matrix and a rank one matrix; thus, the associated linear system of equations can be solved relatively easily even for large n, and a tight upper bound on the nullity of $K(\sigma)$ is obtained simply by counting the zero entries in σ.

We study the nonlinear T^{n+1}-orbital stability of the relative equilibria using the reduced energy momentum method of Simo et al. [1991a]. Relative equilibria are characterized as configurations that are critical points of appropriate functions on the configuration manifold; these functions are simply the composition $L_\xi = L \circ \xi_Q$ of the Lagrangian L with the infinitesimal generator of the steady motion, i.e., the vector field on the configuration manifold associated to the infinitesimal group motion. Formally stable relative equilibria are those for which a particular matrix constructed from the second variation of L_ξ is positive semidefinite with nullvectors consisting only of momentum-preserving infinitesimal group motions. The stability matrices for the coupled top systems are largely determined by the matrix $K(\sigma)$ appearing in the determination of the relative equilibria.

The relative equilibria of the system under investigation can be grouped in three classes according to their symmetries: sleeping configurations, in which all tops are vertical; planar configurations, in which the axes of symmetry of all of the tops lie in a common vertical plane; and nonplanar configurations. This classification is relevant to both the equilibrium and stability analyses. The structure of the matrix $K(\sigma)$ is of particular interest in the case of nonplanar configurations. Nonplanar configurations are possible only if at least one component of the vector σ equals zero, in which case the

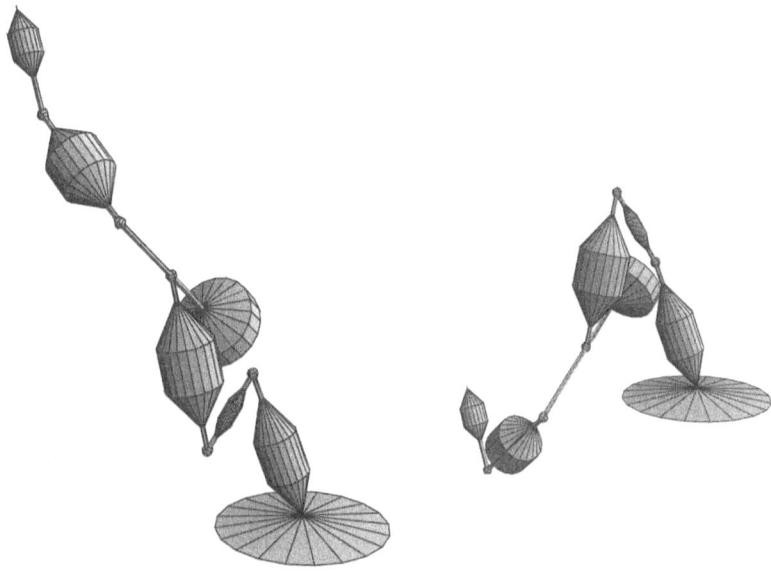

Fig. 1. Two stable relative equilibria hinged at the fourth top. These two relative equilibria have the same horizontal projections, but different vertical projections and spin rates.

matrix $K(\sigma)$ has an "overlapping block" structure that allows the decomposition of the homogeneous system of equations determined by $K(\sigma)$ into three subsystems, each of which is again the sum of a triangular matrix and a rank one matrix. If σ has precisely one zero component, then the nonplanar relative equilibria determined by $K(\sigma)$ can be viewed as two planar configurations "hinged" at some intermediate top. (See Figure 1.) The formal stability test for hinged relative equilibria simplifies to testing two subblocks of $K(\sigma)$ — one determined by the tops before the hinge and the other determined by the tops after the hinge — for positive definiteness and determining the sign of the product of two scalar parameters associated to the hinge top. The latter condition has a simple physical interpretation that the hinge top acts as a counterbalance; in particular, the point of attachment of the hinge top to the previous top must lie between the center of mass of the hinge top and the point of attachment of the subsequent top. (See Figures 1 and 2.) We shall see that the stability of the relative equilibria in Figure 1 can be determined essentially by inspection; the only quantitative information needed is that the hinge top counterbalances the final two tops.

For the most general classes of relative equilibria, the identification of sharp stability conditions — i.e., conditions that are both necessary and sufficient — seems to require a detailed analysis of the particular matrices at hand. Such analyses can readily be carried out numerically, and we briefly discuss some techniques that increase the efficiency of the numerical treatment. For example, for a sleeping configuration, in which all of the tops are vertical, the generator of the steady motion is not unique; a sleeping relative equilibrium is stable if at least one of the associated matrices is definite. While sharp

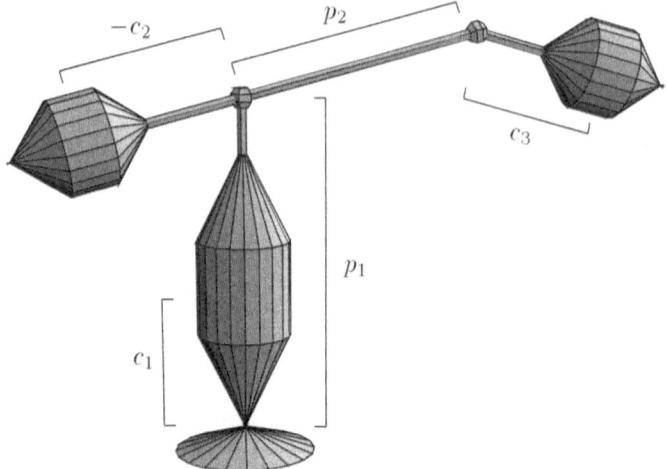

Fig. 2. A representative three-top configuration. Note that $x_2 = m_2 c_2 + m_3 p_2$ may be positive, negative, or zero, depending on the relative masses of the second and third tops.

stability conditions may be difficult to obtain analytically for sleeping configurations, we find simple but nontrivial sufficient conditions for nonlinear stability. By comparing these conditions to some equally simple necessary conditions for formal stability, one can restrict a numerical stability test to the relatively small regions remaining in which the formal stability is unknown.

We conclude this introduction by mentioning one more motivation for the present study. These systems serve as prototypes in the development of general techniques for the analysis of systems of coupled deformable bodies. Relative equilibrium conditions are typically obtained as systems of coupled nonlinear equations that can be solved analytically only in special cases or for low dimensional systems. Manipulation of the equations into a more convenient form appears to be an essential preliminary to a detailed analysis of the relative equilibria of most systems. The equilibrium and stability results derived here can be incorporated into a screening procedure for steady motions of systems of deformable axisymmetric bodies coupled by universal joints and moving under the influence of gravity. Such a screening procedure is motivated by Simo's observation that a relative equilibrium of a general frame invariant system corresponds to a relative equilibrium of the associated rigid-body system obtained by "locking" the system in its equilibrium configuration; in addition, a formally stable relative equilibrium of the general system corresponds to a stable relative equilibrium of the locked system. Thus, when searching for (stable) relative equilibria, one need consider only configuration/generator pairs forming (stable) relative equilibria of the corresponding locked system. Such pairs often possess additional structure that simplifies the analysis of the remaining equilibrium and stability conditions. See, e.g., Simo et al. [1989], Lewis and Simo [1990], Simo et al. [1991a, b], and Lewis [1993].

2. Dynamics

2.1. Equations of Motion

The system under consideration consists of n coupled rigid bodies moving under the influence of gravity. We assume that each body is a Lagrange top, i.e., each body is axisymmetric, with the center of mass on the axis of symmetry, and that the tops are attached to one another in sequence by universal joints on their axes of symmetry. The first top is assumed to have one point on its axis of symmetry fixed in space. The double eigenvalue of the inertia tensor of the i-th top is denoted by I_i; the remaining eigenvalue is denoted I_i'. Each configuration can be identified with a point $\Lambda := (\Lambda_1, \ldots, \Lambda_n) \in (SO(3))^n$ by selecting a reference spatial position for each top and denoting by $\Lambda_i \in SO(3)$ the rotation matrix taking the i-th top from its reference position to its position in the configuration.

The Lagrangian for the coupled top system is

$$L(\dot{\Lambda}_1, \dot{\Lambda}_2, \ldots, \dot{\Lambda}_n)$$
$$= \sum_{i=1}^{n} \int_{\mathcal{B}_i} \rho_i(X) \left(\tfrac{1}{2} \left| \dot{\Lambda}_i X + \sum_{j<i} \dot{\Lambda}_j p_j \right|^2 - g \cdot \left(\Lambda_i X + \sum_{j<i} \Lambda_j p_j \right) \right) dX. \quad (1)$$

Here g denotes the gravitational vector and p_i denotes the vector from the point at which the i-th top is attached to the $(i-1)$-th top to the point at which it is attached to the $(i+1)$-th top. The first top is assumed to have a point fixed in space — p_1 is the vector from this fixed point to the point of attachment of the second top. The coordinates X within each top are taken relative to the point of attachment to the previous top.

The Lagrangian (1) can be expressed in a more convenient form using the spatial angular velocities and an appropriate choice of reference positions. If we define the spatial angular velocities $\omega_1, \ldots, \omega_n \in \mathbb{R}^3$ through the relations

$$\widehat{\omega}_i \Lambda_i y := \omega_i \times (\Lambda_i y) = \dot{\Lambda}_i y, \qquad \text{for all } y \in \mathbb{R}^3, \quad (2)$$

then the point $(\dot{\Lambda}_1, \ldots, \dot{\Lambda}_n) \in (TSO(3))^n \approx T(SO(3)^n)$ can be identified with the pair $(\Lambda, \Omega) \in SO(3)^n \times \mathbb{R}^{3n}$ with components $\Lambda = (\Lambda_1, \ldots, \Lambda_n)$ and $\Omega = (\omega_1, \ldots, \omega_n)$. We denote the mass of the i-th top by m_i and let $\mathbf{c}_i$ denote the vector from the i-th top's point of attachment to the $(i-1)$-th top to the i-th top's center of mass. For convenience, we select an inertial frame $\{e_1, e_2, e_3\}$ and reference positions such that $g = g e_3$ and $p_i = p_i e_3$ for $p_i \geq 0$, $i = 1, \ldots, n-1$. We introduce the additional scalar constants $p_n := 0$,

$$c_i := \langle \mathbf{c}_i, e_3 \rangle, \qquad x_i := m_i c_i + p_i \sum_{i<k} m_k, \qquad d_i := I_i' - I_i - m_i c_i p_i,$$

$i = 1, \ldots, n$, and

$$\kappa_{ij} := d_i \delta_{ij} + p_{\min(i,j)} x_{\max(i,j)},$$

for $i, j = 1, \ldots, n$, and the current spatial inertia tensor $\mathcal{I}: SO(3)^n \to L(\mathbb{R}^{3n})$, with ij-th three by three block

$$\mathcal{I}(\Lambda)_{ij} = (I_i \delta_{ij} + \kappa_{ij} \langle s_i, s_j \rangle) \mathbf{1} - \kappa_{ij} s_i s_j^T,$$

where the unit vector $s_i := \Lambda_i e_3$ specifies the current orientation of the axis of symmetry of the i-th top with respect to the inertial frame $\{e_1, e_2, e_3\}$. Using this notation, the Lagrangian (1) can be expressed as a function of (Λ, Ω) in the concise form

$$L(\Lambda, \Omega) = \tfrac{1}{2}\Omega \cdot \mathcal{I}(\Lambda)\Omega - \langle g, \sum_{i=1}^{n} x_i \Lambda_i e_3 \rangle$$

$$= \sum_{i=1}^{n} \left(\tfrac{l_i}{2} |\omega_i|^2 - x_i \langle g, s_i \rangle \right) + \tfrac{1}{2} \sum_{i,j=1}^{n} \kappa_{ij} \langle \omega_i \times s_i, \omega_j \times s_j \rangle. \tag{3}$$

Note that we use the notation $\langle \, , \, \rangle$ to denote the Euclidean inner product in the "physical space" $\mathbb{R}^3$, while the natural pairing of an element of a vector space with an element of the dual space and the Euclidean inner product on $\mathbb{R}^n$ are both indicated with a dot.

The Euler-Lagrange equations determined by (3) are

$$\dot{\Lambda}_i = \hat{\omega}_i \Lambda_i \qquad \text{and} \qquad I_i \dot{\omega}_i + s_i \times \left(x_i g + \sum_{j=1}^{n} \kappa_{ij} \ddot{s}_j \right) = 0, \tag{4}$$

for $i = 1, \dots, n$, where

$$\ddot{s}_i = \dot{\omega}_i \times s_i + \omega_i \times (\omega_i \times s_i),$$

as follows from differentiation of $s_i = \Lambda_i e_3$.

2.2. Symmetries and Relative Equilibria

The Euler-Lagrange equations (4) are equivariant with respect to the action of the torus $T^{n+1} = (S^1)^{n+1}$ on $T(SO(3))^n$, where the i-th S^1 acts by body rotation of the i-th top about its axis of symmetry and the final S^1 acts by simultaneous spatial rotations of all of the bodies about the vertical axis. Specifically, if we associate to $\theta \in S^1$ the rotation matrix $R_\theta = \exp(\theta \, e_3) \in SO(3)$ determining a (clockwise) rotation through the angle θ about the vertical axis, then

$$(\theta_1, \dots, \theta_n, \varphi) \cdot (\Lambda, \Omega) = ((R_\varphi \Lambda_1 R_{-\theta_1}, \dots, R_\varphi \Lambda_n R_{-\theta_n}), (R_\varphi \omega_1, \dots, R_\varphi \omega_n)). \tag{5}$$

To obtain sharp results, it is necessary to take the full symmetry group into account when analysing the system. To keep the discussion of the symmetries as simple as possible, we now introduce nondegeneracy assumptions that guarantee that the symmetry group is precisely T^{n+1}. If $p_i = c_i = 0$ for some index i, then dynamics of the i-th top relative to its point of attachment is not influenced by the motion of the rest of the system, and its only influence on the other tops is to add its mass to that of the $(i-1)$-th top. In this situation the i-th top acts as a free rigid body and can be arbitrarily rotated about its point of attachment without changing the energy or essentially altering the evolution of the system. If $p_1 = 0$, then the first top acts as a heavy top moving independently of the other $n-1$ tops. We restrict our attention to the case in which the symmetry group is precisely T^{n+1} by imposing the nondegeneracy hypotheses $p_1 \neq 0$, $c_n \neq 0$, and $p_i + |c_i| \neq 0$ for $i = 2, \dots, n-1$.

Remark. Equation (4) implies that $\langle \omega_i, s_i \rangle$ is constant for $i = 1, \dots, n$. Thus the "non-trivial" dynamics of the system are those of the axes of symmetry $s_1, \dots, s_n$. The dynamical system on $(S^2)^n$ determined by the equations

$$\mathbb{P}_{s_i} \left(I_i (\ddot{s}_i + \eta_i \dot{s}_i \times s_i) + x_i g + \sum_{i=1}^{n} \kappa_{ij} \ddot{s}_j \right) = 0,$$

for $i = 1, \ldots, n$, obtained by taking the cross product of the left-hand side of (4) with s_i and setting $\eta_i := \langle \omega_i(0), s_i(0) \rangle$, is equivalent to the system determined by symplectic reduction of the Hamiltonian version of (4) with respect to the action of the group T^n of body rotations. (See, for example, Abraham and Marsden [1978].) Here and throughout the paper, $\mathbb{P}_y$ denotes orthogonal projection onto the plane perpendicular to the vector $y \in \mathbb{R}^3$.

Relative equilibria are steady motions generated by one-parameter subgroups of the symmetry group of the system. For a system of n heavy Lagrange tops satisfying the nondegeneracy conditions given above, the possible equilibrium motions consist of a steady rotation of each top about its axis of symmetry composed with an overall steady spatial rotation about the axis of gravity. The generator of such a relative equilibrium consists of a pair $\zeta \in \mathbb{R}^n$ and $\xi \in \mathbb{R}$, where ζ_i is the rate of spin of the i-th top about its axis of symmetry and ξ is the rate of overall spatial rotation. The motion determined by such a generator is

$$(\Lambda(t), \Omega(t)) = (e^{it\zeta_1}, \ldots, e^{it\zeta_n}, e^{it\xi}) \cdot (\Lambda(0), \Omega(0)), \tag{6}$$

where the action of T^{n+1} on $SO(3)^n \times \mathbb{R}^{3n}$ is given by (5).

Differentiating (6) with respect to t and comparing the resulting equation for $\dot{\Lambda}$ to (2), we see that if (Λ, Ω) is a relative equilibrium, then the solution curve of the Euler-Lagrange equations with initial condition (Λ, Ω) must satisfy

$$\omega_i(t) = \xi\, e_3 - \zeta_i s_i(t), \tag{7}$$

for $i = 1, \ldots, n$ and all t. Differentiating (7) with respect to t yields $\dot{\omega}_i = -\zeta_i \dot{s}_i = \zeta_i \xi\, s_i \times e_3$. Substituting this expression into (4) yields the remaining relative equilibrium conditions,

$$s_i(t) \times \left((I_i \zeta_i \xi + g\, x_i)\, e_3 - \xi^2 \mathbb{P}_{e_3} \sum_{j=1}^n \kappa_{ij} s_j(t) \right) = 0, \tag{8}$$

for $i = 1, \ldots, n$. Note that if the relative equilibrium conditions (7) and (8) are satisfied at any time, then they are satisfied at all times.

Remark. The relative equilibrium conditions also have a variational characterization. In fact, the relative equilibrium conditions can be derived directly from the Lagrangian (3) without making use of the Euler-Lagrange equations (4). The variational approach, which is central to the nonlinear stability analysis, is discussed in the appendix.

We shall see that the relative equilibrium conditions for the coupled top systems determine a system of homogeneous linear equations on the projections of the spatial axes of symmetry $(s_1, \ldots, s_n)$ into the horizontal plane. All coupling between the tops is encoded in this linear system; the remaining n nonlinear scalar equations are easily solved. Thus the determination of the relative equilibria is essentially reduced to linear algebra.

If $\xi = 0$, then (8) immediately yields the relative equilibrium conditions that $x_i s_i$ be parallel to g for $i = 1, \ldots, n$. To study relative equilibria with generator (ζ, ξ), $\xi \neq 0$, we observe that if we temporarily ignore the constraints $|s_i| = 1$, then the system of equations (8) consists of two systems of n homogeneous linear equations each and n uncoupled nonlinear scalar equations. Specifically, let h_1, h_2, and $v \in \mathbb{R}^n$ denote the vectors with components

$$h_{1i} := \langle s_i, e_1 \rangle, \qquad h_{2i} := \langle s_i, e_2 \rangle, \qquad \text{and} \qquad v_i := \langle s_i, e_3 \rangle.$$

Define $K: \mathbb{R}^n \to \mathrm{Sym}(n)$ by

$$K(\sigma)_{ij} := \sigma_i \delta_{ij} - p_{\min(i,j)} x_{\max(i,j)}. \tag{9}$$

If $\xi \neq 0$, then (8) is satisfied if and only if there exists $\sigma \in \mathbb{R}^n$ such that

$$K(\sigma)h_1 = K(\sigma)h_2 = 0, \tag{10}$$

and

$$\xi^2(\sigma_i + d_i)v_i + \xi I_i \zeta_i + g x_i = 0, \qquad \text{for } i = 1, \dots, n. \tag{11}$$

We can identify relative equilibria by first finding triplets (σ, h_1', h_2') satisfying (10) and then, taking advantage of the homogeneity of (10), incorporating the constraints $|s_i| = 1$ by restricting our attention to pairs (h_1, h_2) of vectors in $\mathrm{span}\left[h_1', h_2'\right]$ such that $h_{1i}^2 + h_{2i}^2 \leq 1$, $i = 1, \dots, n$. Setting $v_i^{\pm} = \pm\sqrt{1 - h_{1i}^2 - h_{2i}^2}$ and taking (11) as defining ζ as a function of σ, v, and ξ, which is arbitrary other than being nonzero, we obtain a family of relative equilibria with configurations $\Lambda = (\Lambda_1, \dots, \Lambda_n)$ for any rotation matrices Λ_i satisfying $\Lambda_i e_3 = s_i = (h_{1i}, h_{2i}, v_i^{\pm})$. Thus we obtain the following characterization of relative equilibria, which was originally derived in a slightly different form by Vassileva and Lilov [1987].

Proposition 1. *Relative equilibria with spatial axes of symmetry $s_1, \dots, s_n$ and generator (ζ, ξ), $\xi \neq 0$, exist if and only if there exists $\sigma \in \mathbb{R}^n$ such that $\ker[K(\sigma)] \supset \mathrm{span}[h_1, h_2]$, where $h_{ki} = \langle s_i, e_k \rangle$. If the i-th top is tilted out of the vertical, so that $h_{ki} \neq 0$ for $k = 1$ or 2, then*

$$\sigma_i = p_i x_i + \frac{1}{h_{ki}} \sum_{j \neq i} \kappa_{ij} h_{kj}. \tag{12}$$

If the i-th top is vertical, then σ_i is arbitrary.

If the spatial component ξ of the generator is nonzero, then ζ is related to σ by

$$\zeta_i = -\frac{\xi^2(\sigma_i + d_i)v_i + g x_i}{I_i \xi}, \qquad \text{for } i = 1, \dots, n. \tag{13}$$

If $\xi = 0$, then $x_i s_i$ is parallel to g, $i = 1, \dots, n$ and ζ is arbitrary.

2.3. Solution of the Relative Equilibrium Equations

Equation (12) implies that a relative equilibrium containing no vertical tops determines a unique vector σ. The homogeneity of equation (10) for h_1 and h_2 implies that an 'accordian' deformation of a relative equilibrium — that is, a deformation determined by rescaling h_1 and/or h_2 — yields a new relative equilibrium. (Any admissable rescaling must respect the constraints $|s_i| = 1, i = 1, \ldots, n$.) Additionally, any configuration that determines a relative equilibrium for some nonzero value of the spatial rotation rate ξ also determines a relative equilibrium for *any* nonzero value of ξ. In either case, the spin rates of the tilted tops change in accordance with equation (13).

The most obvious relative equilibria are the sleeping configurations, in which all tops are vertical. Sleeping relative equilibria are possible at any spin rates for any collection of tops, since the horizontal projections h_1 and h_2 are zero and hence trivially satisfy the kernel condition (10) for any matrix $K(\sigma)$. Sleeping relative equilibria have nontrivial isotropy — the family $(\zeta + \xi v, \xi) \in \mathbb{R}^{n+1}$ of algebra elements parametrized by ξ all generate the same sleeping motion with vertical projections v and spatial angular velocities $\omega_i = -\zeta_i s_i, i = 1, \ldots, n$. This nonuniqueness of the generators plays an essential role in the stability analysis.

A relatively simple class of relative equilibria consists of configurations in which all of the symmetry axes $s_1, \ldots, s_n$ lie in the same vertical plane, corresponding to linear dependence of the horizontal projections h_1 and h_2. We shall refer to such relative equilibria as planar relative equilibria. Any vector σ such that $K(\sigma)$ is singular determines a family of planar relative equilibria as follows: Let B_σ denote the unit ball in $\ker[K(\sigma)]$ with respect to the max norm $\|v\|_\infty := \max_{1 \leq i \leq n} |v_i|$. Any nonzero element h of B_σ determines a 2^m family of planar relative equilibria modulo overall orientation and spatial rotation rate ξ, where m denotes the number of components h_i of h satisfying $|h_i| < 1$. Each such component corresponds to a top with spatial symmetry axis s_i having a nonzero vertical component v_i, which can be chosen to be either positive or negative.

Any configuration in which all n tops are tilted out of the vertical determines a unique relative equilibrium (modulo the spatial rotation rate ξ and overall spatial orientation), with spin rates ζ given by (13). Relative equilibria consisting of both tilted and vertical tops exist; however, arbitrary configurations containing both tilted and vertical tops need not correspond to relative equilibria. The i-th top is vertical if and only if

$$0 = \sum_{j \neq i} \kappa_{ij} h_{kj} = -p_i \sum_{j > i} x_j h_{kj} - x_i \sum_{j < i} p_j h_{kj},$$

for $k = 1, 2$. For example, for $n = 3$ and $p_2 x_2 \neq 0$, the possible horizontal projections corresponding to combinations of tilted and vertical tops are

(i) $h_i \in \text{span}[(0, x_3, -x_2)]$, for $\sigma = (c, 0, \sigma_3^*),\}$

(ii) $h_i \in \text{span}\left[\left(\frac{x_1}{x_2}, 0, -\frac{p_1}{p_2}\right)\right]$, for $\sigma = (\sigma_1^*, c, \sigma_3^*)$,

(iii) $h_i \in \text{span}[(p_2, -p_1, 0)]$, for $\sigma = (\sigma_1^*, 0, c)$,

where

$$\sigma_i^* := p_i x_i - \begin{cases} \dfrac{p_i^2 x_{i+1}}{p_{i+1}}, & i \leq 2, \\ \dfrac{x_i^2 p_{i-1}}{x_{i-1}}, & i > 2, \end{cases} \tag{14}$$

and $c \in \mathbb{R}$ is arbitrary. As the following simple corollary of Proposition 1 shows, there are no multi-top relative equilibria containing only one tilted top.

Corollary 1. *A multi-top system has no relative equilibria in which the axes of symmetry of all but one top lie in a common vertical plane.*

Proof. If $n > 1$, then for every index $i = 1, \ldots, n$ the first and last components of the image

$$K(\sigma)e_i = -(p_1 x_i, \ldots, p_{i-1} x_i, p_i x_i - \sigma_i, p_i x_{i+1}, \ldots, p_i x_n)$$

of the i-th standard basis vector e_i of $\mathbb{R}^n$ cannot simultaneously equal zero, given the nondegeneracy assumptions that p_1 and x_n are nonzero and at least one of p_i and x_i is nonzero. Hence $\ker[K(\sigma)]$ contains no vectors with a single nonzero component. This immediately implies that there are no relative equilibria containing a single tilted top.

If we consider a configuration for which the axes of symmetry of all but the i-th top lie in a vertical plane P, the invariance of the system under spatial rotations implies that we can assume without loss of generality that P is the e_1–e_3 plane. Then the only nonzero component of h_2 is the i-th component. Our previous argument shows that this contradicts the equilibrium condition $h_2 \in \ker[K(\sigma)]$.   □

Any two linearly independent vectors h_1 and h_2 in $\ker[K(\sigma)]$ determine a family of nonplanar relative equilibria. However, given two arbitrary vectors h_1 and h_2, there need not be a vector σ such that $\operatorname{span}[h_1, h_2] \subset \ker[K(\sigma)]$. If both h_1 and h_2 are elements of $\ker[K(\sigma)]$, then (13) implies that

$$\sum_{j \neq i} \kappa_{ij}(h_{1i} h_{2j} - h_{2i} h_{1j}) = 0, \qquad \text{for } i = 1, \ldots, n; \tag{15}$$

if h_1 and h_2 are linearly independent, (15) imposes nontrivial relations between the off-diagonal entries of $K(\sigma)$ and the vectors h_1 and h_2. For example, all three-top nonplanar relative equilibria have horizontal projections h_1 and h_2 satisfying

$$h_1, h_2 \in \ker[K(\sigma_1^*, 0, \sigma_3^*)] = \operatorname{span}[(p_2, -p_1, 0), (0, x_3, -x_2)]. \tag{16}$$

Such equilibria exist if and only if $p_2 x_2 \neq 0$.

By exploiting the structure of the matrix $K(\sigma)$, we can identify classes of nonplanar relative equilibria for arbitrary numbers of tops. The matrix $K(\sigma)$ is the sum of a triangular matrix and a rank one matrix. Specifically, $\tilde{K}(\sigma) := K(\sigma) + px^T$ is lower triangular, with ij-th entry

$$\tilde{K}(\sigma)_{ij} = \begin{cases} 0, & i < j \\ \sigma_i, & i = j \\ p_i x_j - x_i p_j, & i > j \end{cases}. \tag{17}$$

Many properties of systems of equations involving the sum of a convenient matrix, such as a triangular matrix, and a rank one matrix have been analysed in detail. (See, e.g., Ciarlet [1989] or Golub and Van Loan [1989].) The following property is particularly relevant to our analysis: Given an m-by-n matrix M, $u \in \mathbb{R}^m$ and $v \in \mathbb{R}^n$, let w denote a solution, if any exists, of the equations $Mw = u$ and $v \cdot w = 1$; set $w = 0$ otherwise. Then

$$\ker[M - uv^T] = \operatorname{span}[w] + (\ker[M] \cap v^\perp), \tag{18}$$

where $v^\perp$ denotes the orthogonal complement to v. An immediate consequence of (18) is the estimate

$$\left| \text{null} \left[M - uv^T \right] - \text{null} \left[M \right] \right| \leq 1, \tag{19}$$

where null $[M]$ denotes the nullity of M. Thus if v denotes the number of zero entries in the vector σ, then null $[K(\sigma)] \leq v + 1$. In fact, as the following proposition shows, at least one intermediate entry σ_ℓ, $1 < \ell < n$, of σ equals zero if ker $[K(\sigma)]$ is multidimensional.

Proposition 2. *If $\sigma_2 \cdots \sigma_{n-1} \neq 0$, then* null $[K(\sigma)] \leq 1$.

Proof. The inequality (19) implies that null $[K(\sigma)] \leq 1$ if $\sigma_1 \cdots \sigma_n \neq 0$; thus it suffices to consider the case $\sigma_1 \sigma_n = 0$. To facilitate the discussion, we introduce the following notation for the projections of vectors to subspaces of $\mathbb{R}^n$ and the restrictions of the matrices $K(\sigma)$ and $\tilde{K}(\sigma)$ to those subspaces. Given $1 \leq \ell \leq n$ and vectors $w, \sigma \in \mathbb{R}^n$, define the vectors $w_\ell \in \mathbb{R}^{\ell-1}$ and $w^\ell \in \mathbb{R}^{n-\ell}$ by

$$w_\ell := (w_1, \ldots, w_{\ell-1}) \qquad \text{and} \qquad w^\ell = (w_{\ell+1}, \ldots, w_n); \tag{20}$$

define the $(\ell - 1) \times (\ell - 1)$ matrix $K_\ell(\sigma)$ by

$$K_\ell(\sigma)_{ij} := K(\sigma)_{ij}, \qquad 1 \leq i, j < \ell, \tag{21}$$

and the $(n - \ell) \times (n - \ell)$ matrix $K^\ell(\sigma)$ by

$$K^\ell(\sigma)_{ij} := K(\sigma)_{(\ell+i)(\ell+j)}, \qquad 1 \leq i, j \leq n - \ell. \tag{22}$$

The lower triangular matrices $\tilde{K}_k(\sigma)$ and $\tilde{K}^\ell(\sigma)$ are defined analogously to $K_k(\sigma)$ and $K^\ell(\sigma)$.

If $\sigma_1 = 0$, then the first row of $\tilde{K}(\sigma)$ is identically zero; hence, $\tilde{K}(\sigma) = K(\sigma) + px^T$ and the nondegeneracy assumption $p_1 \neq 0$ imply that $\alpha \in \text{ker} [K(\sigma)]$ if and only if $x \cdot \alpha = 0$ and $\alpha \in \text{ker} \left[\tilde{K}(\sigma) \right]$. The latter condition can be written in the form

$$\tilde{K}^1(\sigma)\alpha^1 = \alpha_1(x_1 p^1 - p_1 x^1), \tag{23}$$

using (17). If $\sigma_n \neq 0$, then the matrix $\tilde{K}^1(\sigma)$ is lower triangular with nonzero entries on the diagonal and hence is invertible. Thus null $[K(\sigma)]$ is at most one and is nonzero only if the solutions of (23) are orthogonal to x. If $\sigma_1 \neq 0 = \sigma_n$, then the last column of $\tilde{K}(\sigma)$ is identically zero. This case is treated analogously to the previous one, using $K(\sigma) = \tilde{K}(\sigma)^T + xp^T$.

If $\sigma_1 = 0 = \sigma_n$, then $\alpha \in \text{ker} [K(\sigma)]$ satisfies $x \cdot \alpha = 0$ and (23), as before. The additional condition $\sigma_n = 0$ now implies that the final column of $\tilde{K}^1(\sigma)$ is identically zero. Hence (23) is satisfied only if

$$\tilde{K}_n^1(\sigma)\alpha_n^1 = \alpha_1(x_1 p_n^1 - p_1 x_n^1),$$

where the $(n-2) \times (n-2)$ matrix $\tilde{K}_n^1(\sigma)$ is obtained by dropping the first and last rows and columns from $\tilde{K}(\sigma)$, $\alpha_n^1 := (\alpha_2, \ldots, \alpha_{n-1})$, etc. The matrix $\tilde{K}_n^1(\sigma)$ is lower triangular with nonzero diagonal entries and hence is invertible. Hence the condition $x \cdot \alpha = 0$ and the nondegeneracy assumption $x_n \neq 0$ suffice to guarantee that null $[K(\sigma)] \leq 1$. $\square$

Proposition 2 implies that $K(\sigma)$ determines nonplanar relative equilibria only if at least one intermediate entry of σ is zero. We now consider the structure of the relative equilibria determined by vectors σ with precisely one intermediate zero entry. The following proposition shows that if $\sigma_\ell = 0$ for some index $1 < \ell < n$ and $\sigma_i \neq 0$ for $i \neq \ell$, then any triplet (σ, h_1, h_2) with linearly independent h_1 and $h_2 \in \ker[K(\sigma)]$ determines a relative equilibrium that is hinged at the ℓ-th top, i.e., the axes of symmetry of the first $\ell - 1$ tops all lie in a common vertical plane, while those of final $n - \ell$ tops all lie in some other vertical plane. (These planes rotate steadily about the vertical axis as the relative equilibrium precesses.) If $p_\ell x_\ell \neq 0$, then the axis of symmetry of the ℓ-th top is not contained in either of these planes. Figure 1 illustrates two representative six-top relative equilibrium hinged at the fourth top; note that the two configurations have the same horizontal projections.

Proposition 3. *If $\sigma_1 \cdots \sigma_k \neq 0$, then the axes of symmetry of the first k tops all lie in a common vertical plane; if $\sigma_\ell \cdots \sigma_n \neq 0$, then the axes of symmetry of the final ℓ tops all lie in a common vertical plane. In particular, if precisely one component of σ equals zero, then all nonplanar relative equilibria determined by $K(\sigma)$ are hinged at the ℓ-th top.*

Proof. If the relative equilibrium is planar or sleeping, then by definition the axes of symmetry of the tops all lie in a common vertical plane; hence we need only consider nonplanar relative equilibria. There are integers $j > 1$ and $j' < n$ such that the axes of symmetry of the first $j - 1$ tops all lie in a vertical plane P and those of the final $n - j'$ tops all lie in a vertical plane P', while the axis of symmetry of the j-th top lies outside P and that of the j'-th top lies outside P'. By taking appropriate linear combinations of the horizontal projections h_1 and h_2, we can construct vectors α and α' in $\ker[K(\sigma)]$ such that $\alpha_i = 0$ for $i < j$, $\alpha'_i = 0$ for $i > j'$, and $\alpha_j \alpha'_{j'} \neq 0$.

Since p_1 and α_j are both nonzero,

$$0 = (K(\sigma)\alpha)_1 = -p_1 x \cdot \alpha \qquad \text{and} \qquad 0 = (K(\sigma)\alpha)_j = \sigma_j \alpha_j - p_j x \cdot \alpha$$

imply that $\sigma_j = 0$. Hence the assumption $\sigma_1 \cdots \sigma_k \neq 0$ implies that $j > k$; i.e., the axes of symmetry of the first k tops lie in the plane P. Analogously, since x_n and $\alpha'_{j'}$ are nonzero,

$$0 = (K(\sigma)\alpha')_n = -x_n p \cdot \alpha' \qquad \text{and} \qquad 0 = (K(\sigma)\alpha)_{j'} = \sigma_{j'} \alpha_{j'} - x_{j'} p \cdot \alpha'$$

imply that $\sigma_{j'} = 0$ and hence $\ell < j'$.

If σ_ℓ is the only zero component of σ, then the bound (19) on $K(\sigma)$ implies that $\ker[K(\sigma)] = \mathrm{span}[h_1, h_2]$; in this case, all of the nonplanar relative equilibria are hinged. $\square$

As was previously noted, any planar configuration containing no vertical tops determines a relative equilibrium. While the equilibrium conditions for nonplanar configurations are more delicate than those for planar configurations, nonplanar relative equilibria — specifically, hinged relative equilibria — are easily constructed. In fact, the following

Proposition shows that, analogous to the situation for planar equilibria, hinged equilibria exist for any two vertical planes and any tilted, sufficiently near to vertical positions of the tops within those planes. The key to the construction of the hinged relative equilibria is the observation that if $\sigma_\ell = 0$, then the matrix $K(\sigma)$ has the "overlapping block structure"

$$K(\sigma) = \begin{pmatrix} K_\ell(\sigma) & -p_{\ell+1}\left(x^{\ell-1}\right)^T \\ -x^{\ell-1}\left(p_{\ell+1}\right)^T & K^\ell(\sigma) \end{pmatrix}. \tag{24}$$

The components of the matrix (24) are defined by (20), (21), and (22). The lower left-hand entry of the upper off-diagonal block and the upper right-hand entry of the lower off-diagonal block of (24) coincide. This block structure makes it possible to decompose the determination of the kernel of $K(\sigma)$, or the construction of a vector σ such that $K(\sigma)$ has a specified kernel, into three smaller problems which retain the desirable properties of the original.

If $\sigma_k = \cdots = \sigma_\ell = 0$ and $p_k x_\ell \neq 0$, then the homogeneous system of equations $K(\sigma)\alpha = 0$ can be decomposed into three smaller systems of homogeneous equations — one involving only the head α_k, another involving only the tail α^ℓ, and a third system involving all components of α.

Lemma 1. *Let k and ℓ, $1 \leq k \leq \ell \leq n$, be indices such that $\sigma_i = 0$ for $k \leq i \leq \ell$. If $p_k x_\ell \neq 0$, then $\alpha \in \ker[K(\sigma)]$ if and only if*

(i) $\alpha_k \in \ker\left[K_k(\sigma) + \frac{x_k}{p_k}p_k p_k^T\right]$,

(ii) $\alpha^\ell \in \ker\left[K^\ell(\sigma) + \frac{p_\ell}{x_\ell}x^\ell(x^\ell)^T\right]$,

(iii) $x_i p_{i+1} \cdot \alpha_{i+1} + p_i x^i \cdot \alpha^i = 0$ for $k \leq i \leq \ell$.

Proof. The equalities $\sigma_k = \cdots = \sigma_\ell = 0$ imply that

$$(K(\sigma)\alpha)_i = \begin{cases} (K_k(\sigma)\alpha_k)_i - p_i x^{k-1} \cdot \alpha^{k-1}, & i < k \\ -x_i p_{i+1} \cdot \alpha_{i+1} - p_i x^i \cdot \alpha^i, & k \leq i \leq \ell \\ (K^\ell(\sigma)\alpha^\ell)_i - x_i p_{\ell+1} \cdot \alpha_{\ell+1}, & i > \ell \end{cases}.$$

Thus, $K(\sigma)\alpha = 0$ only if

$$K_k(\sigma)\alpha_k = (x^{k-1} \cdot \alpha^{k-1})p_k \qquad \text{and} \qquad K^\ell(\sigma)\alpha^\ell = (p_{\ell+1} \cdot \alpha_{\ell+1})x^\ell.$$

If $p_k \neq 0$ and condition (iii) holds, then $x^{k-1} \cdot \alpha^{k-1} = \frac{x_k}{p_k}p_k \cdot \alpha_k$, and hence

$$K_k(\sigma)\alpha_k - (x^{k-1} \cdot \alpha^{k-1})p_k = \left(K_k(\sigma) + \frac{x_k}{p_k}p_k p_k^T\right)\alpha_k.$$

Analogously, if $x_\ell \neq 0$ and condition (iii) holds, then

$$K^\ell(\sigma)\alpha^\ell - (p_{\ell+1} \cdot \alpha_{\ell+1})x^\ell = \left(K_\ell(\sigma) + \frac{p_\ell}{x_\ell}x^\ell(x^\ell)^T\right)\alpha^\ell. \qquad \square$$

The two decompositions of $K(\sigma)$ — triangular plus rank one and "overlapping blocks" — cooperate, and so the matrices appearing in Lemma 1 are again of the form triangular plus rank one. Specifically,

$$K_k(\sigma) + \tfrac{x_k}{p_k} p_k p_k^T = \tilde{K}_k(\sigma) + p_k \left(\tfrac{x_k}{p_k} p_k - x_k \right)^T,$$

and

$$K^\ell(\sigma) + \tfrac{p_\ell}{x_\ell} x^\ell (x^\ell)^T = \tilde{K}^\ell(\sigma)^T + x^\ell \left(\tfrac{p_\ell}{x_\ell} x^\ell - p^\ell \right)^T.$$

Thus, the kernels of these matrices can be determined using the computationally convenient formula (18).

The decomposition of the kernel condition given in Lemma 1 suggests the following strategy for identifying σ such that $\ker[K(\sigma)]$ is at least two dimensional: Given an index ℓ between 1 and n satisfying $p_\ell x_\ell \neq 0$, select vectors α and α' such that

(i) $\alpha_i \neq 0$ and $\alpha_i' = 0$ for $i < \ell$,
(ii) $\alpha_i = 0$ and $\alpha_i' \neq 0$ for $i > \ell$,
(iii) $\alpha \cdot p = 0 = \alpha' \cdot x$.

σ is then determined as a function of α and α' by the equations

$$K(\sigma)\alpha = K(\sigma)\alpha' = 0.$$

Any triplet (σ, h_1, h_2) with h_1 and $h_2 \in \mathrm{span}\left[\alpha, \alpha'\right]$ and $h_{1i}^2 + h_{2i}^2 \leq 1$ for $i = 1, \ldots, n$ determines a relative equilibrium hinged at the ℓ-th top. If σ_ℓ is the only zero component of σ, then the bound (19) on $K(\sigma)$ implies that $\ker[K(\sigma)] = \mathrm{span}[h_1, h_2]$; in this case, all of the nonplanar relative equilibria are hinged. The strategy outlined above is used in the following Proposition to show that hinged equilibria exist for any two vertical planes and any slightly tilted positions of the tops within those planes.

Proposition 4. *If $p_\ell x_\ell \neq 0$ for some intermediate index $1 < \ell < n$, then for any distinct vertical planes P and P' there are relative equilibria such that the axes of symmetry of the first $\ell - 1$ tops lie in the plane P and the axes of symmetry of the final $n - \ell$ tops lie in the plane P'. All but the ℓ-th top can be tilted at sufficiently small nonzero — but otherwise arbitrary — angles out of the vertical; the position of the ℓ-th top is determined up to the sign of its vertical component by the positions of the others.*

Proof. Let $s_1, \ldots, s_{\ell-1} \in P$ and $s_{\ell+1}, \ldots, s_n \in P'$ be given. Let u and $u' \in \mathbb{R}^3$ be vectors such that $\{u, e_3\}$ is an orthonormal basis for P and $\{u', e_3\}$ is an orthonormal basis for P', and set

$$\alpha = (\langle s_1, u \rangle, \ldots, \langle s_{\ell-1}, u \rangle, \alpha_\ell, 0, \ldots, 0),$$

and

$$\alpha' = (0, \ldots, 0, \alpha_\ell', \langle s_{\ell+1}, u' \rangle, \ldots, \langle s_n, u' \rangle),$$

where α_ℓ and α_ℓ' are as yet undetermined scalars.

We use Lemma 1 to find σ, α_ℓ, and α'_ℓ such that α and α' are nullvectors of $K(\sigma)$. The equation

$$\left(K_\ell(\sigma) + \tfrac{x_\ell}{p_\ell} p_\ell p_\ell^T \right) \alpha_\ell = 0 \tag{25}$$

can be viewed as a system of linear equations for σ_ℓ as a function of α_ℓ. The assumption that none of the first $\ell - 1$ tops are vertical, and hence that $\alpha_1, \ldots, \alpha_{\ell-1}$ are all nonzero, guarantees the existence and uniqueness of the vector σ_ℓ satisfying (25). The second condition of Lemma 1 is trivially satisfied by α, since $\alpha^\ell = 0$. The third condition is satisfied if we set $\alpha_\ell = -(p_\ell \cdot \alpha_\ell)/p_\ell$. Analogously, σ^ℓ and α'_ℓ are uniquely determined by $(\alpha')^\ell$ and the second and third conditions of Lemma 1, while the first condition of the lemma is trivially satisfied. Specifically,

$$\sigma_i = \begin{cases} p_i x_i + \tfrac{p_i}{\alpha_i}\left(x_\ell - \tfrac{x_\ell}{p_\ell} p_\ell \right) \alpha_\ell, & i < j \\ 0, & i = j \\ p_i x_i + \tfrac{x_i}{\alpha_i}\left(p^\ell - \tfrac{p_\ell}{x_\ell} x^\ell \right) (\alpha)^\ell, & i > j \end{cases} \tag{26}$$

and $\alpha'_\ell = -(x^\ell \cdot (\alpha')^\ell)/x_\ell$.

If $s_1, \ldots, s_{\ell-1}$ and $s_{\ell+1}, \ldots, s_n$ are taken sufficiently near to vertical, then $\alpha_\ell^2 + (\alpha'_\ell)^2 \leq 1$, and we can set

$$s_\ell^{\pm} := \left(\langle \alpha_\ell u + \alpha'_\ell u', e_1 \rangle, \langle \alpha_\ell u + \alpha'_\ell u', e_2 \rangle, \pm\sqrt{1 - \alpha_\ell^2 - (\alpha'_\ell)^2} \right).$$

The top positions $s_1, \ldots, s_{\ell-1}, s_\ell^{\pm}, s_{\ell+1}, \ldots, s_n$ determine a hinged relative equilibrium with associated σ given by (26). □

Remark. If $p_\ell x_\ell = 0$, then relative equilibria hinged at the ℓ-th top exist if and only if the horizontal projections h_1 and h_2 satisfy $(h_i)_{\ell+1} \cdot p_{\ell+1} = 0 = (h_i)^{\ell-1} \cdot x^{\ell-1}$ for $i = 1$, 2. In this case, the position of the ℓ-th top is *not* uniquely determined by the positions of the others. If $p_\ell = 0$, then the equation determining σ_ℓ is replaced by

$$(u \cdot e_i) K_\ell(\sigma)\alpha_\ell = ((x_\ell \alpha_\ell u + (\alpha'_\ell + \alpha' \cdot x^\ell)u') \cdot e_i) p_\ell,$$

for $i = 1$, 2. Regrouping these equations, we find that they are equivalent to $\alpha'_\ell = -x^\ell \cdot (\alpha')^\ell/x_\ell$, as before, and $K_\ell(\sigma)\alpha = \alpha_\ell x_\ell p_\ell$. Thus α_ℓ is arbitrary (up to the restriction that $|(h_{1\ell}, h_{2\ell})| \leq 1$) and σ_ℓ is uniquely determined as a function of α_ℓ. The remaining components of σ are determined by the equations $\sigma_\ell = 0$ and $K^\ell(\sigma)(\alpha')^\ell = 0$. The case $x_\ell = 0$ is treated analogously.

As was discussed previously, for $n \leq 2$, all relative equilibria are planar, while for $n = 3$, $\ker\left[K(\sigma_1^*, 0, \sigma_3^*) \right] = \text{span}\left[(p_2, -p_1, 0), (0, x_3, -x_2) \right]$ is the only multi-dimensional kernel. In the corollary below, we consider a system of four tops. For simplicity, we consider only the case $p_2 x_2 p_3 x_3 \neq 0$; the cases in which $p_2 x_2 p_3 x_3 = 0$ can be analysed analogously. We see that for a generic four-top system, i.e., one satisfying $p_2 x_2 p_3 x_3 \neq 0$ and $\frac{p_3}{x_3} \neq \frac{p_2}{x_2}$, all nonplanar relative equilibria are hinged. If $\frac{p_3}{x_3} = \frac{p_2}{x_2}$, then there are relative equilibria such that no two of the axes of symmetry lie in a common vertical plane.

Corollary 2. *If $n = 4$ and $p_3 x_3 \neq \frac{p_2}{x_2}$, there are precisely two families of vectors σ leading to multi-dimensional kernels: If*

$$(\sigma_4 - \sigma_4^*)(\sigma_3 - \sigma_3^*) = \sigma_3 \left(\frac{p_3}{x_3} - \frac{p_2}{x_2} \right) x_4^2, \tag{27}$$

then

$$\ker \left[K(\sigma_1^*, 0, \sigma_3, \sigma_4) \right] \supset \operatorname{span} \left[\left(\frac{1}{p_1}, -\frac{1}{p_2}, 0, 0 \right), \left(0, \frac{\sigma_3}{x_2}, -\frac{\sigma_3^*}{x_3}, -\frac{\sigma_3 - \sigma_3^*}{x_4} \right) \right], \tag{28}$$

for any values of σ_3 and σ_4. If

$$(\sigma_1 - \sigma_1^*)(\sigma_2 - \sigma_2^*) = \sigma_2 \left(\frac{p_3}{x_3} - \frac{p_2}{x_2} \right) p_1^2, \tag{29}$$

then

$$\ker \left[K(\sigma_1, \sigma_2, 0, \sigma_4^*) \right] \supset \operatorname{span} \left[\left(\frac{\sigma_2 - \sigma_2^*}{p_1}, \frac{\sigma_2^*}{p_2}, -\frac{\sigma_2}{p_3}, 0 \right), \left(0, 0, \frac{1}{x_3}, -\frac{1}{x_4} \right) \right], \tag{30}$$

for any values of σ_1 and σ_2. The scalars σ_i^* are given by (14).

If $\frac{p_2}{x_2} = \frac{p_3}{x_3}$, then $\sigma_2^* = \sigma_3^* = 0$ and the vectors $\left(0, \frac{1}{p_2}, -\frac{1}{p_3}, 0 \right)$ and $\left(0, \frac{1}{x_2}, -\frac{1}{x_3}, 0 \right)$ are parallel. In this case,

$$\ker \left[K(\sigma_1^*, 0, 0, \sigma_4) \right] \supset \operatorname{span} \left[\left(\frac{1}{p_1}, -\frac{1}{p_2}, 0, 0 \right), \left(0, \frac{1}{p_2}, -\frac{1}{p_3}, 0 \right) \right], \tag{31}$$

$$\ker \left[K(\sigma_1, 0, 0, \sigma_4^*) \right] \supset \operatorname{span} \left[\left(0, \frac{1}{x_2}, -\frac{1}{x_3}, 0 \right), \left(0, 0, \frac{1}{x_3}, -\frac{1}{x_4} \right) \right], \tag{32}$$

and the families (28), (30), (31), and (32) have the common member $(\sigma_1^*, 0, 0, \sigma_4^*)$, with null $\left[K(\sigma_1^*, 0, 0, \sigma_4^*) \right] = 3$.

Proof. Proposition 2 implies that multi-dimensional kernels are possible only if $\sigma_2 \sigma_3 = 0$.

If $\sigma_2 = 0$, then Lemma 1 implies that $\alpha \in \ker[K(\sigma)]$ if and only if

(i) $(\sigma_1 - \sigma_1^*)\alpha_1 = 0$,
(ii) $(\alpha_3, \alpha_4) \in \ker[M(\sigma_3, \sigma_4)]$, where

$$M(\sigma_3, \sigma_4) := K^2(\sigma) + \frac{p_2}{x_2} x^2 (x^2)^T = \operatorname{diag} \left[\sigma_3, \sigma_4 - \sigma_4^* \right] + \left(\frac{p_2}{x_2} - \frac{p_3}{x_3} \right) x_2 (x_2)^T,$$

(iii) $\alpha_2 = - \left(\frac{p_1}{p_2} \alpha_1 + \frac{x_3}{x_2} \alpha_3 + \frac{x_4}{x_2} \alpha_4 \right)$.

If $\frac{p_2}{x_2} \neq \frac{p_3}{x_3}$, then null $[M(\sigma_3, \sigma_4)] \leq 1$. Hence in this case $K(\sigma)$ has a multi-dimensional kernel if and only if $\sigma_1 = \sigma_1^*$ and $\det M(\sigma_3, \sigma_4) = 0$. The latter condition is equivalent to (27). If $\frac{p_2}{x_2} = \frac{p_3}{x_3}$, then $M(0, \sigma_4^*)$ is identically zero, leading to (32).

The case $\sigma_3 = 0$ is handled analogously, noting that Lemma 1 implies that if $\alpha \in \ker[K(\sigma)]$, then

(i) $(\sigma_4 - \sigma_4^*)\alpha_4 = 0$,
(ii) $(\alpha_1, \alpha_2) \in \ker[N(\sigma_1, \sigma_2)]$, where

$$N(\sigma_1, \sigma_2) := K_3(\sigma) + \frac{x_3}{p_3} p_3 p_3^T = \operatorname{diag} \left[\sigma_1 - \sigma_1^*, \sigma_2 \right] + \left(\frac{x_3}{p_3} - \frac{x_2}{p_2} \right) p_3 p_3^T,$$

(iii) $\alpha_3 = - \left(\frac{p_1}{p_3} \alpha_1 + \frac{p_2}{p_3} \alpha_2 + \frac{x_4}{x_3} \alpha_4 \right)$. $\square$

3. Stability

3.1. Formal and Orbital Stability

When we speak of stability, we shall be concerned ultimately with orbital nonlinear stability — a relative equilibrium v_e is orbitally, nonlinearly stable with respect to a subgroup $\tilde{G}$ of the symmetry group G of the system and an $\tilde{G}$ invariant metric ρ if, given $\epsilon > 0$, there exists $\delta > 0$ such that for any v_0 satisfying $\rho(v_0, v_e) < \delta$, there exists a curve $g(t)$ in $\tilde{G}$ such that the solution curve $v(t)$ with initial condition $v(0) = v_0$ satisfies $\rho(g(t) \cdot v(t), v_e) < \epsilon$. For Hamiltonian or Lagrangian systems with symmetry, nonlinear orbital stability is typically demonstrated using some combination of conserved quantities as a Lyapunov function modulo an appropriate subgroup of the symmetry group.

A traditional choice of conserved quantity, which we shall use here, is the energy-momentum function. If the phase space is the tangent bundle of some manifold Q and the action of the symmetry group G on the phase space is the lift of an action on Q, then a relative equilibrium with generator ξ is a critical point of the energy-momentum function $E - J_\xi$, where $E: P \to \mathbb{R}$ denotes the total energy of the system with Lagrangian L and $J_\xi: P \to \mathbb{R}$ denotes the conserved quantity associated to ξ given by Noether's theorem. Specifically, if $v \in T_q Q$, then

$$(E - J_\xi)(v) = \tfrac{d}{d\epsilon}\big|_{\epsilon=0} L(v + \epsilon(v - \xi_Q(q))) - L(v).$$

(See, e.g., Abraham and Marsden [1978] or Arnold [1978] for detailed treatments of this material.) If the relative equilibrium v_e is a local extremal of $E - J_\xi$ modulo the subgroup G_μ of G preserving the equilibrium value μ of the momentum map, i.e., the full collection of conserved quantities given by Noether's Theorem, then it is nonlinearly G_μ-orbitally stable. This observation motivates the definition of *formal stability*: A relative equilibrium v_e is said to be formally stable if the restriction of the second variation of the energy-momentum function $E - J_\xi$ determined by a generator ξ of v_e to the kernel of the linearization at v_e of the momentum map J is semidefinite, with kernel equal to the tangent space to the orbit of v_e under the subgroup G_μ. Under appropriate technical hypotheses, which are satisfied in the case at hand, formal stability implies nonlinear orbital stability with respect to the subgroup G_μ. See, e.g., Patrick [1992], Lerman [1997], or Ortega and Ratiu [1997].[2] Since the symmetry group T^{n+1} of a nondegenerate system of Lagrange tops is abelian, formal stability implies nonlinear stability modulo T^{n+1} for these systems.

---

[2] Patrick's result assumes that the group action is free; it can be applied to all but the sleeping relative equilibria, since T^{n+1} acts freely on a neighborhood of the group orbit of a tilted relative equilibrium. For the sleeping relative equilibria we can make use of an extension of Patrick's result to the case in which the action is not free, but the isotropy subgroup G_μ of the equilibrium momentum is a subgroup of the isotropy subgroup of the generator ξ; in this case, the map Ψ constructed by Patrick [1992] can be replaced by the constant ξ. In particular, we note that Corollary 4.1 in Lewis [1992], which addresses the relationship between formal and orbital stability of relative equilibria with isotropy, is incorrect; however, if the assumption that the bundle $G_\mu \to (G_\mu \cap G_{v_e})$ is trivial is replaced with the assumption that $G_\mu \subset G_\xi$, then the resulting assertion is valid and can be applied to the case at hand to demonstrate nonlinear G_μ-orbital stability. For more general results, see Ortega and Ratiu [1997].

Formal instability — i.e., failure of the energy-momentum function to serve as a Lyapunov function (modulo the isotropy subgroup) — does not imply that the relative equilibrium is nonlinearly unstable. However, relatively little is known about strategies for the systematic construction of Lyapunov functions for formally unstable relative equilibria. Hence our goal is not a complete analysis of the nonlinear stability conditions, but rather the derivation of some simple sufficient conditions for formal stability. Necessary conditions for formal stability can be derived easily, but should probably not be viewed as having particular significance in their own right; rather, they can be used to (partially) bound the region containing formally stable relative equilibria and thus to facilitate the search for such equilibria. Additionally, the terminology "formally unstable" allows us to refer succinctly to cases in which the formal stability analysis is known to fail, as distinct from those in which the definiteness of the formal stability matrix cannot be immediately determined.

Our grouping of the relative equilibria in the three categories of sleeping configurations, tilted planar configurations, and nonplanar configurations is pertinent to the stability analysis. We incorporate the particular features of each of the three categories into the general expression for the stability two form and then analyse the stability of some particular classes of relative equilibria from each of the categories. We find that if all of the tops are sleeping, then the test for formal stability involves a one-parameter family of $n \times n$ matrices; if any member of this family is positive-definite, then the equilibrium is formally, and hence nonlinearly, stable. In the case of two sleeping tops, we shall explicitly compare the regions determined by two easily derived and computed stability conditions — one necessary, the other sufficient — with the actual region of formal stability. The $n \times n$ matrix $K(\sigma)$ used to determine the relative equilibrium also determines the formal stability of the equilibrium if it is planar. A planar relative equilibrium configuration is formally stable if and only if $K(\sigma)$ is positive semidefinite, with null $[K(\sigma)] \leq 1$. If the equilibrium configuration is nonplanar, then a $2n \times 2n$ matrix with a simple block structure determines formal stability.

Theorem 1. *The formal stability conditions for a relative equilibrium fall into one of the following four categories:*

(i) *A sleeping relative equilibrium, i.e., a relative equilibrium consisting entirely of vertical tops, that is generated by $(\zeta, 0)$ is formally stable if and only if there exists $\xi \in \mathbb{R}$ such that the $n \times n$ matrix $S(\xi)$ with ij-th entry*

$$(S(\xi))_{ij} = (\xi I_i(\zeta_i v_i + \xi) + g x_i v_i)\delta_{ij} + \xi^2 \kappa_{ij} \qquad (33)$$

is negative definite.
(ii) *A tilted relative equilibrium that does not precess is formally unstable.*
(iii) *A planar, tilted, precessing relative equilibrium is formally stable if and only if the associated matrix $K(\sigma)$ is positive semidefinite with nullity one.*
(iv) *If a relative equilibrium is nonplanar and precessing, then let (r_i, θ_i) denote the polar coordinates of the horizontal projections (h_{1i}, h_{2i}) of the axes of symmetry*

of the tops, with arbitrary θ_i if $r_i = 0$. Define the vectors Υ_1 and Υ_2 by

$$\Upsilon_{1i} := r_i^2(\sigma_i + I_i + d_i) \qquad \text{and} \qquad \Upsilon_{2i} := \frac{(I_i\xi^2 - gx_i)r_i}{\xi^2|\Upsilon_1|}, \tag{34}$$

and define the $n \times n$ matrices A, B, and C by

$$\begin{aligned}
a_{ij} &= K(\sigma)_{ij}\cos(\theta_i - \theta_j), \\
b_{ij} &= K(\sigma)_{ij}v_i\sin(\theta_i - \theta_j), \\
c_{ij} &= K(\sigma)_{ij}v_iv_j\cos(\theta_i - \theta_j) + \Upsilon_{1i}\delta_{ij} + \Upsilon_{2i}\Upsilon_{2j}.
\end{aligned} \tag{35}$$

The relative equilibrium is formally stable if and only if the $2n \times 2n$ matrix S_e with block structure

$$S_e = \begin{pmatrix} A & B^T \\ B & C \end{pmatrix} \tag{36}$$

is positive semidefinite, with $\ker[S_e] = \operatorname{span}[(r, 0)]$.

The proof of Theorem 1 is a standard, if somewhat complicated, application of the reduced energy-momentum method (see Simo et al. [1991] or Lewis [1992]) and is given in the appendix. In the following sections, we focus on the identification of some simple classes of stable relative equilibria from the condition that the matrices specified in Theorem 1 be definite or semidefinite.

Remark. Some of the stability conditions given in Theorem 1 appear in Vassileva and Lilov [1987]; the others can be derived from the stability conditions given in that paper. The proof given here, which was derived independently, is included for the sake of completeness.

3.2. Sleeping Configurations

Theorem 1 states that for sleeping relative equilibria, in which all the axes of symmetry are vertical, existence of $\xi \in \mathbb{R}$ such that the matrix $S(\xi)$ given in (33) is negative-definite is a necessary and sufficient condition for formal stability. Even for relatively small values of n, the inequalities determining definiteness appear to be unwieldy; the invariants of $S(\xi)$ are polynomial of order up to $2n$ in ξ. It is relatively easy to test numerically if a particular equilibrium is formally stable, but approximately determining the region of formal stability by numerically analysing every equilibrium on a given mesh can be time-consuming if the mesh is fine and the region to be studied is large. Incomplete stability conditions — i.e., conditions that are necessary or sufficient, but not both — can be used to reduce the size of the region to be tested, since regions that are "obviously" stable or unstable need not be included in the mesh. Identifying parameter values for which inequalities involving quadratic polynomials in ξ can be satisfied simultaneously is relatively easy. The necessary conditions for formal stability that the diagonal entries of $S(\xi)$ and the terms $(e_i \pm e_j) \cdot S(\xi)(e_i \pm e_j)$, where e_i denotes the i-th Euclidean basis vector in $\mathbb{R}^n$, be negative determine regions in parameter space corresponding to formally unstable relative equilibria. On the other hand, the sufficient stability condition that $S(\xi)$ be strictly diagonally dominant with negative diagonal entries determines regions corresponding to stable relative equilibria.

Proposition 5. *Let (Λ, Ω) be a sleeping configuration, i.e., $e_3 \times s_i = e_3 \times \omega_i = 0$ for $i = 1, \ldots, n$. Define the scalar constants*

(i) $\mu_i := I_i \langle \omega_i, s_i \rangle$,

(ii) $\gamma_i := g\, x_i v_i = x_i \langle g, s_i \rangle$,

(iii) $J_i := I_i + \kappa_{ii} = I'_i + p_i^2 \sum_{i<j} m_j$,

(iv) $\tilde{J}_i := J_i + \sum_{j \neq i} |\kappa_{ij}| = I'_i + p_i \sum_{i<j}(p_i m_j + |x_j|) + |x_i| \sum_{j<i} p_j$.

Satisfaction of the inequalities

$$\mu_i^2 > 4\gamma_i J_i, \tag{37}$$

$$\sqrt{\left(\frac{\mu_i}{J_i}\right)^2 - \frac{4\gamma_i}{J_i}} + \sqrt{\left(\frac{\mu_j}{J_j}\right)^2 - \frac{4\gamma_j}{J_j}} > \left|\frac{\mu_i}{J_i} - \frac{\mu_j}{J_j}\right|, \tag{38}$$

and

$$(\mu_i + \mu_j)^2 > 4(\gamma_i + \gamma_j)(J_i + J_j + 2|\kappa_{ij}|), \tag{39}$$

for all i and j, are necessary conditions of formal stability. In particular, if a sleeping configuration is formally stable, then all "upright" tops, that is, tops with $\gamma_i > 0$, rotate in the same direction.

If

$$\mu_i^2 > 4\gamma_i \tilde{J}_i, \tag{40}$$

and

$$\sqrt{\left(\frac{\mu_i}{\tilde{J}_i}\right)^2 - \frac{4\gamma_i}{\tilde{J}_i}} + \sqrt{\left(\frac{\mu_j}{\tilde{J}_j}\right)^2 - \frac{4\gamma_j}{\tilde{J}_j}} > \left|\frac{\mu_i}{\tilde{J}_i} - \frac{\mu_j}{\tilde{J}_j}\right|, \tag{41}$$

for all i and j, then the relative equilibrium is nonlinearly, orbitally stable.

Remark. The inequalities (37) and (40) have the form of the familiar stability condition for a single top, with the moment of inertia I'_i replaced by J_i or $\tilde{J}_i$.

Proof. Theorem 1 states that a sleeping relative equilibrium is formally stable if and only if the matrix $S(\xi)$ given by (33) is negative-definite. The condition that the diagonal entries of $S(\xi)$ be negative yields the necessary formal stability conditions

$$J_i \xi^2 + \mu_i \xi + \gamma_i < 0, \tag{42}$$

for $i = 1, \ldots, n$. Since J_i is always positive, (42) is satisfied for some index i if and only if the corresponding inequality (37) is satisfied, in which case any choice of ξ in the open interval (ξ_i^-, ξ_i^+), where

$$\xi_i^\pm := -\frac{\mu_i}{2J_i} \pm \sqrt{\left(\frac{\mu_i}{2J_i}\right)^2 - \frac{\gamma_i}{J_i}},$$

satisfies (42) for fixed index i. There exists ξ satisfying (42) for all $i = 1, \ldots, n$ if and only if the intersection of all the intervals (ξ_i^-, ξ_i^+) is non-empty, i.e.,

$$\min_{1 \le i \le n} \xi_i^+ > \max_{1 \le i \le n} \xi_i^-. \tag{43}$$

Regrouping terms, we see that (43) is equivalent to the condition that the inequalities (38) hold for all i and j. The inequalities (39) follow from the necessary formal stability condition that ξ can be chosen such that

$$0 > (e_i \pm e_j) \cdot S(\xi)(e_i \pm e_j) = (J_i + J_j \mp 2\kappa_{ij})\xi^2 + (\mu_i + \mu_j)\xi + (\gamma_i + \gamma_j). \tag{44}$$

To show that all tops with $\gamma_i > 0$ must rotate in the same direction, we note that if there are indices i and j such that γ_i, γ_j, and μ_i are all positive, but μ_j is negative, then $\xi_i^+ < 0 < \xi_j^-$ and thus (43) is violated.

To obtain sufficient conditions for stability, we make use of the concept of diagonal dominance. An $n \times n$ matrix M with entries m_{ij} is said to be diagonally dominant if $|m_{ii}| \ge \sum_{j \ne i} |m_{ij}|$ for $i = 1, \ldots, n$; M is strictly diagonally dominant if all of the inequalities are strict. If M is diagonally dominant and $m_{ii} < 0$ for $i = 1, \ldots, n$, then it follows from the Gershgorin circle theorem that M is negative semidefinite; if the diagonal dominance is strict, then M is negative-definite. (See, e.g., Ciarlet [1989], Exercise 1.1–5, or Golub and Van Loan [1989], Theorem 7.2.1.) In particular, if the inequalities (40) and (41) hold for all i and j, then there exists ξ such that

$$-(S(\xi))_{ii} > \sum_{j \ne i} |(S(\xi))_{ij}| = \xi^2 \sum_{j \ne i} |\kappa_{ij}| \qquad \text{for } i = 1, \ldots, n, \tag{45}$$

and hence $S(\xi)$ is negative-definite.   □

Proposition 5 provides a set of necessary conditions for formal stability and a set of sufficient conditions for formal, and hence orbital nonlinear, stability. These stability conditions can be viewed as dividing the "angular momentum space" $\mathbb{R}^n$ for sleeping relative equilibria into regions determined by the formal stability, formal instability, or undetermined formal character of the associated relative equilibria. To develop our intuition about these regions, we now fix a pair (i, j) of indices and characterize the regions in the μ_i–μ_j plane in which the necessary formal stability conditions (37) and (38) are satisfied.

Define the scalar constants

$$R_{ij} := 2(J_i \gamma_j + J_j \gamma_i), \qquad \chi_k := 2\sqrt{J_k \gamma_k}, \qquad \text{and} \qquad \chi_k' := \frac{R_{ij}}{\chi_k},$$

for $k = i, j$. If χ_i and χ_j are real, then the inequalities (37) are satisfied for i and j outside the region containing the origin and bounded by the lines $\mu_i = \pm \chi_i$ and $\mu_j = \pm \chi_j$; if either χ_i or χ_j is imaginary, the corresponding inequality is satisfied everywhere. The equalities corresponding to (38) are satisfied on the intersections of the graphs of the functions

$$T_{ij}^\pm(\mu_i) := \tfrac{1}{2}\left(\mu_i \left(\frac{J_j}{J_i} + \frac{\gamma_j}{\gamma_i} \right) \pm \left| \frac{J_j}{J_i} - \frac{\gamma_j}{\gamma_i} \right| \sqrt{\mu_i^2 - 4\gamma_i J_i} \right), \tag{46}$$

with the regions in which the inequalities $|\mu_i| > \chi_i$, $|\mu_j| > \chi_j$, and $\mu_i \mu_j < R_{ij}$ are satisfied. The curves $\mu_j = T_{ij}^{\pm}(\mu_i)$ and $\mu_i \mu_j = R_{ij}$ intersect at the points $\pm(\chi_i, \chi_i')$ and $\pm(\chi_j', \chi_j)$.

There are three possible situations, determined by the signs of γ_i and γ_j:

- If γ_i and γ_j are both positive, then (37) and (38) are satisfied precisely in the regions

$$
S_{ij} := ((\chi_i, \infty) \times (\chi_j, \infty)) \setminus \left\{ (\mu_i, \mu_j): \chi_j' > \mu_i > \chi_i \text{ and } T_{ij}^-(\mu_i) > \mu_j > \chi_j \right\}
$$

and $-S_{ij} := \left\{ (-\mu_i, -\mu_j): (\mu_i, \mu_j) \in S_{ij} \right\}$. See Figure 3.a.
- If $\gamma_i \gamma_j < 0$ with, say, γ_i negative, then the regions S_{ij} and $-S_{ij}$ in which (37) and (38) are satisfied are given by

$$
S_{ij} := ((\chi_j', \infty) \times (\chi_j, \infty)) \cup \left\{ (\mu_i, \mu_j): \chi_j' > \mu_i \quad \text{and} \quad \mu_j > T_{ij}^-(\mu_i) \right\}
$$

and $-S_{ij} := \left\{ (-\mu_i, -\mu_j): (\mu_i, \mu_j) \in S_{ij} \right\}$. See Figures 3.b and 3.c.
- If both γ_i and γ_j are negative, then $\mu_i T_{ij}^{\pm}(\mu_i) > 2 R_{ij}$ for all μ_i and (37) and (38) are satisfied for all μ_i and μ_j. See Figure 3.d.

To determine the regions in which the inequalities (40) and (41) are satisfied, we can carry out an analogous analysis using $\tilde{J}_i$ and $\tilde{J}_j$ in place of J_i and J_j, yielding regions $\pm \tilde{S}_{ij}$. If $\gamma_i + \gamma_j > 0$, then the lines

$$
\pm L_{ij} := \left\{ (\mu_i, \mu_j): \mu_i + \mu_j = \pm 2\sqrt{(\gamma_i + \gamma_j)(J_i + J_j + 2|\kappa_{ij}|)} \right\}
$$

form the boundaries of the regions in which (39) is satisfied. If $\gamma_i + \gamma_j < 0$, then (39) is satisfied for all values of μ_i and μ_j.

If there are only two tops, then formal stability is unknown only for momentum pairs (μ_1, μ_2) in the regions $\pm(S_{12} \setminus \tilde{S}_{12})$ lying outside the strip S_{12}' bounded by L_{12} and $-L_{12}$. In Figure 4, we consider the stability of upright sleeping configurations for some sample pairs of tops. Since a sleeping relative equilibrium with momenta $(-\mu_1, -\mu_2)$ is stable if and only if the corresponding relative equilibrium with momenta (μ_1, μ_2) is stable, we display only the relevant portions of the upper right-hand quadrant or upper half-plane. We illustrate the boundaries of the regions S_{12} and $\tilde{S}_{12}$ and the approximate boundary of the region of formal stability, which is computed by testing the stability of points on an appropriate mesh. In addition, we plot the line L_{12}, which is tangent to the boundary of $\tilde{S}_{12}$ at (χ_1'', χ_2''), where

$$
\chi_k'' := \tilde{J}_k \sqrt{\frac{\gamma_1 + \gamma_2}{\tilde{J}_1 + \tilde{J}_2}} + \gamma_k \sqrt{\frac{\tilde{J}_1 + \tilde{J}_2}{\gamma_1 + \gamma_2}},
$$

for $k = 1, 2$. Pairs in the strip S_{12}' are necessarily formally unstable, while pairs in $\pm \tilde{S}_{12}$ are formally stable; hence, the boundary of the region of formal stability must pass

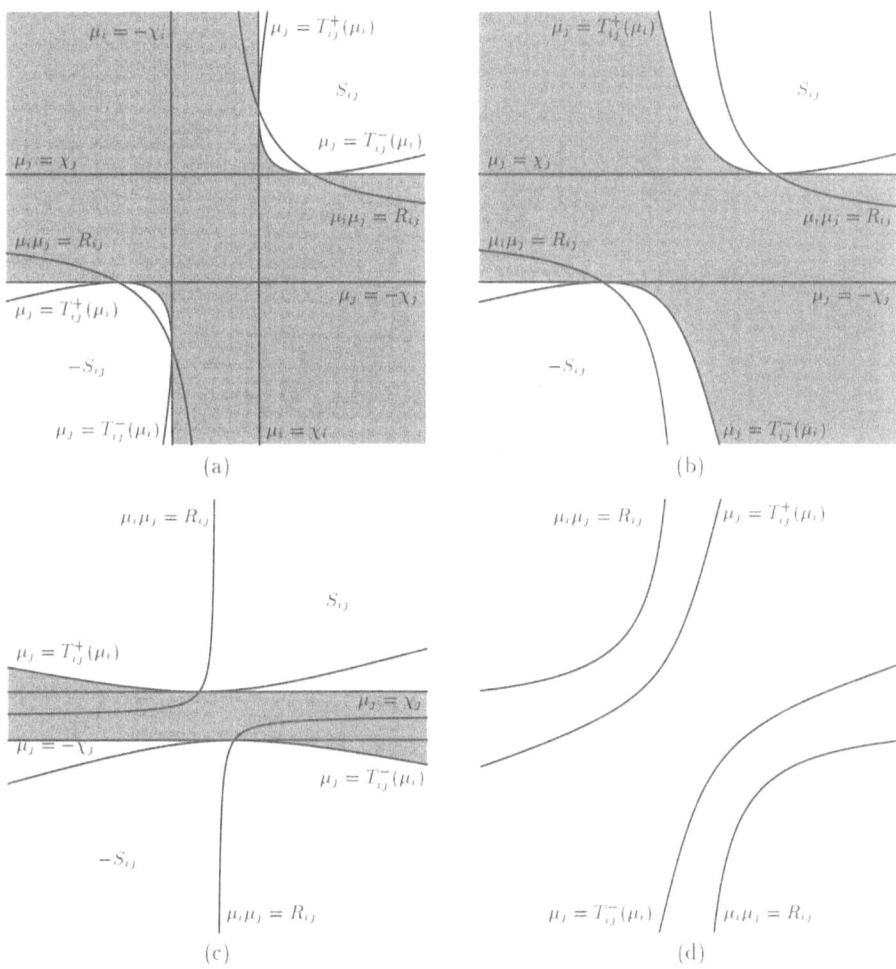

Fig. 3. Representative regions determined by the inequalities (37) and (38), which are satisfied in the white regions and violated in the shaded regions. (a) γ_i and γ_j both positive; (b) $\gamma_j > \gamma_i + \gamma_j > 0$; (c) $\gamma_j > 0 > \gamma_i + \gamma_j$; (d) γ_i and γ_j both negative.

through (χ_1'', χ_2''). (The line L_{12} — respectively $-L_{12}$ — may intersect the boundary of S_{12} — respectively $-S_{12}$ — in either the curve $\mu_2 = T_{12}^{\pm}(\mu_1)$ or the vertical or horizontal portions of the boundary, depending on the parameter values.) The boundary of the region of formal stability asymptotically approaches the boundary of the region S_{12} as μ_1 or μ_2 approaches infinity; if γ_1 and γ_2 have opposite signs, then the boundaries of the regions S_{12} and $\tilde{S}_{12}$ asymptotically approach a line with slope γ_2/γ_1 as (μ_1, μ_2) approaches $(-\infty, \infty)$.

Remark. Note that if $\tilde{J}_i\gamma_j = \tilde{J}_j\gamma_i$, then the curved portion of the boundary of $\pm\tilde{S}_{ij}$ is trivial. In particular, if there are only two tops and $\tilde{J}_1\gamma_2 = \tilde{J}_2\gamma_1$, then the vertical and horizontal boundaries of $\tilde{S}_{12}$ and L_{12} (respectively, $-\tilde{S}_{12}$ and $-L_{12}$) all intersect.

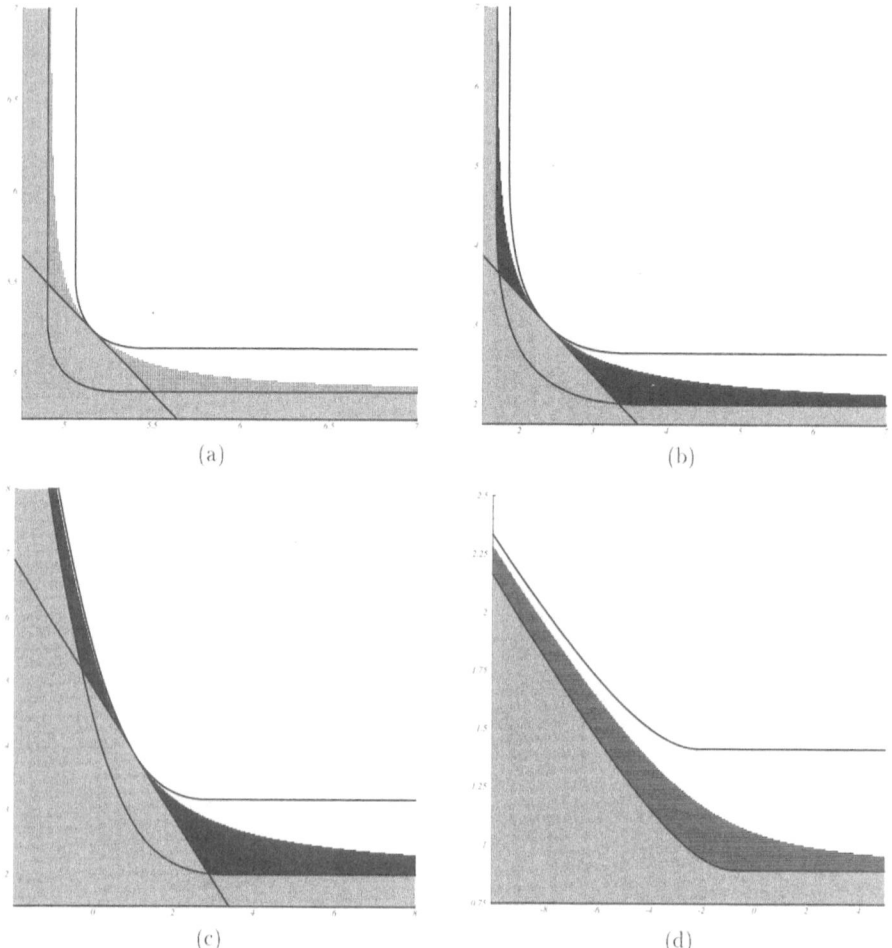

Fig. 4. Representative formal stability diagrams for sleeping relative equilibria consisting of two upright tops. The upper right hand region bounded by the upper smooth elbow curve consists of momenta associated to "obviously" stable relative equilibria satisfying (40) and (41). The light grey region consists of momenta associated to "obviously" formally unstable equilibria that violate at least one of (37), (38), or (39). The dots in the region between the two elbows indicate numerically determined formally unstable points.

For more than two tops, a region of momenta associated to stable equilibria is given by

$$\bigcap_{i<j} \Pi_{ij}^{-1}(\tilde{S}_{ij} \cup (-\tilde{S}_{ij})),$$

where $\Pi_{ij}: \mathbb{R}^n \to \mathbb{R}^2$ denotes projection onto the ij coordinate plane, while a region of momenta of formally unstable relative equilibria is given by

$$\bigcup_{i<j} \Pi_{ij}^{-1}((S_{ij} \cup (-S_{ij}))^c \cup S'_{ij}),$$

where ^{c} denotes the complement in the ij plane.

3.3. Tilted Planar Configurations

Configurations in which all of the tops lie in the same vertical plane, with at least one top tilted out of the vertical, are in some regards the simplest to analyse. As was shown in Theorem 1, a tilted planar relative equilibrium is formally stable if and only if $K(\sigma)$ is positive semidefinite with a one-dimensional kernel. Note that an "accordian" deformation corresponding to a rescaling of the horizontal projections of the axes of symmetry does not alter the stability of the configuration. This can be interpreted as a generalization of the well-known fact that all tilted equilibria of a single Lagrange top are stable, since all such equilibria can be viewed as accordian deformations of any nonsleeping relative equilibrium.

For large values of n and arbitrary vectors σ, the determination of the index of $K(\sigma)$ appears to be a formidable task. However, for particular classes of planar equilibria, formal stability can be verified easily for arbitrary n. As an example, we present a simple sufficient condition for the stability of configurations without foldbacks of top-heavy tops. We say that the i-th top is top-heavy if $x_i \geq 0$, i.e., if the cumulative center of mass and the point of attachment of the i-th top to the $(i+1)$-th top both lie to the same side of the point of attachment of the i-th top to the $(i-1)$-th top on the axis of symmetry. For example, the first and third tops in Figure 1 are clearly top-heavy, while the middle top may or may not be top-heavy, depending on the relative masses of the second and third tops. We say that a planar configuration has no foldbacks if all tops lie on or to the same side of the vertical axis. More precisely, if $h_1 = \cos\theta\, h$ and $h_2 = \sin\theta\, h$ for some angle θ and vector $h \in \mathbb{R}^n$, then the planar configurations with horizontal projections h_1 and h_2 are said to be without foldbacks if $h_i h_j \geq 0$ for $1 \leq i < j \leq n$. If the upper $n-1$ tops are top-heavy and the configuration has no foldbacks, then the equilibrium is stable. If some of the tops are vertical, then the equilibrium need not be stable (the spin rates of the vertical tops are arbitrary), but a stable relative equilibrium can be obtained by adjusting the spin rates of the vertical tops. The proof makes use of diagonal dominance and our previous estimate of null $[K(\sigma)]$.

Proposition 6. *If $x_i \geq 0$ for $i = 2, \ldots, n$ and $h \in \ker[K(\sigma)]$ satisfies $h_i h_j \geq 0$ for $1 \leq i < j \leq n$, then there exists a vector σ' such that the planar relative equilibria determined by σ' and h are nonlinearly orbitally stable and $(\sigma_i' - \sigma_i)h_i = 0$ for $i = 1, \ldots, n$.*

Proof. We can assume without loss of generality that $h_i \geq 0$ for $i = 1, \ldots, n$. Define $\tilde{h} \in \mathbb{R}^n$ by

$$\tilde{h}_i := \begin{cases} h_i, & h_i \neq 0 \\ a_i, & h_i = 0 \end{cases},$$

for some positive constants a_i. Since all of the entries of $\tilde{h}$ are positive, equation (12) uniquely determines the vector $\tilde{\sigma}$ satisfying $K(\tilde{\sigma})\tilde{h} = 0$. The nondegeneracy hypotheses $p_1 x_n \neq 0$ and $p_i + |x_i| \neq 0$ for $i = 1, \ldots, n$ imply that

$$\tilde{\sigma}_i - p_i x_i = \frac{1}{\tilde{h}_i} \sum_{j \neq i} p_{\min(i,j)} x_{\max(i,j)} \tilde{h}_j > 0, \tag{47}$$

for $i = 1, \ldots, n$. If we define $K_{\tilde{h}} := \text{diag}\left[\tilde{h}\right] K(\tilde{\sigma}) \text{diag}\left[\tilde{h}\right]$, then

$$(K_{\tilde{h}})_{ii} = (\sigma_i - p_i x_i)\tilde{h}_i^2 = \tilde{h}_i \sum_{j \neq i} p_{\min(i,j)} x_{\max(i,j)}\tilde{h}_j = \sum_{j \neq i} \left|(K_{\tilde{h}})_{ij}\right|,$$

for $i = 1, \ldots, n$. Thus $K_{\tilde{h}}$ is diagonally dominant with positive entries on the diagonal and hence must be positive semidefinite. Since $\text{diag}[\tilde{h}]$ is invertible, $K(\tilde{\sigma})$ must be positive semidefinite as well. The inequalities (47) imply that

$$\tilde{\sigma}_i > p_i x_i \geq 0,$$

for $i = 2, \ldots, n$; hence, Proposition 2 implies that $\text{null}\left[K(\tilde{\sigma})\right] \leq 1$. Since $\tilde{h} \in \ker\left[K(\tilde{\sigma})\right]$, it must be the case that the eigenvalues $\tilde{\lambda}_1, \ldots, \tilde{\lambda}_n$ of $K(\tilde{\sigma})$ satisfy

$$0 = \tilde{\lambda}_1 < \tilde{\lambda}_2 \leq \cdots \leq \tilde{\lambda}_n.$$

If h_i is strictly positive for $i = 1, \ldots, n$, then $\tilde{h} = h$ and $\tilde{\sigma} = \sigma$; thus, in this case we are done. Otherwise, define $\sigma' \in \mathbb{R}^n$ by

$$\sigma_i' := \begin{cases} \sigma_i, & h_i \neq 0 \\ \tilde{\sigma}_i, & h_i = 0 \end{cases}.$$

$K(\sigma')h = K(\sigma)h = 0$, since the i-th column of $K(\sigma')$ differs from the i-th column of $K(\sigma)$ only if $h_i = 0$. If $\{\iota_1, \ldots, \iota_n\}$ is a permutation of $\{1, \ldots, n\}$ such that $h_{\iota_1} = \cdots = h_{\iota_k} = 0$ for some index $k \geq 1$ and $h_{\iota_{k+1}}, \ldots, h_{\iota_n}$ are positive, then the eigenvalues $\lambda_1', \ldots, \lambda_n'$ of the symmetric matrix $K(\sigma')$ satisfy

$$\left|\tilde{\lambda}_i - \lambda_i'\right| \leq \left\|\tilde{\sigma} - \sigma'\right\|_\infty = \max_{k < i \leq n} \left| h_{\iota_i} \sum_{j=1}^k p_{\min(\iota_i,\iota_j)} x_{\max(\iota_i,\iota_j)} a_{\iota_j} \right|.$$

(See, e.g., Ciarlet [1989], Theorem 2.3-2.) Hence, by taking the a_i's to be sufficiently small, we can guarantee that $\lambda_2', \ldots, \lambda_n'$ are positive. Since $K(\sigma')$ has a nontrivial kernel, $\lambda_1 = 0$ with eigenvector h. Thus Theorem 1 implies that the relative equilibria generated by σ' and h are nonlinearly orbitally stable. $\square$

Remark. Stability of planar relative equilibria in which $h_i h_j > 0$ for all i and j — i.e., all tops are tilted to the same side of the vertical — and $p_i x_i > 0$ for $i = 2, \ldots, n$ was proved by Vassileva and Lilov [1987].

We now consider general two- and three-top planar relative equilibria. We shall see that all formally stable two-top relative equilibria satisfy the hypotheses of Proposition 6, but there are stable three-top relative equilibria with foldbacks.

Proposition 7. *Tilted two-top configurations are formally stable if and only if $h_1 h_2 > 0$.*
   Planar three-top configurations in which all tops are tilted are formally stable if and only if

$$\tilde{p}_1 \tilde{x}_2 + \tilde{p}_2 \tilde{x}_3 + \min\{\tilde{p}_1 \tilde{x}_3, \tilde{p}_2 \tilde{x}_2\} > 0, \tag{48}$$

where $\tilde{p}_i := h_i p_i$ and $\tilde{x}_i := h_i x_i$. Planar three-top configurations in which the i-th top is vertical are formally stable if and only if $p_2 x_2$ is negative and

$$\sigma = (\sigma_1^*, 0, \sigma_3^*) + c(\delta_{i1}, \delta_{i2}, \delta_{i3}), \qquad \text{for some } c > 0, \tag{49}$$

where σ_1^* and σ_3^* are given by (14).

Proof. For $n = 2$, (12) implies that if $h \neq 0$, then

$$K(\sigma) = p_1 x_2 \begin{pmatrix} \frac{h_2}{h_1} & -1 \\ -1 & \frac{h_1}{h_2} \end{pmatrix}.$$

Since the reference orientations were chosen such that $p_1 x_2 > 0$, $K(\sigma)$ is positive semidefinite if and only if $h_1 h_2 > 0$.

When considering the case $n = 3$, it is convenient to make use of the fact that if all of the tops are tilted, then the signature of $K(\sigma)$ is equal to the signature of $K_h = \text{diag}\,[h]\,K(\sigma)\text{diag}\,[h]$ and hence the stability conditions can be derived from the characteristic polynomial

$$\text{char}\,[K_h] = -\lambda \left(\lambda^2 - 2\lambda \left(\tilde{p}_1 \tilde{x}_2 + (\tilde{p}_1 + \tilde{p}_2)\tilde{x}_3 \right) + (\tilde{p}_1 + \tilde{p}_2)\tilde{x}_2 + \tilde{p}_2 \tilde{x}_3 \right),$$

which has two positive roots if and only if (48) holds. If the i-th top is sleeping, then, as was discussed in Section 2.3, σ has the form (49) for some constant c (not necessarily positive). The result follows directly from the characteristic polynomials for each of the cases $i = 1, 2, 3$. $\qquad\square$

3.4. Nonplanar Configurations

We have seen that a vector σ for which $K(\sigma)$ is positive-definite determines a nonlinearly stable sleeping configuration, while a vector σ for which $K(\sigma)$ is positive semidefinite with nullity one determines a one-parameter family of nonlinearly stable sleeping configurations. For nonplanar equilibrium configurations, the kernel of $K(\sigma)$ is necessarily multi-dimensional, while the kernel of the stability matrix $\mathcal{S}_e$ is one-dimensional if the equilibrium is formally stable. Nonetheless, positive semidefiniteness of $K(\sigma)$ may still imply stability, provided that an additional condition is satisfied. Specifically, if $K(\sigma)$ is positive semidefinite with nullity two and the axes of symmetry of the tops do not lie in any two vertical planes, then the configuration is stable.

Proposition 8. *If there are vertical planes P and P' such that the axes of symmetry of the tops in a nonplanar relative equilibrium satisfy $s_i(0) \in P \cup P'$ for $i = 1, \ldots, n$, then the relative equilibrium is formally unstable.*

If $K(\sigma)$ is positive semidefinite with nullity two and the biplanar condition given above fails for some relative equilibrium determined by $K(\sigma)$, then all nonplanar relative equilibria determined by $K(\sigma)$ are stable.

Proof. Theorem 1 implies that a relative equilibrium determined by the triplet $(K(\sigma), h_1, h_2)$ is formally stable if and only if the matrix $\mathcal{S}_e$ given in (36) is positive semidefinite, with ker $[S_e] = \text{span}[(r, 0)]$; here $(h_{1i}, h_{2i}) = r_i(\cos\theta_i, \sin\theta_i)$, as in Theorem 1. If we define the linear maps $Y: \mathbb{R}^n \times \mathbb{R}^n \to \mathbb{R}^n$ and $\tilde{Y}: \mathbb{R}^n \times \mathbb{R}^n \to \mathbb{R}^n$ by

$$\begin{pmatrix} Y(y, \tilde{y})_i \\ \tilde{Y}(y, \tilde{y})_i \end{pmatrix} := \begin{pmatrix} \cos\theta_i & \sin\theta_i \\ -\sin\theta_i & \cos\theta_i \end{pmatrix} \begin{pmatrix} y_i \\ v_i \tilde{y}_i \end{pmatrix},$$

then

$$(y, \tilde{y}) \cdot S_e(y, \tilde{y}) = Y(y, \tilde{y}) \cdot K(\sigma)Y(y, \tilde{y}) + \tilde{Y}(y, \tilde{y}) \cdot K(\sigma)\tilde{Y}(y, \tilde{y})$$
$$+ \tilde{y} \cdot \text{diag}[\Upsilon_1]\tilde{y} + (\Upsilon_2 \cdot \tilde{y})^2.$$

We first consider the biplanar case. We can assume without loss of generality that one of the planes coincides with the e_1–e_3 plane, so that there is a permutation $\iota_1, \ldots, \iota_n$ of $1, \ldots, n$ and an index $\ell < n$ such that $h_{2\iota_j} = 0$ for $j \leq \ell$ and $h_{2\iota_j} \neq 0$ for $j > \ell$. The vector y with components

$$y_{\iota_j} = \begin{cases} h_{1\iota_j}, & j \leq \ell \\ 0, & j > \ell \end{cases},$$

satisfies

$$Y(y, 0) = h_1 - \frac{h_{1\iota_n}}{h_{2\iota_n}}h_2 \in \text{ker}[K(\sigma)] \qquad \text{and} \qquad \tilde{Y}(y, 0) = 0.$$

Thus $(y, 0) \in \text{ker}[S_e]$, but $y \notin \text{span}[r]$, and the relative equilibrium is formally unstable.

We now turn to the case in which $K(\sigma)$ is positive semidefinite with null$[K(\sigma)] = 2$ and the equilibrium configuration is not biplanar. Positive semidefiniteness of $K(\sigma)$ implies that $\sigma_i \geq p_i x_i$ for $i = 1, \ldots, n$. Hence,

$$\Upsilon_{1i} = r_i^2(\sigma_i + I_i + d_i) = r_i^2 \left(\sigma_i - p_i x_i + I_i' + p_i^2 \sum_{j>i} m_j \right) \geq 0,$$

with equality if and only if $r_i = 0$, for all i. If $r_i = 0$, then $\Upsilon_{2i} = 0$ as well. Hence if $K(\sigma)$ is positive semidefinite, then S_e is also positive semidefinite, with

$$\text{ker}[S_e] = \mathcal{K}_\mathcal{O} := \left\{ (y, \tilde{y}): K(\sigma)Y(y, \tilde{y}) = K(\sigma)\tilde{Y}(y, \tilde{y}) = \text{diag}[r]\tilde{y} = 0 \right\}.$$

We now show that if $K(\sigma) = 2$, then $\mathcal{K}_\mathcal{O} = \text{span}[(r, 0)]$ if and only if the equilibrium configuration is not biplanar.

If $Y(y, \tilde{y})$ and $\tilde{Y}(y, \tilde{y})$ lie in ker$[K(\sigma)] = \text{span}[h_1, h_2]$, then there exist scalars α, β, γ, and δ such that

$$\begin{pmatrix} Y(y, \tilde{y}) \\ \tilde{Y}(y, \tilde{y}) \end{pmatrix} = \begin{pmatrix} \alpha + \gamma & \beta - \delta \\ \beta + \delta & \gamma - \alpha \end{pmatrix} \begin{pmatrix} h_1 \\ h_2 \end{pmatrix}. \tag{50}$$

Solving (50) for y and $\tilde{y}$, we find that

$$\begin{pmatrix} y_i \\ v_i \tilde{y}_i \end{pmatrix} = r_i \left(\begin{pmatrix} \alpha \\ \beta \end{pmatrix} + \begin{pmatrix} \cos 2\theta_i & -\sin 2\theta_i \\ \sin 2\theta_i & \cos 2\theta_i \end{pmatrix} \begin{pmatrix} \gamma \\ \delta \end{pmatrix} \right), \tag{51}$$

for $i = 1, \ldots, n$. If $v_i = 0$, then (51) implies that (β, γ, δ) is orthogonal to $(1, \sin 2\theta_{\iota_1}, \cos 2\theta_{\iota_1})$, since $v_i^2 + r_i^2 = 1$ implies that $r_i = 1$ in this case. If $1 > v_i > 0$, then

$$r_i \tilde{y}_i = \frac{r_i^2}{v_i} (1, \sin 2\theta_{\iota_1}, \cos 2\theta_{\iota_1}) \cdot (\beta, \gamma, \delta)$$

is zero if and only if the same orthogonality condition is satisfied. Thus, a vector $\tilde{y}$ satisfies both (51) and diag $[r]\tilde{y} = 0$ if and only if $(\beta, \gamma, \delta) \in \ker[M(\iota_1, \iota_2, \iota_3)]$, where

$$M(\iota_1, \iota_2, \iota_3) := \begin{pmatrix} 1 & \sin 2\theta_{\iota_1} & \cos 2\theta_{\iota_1} \\ 1 & \sin 2\theta_{\iota_2} & \cos 2\theta_{\iota_2} \\ 1 & \sin 2\theta_{\iota_3} & \cos 2\theta_{\iota_3} \end{pmatrix},$$

for any triplet $(\iota_1, \iota_2, \iota_3)$ of indices such that $r_{\iota_1} r_{\iota_2} r_{\iota_3} \neq 0$. If the configuration is not biplanar, i.e., there is at least one triplet such that $(h_{1\iota_1}, h_{2\iota_1})$ is parallel neither to $(h_{1\iota_2}, h_{2\iota_2})$ nor to $(h_{1\iota_3}, h_{2\iota_3})$, then

$$\det M(\iota_1, \iota_2, \iota_3) = -4 \sin(\theta_{\iota_1} - \theta_{\iota_2}) \sin(\theta_{\iota_2} - \theta_{\iota_3}) \sin(\theta_{\iota_3} - \theta_{\iota_1})$$

is nonzero and $\mathcal{K}_{\mathcal{O}} = \text{span}[(r, 0)]$.

The biplanar condition is invariant under invertible linear transformations of the horizontal projections; thus, if the conditions of the lemma are satisfied for any nonplanar configuration associated to $K(\sigma)$, with horizontal projections h_1 and h_2, then they are satisfied for any other nonplanar configuration determined by the same matrix $K(\sigma)$ and horizontal projections h_1' and $h_2' \in \text{span}[h_1, h_2]$. □

The overlapping-block structure (24) of $K(\sigma)$ can be used to simplify verification of the semidefiniteness condition of Proposition 8 for hinged equilibria. In the following corollary we show that $K(\sigma)$ is positive semidefinite with nullity two if $p_\ell x_\ell$ is negative and the diagonal subblocks of (24) are both positive semidefinite with nullity one.

Corollary 3. *If*

(i) $\sigma_\ell = 0$,
(ii) $p_\ell x_\ell < 0$,
(iii) $K_\ell(\sigma)|_{p_\ell^\perp}$ *and* $K^\ell(\sigma)|_{(x^\ell)^\perp}$ *are positive-definite,*

then all nonplanar relative equilibria determined by σ *are stable hinged relative equilibria.*

A nonplanar relative equilibrium hinged at the ℓ-th top, and hence satisfying condition i, is formally stable only if conditions (ii) and (iii) are also satisfied.

Proof. Assume that (i)–(iii) are satisfied for some vector σ such that there are nonplanar relative equilibria associated to $K(\sigma)$. Lemma 1 implies that $\ker[K_\ell(\sigma) + \frac{x_\ell}{p_\ell} p_\ell p_\ell^T]$ and $\ker[K^\ell(\sigma) + \frac{p_\ell}{x_\ell}(x^\ell)(x^\ell)^T]$ are nontrivial, while positive-definiteness of $K_\ell(\sigma)|_{p_\ell^\perp}$ and $K^\ell(\sigma)|_{(x^\ell)^\perp}$ implies that $\ker[K_\ell(\sigma)] \cap p_\ell^\perp$ and $\ker[K^\ell(\sigma)] \cap (x^\ell)^\perp$ are both trivial. Hence, (18) guarantees the existence of $u \in \mathbb{R}^{\ell-1}$ satisfying

$$K_\ell(\sigma)u = p_\ell \quad \text{and} \quad p_\ell \cdot u = -\frac{p_\ell}{x_\ell},$$

and $u' \in \mathbb{R}^{n-\ell}$ satisfying

$$K^{\ell}(\sigma)u' = x^{\ell} \qquad \text{and} \qquad x^{\ell} \cdot u' = -\frac{x_{\ell}}{p_{\ell}}.$$

Hence, given an arbitrary vector $y \in \mathbb{R}^n$, there are vectors $c \in \mathbb{R}^3$, $w \in p_{\ell}^{\perp}$, and $w' \in (x^{\ell})^{\perp}$ such that

$$y = (c_1 u + w, c_2, c_3 u' + w'),$$

and hence

$$y \cdot K(\sigma)y = w \cdot K_{\ell}(\sigma)w + w' \cdot K^{\ell}(\sigma)w' - \frac{(\chi \cdot c)^2}{p_{\ell} x_{\ell}}, \qquad (52)$$

where $\chi = (p_{\ell}, -p_{\ell}x_{\ell}, x_{\ell})$. Assumptions (ii) and (iii) imply that $K(\sigma)$ is positive semidefinite, with

$$\ker[K(\sigma)] = \mathrm{span}\left[\left(u, \frac{1}{x_{\ell}}, 0\right), \left(0, \frac{1}{p_{\ell}}, u'\right)\right]. \qquad (53)$$

The structure of the basis vectors given in (53) guarantees that all nonplanar relative equilibria are hinged at the ℓ-th top.

We now show that axes of symmetry $s_1, \ldots, s_n$ with linearly independent horizontal projections h_1 and h_2 in $\ker[K(\sigma)]$ are not contained in two vertical planes. Equation (53) implies that

$$h_i = \left(\alpha_i u, \frac{\alpha_i}{x_{\ell}} + \frac{\alpha_i'}{p_{\ell}}, \alpha_i' u'\right)$$

for some constants α_i and α_i', $i = 1, 2$. Since u and u' each contain at least one nonzero component, the biplanar condition is satisfied only if $\alpha_1\alpha_2' = \alpha_2\alpha_1'$, i.e., if the equilibrium is planar. Since the equilibrium is assumed to be nonplanar, the vertical planes containing the axes of symmetry of the first $\ell - 1$ tops, the hinge top, and the last $n - \ell$ tops must be distinct. Thus, the assumptions of Proposition 8 are satisfied and the relative equilibrium is stable.

We now show that conditions (ii) and (iii) are necessary for formal stability of a hinged relative equilibrium. Theorem 1 states that a relative equilibrium determined by the triplet $(K(\sigma), h_1, h_2)$ is formally stable if and only if the matrix S_e given by (36) is positive semidefinite, with $\ker[S_e] = \mathrm{span}[(r, 0)]$, where $(h_{1i}, h_{2i}) = r_i(\cos\theta_i, \sin\theta_i)$, $i = 1, \ldots, n$. In particular, if the relative equilibrium is formally stable, then the upper left-hand block A of S_e is positive semidefinite with kernel r.

The axes of symmetry of the first $\ell - 1$ tops lie in a common vertical plane, while those of the last $n - \ell$ tops lie in some other vertical plane; hence, for appropriate assignments of the angles of any sleeping tops, the vector $f \in \mathbb{R}^n$ given by

$$f_i := \begin{cases} \cos(\theta_i - \theta_1), & i < \ell \\ 1, & i = \ell \\ \cos(\theta_i - \theta_n), & i > \ell \end{cases},$$

satisfies $f_i = \pm 1$, for $i = 1, \ldots, n$, and $\cos(\theta_i - \theta_j) = f_i f_j$, for $1 \leq i, j < \ell$ and for

$\ell < i, j \leq n$. Equation (35) thus implies that A has the overlapping block structure

$$A = \text{diag } [\mathbf{f}] \begin{pmatrix} K_\ell(\sigma) & Q \\ Q^T & K^\ell(\sigma) \end{pmatrix} \text{diag } [\mathbf{f}],$$

for some matrix Q.

By assumption, at least one of the first $\ell - 1$ tops and one of the last $n - \ell$ tops is tilted out of the vertical, i.e., there are indices $j < \ell$ and $j' > \ell$ such that r_j and $r_{j'}$ are nonzero. (If not, Corollary 1 would imply that the relative equilibrium was planar or sleeping.) Hence, positivity of $-p_\ell x_\ell = A_{\ell\ell}$ and positive-definiteness of the upper right-hand $(\ell - 1) \times (\ell - 1)$ block and lower left-hand $(n - \ell) \times (n - \ell)$ block of A are all necessary conditions of formal stability. Since the entries of the vector $\mathbf{f}$ are all nonzero, these blocks are positive-definite if and only if $K_\ell(\sigma)$ and $K^\ell(\sigma)$ are positive-definite. $\qquad \square$

Proposition 3 allows us to formulate an analog of the stability result for unfolded top-heavy planar relative equilibria given in Corollary 6 that is applicable to hinged configurations in which all but the hinge top are top-heavy and no foldbacks occur within the upper and lower vertical planes. Using this result, one can essentially determine by inspection that the six-top relative equilibria shown in Figure 1 are stable. The only quantitative information needed is the sign of x_4; if the fourth top is sufficiently massive and set sufficiently far back to counterbalance the final two tops, then the relative equilibria are stable.

Corollary 4. *If there is an index $1 < \ell < n$ such that $p_\ell x_\ell < 0$ and $x_i \geq 0$ for $i \neq \ell$, then a relative equilibrium with horizontal projections $\mathbf{h}_1$ and $\mathbf{h}_2$ with polar coordinates*

$$(h_{1i}, h_{2i}) = \begin{cases} r_i(\cos\phi, \sin\phi), & i < \ell \\ r_i(\cos\psi, \sin\psi), & i > \ell \end{cases}, \qquad (54)$$

for some angles ϕ and ψ, $(\phi - \psi)/\pi \notin \mathbb{N}$, is either nonlinearly orbitally stable or differs from a nonlinearly orbitally stable relative equilibrium only in the spin rates of the vertical tops.

Proof. Analogous to the proof of Corollary 6, we define the vector $\tilde{r} \in \mathbb{R}^n$ by

$$\tilde{r}_i := \begin{cases} r_i, & r_i \neq 0 \\ a_i, & r_i = 0 \end{cases},$$

for some positive constants a_i. Proposition 1 implies that the vector $\tilde{\sigma}$ with components

$$\tilde{\sigma}_i = \begin{cases} p_i x_i + \frac{p_i}{\tilde{r}_i}\left(x_\ell - \frac{x_\ell}{p_\ell}p_\ell\right) \cdot \tilde{r}_\ell, & i < \ell \\ 0, & i = \ell \\ p_i x_i + \frac{x_i}{\tilde{r}_i}\left(p^\ell - \frac{p_\ell}{x_\ell}x^\ell\right) \cdot \tilde{r}^\ell, & i > \ell \end{cases},$$

and the vectors $\tilde{\mathbf{h}}_1$ and $\tilde{\mathbf{h}}_2$ determined by (54), with r_i replaced by $\tilde{r}_i$, and

$$\tilde{h}_{i\ell} = -\frac{p_\ell \cdot (h_i)_\ell}{p_\ell} - \frac{x^\ell \cdot (h_i)^\ell}{x_\ell}$$

determine a nonplanar relative equilibrium. The inequalities $p_\ell x_\ell < 0, x_i \geq 0$ for $i \neq \ell$, imply that the matrices

$$K' := \text{diag}\left[\tilde{r}_\ell\right] K_\ell(\tilde{\sigma})\text{diag}\left[\tilde{r}_\ell\right] \qquad \text{and} \qquad K'' := \text{diag}\left[\tilde{r}^\ell\right] K^\ell(\tilde{\sigma})\text{diag}\left[\tilde{r}^\ell\right]$$

satisfy

$$K'_{ii} = \tilde{r}_i p_i \left(x_\ell - \tfrac{x_\ell}{p_\ell}p_\ell\right) \cdot \tilde{r}_\ell > \tilde{r}_i p_i x_\ell \cdot \tilde{r}_\ell = \sum_{j \neq i} |K'_{ij}| \tag{55}$$

and

$$K''_{ii} = \tilde{r}_i x_i \left(p^\ell - \tfrac{p_\ell}{x_\ell}x^\ell\right) \cdot \tilde{r}^\ell > \tilde{r}_i x_i p^\ell \cdot \tilde{r}^\ell = \sum_{j \neq i} |K''_{ij}|. \tag{56}$$

Equations (55) and (56) imply that K' and K'' are strictly diagonally dominant with positive diagonal entries. Hence, they and the matrices $K_\ell(\tilde{\sigma})$ and $K^\ell(\tilde{\sigma})$ are positive-definite.

If all of the tops are tilted, then $\tilde{r} = r$ and $\tilde{\sigma} = \sigma$; in this case we have established stability of the relative equilibrium. If the equilibrium contains any vertical tops, then we define $\sigma' \in \mathbb{R}^n$ by

$$\sigma'_i := \begin{cases} \sigma_i, & r_i \neq 0 \\ \tilde{\sigma}_i, & r_i = 0 \end{cases},$$

as before. The triplet (σ', h_1, h_2) determines relative equilibria that differs from the originals only in the spin rates of the vertical tops. By choosing the a_i's, if there are any, to be sufficiently small, we can assure that the eigenvalues of $K(\sigma')_\ell$ and $K(\sigma')^\ell$ are close enough to those of $K(\tilde{\sigma})_\ell$ and $K(\tilde{\sigma})^\ell$ that they are also positive. Stability of any relative equilibrium determined by (σ', h_1, h_2) now follows immediately from Proposition 3.   □

The formal stability condition on $p_\ell x_\ell$ appearing in Corollaries 3 and 4 is common to all nonplanar relative equilibria. As the following simple proposition shows, the ℓ-th top must serve as a "counterbalance," with $p_\ell x_\ell < 0$, if some tilted relative equilibrium with associated vector σ satisfying $\sigma_\ell = 0$ is formally stable. In particular, a collection of tops in which the center of mass of each top lies between its point of attachment to the preceding top and its point of attachment to the following top has no formally stable nonplanar relative equilibria. (For example, the set of tops shown Figure 1 can maintain stable nonplanar steady motions only if the middle top is sufficiently massive or set sufficiently far back from the point of attachment to the first top.)

Proposition 9. *If a tilted configuration is formally stable, then $\sigma_i > p_i x_i$ for $i = 1, \ldots, n$. In particular, if $x_i \geq 0$ for $i = 2, \ldots, n - 1$, then all nonplanar relative equilibria are formally unstable.*

Proof. Let e_i denote the i-th Euclidean basis vector. Positive semidefiniteness of $K(\sigma)$ (respectively S_e) implies that

$$\sigma_i - p_i x_i = e_i \cdot K(\sigma)e_i \geq 0, \tag{57}$$

for $i = 1, \ldots, n$, since

$$(e_i, 0) \cdot S_e(e_i, 0) = e_i \cdot K(\sigma)e_i$$

in the nonplanar case.

Corollary 1 implies that no column of $\ker [K(\sigma)]$ is identically zero. Hence, no Euclidean basis vector is a nullvector of $K(\sigma)$ and the inequality (57) must be strict for $i = 1, \ldots, n$.

Proposition 2 implies that there are nonplanar relative equilibria associated to $K(\sigma)$ only if $\sigma_\ell = 0$ for at least one index $1 < \ell < n$. Hence, if $p_\ell x_\ell \geq 0 = \sigma_\ell$, then all tilted relative equilibria determined by $K(\sigma)$ are formally unstable. $\square$

As an application of Corollary 3, we determine the formal stability of the nonplanar relative equilibria for $n = 3$ and for $n = 4$, $p_2 x_2 p_3 x_3 \neq 0$. We shall see that in these cases the only formally stable relative equilibria are those satisfying the assumptions of Corollary 4.

Corollary 5.

(i) *The nonplanar three-top relative equilibria determined by (16) are formally stable if and only if $p_2 x_2$ is negative.*

(ii) *If $\frac{p_2}{x_2} \neq \frac{p_3}{x_2}$ and $p_2 x_2 p_3 x_3 \neq 0$, then the nonplanar four-top relative equilibria determined by equations (27)–(28) are formally stable if and only if $0 > p_3 x_3$ and $\sigma_2 > \sigma_2^*$, while those determined by (29)–(30) are formally stable if and only if $0 > p_2 x_2$ and $\sigma_3 > \sigma_3^*$.*

If $\frac{p_2}{x_2} = \frac{p_3}{x_3}$ and $p_2 x_2 p_3 x_3 \neq 0$, then four-top nonplanar relative equilibria are formally stable if and only if $0 > p_2 x_2$ and $\sigma_i > \sigma_i^*$ for some index i.

Proof. We first consider the case $n = 3$. Both $p_2^\perp$ and $(x^2)^\perp$ are trivial; hence, Proposition 3 implies that a nonplanar relative equilibrium is formally stable if and only if $p_2 x_2$ is negative.

We now consider the case $n = 4$, $\ell = 3$, and $p_2 x_2 p_3 x_3 \neq 0$, corresponding to (29)–(30). Condition (iii) of Corollary 3 states that $0 > p_3 x_3$ is a necessary formal stability condition in this case.

$$\{0\} = (x^3)^\perp \subset \mathbb{R} \qquad \text{and} \qquad \text{span}\,[(p_2, -p_1)] = p_3^\perp \subset \mathbb{R}^2$$

imply that $K^3(\sigma)|_{(x^3)^\perp}$ is trivially definite, while $K_3(\sigma)|_{p_3^\perp} = \text{diag}\left[\sigma_1 - \sigma_1^*, \sigma_2\right]|_{p_3^\perp}$ is positive-definite if and only if

$$
(p_2, -p_1) \cdot K_3(\sigma)(p_2, -p_1) = (\sigma_1 - \sigma_1^*)p_2^2 + \sigma_2 p_1^2
$$
$$
= \begin{cases} (\sigma_1 - \sigma_1^*)p_2^2, & \text{if } \sigma_2 = \sigma_2^* = 0, \\[2mm] \dfrac{(\sigma_2 p_1)^2}{\sigma_2 - \sigma_2^*}, & \text{otherwise,} \end{cases}
$$

is positive. Recall that $\sigma_2^* = \sigma_3^* = 0$ if $\frac{p_2}{x_2} = \frac{p_3}{x_3}$.

The case $n = 4$, $\ell = 2$, with σ given by (27)–(28), is treated analogously. In this case, $p_2^\perp$ is trivial, $(x^2)^\perp = \text{span}\,[(x_4, -x_3)]$, and

$$(x_4, -x_3) \cdot K^2(\sigma)(x_4, -x_3) = \sigma_3 x_4^2 + (\sigma_4 - \sigma_4^*)x_3^2$$

$$= \begin{cases} (\sigma_4 - \sigma_4^*)x_3^2, & \text{if } \sigma_3 = \sigma_3^* = 0 \\[2mm] \dfrac{(\sigma_3 x_4)^2}{\sigma_3 - \sigma_3^*}, & \text{otherwise} \end{cases}$$

The previous argument generally does not apply if $p_2 x_3 = p_3 x_2$, $\sigma_1 = \sigma_1^*$, and $\sigma_4 = \sigma_4^*$, since none of the axes of symmetry need be parallel in this case. However, a straightforward calculation shows that the formal stability matrix S_e, given by (36), of a nonplanar relative equilibrium with $\sigma = (\sigma_1^*, 0, 0, \sigma_4^*)$ has a two-dimensional kernel. Hence, these relative equilibria are always formally unstable.   □

Appendix: Derivation of the Formal Stability Conditions

We briefly describe the formal stability analysis for a Lagrangian system that is equivariant with respect to the lift of an action on the configuration manifold (i.e., the action of an element g of the symmetry group on a vector v tangent to a given curve $\gamma(t)$ is simply $g \cdot v = \frac{d}{d\epsilon}\big|_{\epsilon=0}\, g \cdot \gamma(t)$). Formal stability is defined in terms of the energy-momentum function. However, it has been shown by Smale [1970] that for simple mechanical systems with lifted symmetry group actions, the formal stability test reduces to testing for positive semidefiniteness a bilinear form on the tangent space of the configuration manifold at the equilibrium configuration. This approach has been further developed and generalized by Simo et al. [1991a] and Lewis [1992] to a larger class of systems, including systems for which the symmetry group does not act freely. This modified approach, which is known as the reduced energy-momentum (REM) method, can be summarized as follows: The formal stability of such a relative equilibrium is determined by a two form constructed from variations of a scalar function of the configuration variables and the elements of the algebra of the symmetry group. For simple mechanical systems, i.e., systems with a freely acting symmetry group and Lagrangian of the form "kinetic minus potential energy," this form coincides with the second variation of Smale's amended potential. Positive semidefiniteness of this two form, with kernel consisting of infinitesimal group motions preserving the momentum map, implies formal stability. Note that the symmetry group of the coupled Lagrange top systems does not act freely; the amended potential is not well defined at a sleeping configuration. In many examples it appears to be advantageous to avoid explicit computation of the amended potential even when it is well-defined. The amended potential involves the inverse of the locked inertia tensor, which determines a configuration-dependent inner product on the algebra of the symmetry group if the group action is free; in the REM method the inverse — or generalized inverse if the equilibrium has nontrivial isotropy — is only computed at the equilibrium configuration. Since the relative equilibrium conditions often impose additional structure on the locked inertia tensor at an equilibrium configuration, the REM approach can lead to noticeably simpler calculations than those involved in the direct calculation of the second variation of the amended potential, or in the explicit calculation of vectors tangent to the level set of the momentum map, which is the strategy used in Vassileva and Lilov [1987].

We first outline the formal stability analysis for a relative equilibrium of a general Lagrangian system as described above, then carry out the relevant constructions for the system of n heavy Lagrange tops. The configuration q_e of a relative equilibrium with generator $\xi \in \mathcal{G}$ is a critical point of the locked Lagrangian $L_\xi := L \circ \xi_Q : Q \to \mathbb{R}$, where the infinitesimal generator $\xi_Q : Q \to TQ$ is given by $\xi_Q(q) := \frac{d}{d\epsilon}\big|_{\epsilon=0} \exp(\epsilon \, \xi) \cdot q$. The locked momentum map $\mathbb{I}_\xi : Q \to \mathcal{G}$ is the first variation of the locked Lagrangian with respect to the algebra element. The locked inertia tensor $\mathbb{I}_e$ associated to a relative equilibrium is defined as the second variation of the locked Lagrangian with respect to the algebra elements, evaluated at the equilibrium configuration q_e and generator ξ. Specifically,

$$\mathbb{I}_\xi(q) \cdot \eta := \frac{d}{d\epsilon}\Big|_{\epsilon=0} L_{(\xi+\epsilon\eta)}(q) \qquad \text{and} \qquad \mathbb{I}_e \eta := \frac{d}{d\epsilon}\Big|_{\epsilon=0} \mathbb{I}_{(\xi+\epsilon\eta)}(q_e).$$

Note that even if the locked inertia tensor is not invertible, the pairing $[\mathbb{I}_e]^{-1} : \text{range}[\mathbb{I}_e] \times \text{range}[\mathbb{I}_e] \to \mathbb{R}$ defined by

$$[\mathbb{I}_e]^{-1}(\mu, \nu) := \mu \cdot \zeta \qquad \text{for some } \zeta \text{ satisfying } \mathbb{I}_e \zeta = \nu$$

is still well-defined. If the Lagrangian L is of the form "kinetic minus potential energy," then for arbitrary $\eta \in \mathcal{G}$, $\eta \cdot \mathbb{I}_e \eta$ is the kinetic energy of the infinitesimal group motion $\eta_Q(q_e)$ generated by η. In this case, $\mathbb{I}_e$ is positive semidefinite, with $\ker[\mathbb{I}_e] = \mathcal{G}_{q_e} = \{\eta \in \mathcal{G} : \eta_Q(q_e) = 0\}$.

The stability form has two major components: the second variation $D^2 L_\xi(q_e)$ of the locked Lagrangian and the momentum correction term, which can be expressed in terms of the first variation of the locked momentum map and the generalized inverse of the locked inertia tensor. We set

$$\mathcal{Q}_e := \left\{ \delta q \in T_{q_e}Q : D\mathbb{I}(q_e) \cdot \delta q \in \text{range}[\mathbb{I}_e] \right\},$$

and define the stability two form $\mathcal{B}_e : \mathcal{Q}_e \times \mathcal{Q}_e \to \mathbb{R}$ by

$$\mathcal{B}_e(\delta q, \Delta q) := [\mathbb{I}_e]^{-1}(D\mathbb{I}(q_e) \cdot \delta q, D\mathbb{I}(q_e) \cdot \Delta q) - D^2 L_\xi(q_e)(\delta q, \Delta q). \tag{58}$$

If $\mathcal{B}_e$ is positive semidefinite, with $\mathcal{B}_e(\delta q, \delta q) = 0$ if and only if $\delta q = \eta_Q(q_e)$ for some η such that, $\text{ad}^*_\eta \mathbb{I}_\xi(q_e) = 0$, then the relative equilibrium with configuration q_e and generator ξ is formally stable.

We now apply the analysis outlined above to the system of stacked tops. We note that, since $(S^1)^{n+1}$ does not act freely on $(SO(3))^n$, we cannot simply use Smale's amended potential. Even if we were to treat the sleeping configurations, with nontrivial isotropy, and tilted configurations, with trivial isotropy, separately, we believe that it is more convenient to carry out the necessary calculations using the locked Lagrangian and the associated stability form than it would be to directly work with the amended potential. A steady motion with spatial angular velocities $\omega_i = \xi \, e_3 - \zeta_i s_i$, $i = 1, \dots, n$ satisfying (10) and (11) is a relative equilibrium with generator (ζ, ξ). The locked Lagrangian is obtained by substituting $\omega_i = \xi e_3 - \zeta_i s_i$ into the Lagrangian (3), yielding

$$L_{(\zeta,\xi)}(\Lambda) = \tilde{L}_{(\zeta,\xi)}(s_1, \ldots, s_n)$$

$$:= \sum_{i=1}^{n} \left(\tfrac{I_i}{2}\left(\xi^2 + \zeta_i^2\right) - (\xi \zeta_i I_i + g\, x_i)\langle e_3, s_i\rangle\right) + \tfrac{\xi^2}{2} \sum_{i,j=1}^{n} \kappa_{ij} \langle s_i, \mathbb{P}_g s_j\rangle,$$

where, as before, $s_i = \Lambda_i e_3$ denotes the axis of symmetry of the i-th top. It follows immediately from the $(S^1)^{n+1}$ invariance of $L_{(\zeta,\xi)}$ that Λ is a critical point of $L_{(\zeta,\xi)}$ if and only if $(s_1, \ldots, s_n)$ is a critical point of $\tilde{L}_{(\zeta,\xi)}: (S^2)^n \to \mathbb{R}$. A standard Lagrange multiplier argument implies that $D\tilde{L}_{(\zeta,\xi)}(s_1, \ldots, s_n) = 0$ if and only if there exists $\lambda \in \mathbb{R}^n$ such that

$$\xi^2 \mathbb{P}_g \sum_{j=1}^{n} \kappa_{ij} s_j - (\xi I_i \zeta_i + g\, x_i)e_3 = \lambda_i s_i, \qquad \text{for } i = 1, \ldots, n. \qquad (59)$$

The scalars σ_i appearing in (59) are related to the corresponding Lagrange multipliers λ_i in (59) by $\lambda_i = \xi^2(\sigma_i + d_i)$. Thus, equations (10) and (11) are equivalent to the condition that $\Lambda = (\Lambda_1, \ldots, \Lambda_n)$ is a critical point of the locked Lagrangian.

As the first step in proving Theorem 1, we determine the stability form $\mathcal{B}_e$ given by (58). Specializing the constructions outlined above to the case of n Lagrange tops, we obtain the following:

Lemma 2. *Let Λ_e denote the configuration of a relative equilibrium with generator (ζ, ξ). If $\xi = 0$ or the configuration is sleeping, then the ij-th three-by-three block of the stability form $\mathcal{B}_e$ is given by*

$$(\mathcal{B}_e)_{ij} = ((\lambda_i + \xi^2 \kappa_{ii})\delta_{ij} - \xi^2 \kappa_{ij} v_i v_j)\mathbb{P}_g,$$

where λ_i is given by (59). Otherwise, $\mathcal{B}_e$ has the block structure

$$\xi^{-2}(\mathcal{B}_e)_{ij} = b_i b_j^T + (\sigma_i + I_i + \kappa_{ii})\delta_{ij} t_i t_i^T + \mathcal{K}_{ij}(\sigma), \qquad (60)$$

where σ is given by (10) and

- $t_i = e_3 \times s_i$ *and* $r_i := |t_i|$ *for* $i = 1, \ldots, n$,
- $b_i := \begin{cases} 0, & \xi\, r_i = 0 \\ \frac{\gamma_{2i}}{r_i} t_i, & \text{otherwise} \end{cases}$,
- $\mathcal{K}_{ij} := K(\sigma)_{ij}\left(\langle s_i, s_j\rangle \mathbf{1} - s_j s_i^T - t_i t_j^T\right) \in L(3)$.

Proof. We first compute the second variation of the locked Lagrangian $L_{(\zeta,\xi)}$. A variation $\delta\Lambda$ of a rotation matrix Λ can be written in the form $\delta\Lambda = \widehat{\delta\theta}\Lambda$ for some vector $\delta\theta \in \mathbb{R}^3$ determined by the relationship

$$\widehat{\delta\theta}y := \delta\theta \times y = \delta\Lambda\Lambda^T y,$$

for all $y \in \mathbb{R}^3$. In the case at hand, a variation $\delta\Lambda := (\widehat{\delta\theta}_1 \Lambda_1, \ldots, \widehat{\delta\theta}_n \Lambda_n)$ corresponds to the variation $(\delta\theta_1 \times s_1, \ldots, \delta\theta_n \times s_n)$ of the spatial representations of the axes of symmetry of the tops. Setting $\Delta\Lambda := \left(\widehat{\Delta\theta}_1\Lambda_1, \ldots, \widehat{\Delta\theta}_n\Lambda_n\right)$, we find that

$$
\begin{aligned}
D^2 &L_{(\zeta,\xi)}(\Lambda_e)(\delta\Lambda, \Delta\Lambda) \\
&= \sum_{i=1}^{n}\Bigl((-(\xi\zeta_i + g\,x_i))\langle\delta\theta_i \times (\Delta\theta_i \times s_i), e_3\rangle \\
&\qquad\qquad + \xi^2 \sum_{j=1}^{n} \kappa_{ij}\left(\langle\delta\theta_i \times (\Delta\theta_i \times s_i), \mathbb{P}_g s_j\rangle + \langle\delta\theta_i \times s_i, \mathbb{P}_g(\Delta\theta_i \times s_j)\rangle\right)\Bigr) \qquad (61) \\
&= \sum_{i=1}^{n}\Bigl(-\lambda_i\langle\delta\theta_i, \mathbb{P}_{s_i}\Delta\theta_i\rangle \\
&\qquad\qquad + \xi^2 \sum_{j=1}^{n} \kappa_{ij}\left(\langle s_i, s_j\rangle\langle\delta\theta_i, \Delta\theta_j\rangle - \langle s_j, \delta\theta_i\rangle\langle s_i, \Delta\theta_j\rangle - \langle t_i, \delta\theta_i\rangle\langle t_j, \Delta\theta_j\rangle\right)\Bigr),
\end{aligned}
$$

which follows from the equilibrium conditions (8) and $\mathbb{P}_{s_i} = \mathbf{1} - s_i s_i^T$.

The momentum correction term is determined as follows: Straightforward calculations show that the first variation of the locked momentum map $\mathbb{I}_{(\zeta,\xi)}\colon (SO(3))^n \to \mathbb{R}^{n+1}$ satisfies

$$
\begin{aligned}
\left(D\mathbb{I}_{(\zeta,\xi)}(\Lambda_e)\cdot\delta\Lambda\right)\cdot(\tilde{\zeta}, \tilde{\xi}) &= \sum_{i=1}^{n}\left\langle -(\tilde{\xi}\zeta_i + \xi\tilde{\zeta}_i)I_i\,e_3 + \xi\tilde{\xi}\,\mathbb{P}_{e_3}\sum_{j\neq i}\kappa_{ij}s_j, \delta\theta_i \times s_i\right\rangle \\
&= \sum_{i=1}^{n}(\xi I_i\tilde{\zeta}_i - \tilde{\xi}\,g\,x_i)\langle t_i, \delta\theta_i\rangle \qquad (62)
\end{aligned}
$$

at a critical point Λ_e of $L_{(\zeta,\xi)}$. In other words,

$$
D\mathbb{I}_{(\zeta,\xi)}(\Lambda_e)\cdot\delta\Lambda = \left(\xi\left(I_1\langle t_1, \delta\theta_1\rangle, \ldots, I_n\langle t_n, \delta\theta_n\rangle\right), -g\sum_{i=1}^{n}x_i\langle t_i, \delta\theta_i\rangle\right). \qquad (63)
$$

The first variation of the locked momentum map is identically zero if the configuration Λ_e is sleeping, since in this case $t_i = e_3 \times s_i = 0$ for all i. Hence, the momentum correction term is trivial for sleeping configurations and the space of admissible variations is the entire tangent space $T_{\Lambda_e}(SO(3))^n$. Similarly, if $\xi = 0$, then the relative equilibrium conditions imply that $x_i s_i$ is parallel to g, and hence $x_i t_i = 0$, for $i = 1, \ldots, n$. Hence, the momentum correction term is identically zero and $\mathcal{B}_e = -D^2 L_{(\zeta,\xi)}(\Lambda_e)$ unless $\xi\,t_i$ is nonzero for at least one index i.

The locked inertia tensor $\mathbb{I}_e$ satisfies

$$
\mathbb{I}_e = \mathrm{diag}\,[\mathbf{I}, \nu + w\cdot v] - (0, 1)(w, 0)^T - (w, 0)(0, 1)^T,
$$

where $w := \mathrm{diag}\,[\mathbf{I}]\,v = (I_1 v_1, \ldots, I_n v_n)$ and $\nu = (v, 1)\cdot\mathbb{I}_e(v, 1)$ is the kinetic energy of the infinitesimal group motion generated by $(v, 1)$. Since the kinetic energy is positive-definite, $\nu \geq 0$, with $\nu = 0$ if and only if $(v, 1)$ lies in the isotropy subalgebra $\mathcal{G}_{\Lambda_e}$ of Λ_e. If at least one top is tilted, then $\mathcal{G}_{\Lambda_e}$ is trivial and the locked inertia tensor $\mathbb{I}_e$ is invertible, with inverse

$$
\mathbb{I}_e^{-1} = \mathrm{diag}\,\left[I_1^{-1}, \ldots, I_n^{-1}, 0\right] + \tfrac{1}{\nu}(v, 1)(v, 1)^T \qquad (64)
$$

configurations. It follows that the space $\mathcal{Q}_e$ of admissible variations is again the entire tangent space $T_{\Lambda_e}(SO(3))^n$. Combining (63) and (64) yields the momentum correction term

$$[\mathbb{I}_e]^{-1}(D\mathbb{I}_{(\zeta,\xi)}(\Lambda_e), D\mathbb{I}_{(\zeta,\xi)}(\Lambda_e)) = \xi^2(\beta\beta^T + \mathrm{diag}\left[I_1 t_1 t_1^T, \ldots, I_n t_n t_n^T\right]), \quad (65)$$

where $\beta := (b_1, \ldots, b_n) \in \mathbb{R}^{3n}$. Combining (61) and (65) yields (60). $\qquad\square$

Proof of Theorem 1. A relative equilibrium is formally stable if and only if the associated two form $\mathcal{B}_e$ is positive semidefinite with kernel equal to the space of momentum-preserving group motions. The action of the symmetry group $(S^1)^{n+1}$ for the system of heavy Lagrange tops is abelian and hence preserves the equivariant momentum map associated to the action of $(S^1)^{n+1}$ on $TSO(3)^n$. Thus, a relative equilibrium Λ_e with unique generator (ζ, ξ) is formally stable if and only if $\mathcal{B}_e(\delta\Lambda, \delta\Lambda) \geq 0$ for all $\delta\Lambda \in T_{\Lambda_e}SO(3)^n$, with equality only if $\delta\Lambda$ is generated by an infinitesimal group motion. If the isotropy subalgebra of the equilibrium is nontrivial, as is the case for sleeping configurations, then the relative equilibrium is formally stable if and only if $\mathcal{B}_e$ is positive semidefinite, with zero set consisting of infinitesimal group motions, for at least one generator of the motion. If we set $\zeta_i = \langle \omega_i, s_i \rangle$, then for any value of ξ, $(\zeta + \xi v, \xi)$ generates the same sleeping relative equilibrium. Any one of the generators $(\zeta + \xi v, \xi)$ can be used in the stability analysis; to obtain sharp formal stability conditions, we need to make use of this flexibility.

We represent the stability form with respect to a convenient basis for a complement to the space of infinitesimal rotations of the tops about their axes of symmetry. (This complement can be identified in a natural way with the tangent bundle of the manifold $(S^2)^n$ of possible spatial positions of the axes of symmetry of the tops.) An infinitesimal body rotation of the i-th top is determined by the tangent vector $\delta\Lambda$ with $\delta\Lambda_j = \delta_{ij}\hat{s}_i\Lambda_i$, where ˇdenotes the isomorphism between $\mathbb{R}^3$ and three by three skew-symmetric matrices determined by the cross product. If we set $r_i := |t_i|$, then a complement to the subspace of infinitesimal body rotations is spanned by the orthogonal $3n$-vectors

$$u_i := \begin{cases} \frac{1}{r_i}(0, \ldots, 0, \hat{e}_3\Lambda_i, 0, \ldots, 0), & r_i \neq 0 \\ (0, \ldots, 0, v_i(\cos\theta_i\hat{e}_1 - \sin\theta_i\hat{e}_2)\Lambda_i, 0, \ldots, 0), & r_i = 0 \end{cases},$$

and

$$\tilde{u}_i := \begin{cases} -\frac{1}{r_i}(0, \ldots, 0, \hat{t}_i\Lambda_i, 0, \ldots, 0), & r_i \neq 0 \\ (0, \ldots, 0, (\sin\theta_i\hat{e}_1 + \cos\theta_i\hat{e}_2), \Lambda_i, 0, \ldots, 0), & r_i = 0 \end{cases},$$

for $i = 1, \ldots, n$. This space can be identified in a natural way with the tangent space of $(S^2)^n$. (Note that the vector u_i is not a unit vector if $0 < r_i < 1$; this convention is chosen so as to simplify some of the relevant identities.)

Lemma 2 implies that if Λ_e is a sleeping configuration, with $s_i = \pm e_3$ for $i = 1, \ldots, n$ and a one-parameter family of generators $(\zeta + \xi v, \xi)$, then for any choice of ξ there is an associated stability form

$$\mathcal{B}_e(u_i, u_j) = \mathcal{B}_e(\tilde{u}_i, \tilde{u}_j) = \lambda_i\delta_{ij} - \xi^2\kappa_{ij}v_iv_j \quad \text{and} \quad \mathcal{B}_e(u_i, \tilde{u}_j) = 0,$$

for all i and j, where

$$-\lambda_i = v_i(\xi\, I_i\, (\zeta_i + \xi\, v_i) + g\, x_i) = \xi\, I_i\, (\zeta_i\, v_i + \xi) + g\, x_i\, v_i,$$

for $i = 1, \ldots, n$, as follows from (59). Thus the restriction of $\mathcal{B}_e$ to span $[u_1, \ldots, u_n, \tilde{u}_1,$ $\ldots, \tilde{u}_n]$ block diagonalizes into two identical blocks $-\mathcal{S}(\xi)$, where $\mathcal{S}(\xi)$ is given by (33). In this case, the space of infinitesimal spatial rotations of Λ_e is contained in the space of infinitesimal body rotations, i.e., the complement to span $\big[u_1, \ldots, u_n, \tilde{u}_1, \ldots, \tilde{u}_n\big]$; hence the relative-equilibrium is formally stable if and only if $\mathcal{S}(\xi)$ is negative definite for some choice of ξ.

Straightforward calculations show that, for tilted configurations,

$$\mathcal{B}_e(u_i, u_j) = a_{ij}, \qquad \mathcal{B}_e(\tilde{u}_i, u_j) = b_{ij}, \qquad \text{and} \qquad \mathcal{B}_e(\tilde{u}_i, \tilde{u}_j) = c_{ij},$$

where a_{ij}, b_{ij}, and c_{ij} are given by (35). Thus, the representation of the restriction of the stability form $\mathcal{B}_e$ to span $\big[u_1, \ldots, u_n, \tilde{u}_1, \ldots, \tilde{u}_n\big]$ is the matrix $\mathcal{S}_e$ given in (36). A unit speed infinitesimal spatial rotation of the configuration corresponds to the variation $\sum_{i=1}^n r_i u_i$; thus $\mathcal{B}_e$ is positive semidefinite with ker $[\mathcal{B}_e]$ consisting precisely of infinitesimal group motions if and only if $\mathcal{S}_e$ is positive semidefinite with ker $[\mathcal{S}_e] = \text{span}\,[(r, 0)]$.

The stability condition (iii) for planar configurations is derived from the general condition (iv), as follows. If all of the tops lie in the same vertical plane, i.e., $\theta_i = \theta_j$ modulo π for all i and j, then block A of the stability matrix $\mathcal{S}_e$ equals $K(\sigma)$ and the off-diagonal block B is identically zero. Hence, the general stability condition that $\mathcal{S}_e$ be positive semidefinite, with ker $[\mathcal{S}_e] = \text{span}\,[(r, 0)]$, simplifies to the conditions that $K(\sigma)$ be positive semidefinite, with kernel span $[r]$, and C be positive definite. Observe that

$$C = \text{diag}\,[v]\,K(\sigma)\,\text{diag}\,[v] + \text{diag}\,[\Upsilon_1] + \Upsilon_2\Upsilon_2^T.$$

As is shown in the proof of Proposition 8, positive semidefiniteness of $K(\sigma)$ implies that the entries of Υ_1 are nonnegative, with $\Upsilon_{1i} = 0$ if and only if $r_i = 0$. Thus, if diag $[r]\,u \neq 0$, then

$$u \cdot Cu \geq u \cdot \text{diag}\,[\Upsilon_1]\,u > 0.$$

If diag $[v]\,u \notin \text{span}\,[r]$, then positive semidefiniteness of $K(\sigma)$ implies that

$$u \cdot Cu \geq (\text{diag}\,[v]\,u) \cdot K(\sigma)(\text{diag}\,[v]\,u) > 0.$$

If diag $[r]\,u = 0$ and diag $[v]\,u \in \text{span}\,[r]$, then $r_i^2 + v_i^2 = 1$ for all i implies $u = 0$. Thus, if $K(\sigma)$ is positive semidefinite with ker $[K(\sigma)] = \text{span}\,[r]$, then C is positive-definite and the relative equilibrium is stable.

Finally, we show that non-sleeping relative equilibria with trivial spatial generator are formally unstable. If $\xi = 0$, then the momentum correction term is identically zero, as noted above, and

$$\mathcal{S}_e = -g\,\text{diag}\,[x_1v_1, \ldots, x_nv_n, x_1v_1, \ldots, x_nv_n].$$

The equilibrium conditions imply that if $\xi = 0$, then $x_i r_i = 0$ for $i = 1, \ldots, n$; hence if any top is tilted out of the vertical, i.e. if $r_i \neq 0$ for some index i, then $x_i = 0$ and ker $[\mathcal{S}_e]$ is multi-dimensional. $\qquad\qquad\square$

References

[1] Abraham R. and Marsden J. E. (1978) **Foundations of Mechanics**, second edition. Reading, MA: Benjamin/Cummings Publishing Company.

[1a] Arnold V. I. (1989) **Mathematical Methods of Classical Mechanics**, second edition. New York, NY: Springer-Verlag Inc.

[2] Ciarlet P. (1989) **Introduction to Numerical Linear Algebra and Optimization**. Cambridge, UK: Cambridge University Press.

[3] Golub G. and Van Loan C. (1989) **Matrix Computations**, second edition. Baltimore and London: The Johns Hopkins University Press.

[4] Klein F. and Sommerfeld A. (1910) **Theorie des Kreisels**. Leipzig: Teubner.

[5] Lerman E. (1997) Relative equilibria at singular values of the moment map. (Preprint.)

[6] Lewis D. (1992) Lagrangian block diagonalization. *J. Dyn. Diff. Eqs.* **4**, 1–41.

[7] Lewis D. (1993) Bifurcation of liquid drops. *Nonlinearity* **6**, 491–522.

[8] Lewis D. and Simo J.-C. (1990) Nonlinear stability of rotating pseudo-rigid bodies. *Proc. Royal Soc. London A.* **427**, 281–319.

[9] Ortega J.-P. and Ratiu T. (1997) Stability of relative equilibria: Symplectic block diagonalization. (Preprint.)

[10] Patrick G. (1989) The dynamics of two coupled rigid bodies in three-space. *Contemp. Math.* **97**, Dynamics and control of multibody systems. 315–335.

[11] Patrick G. (1992) Relative equilibria in Hamiltonian systems: The dynamic interpretation of nonlinear stability on a reduced phase space. *J. Geometry Phys.* **9**, 111–119.

[12] Routh E. J. (1905) **Advanced Rigid Dynamics**. London: McMillan and Co. Reprinted by Dover, 1960.

[13] Simo J.-C., Marsden J. E., Lewis D., and Posbergh T. (1989) Block diagonalization and the energy-momentum method. *Contemp. Math.* **97**, 297–314.

[14] Simo J.-C., Lewis D., and Marsden J. E. (1991a) The stability of relative equilibria. Part I: The reduced energy-momentum method. *Arch. Rat. Mech. Anal.* **115**, 15–59.

[15] Simo J.-C., Marsden J. E., and Posbergh T. (1991b) The stability of relative equilibria. Part II: The reduced energy-momentum method. *Arch. Rat. Mech. Anal.* **115**, 61–100.

[16] Smale S. (1970) Topology and Mechanics. I. *Inventiones Math.* **10**, 305–331.

[17] Vassileva N. and Lilov L. (1987) A stationary motion stability analysis of Lagrange gyroscopic systems with a tree-like structure. (Russian) *Teoret. Prilozhna Mekh.* **18**, 17–26.

[18] Wittenburg J. (1977) **Dynamics of Systems of Rigid Bodies**. Stuttgart: B. G. Teubner.

On the Bifurcation and Stability of Rigidly Rotating Inviscid Liquid Bridges

H.-P. Kruse and J. Scheurle*
Zentrum Mathematik, TU München, Arcisstrasse 21, D-80290 München, Germany

Received June 21, 1996; revision received October 2, 1997, and accepted for publication October 9, 1997
Communicated by Jerrold Marsden

This paper is dedicated to the memory of Juan-Carlos Simo

Summary. We consider a mathematical model that describes the motion of an ideal fluid of finite volume that forms a bridge between two fixed parallel plates. Most importantly, this model includes capillarity effects at the plates and surface tension at the free surface of the liquid bridge. We point out that the liquid can stick to the plates due to the inner pressure even in the absence of adhesion forces. We use both the Hamiltonian structure and the symmetry group of this model to perform a bifurcation and stability analysis for relative equilibrium solutions. Starting from rigidly rotating, circularly cylindrical fluid bridges, which exist for arbitrary values of the angular velocity and vanishing adhesion forces, we find various symmetry-breaking bifurcations and prove corresponding stability results. Either the angular velocity or the angular momentum can be used as a bifurcation parameter. This analysis reduces to find critical points and corresponding definiteness properties of a potential function involving the respective bifurcation parameter.

1. Introduction

We consider the motion of an ideal, i.e., incompressible and inviscid, fluid of finite volume between two parallel flat plates. The plates are assumed to be at rest. We only take into account surface tension along the free surface of the fluid and adhesion forces along the surfaces of contact between the fluid and the plates. The influence of other forces such as gravity is neglected (cf. Concus and Finn [4]). Also, the complete separation of the fluid from one of the plates will be excluded, i.e., throughout the motion, the fluid forms

---

* Supported by the DFG under the Contract Sch 233/3-1.

a bridge between the plates. This assumption seems to be reasonable if adhesion forces attract the fluid to the plates. We argue that it is also acceptable even if no adhesion forces act, provided that the fluid volume is sufficiently large. In this case the internal pressure prevents the separation of the fluid bridge from the plates. A specific type of motion that we are going to study in the case of zero adhesion forces will be rigid rotations of circular liquid cylinders connecting the two plates. These represent solutions of the underlying equations of motion for arbitrary large fluid volumes and arbitrary large angular velocities. We use a mathematical model for this problem that has been proposed in Kruse [10] (see also Kruse, Marsden, and Scheurle [13]). It is based on Euler's equation for the motion of an ideal fluid. In particular, we prescribe a fixed angle of contact between the fluid and the plates. Kruse [10] showed that the corresponding equations of motion have a Hamiltonian structure in the sense of mechanics on Poisson manifolds which generalizes the Hamiltonian structure derived by Lewis, Marsden, Montgomery, and Ratiu [17] for free boundary problems in fluid mechanics without capillarity effects (cf. also Arnold [2]). The fixed angle of contact condition can thus be derived from a variational principle. This condition agrees with the one for the static case already derived by Gauss (see Finn [8]). However, its validity is still a point of controversy in the literature if the liquid is not at rest. See Fermigier and Jenffer [7] and Thompson and Robbins [29] [30] for recent work that provides confirmation from a physical point of view. Furthermore, that model possesses a symmetry given by the S^1 action on Eulerian phase space, which is induced by the action from the left on Lagrangian phase space of rotations around a fixed axis perpendicular to the plates. In the present paper we only use the Eulerian or spatial formulation of fluid mechanics and work with that symmetry. Note that in her study of planar liquid drops, Lewis [16] uses the Lagrangian or material description and takes into account the full symmetry group of the system, which includes the action of the group of volume-preserving diffeomorphisms of the reference configuration of the drop.

Rigidly rotating liquid bridges are relative equilibrium solutions of the equations of motion that rotate uniformly in a fixed shape around an axis perpendicular to the plates. Based on general theory for relative equilibria of Hamiltonian systems with symmetry, the existence of this type of solutions, in the shape of surfaces of revolution around the axis of rotation, has been established for any size of the adhesion forces and small angular velocities in [10] and [13]. Furthermore, for these solutions and variations of their shape with respect to both the axial as well as the angular direction, in Kruse [11] specific stability conditions have been derived using ideas related to the so-called reduced energy momentum method (cf. Lewis [15], [16], Lewis and Simo [19], Marsden, Simo, Lewis, and Posbergh [21], Simo, Lewis, and Marsden [25], Simo, Posbergh, and Marsden [26], [27]). In particular, for circularly cylindrical liquid bridges that exist for any angular velocity and volume, an explicit stability criterium involving only physical data came out of this. These results extend work of Concus and Finn [3] and Vogel [31], [32] on static capillarity problems (see also Finn [8]).

In what follows, we continue our earlier studies of the liquid bridge problem by studying bifurcations of relative equilibria from the family of explicitly known solutions that represent rigidly rotating, circularly cylindrical liquid bridges. There are two natural choices for the bifurcation parameter—one being the angular velocity, the other being the angular momentum—both of which we are going to exploit. Using the angular velocity corresponds to characterizing relative equilibria as critical points of the so-called augmented potential, and using the angular momentum corresponds to characterizing relative equilibria as critical points of the so-called amended potential. In both cases, the volume constraint has to be taken into account.

Both these potential energy functions also provide sufficient conditions for the (formal) stability of relative equilibria. Indeed, positive definiteness of their second variation transversal to the orbit of the symmetry group at the corresponding critical point implies that the so-called energy-momentum function is a Lyapunov function for the problem subject to the volume constraint. We emphasize that energy-based stability criteria just provide sufficient conditions, and not sufficient as well as necessary conditions for non-linear stability in general (see Oh [22] and Marsden [20, Chapter 10] for sufficient and necessary conditions). In this connection it is to be noted that the amended potential yields a weaker stability condition than the augmented potential, provided that it is well defined (cf. Marsden [20]). In fact, while for the basic cylindrical solutions considered here, the stability criteria corresponding to those two potentials are equivalent, it turns out that this is not true anymore for certain bifurcating branches of solutions.

Of course, the fact that the Hamiltonian system under consideration is infinite-dimensional makes a rigorous stability analysis subtle in terms of questions such as existence and uniqueness of solutions with prescribed initial data. Also, positive definiteness of a functional at a critical point does not necessarily imply that the critical point is a local minimum. These questions will not be addressed here. In this regard, our stability analysis is formal.

In this paper we show that the stability results obtained for the circularly cylindrical liquid bridges in [11] are sharp in the sense that at parameter values where they become violated, there occurs a bifurcation to liquid bridges of different shapes. Depending on the physical data, there turn out to be two qualitatively different cases. In one case, the second variations of the potential functions mentioned above change from positive definite to indefinite by a simple critical eigenvalue crossing through zero, and the corresponding critical eigenmode is independent of the angular direction with respect to the basic cylinder solution. In this case, noncylindrical rotationally symmetric relative equilibrium solutions bifurcate at the critical point. In the second case, the critical eigenvalue is of double multiplicity and the corresponding eigenmodes are independent of the axial cylinder direction. This leads to the bifurcation of noncircular cylindrical relative equilibrium solutions. In fact, this second case corresponds to the special situation of planar fluid drops that have been considered by Lewis, Marsden, and Ratiu [18] and Lewis [14]. Therefore, we restrict the analysis to axisymmetric solutions in the present paper and just state some results otherwise. In fact, we note that the analysis required by a critical eigenvalue of multiplicity two is quite involved, in particular, as far as stability is concerned. In turn, in the nonaxisymmetric case, the results are not by far complete at present.

It turns out that we can use standard tools of bifurcation theory in the axisymmetric case. In fact, we just have to study the bifurcation structure of critical points of the potential functions near a simple eigenvalue. In particular, we can use the standard exchange of stability principle to determine the definiteness properties of the second variations of the potential functions. By definition, the second variations along the basic solution lose their definiteness at the critical point. Moreover, the second variations along the bifurcating branch of solutions are positive definite if the bifurcation is supercritical and indefinite if the bifurcation is subcritical with respect to the respective bifurcation parameter. Hence, these properties follow solely from the direction of bifurcation at the critical point. Here is where the choice of the potential function makes a difference. Namely, there exist ranges of the physical data where the bifurcation is supercritical with respect to the angular momentum parameter but subcritical with respect to the angular velocity parameter. Thus, the energetic stability test for the bifurcating solutions succeeds using the amended potential, but fails using the augmented potential. This shows that using the amended potential leads to weaker stability conditions than using the augmented potential, because positive definiteness of the second variation of the augmented potential always implies positive definiteness of the second variation of the amended potential. The latter fact has been proved by Simo, Lewis, and Marsden in [25]. We also note, as a consequence, that bistability of relative equilibria for a fixed value of the angular velocity is possible in the present case.

The paper is organized as follows. In Section 2 we state the governing equations of motion and describe their Hamiltonian structure. In Section 3 we briefly summarize known results about the existence and stability of relative equilibrium solutions, in particular, concerning rotationally symmetric ones. Finally, in Section 4 we perform the stability and bifurcation analysis for the circularly cylindrical relative equilibrium solutions as described above.

2. The Equations of Motion

We assume that at any instant of time t, the free surface of the fluid bridge is given as the graph of a real-valued function

$$r = \Sigma(\varphi, z; t) \qquad (r > 0, 0 \le \varphi < 2\pi, 0 \le z \le h), \qquad (2.1)$$

where r, φ, z are cylindrical coordinates in the Euclidean 3-space with origin at one plate and the z-axis perpendicular to the plates; h denotes the distance of the two plates. Note that this assumption excludes the complete separation of the fluid bridge from one of the plates. Then we have the following equations for spatial representations

$$v = v(r, \varphi, z; t), \qquad p = p(r, \varphi, z; t), \qquad \Sigma = \Sigma(\varphi, z; t),$$

of the velocity field, the pressure field, and the free surface of the fluid:

$$
\begin{aligned}
\frac{\partial v}{\partial t} + (v \cdot \nabla)v &= -\nabla p & &\text{in } D_\Sigma, \\[4pt]
\nabla \cdot v &= 0 & &\text{in } D_\Sigma, \\[4pt]
\frac{\partial \Sigma}{\partial t} &= \frac{v \cdot n}{e_r \cdot n} & &\text{on } r = \Sigma, \\[4pt]
p &= \tau \kappa & &\text{on } r = \Sigma, \\[4pt]
v \cdot n &= 0 & &\text{on } \Sigma_1 \cup \Sigma_2, \\[4pt]
\cos \gamma_j &= \frac{\sigma_j}{\tau} & &\text{on } c_j \ (j = 1, 2).
\end{aligned}
\qquad (2.2)
$$

Here, ∇ is the Nabla operator; at any instant of time, D_Σ denotes the region between the plates which is occupied by the fluid; for $j = 1$ and $j = 2$, Σ_j is the region in the j-th plate P_j which is wetted by the fluid, and c_j denotes the boundary curve of Σ_j, i.e., the curve of intersection of the free surface Σ with the plate P_j. The outer unit normal with respect to D_Σ is always denoted by n. The first two equations are just Euler's equations for ideal fluids, i.e., the balance equation for linear momentum and the continuity equation which models incompressibility of the fluid. The third equation constitutes a kinematic condition for the evolution of the free surface. Here and subsequently, e_r denotes the unit vector in radial direction. Along the free surface, the pressure is supposed to be balanced by surface tension. In the fourth equation, κ denotes the mean curvature of the free surface, and $\tau > 0$ is the material constant of surface tension. As usual, along the rigid walls, we assume slip boundary conditions given by the fifth equation. Finally, γ_j denotes the angle of contact of the fluid with the plate P_j, i.e., the angle between the outer unit normal n_j of Σ_j inside P_j and the outer unit conormal of the free surface $r = \Sigma$ along the curve c_j. As explained in the introduction, it is assumed to be constant and given by the sixth equation in (2.2), where σ_j denotes the adhesion coefficient with respect to the plate P_j, which is another material constant. For simplicity, we have set the fluid density equal to one here.

As indicated in Section 1, these equations are Hamiltonian in the sense of mechanics on Poisson manifolds. The *Hamiltonian function* is given by

$$H(\Sigma, v) = \frac{1}{2} \int_{D_\Sigma} \|v\|^2 \, dV + \tau \int_\Sigma dA - \sum_{j=1,2} \sigma_j \int_{\Sigma_j} dA, \qquad (2.3)$$

where the volume integral describes the kinetic energy of the fluid drop, $\| \cdot \|$ is the Euclidean norm, and the surface integrals describe the potential energies. The dynamic variables are the free boundary Σ and the spatial velocity field v, a divergence-free smooth vector field in the region D_Σ bounded by Σ and the plates P_j. Also, v is supposed to satisfy the slip boundary condition along Σ_1 and Σ_2. The surface Σ is represented by a sufficiently smooth function $\Sigma(\varphi, z)$ as in (2.1). We assume that the region enclosed by Σ, Σ_1, and Σ_2 has a fixed prescribed volume. The Eulerian phase space $\mathcal{N}$ can be identified with all such pairs (Σ, v) (cf. Kruse, Marsden, and Scheurle [13]). Variations of Σ and v are denoted by $\delta\Sigma$ and δv, respectively. Following Kruse [10], we are going to introduce a Poisson bracket for functions $F, G: \mathcal{N} \to \mathbb{R}$, which possess functional derivatives defined as follows.

We say that such a function F has a *functional derivative with respect to* Σ *at* $(\Sigma, v) \in \mathcal{N}$, if there exist maps $\frac{\delta F}{\delta \Sigma}(\Sigma, v): \Sigma \to \mathbb{R}$ and $\frac{\delta_j F}{\delta \Sigma}(\Sigma, v): c_j \to \mathbb{R}$ $(j = 1, 2)$, such that

$$\frac{d}{d\varepsilon}\bigg|_{\varepsilon=0} F(\Sigma_\varepsilon, v) = \int_\Sigma \frac{\delta F}{\delta \Sigma}(\Sigma, v)\delta\Sigma \, dA + \sum_{j=1,2} \int_{c_j} \frac{\delta_j F}{\delta \Sigma}(\Sigma, v)\delta\Sigma \, ds$$

holds for any curve $\varepsilon \mapsto \Sigma_\varepsilon$ of admissible surfaces with $\Sigma_0 = \Sigma$ and $\frac{d}{d\varepsilon}\big|_{\varepsilon=0} \Sigma_\varepsilon = \delta\Sigma$. Here c_j denotes the curve of intersection of Σ with the plate P_j as above; for path integrals, the element of integration is denoted by ds. Similarly, we say that a function F has a *functional derivative with respect to* v *at* $(\Sigma, v) \in \mathcal{N}$, if there exists a divergence-free vector field $\frac{\delta F}{\delta v}(\Sigma, v)$ in D_Σ, the normal component of which vanishes along Σ_1

and Σ_2, such that

$$\frac{d}{d\varepsilon}\bigg|_{\varepsilon=0} F(\Sigma, v_\varepsilon) = \int\limits_{D_\Sigma} \frac{\delta F}{\delta v}(\Sigma, v) \cdot \delta v\, dV$$

holds for any curve $\varepsilon \mapsto v_\varepsilon$ of admissible vector fields with $v_0 = v$ and $\frac{d}{d\varepsilon}\big|_{\varepsilon=0} v_\varepsilon = \delta v$.

Let $\mathcal{D}$ be the set of all functions $F: \mathcal{N} \to \mathbb{R}$, which have functional derivatives as defined above at any point $(\Sigma, v) \in \mathcal{N}$. We have $H \in \mathcal{D}$. In fact,

$$\frac{\delta H}{\delta \Sigma}(\Sigma, v) = \left(\frac{1}{2}\|v\|^2 + \tau\kappa\right) e_r \cdot n,$$

$$\frac{\delta_j H}{\delta \Sigma}(\Sigma, v) = (\tau \cos \gamma_j - \sigma_j) e_r \cdot n_j \quad (j = 1, 2), \tag{2.4}$$

$$\frac{\delta H}{\delta v}(\Sigma, v) = v.$$

We now define a Poisson bracket on $\mathcal{N}$ as follows. For functions $F, G \in \mathcal{D}$, we set

$$\{F, G\} = \int\limits_{D_\Sigma} (\nabla \times v) \cdot \left(\frac{\delta F}{\delta v} \times \frac{\delta G}{\delta v}\right) dV$$

$$+ \int\limits_{\Sigma} \left[\frac{\delta F}{\delta \Sigma}\left(\frac{\delta G}{\delta v} \cdot n\right) - \frac{\delta G}{\delta \Sigma}\left(\frac{\delta F}{\delta v} \cdot n\right)\right] \frac{1}{e_r \cdot n}\, dA$$

$$+ \sum_{j=1,2} \int\limits_{c_j} \left[\frac{\delta_j F}{\delta \Sigma}\left(\frac{\delta G}{\delta v} \cdot n\right) - \frac{\delta_j G}{\delta \Sigma}\left(\frac{\delta F}{\delta v} \cdot n\right)\right] \frac{1}{e_r \cdot n}\, ds.$$

With this Poisson bracket, using the divergence theorem, it is not hard to show that for any solution (v, p, Σ) of the basic equation (2.2), the relation

$$\dot{F} = \{F, H\} \qquad \text{for all } F \in \mathcal{D} \tag{2.5}$$

is satisfied along the curve $t \mapsto (\Sigma, v)$ in $\mathcal{N}$. Conversely, given any such curve for which this relation is satisfied, one can construct a pressure field p in D_Σ, such that (v, p, Σ) is a solution of (2.2) (see Kruse [10]). The pressure field satisfies the following boundary value problem, where Δ is the Laplace operator:

$$\begin{array}{lll} \Delta p = -\nabla \cdot ((v \cdot \nabla)v) & \text{in } D_\Sigma, & \\ p = \tau\kappa & \text{on } \Sigma, & \tag{2.6} \\ \nabla p \cdot n = -((v \cdot \nabla)v) \cdot n & \text{on } \Sigma_j, & (j = 1, 2). \end{array}$$

In that sense, the equations in (2.2) are equivalent to the so-called Hamiltonian equation (2.5).

We remark that the question whether the Poisson bracket introduced above satisfies Jacobi's identity and Leibniz's rule, respectively, has not been settled yet. Usually, both of these identities are part of the notion of a Poisson bracket. However, we will not make use of them in the following analysis. On the other hand, the facts that the bracket is bilinear and anticommutative play a role.

3. Rigidly Rotating Liquid Bridges

Now we look for special solutions of (2.2) for which the fluid bridges rigidly rotate around the z-axis with constant angular velocity ω, in the shape of a free surface of the form

$$\Sigma_f: r = f(\varphi, z) \qquad (0 \le \varphi < 2\pi, \ 0 \le z \le h).$$

These are so-called relative equilibrium solutions. It is straightforward to check that for this kind of solutions, the basic equations (2.2) reduce to the following boundary value problem for the function $f = f(\varphi, z)$:

$$\tau \kappa_f - \tfrac{1}{2}\omega^2 f^2 = c,$$
$$\cos \gamma_j = \tfrac{\sigma_j}{\tau} \quad \text{on} \quad c_j \ (j = 1, 2). \tag{3.1}$$

Here, κ_f denotes the mean curvature of the free surface Σ_f, and c is an arbitrary real constant which is related to the pressure field as follows:

$$p = c + \frac{1}{2}\omega^2 r^2. \tag{3.2}$$

As indicated in the introduction, we are primarily dealing with rotationally symmetric relative equilibria in the present paper. In this case, $f = f(z)$ is just a function of z, and (3.1) becomes the following two-point boundary value problem for a nonlinear second-order ordinary differential equation:

$$\tau \kappa_f - \frac{1}{2}\omega^2 f^2 = c,$$

$$f'(0) = \frac{-\sigma_1}{\sqrt{\tau^2 - \sigma_1^2}} =: \rho_1, \tag{3.3}$$

$$f'(h) = \frac{\sigma_2}{\sqrt{\tau^2 - \sigma_2^2}} =: \rho_2,$$

with

$$\kappa_f = \frac{1}{f(1 + (f')^2)^{1/2}} - \frac{f''}{(1 + (f')^2)^{3/2}}.$$

Here, f' and f'' denote derivatives with respect to z. For later use, we also note that the volume of the liquid bridge is given by

$$\text{vol}(D_{\Sigma_f}) = \pi \int_0^h f^2 \, dz \tag{3.4}$$

in this case.

The following theorem ensures the existence of many solutions for this problem. See Kruse, Marsden, and Scheurle [13] for a proof.

Theorem 3.1. *Let the constants h, τ, ρ_1, and ρ_2, as well as a number $K > 0$ be given. Then there exists a constant ω_0 such that for all $\omega \in [-\omega_0, \omega_0]$, (3.3) has a solution (f, c) with vol $(D_{\Sigma_f}) < K$.*

Furthermore, suppose that no adhesion forces act along the plates, i.e., $\sigma_1 = \sigma_2 = 0$ holds. Then, apart from Theorem 3.1, for any ω and any $d > 0$,

$$f(z) = d \tag{3.5}$$

is obviously a solution of (3.3) with the appropriate constant $c = \frac{\tau}{d} - \frac{1}{2}\omega^2 d^2$. Hence, as already pointed out before, in that case solutions of (2.2) exist which represent liquid bridges in the shape of circular cylinders of any volume rotating with any angular velocity. For the stability and bifurcation analysis in the next section, these circularly cylindrical relative equilibria are considered to be the basic solutions. This means that these are the known solutions, and we are searching for solutions branching from these solutions where they become unstable for the first time as ω increases. They are stable as long as h/d and ω are sufficiently small. Moreover, inside parameter regions, where the primary instability is caused by rotationally symmetric variations of the cylindrical surface shape, there occurs a bifurcation of noncylindrical, rotationally symmetric relative equilibria which in turn may or may not be stable themselves. Another kind of primary criticality is analogous to the case of a circular planar liquid drop studied by Lewis, Marsden, and Ratiu [18] and leads to a bifurcation of cylindrical relative equilibria that are not rotationally symmetric anymore, i.e., the function f in (3.1) is just a function of φ.

In the remaining part of the present section, we are going to summarize some general theory concerning the characterization and stability of relative equilibria in order to set the stage for the analysis in Section 4. The reader is referred to one of the standard books such as Abraham and Marsden [1] for more information (cf. also Marsden [20] and the original papers of Simo, Lewis, and Marsden [25], as well as Simo, Posbergh, and Marsden [26], [27]).

Let $(P, \{ , \})$ be a *Poisson manifold* and let $\mathcal{F}(P)$ denote the set of smooth real-valued functions on P, i.e., we have the *Poisson bracket* operation

$$\{ , \}: \mathcal{F}(P) \times \mathcal{F}(P) \rightarrow \mathcal{F}(P),$$

which is usually supposed to be bilinear and anticommutative, and to satisfy Jacobi's identity as well as Leibniz's rule. Define X_F to be the unique vector field on P satisfying

$$dG \cdot X_F = \{G, F\} \quad \text{for all } G \in \mathcal{F}(P), \tag{3.6}$$

where dG denotes the differential of G. X_F is called the *Hamiltonian vector field* corresponding to F.

Furthermore, we assume that a Lie group G acts on P (by Poisson maps) and we denote the infinitesimal generator on P of an element ξ in the Lie algebra g corresponding to G by ξ_P. The vector field ξ_P is obtained at z by differentiating the one-parameter group $\exp(\varepsilon\xi)$ generated by ξ with respect to ε at $\varepsilon = 0$:

$$\xi_P(z) = \frac{d}{d\varepsilon}[\exp(\varepsilon\xi) \cdot z]\Big|_{\varepsilon=0}. \tag{3.7}$$

Definition 3.1. *A map* $J\colon P \to g^*$ *is called a momentum map if* $X_{\langle J,\xi\rangle} = \xi_P$ *for each* $\xi \in g$, *where* $\langle J, \xi\rangle(z) = \langle J(z), \xi\rangle$ *and* $\langle\cdot,\cdot\rangle$ *denotes the pairing between g and its dual* g^*.

Now we consider Hamilton's equation

$$\dot{F} = \{F, H\} \qquad \text{for all } F \text{ in } \mathcal{F}(P), \tag{3.8}$$

with some Hamiltonian function $H \in \mathcal{F}(P)$ that is G-invariant, i.e.,

$$H(\gamma z) = H(z) \qquad \text{for all } z \in P \text{ and } \gamma \in G. \tag{3.9}$$

In this case we call equation (3.8) *G-symmetric*. It is an immediate consequence of the anticommutativity of the Poisson bracket that H is conserved along the solution curves of (3.8). Moreover, the same is true for the function $\langle J, \xi\rangle$ in the symmetric case. In fact, differentiating (3.9) with respect to γ in the direction of $\xi \in g$ for some $z \in P$ gives $dH(z) \cdot \xi_P(z) = 0$, and so $\{H, \langle J, \xi\rangle\} = 0$. Thus, the claim follows by (3.8). This is actually Noether's theorem.

Definition 3.2. *A point* $z_e \in P$ *is called a relative equilibrium of equation (3.8) if*

$$z_e(t) = \exp(t\xi) \cdot z_e \qquad (t \in \mathbb{R}), \tag{3.10}$$

is a solution curve of (3.8) for some $\xi \in g$.

Hence, the dynamic orbit of a relative equilibrium $z_e \in P$ is the orbit of a one-parameter subgroup of G. Of course, any genuine equilibrium is also a relative equilibrium. But there are much more sophisticated examples, such as quasiperiodic orbits (cf. Scheurle [24]).

Next we present an equivalent variational characterization of a relative equilibrium. See, for example, Marsden [20] for other equivalent definitions.

Proposition 3.1. *Let* $z_e \in P$ *and set* $\mu = J(z_e)$. *Then* z_e *is a relative equilibrium of (3.8) if and only if* z_e *is a critical point of the so-called augmented Hamiltonian (energy-momentum function)*

$$H_\xi(z) = H(z) - \langle J(z) - \mu, \xi\rangle.$$

Sketch of the proof. We take any function $F \in \mathcal{F}(P)$ and evaluate it along $z_e(t)$ given by (3.10). It follows that $\dot{F} = \{F, H\}$ holds along $z_e(t)$ if and only if $dH_\xi(z_e) = 0$. Indeed, we have $\dot{F} = dF\xi_P = \{F, \langle J, \xi\rangle\} = -d\langle J, \xi\rangle \cdot X_F$, and $\{F, H\} = -dH \cdot X_F$. Using $\exp(\varepsilon\xi)$-invariance of both the functions $\langle J, \xi\rangle$ and H, the claim follows.

Since H and $\langle J, \xi\rangle$ are constant along solution curves of equation (3.8), the augmented Hamiltonian H_ξ is also conserved. Thus, if the second variation $d^2 H_\xi(z_e)$ of H_ξ at a relative equilibrium z_e has certain definiteness properties, then one uses H_ξ as a kind of Liapunov function to ascertain stability. This is the general technique of the *energy-momentum method* to the development of which the late Juan-Carlos Simo made essential

contributions. Of course, definiteness in this context has to be properly interpreted to take
into account the conservation of the momentum and the fact that $d^2 H_\xi(z_e)$ may have zero
eigenvalues due to its invariance under a subgroup of the symmetry group. Therefore, the
second variation is tested for definiteness only on a particular subspace of variations of
z_e. The variations δz must satisfy the linearized momentum constraint $\delta z \in \ker[DJ(z_e)]$,
and must not lie in symmetry directions. In turn, stability is to be understood in the sense
of orbital stability with respect to group orbits. For example, a drift along group orbits
of nearby solutions is not excluded by this notion of stability (cf. Patrick [23]). Also, as
mentioned in the introduction, for infinite-dimensional systems additional hypotheses
are necessary in order to justify the use of H_ξ as a Liapunov function. From the practical
point of view, a helpful simplification of the energy-momentum method can be achieved
in the context of so-called simple mechanical systems. In simple mechanical systems the
Hamiltonian H may be written as the sum of the kinetic and potential energies (cf. Smale
[28]). Evidently, the Hamiltonian in (2.3) is of this form. In this case, the bifurcation and
stability analysis for relative equilibria can be restricted from the full phase space to the
"configuration manifold" by means of either the augmented potential energy function
V_ξ or the amended potential energy function V_μ. This possibility is also an important
feature of the so-called *reduced energy momentum method* as developed by Lewis [15],
[16], Lewis and Simo [19], Marsden, Simo, Lewis, and Posbergh [21], Simo, Lewis,
and Marsden [25], and Simo, Posbergh, and Marsden [26], [27] for relative equilibria
on cotangent bundles and rather general symmetry groups. We conclude this section by
defining V_ξ and V_μ in the present context and showing their relationship to H_ξ.

Subsequently we use the notation of Section 2 and consider $(\mathcal{N}, \{\ ,\ \})$ to be the
underlying Poisson manifold. Although this choice does not completely fulfill the requirements of the general theory—in particular, as pointed out in Section 2, it is not
known whether Jacobi's identity is valid for the bracket $\{\ ,\ \}$—it still makes sense to use
the relevant part of the theory. Obviously, (2.2) possesses a rotational symmetry about
the z-axis. The corresponding momentum map $J = J(\Sigma, v)$ assigns to each fluid state
$(\Sigma, v) \in \mathcal{N}$ the corresponding classical angular momentum about the z-axis that is given
by

$$J = J(\Sigma, v) = \int_{x \in D_\Sigma} (x \times v) \cdot e_z \, dV. \tag{3.11}$$

Also, we introduce the moment of inertia about the z-axis corresponding to any fluid
volume D_Σ,

$$I = I(\Sigma) = \int_{D_\Sigma} r^2 \, dV. \tag{3.12}$$

Now consider a relative equilibrium of (2.2) given by a function $f = f(\varphi, z)$ and
an angular velocity value $\omega \in \mathbb{R}$ according to (3.1). Denote the free surface of the
corresponding fluid bridge by Σ_f and the velocity field corresponding to ω by v_ω. Thus,
$(\Sigma_f, v_\omega) \in \mathcal{N}$ is a critical point of the augmented Hamiltonian

$$H_\omega = H - \omega(J - \mu), \tag{3.13}$$

where H is defined as in (2.3) and $\mu = J(\Sigma_f, v_\omega)$. Conversely, any critical point of H_ω
in $\mathcal{N}$ for some ω determines a relative equilibrium of (2.2). Due to the special form of H,

which is characteristic for a simple mechanical system, H_ω can be rewritten as follows:

$$H_\omega = K_\omega + V_\omega + \omega\mu, \qquad (3.14)$$

where

$$K_\omega = \frac{1}{2} \int_{D_\Sigma} \|v - v_\omega\|^2 \, dV$$

is the *augmented kinetic energy*, and the *augmented potential function* $V_\omega: \mathcal{N} \to \mathbb{R}$ is defined by

$$V_\omega = V_p - \frac{1}{2} I \omega^2,$$

where V_p is the sum of the potential energies in H. Since K_ω attains its minimal value at (Σ_f, v_ω), it follows that (Σ_f, v_ω) is a critical point of H_ω if and only if Σ_f is a critical point of V_ω for some ω. Moreover, if Σ_f is a local minimum of V_ω, then (Σ_f, v_ω) clearly is a local minimum of H_ω. Also, definiteness, interpreted as described above, of $d^2 V_\omega(\Sigma_f)$ implies definiteness of $d^2 H_\omega(\Sigma_f, v_\omega)$.

There is another way to rearrange the terms in H_ω. Namely, for $\omega = \frac{\mu}{I}$ we have

$$H_\omega = K_\mu + V_\mu, \qquad (3.15)$$

where

$$K_\mu = \frac{1}{2} \int_{D_\Sigma} \|v - \alpha_{(\Sigma,\mu)}\|^2 \, dV$$

is the *amended kinetic energy*. Here the vector field $\alpha_{(\Sigma,\mu)}$ is given by

$$\alpha_{(\Sigma,\mu)}(x) = \frac{\mu}{I}(e_z \times x),$$

for all $x \in D_\Sigma$. Furthermore, $V_\mu: \mathcal{N} \to \mathbb{R}$ denotes the *amended potential function*, which is defined as follows:

$$V_\mu = V_p + \frac{1}{2} \frac{\mu^2}{I}.$$

Since $\alpha_{(\Sigma,\mu)} = v_\omega$ with $\omega = \frac{\mu}{I}$, the same remarks are in order as in the previous case. In particular, definiteness of $d^2 V_\mu(\Sigma_f)$ implies definiteness of $d^2 H_\omega(\Sigma_f, v_\omega)$. Finally, in this section we note that for $\Sigma = \Sigma_f$,

$$d^2 V_\mu = d^2 V_\omega + \frac{\omega^2}{I} \langle dI\cdot, dI\cdot\rangle$$

holds and the correction term $\frac{\omega^2}{I} \langle dI\cdot, dI\cdot\rangle$ is nonnegative (cf. Simo, Lewis, and Marsden [25]). Thus, if $d^2 V_\omega$ is positive definite for $\Sigma = \Sigma_f$, then so is $d^2 V_\mu$, but not necessarily conversely.   □

4. Bifurcation Analysis

In this section we analyze bifurcations of relative equilibria of (2.2) from the basic branch of relative equilibria that represent circularly cylindrical liquid bridges. As indicated before, we concentrate on the rotationally symmetric case here.

According to (3.14), relative equilibria with angular velocity ω can be characterized as critical points of the augmented potential V_ω for any given angular velocity $\omega \in \mathbb{R}$. In the rotationally symmetric case, the free surface Σ_f is a surface of revolution around the z-axis and is given by a function $f = f(z)$, $0 \leq z \leq h$. Here,

$$V_\omega(\Sigma_f) = 2\pi\tau \int_0^h f\sqrt{1 + (f')^2} \, dz - \sigma_1\pi f^2(h) - \sigma_2\pi f^2(0)$$
$$-\frac{\pi}{4}\omega^2 \int_0^h f^4 \, dz. \tag{4.1}$$

In this section we fulfill the condition of a prescribed fluid volume by introducing a Lagrange multiplier $c \in \mathbb{R}$, i.e., we look for critical points (f, c) of the Lagrange functional

$$L_\omega(\Sigma_f) = V_\omega(\Sigma_f) - c \, \text{vol} \, (D_{\Sigma_f}), \tag{4.2}$$

such that

$$\text{vol} \, (D_{\Sigma_f}) = S, \tag{4.3}$$

where vol (D_{Σ_f}) is defined as in (3.4), and S is some constant. Note that c in (4.2) corresponds to the constant c in (3.1).

Since we study bifurcations from circularly cylindrical liquid bridges, we set $S = \pi h d^2$, where d is the radius of the corresponding cylinder, and in addition we impose the boundary conditions

$$f'(0) = f'(h) = 0. \tag{4.4}$$

In the sense of the calculus of variations, these are the natural boundary conditions for the Euler–Lagrange equation corresponding to the described variational principle with $\sigma_2 = \sigma_1 = 0$. This Euler–Lagrange equation evidently is equivalent to the differential equation in (3.3). It differs only by an overall factor $2\pi f$, which is assumed to be positive. Thus, critical points are determined by the boundary-value problem (3.3), with $\rho_2 = \rho_1 = 0$ together with the above volume constraint.

To write this problem as an operator equation, we introduce the following operator for any $\omega \in \mathbb{R}$:

$$\Phi_\omega : \{f \in W_2^2(0, h) \mid f'(0) = f'(h) = 0\} \times \mathbb{R} \to L_2(0, h) \times \mathbb{R};$$

$$(f, c) \mapsto \left(\tau\kappa_f - \frac{1}{2}\omega^2 f^2 - c, \quad \pi \int_0^h f^2 \, dz - S\right). \tag{4.5}$$

Here $W_2^2(0, h)$ and $L_2(0, h)$ are the standard Sobolev spaces of functions $f = f(z): [0, h] \to \mathbb{R}$, and κ_f is defined as in (3.3). Thus, we have to solve the operator equation

$$\Phi_\omega(f, c) = (0, 0). \tag{4.6}$$

Moreover, again up to the positive factor $2\pi f$, the linear operator associated to the second variation $d^2 L_\omega(\Sigma_f)$ of the Lagrange functional along the constraint manifold at a critical point (f, c) is given by the linearization with respect to f of the first component of Φ_ω, i.e., by

$$A_\omega(f) \quad : \quad \{g \in W_2^2(0, h) \mid g'(0) = g'(h) = 0\} \to L_2(0, h);$$

$$g \mapsto -\frac{\tau g}{f^2(1 + (f')^2)^{1/2}} - \frac{\tau f' g'}{f(1 + (f')^2)^{3/2}} \tag{4.7}$$
$$-\frac{\tau g''}{(1 + (f')^2)^{3/2}} + \frac{3\tau f'' f' g'}{(1 + (f')^2)^{5/2}} - \omega^2 fg,$$

together with the linearized volume constraint

$$\int_0^h fg \, dz = 0. \tag{4.8}$$

We remark that for rotationally symmetric liquid bridges, any variation of the shape given by some $g \in W_2^2(0, h)$ is transversal to the orbit of the symmetry group. Hence, $d^2 L_\omega(\Sigma_f)$ is positive definite if and only if its eigenvalues are positive, i.e., if and only if the eigenvalues of the linear operator $A_\omega(f)$ are positive. Note that $A_\omega(f)$ is a closed operator with a compact resolvent in $L_2(0, h)$.

Obviously,

$$f \equiv d, \qquad c = \frac{\tau}{d} - \frac{1}{2}\omega^2 d^2 \tag{4.9}$$

is a solution of equation (4.6) for all values of ω. It represents the basic branch of relative equilibria. The corresponding operator $A_\omega(d)$ is given by

$$A_\omega(d)g = -\frac{\tau}{d^2}g - \tau g'' - \omega^2 dg.$$

It has simple eigenvalues $\lambda_l = -\frac{\tau}{d^2} + \frac{\tau \pi^2 l^2}{h^2} - \omega^2 d$ ($l = 1, 2, 3, \ldots$), where $\cos\left(\frac{\pi l}{h}z\right)$ are corresponding eigenfunctions. Thus, all eigenvalues are positive if and only if ω satisfies the inequality

$$\frac{h^2}{\pi^2 d^2} + \frac{\omega^2 h^2 d}{\pi^2 \tau} < 1. \tag{4.10}$$

In other words, this condition implies stability of circularly cylindrical liquid bridges with respect to rotationally symmetric variations of the shape of the free surface such that the contact angle remains equal to π at each plate. In particular, (4.10) is satisfied for sufficiently small values of ω provided that

$$\eta = \frac{h}{d} < \pi \tag{4.11}$$

holds for the *aspect ratio* η. In other words, for fixed h, the fluid volume S must be sufficiently large. Subsequently, we assume (4.11) to be true (cf. Finn and Vogel [9], Zhou [33]).

Let us now consider a critical value $\omega = \omega_o$ such that the left-hand side of the inequality in (4.10) becomes equal to 1. Then we have the following bifurcation result.

Theorem 4.1. *For $\omega = \omega_0$, the solution in (4.9) is a bifurcation point of equation (4.6). There, the basic solution branch undergoes a subcritical pitchfork bifurcation. The bifurcating solutions exist locally for $\omega^2 < \omega_0^2$ and represent noncylindrical, rotationally symmetric liquid bridges.*

Proof. We use the standard theory for bifurcation problems with a one-dimensional null space (Crandall and Rabinowitz [5]). In fact, it is straightforward to verify the corresponding assumptions. In particular, the derivative $\mathcal{A} := D_{(f,c)}\Phi_{\omega_0}(d, c_0)$ is a Fredholm operator with index zero, where (d, c_0) denotes the basic solution given by (4.9) with $\omega = \omega_0$. Furthermore, the kernel of this operator is one-dimensional and spanned by $\varphi := \left(\cos\left(\frac{\pi}{h}z\right), 0\right)$. Let $\mathcal{B} := D^2_{(\omega,(f,c))}\Phi_{\omega_0}(d, c_0)$. Then it follows that $\mathcal{B}(1, \varphi) \notin$ range $\mathcal{A}$. Hence, $\omega = \omega_0$, $f \equiv d$, $c = c_0$ is a bifurcation point of (4.6). In order to determine the type of the corresponding bifurcation, one has to compute further derivatives of Φ_{ω_0} at (d, c_0). One concludes that the second derivative in the direction of φ is contained in range $\mathcal{A}$, while the third derivative in this direction is not. Thus, one has a pitchfork bifurcation. Finally, one shows that the components of the latter derivative and of $\mathcal{B}(1, \varphi)$ in the direction of the subspace spanned by φ, which is complementary to range $\mathcal{A}$, are parallel for $\omega_0 > 0$ and antiparallel for $\omega_0 < 0$. Hence, the bifurcating solutions exist for $\omega^2 < \omega_0^2$ in both cases, i.e., the bifurcation is subcritical. To first order, as $\omega \to \omega_0$, the bifurcating solutions are given by $(f, c) \sim (d, c_0) \pm \alpha(\omega_0^2 - \omega^2)^{1/2}\varphi$, where α is some positive constant.

Next we perform the same bifurcation analysis based on the amended potential V_μ. In the rotational symmetric case with $\sigma_1 = \sigma_2 = 0$, this is given by

$$V_\mu(\Sigma_f) = 2\pi\tau \int\limits_0^h f\sqrt{1 + (f')^2}\, dz + \frac{\mu^2}{\pi}\left(\int\limits_0^h f^4 dz\right)^{-1}. \tag{4.12}$$

As in the previous analysis, we introduce the volume constraint (4.3) and consider the Lagrange functional

$$L_\mu(\Sigma_f) = V_\mu(\Sigma_f) - c\,\mathrm{vol}\,(D_{\Sigma_f}). \tag{4.13}$$

This time the corresponding Euler–Lagrange equation reads as follows after division by $2\pi f$:

$$\tau\kappa_f - \frac{2\mu^2}{\pi^2}f^2\left(\int\limits_0^h f^4\,dz\right)^{-2} - c = 0. \tag{4.14}$$

This is complemented by the boundary conditions (4.4) again. Here critical points are determined as solutions of the operator equation

$$\Psi_\mu(f, c) = (0, 0), \tag{4.15}$$

where Ψ_μ denotes the following operator for any $\mu \in \mathbb{R}$:

$$\Psi_\mu : \{f \in W_2^2(0, h) | f'(0) = f'(h) = 0\} \times \mathbb{R} \to L_2(0, h) \times \mathbb{R};$$

$$(f, c) \mapsto \left(\tau \kappa_f - \frac{2\mu^2}{\pi^2} f^2 \left(\int_0^h f^4 dz \right)^{-2} - c, \ \pi \int_0^h f^2 dz - S \right). \quad (4.16)$$

Analogously to $A_\omega(f)$ in (4.7), we now obtain the operator

$$A_\mu(f) : \{g \in W_2^2(0, h) \mid g'(0) = g'(h) = 0\} \to L_2(0, h);$$

$$g \mapsto -\frac{\tau g}{f^2(1 + (f')^2)^{1/2}} - \frac{\tau f' g'}{f(1 + (f')^2)^{3/2}} - \frac{\tau g''}{(1 + (f')^2)^{3/2}}$$

$$+ \frac{3\tau f'' f' g'}{(1 + (f')^2)^{5/2}} + \frac{16\mu^2}{\pi^2} \left(\int_0^h f^4 dz \right)^{-3} \left(\int_0^h f^3 g \, dz \right) f^2$$

$$- \frac{4\mu^2}{\pi^2} \left(\int_0^h f^4 dz \right)^{-2} fg, \quad (4.17)$$

together with the linearized volume constraint (4.8). This operator is again closed and has a compact resolvent in $L^2(0, h)$. Therefore, $d^2 L_\mu(\Sigma_f)$ is positive definite at a critical point (f, c) under the constraints mentioned, if and only if the eigenvalues of $A_\mu(f)$ are positive.

In terms of equation (4.15), the basic solution branch is given by

$$f \equiv d, \qquad c = \frac{\tau}{d} - \frac{2\mu^2}{\pi^2 h^2 d^6} \qquad (\mu \in \mathbb{R}). \quad (4.18)$$

Along this solution branch, we have $A_\mu(d)g = -\frac{\tau}{d^2}g - \tau g'' - \frac{4\mu^2}{\pi^2 h^2 d^7} g$ for any admissible g. The eigenvalues of this operator are simple and given by $\tilde{\lambda}_l = -\frac{\tau}{d^2} + \frac{\tau \pi^2 l^2}{h^2} - \frac{4\mu^2}{\pi^2 h^2 d^7}$ ($l = 1, 2, \ldots$); corresponding eigenfunctions are $\cos\left(\frac{\pi l}{h} z\right)$. Note that $\tilde{\lambda}_l = \lambda_l$ holds, where λ_l are the eigenvalues of $A_\omega(d)$, if and only if

$$\mu = \frac{\pi}{2} hd^4 \omega = I(\Sigma_d) \omega. \quad (4.19)$$

Recall that $I(\Sigma_d)$ denotes the moment of inertia about the z-axis of the fluid volume D_{Σ_d}; see (3.12) for the general definition. Of course, the relation (4.19) between the angular momentum μ and the angular velocity ω is satisfied for relative equilibria. Hence, in the present case the stability test with V_μ leads to the same explicit stability condition (4.10) as with V_ω. The second variation of L_μ becomes indefinite along the basic solution branch (4.18) as μ^2 increases through μ_0^2. Here ω_0 is defined as in Theorem 4.1, and

$$\mu_0 = I(\Sigma_d)\omega_0 \quad (4.20)$$

denotes the corresponding critical value of μ, where the left-hand side of inequality (4.10) becomes equal to 1. It remains to study the associated bifurcation in terms of V_μ rather than V_ω as in Theorem 4.1. As pointed out before, there appears to be a difference concerning the direction of bifurcation. $\square$

Theorem 4.2. *For* $\mu = \mu_0$, *the solution in (4.18) is a bifurcation point of equation (4.15). There, the basic solution branch undergoes a pitchfork bifurcation. Let*

$$0 < \eta_\pm = \frac{147 \pm \sqrt{15777}}{324} \pi^2 < \pi^2,$$

and let $\eta = \frac{h}{d}$ *be the aspect ratio as in (4.11). Then the bifurcation is subcritical for* $\eta^2 \in (0, \eta_-) \cup (\eta_+, \pi^2)$, *but supercritical for* $\eta^2 \in (\eta_-, \eta_+)$. *Thus, the bifurcating solutions exist locally for* $\mu^2 < \mu_0^2$ *in the first case and for* $\mu^2 > \mu_0^2$ *in the second case. In particular, they are stable with respect to axisymmetric perturbations in the second case.*

Sketch of the proof. Except for the assertion concerning stability, this theorem can be proved following exactly the lines of the proof of Theorem 4.1, with the obvious modification in the supercritical case. Therefore we just establish the stability claim here. To this end we use the well-known principle of exchange of stability for bifurcation problems with a one-dimensional null space (Crandall and Rabinowitz [6]). It tells here that the critical eigenvalue of $A_\mu(f)$ along the bifurcating solution branch becomes negative in the subcritical case, but positive in the supercritical case, slightly away from the bifurcation point. Here we use the fact that the critical eigenvalue of $A_\mu(f)$ coincides with the critical eigenvalue of $D_{(f,c)}\Psi_\mu(f, c)$ in this situation. Thus, the claim follows from the corresponding definiteness properties of $d^2L_\mu(\Sigma_f)$.

By local uniqueness, the bifurcating solutions in the Theorems 4.1 and 4.2 coincide, where ω and μ are related according to (4.19). Analogously to the previous proof, it follows that $d^2L_\omega(\Sigma_f)$ is always indefinite there, since the bifurcation is subcritical in terms of ω for all η. Hence, for $\eta^2 \in (\eta_-, \eta_+)$ we have the situation that the second variation corresponding to L_ω is indefinite at a relative equilibrium, although the latter is stable.

Finally, we point out that Kruse [11] has studied the stability of circularly cylindrical liquid bridges allowing also disturbances of the free surface that are not rotationally symmetric. He has used L_ω and found an additional condition, namely

$$\omega^2 < \frac{3\tau}{d^3}, \tag{4.21}$$

which, together with (4.10), is sufficient to guarantee stability in this broader sense. It is obvious that the stability result in Theorem 4.2 remains valid even in this sense, if (4.21) is still satisfied while ω^2 increases through the critical value ω_0^2. One can easily verify that the latter is indeed the case for $\eta^2 > \frac{\pi^2}{4}$. Note that $\eta_- < \frac{\pi^2}{4} < \eta_+$. On the other hand, for $0 < \eta^2 < \frac{\pi^2}{4}$, the second variation of L_ω first becomes indefinite when ω^2 increases through $\bar{\omega}^2 = \frac{3\tau}{d^3}$, i.e., by violation of (4.21). Here the critical eigenvalue is double, which makes the analysis more difficult than in the simple case. In fact, here the basic solution appears to stay spectrally stable beyond that criticality (cf. Kruse, Mahalov, and Marsden [12]). Nevertheless, there occurs a bifurcation of relative equilibrium solutions that are cylindrical, but not rotationally symmetric. They have just a $\mathbb{Z}_2 \times \mathbb{Z}_2$-symmetry. In terms of ω, this bifurcation is always subcritical; in terms of μ, it is always supercritical. These results can be proved by means of an equivariant branching lemma following Lewis, Marsden, and Ratiu [18]. The question of stability has not yet been settled here. $\square$

Acknowledgments

We thank one of the referees for his comments, which helped us to clarify our statements in a number of places in the text.

References

1. Abraham R. and Marsden J. E., *Foundations of Mechanics*, 2nd ed., Addison-Wesley Publishing Co., Reading, Mass., 1978.

2. Arnold V. I., Sur la géométrie différentielle des groupes de Lie de dimension infinie et ses applications à l'hydrodynamique des fluides parfaits, *Ann. Inst. Fourier*, Grenoble 16(1) (1966), 319–361.

3. Concus P. and Finn R., The shape of a pendent liquid drop, *Philos. Trans. Roy. Soc. London Ser. A*, 292 (1979), 307–340.

4. Concus P. and Finn R., Capillary surfaces in microgravity, in *Low-Gravity Fluid Mechanics and Transport Phenomena*, J. N. Koster and R. L. Sani, eds., Progress in Astronautics and Aeronautics 130, AIAA, Washington, DC, 1990, 183–206.

5. Crandall M. G. and Rabinowitz P. H., Bifurcation from simple eigenvalues, *J. Funct. Anal.* 8 (1971), 321–340.

6. Crandall M. G. and Rabinowitz P. H., Bifurcation, perturbation of simple eigenvalues and linearized stability, *Arch. Ratl. Mech. Anal.* 52 (1973), 161–180.

7. Fermigier M. and Jenffer, P., An experimental investigation of the dynamic contact angle in liquid-liquid systems, *J. Coloid Interface Sci.* 146(1) (1991), 226–241.

8. Finn R., *Equilibrium Capillary Surfaces*, Springer-Verlag, New York, 1986.

9. Finn R. and Vogel T. I., On the volume infimum for liquid bridges, *Z. Anal. Anw.* 11 (1992), 3–23.

10. Kruse H.-P., Flüssigkeitstropfen zwischen parallelen Platten: Hamiltonsche Struktur, Existenz von Lösungen und Stabilität, Doctoral thesis, Universität Hamburg, 1992.

11. Kruse H.-P., The Hamiltonian structure of the equations of motion of a liquid drop trapped between two plates, Preprint, 1994.

12. Kruse H.-P., Mahalov A., and Marsden J. E., On the three-dimensional instabilities of rotating liquid bridges, Preprint 1995.

13. Kruse H.-P., Marsden J. E., and Scheurle J., On uniformly rotating fluid drops trapped between two parallel plates, *Lect. Appl. Math.* 29 (1993), 307–317.

14. Lewis D., Nonlinear stability of a rotating planar liquid drop, *Arch. Ratl. Mech. Anal.* 106 (1989), 287–333.

15. Lewis D., Lagrangian Block Diagonalization, *J. Dyn. Diff. Eq.* 4(1) (1992), 1–41.

16. Lewis D., Bifurcations of liquid drops, *Nonlinearity* 6 (1993), 491–522.

17. Lewis D., Marsden J. E., Montgomery R., and Ratiu T. S., The Hamiltonian structure for dynamic free boundary problems, *Physica D* 18 (1986), 391–404.

18. Lewis D., Marsden J. E., and Ratiu T. S., Stability and bifurcation of a rotating planar liquid drop, *J. Math. Phys.* 28 (1987), 2500–2515.

19. Lewis, D. and Simo J.-C., Nonlinear stability of rotating pseudo-rigid bodies, *Proc. Roy. Soc. London A* 427 (1990), 281–319.
20. Marsden J. E., Lectures on Mechanics, *London Math. Soc. Lect. Note Series*, Vol. 174, Cambridge Univ. Press, Cambridge, 1992.
21. Marsden J. E., Simo J.-C., Lewis D., and Posbergh T. A., A block diagonalization theorem in the energy momentum method, *Cont. Math. AMS* 97 (1989), 297–313.
22. Oh, Y. G., A stability criterion for Hamiltonian systems with symmetry, *Geom. Phys.* 4 (1987), 163–182.
23. Patrick G. W., Relative equilibria of Hamiltonian systems with symmetry: Linearization, smoothness, and drift, *J. Nonl. Sci.* 5(5) (1995), 373–418.
24. Scheurle J., Some aspects of successive bifurcations in the Couette-Taylor problem, in *Pattern Formation: Symmetry Methods and Applications*, J. Chadam, M. Golubitsky, W. F. Langford, and B. Wetton, eds., Fields Inst. Comm. 5 (1996), 335–345.
25. Simo J.-C., Lewis D., and Marsden J. E., Stability of relative equilibria. Part I: The reduced energy momentum method, *Arch. Ratl. Mech. Anal.* 115 (1991), 15–59.
26. Simo J.-C., Posbergh T. A., and Marsden J. E., Stability of coupled rigid bodies and geometrically exact rods: Block diagonalization and the energy-momentum method, *Phys. Reports* 193 (1990), 279–360.
27. Simo J.-C., Posbergh T. A., and Marsden J. E., Stability of relative equilibria. Part II: Application to nonlinear elasticity, *Arch. Ratl. Mech. Anal.* 115 (1991), 61–100.
28. Smale S., Topology and mechanics, *Inv. Math.* 10 (1970), 305–331; 11 (1970), 45–64.
29. Thompson P. and Robbins M., Simulations of contact-line motion: Slip and the dynamic contact angle, *Phys. Rev. Lett.* 63(7) (1989), 766–769.
30. Thompson P. and Robbins M., To slip or not to slip?, *Physics World*, November 1990, 35–38.
31. Vogel T. I., Stability of a liquid drop trapped between two parallel planes, *SIAM J. Appl. Math.* 47 (1987), 516–525.
32. Vogel T. I., Stability of a liquid drop trapped between two planes II: General contact angles, *SIAM J. Appl. Math.* 49 (1989), 1009–1028.
33. Zhou L., On the volume infimum for liquid bridges, *Z. Anal. Anw.* 12 (1993), 629–642.